AF768413

Brain and Values:
Is a Biological Science of Values Possible

Edited by:

Karl H. Pribram
Stanford University &
Radford University

Proceedings of the Fifth Appalachian Conference on Behavioral Neurodynamics

LEA LAWRENCE ERLBAUM ASSOCIATES, PUBLISHERS
1998 Mahwah, New Jersey London

BRAIN AND VALUES

Lawrence Erlbaum Associates, Inc., Publishers
10 Industrial Avenue
Mahwah, New Jersey 07430

Library of Congress Cataloging-in-Publication Data

Brain and values / edited by Karl H. Pribram
p. cm.
Includes bibliographical references and index.
ISBN 0-8058-3154-1 (pbk. : alk. paper)
1. Neuropsychology--Congresses. 2. Brain--Congresses. 3. Values--Congresses. I. Pribram, Karl H., 1919- .
[DNLM: 1. Brain--physiology congresses. 2. Nerve Net--physiology congresses. 3. Mental Processes--physiology congresses. 4. Models, Neurological congresses. 5. Neuropsychology congresses. WL 300 B81232 1998]
QP360.B695 1998
612.8'2--dc21
DNLM/DLC
for Library of Congress 98-17067
CIP

Books published by Lawrence Erlbaum Associates are printed on acid-free paper, and their bindings are chosen for strength and durability.

Printed in the United States of America

10 9 8 7 6 5 4 3 2 1

TABLE OF CONTENTS

Table of Contents

Utility

Creativity

Afterword

Acknowledgments

As for the first four Proceedings, I am deeply grateful to Sam Leven who has made these Proceedings possible. Also Walter and Dorothy Morris have steadily provided financial and emotional support and are specifically responsible for making this publication possible. Interest of the International Neural Network Society has lent a special impetus to the Proceedings.

Additionally, I acknowledge and thank Terry Lynn Hayes, Shelli D. Meade and Lynn Doody for their efforts in preparing the manuscript for publication.

KEYNOTE

Keynote

Successful Intelligence: An Expanded Approach to Understanding Intelligence

Robert J. Sternberg
Department of Psychology
Yale University
Box 208205
New Haven, CT 06520

Jack points out to his pal, Tom, the boy Jack has identified as the stupidest in his class, Irvin. Jack, who considers himself the smartest in the class, likes to make fun of Irvin.

"You want to see what 'stupid' means, Tom? Watch this."
"Hey, Irvin. Here are two coins. Take whichever one you want. It's yours."
Irvin looks at the two coins, a nickel and a dime. He stares at them for a while, and then selects the larger coin, the nickel.
"Go ahead, Irv, take it, it's yours," laughs Jack.
Irvin takes the coin and walks away.

An adult, who has been watching the transaction from a distance, walks up to Irvin and gently points out that the dime is worth more than the nickel, despite the dime's being smaller, and that Irvin has just cost himself 5 cents.
"Oh, I know that," replies Irvin, "but if I picked the dime, Jack never would ask me to choose between the two coins again, and this way, he'll keep asking me again and again. I've already collected over a dollar from him, and all I have to do is keep choosing the nickel."

This apocryphal story points out something we already intuitively know--that someone can be slow in school but think well outside it, and vice versa. The well-rehearsed question, "How can someone so smart be so dumb?", reminds us of how someone can be a good or a bad thinker, however well he or she may do in a school setting.

Introduction

The Nature of Successful Intelligence

What is successful intelligence? How does it shed a different light on the construct of intelligence? I consider these questions here.

Three Ways of Thinking

Two boys are walking in a forest. They are quite different. The first boy's teachers think he is smart, his parents think he is smart, and as a result, he thinks he is smart. He has good test scores, good grades, and other good paper credentials that will get him far in his scholastic life. Few people consider the second boy smart. His test scores are nothing great, his grades aren't so good, and his other paper credentials are, in general, marginal. At best, people would call him shrewd or street smart. As the two boys walk along in the forest, they encounter a problem, namely, a huge, furious, hungry-looking grizzly bear. It is charging straight at them. The first boy, calculating that the grizzly bear will overtake them in 17.3 seconds, panics. In this state, he looks at the second boy, who is calmly taking off his hiking boots and putting on his jogging shoes.

The first boy says to the second boy, "You must be crazy. There is no way we are going to outrun that grizzly bear!"

The second boy replies, "That's true. But all I have to do is outrun you!"

In the preceding story, both boys are smart, but they are smart in different ways. Let's discuss different ways of being smart and different ways of using the smarts we have. All tests measure only a small part of intelligence. Many people with modest test scores are nevertheless highly intelligent.

To be successfully intelligent is to have the ability to succeed in what you or others value for you in life. It requires abilities to think well in one or more of three different ways: analytic, creative, and practical (Sternberg, 1985a, 1988, 1996). One of the missing elements of our system of education is that typically only one of these ways is valued in tests and in the classroom. Yet no one way is any better than either of the others, and, ironically, the style of intelligence that schools most readily recognize as smart may well be less useful than the others to many students in their adult lives.

1. Critical-analytic intelligence. Alice (a real student, but with her name changed) was the teacher's dream. She scored high on tests, performed well in class, and in general, did everything that a teacher would expect a bright student to do. The result was that Alice was always considered to be at or near the top of her class. Her high test scores were accepted as a valid indicant of her ability to do outstanding work throughout her academic career. Yet by the time Alice was finished with graduate school in psychology, she was performing at a very modest level. About 70 to 80 percent of her classmates were doing better. Alice-types occur at all levels of schooling.

For example, consider Ben. Ben, at just five years old, started school early. In October of the first grade, he tested at a third-grade level in oral reading and could decode almost any word. His general reading comprehension also was excellent. When asked to write something, he produced a coherent story a page long that included commas and quotation marks. He also requested permission to look up the spelling of some of the more difficult words in a dictionary--which he accomplished on his own when permission was granted!

According to all accounts, this boy was not in any kind of instructional program in preschool; he was simply fascinated by books and words, and he spent a lot of time on his own trying to figure out how to read and spell. Here's a child almost everybody recognized as smart. There was just one problem: Ben's story was uninspired, and so was the rest of his writing. It was technically excellent, but dull. Asked to write creatively, Ben seemed at sea, and generally repeated with little change plots from stories he had read in school or on his own.

The question that naturally arises is, what went wrong with Alice, and might with Billy? The answer, quite simply, is that while Alice was excellent at remembering and analyzing other people's ideas, she was not very good at coming up with ideas of her own. Consequently, she faltered in advanced schooling, where (as in life) it is necessary to come up with ideas of one's own. Similarly, although Billy's academic skills certainly are impressive, he will require more than just a high level of academic achievement in order to do well in advanced schooling and in adult life.

If we think about schooling as preparation for the world of work, we need to be concerned about whether schooling requires and develops creative thinking, because for Alice, and eventually for Ben, to stay competitive in most jobs, it is and will be necessary for them to come up with their own ideas.

In science, for example, you need more than the ability to memorize facts in a book or to solve ready-made problems at the ends of chapters. The practice of science requires the ability to come up with creative, significant ideas that make a difference to the field, and ultimately, to the world. People who generate important scientific ideas are not necessarily those who are best at memorizing facts or solving textbook problems. Indeed, they may be people who don't particularly like to do these things and who therefore do not show themselves at their best in school settings.

The same dynamic applies in other occupations. Consider writing or art. It is one thing to succeed in writing good essays when told what to write about, or to draw nice pictures when told what the pictures should show. It is quite another thing to come up with one's own ideas for stories or pictures. For example, outside a classroom, I once observed roughly two dozen pictures of children's houses. They were nice pictures of houses, but it was clear that the teacher had told the students what to draw--it did not seem likely that 24 children had independently decided to draw pictures of houses. In the real world of art and writing, though, someone is not always there to tell the artist or writer what their composition should deal with. Indeed, creative writers and artists are, almost by definition, people who come up with their own imaginative ideas.

The problem with the current situation is, of course, that telling students what to do is often unrealistic with respect to what the students will be required to do later. Educators need to stop always formulating problems for students and instead sometimes urge the students to formulate problems for themselves.

One could argue that most students will not become scientists or writers or artists, but the situation is no different in a very pragmatic occupation like business. Many of the executives interviewed during our studies of practical intelligence complained that one could hire a top-level graduate of a business school and get someone who might be good at analyzing textbook cases, but who was unable to come up with innovative ideas for new business products or services, rearrangements that create more shelf space, or ways to stay competitive with similar industries in other countries.

The point, of course, is that there are large gaps between the kind of performance needed for success in a business setting and the kind of performance needed for success in schools, even schools that are supposed to be quite practical in training students for the world of business. Thus we often end up with adults who are unable to do what is expected of them.

This same problem even afflicts the study of education itself. It is one thing to get A's in education courses and quite another to succeed when called upon to be innovative in a classroom setting. We know from experience just how challenging classroom situations can be. For example, several years ago I was giving a lecture at the University of Puerto Rico, and found myself confronting a serious classroom management problem, namely, the professors of education in the audience just weren't listening. For whatever reasons, they had decided to tune out, and they were walking around the room and in and out of the room, and generally not being very attentive.

I tried the standard, uncreative techniques one learns in the course of training to be a teacher. I tried lowering my voice in the hope that these professors would then lower their voices so that they would be able to hear me. Of course, he was assuming that they wanted to hear, an assumption that proved false. Instead, they appeared to be grateful that I had lowered my voice so that they could hear themselves better. I then tried asking them to be quiet, but that didn't work, either. Finally, after I had given up, a woman in the audience shot up and said something in rapid-fire Spanish. After that, you could have heard a pin drop, and the audience remained silent and attentive for the rest of the session.

What did she say? She had capitalized on her understanding that Puerto Rico is a shame culture, not a guilt culture. My attempts to make the audience feel guilty might have worked in the mainland United States, but were ineffective in Puerto Rico. In contrast, the woman pointed out to the audience that, if they continued to be noisy, I would leave with a poor impression of the University of Puerto Rico. I would then report this bad impression to others. She said that the audience had no right to convey a bad impression and thereby to cast shame on the university. This appeal was effective in achieving the behavioral change that I had unsuccessfully sought.

Why is it that the students whom we consider to be bright tend to be bright like Alice? In other words, why are they so often test-smart, but not necessarily smart in other ways?

Children are not born to be smart only in this limited way; rather, we shape them to be. Our system of education, in essence, creates Alices by continually reinforcing or rewarding students for being test-smart. Indeed, the main lesson that students learn is that it pays to be smart like Alice. As rewards, you receive good grades, good placements in high tracks, awards, and later, impressive college admissions.

There is an irony in our scholastic reward system. In school, children are continually reinforced for being smart like Alice, but after the children leave school, the world changes. IQ is no longer enough. It is not that analytic smarts stop mattering, but they certainly matter less. So people come to be intermittently reinforced rather than continually reinforced. That is to say, they are rewarded some times for their IQ-like smarts, but not other times. And here is the irony: Psychologists have found (see Sternberg, 1995) that intermittent reinforcement (rewarding behavior some but not all of the time) actually strengthens the behavior. In other words, when IQ stops mattering so much, and people start to be intermittently reinforced for it, they even more strongly start relying on it, even though this response is inappropriate.

Consider an analogy, here of intelligence as it applies to relationships (Sternberg, 1987). It sometimes comes to pass in relationships that one partner realizes that the relationship just isn't working. Suppose, for example, that a woman is with a guy who keeps beating on her, or cheating on her. She eventually decides it's just not worth the pain. So she tells the guy that the relationship is through--kaput. In the old times, we used to call such a message "dumping the guy," but today, in the modern world, we call it "restructuring a relationship." So the woman tells the guy that sure, she will still be friends with him and supportive of him and so on and so forth; she just doesn't want to have an intimate relationship with him anymore.

What often happens, curiously, is that after receiving this relatively straightforward message, the guy tries to get the woman back. And he keeps trying to the point that it is annoying. The woman may find it totally beyond her grasp why he keeps trying, especially when his behavior was so outrageous before. Sometimes women actually get suckered into taking the guys back, at which time they almost immediately recommence their old pattern of behavior. But why do guys do this, when they are being given the signal that the relationship is over?

The reason they do so is because of intermittent reinforcement. By being nice to the guy, and maintaining the friendship, the woman is inadvertently giving the guy a mixed message: He is being given some reward. And intermittent reinforcement strengthens old patterns of behavior. So the guy tries to maintain the relationship. The nicest thing the woman could do, actually, is to move as far away as possible, or act as cold as possible. Trying to maintain the friendship makes it harder on the guy, and on her.

You can see the effects of intermittent reinforcement on the utilization of abilities in a variety of ways. One way is to realize that those with the very highest IQs are generally not those who are most successful in life. There is a small but positive correlation between IQ and various kinds of measured success, but at the top, the relationship weakens. People with extremely high IQs often don't achieve as great success. Why? Because they try to overcapitalize on their IQ points.

Again, you can see the result of such overcapitalization in relationships. If you are ever in a relationship with someone and get into an argument, and your partner goes to pains to dazzle you with his or her impeccable logic and powers of abstract reasoning, pointing out at the same time your own stupidity or emotionality, you know that the technique of logical wizardry is not a successful way of resolving conflicts. It tends to make them worse. Someone who tries to use his or her IQ points where they just don't fit worsens rather than improves a situation.

You can also read the book, Whatever Happened to the Quiz Kids?, by Ruth Duskin Feldman (1982). The book is interesting, in that it tells what happened to these extremely bright youngsters when they became adults. The youngsters, who were first on the radio and then on the television show, The Quiz Kids, had extremely high IQs, in many cases, over 160. But their adult lives were, for the most part, rather ordinary, and none of them actually turned out to be a stellar success. You can read about example after example of people trying to overcapitalize on their IQs.

One indication that schools mold students into Alices comes from Joe Glick's study of the Kpelle tribe (Cole, Gay, Glick & Sharp, 1971). Glick asked adult members of the tribe to sort terms into categories. For example, they might be asked to sort names of fruits (apple, orange, grapefruit) or names of vegetables (celery, lettuce, broccoli) or names of vehicles (bus, boat, car). Glick found that the Kpelle sorted functionally. For example, they would sort "apple" with "eat" and "car" with "gas" because people eat apples and cars use gas.

In our culture, only young children sort functionally. The Kpelle's functional kind of sorting behavior is considered stupid when it is done by an adult. Older children and adults are expected to sort taxonomically (putting fruits together) or hierarchically (putting the word "fruit" over the names of different fruits and then perhaps putting the word "food" over the whole lot).

Glick tried, without initial success, to get the Kpelle to sort in other ways. When he was about to conclude that they simply didn't have the mental ability to do things any differently, he decided as a last resort to ask them how a stupid person would do the task. At this point, they sorted taxonomically, and with no trouble at all. Why would the Kpelle consider taxonomic sorting stupid? The answer is that the Kpelle do not grow up in our educational system and--even more important--do not take our tests. In everyday life, we tend to think functionally. We think of eating apples or using gas in cars. We learn to think taxonomically in school, but for the most part this kind of thinking remains limited to artificial settings. A problem arises, therefore, when advanced students or career aspirants have to start thinking in ways that they have not been conditioned in school to think, that is, when they need to start turning out their own ideas rather than reciting back or analyzing ideas that other people have had.

We need to take a broad view of intelligence, recognizing that what is viewed as smart in one place may not be in another. for example, Elena Grigorenko, Catherine Nokes, Paul Wenzel Geissler, and myself are currently collaborating in a study of conceptions of intelligence in a rural Kenyan village. The children in the village know large numbers of natural herbal medicines they can use to treat parasitic infections. This information is adaptive and intelligent in their context. In the U.S., it would not be. But if we translated a Kenyan "IQ test" based on this knowledge into English, the U.S. children would really look quite stupid on the test. Thus, we need to think broadly about the skills that constitute intelligence.

Because practical as well as creative kinds of skills have not been actively encouraged or selected for, students tend not to develop them. In this respect, then, our schools essentially mislead and misprepare students by developing and rewarding a set of skills that will be important in later life, but much less important than they are in school.

I want to emphasize that the kinds of analytical abilities shown by Alice and Ben do matter in later life. In life, you need to be able to analyze problems. It's very hard to get through life without analytic abilities. At work, when things go wrong, you need to be able to analyze just what is wrong. Similarly, in personal relationships, misanalyzing the source of, say, an argument can be quite a disaster. Couples who repeatedly fight about the same thing often continue to fight because what they are fighting about isn't really the underlying source of friction. Thus, even if they resolve the fight, they don't resolve the problem that caused the fight, resulting in their continuing to argue. Clearly, analytical abilities are important.

Let's return, for a moment, to the story of the two boys in the forest. The one who survived was the one who was not particularly skilled in analytical abilities. But what eventually happened to him? Well, he acquired a phobia of forests. He decided that the only way to fight his phobia was to learn antiphobic techniques, like deep relaxation, heavy breathing, self-hypnosis, and the like. So he learned them, and then returned to the forest to test them out. They worked; he found himself totally relaxed. As he was relaxing in the forest, however, he learned a valuable lesson, albeit, too late. The lesson was that whereas lightning never strikes twice in the same place, grizzly bears do. The grizzly bear, which had not eaten since the last boy, now came charging again. The boy, petrified, first thought that maybe he could outrun the other boy. Then he remembered that that boy was dead. Next he thought about trying to climb up a tree, but clearly, it was too late. He didn't know exactly how many seconds he had left, but there weren't many. Things were looking bad.

Finally, he decided in desperation that the only course of action left was to get down on his knees and start to pray. So he did: He prayed to the Lord that the grizzly bear would become a good, religious being like himself. This course of action may sound like a last resort, but in fact, it wasn't totally misguided. Just as the grizzly bear was about to attack the boy and eat him alive, it stopped dead in its tracks and it got down to pray. But its prayer illustrated the importance of analytic powers, or the lack thereof. The boy had not specified any religion in particular: He had not adequately analyzed the problem. The grizzly bear prayed, "I thank thee, oh Lord, for the offering I am about to receive," and that was the end of that boy too. So you see, analytic abilities do matter, and not just in forests populated by grizzly bears!

2. Creative-synthetic intelligence. A student whom I shall call Barbara (not her real name) exhibits a second way of being successfully intelligent. Barbara's grades were good, although by no means spectacular. Her undergraduate teachers thought that she was just terrific, despite the fact that her test scores were very weak. Moreover, Barbara included a portfolio of her work with her application, and I was tremendously impressed with the work she had done. It occurred to me at the time that her chances of getting a job at Yale as an assistant professor (for which you need a PhD) would actually have been quite a bit higher than her chances of being admitted to graduate school, because getting a job does not require a standardized ability test, whereas getting into graduate school does.

When Barbara applied to our graduate program in psychology, I figured that she would, despite the scores, be admitted. For one thing, we claim to care about creativity, and Barbara had amply demonstrated her creativity in her work. What better predictor of future creative work could one find than past creative work? Actually, none. Certainly, a test like the Graduate Record Examination, which does not measure creativity, will not be as good a predictor of creative work as the creative work itself. In general, the best predictor of any kind of behavior in the future is the same kind of behavior in the past. This principle applies universally. For example, if you are going out with someone who is verbally abusive, but who promises to stop being verbally abusive after the marriage, the best predictor of what the person will do after the marriage is what he or she has done before the marriage, just as the best predictor of what the person will say after the marriage is what he or she has said before the marriage.

At the time, I was the Director of Graduate Studies for the Psychology Department, so I figured that the other members of the admissions committee more or less would have to listen to me. and besides, I was the expert on the committee with regard to abilities and how to test them. The committee did not, in fact, just blow off Barbara's application. They considered it quite seriously, spending close to a half-hour considering it. They discussed it at great length. There was a distressing aspect to this discussion. It became obvious to me during the discussion that the test scores were coloring everything else about the application. In other words, having seen the low test scores, committee members were looking to interpret other aspects of Barbara's admissions file as supporting the test scores. This pattern of behavior is not unusual.

Work by Solomon Asch (1946) has shown that certain traits of a person are what he referred to as central traits. These traits are the ones around which people organize other information. And the pseudo-quantitative precision of test scores almost guarantees that they will be interpreted as central traits. So what was happening was that committee members were using the central trait of intelligence, as they gleaned it from the test scores, to interpret other information. Thus, for example, if a comment in a letter of recommendation could be interpreted either somewhat more favorably or somewhat more negatively, the members of the committee would interpret it somewhat more negatively to make the comment consistent with the test scores.

By the end of the discussion, the members of the committee felt very good about the decision they were making. In fact, Barbara was rejected by an almost unanimous vote. In fact, I was the only person who voted to admit her. Even though Barbara had included a portfolio of her work, which demonstrated a high degree of competence, the other admissions people made their decisions largely on the basis of her test scores. In other words, they had more confidence in fallible and often weak predictors of creative work than they had in the work itself. We often see this odd situation in education today. The predictor of the performance has become more important than the performance itself! Barbaras don't occur just at the graduate level.

Consider Jeannie as another good example of creative abilities. At age 6, Jeannie had been a good solid student--above average in most areas, but not the kind of exceptional early reader and writer that Ben was. What teachers repeatedly commented on was that Jeannie was "very creative," "an independent thinker," "has an unusual way of thinking," etc. Some of her art work was even selected to represent her school at a regional art show. But nowhere did Jeannie's creativity get recognized in her report cards, because there was no entry, among all the many checkoff boxes, for creative ability or performance, neither in Jeannie's school nor any other I am aware of. It's odd, isn't it? The report cards that elementary schools typically offer has multiple boxes to measure basic academic competencies and interpersonal behavior, but creative behavior is totally unrecognized. If anything, teachers tend to penalize it, because creative children can so often be perceived as disruptive.

Jeannie's creative ability was not limited to art. For instance, Jeannie's favorite subject in school was science, and she frequently initiated her own science projects at home, such as growing bacteria, gathering insect specimens outdoors, and so on. Once she used an old deflated balloon and a diagram of the human digestive tract from a book to make her own "model" of a stomach and intestines. She once saw a children's game on television that interested her; instead of pestering her parents to buy it for her (which she knew would be futile), she constructed her own cardboard version of it. Thus, she was really good at generating interesting ideas and at carrying them out independently.

Some of Jeannie's teachers truly appreciated her creative abilities. But overall, her abilities were less valued that were those of Ben. Some people even viewed her negatively, because she tended to have strong preferences for the kinds of problems she wanted to work on and for how she wanted to do things. Sadly, parents often find themselves having to encourage their children to suppress their creative instincts, less the schools regard the children as misfits in need of being put in their place.

Why has Barbaras' future been jeopardized, and why is Jeannie's at risk as well? Why do we pay more attention to predictors than to performance? And in general, why do we pay so much attention to results on tests of IQ and related abilities?

Most of the time, the Barbaras of the world simply get shut out of the system. Basically, they're screwed. For example, Barbara was rejected by our graduate program because of her GREs. If she also applied to other competitive programs, chances are she would be rejected by those programs as well, for exactly the same reason. Their admissions committees, too, might pay lip service to her portfolio and letters of recommendation--prior to giving her the axe. Or they might never even get to this point. Many graduate programs prescreen their applicants so that those, like Barbara, who have low scores, never even get their applications reviewed by anyone except a secretary who shunts their applications aside.

So Barbara can pretty much forget about admission to a competitive program in psychology. Unfortunately, she can also pretty much forget about admission to any other kinds of graduate program because all of the programs require the same test. If Barbara wants to go to law school, she can pretty much forget about that too, because law schools require the Law School Admission Test (LSAT), which is very similar to and highly related to the GRE. If Barbara wants to go to medical school, she will have to take the Medical College Admission Test (MCAT), and almost certainly won't do very well on that either. And if she decides to pursue something more practical, such as business school, the BMAT will almost assuredly keep her out. In other words, Barbara's in big trouble, no matter what she wants to do, just so long as whatever it is requires an admittance test.

Barbara had a piece of good luck, but one that very very few of those who test poorly have. She got another chance. I decided to hire Barbara as a research associate because I believed that she showed much better potential than the test scores indicated. Pretty selfless of me! Actually, it's so hard to be totally selfless these days. At the time, Barbara was working for one of my main competitors in the field, and I figured that it would be better to have Barbara in my camp than in his. And I was not disappointed. Her work as a research associate was highly creative and innovative. Two years later, she was admitted as the top pick into our program. But do you suppose that Barbara's case changed the system? If anything, people's reaction was to regard Barbara as an odd exception to a sound rule. We need to open up our thinking and reappraise our educational superstitions if we are ever to change the way our educational system functions. Although, after many years of effort, things are loosening up, we still have far to go.

I gave myself a lot of credit for my efforts to get Barbara admitted, and I deserve that credit. She was an outstanding student, after all. But lest it sound like I brag too much, I'll tell you the sequel. As in the case of the two boys in the forest, not to mention the innumerable Rocky movies, sometimes the sequel is not as impressive as the original story. Barbara, having been singularly successful in graduate school, got two job offers, one from an enormously prestigious institution and the other from an institution of somewhat lesser prestige. The problem was that her creative work was of a kind that the less prestigious institution seemed more likely to value. The more prestigious institution seemed then, and to me still seems now, to value Alice-like more than Barbara-like abilities. Stupidly, I encouraged Barbara to take the more prestigious job, telling her she would always wonder whether she could have made it there if she didn't in fact go there. Really dumb advice.

Barbara accepted my advice, and took the more prestigious job. In fact, they did not particularly value what she had to offer. To the extent that they valued creative intelligence, it wasn't the kind she had, but rather a kind that is much more molecular and detail-oriented. Barbara just didn't work at a micro-detail level. The result was that Barbara was not perceived as particularly able there, and eventually left, her career off to what was almost certainly a poorer start than it would have had at the other institution. The lesson, of course, is that creative intelligence is valuable only to the extent that the environment in which it is displayed values it.

3. Practical-contextual intelligence. When Celia (not her real name) applied to our graduate program in psychology, she had grades that were good but not great, test scores that were good but not great, and letters of recommendation that were good but not great. In fact, just about everything in her application seemed to be good but not great. Naturally, we admitted Celia because every program needs people who are good but not great. Indeed, in our program, her work proved to be exactly what we had predicted: good, but not great--so we figured we hit it on that one.

But what a surprise Celia gave us when it came to getting a job! Everyone wanted to hire Celia! That raised an intriguing question. Why would someone who lacked Alice's analytic ability and Barbara's creative ability do so spectacularly well in the job market?

The answer was actually very simple. Celia was something like the second boy in the opening bear-in-the-forest story. She had an abundance of practical intelligence, or simple common sense. She could go into an environment, figure out what she needed to do to thrive there, and then do it.

For example, Celia knew how to interview effectively, how to interact well with other students, how to get her work done. She also was aware of what kinds of things do and don't work. In other words, she was street smart in an academic setting. She knew something that is true, though seldom acknowledged: that in school, as well as life, one needs a certain amount of practical smartness in order to adapt to the environment. These talents can be seen in younger children as well.

For instance, a television news story discussed a child whose mother was epileptic. The child was about five or six years old. The mother had had a seizure just as she was about to get into the shower, and she was lying unconscious in the tub with scalding water pouring over her. There was no one else in the house but the little boy, who called 911, gave the dispatcher directions to his house, and so on. But the thing that really impressed us is that, before calling 911, the boy turned on the cold-water tap full force to keep his mother from being even more badly burned. He had also tried to turn off the hot-water tap, but it was too hot for him to turn it off.

Another example involves a child whom I'll call Dolores. Dolores was ten years old, in the fifth grade, and she, along with a group of three other fifth-graders, needed help in basic academic subjects. Dolores's IQ was somewhere in the 70s, a good 20 to 30 points below the IQs of the three other children in her group. In fact, Dolores had once been placed in a class for retarded children. (This was in the early to middle 1970s, when children with "borderlinc" IQs ended up in mental retardation placements much more often than they do today.) Indeed, Dolores did seem to have a significantly harder time than did the other children in her group in acquiring academic skills and with analytical types of tasks (e.g., with higher level reading comprehension questions or problem solving in math).

However, when the four children were "mainstreamed" into a regular classroom full-time, surprisingly, it was Dolores who was most successful. Dolores definitely had "practical intelligence"--she knew how to approach the teacher to ask for help, how to function independently in a regular-classroom setting, how to get along with other children, and so on. Dolores was one of those few children who actually elicit constructive criticism about how she can "make it" in a regular classroom. And she wasn't just sitting there being compliant but not learning anything. Rather, she made the best possible use of the abilities she had, whereas the other children in her group, despite their higher IQs, often failed to mobilize their abilities effectively without the constant adult direction and supervision available in a special-education setting.

Sometimes so-called mentally retarded people have tremendous practical skills. They have to in order to get through life. For example, Edgerton (1967) describes a mentally retarded man who was unable to tell time. But the poor man knew that, in order to get through life, he somehow needed a way of knowing what time it was. So he would walk around wearing a broken watch. When he needed to know the time, he would ask someone on the street to tell him the time, pointing out that his own watch was broken. Not bad for someone labeled as retarded.

Whereas almost everyone would accept Alice and Ben as smart, and many people would regard Barbara and Jeannie as smart (albeit in their own ways), few people would think of Celia and certainly not of Dolores as smart. They might concede that Celia and Dolores have common sense, but would not see that as part of intelligence. They might even say that Celia and Dolores are manipulative and reject the idea that being manipulative is an element of intelligence.

Not so. The kind of practical intelligence that Celia and Dolores exhibited is every bit as important as Alice's analytic or Barbara's synthetic intelligence. The reason is that different situations call for different kinds of intelligence. Furthermore, if we value only one kind of intelligence in school, we will seriously underestimate a lot of students. We will peg them as much less intelligent than they really are.

This tendency to undervalue certain forms of intelligence became apparent in our own research in California. We compared conceptions of intelligence among parents from different ethnic groups (Okagaki & Sternberg, 1993). We found that the more parents emphasize in their conception of intelligence social competence skills, such as getting along with peers and helping out the family, the less bright their own children look according to the standards of the schools. In other words, the mismatch between what the parents emphasized in their environment and what the schools required in their environment resulted in kids who might be quite competent in the home and community setting, but who would be judged as intellectually lacking in the school.

More recently, we have completed a study investigating conceptions of intelligence in Taiwan (Yang & Sternberg, 1997). We found that the Taiwanese conception of intelligence includes a broad cognitive factor, of the kind found in conventional theories of intelligence, but also includes interpersonal, intrapersonal, self-assertion, and self-abnegation factors. Clearly, the concept of intelligence in Taiwan is quite a bit richer than one of just a general intellectual ability.

Along a similar line of inquiry, Shirley Heath (1983) compared the language behavior of children in three communities:

--Trackton, a lower social class black community
--Roadville, a lower social class white community
--Gateway, a middle social class white community.

Heath found that the children from Trackton performed quite a bit worse than the children from Roadville or Gateway as soon as the children started school, but that the idea of how smart children are in school may be largely dependent on the match between parental and school conceptions of intelligence. The children in Trackton, therefore, might actually have been no less intelligent than the children in Roadville or Gateway.

For example, parents in Trackton were found to emphasize the importance of nonverbal communication. To communicate successfully in Trackton, it was necessary to be very adept at nonverbal cues, both with respect to understanding them and with respect to transmitting them. In Roadville and Gateway, on the other hand, more emphasis was placed on verbal skills, an emphasis that was a better match to the demands of the school. As a result, children from Roadville and Gateway appeared smarter than did children from Trackton, but they may actually have been no smarter. Once again, the middle class (especially the white middle class) benefited from the match between school values and home and community values.

It is quite plausible to argue that white middle-class culture undervalues the importance of nonverbal communication. For example, many boring teachers or professors can go on being boring year after year, precisely because they ignore the nonverbal communication of the audience. None of the students have the courage to risk an F by telling the teachers or professors that they are boring. Furthermore, this behavior certainly would not be practically intelligent! If the educators were to pay attention to nonverbal cues, they might well realize their failure to command attention and might even do something about it.

Sensitivity to nonverbal communication can be a key to success in an interview setting also. Information as to how well an interview is going is almost exclusively nonverbal. Interviewers know that they are not supposed to reveal their feelings about the person being interviewed. At times, there might be nothing they would rather say than, "Please leave now. I know that we have another 25 minutes left in the interview, but we both know that you are wasting my time and I am wasting yours." The interviewer may feel that way, but certainly can't say so. Nevertheless, the interviewer's feelings are likely to leak out nonverbally. If the applicant is sensitive to the nonverbal communication, he or she at least has a chance of changing the way the interview is going.

In short, the ways of Trackton have something to teach us all. A child from Roadville or Gateway would look as stupid in Trackton as a child from Trackton would look in Roadville or Gateway. We need to acknowledge the multiple styles of intelligence.

It is interesting as well to compare Roadville versus Gateway. When the children from these two communities start school, they look roughly comparable. Within a few years, however, the white middle-class children from Gateway are doing better than the white lower-class children from Roadville. What happened? Do the Roadville children have some kind of "inherited cumulative deficit," as some would have us believe? We believe that the explanation is a lot simpler: that the views about the nature of education and intelligence that are commonly held in Roadville render the children there less smart appearing in school.

For example, parents in Roadville are more likely to believe that their role as teachers stops when the children start school; at that point, parents stop intervening in their children's education. They, like many low-SES parents, have limited educations themselves, and may feel intimidated by the school, or ill-equipped to help especially as their children get older (e.g., Snow, Barnes, Chandler, Goodman, & Hemphill, 1991). Gateway parents, on the other hand, continue to intervene, to the advantage of their children.

Moreover, parents from Roadville emphasize memory in their concept of intelligence, whereas parents from Gateway emphasize reasoning. As the years go by and reasoning becomes more important, the children from Gateway become progressively more advantaged.

In sum, context matters. We cannot consider intelligence in isolation from context. To do so may lead to seriously erroneous conclusions about children's capacity to learn. All the investigations described above suggest that practical intelligence matters, but unfortunately it is not what our tests measure, nor is it sufficiently emphasized in schools. Educators need to begin to consider not only the intelligence of Alice, but also that of Barbara and Celia.

Of course, although people usually have a preferred style of intelligence, they do not use only one style exclusively. In everyone, there is come combination of analytic, creative, and practical intelligence. We need to foster all these aspects of intelligence, not to favor just one. In addition, we need to recognize that people who are really smart in their lives are people who figure out (a) what it is they are good at, (b) what it is they are not good at, and (c) what they can do to make the most of their strengths while remediating or compensating for their weaknesses.

In other words, the most functionally intelligent people are not necessarily the ones with the greatest degree of successful intelligence in any of its three styles. Being smart in everyday life means making the most of what one has, not conforming to any preset stereotypical pattern of what others may consider smart. It is this view of intelligence and of styles of intelligence that we need to adopt in order to obtain the most from our students and ourselves. Being academically smart, like Alice and Ben, is important in school and even, to an extent, in later life; but there is more to intelligence than what tests measure!

The Mental Processes Underlying Successful Intelligence

A distinguished historian arrived in a city where he was scheduled to give a talk on memory to the faculty and students in a small college. The historian was exhausted because this was the last in a long string of engagements, and he knew he was incapable of giving a good talk at this point. As he was riding in a taxi to his destination, he had an idea. He offered his taxi driver $50 if the driver would give the talk for him. he didn't know anyone at the college, and he assured his driver that because the talk was completely written out, the driver didn't need to know anything at all about the topic of the talk--the allies' initial failure to recognize that Hitler's ambitions would have no limit. All the driver had to do was to deliver the talk. Moreover, the historian assured the driver, he, the historian, would be in the audience just in case something went wrong. The driver could always rely on the historian to bail him out in case things got tough.

Business had been slow, so the taxi driver decided that it was worth going through with the deal for the $50. He gave the talk, and it went flawlessly. Then the question-and-answer period started, and the taxi driver found himself unable to answer even the first, simple question about his talk that someone from the audience asked. He knew the answer was in the talk he had given, but he had been so busy concentrating on the talk that he couldn't remember a thing he had said.

Thinking quickly, the driver looked the questioner straight in the eye, and responded: "Why, that question is so easy, even my taxi driver could answer it, and he just happens to be in the audience."

The story of the historian and the taxi driver illustrates how a person who ably--and quickly--sizes up a situation and defines it properly, can get him or herself out of some pretty hot water. What kinds of skills did the taxi driver use in thinking about his situation, and what kinds of skills do people do when they think intelligently in their everyday lives?

In my theory of successful intelligence, the same basic set of thinking skills underlies all three aspects of intelligence--analytical, creative, and practical. Analytical people are particularly adept at applying these skills to familiar and often academic kinds of problems. Creative people are particularly adept at applying them to relatively novel problems. Practical people are particularly adept at applying them to everyday problems.

After decades of bickering, many research psychologists seem to be reaching at least a partial consensus as to what the thinking skills underlying intelligence are. My own version of this view (Sternberg, 1985, 1997), which is similar in some ways to that of others (e.g., Butterfield, Wambold, & Belmont, 1973; Campione & Brown, 1979; Carroll, 1981, 1993; Hunt, 1978; Pellegrino & Glaser, 1980; Snow, 1979), is that good thinking consists of a set of developed thinking and learning skills used in academic and everyday problem solving. Some of these skills are pretty much "cognitive skills," but others are more social-practical, and could better be seen even as being in the realm of what Salovey and Mayer (1990) and Goleman (1995) have called "emotional intelligence." All of these skills are developed through experience, although individuals may differ in terms of how easy it is potentially for them to develop the skills (see Sternberg & Grigorenko, 1997). I'll consider here first the more cognitive, and then the more affective skills.

Cognitive Skills

1. Problem identification. Here, you need both to recognize that you have a problem and to define what the problem is. Consider, for example, a student assigned to write a social studies paper on a topic of his or her choice. The quality of the outcome in large part depends on the choice of topic; some topics will not yield an interesting paper regardless of what one does with them. Although no topic guarantees a good paper, some seem to preclude a good one.

The ability to identify problems is measured indirectly by ability tests, but in an academic way. Distractors on intelligence and other tests are frequently the right answers to the wrong problems. On arithmetic problem-solving tests, distractors are often correct answers to sub-problems of the full problem, and thus might be correct outcomes for intermediate stages of problem solution. The idea, quite bluntly, is to trick test-takers into thinking that one of the incorrect answers is correct, when in fact it is only a partial solution to the problem. Returning to Alice (mentioned earlier), note that she is someone who may be quite adept at solving problems, but who is not necessarily adept at identifying good problems to solve.

The ability to define problems is of key importance in everyday settings, not just in academic ones. Back in 1974, Detroit got killed when it thought that the royal road to profits was to make larger and larger, more and more expensive cars. When the gas crunch hit, people turned to smaller cars, and voila, the Japanese had them waiting for U.S. consumers to buy. Detroit doesn't always seem to learn rapidly. Auto manufacturers have groaned and whined for years about being excluded from Japanese markets. Their cars have been, in the past, a flop in the Japanese markets. But it has been disingenuous to put the blame squarely on the Japanese government. In Japan, people drive on the right side of the road, as in England, and yet Detroit has been trying to sell cars with the driver's gear (steering wheel, brakes, gas pedal, etc.) on the left side of the car. Little wonder the Japanese haven't been buying. Japanese manufacturers, in contrast, have never tried to sell in the U.S. the cars they sell in Japan with the driver's gear on the right side. When the newspapers trumpeted, late in 1995, that Detroit had finally started adjusting their cars, it was time for a celebration, but one that they should have had twenty years ago.

Politicians are notorious for misidentifying problems. Back in 1972, during the hellish days of Watergate, the Nixon administration made a big mistake, and then made an even bigger one. The big mistake was to send men to burglarize the office of the Democratic National Party in the Watergate Hotel, for reasons about which there continues to be speculation but no real certainty. The burglars had the bad fortune to be caught, resulting in an enormous scandal. But the second mistake of the administration was even worse. They instituted a massive coverup, which proved to be disastrous. Machiavelli warned that if you have good news to give the people, dribble it out to them slowly, but if you have bad news to give the people, give it to them all at once. As the coverup unraveled, there was plenty of bad news to the give the people, but they got it in painfully slow bits, exactly counter to the advice of Machiavelli. In identifying the problem as one of coverup rather than of coming clean, they lost it all, and Nixon eventually resigned.

Democrats are not immune to the same problem, as the recent Whitewater coverup has shown. Moreover, when the Democrats lost the midterm elections of 1994, the message of the populace to Clinton and the other Democrats seemed to be clear to all: Take a stand, a stand by it. People were tired of the equivocation, tired of positions that scarcely lasted a day. A going joke was that if you didn't agree with one of Clinton's positions, you shouldn't worry about it, because whatever your position was, it would certainly come to be his, sooner or later.

Businesses can have the same problem. The A. H. Robins Company tried to hide from the public the enormous damage done by an intrauterine device (IUD), the Dalkon Shield, and as a result, when the damage became known, the company was forced eventually to go into bankruptcy. The makers of silicone breast implants--Dow Corning--has recently found itself suffering the same fate. Not all businesses misidentify problems.

Some years back, Johnson and Johnson found itself in a most unpleasant position: Someone was lacing extra-strength Tylenol with the permanent cure to headaches: cyanide. Take a couple of these capsules, and you'd never have a headache again. In fact, Johnson and Johnson was in no way responsible for what was quickly becoming a health as well as a public-relations fiasco. The company chose not to go the cover-up route. It immediately recalled all of its extra-strength Tylenol, and then, other Tylenol products as well. Some said that the company had made a fatal mistake--that the Tylenol brand was finished as an analgesic. But it quickly came back as an enormously popular analgesic. The company had, in fact, identified the problem correctly. More recently, Intel, after several attempts to minimize a problem with their Pentium chip, realized that their refusal to acknowledge the problem was backfiring, and offered to replace all Pentium chips that were returned to them.

Of course, we all need to be careful about how we identify problems. For example, a friend of mine was interested in opening a franchise for home-security products. The franchiser made him an offer that it suggested he couldn't refuse: a large territory in northern Vermont with absolutely no competition to cut into his sales. My friend had the good sense to refuse the generous offer. He knew that northern Vermont would have no competition because there aren't many people and there is very little crime. He identified the problem he would face was not one of competition, but one of having customers in the first place.

2. Process selection. In order to solve a problem successfully, one must select or discover a set of appropriate processes for solving that problem. Consider the steps in writing a research paper on, say, the declining role of the United States as "the world's policeman." One needs to identify possible relevant sources of information, discarding sources that are irrelevant; seek those sources in libraries or elsewhere; cull from those sources information that is relevant, and ignore information that is irrelevant; evaluate the credibility of the various sources; and so on.

Intelligence tests measure process selection ability, but again, only indirectly and in an academic kind of environment. In order to solve a test problem, an individual must select processes that will yield a correct answer. Except for rare cases, the tests do not separate the ability to select a set of processes from the ability to execute them. One of these rare exceptions is the kind of arithmetic problem-solving test that asks examinees what operations (addition, subtraction, multiplication, division) they would use to solve problems, but does not require them to execute the operations. A potential problem with this kind of test is that individuals are not always aware of their own process-selection practices; forcing them to bring these practices out into the open may change the nature of the problem solving required by the test.

Selecting a process by which to accomplish a goal is as important in everyday life as it is in the academic environment, but the people who are good at it in the one setting often are not in another. For example, suppose you identify a problem you are facing as one of wanting greater closeness with a partner who seems to be keeping his or her distance. It's a tremendously common problem in close relationships. Most people immediately start to employ the same process: They start to bring subtle or not so subtle pressure to bear on the partner so as to bring him or her closer. For example, they might complain that the partner is too distant or uncaring, or they might try to make the partner feel guilty for keeping such a distance when they are themselves go giving of themselves. Yes, it's a common solution process, and a bad one. It usually ends up backfiring, driving the partner further away. A more effective process for achieving the goal is, paradoxically, to create greater distance (Sternberg, 1987). As the level of distance becomes greater than that comfortable to the partner, the partner is placed in the position of trying to achieve greater closeness, and usually then starts to do so in order to reestablish equilibrium.

An interesting problem that requires one to select a process of solution is called prisoner's dilemma (Luce & Raiffa, 1957). The problem goes like this: Suppose that two men have been arrested and are charged with pulling off a bank heist. Each man is found to have an unregistered firearm at the time of arrest. The police do not have conclusive evidence that the two men actually robbed the bank. The police need a confession from at least one of the men so that the district attorney can win the case. The distinct attorney hatches a Machiavellian plan in order to wring out a confession from at least one prisoner.

She separates the two prisoners, preventing them from communicating with each other. The district attorney tells each of the accused that what happens to him will depend not only on what he does, but also on what the other prisoner does. If neither prisoner confesses, the district attorney will be unable to prove the bank-robbery charge, but she will prosecute both of them on the charge of illegal possession of a firearm, and each prisoner will be jailed for a term of one year. The district attorney informs the prisoners that if they both confess, she will recommend an intermediate-length term for each of them: 10 years in prison. In addition, however, the district attorney offers each prisoner immunity from prosecution if he is the only one to confess. In this case, the prisoner who confesses will go scot-free, whereas the prisoner who does not confess will be slapped with the maximum possible sentence--20 years.

The prisoner's dilemma presents a complicated situation for the prisoners. The greatest individual payoff is for being the only one to pull a double-cross, thereby enabling oneself to go scot-free. However, if both prisoners double-cross each other, they will both go to jail for a decade. The greatest loss is for being the only one not to double-cross the other, and thereby having to spend the maximum term in prison. The best deal for the two partners together is for each prisoner to remain silent--a potentially costly strategy if one partner is less altruistic than the other.

Psychologists were originally interested in the prisoners' dilemma in and of itself, because it illustrated different processes people could use to deal with a difficult decision-making problem. For example, they could use the minimax loss rule, by which they make a choice that minimizes their maximum loss. Using this rule, the prisoner in the prisoner's dilemma will confess because he wants to minimize the possibility of going to jail for 20 years, the maximum loss.

But the prisoner's dilemma is also interesting when one considers a generalized example where one has to continue to interact with another person over a period of time. Suppose that two people's interests roughly coincide to those in the prisoner's dilemma, but that one will repeatedly have to interact with the other person. What is the best process of decision making, given repeated decisions in a negotiation. For example, in a single negotiation, your best bet may seem to be to try to screw the other individual, but then you will have to deal with his or her wrath in the next negotiation, and so that bet may not be so great after all.

Simulations via computer show that the best decision-making process is actually a very simple one, called "tit-for-tat". In the first negotiation, you give the individual (or group) with whom you are negotiating the benefit of the doubt (e.g., in the prisoner's dilemma, you keep quiet). After that, you do exactly what the negotiating partner does. If the partner acts favorably toward you, you respond favorably; if the partner acts adversely, you act adversely. This simple process, it turns out, maximizes your payoff.

3. Representation of information. In most tasks requiring intelligent performance, the individual must represent information in a useful way, both internally (in one's head) and externally (on paper). A student collecting information for the paper on the U.S. as the world's policeman might organize his or her notes by authors of books and articles, or by topics. The latter organization is usually more effective, although probably less widely used. Similarly, encoding the information by topic will be much more useful in later recall than encoding the information by author.

Representation of information is important in many life pursuits, for example, negotiations of any kind. Many Israelis viewed Yassir Araft and the PLO as murderers and liars whose word could not be trusted in the slightest. Of course, many Palestinians shared a similar view with respect to the Israelis. The correctness of their views could be debated, but as long as each side held this kind of representation of the other, negotiations would be an exercise in futility. In such negotiations, it is the representation of the other side, rather than what the other side is actually like, that matters. What sense is there in negotiating a peace agreement with people who you are certain will break the agreement at their earliest convenience?

It is tempting to point one's finger at others as holding these simplified forms of representation, but the truth is that we in the United States are no different, and often even more simplistic. Early in the Vietnam War, we held a very simplified representation of what was happening: The U.S. was fighting the Communist threat, which had extended even to southeast Asia.

This kind of representation is nothing new. I write these words just a few blocks from the Alamo, where one is able to hear a stirring film entitled, "The Price of Freedom." Very touching, but perhaps a bit one-sided. From the Mexican point of view, the U.S. stole roughly half of its country, and needless to say, Native Americans tend to see the pioneers' conquest of their lands in a way that is somewhat different from the way it was presented to us in school, at least in those years during the 1960s when I, like other students, learned to view Native Americans as an unacceptable blockade to progress and civilization. What was left out was the matter of whose progress and civilization.

Bad representation of information perhaps reaches its extreme in neurologically impaired patients:

There was the hint of a smile on his face. He also appeared to have decided that the examination was over and started to look around for his hat. He reached out his hand and took hold of his wife's head, tried to lift it off, and put it on. He had apparently mistaken his wife for a hat! His wife looked as if she was used to such things (Sacks, 1985).

4. Strategy formation. Selection of processes and a representation for information must be accompanied by the formulation of a strategy for sequencing processes in the order they act on the representation. Ineffective sequencing of steps can result not only in wasted time and effort, but in a poor product. For example, students often try to write introductions to papers before their research is completed, figuring that although the research may affect the main body of the paper, it should have little or no effect on a section that merely describes the goals and motivations behind the paper. But as experienced authors know, goals and motivations often change as a project progresses, and sometimes the resultant paper isn't anything like the paper one originally intended to write.

Strategy formation is a key part of intelligence, a fact illustrated, like so many other facts, by the O.J. Simpson trial. The defense decided early on to rely heavily on a jury-selection consultant, who would help them decide which potential jurors would be likely to arrive at a "not guilty" verdict after the trial was over. The prosecution hired a jury consultant, but decided not to rely on him. This decision proved to be disastrous, especially because the prosecution's own consultant viewed the jury that was selected as heavily pro-defense. It is now a matter of history that the defense strategy paid off. One could argue, probably correctly, that the not-guilty verdict was in large measure determined before the trial even started.

Strategy formation in business is in large part what determines whether a business will succeed or fail. From the beginning, IBM and Apple decided on two completely different strategies with respect to their operating systems as means of maximizing profits. IBM went for an open system, which others could and did copy. Of course, this strategy encouraged clones, but it also resulted in enormous amounts of software being created for IBM and IBM compatible computers, with the result that IBM's fortunes became tied to the success of their marketing finesse, which was not always, perhaps, the greatest. But Apple's decision to go for a proprietary operating system and jealously guarded their rights to it. The result: relatively little software and a declining market share.

Bad strategies can be costly. Just recently, a trader in the United States for the Daiwa Bank made some bad decisions, and lost $400,000. He decided that he needed to rethink his strategies and gain back the money. Possibly in collusion with high officials of the bank, he kept secret his bad trades, and tried to recoup his losses. He never did, and ultimately he was discovered, by which time the bank's losses reached into the billions. Definitely bad strategy on the trader's part, not to mention the bank's.

5. Allocation of resources. Virtually all tasks can be allocated only limited amounts of time and other resources. An important decision to make in performing a task is how to allocate time to the various components in order to optimize performance. Poor time allocation may turn a potentially excellent product into a mediocre one. Students commonly allow insufficient time for actually writing a paper. They spend a great deal of time doing research, but then find they don't have

enough time to write the kind of paper they could, on the basis on their research, if they had more time. As a result, the final product does not well represent the work that went into it.

We have studied time allocation in the laboratory in order to study its role in intelligence (Sternberg, 1981). In one study, I had participants solve difficult verbal analogy problems that required substantial amounts of reasoning in order to complete them. Participants were timed while they solved the problems. I used mathematical-modeling techniques to extract two planning scores for each participant, called global planning and local planning. Global planning was up-front planning, the planning the participants did before they even started solving a given problem. Global planning is used to get the lay of the land, to get a sense of a problem before one actually starts solving it. Local planning was planning done in the course of problem solving. It is used to micromanage a task as one goes through it. I found that better reasoners generally allocated more time to global planning than did poorer reasoners: In other words, they actually spent more time on global planning than did the poorer reasoners. Poorer reasoners, in contrast, spent more time on local planning. In other words, the main difference between the better and poorer reasoners was not well captured simply by a score on a test, but by a qualitative difference in strategy, namely, how they allocated time between global and local planning.

In another study, Richard Wagner and I studied how people allocate time when they read. We had noticed what we believed to be a serious flaw in the kinds of tests used to measure reading comprehension, for example, on the SAT or any of the other tests that require such comprehension. Such tests seemed to us to give a relatively distorted view of what it means to be a good reader.

In a reading test, one reads all the passages for maximum comprehension, because one knows that the tester can and probably will ask almost anything about the passage one has read. There are usually some general questions following each reading passage, but also some picky detail questions. But in real life, whether in school, on the job, or in just reading for pleasure, one doesn't read this way. When you read a newspaper or a magazine in your free time, you rarely pay attention to details. When you read instructions for how to put a piece of equipment together, you may read the instructions and pay attention even to the small details. When you read as a student, you allocate your time as a function of how you will be tested: For example, you may read a history assignment differently depending on whether you will be tested, say, with general essays or a picky multiple-choice test.

So we had participants read different kinds of reading passages with different purposes in mind. Sometimes we had the participants read for the main idea, other times for the gist of the passage, still other times for details, and yet other times for analysis and evaluation. We then looked at how the participants allocated their time. We found that good readers allocated their time differentially, depending upon the purpose for which they were reading a given passage. Average and poor readers, however, either showed no differential time allocation across types of passages, or arbitrary time allocation. Their system did not correspond to the assignment they were given. In short, time allocation is a crucial part of good reading, and is not well measured by the reading tests.

Time allocation is also important on the job. Jimmy Carter, as President, was notorious for his inability or unwillingness to allocate his time effectively. He would get involved in minutiae and details that he would have better left to subordinates. But he seemed to be unable or unwilling adequately to delegate authority, and thus ended up tending to these kinds of trivia that would have best been left to others. At the same time, higher level decision making was often delayed or insufficiently thought through, as in the sunning failure of the rescue mission of soldiers trapped in Iran.

Carter's presidency is generally not considered to have been among the more successful ones. But as a post-presidential statesman, Carter has been considerably more successful, perhaps in part because Carter has been more of a one-man show, so that his allocations of resources have mattered less than they did during his presidency.

6. Solution monitoring. As individuals proceed through a task, they must keep track of what they have already done, what they are currently doing, and what remains to be done. They must also check that their skills have been applied to the task in a way that is bringing them closer to solution. In writing a research paper, it is important to keep track of sources that were already consulted, so as not to waste time reconsulting them. One further needs to keep track of what kinds of information have been collected and what kinds still need to be collected.

Monitoring lapses are responsible for some of the more grievous failures we encounter in our personal lives. In studying personal relationships, for example (Sternberg, 1986), I have been struck by the number of people who have reported that they now can see that what they find to be intolerable flaws in their partners were not hidden before they tied the knot. On the contrary, they were often quite noticeable, and other people may even have pointed them out. The evidence was there; they chose not to see it, or at best, they chose not to monitor what they were seeing.

For example, an individual may notice that violence or disloyalty is an entrenched aspect of their partner's story about relationships; yet they will fail to think through the implications of this fact for their long-term happiness in the relationship (Sternberg, 1995, 1996a). The result is that they get stuck living with it.

Monitoring failures are costly at the organizational level as well. Sometimes, the failures are so dramatic that they are hard to accept at face value. For example, the Central Intelligence Agency is supposed to specialize in monitoring information, including well-hidden information. This fact makes all the more awesome the agency's failure to detect the treason of Aldrich Ames, whose treachery was hardly concealed. Ames was living in a lifestyle that was publicly incommensurate with the salary he was taking home from the CIA. The agency's failure adequately to monitor this behavior, despite many clues to it, resulted in the worst scandal of betrayal in the agency's history.

7. Evaluating solutions. This step involves sensitivity to feedback and translating the feedback into an action plan. In performing a task, there are often various sources of internal and external feedback. Internal feedback derives from one's own perceptions of how well task performance is going, external feedback from other people's perceptions. Sensitivity to feedback is a major determinant of a person's potential to improve his or her work. This ability is probably at least as relevant for future as for present task performance.

Our educational system would improve immediately and substantially if evaluation were taken more seriously in our schools. Relatively few of the programs that schools implement are ever seriously evaluated; the large majority receive no formal evaluation at all (Sternberg & Bhana, 1986). Today, many school-reform efforts are being made, but if the past is any key to the future, few of them will be evaluated.

Evaluation is also important in one's personal life, of course. People are often loathe to evaluate things when their own personal interests may be threatened. For example, many people stay for long periods of time in failing relationships. After these relationships end, they report that during the relationship, they were reluctant to ever really seriously evaluate how the relationship was going, because they found doing so too threatening. At times they may have thought about the issue, but they dropped it quickly because it was too painful. Had they seriously evaluated the relationship earlier, they might have left a long time before they did. Of course, emotional as well as cognitive skills are needed adequately to make such an evaluation, and it is to such skills that I turn next.

Conclusion

Successful intelligence is one's ability to succeed in life, as measured either by one's own standards or the standards of others. Successful intelligence is much broader as a construct than is intelligence as it is traditionally defined. If what we care about is what people get out of life, and give to others through it, then we might choose to concentrate on successful intelligence, rather than the narrow kind of intelligence measured by conventional tests of abilities.

Author Note

The work reported herein was supported under the Javits Act program (Grant #R206R50001) as administered by the Office of Educational Research and Improvement, U.S. Department of Education. The findings and opinions expressed in this report do not reflect the positions or policies of the Office of Educational Research and Improvement or the U.S. Department of Education.

References

Asch, S. (1946). Forming impressions of personality. Journal of Abnormal and Social Psychology. 41, 258-290.

Butterfield, E.C., & Wambold, C., & Belmont, J.M. (1973). On the theory and practice of improving short-term memory. American Journal of Mental Deficiency, 77, 654-669.

Campione, J.C., & Brown, A.L. (1979). Toward a theory of intelligence: Contributions from research with retarded children. In R.J. Sternberg & D.K. Detterman (Eds.), Human intelligence: Perspectives on its theory and measurement. Norwood, NJ: Ablex.

Carroll, J. B. (1993). Human cognitive abilities. New York: Cambridge University Press.

Cole, M., Gay, J., Glick, J., & Sharp D. W. (1971). The cultural context of learning and thinking. New York: Basic Books.

Edgerton, R. (1967). The cloak of competence. Berkeley, CA: University of California Press.

Goleman, S. (1995). Emotional intelligence. Bantam, CT: Bantam.

Heath, S. B. (1983). Ways with words. New York: Cambridge University Press.

Hunt, E. B., (1978). Mechanics of verbal ability. Psychological Review, 85, 109-30.

Luce, R. D., & Raifa, H. (1957). Games and decisions. New York: Wiley.

Okagaki, L., & Sternberg, R. J. (1993). Parental beliefs and children's school performance. Child Development, 64(1), 36-56.

Pellegrino, J. W., & Glaser, R. (1980). Components of inductive reasoning. In R. E. Snow, P-A Federico, & W. E. Montague (Eds), Aptitude, learning, and instruction Vol (1): Cognitive process analyses of aptitude. Hillsdale, NJ: Lawrence Erlbaum Associates.

Sacks, O. (1985). The man who mistook his wife for a hat. New York: HarperCollins

Salovey, P., & Mayer, J. D. (1990). Emotional Intelligence. Imagination, Cognition and Personality, 9(3), 185-211.

Snow, R. E. (1979). Theory and method for research on aptitude processes. In R. J. Sternberg & D. K. Detterman (Eds.), Human intelligence: Perspectives on its theory and measurement (pp. 105-137). Norwood, NJ: Ablex.

Snow, C. E., Barnes, W. S., Chandler, J., Goodman, J. F., & Hemphill, L. (1991). Unfulfilled expectations: Home and school influences on literacy. Cambridge, MA: Harvard University Press.

Sternberg, R. J. (1981). Intelligence and nonentrenchment. Journal of Educational Psychology, 73, 1-16.

Sternberg, R. J. (1985). Beyond IQ: A triarchic theory of human intelligence. New York: Cambridge University Press.

Sternberg, R. J. (1985a). Implicit theories of intelligence, creativity, and wisdom. Journal of Personality and Social Psychology, 49, 607-627.

Sternberg, R. J. (1986). Intelligence applied: Understanding and increasing your intellectual skills. San Diego: Harcourt Brace Jovanovich.

Sternberg, R. J. (1987). Most vocabulary is learned from context. In M. G. McKeown & M. E. Curtis (Eds.), The nature of vocabulary acquisition (pp. 89-105). Hillsdale, NJ: Lawrence Erlbaum.

Sternberg, R. J. (1995). In search of the human mind. Orlando: Harcourt Brace College Publishers.

Sternberg, R. J. (1996). Successful intelligence. New York: Simon & Schuster.

Sternberg, R. J. (1996a). Love stories. Personal Relationships, 3, 1359-1379.

Sternberg, R. J. (1997). Thinking styles. New York: Cambridge University Press.

Sternberg, R. J., & Bhana, K. (1986). Synthesis of research on the effectiveness of intellectual skills programs: Snake-oil remedies or miracle cures? Educational Leadership, 44(2), 60-67.

Sternberg, R. J., & Grigorenko, E. L. (Eds.), (1997). Intelligence: Heredity and environment. New York: Cambridge University Press.

Yang, S. Y. & Sternberg, R. J. (1997). Taiwanese conceptions of intelligence. Unpublished manuscript, Yale University, New Haven, CT.

FOREWORD

Foreword

Ruminations on Sex and Death; Memory and Value

Karl H. Pribram
Center for Brain Research and Informational Sciences
Radford University
Peters Hall Laboratory/Adams Street
Radford, VA 24142
Telephone: 540-831-6108 / Fax: 540-831-6236
E-Mail: kpribram@runet.edu

When Appalachian V was over I felt deep disappointment: More that half the participants had addressed learning and memory -- not, as I had expected, a scientific analysis of how values become organized by the human brain. Over the year that I've been trying to formulate the *Proceedings* for publication, however, I have begun to wonder whether, perhaps, some deeper intuition, probably unconscious, had led these participants to do what they had done.

A clue as to what that intuition might be is given by Bruce MacLennon's introductory presentation. As he states, there is a critical connection between human values and the process of learning and remembering. We value food, shelter, comfort, excitement, challenge, companionship -- and we can find both biological roots and brain mnemic processes for all these evaluations.

At the Second Tucson Conference on Consciousness, the most telling paper in my own view was not presented by a scientist or philosopher -- but by a lawyer. He noted that should the reductionist program (espoused by the Churchlands, Dan Dennet and Francis Crick) succeed, the law would be in serious difficulty. If, indeed, rape and murder were "nothing but" brain processes, where would accountability have gone? My colleagues and I do not have this difficulty: the presentations that made up Appalachian III stated over and over that there is a category error involved in such reductionism. Scale (or level) of organization and the facts of emergence assure that the whole is not only greater than the sum of its parts but is different from such a sum.

Still, there is more to an answer regarding the dilemma of human accountability than the level (social and cultural) at which it operates. How can humans justify wars in the name of spiritual guidance, how do they justify bonus payments of millions of dollars to themselves, while putting their employees out of work? Just what <u>is</u> going on in their brains that allows justification?

One obvious answer is that humans learn and then remember (and act accordingly) what their culture has taught them. Thus, as intuited by the conferees of Appalachian V, learning and remembering are critical components of values -- and I have organized their presentations under the heading "Preferences."

But where do preferences originate? There is an abundance of evidence that the preferences we hold are rooted in collective memories of experiences of death. Without memory, we would not be concerned with dying. The death of intimates would fade and anticipation of our own death would not exist (Edelman's prescient book-title *The Remembered Present* would have been useful in making this point).

But death is biologically linked to another process which was called to our attention recently by the cloning of a sheep: sex. Cloning circumvents death. Sexual reproduction is responsible for death.

In traditional societies and our own mythology and religious practices, this juxtaposition of the sexual and the sacred has often been made explicit. The sacred deals with how we handle death; the sexual, how we circumvent it. The temples of the classical period usually dealt with both. In the Proceedings, I have grouped brain processes aimed at circumventing death under the rubric "Utility."

A. Earl Walker, head of neurosurgery at the University of Chicago when I was there, had made major contributions to our understanding of thalamocortical anatomy. His teaching to us was that if one wanted to know how a particular brain system is organized, first look at its input from (and output to) the periphery. Biological needs are surely important roots of values, and one set of brain systems, the frontolimbic formations, receive their input from -- and control the output to -- the visceral and autonomically regulated peripheral structures.

Autonomic regulation takes two forms, sympathetic and parasympathetic. Essentially, the sympathetic system activates emergency responses (fight and flight reactions) in the circumvention of death. The parasympathetic system regulates longer range circumvention by processes involved in daily maintenance and the maintenance of the species (feeding and sex).

Of interest is the fact that the parasympathetic system has two divisions -- a cranial and a sacral. Thus, when autonomic system involvement shifts from emergency responses to daily maintenance responses, sacral and cranial (brain) regulation takes over.

There is considerable misunderstanding regarding the nature of these maintenance functions. For instance, sexuality and feeding are usually conceived, in our society, as base, simple and mostly a function of our viscera. This is not so. It is the brain that regulates sexuality; gonadal hormones initiate, in embryo, the changes in the brain that are expressed after puberty. Expression of sexuality critically involves the cerebral cortex. A study by Frank Beach, using cats, showed that decortication totally abolished female sexual behavior. A parallel case can be made for the functions of the "olfactory" cortex in making life pleasant for gourmets. (Contrast this with the fact that both classical and operant conditioning survive decortication.)

As developed in my paper in this volume, maintenance is based on a set of processes that can be described in terms of their utility to the organism. Perceived utility leads to cultural preferences which, by way of learning and remembering provide guides to human social transactions. These, more often than not, lead to creative agency. As noted in the final passages of my 1971 book, *Languages of the Brain*:

> Man's brain is so constructed that piece by piece he apprehends the whole through the operations of coding and recoding. Languages, verbal (linguistic) and nonverbal (cultural), are constituted of these pieces. When, because of linguistic and cultural affluence, the means-ends reversal occurs, these languages begin to live lives of their own. Thus complexity is compounded and the original organization can easily be lost sight of. Biological processes have, however, built-in renewal mechanisms. When the linguistic and cultural structures become too cumbersome or conflict with each other, they are often degraded, pruned back to their more essential roots. Clearer vision is then attained of the basic organization which gave rise to the process originally; historical comparison can be made between the primitive and the sophisticated version of the language or culture.

Thus, gradually, wisdom is attained in the regulation of human affairs. In contrast to the cries of woe that are increasingly heard as we approach the new millennium, the biologist immersed in the study of brain process faces social issues hopefully. The power of this peculiar biological organ, the brain, especially in man, is only beginning to be fathomed. True, we must get on with the job before some of the cultural-language structures that have suffered the means-ends reversal overwhelm their creators. But the evidence suggests that remedial language-cultures will quickly be formed by those same sorts of brains that initiated the original. Thus wisdom is recurrently achieved. The biological process does not cease. Men's brains, through Image and Act, will create and communicate continuously, constructing languages – the regulators of human affairs. (384-385)

INTRODUCTION

1

Mixing Memory and Desire: Want and Will in Neural Modeling

Bruce J. MacLennan
Computer Science Department
University of Tennessee
Knoxville, TN 37996-1301
Telephone: 423-974-5067 / Fax: 423-974-4404
E-Mail: maclennan@cs.utk.edu

Abstract

Values are critical for intelligent behavior, since values determine interests, and interests determine relevance. Therefore we address relevance and its role in intelligent behavior in animals and machines. Animals avoid exhaustive enumeration of possibilities by focusing on relevant aspects of the environment, which emerge into the (cognitive) foreground, while suppressing irrelevant aspects, which submerge into the background. Nevertheless, the background is not invisible, and aspects of it can pop into the foreground if background processing deems them potentially relevant.

This illuminates the differences between representation in natural intelligence and (traditional) artificial intelligence. Traditionally artificial intelligence has started with simple, primitive features, and attempted to construct from them a representation of the environment. If too few features are used, then the processing is imprecise and crude. However, if sufficient features are used to permit precise processing in all contexts, then the system is defeated by the combinatorial explosion of features. In natural intelligence, in contrast, we begin with a nervous system that can process in real-time the "concrete space" represented by the interface between the animal's nervous system and its environment. The separation of foreground from background then serves to increase the efficiency of this process. Instead of trying to construct the concrete world from abstract predicates, the brain projects the very high-dimensional concrete world into lower dimensional subspaces; this projection is context-sensitive and rapidly adaptable. Therefore it is not vulnerable to the combinatorial explosion.

We consider the connection between these ideas and the concepts of intentionality, as discussed by Brentano and Husserl, and information, as quantified by Shannon and Weaver. In particular, the Shannon-Weaver measure ignores relevance, which is essential to biological information. Further, Brentano and Husserl characterized intentionality in terms of the "directedness of consciousness," which can be explained as a decrease in the entropy (disorder) in the probability of processing, which is produced by the separation of foreground from background.

Essential to these ideas are questions of how contexts are switched, which defines cognitive/behavioral episodes, and how new contexts are created, which allows the efficiency of foreground/background processing to be extended to new behaviors and cognitive domains.

Next we consider mathematical characterizations of the foreground/background distinction, which we treat as a dynamic separation of the concrete space into (approximately) orthogonal subspaces, which are processed differently. Background processing is characterized by large receptive fields which project into a space of relatively low dimension to accomplish rough categorization of a novel stimulus and its approximate location. Such background processing is partly innate and partly learned, and we discuss possible correlational (Hebbian) learning mechanisms.

Foreground processing is characterized by small receptive fields which project into a space of comparatively high dimension to accomplish precise categorization and localization of the stimuli relevant to the context. We also consider mathematical models of valences and affordances, which are an aspect of the foreground. Cells processing foreground information have no fixed meaning (i.e., their meaning is contextual), so it is necessary to explain how the processing accomplished by foreground neurons can be made relative to the context. Thus we consider the properties of several simple mathematical models of how the contextual representation controls foreground processing.

We show how simple correlational processes accomplish the contextual separation of foreground from background on the basis of differential reinforcement. That is, these processes account for the contextual separation of the concrete space into disjoint subspaces corresponding to the foreground and background.

Since an episode may comprise the activation of several contexts (at varying levels of activity) we consider models, suggested by quantum mechanics, of foreground processing in superposition. That is, the contextual state may be a weighted superposition of several pure contexts, with a corresponding superposition of the foreground representations and the processes operating on them. This leads us to a consideration of the nature and origin of contexts. Although some contexts are innate, many are learned. We discuss a mathematical model of contexts which allows a context to split into several contexts, agglutinate from several contexts, or to constellate out of relatively acontextual processing. Finally, we consider the acontextual processing which occurs when the current context is no longer relevant, and may trigger the switch to another context or the formation of a new context. We relate this to the situation known as "breakdown" in phenomenology.

1. Relevance and Intelligence

A sense of *relevance* is one of the central components of intelligence. As Dreyfus (1979; Dreyfus & Dreyfus 1986) has shown in his critiques of traditional approaches to artificial intelligence, experts have a well-tuned sense of relevance that saves them from having to consider irrelevant aspects of the situation at hand. For example, unlike the chess program, which must consider every possible move, if only to reject it on the basis of heuristics, the chess master sees only the relevant possibilities. Instead of wasting time on irrelevant possibilities, as the program does, the chess master can devote greater cognitive resources to possibilities that are genuinely important.

However, a well-developed sense of relevance is not peculiar to specialized activities, such as playing chess; rather it is characteristic of expert (i.e., skillful) behavior throughout the animal kingdom. The cat stalking a bird is guided by relevance no less than the chess expert stalking an opponent's king. Relevance then is a common denominator of intelligence, since it guides effective allocation of cognitive resources.

Nevertheless, until recently (e.g. Sperber & Wilson 1986), relevance has been the abandoned child of informatics. Indeed the measure of information enshrined in Shannon's information theory (Shannon 1948; Shannon & Weaver 1949) ignores relevance. As Hamming (1980, 103) has remarked, according to Shannon's measure, the book with the most information is a book of random numbers (since it is completely unpredictable and therefore has no redundancy). This paradox arises from the divergence between our ordinary understanding of information and Shannon's formalization of it: ordinarily, for something to be informative, it must inform us of something relevant (actually or potentially).

One cannot understand relevance without understanding *context*, for aspects of the environment that are relevant in one behavioral context are not in another. Therefore, a significant component of intelligence is being able to determine the appropriate context and to switch between contexts quickly. When an aggressive dog appears nearby, the intelligent cat will switch contexts, abandon its pursuit of the bird, and flee the potential predator. When a fire alarm sounds, the intelligent chess player abandons the game and evacuates with everyone else. Different aspects of the environment, such as "Exit" signs, will become salient, while previously relevant chess configurations are ignored.

The phenomenologist philosophers, such as Husserl and Heidegger, have done much to improve our understanding of relevance and context (see, e.g., Dreyfus 1982, 17-21; Dreyfus 1991, 118-20). For example, by stressing that we are "always already in a situation" (Dreyfus 1979, 53) they remind us that perception and cognition mediated by a context-dependent sense of relevance is the norm rather than an exception. Indeed, comparatively context-free cognition occurs only in "breakdown" situations (Dreyfus 1991, 70ff), when we cannot readily determine a context appropriate to the situation.

The role of relevance and context can be understood in terms of the information-theoretic concept of entropy. If we think of a probability distribution p_x over the myriad concrete components x of the situation, or equivalently, if p_x represents the relative allocation of cognitive resources to x, then the *entropy* of this distribution, $H\{p\}$, defined,

$$H\{p\} = - {}_x\, p_x \log p_x,$$

measures the "structure of relevance." The entropy is maximized (representing maximum disorder and minimum structure) when the distribution is uniform, that is, when all x get equal attention (as in context-free "breakdown"). The entropy is lower (representing greater structure) to the extent that attention is focused on some x to the exclusion of others. A lower entropy distribution gives a more definite content to consciousness, by directing it toward some aspects of the world at the expense of others (MacLennan 1988, 172-3). For the phenomenologists, such "directedness" is an essential characteristic of consciousness and intentionality (Brentano 1925; Husserl 1931, §84).

Based on the foregoing discussion we can understand the critical role played by values in intelligent behavior. The term "value" has many senses, even in philosophy (Frankena 1967). I will use *value* to refer to anything that an organism *values* (positively or negatively), that is, anything that it tends to seek or avoid. Thus, common positive values include food and sex, negative ones include pain and injury. Values define *potential interests*, which become *actual interests* in an appropriate context. Thus food is normally a potential interest until the animal becomes hungry, whereupon it becomes an actual interest. Context-sensitive processes in the animal's nervous system then direct its attention to aspects of the environment relevant to its current interests. Jung (1978, 32-3) has stressed the central role of value:

> The function of value — feeling — is an integral part of our conscious orientation and ought not to be missing in a psychological judgement of any scope, otherwise the model we are trying to build of the real process will be incomplete. Every psychic process has a value quality attached to it, namely its feeling-tone. This indicates the degree to which the subject is *affected* by the process or how much it means to him (in so far as the process reaches consciousness at all). It is through the "affect" that the whole subject becomes involved and so comes to feel the whole weight of reality.

In the following sections I will consider processes that might subserve contextual processing by identifying the appropriate context and using it to control the extraction and processing of relevant information.

2. Function of Contextual Processing

2.1. Function

In this paper I will often take a robotics perspective on the problems of relevance, context and values. That is, I will be asking the question: How *ought* contextual processing to operate in an intelligent autonomous robot? In this way, we will come to the biological functions of values and context from first principles, which will illuminate their role in natural animal intelligence. (See Kilmer, McCulloch & Blum 1969, especially §9, for an earlier parallel treatment of natural and artificial contextual processing systems.)

It appears to me that the principal role of contextual processing is the efficient allocation of neural resources (cognitive processing power). The nervous system is metabolically expensive, and so each neuron must earn its keep. Contextual processing is more efficient by devoting greater resources to relevant aspects of the world than to irrelevant aspects; allocation is *biased* toward relevant aspects. In this way, for fixed total resources, relevant aspects may be processed more extensively than would be possible with an unbiased allocation.

The disadvantage of contextual processing is a consequence of this same bias, for the implicit judgment of relevance is historical, deriving from the history of the individual and its species. Thus there is a danger that the organism or robot will be ill-equipped to handle the future, that it will be insensitive to previously irrelevant aspects that are now relevant, that it will be confined within the contextual boundaries of the past.

Therefore, if a robot or animal is to be adaptive in the face of a future that differs from the past, it cannot ignore irrelevant aspects of its environment; they must be processed, albeit with a lesser allocation of resources. Further, it must be capable of detecting situations that do not fit into known contexts, and be able to construct new contexts capturing the "structure of relevance" in such situations.

Since contextual processing focuses cognition on restricted aspects of the situation, to the partial neglect of other aspects, I will refer to the disadvantages of contextual processing as the problem of becoming "single-minded."

2.2. Means

We can identify a number of means by which contextual processing accomplishes its function. First, depending on the context, certain aspects of the situation are projected into the foreground, where more resources are devoted to them, while the remainder (or the whole situation) is projected into the background, where it is processed with a lower resource investment and in a relatively unbiased (context-free) way. That is, currently irrelevant aspects are in the background, but not "invisible." Further, the variable content of the foreground implies that the fixed neural resources devoted to foreground processing must be capable of performing different functions depending on the context.

Contextual processing, by directing attention toward some aspects, directs it away from others; therefore, management of the context becomes a critical task. First, it is important to recognize when a new context should be activated, either on the basis of foreground processing or, even more importantly, on the basis of background processing. This is, in effect, a categorization problem: classifying the current situation as a cognitive context. (This view is especially consistent with Ruth Garrett Millikan's characterization, in her Presidential Address to the American Philosophical Association, of categorization as a process of *reindentification*; see also Millikan 1984, ch. 16.)

Next, since several contexts may be demanding attention, it is important to have a representation of the urgency or importance of these competing demands, so that resources may be allocated most effectively. For example, hunger (for an animal) and low battery power (for a robot) are gradually changing conditions that we expect to make progressively more insistent demands; they can be set aside for a time, but eventually must be handled. (In psychological terms, this is the *dynamical* aspect of contextual processing.) Finally, for flexibility of behavior, it is important that both the "context triggers" and their dynamical strength be adaptable through learning.

Indeed, adaptiveness requires some means of assessing the success of contextual processing, which is provided by various kinds of reinforcement. Often this is the elimination of the conditions (e.g. hunger) that activated the context. However, just as the triggering conditions can be learned, so can the reinforcing conditions, which is important to the successful pursuit of long-term activities, for which direct (somatic) rewards may be delayed. Further, reinforcement must be contextual, otherwise it is difficult to relate it to its preconditions. This is easiest when the reinforcement occurs in the same *episode* (interval of constant context) that led to it, but often the reinforcement is delayed. In these cases the reinforcement must include information sufficient to cause reactivation of the context to which it applies. (Of course, I am not claiming these processes should be expected to work perfectly in either animal or machine; I am trying to lay down conditions that should be satisfied to have a higher probability of success.) One of primary functions of reinforcement is in determining relevance; aspects that have appeared in an episode leading to reinforcement will be *relevated* (raised in relevance, from Latin *relevare*, to lift up;

Pribram 1991, 230). Beyond this, reinforcement is used to tune foreground processing to increase the probability of achievement of positive reinforcement, or avoidance of negative reinforcement.

Finally, we must "expect the unexpected"; that is, there must be some means of handling situations that do not fit into a recognized context. This could be known context arising in a new way (i.e., without its known trigger conditions), or it could be an entirely new context. Without knowing the context, which tells us how to bias processing, the situation must be processed in a relatively unbiased way, which is less efficient, since it doesn't allow resources to be diverted from irrelevant aspects. Since this *acontextual processing mode* is relatively inefficient, it is important to activate an appropriate context as quickly as possible. Therefore, most of the cognitive effort will be devoted to either categorizing the novel situation as a known context, or in constructing a new context (a new contextual category) to accommodate it.

3. General Framework

Having outlined the general functions of contextual processing and the means to accomplish them, we can turn to a more specific model (Fig, 1).

3.1. Concretum

We begin with an abstraction, the *concretum*, which is defined as the space of all inputs to contextual processing, including all sensory and somatic inputs as well as feedback from other brain areas. More precisely, if we define contextual processing to include all the processes that are not context-free, then the concretum comprises all the neural processes on which contextual processing depends. In particular, it includes *protocritical signals* such as comfort/discomfort (Pribram 1991, 200-14). However, for the present discussion, it is not important to decide exactly what the concretum includes; it suffices to observe that it is of very high dimension.

I will use W to stand for the concretum, since it is, so far as the organism is concerned, the concrete *world*. Mathematically W is a set comprising all the possible states y of the concretum. The state $y(t)$ at a given time represents the instantaneous activity of the neural tissue corresponding to the concretum; we will generally think of this "instant" as having some finite duration, in accord with the Gabor Uncertainty Principle. That is, $y(t)$ is extended in time as well as space.

3.2. Primacy of the Concrete

It is necessary to say a few words about the "primacy of the concrete," a lesson we learn from phenomenology. Traditionally AI has tried to construct a representation of the world by a combination of context-free features. If too few features are used, the representation is crude and incomplete, but the use of an increasing number of features eventually leads to a combinatorial explosion of possibilities. The traditional AI program is ultimately defeated by the Scylla and Charybdis of either processing all the possibilities (context-free processing) or risking missing the unexpected (single-mindedness).

Biological intelligence does not have this shortcoming, and it is important to consider why. Biological intelligence begins with the concrete world with whatever resolution and dimensions it is represented by its nervous system. For example, the human brain has millions of sensory inputs, which suggests the dimension of our concrete world. For biological intelligence, we begin with the assumption that the organism can process its concretum in real time (for otherwise the species would not survive). Contextual processing improves the efficiency of this process by a context-dependent projection of some aspects of the concretum into a subspace where they can be processed with greater precision. Thus, instead of the *context-free features* of traditional AI, the nervous system improves its performance by means of *context-sensitive aspects*.

At root, we are really dealing with the gap between the continuous and the discrete. We can divide a continuum into smaller continua, some of which may be magnified; that is what biological intelligence accomplishes through context-sensitive projection of aspects of the concretum. On the other hand, a continuum cannot be constructed from any finite number of discrete elements; that is the impossibility that traditional AI tried to accomplish. As Zeno observed 2500 years ago in his *Dichotomy*, "There is no motion, because what moves must arrive at the middle of its course before it reaches the end" (Aris. *Phys.* 239b11). We apparently cannot get from A to B in any finite number of discrete steps, each of which goes half the remaining way. So also, no increasing number of more finely divided abstract context-free features will capture the concreteness of the world. Yet in everyday life we go from A to B all the time; continuous motion gets us from A to B and through all the points in between. A posteriori we can divide the motion into parts, but the continuous motion is primary, and its division into discrete parts secondary and approximate.

3.3. Context and the Foreground & Background Projections

First we postulate a space X of possible context representations c; I will consider later (section 5) what they may "look like." We postulate two additional spaces F and G representing possible states of the foreground and background, and corresponding context-dependent projections *f*: X ¥ *W* Æ F, *g*: X ¥ *W* Æ G (i.e., *figure* and *ground*). Thus, f = *f*(c, y) is the foreground representation in context c of the concrete state y, and g = *g*(c, y) is the corresponding background.[1] Note that *f* and *g* are "projections" in a general sense: so *g* represents the state of the concretum in a manner that is unbiased with respect to context, but optimized for the detection of potential context changes. Conversely, *f* projects the concrete state in a manner that is specialized to the context at that time; that is, it is tuned to information relevant to that context. This implies that neurons in the foreground projection may be used differently in different contexts, which is supported by evidence of place cells in rat hippocampus (O'Keefe 1986, 82-4; Pribram 1991, 233-4).

3.4. Control of Foreground Processing

Since the primary goal of contextual processing is to allocate neural resources to different processing tasks in different contexts, we define the foreground process to be a function of context as well as the foreground representation, *F*: X ¥ F Æ $_{F}$. The output space of foreground processing $_{F}$ will be left unspecified for the time being, but it must include all the outputs of attentive cognition, including those controlling motor behavior. Similarly, *G*: X ¥ G Æ $_{G}$ is the background process with its output in $_{G}$. Notice that the foreground and background processes *F*[c, *f*(c, y)] and *G*[c, *g*(c, y)] might be computed directly without explicit representation in the brain of the projections f = *f*(c, y) and g = *g*(c, y), although it is conceptually simpler to considerthem two-step processes.

3.5. Context Activation

As noted previously, the primary function of background processing is the switching of contexts, thus avoiding "single mindedness." However, attentive foreground processing can also detect conditions for the activation of contexts. Therefore we define an "arbitration function" *A*: $_{F}$ ¥ $_{G}$ Æ X which takes the results of foreground and background processing and determines the current context. This arbitration is based on the strength with which contexts are activated, which is one of the affective components of the outputs of both foreground and background contextual categorization processes. In our model, the context representations are normalized ||c|| = 1, so that the magnitude of the output *a*c of the contextual categorization process reflects the strength of its activation, ||*a*c|| = |*a*|.

3.6. Reinforcement & Adaptation

Finally, there are adaptive processes that, on the basis of reinforcement, adjust the foreground and (perhaps) background projections and foreground and background processing. This is because (1) the foreground aspects are relevated on the basis of their role in achieving positive reinforcement or in avoiding negative reinforcement, (2) foreground processing must be adapted to accomplish its purpose more effectively, and (3) background processing needs to adapt

[1] I think it is likely that the background projection is independent of the context, but the current formulation is more general and permits, for example, subtraction of the foreground information from the background.

to identify context switches more efficiently and reliably. This requires determining both the context to be activated and the strength of its activation.

Since they are adaptive, the foreground and background projections and processes are all repositories of "memory." This memory mediates the activation of contexts, in particular, the strength of their activation and so their potential for commanding the animal's behavior; it also mediates reinforcement, much of which is learned, though it be based on a protocritical core. For example, increasing hunger corresponds to increasing activation of a food-seeking context which causes restaurants, refrigerators, etc. to become more salient. Nevertheless, less tangible learned reinforcements, both current and future expected, may raise the activation of the doing-mathematics context, sufficient to keep us working on the derivation rather than heading for the refrigerator.

In effect, the input to the reinforcement process corresponds to the evaluative dimension (pleasant/unpleasant, etc., thus what it tends to seek or avoid), whereas the input to arbitrator corresponds to the intentional dimension (important/unimportant, thus what grabs the animal's attention). In terms adapted from Miller, Galanter & Pribram (1960, 62), we can say that memory — in the broad sense, both individual and species (cf. Jung 1978, 27-9) — is used to evaluate the current image, which in turn activates a context by which current value may be improved. *Motive* results from a combination of *value* (implicit in the evaluation) and *intention* (implicit in the context).

4. Contextual Processing

There are several ways that context can control foreground processing, thereby allowing neural resources to be used differently in different contexts. I will describe two, both based on bilinear (product forming) synapses.

4.1. Inner-product Model

Consider a single neuron in foreground processing and the inner product its dendritic net forms with the foreground representation f (including here the results computed by other foreground processing neurons) before it is passed through any nonlinear activation function. For contextual processing, we would like the possibility of a different inner product ·f, p_kÒ = $f^T p_k$ in each context c_k. (Recall that fields such as f and c_k. are taken to be temporally as well as spatially extended, therefore the *characteristic pattern* p_k represents the spatiotemporal pattern to which the neuron is tuned by the context. Further, apparent matrix expressions such as $f^T p_k$ involve spatiotemporal inner products.) This context-dependent process is accomplished by the bilinear foreground computation $F(c, f) = f^T Mc$ where $M = \ _j \ p_j c_j^T$ (a linear superposition of spatiotemporal outer products). The inner product will be exact if contextual representations are orthonormal; it will be inexact to the extent they are not. Reinforcement can lead to this kind of contextual processing through simple correlational learning, for example,

$$M \ \mu \ \pm fc^T.$$

(The sign of the change is + or - depending on whether the reinforcement is positive or negative.)

4.2. Correlation Model

Pribram (1991, 233-8) reviews O'Keefe's (1986, 82-4) observation that hippocampal function resembles holographic reconstruction with the theta-rhythm serving as a reference beam. Therefore I will briefly consider how the context can control foreground processing through a holographic (i.e. convolutional or correlational) mechanism.

To begin, we define the correlation c*y of two spatiotemporally extended fields, c and y. The j-th spatial component of the correlation at time t is given by a superposition of temporal correlations:

$$(c * y)_j(t) = \sum_k (c_{k-j} * y_k)(t).$$

These correlations can be computed by linear dendritic nets that have impulse responses $h_k(t) = \mathrm{y}_k(-t)$. Expanding the temporal correlations by integration over the temporal extension T, we get:

$$(\mathrm{c} * \mathrm{y})_j(t) \;=\; {}_k\, \acute{\mathrm{U}}_T\, \mathrm{c}_{k\text{-}j}(s\text{-}t)\;\; \mathrm{y}_k(s)\; \mathrm{d}s.$$

Again, we would like, in the ideal case, the linear input to a foreground neuron to be ·f, $\mathbf{p}_j$Ò in context c_j. If the "hologram" h = $_k\,\mathrm{c}_k * \mathbf{p}_k$ is a superposition of correlations between contexts and patterns, and if p = c_j * f is the correlation of context c_j with input f, then we can show that

$$\cdot\mathrm{h},\, \mathrm{p}\grave{\mathrm{O}} \;\; \mathrm{e} \;\; \cdot\mathrm{a}_j\, f\, \mathbf{p}_j,\, \mathrm{f}\grave{\mathrm{O}} \;+\;\; {}_{j\,k}\cdot\mathrm{k}_{jk} f\, \mathbf{p}_k,\, \mathrm{f}\grave{\mathrm{O}},$$

where f stands for convolution, $\mathrm{a}_j = \mathrm{c}_j * \mathrm{c}_j$ is the autocorrelation of context c_j, and $\mathrm{k}_{jk} = \mathrm{c}_j * \mathrm{c}_k$ is the cross-correlation of contexts c_j and c_k. The effect of the first convolution is to "blur" the characteristic pattern $\mathbf{p}_j$ by the autocorrelation field a_j. In the ideal case, a_j = d (a Dirac delta function), $\mathrm{k}_{jk} = 0$, and we have exact selection of the foreground computation, ·h, pÒ e ·f, $\mathbf{p}_j$Ò. The inner product can be computed in two different ways:

$$\cdot\mathrm{h},\, \mathrm{p}\grave{\mathrm{O}} \;\; \mathrm{e} \;\; \cdot\mathrm{h},\, \mathrm{c} * \mathrm{f}\grave{\mathrm{O}} \;\; \mathrm{e} \;\; \cdot\mathrm{c} * \mathrm{h},\, \mathrm{f}\grave{\mathrm{O}}.$$

In the first case, ·h, c * fÒ, the context and input are allowed to interfere spatiotemporally, and an inner product is formed with the "hologram" h. In the second case, ·c * h, fÒ, the context is allowed to interfere spatiotemporally with the "hologram," and an inner product is formed between the result and the input. The "hologram" results from positive or negative reinforcement of the interference pattern (correlation) between the current context and the reinforced input,

$$\mathrm{h} \;\; \mu \;\; \pm \mathrm{c} * \mathrm{f}.$$

4.3. The Foreground Projection

Pribram (1991, Apps. C - F) defines a context as a complete orthonormal system (CONS) defined by the eigenvectors of the dendritic microprocesses defined by the neural wave equation; they are the stable states that constitute an episode. In effect, the context, as CONS, defines the projectionof the state onto a set of axes represented by the eigenvectors. Our model includes Pribram's as a special case by abstracting away from the details of the neural wave equation (and by taking what amounts to time-windowed or wavelet view).

In our model the foreground projection, f = f(c, y), is controlled in the same ways as foreground processing. If we take projection to be linear process, then $\mathrm{f}_i = \mathrm{y}^{\mathrm{T}} \mathrm{M}_i \mathrm{c}$ for the inner-product encoding, and f_i = ·h_i, c * yÒ e ·c * h_i, yÒ for the correlation encoding. Based on these equations we can define the impulse response or characteristic pattern of the dendritic net of a foreground neuron, $h_i = \mathrm{M}_i\,\mathrm{c}$ or $h_i = \mathrm{c} * \mathrm{h}_i$, respectively. Thus, keyed by the context, each foreground neuron f_i receives some linear combination of spatiotemporal "microfeatures" (wavelets) from the concretum. If the neuron is unused in that context (i.e. is not a part of a foreground aspect in that context) then its characteristic pattern is the zero field; that is, $h_i = \mathbf{0}$.

Adaptation also takes place in the same ways as in foreground processing, $\mathrm{M}_i \;\mu\; \pm \mathrm{yc}^{\mathrm{T}}$ or $\mathrm{h}_i \;\mu\; \pm \mathrm{c} * \mathrm{y}$. In this case positive or negative reinforcement is strengthening a particular projection into foreground dimension i, up to some maximum. This strength, which corresponds to the relevance of a contextual aspect, can be measured by the norm of the characteristic pattern, $\|h_i\|$. The entropy of a context can be defined in terms of the norm of the total of the characteristic patterns of all foreground neurons, $h^* = \|\,{}_i\;\; h_i\|$. Thus, $h^* = \|(\,{}_i\, \mathrm{M}\;{}_i)\mathrm{c}\|$ or $h^* = \|\mathrm{c} * (\,{}_i\, \mathrm{h}_i)\|$, respectively. The entropy of context c is

$$H_{\mathrm{c}} = (\textstyle\sum_k h^*_k \log h^*_k) \;/\; (\textstyle\sum_k h^*_k)$$

It seems likely to me that relevation is primarily a one-way process; that is, once reinforcement has made an aspect relevant in some context, there is little if anything that can make it irrelevant. That is, *delevation* (from Latin *delevare*, to smooth out) does not take place (or takes place extremely slowly). (From a robotics standpoint, this is what we would want, because relevant aspects need not occur in a every instance of a context.) Delevation is different from an animal learning to ignore an unreinforced foreground aspect (which is still relevant, in spite of being unreinforced). Relevation is adaptation of the foreground projection f; learning to ignore an unreinforced (yet relevant) stimulus occurs through adaptation of the foreground process F.[2]

5. Representation of Contexts

The preceding discussion shows that contextual representations work best if (in the inner-product case) they are orthonormal, or if (in the correlational case) they are strongly autocorrelated (i.e. they are aperiodic) and weakly cross-correlated. In this section I will consider in more detail the implications of orthonormality and its lack.

5.1. Approximate Discreteness

To a first approximation, contexts form a discrete set. For example, it has been estimated that a vertebrate animal has a couple dozen "behavior modes" (contexts), which are discrete and mutually incompatible (Kilmer & al. 1969). Typical behavior modes include eating, drinking, hunting, exploring, fighting, fleeing, mating, nest building, grooming, sleeping, and so forth. There are several reasons, however, for considering this discreteness to be only approximate.

First, we must say what it means, in mathematical terms, for a set to be discrete. A set is discrete (or has the discrete topology) if there are only two possible distances between the elements of the set: they have a distance of 0 if they are the same element, and they have a distance of 1 if they are not the same. That is,

$$d(x, x) = 0, \quad \text{and} \quad d(x, y) = 1 \text{ if } x \neq y.$$

It is easy to show that a set of orthonormal vectors forms a discrete set under the inner-product metric (the metric defined by the L_2 norm):

$$d(u, v) = \| u - v \| / \sqrt{2}$$

$$= \sqrt{[(u - v)^2 / 2]}$$
$$= \sqrt{[\, \|u\|^2 + \|v\|^2 - \langle u, v \rangle \,]}$$
$$= \sqrt{[\, 1 - \langle u, v \rangle \,]}$$
$$= 0 \text{ if } u = v, \text{ or } = 1 \text{ if } u \neq v.$$

Therefore, the set of contexts will be discrete to the extent that the context representations c are orthonormal. This is biologically unrealistic, so in practice we expect the contexts to form "clusters" or "constellations" with some overlap between them.

5.2. Innate and Learned Contexts

Although some vertebrates may have a fixed set of 25 or so behavioral modes, this is certainly not the case for human beings; we acquire many new contexts, such as playing chess, doing mathematics, and delivering papers at conferences. Like innate contexts, learned contexts define a space of relevant aspects of the world. We have already considered how, given an orthonormal representation for a context, the corresponding foreground projection and process can be learned,
but we have not addressed the creation of a contextual representation.

---

[2] Delevation probably takes place during the creation of a context from the unbiased "breakdown" context, since unreinforced aspects must be suppressed and left to background processing.

5.3. Creation of Contexts

Representations for (approximately discrete) contexts can be created by a variety of orthogonalization processes. The existing context representations $c_1, \ldots, c_n$ span a space, and a representation for a new context c_{n+1} can be chosen from the orthogonal complement of that space. Recalling that contextual categorization is a problem of reidentifying a relevance structure, we can see that the creation of a contextual representation requires correlating the "trigger conditions" that will activate the context (as determined by background processing or prior foreground processing) with elements of the concretum that have emerged as potentially relevant. Since the trigger conditions are a priori relevant, they form a kernel around which a relevance structure can develop. Furthermore, new contexts often develop out of "breakdown" situations in which processing is acontextual, that is, attentive but unbiased. Reinforcement in this mode allows a tentative separation of the relevant from the irrelevant.

Suppose therefore that in acontextual processing, the activity of certain foreground neurons $f_{r1}, \ldots, f_{rm}$ has been reinforced, while others $f_{q1}, \ldots, f_{qn}$ have not. We want to create a contextual representation c that relevates $r_1, \ldots, r_n$ but delevates $q_1, \ldots, q_n$. That is, we want $y^T M_{ri}\, c = 1$ and $y^T M_{qj}\, c = 0$. This can be accomplished by picking a (normalized) c that is in the subspace spanned by $M_{ri}{}^T y$ but orthogonal to the subspace spanned by the $M_{qj}{}^T y$. Further relevation can then take place by adjusting the M_i on the basis of reinforced foreground processing in context c.

5.4. Superposition of Contexts

Introspection suggests that at any given time we can be operating in a limited number of compatible contexts. For example, we can do a mathematical derivation or plan a lecture while we are driving; we can look for a restaurant or a mailbox at the same time. The model of contextual processing that we have presented is easily extended to accommodate "parallel processing" in multiple contexts. Much as a quantum-mechanical system can be in a linear superposition of pure states, so an organism can be in a superposition of contexts. If we let a_k represent the strength of activation of context k, then the effective context is a linear superposition $c = {}_k a_k c_k$.

Since the foreground projection, as we have defined it, is linear, the foregrounds will also exist in superposition. Of course, they could interfere with each other, but the lack of such interference is part of what makes contexts compatible and therefore amenable to parallel processing in superposition. For example, much of the visual processing involved in the looking-for-restaurant context is compatible with that in the looking-for-mailbox context. Similarly, since the dendritic processes are linear, a superposition of contexts will create a corresponding superposition of dendritic states in foreground processing. Again, to the extent that the processes do not interfere, they can proceed in parallel. This will depend on the extent to which they have compatible, or at least relatively orthogonal, foreground projections.

It seems likely that the interference must limit the extent of parallel processing in superposition. Given a certain level of primary activation by "trigger conditions," competitive or other nonlinear processes may ensure that one context is strongly activated, and at most a couple others are at most weakly activated. In that way, the "noise" contributed by less important contexts would be kept at a manageable level. Furthermore, overt behavior would, for the most part, be under the control of a single context, thus giving the illusion of "mutually incompatible modes of vertebrate behavior" (Kilmer & al. 1969, 279).

References

Brentano, F. (1925). *Psychologie von empirischen Standpunkt.* Meiner.

Dreyfus, H. L. (1979). *What computers can't do.* New York: Harper & Row.

Dreyfus, H. L. (Ed.) (1982). *Husserl, intentionality and cognitive science.* Cambridge: MIT Press.

Dreyfus, H. L. (1991). *Being-in-the-world: A commentary on Heidegger's* Being and Time, *division I.* Cambridge: MIT Press.

Dreyfus, H. L., & Dreyfus, S. E. (1986). *Mind over machine: The power of human intuition and expertise in the era of the computer.* New York: The Free Press.
Frankena, W. K. (1967). Value and valuation. In P. Edwards (Ed.), *Encyclopedia of Philosophy* (Vol. 8, pp. 229-32). New York: Macmillan Publ. Co.
Hamming, R. W. (1980). *Coding and information theory.* Englewood-Cliffs: Prentice-Hall.
Husserl, E. (1931). *Ideas* (transl. by W. R. B. Gibson of *Ideen zu einer reinen Phänomenologie und phänomenologischen Philosophie*, 1913). Allen & Unwin, Macmillan.
Jung, C. G. (1978). *Aion: Researches into the phenomenology of the self*; Collected Works, Vol. 9, Pt. II, 2nd ed. (transl. R. F. C. Hull). Princeton: Princeton Univ. Press.
Kilmer, W. L., McCulloch, W. S., & Blum, J. (1969). A model of the vertebrate central command system. *International Journ. Man-Machine Studies* **1**: 279-309.
MacLennan, B. J. (1988). Logic for the new AI. In James H. Fetzer (Ed.), *Aspects of artificial intelligence* (pp. 163-92). Kluwer Academic Publishers.
Miller, G. A., Galanter, E., & Pribram, K. H. (1960). *Plans and the structure of behavior.* New York: Holt, Rinehart & Winston.
Millikan, R. G. (1984). *Language, thought, and other biological categories: New foundations for realism.* Cambridge: MIT Press.
O'Keefe, J. (1986). Is consciousness the gateway to the hippocampal cognitive map? A speculative essay on the neural basis of mind. In David A. Oakley (Ed.), *Brain and mind* (pp. 59-98). London & New York: Methuen.
Pribram, K. H. (1991). *Brain and perception: Holonomy and structure in figural processing.* Hillsdale: Lawrence Erlbaum.
Shannon, C. E. (1948). The mathematical theory of communication. *The Bell System Technical Journal* **27**: 379-423.
Shannon, C. E., & Weaver, W. (1949). *The mathematical theory of communication.* Urbana: Univ. of Illinois Press.
Sperber, D., & Wilson, D. (1986). *Relevance: Communication and cognition.* Cambridge: Harvard Univ. Press.

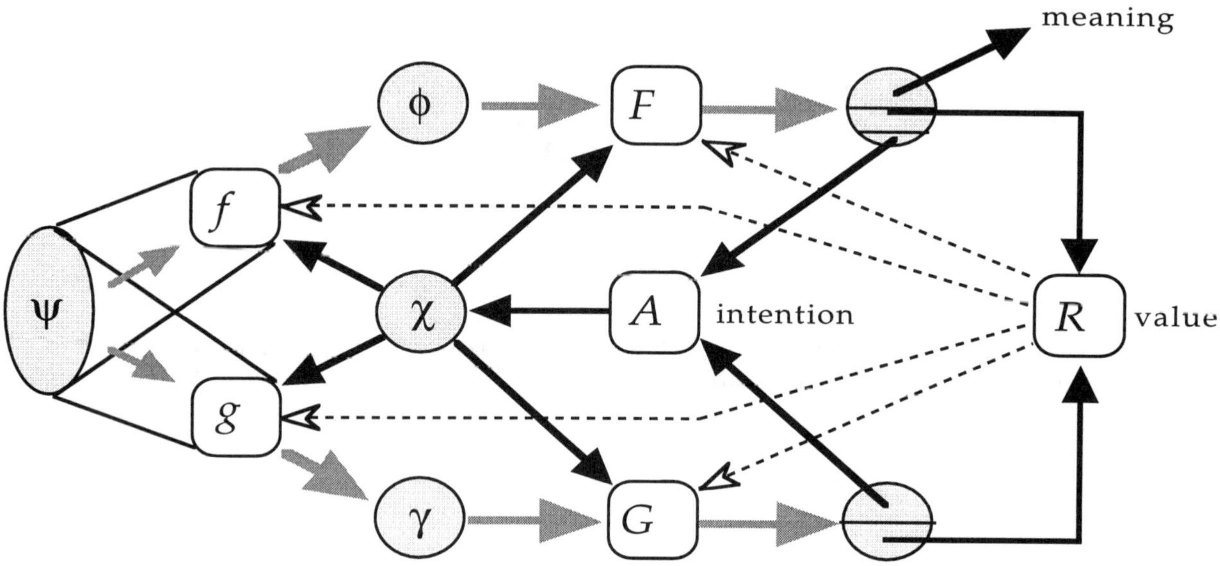

Figure 1

2

On Brain and Value: Utility, Preference, Play and Creativity

Karl H. Pribram
Center for Brain Research and Informational Sciences
Radford University
Box 6977 - R U Station
Radford, VA 24142
Telephone: 540-831-6108 / Fax: 540-831-6236
E-Mail: kpribram@runet.edu
July 28, 1997

Abstract

At the Fifteenth International Congress of Psychology held in Brussels in 1957, I presented evidence from primate studies regarding the biological determinants of values. I stated then:

> "the empirical relations that determine the value of a piece of currency depend, in part, on the utility of that piece of currency for any individual. The currency used in the primate neurobehavioral experiments reported was a food pellet or peanut. Two interrelated classes of variables have been abstracted by economists to determine utility: demand and expectation; two similar classes (need and probability distribution) can be delineated from the experiments reported here -- each of the classes related to a distinct neural mechanism. A still different neural mechanism has been delineated whereby (preferences among) values can be discriminated." (Pribram, 1957, p. 82)

What follows here is an updated version of these findings and analyses, which provide the prologue to and organizational framework for the Proceedings of the Fifth Appalachian Conference on Behavioral Neurodynamics.

Familiarity

Take an often repeated experiment. Five hundred pictures are displayed for observation, then mixed with 500 others portraying somewhat similar subjects. Next the full set of 1000 pictures is displayed, and the observer has only to indicate which pictures are familiar and which are unfamiliar. Most of us perform this task with a remarkable 90% or better score.

Next we encounter a patient who experiences bizarre (to her) feelings of familiarity in places she is sure she has never been. Another patient complains that he has just the opposite experience: He comes home only to feel it, on occasion, to be strange and unfamiliar to him. These feelings of "deja" (already) and "jamais (never) vu" (seen), as they are called, are related to epileptic electrical discharges recorded from the region of the amygdala, an almond shaped basal ganglion (the Greek for almond is amygdala) central to the functions of the limbic systems (limbus means border -- the limbic portions lie on the medial border of the cerebral hemispheres).

Such epileptic seizures, when they are prolonged, can prevent a whole episode of experience from becoming a part of the familiar narrative by which we experience our continuity as selves. Similarly, when the amygdala of both hemispheres of the brain are removed, monkeys no longer perform adequately on tasks comparable to those which test for familiarization in humans (Douglas & Pribram, 1966; 1969; Pribram, Douglas & Pribram, 1969).

Familiarization is an odd process. Confronted by a novel event organisms manifest an orienting reaction. When such an event recurs repeatedly, the orienting reaction subsides (habituates) after some 3-10 repetitions. After removal of the amygdala, however, the orienting reaction continues and continues and continues. Familiarization fails to occur.

The orienting reaction has, in addition to obvious behavioral orienting toward the novel stimulus, visceral components. Heart rate, blood pressure, and respiratory changes occur, as well as changes in brain electrical recordings. Such changes are produced by electrical stimulation of the amygdala and the cortical systems related to it (Kaada, Pribram & Epstein, 1949; Pribram, Lennox & Dunsmore, 1950). After amygdalectomy the visceral components of orienting no longer occur. Neither does familiarization: as noted, behavioral orienting continues (Pribram, Reitz, McNeil & Spevack, 1979).

From these observations we may conclude that familiarization depends on the occurrence of visceral reactions which are processed by the amygdala and related limbic systems of the brain. Further, on the basis of this and other evidence, we can conclude that each familiarization consists of an episode initiated by an orienting reaction to a novel event and terminated by the next experienced novel event.

Novelty as Rearrangement

What then might be experienced as a novel event? Again, experiencing novelty is an odd process. Novelty turns out to be experienced when a rearrangement of, or minor change in, the familiar takes place. Productive invention is based on a thorough familiarity with the inventory at hand. Reinventing the wheel is not innovative. A literary novel is composed of nuances, repeated small changes in the familiar which carry the plot gradually forward. If novelty were entirely unexpected, *Reader's Digest* would not be able to delete whole sections of the manuscript.

An experiment performed in Belgium (Smets, 1973) makes the point succinctly. Humans were shown two types of changes in displays while changes in visceral measurements were recorded. One change in display radically increased or decreased the amount of information; the other merely rearranged the items in the display. Practically no visceral reactions were recorded when the amount of information was changed. By contrast, rearrangement produced marked visceral responses. It is these repeated rearrangements of the familiar that elicit such strong "gut reactions" from music as well. (Pribram, 1982; Pribram et. al, 1966).

Thus, the experience of novelty is produced by rearrangement and change of the familiar. In turn, however, the experiencing of the familiar depends on novelty: Without novelty, the familiar recedes from consciousness. We walk through doors, giving the surrounding walls nary a glance -- unless an earthquake shakes up the familiar into novel configurations.

In more ordinary circumstances there are occasions which also demand this shaking up of the familiar in order to invent, to create novel configurations in our conscious awareness. Fortunately there is another part of the limbic brain, the hippocampus, lying just behind the amygdala, which makes such internal shake-ups possible. Damage to this sea-horse-like structure (hippocampus is Greek for seahorse) in humans makes it impossible for them to resist a practically total wipe out of current experience every time a novel (distracting) event occurs (Pribram 1986; 1991, Lecture 9)

Events as Consequences

Understanding novelty as a change in the familiar is the key to understanding values. Evaluation of an "event" or "eventuality" is achieved by activating (dishabituating) a previously familiar, tacitly known (see Polyani, 1960) aspect of a situation. The novel <u>event</u> is the <u>outcome</u>, the <u>consequence</u> of the process (Bateson, 1976; Konrad & Bagshaw, 1970). In fact the word event is the derived Latin equivalent of outcome: ex (out) - venire (come).

How are events, the consequences of habituation and dishabituation of the novel and familiar achieved? Let us first examine a biological paradox. Our respiratory mechanism is based on the ability of our tissues to metabolize food in the presence of oxygen. Without oxygen we quickly cease to live. However, the neural control over respiration is practically insensitive to oxygen deprivation -- lack of oxygen is reflected only minimally, if at all, in a change in respiratory rate, discomfort or alarm. By contrast, very minute changes in the partial pressure of carbon dioxide -- the end product (with water) of our respiratory metabolism -- are sensed by cells in the brainstem which, when carbon dioxide accumulates, initiate increases in respiratory rate, feelings of discomfort and even alarm.

This illustration is but one example of the fact that biological control mechanisms are often indirect and circuitous. Sexual selection takes place for the pleasure it provides but assures diversification and survival for the species; food is chosen on the basis of taste but accomplishes the maintenance of energy sources for the organism; clothes are donned for adornment but protect against weather, etc. The immediate control over behavior is exerted by a mechanism that is remote from the biological significance that the behavior entails.

Utility and Futility

There is evidence from the results of brain research that this universal biological paradox applies to the effects of the consequences of behavior on the organization of human values as well. Some fifty years ago I noted that monkeys whose brains had been subjected to removal of the amygdala would place all sorts of objects in their mouths, chew on them and if they were chewable would swallow them (Fulton, Pribram, Stevenson & Wall , 1949; Pribram & Bagshaw, 1953). We first wondered if the animal's taste mechanism had been disturbed and showed that their primary taste preferences were intact and that the area of the brain responsible for primary gustatory sensation was located elsewhere (Bagshaw & Pribram, 1953). We next wondered whether some higher order system of preferences had been disrupted by the resections of the amygdala -- that "good taste" had been abolished and gourmet had been turned into gourmand. But careful testing showed that the order of choosing food and non-food objects had not, in fact, been altered. What was preferred before the brain operation continued to be preferred in just the same order following the operation.

Preferences show transitivity -- they are hierarchically arranged -- utilities are not. One can prefer vegetables to meat, but it is difficult to choose one child's wellbeing over another's. For the lesioned monkeys the cutoff point for continuing eating was changed. The monkeys with brain lesions accepted a much wider range of objects before they stopped eating (Wilson, 1959; Weiskrantz

& Wilson, 1958). More elegant experiments demonstrated that chemical stimulations of this part of the brain (the amygdala) though they would not initiate behavior, would regulate the amount of ongoing eating and drinking, the amount varying precisely with the volume of chemical injected (Grossman, 1967; Russell, Singer, Flanagan, Stone & Russell, 1968). In short, this part of the brain quantitatively controls the consequences of behaviors that satisfy one or another biological need once these behaviors are initiated.

The brain processes directly in control of need satisfaction (see Livesey, 1986 for a comprehensive review) lend themselves to classification into three major categories (Pribram & McGuinness, 1975; 1992). The processes just discussed that stop behavior and regulate the amount of the consequential behavior constitutes but one of these categories. Another process (centered on the basal ganglia of the forebrain -- see Pribram & McGuinness, 1975 for review) readies the organism to behave in a certain manner and <u>initiates</u> the consequential behavior. A third process (based on the hippocampal formation), <u>coordinates</u> these stop (familiarization), and go (readiness), processes into a smoothly functioning system to operate within certain limits of tolerance. Coordination takes effort and shifts the type of control from a closed loop feedback homeostatic to a helical open loop feed-forward process. I have detailed elsewhere (Pribram, 1971, 1975a, 1992, 1997) the evidence that feed-forward processes are the basis for intentional behaviors and for that special human ability philosophers call "intentionality". (Intentionality stands in relation to perception as intention stands to behavioral performance. Both intentionality and intention share the characteristic that they are <u>about</u> the external objective world but need not be realized in it.)

When effort reaches limits of tolerance, coordination breaks down and the organism reverts to homeostatic control. Under such circumstances homeostasis may also fail to work efficiently (Ashby 1960), with the result either that the organism experiences, a) upset due to loss of control, or b) obsession, compulsion or boredom due to attempted overcontrol. The management of upset and of overcontrol are continuously faced in the psychiatric clinic and in everyday life by clinical psychologists. Understanding can be framed within the "utility" theory of economic behavior and "futility" theory to deal with the "games people play" such as that proposed by Berne (1961) to deal with upset, compulsion and boredom.

Some Economic Theory

In their influential volume on the theory of games and economic behavior, Von Neumann and Morgenstern outline the elements of a quantitative approach to the problem of value based choices that regulate competitive behavior (Von Neumann & Morgenstern, 1953). In common with other economic theorists, they distinguish behavior that 1) meets the needs and desires of the individual; 2) his estimate of the probability that the occurrence of the need-satisfying behavior will in fact meet a particular need; and 3) the construction of a preference hierarchy from 1 & 2. Needs and desires are measurable in terms of the amount of behavior entailed in their satisfaction, provided a zero point or anchor and a scaling of increments of that behavior become available. Von Neuman and Morgenstern use the illustration of measuring temperature. A zero point must be chosen (e.g., the freezing point of water) and a scale (e.g., adding a portion of a quantity of heat to raise the temperature to the boiling point) developed. Measurement is relative to the zero and scale chosen unless some absolute zero is discerned and agreed upon.

Von Neuman and Morgenstern's theory reflects some of the brain biological facts outlined in the previous section. The experimental analysis of the mathematical principles regarding reinforcing, i.e. consequential events: How these two factors operate and interact was shown by Peter Killeen in his presentation at the Fourth Appalachian Conference on Behavioral Neurodynamics in 1996 and in his earlier presentation (1994) in Behavioral and Brain Sciences:

1) Needs: The evidence for a measurable, filtratable process regulating needs is provided by the carefully chemically titrated satiety process delineated by injections into the amygdala noted above (Russel, Singer, Flanagen, Stone & Russel, 1968).

2) Probabilities: The evidence for a process that estimates the probability that the consequences of behavior will provide satisfaction of a biological need comes from studies comparing the results of food deprivation with those of produced by frontal lobe resection. Using the fixed-interval operant conditioning technique, I showed that monkeys deprived of food would change the rate of responding without changing the distribution of responses across the interval. By contrast, monkeys with resections of the most forward portion of their frontal lobes would fail to distribute their responses probablistically, whereas intact monkeys would (Pribram, 1961).

In terms of human endeavor, we have all observed that students tend to distribute their work between examinations so that maximum activity takes place just before the critical moment when tests are given. We urge them to plan differently -- but the experimental result obtained with monkeys described above suggests that if they were deprived of their frontal cortex, the distribution of activity would cease and the students would come completely unprepared for "the moment of truth". In fact, patients with frontal lobe tumors or excision behave in just this fashion -- unable to plan, to distribute their responses according to an estimate of the probabilities that the responses will be effective.[1]

Games and Play

3) Preferences: But life is not a game, especially not a zero-sum game. Humans are creative and find ever greater opportunities for making their behavior eventful and consequential. Witness Hong Kong, a rocky island (and small peninsula) that has become one of the great ports and trading centers of the world through sheer innovative effort. Life is open-ended, akin to a play: Shakespeare stated it so clearly: "and all the world's a stage".

Games are closed systems where control is exercised by way of feedback processes called homeostasis that operate much as do thermostats. Over the past decades, investigators (Ashby, 1960; Mackay, 1962; Mittelstaedt, 1968; Pribram, 1971 Chapter 5; Pribram & Gill, 1976, Chapter 1) have begun to distinguish between feedback organizations that are error-sensitive (such as simple thermostats) and those whose settings can be controlled. In a thermostat a small wheel is usually attached which alters the gap between two pieces of metal which, when they touch, close the switch that turns off the heat source. Heating the metal also closes the gap -- thus the critical distance between the pieces of metal is controlled by two sources operating in parallel, heat and the dial setting. The dial provides a bias on the setting, thus a range of settings can be programmed around which the ambient temperature (the measure on the amount of heat) will fluctuate.

These biasing mechanisms have become known as information processing mechanisms especially in their parallel distributed processing (PDP) form. The organization of information processing is open loop or helical rather than closed as in the case of feedbacks. Information is processed because pattern matching is involved: The furnace is to be turned on or off, just as in the

---

[1]And we have recently obtained evidence that the frontal part of the brain is still maturing as late as 18 - 21 years of age (Hudspeth & Pribram, 1990; 1992).

case of feedback organization. However, the turning on and off is no longer based solely on error processing, the differencing between a current and an inflexible desired state. Rather, the desired state is programmed by the dial. To the extent that the dial can be turned to a range of settings independent of the current temperature, the mechanism feeds information forward to control the process. As noted, the hippocampal formation makes possible such a shift from a feedback to a feedforward process. A model of how this shift is made has been developed (Pribram 1991, Lecture 9; Pribram 1997). For here the important consideration is that the model in involves an interaction between the hippocampal formation and the cortex of the cerebral convexity to which we turn next.[2]

Preferences and Prejudices

Rather than the frontolimbic parts of the brain considered so far, it is the posterior convexity of the brain which, when injured, affects choices among situations (Pribram, 1969; See review in Pribram, 1991, Lecture 7). Analysis has established that such choices, discriminations used in animal research, are akin to those involving preferences as these describe human choices (see the seminal review by Irwin, 1958). There is thus substantial evidence that the brain systems involved in establishing preferences are separate from those regulating utilities through probability estimates of the satisfaction of biological needs.

How are preferences developed? As human primates we go about our activities estimating the probability that particular needs will be satisfied. For example, after a tennis game we are thirsty and set to drink a specific quantity on the basis of the tissue osmolarity, the "concentration of water" (or of the electrolytes dissolved in the water), which is sensed by the brain. On the basis of prior experience, we either gulp down the required amount of liquid rather indiscriminately, or having on several occasions experienced cramps as a result of drinking too rapidly or because the liquid was too cold, we sip more casually instead, sitting down to share a leisurely afternoon iced tea or relaxing alcoholic beverage.

But it is peculiarly human that we might never have had the experience of cramps induced by too rapid satiation of thirst with inappropriate liquids. We might simply follow the guidelines for appropriate behavior given to us by our care-taking elders or our peers. We then say that we *prefer* the civilized behavior that is defined by sipping drinks at our tennis club. Again, nature's paradox

---

[2]The model involves a content addressable holographic-like matching between current input and stored memory. The model is based on evidence obtained with microelectrode recordings made within the hippocampal system and within the somatosensory cortex of the parietal lobe (Pribram, 1997).

A holographic process is based on the distribution of, and therefore enfolding of, information over and within an extent (a spatial and temporal envelope). In economics, the marketplace is such a holographic structure: each transaction enfolds the values of the whole, which are distributed throughout the extent of the whole. Thus, when one spends a unit of currency -- say a dollar -- the current value of that unit (currency) represents the productivity of Japan, China and other far eastern economies, the adjustments of European nations to a common market and common currency, the appearance of cow disease in England, the Wall Street Stock Market, etc.

In the same fashion, the current valuation of an event occurs within the marketplace of the episode within which the event is generated. Valuation thus depends on the values attributed to the variety of transactions that compose the episode.

shows itself: biology, in terms of brain processes, substitutes civility for direct action to achieve utility.

I believe that this example can be generalized to a description of how human preferences come to be organized. *The experienced utilities (and futilities) of individuals become encoded in culture to be presented to others as preferences.* Preferences are, from time to time, checked against utilities by individuals or by groups of individuals as in the recent revolutionary decade of the 1960's and again in Europe in 1989. Alternatives were explored, and if the situation has changed sufficiently (e.g., the advent of the birth control pill and Michael Gorbachev's decision to withhold troops) new preferences may emerge.

To summarize: Both theory and brain research make a good case for distinguishing between utilities and preferences. Utilities are based on biological needs, quantitative controls over the behavior based on *internal controls* which are subject to probability estimation of their satisfaction based on covariation among encoded experiences. Preferences, by contrast, devolve on a separate and distinct process which involves the ability to discriminate between invariant *situations*. In short, *utilities are state and situation specific and thus context dependent, whereas preferences are situation general hierarchically arranged and context free.*

Ethics and Morals

In ordinary life, experience controls the bias on the homeostatic regulations of the organism. Ordinarily, the eventualities are taken into account, and the response appropriate to the occasion is computed. Unless addicted, we ordinarily also take into account the limits over which control can be exercised before breakdown or futility occurs. Thus, eventual risk and cost-benefit ratio are considered. These considerations are state dependent and thus direct the attentions and intentions of the organism.

However, different situations may induce different states and therefore different utilities. But as situations may resemble one another, identities can be assigned to guide intentions. In most physical situations identifications work well. They are the basis of classifications in terms of attributes. (See review in Pribram, 1991, Lecture 7.)

Nonetheless, identifications can be assigned inappropriately. An experimental demonstration illustrates this point. Subjects are asked to describe the pattern of lights switching on and off in a matrix of possible positions. The descriptions are rated as closer to (warmer) or further from (colder) correct. But both the switching of lights and the ratings are in fact given randomly. Despite this, most subjects come up with ingenious descriptions of the paths of light placements -- some of these descriptions are incredibly complex.

Such misidentifications abound in social situations. Once an identification has been made, the subject defends it against dissolution despite being given additional information (such as "the appearance of the lights was actually random" or "I gave you hints as to colder or warmer randomly"), perhaps because putative correlations were not strained beyond credibility. Thus, preferences (in contrast to utilities) become automatized (Pribram, 1975b). They are not computed anew for each situational episode since they are grounded in the putative invariances obtaining in the situation, not its variety.

There is danger in this. Automatized preferences in social situations are difficult-to-change prejudices.

The distinction between utilities and preferences finds an echo in the psychoanalytic distinction between ethics and morals. Otto Rank (1936) suggested that ethics (based on relations to the mother) are internal feelings based on covariation among events, the eventual consequences of our behaviors. By contrast morals (based on instructions provided by the father) are external given that fit one to society. Gilligan's research (1982) has supported this difference between the development of male rule of governed morality and the more flexible ethical development of females, and Hayek (1988), in his book *The Fatal Conceit: The Errors of Socialism*, has described how societal morals become established.

The moralist does as (s)he should; whereas ethically we explore our wants (utilities and futilities). The moral person uses his moral preferences (and prejudices) automatically -- that is, without analysis, to support his intentions. The ethical person, paradoxically, analyzes each valuation with respect to his or her wants. The moral person invokes God or other ideology to support his preferences in a dispute; the person on the other side of the dispute is equally certain that God supports his/her opposite preferences. The ethical person intuitively feels and therefore recognizes that his/her wants (i.e., internal utility states) are situation specific and context dependent and thus may not be those of the other person. Much of psychoanalytic procedure is devoted to loosening morally given "shoulds" so that ethically determined "wants" can become realized. In social situations, wants, i.e., utilities and futilities, are negotiated by the interplay between a person's own wants and those of others. Such interplay frequently evokes paradox as when former rivals emerge as staunch friends.

Paradox and Wisdom

The results of this biobehavioral analysis return us to the biological paradox described at the outset of the analysis. Just as in the brain's control over respiration, feeding, drinking and sexual behavior, the control of the organism's values is ordinarily accomplished by way of his preferences. These are situationally derived and *appear* to be inviolate because they reflect putative invariances across situations. Man is thus shaken when he finds his preferences dissolving under the onslaught of major situational change. He is forced to relinquish his automatic mechanisms of information processing and rely on the computation of covariances (Pribram 1991, Lecture 10) from one episode to another. Such computations entail risk (the dangers of futility) as we have seen, and are thus likely to be accompanied by upset, and countered by overcontrol leading to dogmatic reassertion of an absolute which has become obsolete, or by relinquishing of all values leading to anomie and anarchy.

Does this mean that there are no universal values? Of course not. It means only that, understandably but paradoxically, we have identified universals with our preferences rather than with our utilities and futilities. The paradox consists of the fact that utilities, since they deal with episode and state dependent variables, *seem* so unreliable. But as long as episodes do not differ catastrophically, and as long as the brain's computational machinery is universal to all mankind, universals *can* be derived from such computations. In fact, as we saw above, preferences are often a cultural representation based on computed universals. Carl Jung (1968) referred to the utilities upon which such universals are established as archetypes.

The paradox penetrates further. Identifications signifying preferences constitute knowledge, especially scientific knowledge. (The terms science is derived from the Latin *sciencia*, to know). Knowledge at any moment is, therefore, finite, bounded by the limits that describe the situations over which the identifications hold.

Computations of covariations among recurring episodes have no such bounds. Variety can be as infinite as possible covariations, as infinite as the reach of the brain processes involved. Changes in biological states and major shifts in situations continuously reset the process as long as life lasts. Resetting removes prior established limits and often results in solutions that appear to be paradoxical and even enigmatic -- just as in mathematics dividing a line of unlimited ("infinite") length results in two lines of unlimited ("infinite") length -- thus one half equaling the whole. Ordinarily we speak of processes that involve such paradoxes as demanding wisdom for their resolution (Pribram 1971, Chapter 20).

Wisdom is at home with paradox. Solomon's apparent cruelty in dividing a baby in two proved the humane solution to a difficult dilemma. The mother's "giving up" her infant paradoxically resulted in her retaining it as is so often the case in interpersonal transactions. G. Spencer Brown in his *Laws of Form* (1972) provides a mathematics for handling paradox by "inventing" an imaginary Boolean descriptor. Spencer Brown addresses the Russell-Whitehead paradox ("I am a liar") but his formulations are applicable to a wider range of problems in the domain of the infinite.

Wisdom is thus based on acknowledged change produced in an unacknowledged, tacit familiar context and are therefore guided by the familiar. Nevertheless such change can take many forms. What makes for successful change? To answer this question we turn again to the part of the brain that is involved in controlling the probabilistic distribution of consequential behavior. As we have seen, this part of the brain, the anterior frontal cortex, integrates the processes that involve the amygdala and hippocampal systems.

The far forward portions of the frontal lobes of the brain serve this controlling function in the following fashion. The lower orbital part of the lobes, seated above the eyeball, are directly connected with the amygdala; the upper part is connected with the hippocampus; while the middle parts connect with the rest of the brain. Experiments performed with monkeys and observations made on humans have shown that the lower parts of the lobe are involved in dealing, on the basis of familiarity, with the propriety of actions to be undertaken. The upper part of the lobe arrange the priority of the actions taken on the basis of their effectiveness. Finally the mid portions of the lobe integrate propriety and priority into a practical course of action on the basis of probability estimates of the success or failure of its consequences. Consequences, sequences of events, outcomes of actions, when they are novel, produce visceral responses until satisfying familiarity is established. A succession of novel events can thus become familiarized as a narrative, a story which serves on subsequent occasions to guide inference as to proprieties, priorities and practicalities. The inferences are not reasoned in any logical sense; rather they simply accrue probable practicality from prior experience. The valuation is intuitive.

Thus, one way of summarizing the results of this neuropsychological analysis is to state that preferences reflect hierarchically organized information processes that lead to knowledge: the identification of finite invariant properties of situations. By contrast, the processing of utilities and futilities spawns wisdom: the processing of paradox among an infinite variety of narratives. In keeping with this rootedness in paradox, wisdom, though in the realm of the infinite, is not unfathomable. Slowly it is yielding the secrets of its structure to scientific knowledge while carefully harboring its contents. For wisdom is displayed in parables in an infinite diversity that paradoxically unifies our humanness.

References

Ashby, W. R. (1960). *Design for a Brain: The Origin of Adaptive Behavior* (2nd Edition). New York: John Wiley & Sons.

Bagshaw, M. H. & Pribram, K. H. (1953). Cortical organization in gustation (Macaca Mulatta). *Journal of Neurophysiology, 16*, 499-508.

Bateson, P. G. R. (1976). Psychology of knowing another side. *New Scientist, Jan.*, pp. 166--167.

Brown, GS (1972) *Laws of Form*. New York: The Julian Press, Inc.

Berne, E. (1961). *Transactional Analysis in Psychotherapy*. New York: Grove Press, Inc.

Douglas, R. J. & Pribram, K. H. (1966) Learning and limbic lesions. *Neuropsychologia, 4,* pp. 197-220.

Douglas, R. J. & Pribram, K. H. (1969) Distraction and habituation in monkeys with limbic lesions. *J. Comp. Physiol. Psychol., 69,* pp. 473-480.

Fulton, J. F., Pribram, K. H., Stevenson, J. A. F., & Wall, P. (1949) Interrelations between orbital gyrus, insula, temporal tip and anterior cingulate gyrus. *Trans. Am. Neurol. Assoc.*, pp. 175-179.

Gilligan, C. (1982) *In a Different Voice*. Harvard Univ. Press.

Grossman, S. P. (1967). *A Textbook of Physiological Psychology*. New York: Wiley.

Hayek, F.A. (1988) *The Fatal Concert: The Errors of Socialism*. W.W. Bartley, III (Ed.), London: Routledge.

Hudspeth, W. J. & Pribram, K. H. (1990). Stages of brain and cognitive maturation. *Journal of Educational Psychology, 82*(4), 881-884.

Hudspeth, W. J. & Pribram, K. H. (1992) Psychophysiological indices of cognitive maturation. *International Journal of Psychophysiology, 12*, 19-29.

Irwin, F. W. (1958). An analysis of the concepts of discrimination and preference. *The American Journal of Psych.*, Callenbach Commemorative Number. Vol. LXXI, No. 1.

Jung, C. G. (1968). *Analytical psychology: Its theory and practice: the Tavistock lectures*. New York: Pantheon Books.

Kaada, B.R., Pribram, K.H. & Epstein, J.A. (1949). Respiratory and vascular responses in monkeys from temporal pole, insular, orbital surface and cingulate gyrus. *Journal of Neurophysiology, 12*, 347-356.

Killeen, P.R. (1994) Mathematical principles of reinforcement. *Behavioral and Brain Sciences*, vol. 17, no. 1, pp. 105-172.

Killeen, P.R. (1996) Mechanics of the animate. In K.H. Pribram and J.S. King (Eds.) *Learning as Self Organization*. Mahwah, NJ: Lawrence Erlbaum Associates, 103-140.

Konrad, K.W. & Bagshaw, M.H. (1970) Effect of novel stimuli on cats reared in a restricted environment. *J. Comp. Physiol. Psychol.*, 70, pp. 157-164.

Livesey, P.J. (1986) *Learning and Emotion, Vol. 1*. Mahwah, NJ: Lawrence Erlbaum Associates, Inc.

Mackay, D. M. (1962). Self-organization in the time domain. In: M.C. Yovits, G. T. Jacobi & G. D. Golstein (Eds.) *Self-Organizing Systems*, (pp. 37-48). Washington, DC: Spartan Books.

Mittelstaedt, H. (1968). Discussion. In: D. P. Kimble (Ed.), *Experience and Capacity* (pp. 46-49). New York: The New York Academy of Sciences, Interdisciplinary Communications Program.

Polyani, M. (1960). *Personal Knowledge: Towards a post-critical philosophy*. Chicago: University of Chicago Press.

Pribram, K.H. (1957) On the neurology of values and the neurology of perception. In *Proceedings of the fifteenth international congress of psychology*. North-Holland Publishing Company: Amsterdam, pp. 77-82, and pp. 230-231.

Pribram, K. H. (1961). A further experimental analysis of the behavioral deficit that follows injury to the primate frontal cortex. *Experimental Neurology, 3*, 432-466.

Pribram, K. H. (1969). The Amnestic Syndromes: Disturbances in Coding? In: G. A. Talland and N. C. Waugh (Eds.), *Pathology of Memory*. New York: Academic Press.

Pribram, K. H. (1971). *Languages of the Brain: Experimental Paradoxes and Principles in Neuropsychology*. Englewood Cliffs, NJ: Prentice-Hall, Inc.

Pribram, K. H. (1975a). Self-consciousness and intentionality. In: G. E. Schwartz & D. Shapiro (Eds.), *Consciousness and Self-Regulation: Advances in Research*. New York: Plenum Press.

Pribram, K. H. (1975b). Modes of central processing in human learning and remembering. In: T. J. Teyler (Ed.), *Brain and Learning* (pp. 147-163). Stamford, CT: Greylock Press.

Pribram, K. H. (1982) Brain mechanisms in music: Prolegomena for a theory of the meaning of meaning. In M. Clynes (Ed.), *Music, Mind, and Brain*. New York: Plenum, pp. 21-35.

Pribram, K. H. (1986) The cognitive revolution and mind/brain issues. *American Psychologist, Vol. 41, No. 5*, pp. 507-520.

Pribram, K. H. & Bagshaw, M. H. (1953). Further analysis of the temporal lobe syndrome utilizing frontotemporal ablations. *Journal of Comp. Neurology, 99*, 347-375.

Pribram, K. H. (1991) *Brain and Perception: Holonomy and Structure in Figural Processing*. New Jersey: Lawrence Erlbaum Associates, Inc.

Pribram, K. H. (1992) Explorations in Experimental Neuropsychology. In: Samson & Adelman (Eds.), *The Neurosciences: Paths of Discovery II* (pp. 307-3250. Cambridge, MA: Birkhäuser Boston Inc.

Pribram, K.H. (1997) What is Mind that the Brain May Order It?. In V. Mandrekar & P.R. Masani (Eds.) *Proceedings of Symposia in Applied Mathematics, Vol. 2: Proceedings of the Norbert Wiener Centenary Congress, 1994*. Providence, RI: American Mathematical Society, pp. 301-329.

Pribram, K. H., Douglas, R. & Pribram, B. J. (1969) The nature of non-limbic learning. *J. Comp. Physiol. Psychol., 69*, pp. 765-772.

Pribram, K. H. & Gill, M. M. (1976). *Freud's "Project for a Scientific Psychology": Preface to Contemporary Cognitive Theory and Neuropsychology*. London: Hutchinson Publishing Group, Ltd.

Pribram, K. H., Lim, H., Poppen, R., & Bagshaw, M. H. (1966) Limbic lesions and the temporal structure of redundancy. *J. Comp. Physiol. Psychol., 61*, pp. 365-373.

Pribram, K. H., Lennox, M. A., & Dunsmore, R. H. (1950) Some connections of the orbito-fronto-temporal limbic and hippocampal areas of Macaca Mulatta. *J. Neurophysiol., 13*, pp. 127-135.

Pribram, K. H. & McGuinness, D. (1992) Attention and para-attentional processing: Event-related brain potentials as tests of a model. In: D. Friedman & G. Bruder (Eds.), *Annals of the New York Academy of Sciences, 658* (pp. 65-92). New York: New York Academy of Sciences.

Pribram, K. H. & McGuinness, D. (1975). Arousal, Activation and effort in the control of attention. *Psychological Review, 82*(2), 116-149.

Pribram, K. H., Reitz, S., McNeil, M. & Spevack, A. A. (1979) The effect of amygdalectomy on orienting and classical conditioning. *Pavlovian J. Biol. Sci., 14,* pp. 203-217.

Rank, O. (1936). *Truth and reality: A life history of the human will.* New York: Knopf.

Russell, R. W., Singer, G., Flanagan, F., Stone, M. & Russell, J. W. (1968). Quantitative relations in amygdala modulation of drinking. *Physiology and Behavior, 3,* 871-875.

Smets, G. (1973) *Aesthetic Judgment and Arousal.* Leuven, Belgium: Leuven University Press.

Spencer Brown, G. (1972). *Laws of Form.* New York, NY: The Julian Press Inc.

Von Neumann, J. & Morgenstern, O. (1953). *Theory of Games and Economic Behavior.* Princeton: Princeton University Press (3rd ed.) Chap II. pp: 60-73; Chap. I, pp. 19-20; 24-28; 39-41.

Weiskrantz, L. & Wilson, W. A. (1958). The effect of ventral rhinencephalic lesions on avoidance thresholds in monkeys. *Journal of Comp. Physiological Psychology, 51,* 167-171.

Wilson, W. (1959). Role of learning, perception and reward in monkeys choice of food. *American Journal of Psychology, 72,* 560-565.

3

Values, Goals and Utility in an Engineering-Based Theory of Mammalian Intelligence

Paul J. Werbos
*National Science Foundation, Room 675**
Arlington, VA 22230
pwerbos@nsf.gov

Introduction and Strategy

Values, goals and utility play a central role both in formal psychology and in everyday life. This paper will try to explain how our understanding of these concepts can be further unified and enriched, based on a new engineering-based theory of mammalian intelligence [1].

The new theory is essentially the latest iteration of an ongoing dialogue between neuroscientists like Karl Pribram and engineers like myself, each trying to grow beyond the usual limitations of these disciplines. A major goal of this effort is to bridge the gap between functional, realistic and intuitive descriptions of the brain and more formal mathematical, computational representations. After we develop a better understanding of how the higher-level functional abilities of the brain could be replicated, by certain types of parallel distributed computational structure, we will then be ready to do a more serious job of matching such structures to the lower-level features of neural circuitry and to suggest new experiments.

The underlying strategy of this approach has been discussed at great length in the past [2-4]. Those prior discussions have elaborated on issues such as the role of complexity, the importance of learning, the choice of mathematical tools, the strategies for experimental validation, and so on. Nevertheless, there are three strategic issues which need to be discussed yet again, because they are a source of concern to many researchers: (1) consciousness versus intelligence; (2) prior information versus learning; (3) neural networks versus computational neuroscience.

Consciousness Versus Intelligence

The new theory tries to replicate the higher-level intelligence of the brains of mammals. This is not the same thing as explaining consciousness in the minds of humans.

Many of us are actually more interested in human consciousness, in principle, than in mammalian intelligence. For example, I have written at length about the various meanings of the word "conscious," about the role of human language and symbolic reasoning, quantum computing, the soul, and so on[5]. However, those topics are beyond the scope of this article.

---

*The views herein are those of the author, not those of NSF, though they are written on government time.

I will argue here that a deeper understanding of mammalian intelligence is actually a prerequisite to a deeper understanding of human consciousness, just as algebra is a prerequisite to calculus and calculus is a prerequisite to understanding gravity. In reality, in everyday life, we have to develop some kind of intuitive understanding of gravity (dropping things and breaking things) before we even learn algebra . In the same way, intuitive studies of language and the soul can be very important to everyday life. But for a deeper understanding of gravity -- and of the higher levels of intelligence -- one must first master the prerequisites.

In fact, the word "intelligence" actually represents one of the meanings of the word "consciousness" in everyday conversation. Sometimes when we refer to systems as possessing "consciousness," we mean that they are intelligent.

Some authors who write about "consciousness" or "intelligent systems" act as if consciousness were some kind of binary variable. They ask: "Is it conscious or is it not?", as if the answer should be "yes" or "no." In my view, the question itself is fundamentally misleading. On the other, some classical behaviorists have treated intelligence as if it were a continuous variable, like the Intelligence Quotient. In the extreme form of that view, snails and humans are treated as if they are qualitatively the same, and different only in their speed of learning.

Based on empirical reality, I would argue that "consciousness" or "intelligence" is actually more like a staircase -- a discrete series of levels, each level qualitatively different from the level before. (Actually, it is a tilty staircase, insofar as there are some quantitative differences within each level, but these are not as important as the qualitative differences between levels.) The most important levels, for our purposes here, correspond roughly to the classes of vertebrates -- fish versus amphibians versus reptiles versus birds versus mammals versus the symbolic-reasoning level of intelligence. These are the forms of intelligence that we actually see in nature.

At a superficial level, the differences between different mammals seem very large. For example, the neocortex of a normal adult monkey contains many more visual areas than does the neocortex of a mouse. However, lower classes of vertebrate do not even possess a neocortex at all. Furthermore, the same underlying principles of learning seem to determine what is actually learned and stored in the various areas in the neocortex. When one area is damaged but the necessary connections are present, other parts of the neocortex can learn to take over the functions of the damaged area. There is a huge literature on "mass action" and "equipotentiality" which has demonstrated this. (This literature originated in the classic work of Lashley, Pribram, Freeman and their colleagues.) For example, neurons in the temporal cortex have learned to take over as edge detectors in some experiments. The new theory to be described here [1] tries to replicate the underlying learning abilities which are essentially the same across all mammals.

Years ago, in comparative psychology, Bitterman [6] demonstrated qualitative differences in the nature of learning across different classes of vertebrates. The details are beyond the scope of this paper, but did contribute to the development of our new theory.

Many researchers in Artificial Intelligence (AI) have argued that "real intelligence" requires the use of symbolic reasoning. One such researcher begins his talks by asking how Einstein reasoned, in order to develop his theories. I would certainly agree that these are excellent examples of higher-order intelligence. However, I would question our chances of building an artificial Einstein before we can build even the crudest version of an artificial mouse. If Einstein and the mouse were unrelated phenomena, it might make sense to aim directly to build an Einstein, since he is more interesting; however, the symbolic reasoning used by Einstein and by other relatively intelligent humans is deeply rooted in the mammalian neocortex and in other brain structures which we share with other mammals. There are good humanistic reasons to try to understand symbolic reasoning better, even now, as I discussed above; however, a deep understanding of the neocortex and of mammalian intelligence in general should permit a much deeper understanding of symbolic reasoning as well in the more distant future.

Prior Information Versus Learning

The new theory to be described here is a general purpose learning design. It can be started out with essentially no concrete knowledge of its environment. Given time, it can develop a very complex understanding of its environment, and a complex plan of action, based entirely on learning from experience. The learning-based approach is now very familiar in the neural network field, where it has led to many practical, empirically-tested applications now being used in technology and in finance. The success of this approach has contributed to a resurgence of interest in learning even within hard-core AI.

On the other hand, some AI researchers still speculate that learning-based designs by themselves could never replicate higher-order intelligence such as what we see in the neocortex. They speculate that higher-order intelligence can only grow in the soil of very extensive, hard-wired prior knowledge which is very concrete and specific in nature. Certainly, in the nervous system, there are many lower-level sensory and motor circuits and reflexes which are very specialized and preprogrammed in nature. But some of these researchers go on to claim that much of the higher-order processing -- such as the entire path up from edge detectors through to object identification -- is preprogrammed in a concrete way. As evidence, they argue that some mammals -- like ungulates -- are actually born fully functional, and able to perform these kinds of cognitive tasks, before they have had any time to learn anything. They even argue that much of the supposed "learning" in human infants may be part of a preprogrammed developmental process, similar to what happens to baby ungulates in the womb, and independent of learning experience.

This paper will not try to evaluate the last few claims, one way or another. But there is a crucial logical point which needs to be considered: even if these claims should be true, even if there is substantial concrete higher-order information born into the mammalian brain, this still does not imply a lack of ability to learn this same information if the occasion should arise.

As an example, even if certain cells in the neocortex are set up as edge detectors, right from birth, this does not imply that neocortical cells lack the ability to learn edge detection, through ordinary learning from experience. In fact, the temporal lobe experiments mentioned in the previous section clearly demonstrate that these cells do possess that learning ability.

There is no real paradox here. In practical engineering applications, one often initializes the weights and connections of an artificial neural network (ANN) to the best available initial guess, based on the available prior information, even though these same weights and connections will then be adapted freely as if they had been the result of learning. There is no special reason to initialize the weights to zero.

One would expect evolution to use the same general sort of opportunistic approach, combining both prior information and a generalized learning ability, to the extent that the machinery of DNA permits the transmission of specific information.

All forms of learning do tend to make some sorts of prior assumptions about the world, implicitly, of a very general theoretical sort [7, ch.10]. The new theory to be described here is no exception. However, it does not require extensive, concrete forms of prior knowledge, of the sort which is popular in the world of expert systems in AI.

Neural Networks Versus Computational Neuroscience

The new theory to be described here is quite different in flavor from the usual differential equation models which are most familiar in neuroscience. It results from a modeling approach very different from conventional computational neuroscience. This section will discuss these differences, somewhat briefly and crudely, in order to help the reader understand how these different approaches are complementary.

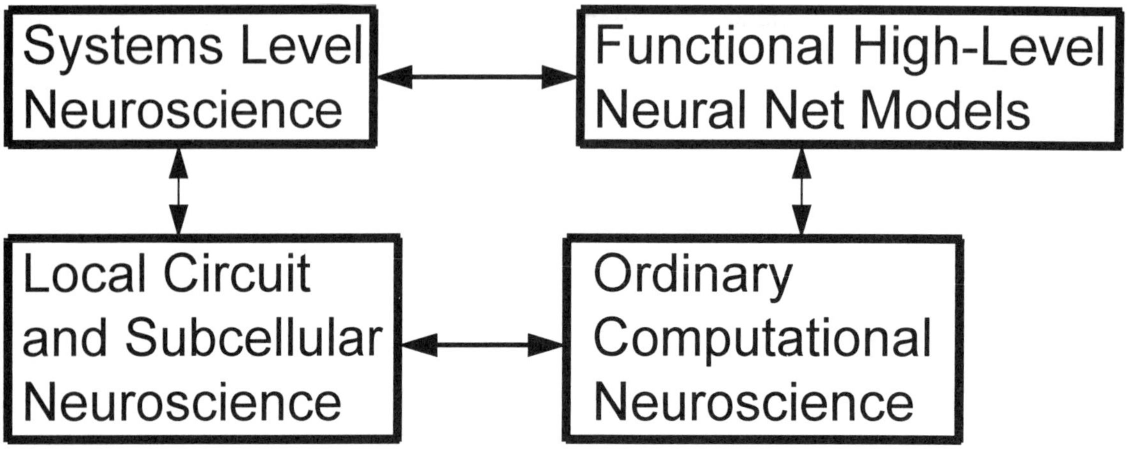

Figure 1. Some Major Strands of Research Addressing Mammalian Brains

Figure 1 provides a crude overview of some important strands of research which try to understand the mammalian brain. Research involving invertebrates, etc., is beyond the scope of this paper. The chapter by James McClelland in this book probably belongs in a new box on the middle right, but that too is beyond the scope of this paper.

On the left side of Figure 1 are two boxes representing two important trends in neuroscience which have had an uneasy coexistence for decades. Systems level neuroscience addresses higher-order issues such as the functional significance of major parts of the brain, the representation of emotions in the brain, the location and dynamics of working memory, the interaction between the neocortex, the limbic system and the basal ganglia, and so on. It attempts to understand fundamental issues such as the nature of human intelligence and motivation, through qualitative analysis of extensive clinical and experimental information. It tries to account for experimental information from lower levels, but the major focus is on understanding the whole system. Karl Pribram has been one of the serious contributors to this field for decades.

Local circuit and subcellular neuroscience operates at a very different level. This area has had a tremendous spurt of growth in recent decades, due to the development of genetic technology and other new technologies which tell us more and more about the molecular events involved in some forms of learning, about connections between neurons, and so on.

Many neuroscientists believe that the main challenge to neuroscience as a whole is to explain the higher-level phenomena in terms of the local circuit and subcellular phenomena. In other words, the challenge is to connect the two levels of analysis more completely.

On the other hand, many other neuroscientists believe that the key challenge to the field is to put the whole field on a more precise, more mathematical footing. Their goal is to become more like physics, in some sense. These neuroscientists have developed more and more differential equation models -- some rooted in extremely detailed physics and chemistry -- which attempt to describe local circuits very precisely. By building up from such models, at ever higher levels of abstraction, they have also built up to models which actually try to replicate meaningful mental functions, such as associative memory or early vision in the adult.

The theory to be described here fits into the upper right box, which is currently the most sparsely populated part of this figure. Rather than modeling the brain from a bottoms-up point of view, it takes a top-down approach. It tries to provide a mathematical articulation or model of the kinds of phenomena which neuroscientists like Pribram have focused on, rather than articulating subcellular phenomena.

In the long term, I agree that the central challenge is to connect the two levels -- to develop detailed neural network models which match the subcellular data and the systems level information, both at the same time. There is a great need for researchers in the upper boxes and in the lower boxes to study each others' work, in order to strengthen the effort to build a bridge between the two levels. On the other hand, there is also a great deal of important work to be done in reinforcing the foundations at either side of the bridge. Particularly in the upper right box, the ideas are so new that considerable effort will be needed to sharpen, validate and consolidate the theory to be described here.

In the upper right box, the focus is on higher-level functionality. The goal is to develop neural network models which replicate higher-order learning-based intelligence -- not just a component of intelligence, but the whole system. Thus the lower right box tends to draw most heavily on chemistry and physics, while the upper right box requires a heavy emphasis on engineering. Please bear in mind that "engineering" does not consistent of a collection of clever monkeys; it is a large, long-standing intellectual endeavor, which has developed rigorous mathematical techniques addressing the issue of functionality -- of understanding what works, what does not, how to develop new designs, and how to understand a wide variety of challenging tasks. Engineering concepts and engineering testbeds provide a very crucial intellectual filter and empirical testing ground for allegedly functional designs -- including designs in the upper right box.

Most models used in engineering or in the artificial neural network (ANN) field do not fit within any of the boxes in Figure 1, because they try to achieve functionality in more limited kinds of tasks. Nevertheless, the theory to be described here was developed within the disciplines of ANNs and of engineering, because of the filters which they provide.

How important are these filters? As of 1995 [8], only one class of neural network model had ever been implemented which met four basic necessary tests for a model of brain-like intelligence: (1) demonstrated engineering functionality, for a completely learning-based system; (2) possession of an "emotional" or "values" component, enabling the system to learn to pursue goals effectively over time; (3) possession and use of an "expectations" component, as needed to address behavioral experiments in classical conditioning; (4) the ability to take physical actions (e.g. control motors or muscles).

These are necessary tests, not sufficient tests, but, when applied together, they rule out all the other models, including those which have emerged from computational neuroscience.

Models in this class have since been adapted and extended to a considerable extent, in order to cope with difficult engineering control applications [9,10]. As a result of subsequent analysis -- both engineering analysis and the study of issues raised by Pribram and others -- these models are now essentially obsolete as models of mammalian intelligence; new, more complete models have been developed. However, the original models have still played a crucial role because: (1) by showing how it is possible to meet the four tests together, they opened the door to the development of more complex models in the same general family; (2) because they embody difficult control concepts in a simple context, they are crucial to our ability to really understand and develop more complex models embodying the same principles; (3) as testable, implemented designs they continue to provide a flow of empirical information about what works and what does not, and about the interface between higher-level models and lower-level components.

The Larger Context: The Role of Values and Utility

The new theory [1] is a theory of intelligence. But the mammalian brain includes more than just intelligence. Even before I began to develop serious designs for intelligent systems, decades ago[2], I started out by defining a working picture of how the intelligence fits in as part of the larger system [2, 11]. This picture is illustrated somewhat crudely in Figure 2.

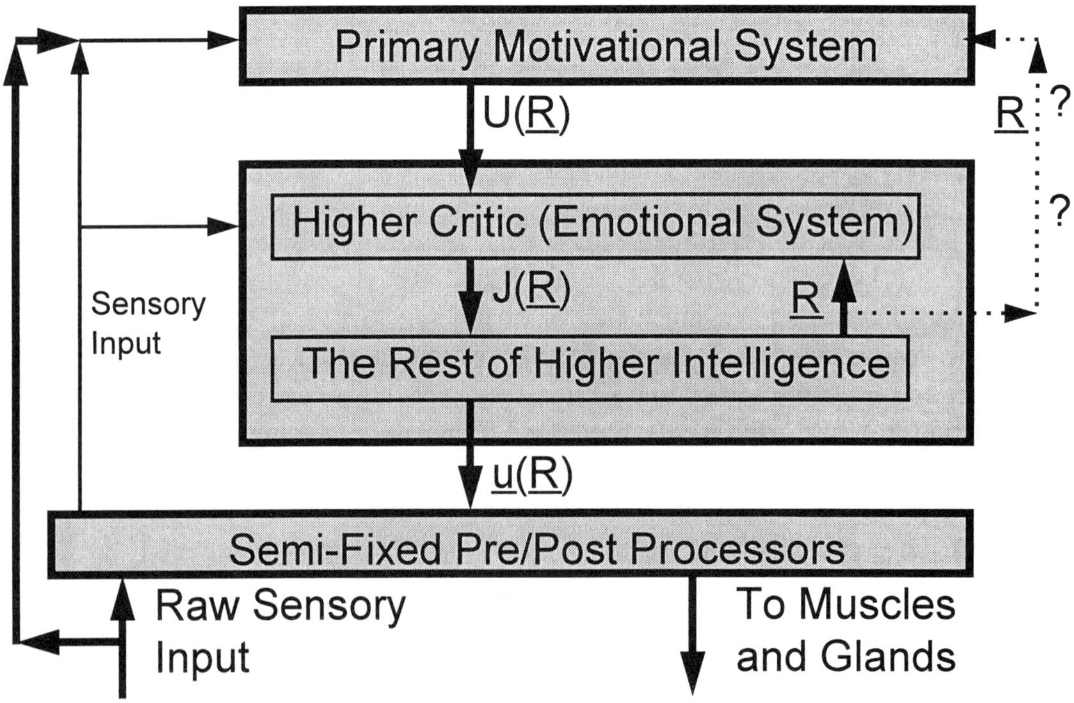

Figure 2. Intuitive Picture of What the Brain Does

In Figure 2, the intelligence proper consists of one big box containing two boxes inside of it -- the higher Critic and "the Rest."

In this picture, the intelligence proper builds up an image or representation of the present state of the world, denoted by the vector **R**. The Primary Motivational System (PMS) provides a utility function,
U(**R**). The job of the intelligent system is to develop a strategy for how to maximize U, over the long-term future, by manipulating the actions which it controls. The choice of actions is represented by another vector, **u**. Below the level of higher-order intelligence -- which is based on truly generalized learning dynamics -- there are more or less fixed preprocessors and postprocessors, such as the retina, the motor neurons in the spine, and even the "motor pools" in the brain stem[12]. These lower-level systems often do contain some kind of adaptation abilities, responsive to higher-level actions (**u**), but their adaptation tends to be somewhat specialized in nature.

Of course, this picture owes a lot to the classical work of Von Neumann and Morgenstern[13],
which in turn was essentially just a mathematical formulation of ideas from John Stuart Mill and Jeremy Bentham. Von Neumann's notion of "utility function" is well known in economics, but somewhat less so in other disciplines. "Utility," in Von Neumann's sense, is not a measure of the usefulness of an object.
Rather, it is a measure used to evaluate the overall state of the entire world, **R**. (But see [3,4] for discussion of marginal utilities and prices.) Utility is a global measure of "success" or "performance," or
any other "figure of merit" which an intelligent system tries to maximize. It represents the most fundamental desires or "needs" of the organism, as discussed in the chapter by Pribram in this book.

All across the social sciences, there have been very rabid debates for and against the idea that people maximize some kind of utility function. In this picture, the organism is not assumed to do a perfect job of maximizing utility; rather, it is assumed to learn to do a better and better job, over time, subject to certain constraints which limit the performance of any real-world computing system. Perhaps the most important constraint is the need for a balance between play or novelty-seeking behavior, versus the danger of becoming stuck forever in a "rut" or in a "local minimum." See [5,14] for more discussion of these issues.

The utility function U(**R**) is not the only measure of value depicted in Figure 2. There is also another function, J(**R**), which is learned by the intelligent system proper. This function J(**R**) provides a kind of immediate strategic assessment of the situation **R**. The network which outputs J is called a "Critic" network.

The need for the Critic network emerges directly from the mathematics, which I will discuss in more detail below. Intuitively, however, the J function corresponds to the learned preferences discussed by Pribram in his chapter in this book. (See [3,4] for more discussion of this intuition.)

Following this picture, there are really four fundamental issues we need to consider when we try to understand "values":

(1) How does the Primary Motivational System work? Where do the basic "needs" come from?

(2) How does the Upper Critic work? Where do our learned hopes and fears and preferences come from?
(3) Are there <u>other</u> measures of value in the "Rest of System," and, if so, how do <u>they</u> work?
(4) Going beyond science, how can we as human beings do a better job of formulating and implementing our values, in light of this information?

Most of the chapters in this book, like most of my own published work, focus on "values" in the sense of question 2. They focus on the effort to understand the higher levels of intelligence as such, including the <u>learned</u> emotional associations in the main part of the limbic system. The new theory [1] addresses questions 2 and 3.

Nevertheless, as Pribram points out, the PMS is more fundamental and more universal, in some sense. It is the foundation on which the rest is built. Better intellectual understanding can logically supersede what we have learned from experience (questions 2 and 3), but it cannot supersede the PMS.

In my own work, I have discussed the PMS somewhat, mainly on a theoretical basis [2,5]. At this conference, Allan Schore gave an impressive talk on the process of <u>imprinting</u>, which is crucial to the dynamics of the PMS. The PMS obviously includes some fixed systems in the hypothalamus and epithalamus, which respond to a wide variety of simple variables like blood sugar, pain, and so on.

As Pribram points out, it often includes simple measurable variables which serve as indicators, in effect, of concepts which are more important to the organism but harder to measure. The PMS also includes some fairly complex social responses, as described by Schore, based in part on something like an associative memory. This is <u>not</u> the kind of associative memory which one uses as part of a utility maximization scheme; rather, it is a special kind of associative memory, with laws of its own, which requires more of an empirical approach to understanding.

The remainder of this paper will describe the new theory of intelligence, which includes a response to questions 2 and 3. For the sake of completeness, I will conclude this section with some thoughts about question 4.

Question 4 really raises the larger questions of ethics and of the concrete goals we pursue in our lives and in our society. In the past [5], I have argued that a sane and sapient person would gradually learn a two-pronged approach to these issues: (1) improved self-understanding, by learning, in effect, what one's <u>own</u> PMS is really asking for (U); (2) a thoroughly systematic, intelligent effort -- supported by a synergistic use of symbolic reasoning and focused intuition -- to go ahead and maximize one's U over the long-term future, over a large playing field. Broadly speaking, this is consistent with the spirit of Pribram's comments in his chapter as well. <u>Both</u> prongs of this approach encourage us to learn to "see" all inputs to our consciousness -- including the emotional inputs -- in the same focused and intelligent way that many of us reserve for visual inputs only.

Certain cultural authorities would consider this approach extremely threatening. When people think for themselves, they may become harder to predict and control. However, these fears indicate a disrespect for human intelligence and a short-sighted effort to maximize the authorities' own power. (Indeed, such authorities have been known to foster illiteracy and the use of drugs in order to advance their allegedly higher ethical values.) As our environment becomes more and more challenging, and more and more dependent on the honest and creative management of complex information, such static, repressive and authoritarian attitudes become less and less consistent with human survival.

Description of The New Theory

Introduction

The new theory can be seen as a hybrid between hierarchical task planning, as practiced in AI, and neural network designs based on reinforcement learning or adaptive critics. In effect, the task planning aspect provides a way to operationalize the classic theory of Miller, Galanter and Pribram[15].

The neural network aspect provides a way to link the system to higher values and emotions, and to the brain. The design as a whole provides a unification of these two different approaches or aspects of higher-level intelligence.

The new theory is a natural outcome of earlier theories and concepts discussed in [3]. The details of the new theory were first reported as part of an international patent application filed on June 4, 1997. The main design-specification part of that disclosure was included in [1]. Links to biology are mentioned in [1], but they are sandwiched between discussions of the engineering details, which are quite complicated. This paper will not reproduce all of those details; instead, it will try to explain the points of greatest importance to biology and psychology.

Even one year ago, I had no intention of devising a theory which is strictly and classically hierarchical in nature. The concept of hierarchy has been used all too often as a quick ad hoc solution in AI, and it often limits the learning capability and flexibility of the resulting design. However, the seminal work of Vernon Brooks on motor control[16] makes it clear that some hierarchical approaches can still be compatible with parts of the biology. In the end, a careful study of the mathematics [1] have me no choice, and it also provided a basis for incremental, neural-style learning in such a system.

In my present view, the new theory fits the mammalian brain in much the same way that certain line drawings can fit an oil painting. Often, before an artist paints an oil painting, he first draws a cartoon-style, black and white sketch of the same scene. The contours are all there, and all the objects are there, but the colors and the fine texture all have yet to be filled in. In the same way, the new theory -- as an engineering design -- actually involves several choices between possible subsystems. New subsystems will be needed, in the future, in order to improve performance and in order to fit the lower-level details of the brain more exactly. Nevertheless, this theory -- unlike its predecessors -- does seem to encompass all the major higher-order structure which needs to be filled in. The work ahead of us is less like running ahead into the wilderness, and more like settling in and developing known territory.

The work leading up to the new theory has already been published in the past. However, a certain degree of condensed review is necessary, to make this paper at least partly self-contained.

Relation to Prior Theories: Concepts and History

The new theory, like its predecessors, is a general purpose learning system. Like its predecessors, it is designed to learn how to develop a strategy of action which will maximize some externally specified utility function, U, over all future times. Expressed as an engineering design, it is a system which learns how to perform optimization over time. It is intended to operate in a generalized manner, such that it could learn to perform "any task," based on learning from experience, without concrete prior knowledge of the task. This family of designs is sometimes referred to as "reinforcement learning systems," but it is more accurate to call them "adaptive critic systems" or "approximate dynamic programming" (ADP).

If the new theory and its predecessors both perform this same function, then what is the difference between them? There are two important differences. First, the new design contains new features which substantially expand the range of tasks which can be learned in a practical way, at a reasonable learning speed. Second, these features provide a more complete fit to global features of the brain.

Decades ago, when this work began, I was impressed by the work of Hebb[17], who seemed to suggest that higher-level intelligence could emerge from the operation of a simple, general learning rule, operating across all neurons in the brain. That simple vision simply did not work out in functional or engineering terms, and it never provided any kind of explanation for the major structural components of the brain. Later, a variety of "reinforcement learning" models or designs were developed by several authors [11,18,19]; however, none of them met the simple four tests for a brain-like system mentioned in the Introduction above and discussed at length in [5]. This section will only discuss designs which do at least meet those minimal tests.

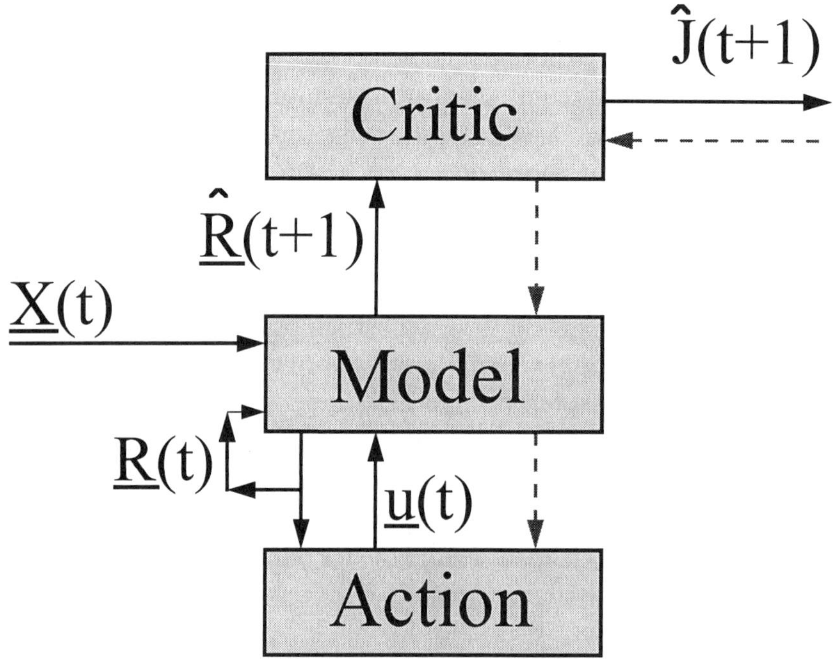

Figure 3. The Earliest "Brain-Like" Control Design

Figure 3 illustrates the simplest design in this family. In this design, there were three major components of the overall "brain." The highest was the emotional system or "Critic," whose task was to estimate the function "J" discussed above. I associated the Critic with the limbic system of the brain. The middle was the "Model" -- a neural network trained to act as a simulation model of the environment of the organism. (Following standard engineering approaches, the Model also calculated an "estimated state vector **R**", i.e. a representation of the current state of the environment.) The Model served, in effect, as an expectations system. I associated it with the cerebro-thalamic system. Finally, the system also contains an Action network -- a network trained to output control signals **u**, which affect muscles, glands and the orientation of sensors.

The mathematics of how to adapt this system were first developed in 1970, first published in 1977[20], and explained more completely in 1992 [7]. The first practical engineering test, in simulation, was published by Jameson in 1993. By 1995, three additional groups -- Santiago and myself, Wunsch and Prokhorov, and Balakrishnan -- had implemented, tested and published more advanced designs of the same general flavor[8]. In a 1977 conference [9], four new groups reported implementations, some with substantial practical significance -- Ford Motor Company, Accurate Automation Corporation, Lendaris and Tang. Dolmatova and myself, working through Scientific Cybernetics, Inc., have developed some new implementations, which will be published elsewhere in 1997. In addition to these nine groups in the United States, one researcher in Sweden -- Landelius [21] -- has also had important results. Even the simple design in Figure 3 is substantially more complex than the underlying designs used in the past in control engineering; however, as we develop more understanding of its practical capabilities relative to earlier control designs [10], the practical use of this family should continue to grow.

In Figure 3, the broken arrows represent a flow of value information, which is crucial to the dynamics of learning in this design. The overall design given in [7, chs. 3 and 13] does not specify what kinds of networks to insert into these boxes. It deliberately gives the user or modeler some flexibility in that regard. However, it does specify how to calculate the flow of value information, for any choice of network, and how to use that information in adapting the Critic and the Action network. (A different chapter -- [7,ch.10] -- discusses the structure and dynamics of learning for Model networks.) Those value calculations turn out to match exactly the rules for the "flow of cathexis" proposed by Freud years ago [2,3]. The same underlying algorithm which I developed for that application is now being used in mostpractical applications of ANNs, both in technology and in finance [2].

Back in 1970, I hoped that this kind of design would be enough to replicate higher-order intelligence. If the three boxes were filled in with simple neural networks, each made up of one class of neuron, then Hebb's dream would almost be fulfilled. Instead of one "general neuron model," one would need three general neuron models, one for each class of neuron. One would also have a basis for understanding the division of the brain into major components [2,3,22].

Over the years, this hope had to be modified. Engineering research told us more and more about how to "fill in the boxes." It told us that we need to fill in the boxes with fairly complex structures, in order to get maximum performance [7]. Instead of three cell types, the research pointed towards a need for boxes within boxes, ultimately leading to dozens and dozens of basic, generalized cell types. But this did not invalidate the overall model, at first. In fact, it led to an improved fit with neuroscience; for example, it began to suggest an explanation for some of the more detailed features of the cerebro-thalamic system[3].

By 1992, however, this study of substructures had led us to a major paradox [7,23]. We discovered that many higher-order learning tasks require us to fill in some of the boxes with a certain class of neural network -- the Simultaneous Recurrent Network (SRN) -- which requires many iterations in order to settle down and calculate its output [24,25]. This slow speed of calculation is inconsistent with the requirement for fast calculation (circa 100-200 cycles per second) for the smooth coordination of bodily movements.

To explain this, I then proposed a "two brains in one" [3] theory of the mammalian brain. According to that theory, the brain contains two entire adaptive critic learning systems, each with its own Critic network. In the upper brain (limbic system and cerebro-thalamic system), the relevant boxes are filled in with these slow but powerful SRN components. Thus the upper brain performs a major cycle of computation over a relatively long time period -- 0.1 to 0.25 seconds, corresponding to alpha or theta rhythms. (Within that period, it goes through many minor iterations, as required by the SRN model and verified by recent multichannel EEG recordings in Pribram's laboratories.) The lower brain (mainly the olive and cerebellum) uses simpler, more feedforward components, which allow it to operate at a faster rate. The lower brain is trained, in effect, to try to make the upper brain happy.

All of these theories had one really large, gaping hole in them: the lack of an explanation for the role of the basal ganglia. Decades ago, this did not seem like such a big hole, because the importance of the basal ganglia was not yet so well understood. But evidence has begun to accumulate [16,26,3,4] showing that the basal ganglia do indeed play a crucial role in the brain. Fortunately, this role also seems to be related to certain functional weaknesses of the earlier theories as well. It seems to involve the organization of time or "tasks" into large blocks or "chunks," which make it easier for the system to do long-term planning. Thus once again, by working to improve the functional capabilities of our theory, we can expect to converge more closely to neuroscience as well. For the complete form of our new theory, we also revisit the treatment of space, which immediately suggests new interpretations of some well-known information about vision in the neocortex.

Relation to Prior Theories: Basics of the Mathematics

The new theory continues to assume that there is a kind of master-slave relation between the "upper brain" and the "lower brain," as in [3]. It replaces the old picture of the upper brain with something far more complex.

The previous picture of the upper brain was still essentially based on Figure 3 (with various elaborations discussed in the engineering literature[5]). Figure 3, in turn, was originally derived from an effort to approximate dynamic programming. Dynamic programming is the only exact and efficient general-purpose algorithm available to calculate the optimal strategy of action over time in an uncertain, noisy, nonlinear environment.

In dynamic programming, the user supplies a utility function U(**X**) and a stochastic model to predict **X**(t+1) as a function of **X**(t) and **u**(t), where **X** denotes the present state of the environment and **u** denotes the actions taken by the controller at time t. (For simplicity I will assume that the state of the world as seen by the senses, **X**, is the same as the actual state of the world, **R**; however, the difference between the two is an important part of the theory.) Our task, in control, is to find the optimal strategy of action **u**(**X**). We do this by solving the Bellman equation:

$$J(\underline{X}(t)) = U(\underline{X}(t)) + \max_{\underline{u}(t)} \langle J((\underline{X}(t+1)) / (1+r) \rangle \quad (1)$$

where the angle brackets denote expected value, and where this "r" is a kind of interest rate which can be set to zero for now. Our task is to find a function J(**X**) which satisfies this equation. According to the theorems of dynamic programming, the optimal strategy **u**(**X**) can be found simply by picking the actions which maximize <J(**X**(t+1))>, exactly as in equation 1.

Pure dynamic programming is not useful even in medium-scale control applications, because it is simply too difficult to find an exact solution to equation 1, in the general case. In approximate dynamic programming (ADP), we try to learn an approximate solution. In the simplest forms of ADP, we approximate the function J by some network J(**X**,W), where "W" represents the set of weights or parameters (e.g., synapse strengths) in the network. The goal is to adapt the weights W, so as to make J(**X**,W) fit equation 1 as closely as possible, in some sense. The equations used to adapt the weights W effectively imply a theory for how the synapse strengths are adapted over time in the corresponding biological neural network.

In the simplest previous theory, the Critic would be adapted as follows. At every other time t, we would remember **X**(t), and then wait until time t+1 before performing adaptation. At time t+1, we would calculate J*=U(**X**(t))+J(**X**(t+1)). Then we would go back to considering **X**(t). We would adjust the weights W so as to make J(**X**(t),W) match more closely to J*. I called this approach Heuristic Dynamic Programming (HDP) when I first published it in 1977 [20]; however, it was rediscovered in 1983 under the name "Temporal Differences"[18].

There are many variations of this approach, but the ones which work all have the same general effect: they update the estimate of J(t) so as to account for information one time period ahead, from time t+1. In effect, after training, the organism "sees into the future" one time period further than it did before learning. But if one time period is a very short interval of time (such as 0.25 second!), this may not be much of an improvement in foresight. Furthermore, it would take considerable training, over many possible states **X**(t), before the system could look ahead one time period further across all of these possible states.

This "foresight horizon problem" is not new. I discussed it, for example, in [27,28]. There are many ways to try to improve the degree of foresight, even for this class of designs; some of these approaches seem quite promising, while others have been proven incorrect [7,28]. Nevertheless, whatever tricks or design modifications one uses, there is no substitute for the simple effort to use "t+T" in place of "t+1," where T is a larger time interval. More precisely, by using "t+T" instead of "t+1," in addition to other design improvements, we can still expect to achieve much greater foresight. The problem lies in how to do this, within the framework of a learning-based neural network design. The new theory provides a solution to this problem. In AI, this problem is commonly referred to as the problem of "temporal chunking."

Another major deficiency of the previous theories concerns the treatment of space. In AI, for example, the problem of "spatial chunking" has received a lot of attention. John Holland has talked about road maps as an example of a compressed representation of space. Albus [29] has discussed the problem of coordinating the actions of diverse robots spread over a large region of space such as a factory floor.

Even in the new theory, the treatment of space is seen as a subsystem issue. In principle, it does not affect the highest-level structure of the mathematical theory. Individual mammals, unlike factory management systems, are not really spread out over a large region of space. Humans and other mammals are not 100% flexible about "doing two things at the same time."

Nevertheless, the treatment of space really does have some pervasive implications. There are some basic principles of symmetry which have been exploited very effectively in the image processing part of the neural network community [30,24,25], but largely ignored in other parts of the community.

The new theory assumes that the brain exploits a similar but more general form of symmetry, which I call "object symmetry"[1]. In essence, object symmetry simply predicts that objects will behave the same way, as a function of their relations to other objects, regardless of where they are physically located or seen. It is extremely important to exploit this principle, when trying to adapt to a very complex environment which contains thousands of significant objects and locations in it, but traditional neural networks do not do so.

In principle, there are three ways to train neural networks to exploit object symmetry. One way is to broadcast connections and weights from a "master copy" of a small neural network; this is not biologically plausible, in my view. Another way involves the use of special tricks to encourage separate but similar regions to learn the same relationships; for example, rapid eye movements may result in many areas in visual cortex scanning the same object, and learning from it. In my view, this kind of parallel learning is an important part of what happens in neocortex; for example, it explains the ability of adults to classify all the letters in a word at the same time, in parallel, in one major cycle time (circa 0.1-0.25 second). On the other hand, such parallel training does not exploit the full theoretical potential implied by symmetry; it requires a lot of learning experience.

The third approach is multiplexing -- the use of the same network, at different times, with different inputs, in order to recognize and predict different objects, accounting for their relations to a few related objects. The literature on dynamic feature binding in visual physiology strongly suggests that the brain does exploit the multiplexing approach, particularly in the regions of temporal cortex which recognize objects and (in humans) link up to the nearby speech centers. A key part of the new theory deals with methods for defining and adapting networks which map back and forth between ordinary vectors $\underline{\mathbf{X}}$ and structures made up of objects and relations obeying object symmetry [1]. This in turn would have substantial implications for the organization of neocortex and hippocampus.

In the brain, this approach highlights the need for the organism to maintain a kind of global "world model" [29] or inventory of objects in associative memory, even as the organism really focuses on one (or a few) objects at any time. Fukushima [9] has presented a theory of how this might be done, by patching together visual images, but his theory does not exploit the links between images and places which are crucial to optimal performance. In our theory, we can use essentially the same sort of architecture, except that visual images are replaced by a cluster of objects and relations. Because such a cluster would include links between specific objects or places, better performance could be expected.

Temporal Chunking: Reasons for a Task-Oriented Approach

Strictly speaking, temporal chunking presents an opportunity, not a problem. The conventional t/(t+1) designs do converge, in theory, to an optimal strategy -- eventually. My claim is that new designs, exploiting temporal chunking, can converge faster, when foresight over long periods of time is required.

In classical AI, there are two common forms of temporal chunking -- multiresolutional or clock-based approaches, and task-based or event-based approaches. This section will try to explain, a bit more precisely, why these approaches can improve foresight, and why the task-based approach is probably preferable.

For simplicity, consider the situation of an environment which can only exist in N possible states, where N is finite. Thus the state of the environment at any time, t, can be represented as some integer i
between 1 and N. Consider the situation where the strategy of action $\underline{\mathbf{u}}(i)$ is fixed, and we are simply trying to learn/approximate the function J(i). For more explanation of this simple case, see [30,31].

In this situation, the utility function U can be represented as a vector with N components, U_1 through U_N, where U_i is simply U(i). Likewise, the function J(i) can be represented as a vector $\underline{\mathbf{J}}$. The model of the environment simply consists of a set of numbers, P_{ij} , representing the probability that the environment will occupy state i at time t+1 if it was in state j at time t. P is just a matrix of transition probabilities, and is called a Markhov chain. Let us define $M=P^T$, the transpose of P.

In this situation, the J function used in the Bellman equation has a simple representation:

$$\underline{\mathbf{J}} = \underline{\mathbf{U}} + M\underline{\mathbf{U}} + M^2\underline{\mathbf{U}} + M^3\underline{\mathbf{U}} + \dots \qquad (2)$$

The first term on the right represents the utility U in the current state (time t), the second term represents the utility in the next state (time t+1), and so on. (Roughly speaking, if $\underline{\mathbf{X}}(t+1)=P\underline{\mathbf{X}}(t)$, then $\underline{\mathbf{U}}^T\underline{\mathbf{X}}(t+1) =$
$\underline{\mathbf{U}}^TP\underline{\mathbf{X}}(t) = (M\underline{\mathbf{U}})^T \underline{\mathbf{X}}(t)$.) $\underline{\mathbf{J}}$ represents the sum of utility over all future time periods, which is what we really want to maximize. From equation 2, we may immediately deduce the Bellman equation for this special case:

$$\underline{\mathbf{J}} = \underline{\mathbf{U}} + M\underline{\mathbf{J}} \qquad (3)$$

The usual adaptive critic methods basically approximate an old update rule discussed by Howard [30]:

$$\underline{\mathbf{J}}^{(n+1)} = \underline{\mathbf{U}} + M\underline{\mathbf{J}}^{(n)} \qquad (4)$$

This states that our n+1-th estimate of the J function is based on the n-th estimate as shown in this equation. If we initialize $\underline{\mathbf{J}}^{(0)}$ to equal $\underline{\mathbf{U}}$, then it is easy to deduce that:

$$\underline{\mathbf{J}}^{(n)} = \underline{\mathbf{U}} + M\underline{\mathbf{U}} + \dots + M^n\underline{\mathbf{U}} \qquad (5)$$

In other words, after n such very large updates (each accounting for every possible state of the environment!), our J function "sees" only n steps into the future.

In matrix algebra, there is a much faster way to calculate the same sort of result. We may write it as:

$$\underline{\mathbf{J}}^{(n)} = (I + M^{2^n}) \dots (I + M^8)(I + M^4)(I + M^2)(I + M)\underline{\mathbf{U}} \qquad (6)$$

(As an example, notice how $(I+M^4)(I+M^2)(I+M) = I+M+M^2+M^3+...+M^7$!)
This sort of update procedure allows us to see more than 2^n time periods into the future after only n updates! Foresight grows exponentially, not linearly. The price is that we need to compute the matrices M^2, M^4, etc. But notice what these matrices represent! $(M^8)_{ij}$, for example, simply represents the probability of ending up in state i at time t+8, if you start out in state j at time t. These matrices are simply predictive models of the environment, jumping ahead 2 time periods, 4 time periods, etc. Thus equation 6 simply represents the classical idea of multiresolutional analysis, based on models that predict over large but fixed time intervals.

In the real world, however, there is a serious problem here. First of all, it is hard to imagine a biological basis for such an arrangement. Second of all, the original matrix M will usually be extremely sparse, but the matrix M^k for large k will tend to be fully connected. If we do not somehow exploit the sparsity of M, then we do not achieve the full potential for cost reduction and efficiency.

In developing the new model, I began by assuming that the states of the environment can be divided up into blocks, A, B, etc. This led to a modified Bellman equation which can be written, crudely, as:

$$\underline{\mathbf{J}}^{A} = \underline{\mathbf{J}}^{A0} + J^{AI}\underline{\mathbf{J}}^{A+}, \qquad (7)$$

where $\underline{\mathbf{J}}^{A}$ represents the J function for states within block A; where $\underline{\mathbf{J}}^{A0}$ represents the payoffs (U) to be expected in the future within block A, before the system next exits that block; where $\underline{\mathbf{J}}^{A+}$ represents the J function for states immediately outside block A, which I call the "postexit" states of A; and where J^{AI} is a kind of connection matrix.

In order to learn J most effectively, one can begin by learning $\underline{\mathbf{J}}^{A0}$ and J^{AI} within each block; this learning is done on a purely local basis, because it does not require knowledge of anything outside the block. Then, after these "local critics" have been adapted, one can use equation 7 to propagate values backwards from the end of the block ($\underline{\mathbf{J}}^{A+}$) to the initial states at the beginning of the block, which also serve as the postexit states of the previous block. In other words, equation 7 can act like equation 4, except that an update jumps over the entire time-history of a block in one direct calculation.

To make full use of this kind of procedure, one needs to develop a hierarchy of blocks within blocks (somewhat analogous to equation 6). One also needs to consider how we have a choice between blocks or tasks, at certain times. In reality, the choices we face are often very discrete in nature. (Consider, for example, the choice of where to place the next stone, in a game of Go.) Because they are discrete choices -- each representing a kind of local optimum -- we need to use a crisp, discrete decision-making system, very different from the incremental and fuzzy value systems which work best at the highest and lowest levels of the brain. (Actually, the new theory does allow for fuzzy and incremental decision making at the middle level, within each decision block.)

The greatest challenge in developing the new theory was to find a smooth and effective approach to approximating this kind of design with neural networks, and to maintain a learning-based approach throughout.

Overview of the Structure

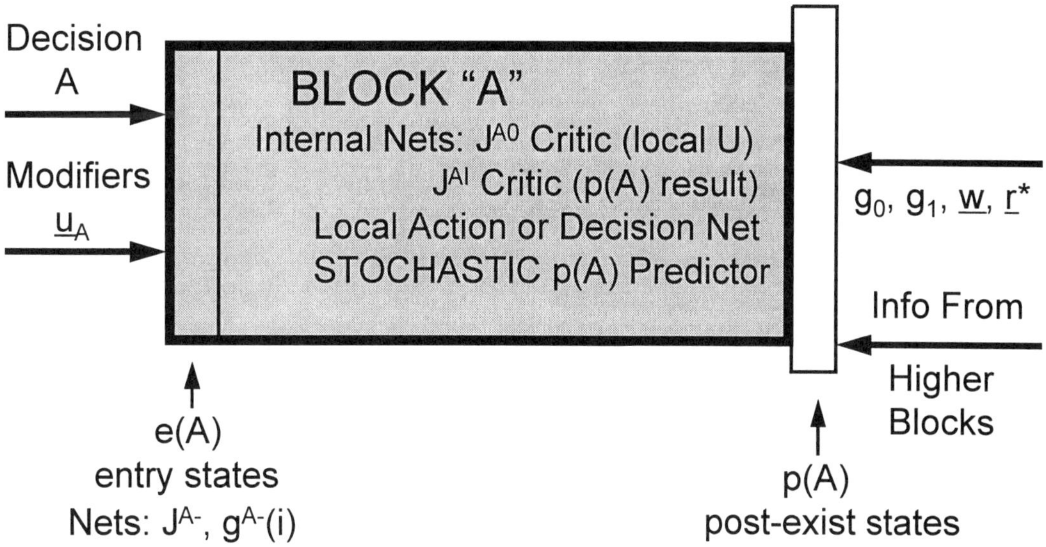

Figure 4. Structure of a Decision Block

According to the new theory, the "middle level" of the brain is made up, in effect, of a library of "decision blocks," illustrated in Figure 4. In principle, each decision block is separate from every other block; however, as in any neural network design, a sharing of hidden nodes and of inputs is allowed.

Each decision block has a "level." When decision block A is activated, it then has the job of deciding which lower-level decision block it should activate. The system is sequential. In other words, at any given time, only one block is activated at any level; however, new decision blocks can be constructed as the attempt to activate two unrelated blocks on the same level at the same time. (Usually, however, it takes some time for the system to learn how to do two specific things at once in a smooth manner.) In the brain, the "level" would probably correspond to the level of the relevant loop in the neocortex-basal-thalamus reverberatory circuit, as described by Brooks [16]. (In theory, however, there could also be a kind of linked-list connection between a decision and its subordinate.)

The trickiest part, in any decision block, is how to modulate the actions in that block, so as to better account for what later decision blocks will want after the current task is completed. This is done by providing the secondary decision information suggested in the figure: "adverbs" $\underline{\mathbf{u}}_A$, a fuzzy goal image **g**, and some sort of condensed representation of longer-term goals. In effect, this information provides a kind of condensed representation of the vector $\underline{\mathbf{J}}^{A+}$ of equation 7. The fuzzy goal image approximates J, in effect, as:

$$g_0 + g_1 \sum w_i (R_i - R_i^*)^2 \qquad (8)$$

This representation may seem a bit arbitrary, at first, but it emerges rather strongly from careful consideration of the difficulty of sharing communications between decision blocks, and of the limitations of possible alternatives [1]. This is a strong theoretical prediction -- that explicit images of goals or of desired states of the world play a central role in the effectiveness of human planning. Perhaps the word "goal" really is another one of those four-letter words representing a hard-wired invariant of human thought.

Within each decision block, as before, there is also a need to adapt local critic networks, representing the $\underline{\mathbf{J}}^{A0}$ and J^{AI} of equation 7. Another network $\underline{\mathbf{J}}^{A+}$ can improve the accuracy, in principle, of the interface between this block and its successors; however, it is not strictly necessary. Yet another network $\underline{\mathbf{J}}^{A-}$ is needed to help decision block A to formulate the goals $\mathbf{g}^{A-}$ which it could pass back to the previous decision blocks.

Another key feature of this design is the need for a kind of decision network within each block, analogous to the action network of Figure 3. At this level, however, the concern about local optima and the need for exploration of alternatives is very great; therefore, the decision networks must be stochastic networks, which propose possible decisions, based on some notion of their expected payoffs. A key part of the new design [1] is an approach to training such stochastic networks, which tries to make them output options based on a Gibbs function of their expected value (in effect, J). This function includes a kind of global "temperature" parameter, which is compatible with discussions by Levine and Leven [32], among others, of global variations in novelty-seeking behavior in humans and other organisms. The assumption here is that the neocortex outputs these possible options, but that the discrete go/no-go choices are made in the basal ganglia (except for "decisions" output directly from the motor cortex to the lower brain). This is also consistent with Grossberg's view that data on the basal ganglia imply that discrete, binary go/no-go decissions are made in that part of the brain. The local critics provide a basis for making these final choices, as a function of the global "temperature."

Finally, the decision blocks must each contain a kind of internal Model, similar in a way to the Model network of Figure 3. This can be useful as part of the system to adapt the other networks, as in Figure 3. But it provides another important capability: the capability to simulate possible decisions and results. This capability for simulation or imagination or "dreaming"[22] is crucial to actually exploiting the benefits of large-scale temporal chunking. If a decision block represents a task which takes T time periods to unfold, then it should be possible to consider T alternative possible decisions or tasks or scenarios during a period when only one actual task is accomplished! In the brain, the multiplexing of different decision levels may reduce this benefit somewhat, but one would still expect that more time is spent considering future possibilities than in tracking present reality.

Another crucial aspect of this design is that the Model within a higher-level decision block must be truly stochastic, like the SEDP design in [7, ch.13]. The reasons for this are discussed further in [1], along with the details of how to adapt these various component networks.

References

[1]P.Werbos, A Brain-Like Design To Learn Optimal Decision Strategies in Complex Environments, in M.Karny, K.Warwick and V.Kurkova, eds, *Dealing with Complexity: A Neural Networks Approach*. Springer, London, 1997.

[2] P.Werbos, *The Roots of Backpropagation: From Ordered Derivatives to Neural Networks and Political Forecasting*, Wiley, 1994.

[3] K.Pribram, ed., *Learning as Self-Organization* , Erlbaum 1996.

[4] K.Pribram, ed., *Origins: Brain and Self-Organization*, Erlbaum, 1994, p.46-52.

[5] P.Werbos, Optimization: A Foundation for understanding consciousness. In D.Levine & W. Elsberry (eds) *Optimality in Biological and Artificial Networks?*, Erlbaum, 1996.

[6] Bitterman ME 1965 The evolution of intelligence *Scientific American* January

[7] . D.White & D.Sofge, eds, *Handbook of Intelligent Control*, Van Nostrand, 1992.

[8] P.Werbos, Optimal neurocontrol: Practical benefits, new results and biological evidence, *Proc. World Cong. on Neural Networks(WCNN95)*,Erlbaum, 1995

[9] *Proc. Int'l Conf. Neural Networks (ICNN97)*, IEEE, 1977.

[10] P. Werbos, Neurocontrollers, in J.Webster, ed, *Encyclopedia of Electronics and Electrical Engineering*, Wiley, forthcoming.

[11] P.Werbos, The elements of intelligence. Cybernetica (Namur), No.3, 1968.

[12] W.Nauta & M.Feirtag, *Fundamental Neuro-anatomy*, W.H.Freeman, 1986.

[13]J.Von Neumann and O.Morgenstern, *The Theory of Games and Economic Behavior*,Princeton NJ: Princeton U. Press, 1953.

[14] P.Werbos, The cytoskeleton: Why it may be crucial to human learning and to neurocontrol, *Nanobiology*, Vol. 1, No.1, 1992.

[15] Miller GA Galanter EH and Pribram K 1960 *Plans and the Structure of Behavior* (New York: Holt, Rinehart and Winston)

[16] Vernon Brooks, *The Neural Basis of Motor Control*, Oxford U. Press, 198_.

[17] D.O.Hebb, *The Organization of Behavior*, Wiley, 1949.

[18] A.Barto, R.Sutton and C.Anderson, Neuronlike adaptive elements that can solve difficult learning control problems, *IEEE Trans. SMC*, Vol. 13, No.5, 1983, p.834-846.

[19] B.Widrow, N.Gupta & S.Maitra, Punish/reward: learning with a Critic in adaptive threshold systems, *IEEE Trans. SMC*, 1973, Vol. 5, p.455-465.

[20] P.Werbos, Advanced forecasting for global crisis warning and models of intelligence, *General Systems Yearbook*, 1977 issue.

[21] T.Landelius, *Reinforcement Learning and Distributed Local Model Synthesis*, Ph.D. thesis and Report No.469, Department of Electrical Engineering, Linkoping U., 58183, Linkoping, Sweden.

[22] P.Werbos, Building and understanding adaptive systems: A statistical/numerical approach to factory automation and brain research, *IEEE Trans. SMC*, Jan./Feb. 1987.

.[23] P.Werbos and A.Pellionisz, Neurocontrol and neurobiology, *Proc. Int'l Joint Conf. Neural Networks (IJCNN92)*, IEEE, 1992.

[24] P.Werbos & X.Z.Pang, Generalized maze navigation: SRN critics solve what feedforward or Hebbian nets cannot.*Proc. Conf. Systems, Man and Cybernetics (SMC)* (Beijing), IEEE, 1996. (An earlier version appeared in WCNN96.)

[25] X.Z.Pang & P.Werbos, Neural network design for J function approximation in dynamic programming, *Math. Modelling and Scientific Computing* (a Principia Scientia journal), special issue on neural nets planned for circa December 1996. See also links on www.glue.umd.edu/~pangxz

[26] J.Houk, J.Davis & D.Beiser (eds), *Models of Information Processing in the Basal Ganglia*, MIT Press, 1995.

[27] W.T.Miller, R.Sutton & P.Werbos (eds), *Neural Networks for Control*, MIT Press,1990, now in paper

[28] P.Werbos, Consistency of HDP applied to a simple reinforcement learning problem, *Neural Networks*, March 1990

[29] J.Albus, Outline of Intelligence, *IEEE Trans. Systems, Man and Cybernetics*, Vol.21, No.2, 1991.

[30] R.Howard, *Dynamic Programming and Markhov Processes*, MIT Press, 1960.

[31]D.Bertsekas, *Dynamic Programming and Optimal Control*, Belmont, MA: Athena Scientific, 1995.

[32] D.S.Levine and S.J.Leven, eds, *Motivation, Emotion, and Gvoal Direction in Neural Networks*, Erlbaum, 1992

PREFERENCE

4

Stimulus Class Formation in Animals

Thomas R. Zentall
Department of Psychology
University of Kentucky
Lexington, KY 40506-0044
Telephone: 606-257-4076/Fax: 606-323-1979
E-Mail: Zentall@pop.uky.edu

Abstract

The simplest kind of stimulus class is one defined simply by a common trained response. Other stimulus classes do not require training of each of the members but they derive their membership from the physical similarity among class members through the process of stimulus generalization. Stimulus classes of greater interest are nonsimilarity based classes for which membership implies properties not explicitly experienced, or so-called *emergent* properties. Nonsimilarity-based or arbitrary classes can develop when two or more arbitrary stimuli become directly associated through their pairing with a common event. These classes appear to develop during matching-to-sample training when two or more samples are associated with the same comparison (i.e., many-to-one matching).

Convergent evidence will be presented for the development of emergent relations between these samples (or for their common coding) in the form of: (1) Transfer: When a subset of these samples is associated with new comparisons there is evidence that the remaining samples are also associated with the new comparisons. (2) Interference/facilitation: When a subset of these samples is associated with new comparisons and a delay is inserted between the samples and new comparisons, interpolation of one of the remaining samples into the delay will interfere with or will facilitate matching accuracy depending on whether the interpolated sample was paired with the same comparison as the current sample in original many-to-one training. (3) Partial versus total reversal: When originally trained sample-comparison associations are reversed, reversals are faster when all of the associations are reversed than when only half of the associations are reversed (and the rest remain relevant). (4) Samples associated with the same comparison are harder to discriminate from each other (in a simple successive discrimination) than are samples associated with different comparisons. (5) The retention functions (matching accuracy as a function of delay inserted between the sample offset and comparison onset) for samples associated with the same comparison are much more similar in slope than those same retention functions when samples are associated with different comparisons.

The relation between this kind of stimulus class formation and formal (Sidman) stimulus equivalence will be discussed. Specifically, direct evidence for the three defining characteristics of stimulus equivalence (reflexivity, symmetry, and transitivity) in animals will be presented. Finally, the relation between stimulus class formation and human language will be addressed.

A stimulus class is defined by any group of stimuli that have some property in common. The simplest kind of stimulus class is one defined by a common response. All of the items in a refrigerator can be thought of as a stimulus class because they share a common location and can each be obtained by making a similar response (opening the door of the refrigerator). Stimulus classes of this type typically derive their membership from specific experience (i.e., having placed something in the refrigerator). Representing the members of a class by a common tag or label can serve as a useful and efficient means of storing, retrieving, and communicating information (e.g., about where the item can be found). But if the only relation among members of such a stimulus class is their physical location, such a class would have few implications for the organization of information beyond what was already known about specific experienced responses.

Stimulus classes of greater interest are those for which membership implies properties not explicitly experienced. In the case of the refrigerator example, if I knew that an novel food object had been placed in the refrigerator, I might conclude that it would likely spoil relatively quickly at room temperature. This would be an inference, or emergent property of the novel substance, based on learning about other objects that are typically placed in a refrigerator. The kind of stimulus class of interest here is that for which untrained properties can be shown to emerge. In other words, they are classes in which members are not just related to the class but are also related to each other, such that, unless explicitly trained otherwise, whatever happens to one member of the class will have an effect on the other members.

Consider for a moment the stimulus class DOG as it might be used by humans. Not only might this class contain experiences with a wide variety of dog breeds, but it might also contain pictures of dogs, the sound of the spoken word "dog," and the written word "DOG." The value of such as stimulus class is in our ability as humans to use information about one member of the class to make inferences about other members of the class. If you saw a dog bark, you might assume (whether correctly or not) that other dogs bark as well. Similarly, if you were told that there was a widespread outbreak of rabies among dogs, that might change your behavior in the presence of dogs that you encountered (other members of that class).

Perhaps the most important benefit to an organism of emergent relations among members of a stimulus class is when the value of one member of the class changes (e.g., is associated with fear or pain) and the organism can react appropriately, not only to that member of the class, but to other members of the class as well. Thus, emergent relations among members of a stimulus class allow a specific learning experience involving a change in value to have wide spread effects on the value of unexperienced events in the class. The capacity for emergent relations allows organisms to extend learning far beyond the events experienced.

In the remained of this chapter I will describe the various emergent relations that can develop among stimuli in a class and will present evidence of these relations. I will also address some of the controversies in the literature concerning emergent relations among stimuli in a class by humans and nonhumans.

Stimulus Classes that Show Emergent Relations

Similarity Based Classes: Stimulus Generalization

The simplest kind of emergent stimulus property is one based on stimulus similarity. Given that an organism has been trained to respond to a particular stimulus, it also will tend to respond to other stimuli in proportion to their similarity to the training stimulus. Thus, there is generally a monotonic relation between the similarity of the test stimulus to the training stimulus and the strength of the response. This function, relating stimulus similarity to response strength, is generally referred to as the *gradient of stimulus generalization.* To the extent that such generalized responding reflects the organisms inability to distinguish between the training stimulus and similar test stimuli (e.g., because the training and test stimuli may share many stimulus elements, Estes, 1950) the phenomenon may be of interest primarily to map an organism's sensory capacity (i.e., determine psychophysical functions for the animal). Generalization gradients have also served as a tool for the assessment of stimulus control (see, e.g., Honig & Urcuioli, 1981). If, for example, one trains a pigeon to respond to a red light in the presence of a 1000 Hz tone, one can ask to what extent the hue and the tone control responding by systematically varying each.

But generalization gradients can have theoretical implications as well. For example, some would argue that if an organism's ability to detect stimulus differences varies as a function prior experience, it suggests the presence of theoretically important psychological processes such as attention (Lashley & Wade, 1946), and perceptual learning (Hall, 1991). Thus, obtaining a flat generalization gradient following a particular kind of experience may suggest that an animal is not attending to stimuli that would otherwise gain control of responding.

Alternatively, stimulus generalization may reflect a process quite different from the inability to detect stimulus differences. Hull (1943) has suggested that animals have a natural tendency to respond to stimuli similar to those experienced in training, even when those stimuli can be readily discriminated from the training stimuli and from each other. Consistent with this hypothesis is the finding that a declining gradient of stimulus generalization does not depend on the animal's prior experience with discriminations involving the tested dimension (Peterson, 1962; Riley & Leuin, 1971). Apparently, the detection of stimulus differences can occur in the absence of prior discrimination experience.

Although stimulus generalization has important implications for general theories of learning (see e.g., Mostofsky, 1965), I will not consider such examples of stimulus classes in assessing the ability of animals to demonstrate emergent relations among stimuli.

Nonsimilarity Based Classes

Similarity based classes can be defined by the relatively high correlation between stimulus similarity and the strength of the response. Nonsimilarity based classes are defined by the relative absence of such a correlation.

Natural Classes

When class membership is defined by a particular element present or absent in a complex polymorphic stimulus display, it is often referred to as a natural class because the constituent elements usually are not under experimental control. Photographic slides of natural scenes are often used in this kind of research and the relevant element defining a correct response is typically the presence versus the absence of a person or thing which itself may vary considerably over exemplars (e.g., Herrnstein & Loveland, 1964). For example, members of a "person present" class may differ in size, age, race, sex, amount and type of clothing, portion of the person that can be seen, and even number of people present. On the other hand, many of the elements present in the "person present" slides are also present in the "person absent" slides (e.g., trees, cars, houses). Thus, what characterizes the exemplars in such a natural present/absent class is the great variability in the nature of the elements constituting class membership as well the great similarity in elements common to both classes. In fact, it has been argued that any attempt to extract a feature common to all the members of the "present" class but absent from the "absence" class must surely fail (Herrnstein, Loveland, & Cable, 1976). Alternatively, given the complexity of the stimulus displays, it can be argued that such a presence/absence class may be definable by the presence or absence of one or more of a small number of elements. Furthermore, any demonstration of transfer to a new set of exemplars may be accounted for in terms of the similarity of the test stimulus to one or more of the training stimuli from the same class (see Astley & Wasserman, 1996; Greene, 1983).

Although difficulties inherent in the use of natural pictures as exemplars in stimulus class experiments with animals prevent making strong conclusions about the results, the variety of published convergent evidence using these presence/absence procedures is impressive. Among the natural concepts studied have been "trees"and "water" (Herrnstein et al., 1976), "pigeons" (Poole & Lander, 1971), "fish" (Herrnstein & de Villiers, 1980), "leaves" (of a particular species of tree, Cerella, 1979), and even "a particular person" (among exemplars involving other individuals, (Herrnstein et al., 1976). What is also impressive in this line of research is the nature of the typical error made by pigeons. Herrnstein et al. noted, for example, that pigeons often made errors to exemplars that humans might assign to the boundaries of the class (e.g., a tree-present response might be made to a picture of a bunch of celery or a picture of an ivy-covered wall and a tree-absent response might be made to a scene involving a distant tree that might not be detected by a human observer).

Preexisting Stimulus Classes

Occasionally, when stimuli have been varied along a specifiable stimulus dimension, preexisting classes have been found in animals. These classes differ from similarity based classes in that over a broad range of stimulus values, there appears to be little change in the strength of the response, however, at the boundary of the class, there is an abrupt change in response strength. An example of such dimensional classes in humans are color names. For example, when presented with a broad range of long-wave-length hues humans typically respond with the color name "red." At some point, as the wave length gets sufficiently short, the name of the hue will suddenly change to "orange."

It is tempting to attribute these language-based categories that humans use so readily to the acquired use of verbal labels (i.e., when a range of wavelengths is consistently referred to as "red," it may take on a common property, see e.g., Lawrence, 1963). However, the similarity of the categories that humans use, over a large number of unrelated and often isolated cultures, suggests that these color categories are probably determined more by our biology than by our experience (Berlin & Kay, 1969; Heider, 1972).

Remarkably, pigeons too show evidence of color naming classes (Wright & Cumming, 1971; Zentall & Edwards, 1984; Zentall, Hogan, Howard, & Moore, 1978; Zentall, Jackson-Smith, & Jagielo, 1990; Zentall, Jackson-Smith, Jagielo, & Nallan, 1986). Wright and Cumming demonstrated color naming classes following matching-to-sample training by comparing the point of subjective indifference for various pairs hue samples. If pigeons treated the color spectrum as a continuum, one would expect that presenting as samples various hues that fell between the two hues used in training would produce a psychophysical choice function with the point of indifference at the psychological "midpoint" between the two training samples. Instead, they found that the point of subjective indifference was relatively independent of the training sample values. Rather, the points of indifference were virtually constant at about 540 nm and 595 nm. Wright and Cumming concluded that pigeons had three color "names": One for all the hues below 540 nm (the reds, oranges, and some of the yellows), one for all the hues between 540 nm and 595 nm (the rest of the yellows and some of the greens), and one for all the hues above 595 nm (the rest of the greens and all of the blues). In training, these pigeons had presumably learned to match the comparison hue to the sample hue. When these pigeons encountered test trials on which there was no matching hue, they responded to the comparison that came from the same color class as the sample.

Converging evidence for color classes has come from research in which pigeons learned two independent two-hue conditional discriminations, red-green, and yellow-blue (Zentall et al., 1978). Following acquisition, one of the hues from one task replaced one of the comparisons from the other task. For example, if in training the pigeon had learned to respond to the red and not the green comparison when the sample was red, test trials might involve a red sample with comparisons that were red and yellow, red and blue, yellow and green, or blue and green. When the comparisons were red and blue or yellow and green, pigeons consistently chose the red or yellow comparison. When the comparisons were red and yellow or blue and green, comparison choice was virtually random. Pigeons acted as if they had coded the red and yellow samples similarly. Thus, when either one of these hues was available as a comparison, it was selected. When both or neither hue was available, comparison choice was inconsistent. Similar results were found with yellow samples when either the yellow or the blue comparison was replaced with either red or green. A similar pattern of results found for the green and blue samples suggested that green and blue samples were coded using a common class. The possibility that the pigeons were merely confusing red and yellow (and blue and green) comparisons was ruled out by training some pigeons on an oddity task (e.g., when the sample was red a response to a green rather than a red comparison was correct) with one pair of hues and on a matching task with the other pair (i.e., yellow and blue). In this case, when the sample came from the presumed red/yellow color class, the correct response should have been to green (if the sample was red) or to yellow (if the sample was yellow). A critical test was a trial on which the sample was red and the comparisons were red and yellow. According to the rule that they had, presumably, learned, the pigeons should have consistently chosen the yellow comparison -- and they did. Had the pigeons simply confused red with yellow comparisons, however, they still should have been confused -- but they were not.

Evidence for the third, yellow-green, color class was found when the original hue pairs in training were red and yellow, and green and blue (Zentall & Edwards, 1984). In this experiment, although the pigeons appeared to code yellow and green samples similarly, there was no evidence for the common coding of red and blue samples.

Evidence has also been found for the presence of preexisting higher-order stimulus classes involving hues and shapes (Zentall et al., 1986). When pigeons learned one matching task involving red and circle and another involving green and plus, they responded consistently whenever the comparisons were one hue and one shape, but they responded inconsistently whenever the comparisons were two hues or two shapes. Again,, the possibility that there was confusion between the two hues and between the two shapes was ruled out by finding similar results when training involved one matching task and one oddity task.

Common Coding Classes

What characterizes the preexisting classes discussed in the previous section is that they were not explicitly trained but were merely demonstrated by the ambiguity of the training conditions (e.g., flowers) and to make a different new response (e.g., peck the bottom right response key) to exemplars of one of the categories from the other category pair (e.g., people). On test trials, the pigeons made the appropriate new responses to exemplars of the two remaining categories (i.e., chairs, top left and cars, bottom right). Thus, the pigeons appear to have learned two arbitrary, superordinate categories based on (but not restricted to) their common association with a particular response location. Although natural stimuli were used, the importance of this study was that two different unrelated "natural classes" formed a new arbitrary class. The Wasserman et al. procedure is actually a special case of many-to-one matching in which following training, one pair of the original samples is associated with new comparisons and in test, the relation is assessed between the remaining samples and the new comparisons (Urcuioli, Zentall, & DeMarse, 1995; Urcuioli, Zentall, Jackson-Smith,& Steirn, 1989, see Table 1). Using this transfer design, evidence for the formation of two arbitrary stimulus classes (or common codes) has been found in matching performance on test trials when pairings consistent with the presumed stimulus classes have been compared with pairings inconsistent with those presumed classes (see Figure 1).

TABLE 1
Common coding: many-to-one transfer design

Group	Phase 1	Phase 2	Test
Consistent	P --> V	P --> R	
	S --> H	S --> G	
	B --> V		B --> R
	W--> H		W--> G
Inconsistent	P --> V	P --> R	
	S --> H	S --> G	
	B --> V		B --> G
	W--> H		W --> R

<u>Note</u>. P = a white line-drawn plus, S = a white line-drawn square, V = three vertical black lines, H = three horizontal black lines, R = red, G = green. Lines were projected on a white background and shapes were projected on a white background. For each trial type the first letter represents the sample and the second the correct comparison. In test, samples from Phase 1 that were not presented in Phase 2 were presented with the comparisons from Phase 2 and reinforced responding was either consistent with or inconsistent with the presumed stimulus classes.

As can be seen in Figure 1, the degree of transfer varies considerably among pigeons. One can ask if the individual differences in transfer are predictable from the pattern of the pigeons' acquisition of many-to-one matching. If the pigeons commonly code hue and line samples associated with a common comparison, perhaps they would also acquire those associations at a relatively similar rate. In fact, consistent with this hypothesis, there was a significant correlation between the ratio of sessions to acquire the hue-sample associations to sessions to acquire the line-sample associations, and the absolute magnitude of the transfer effect. The closer to 1.0 was the ratio, the greater was the transfer (positive or negative).

there was no explicit attempt to train the pigeons to commonly code red and yellow). In the case of the stimulus classes that follow, however, an attempt is made, through explicit training, to have the animals commonly code arbitrary stimuli that should not be members of preexisting classes.

Classes defined by a common history of reinforcement. One technique that has been used to develop arbitrary stimulus classes is to randomly assign stimuli to reinforcement classes A and B (Vaughan, 1988). The stimuli Vaughan used were all photographs of trees. On any day all the members of class A were either S+s or S-s and all the members of class B were the opposite. The defining characteristic of each of two stimulus classes (i.e., the value of members of each class) was allowed to vary from session to session of training, such that the value of each class over sessions was the same. In fact, the only feature that could be used to distinguish between the two classes was the reinforcement conditions that were constant within a training session but varied between sessions. Thus, after training pigeons that responses to stimuli in set A were reinforced and those to stimuli in set B were not, the valence associated with each set was reversed, and then reversed again, repeatedly. After a large number of such reversals, Vaughan found that early in a session (i.e., after a small number of stimuli from each set had been presented), the pigeons could identify the contingencies associated with all of the members of each set and could respond appropriately. Thus, these arbitrarily assigned stimuli had become two functional stimulus classes, in spite of the fact that they shared no more in common (in terms of their physical similarity or their overall reinforcement histories) than they did with each member of the other class -- once the current status of each set was determined, the pigeons could respond appropriately to the remaining members.

There are two points of interest concerning Vaughan's procedure. First it should be noted that the stimulus classes formed are sufficiently abstract that even we humans do not have a simple name for them! The closest we can come to a class name is, "stimuli for which the schedule of reinforcement is the same on any particular day," - or more abstractly, "stimuli that always 'go together'."

On the other hand, as mentioned earlier with regard to the natural stimuli used by Herrnstein and his colleagues (e.g., Herrnstein & Loveland, 1964), although it may seem unlikely, extensive exposure to complex stimulus displays may allow the animal to extract common elements from each of the sets. The richness of the stimuli used in this research may represent both the ultimate strength and weakness of this line of research.

Stimulus classes defined by many-to-one matching procedures. The hallmark of this procedure is the use of matching-to-sample in which two (or more) samples are mapped onto each of two comparisons. The question asked is whether an emergent relation has developed between the two samples as the result of their association with a common comparison. A number of techniques for assessing the emergent relation have been reported and I will consider each in turn.

Transfer of training design. A good example of the use of transfer of training in assessing the emergent relations that develop between samples associated with the same comparison was reported by Wasserman, DeVolder, and Coppage (1992). Wasserman et al. trained pigeons to make a particular response (e.g., peck the top right response key) to all exemplars (photographs) of two different categories (e.g., flowers and chairs) and to make a different response (peck the bottom left response key) to exemplars of two other categories (e.g., people and cars). The pigeons were then trained to make a new response (peck the top left response key) to exemplars of one category (e.g.,

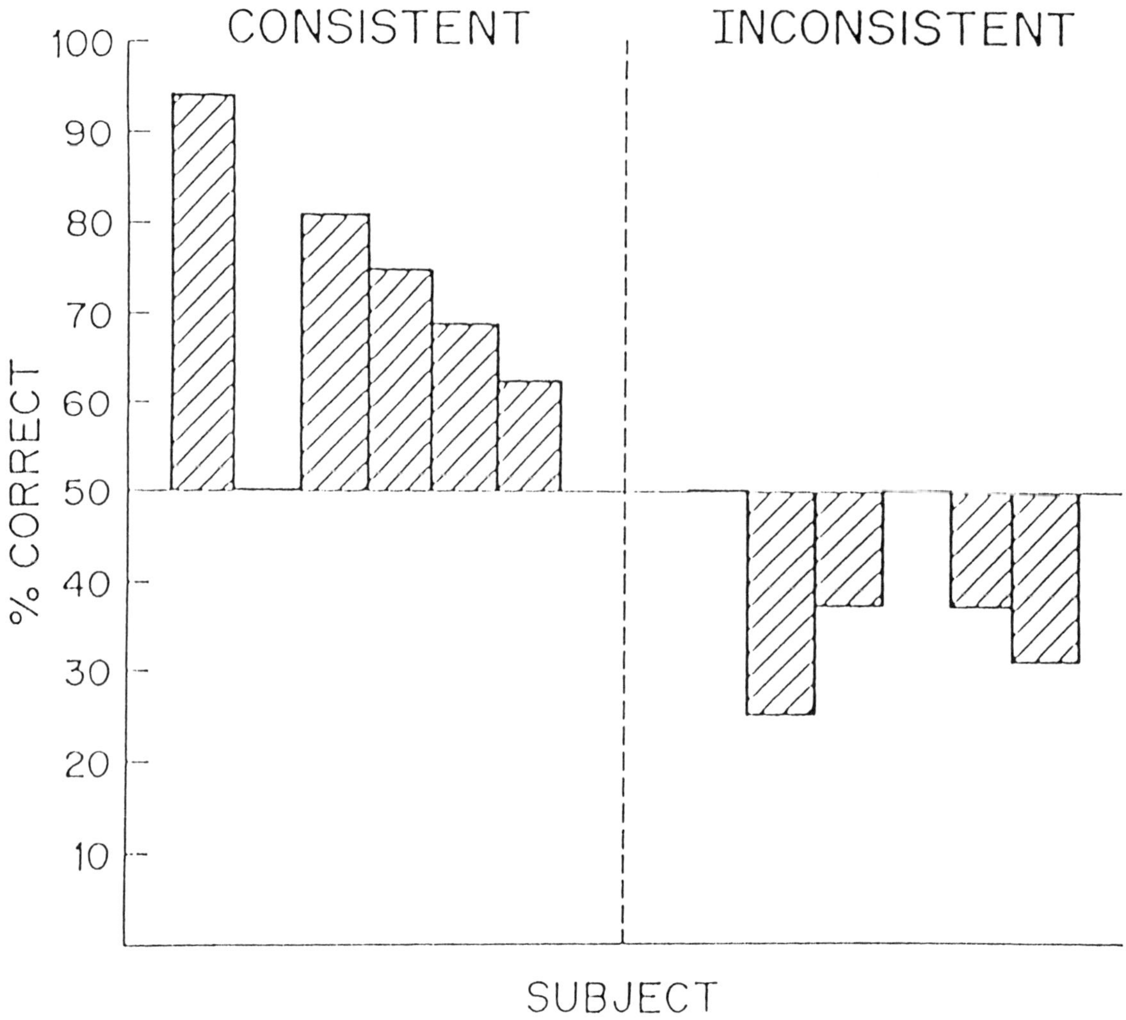

Figure 1. Data from many-to-one matching transfer experiment (Urcuioli et al., 1989, Exp. 1). Pigeons were originally trained on a many-to-one matching task with red and vertical samples mapped on to one comparison and green and horizontal samples mapped onto the other. The red and green samples were then associated with new comparisons and on test trials the remaining samples were presented with the new comparisons and correct responses were either *consistent* or *inconsistent* with the presumed common codes (see Table 1).

Interference/facilitation design. Converging evidence for the formation of arbitrary stimulus classes has been found using a related design in which following training with many-to-one matching and then interim training with one pair of the original samples and new comparisons, one of the remaining samples is inserted into the delay between the sample offset and comparison onset (Zentall, Sherburne, & Urcuioli, 1993). If samples associated with the same comparisons are commonly coded, then the insertion of a commonly coded sample should facilitate matching accuracy, while the insertion of sample having the same code as the alternative sample should interfere with matching accuracy (see Table 2).

TABLE 2
Common coding: many-to-one interference/facilitation design

Phase 1	Phase 2A	Phase 2B	Test
R --> V	R --> C	R -------> C	R ---------> C
G --> H	G --> D	G -------> D	G ---------> D
V --> V			R --- V --> C
H --> H			G --- V --> D
			R --- H --> C
			G --- H --> D

Note. R = red, G = green, V = three vertical white lines, H = three horizontal white lines, C = a white line-drawn circle, D = a white dot. All lines and shapes were projected on a black background. For each trial type the first letter represents the sample and the second the correct comparison. In Phase 2B, a 2-sec delay was inserted between the sample and comparisons. In Phase 3, the delay was filled with either of the two remaining samples on some trials or remained empty on others. Other groups received line-sample training in Phase 2 and were tested with hues inserted during the delay. For half of the pigeons the Phase 1 line-sample/line-comparison task was mismatching rather than matching.

Of particular interest in this study was the fact that pigeons that acquired the original many-to-one matching task quickly (because line-sample associations were acquired at about the same rate as hue-sample associations) showed a different pattern of results from those that acquired the task slowly. For the fast learners, relative to performance on control trials in which no stimulus was inserted into the sample-comparison delay, matching accuracy improved when the presumably commonly-coded sample was inserted into the delay and it suffered when the other sample, associated with the other code, was inserted. For the slow learners, however, relative to matching accuracy on control trials performance was disrupted whenever either remaining sample was inserted into the delay. For the slow learners, it appeared that when the remaining samples were inserted into the delay they were treated as if they were novel stimuli (see Zentall, 1973).

The different pattern of results for fast and slow learners suggests that the fast learners coded samples associated with the same comparisons similarly, but the slow learners did not. Consistent with this interpretation, was the fact that although the pair of associations involving the generally-slower-learned shapes was acquired faster by the fast learners than by the slow learners, the pair of sample-comparison associations involving the generally-faster-learned hues was acquired at the same rate by both fast and slow learners. Thus, it appears that the fast learners may have acquired the original task faster *because* they commonly coding samples associated with the same comparison. Coding the shape samples as hues may have speeded their association with the comparison stimuli. Thus, it may be that the basis for the common codes (or the "names" of the stimulus classes) that appear to result from many-to-one training may be some representation of the samples that are easier to associate with comparisons (i.e., in this case, the hue samples). This issue will come up again in the next section.

Partial versus total reversal design. Another source of support for the notion that many-to-one matching results in the formation of two arbitrary classes, each consisting of samples associated with the same comparison comes from the partial versus reversal design (Zentall, Steirn, Sherburne, & Urcuioli, 1991). If following acquisition of many-to-one matching involving hue and line-orientation samples (one hue and one line orientation associated with each comparison) the line-orientation-comparison associations are reversed, the reversal learning appears to be faster if the hue-comparison associations are reversed as well. The design of this experiment appears in Table 3.

TABLE 3
Common Coding Many-to-One Partial Versus Total Reversal

Phase 1	Phase 2		
All Groups	Group Hue	Group Line	Group Hue-Line
R --> C	R --> D	R --> C	R --> D
G --> D	G --> C	G --> D	G --> C
V --> C	V --> C	V --> D	V --> D
H --> D	H --> D	H --> C	H --> C

Note. The first letter of each pair represents the sample (R = red, G = green, V = vertical, H = Horizontal) and the second letter represents the correct comparison (C = circle, D = dot).

This design can not only tell us that samples associated with the same comparison are commonly represented by animals, but it may also suggest to us what that common code might be. Although identifying the nature of the codes used by pigeons when forming these presumed arbitrary classes may appear to be an impossible task, suggestive evidence comes from certain inconsistencies in the data. The fact that facilitated reversal learning involving line-orientation samples occurs when hue sample-comparison associations are reversed as well, but a similar advantage of a total reversal over a partial reversal is not found when the partial reversal involves the hue samples suggests that hues may form the basis for the common codes (i.e., red and vertical lines may be both coded as red, whereas green and horizontal lines may be both coded as green). This might be the case because if the common representations consist of hues and hue samples are considered, the advantage of the total reversal over partial reversal might be diminished. Considering the hue sample, for both partial and total reversals similar hue associations would have to be reversed. For line-orientation samples, however, the partial reversal should be particularly difficult because the line-orientation sample first would have to be disengaged from the hue samples and then reversed on their own. For the total reversal, however, if line-orientation samples were coded as hues, reversing the hue sample associations should suffice.

The advantage of a total reversal over a partial reversal is not reserved for many-to-one matching training. Similar results have been reported when the training task involved a simple successive discrimination in which responding to two stimuli (e.g., red and vertical lines) was followed by a food outcome and responding to two other stimuli (i.e., green and horizontal lines) was extinguished. When the outcomes associated with only the line orientations were reversed, reacquisition was slower than when the outcomes associated with the hues were also reversed (Zentall et al., 1991, Exp. 2; see also Nakagawa, 1986).

Discrimination design. Findings of color "naming" by pigeons indicated thc presence of stimulus classes independent of simple stimulus generalization because the pigeons responded similarly to stimuli that were spectrally quite distinct, yet they could clearly discriminate between them (Zentall & Edwards, 1984). The defining characteristic of a stimulus class is that responding is relatively flat over a broad range of stimulus values (constituting the class) and then drops off rapidly at the boundary of the class.

For the presumed arbitrary classes resulting from many-to-one matching, it should be the case that the training procedures responsible for producing the stimulus classes should alter the discriminability of the stimuli involved (see Lawrence, 1949). Thus, if sample stimuli associated with the same comparison form a stimulus class, those stimuli should be more difficult to discriminate from each other than samples associated with different comparisons. Recent data support this hypothesis. After training pigeons with many-to-one matching, Kaiser, Sherburne, and Zentall, (1996), transferred then to a simple successive discrimination involving the four samples (reposes were reinforced in the presence of two of those samples and extinguished in the presence of the two others). When the two S+s were samples associated with one of the comparisons (and the S-s were associated with the other comparison) acquisition of the successive discrimination was significantly faster than when the two S+s were samples associated with different comparisons (and the S-s were also associated with different comparisons). Thus, the presumed common coding of samples associated with the same comparison causes them to be more difficult to discriminate than samples associated with different comparisons. Again, the ratio of sessions to acquire the hue- versus line-sample associations was a good predictor of acquisition of the successive discrimination A significant positive correlation was found for the consistent group and a significant negative correlation was found for the inconsistent group.

Slope of the retention functions. The insertion of delays of various durations between the offset of the sample and the onset of the comparisons has been used as a means of assessing memory for the sample in delayed matching by pigeons (Blough, 1959; Roberts, 1972). The rate of forgetting or slope of the retention function can be affected by several factors including the nature of the samples. For example, it has been reported consistently that hue samples are remembered longer (or retained better) than line-orientation samples (Farthing, Wagner, Gilmour, & Waxman, 1977; Zentall, Urcuioli, Jagielo, & Jackson-Smith, 1989). The simplest explanation of this effect is that the line orientation samples are less discriminable than the hue samples, and the passage of time allows for the memory traces of the line-orientation samples to be more confused with each other than the memory traces of the hue samples.

With regard to common coding effects, it is noteworthy that when training involves the many-to-one mapping of samples onto comparisons and one hue and one line orientation are each mapped onto one of the comparisons, the differences in slope of the retention functions between hues and line orientations virtually disappear, whereas when those same stimuli are involved in other mappings (one-to-one or one-to-many) the hues are remembered consistently better (Urcuioli & Zentall, 1986; Zentall, Urcuioli, Jagielo, & Jackson-Smith, 1989). This finding is consistent with the hypothesis that in many-to-one matching, samples associated with the same comparison are commonly coded because if, for example, red samples and vertical-line samples are coded or represented similarly by the pigeon, one would expect the retention functions produced by the memories to be the similar.

Thus, retention functions can be used to suggest that stimuli associated with the same comparison are commonly coded because their retention functions, which would typically show quite different slopes (when they are associated with the different comparison), show similar slopes when they are associated with the same comparison. But retention functions can also be used to help identify the nature of the common code. To understand the logic of this reasoning I will have to digress briefly. When stimuli from the same dimension (e.g. hues) are used as samples the resulting retention functions for individual samples (e.g., red vs. green) are generally quite similar. When samples involve the presence versus the absence of a particular attribute, however, parallel retention functions are generally not obtained. Samples for which asymmetrical retention functions generally have been obtained are the presence versus the absence of food, the presence versus the absence of pecking (pigeons are required to peck at one stimulus and to refrain from pecking the other), and the presence versus the absence of a stimulus (if there was a sample, one comparison is correct, if there was no sample, the other is correct). In general, with increasing delay, matching accuracy on absence sample trials is relatively flat and high, whereas matching accuracy on presence sample trials starts off at a high level at short delays but declines rapidly and often falls below chance. The retention-function asymmetries have important implications for the nature of the sample code but for purposes here it is sufficient to note that some sample pairs yield parallel retention functions whereas others do not. This symmetry-versus-asymmetry retention-function difference depending on the nature of the samples can be used to track the nature of the underlying sample representation. What would happen, for example, if two samples that typically yield parallel retention functions were combined in a many-to-one matching task with two samples that typically yield asymmetrical retention functions. If samples associated with the same comparison are commonly coded, one might expect that the two pairs of retention functions would tend to conform to one another (either both become symmetrical or both become asymmetrical). Furthermore, if common retention functions were obtained, the nature of the common sample codes could be inferred from whether normally symmetrical functions became asymmetrical, or the reverse. To be specific, if pigeons are trained to match red and green samples and food and no-food samples to shape comparisons (e.g., red and food samples to one shape comparison, and green and horizontal samples to the other) would the hue samples take on the asymmetry of the food/no-food samples or the reverse? Or would the samples maintain their original "personalities" in spite of the many-to-one associations?

Zentall, Sherburne, and Urcuioli (1995) found that following many-to-one training (with no delay), the delay functions associated with food and no-food samples retained their asymmetry. On the other hand, the red and green samples which normally produce parallel retention functions took on the asymmetrical retention functions of the food/no-food samples (see Figure 2). Thus, in the example given above, retention functions obtained following green samples were high and flat, whereas retention functions following red comparisons were steeply declining. Given that the hue sample retention functions, but not the food/no-food sample functions, were altered by the many-to-one matching experience, it appears that under the present conditions, red samples were coded as food, and green samples were coded as no food (see also, Urcuioli, DeMarse, & Zentall, 1994).

Under somewhat different conditions, Grant and Spetch (1993) found that many-to-one training resulted in parallel retention functions following samples that typically yield asymmetrical retention functions, if the other samples were hues and initial training was with the hue samples alone. Thus, these experiments suggest that representations of one of the two pairs of samples (perhaps the easier to discriminate or associate with the comparisons, or the one that is trained first) may serve as the basis for the presumed common codes.

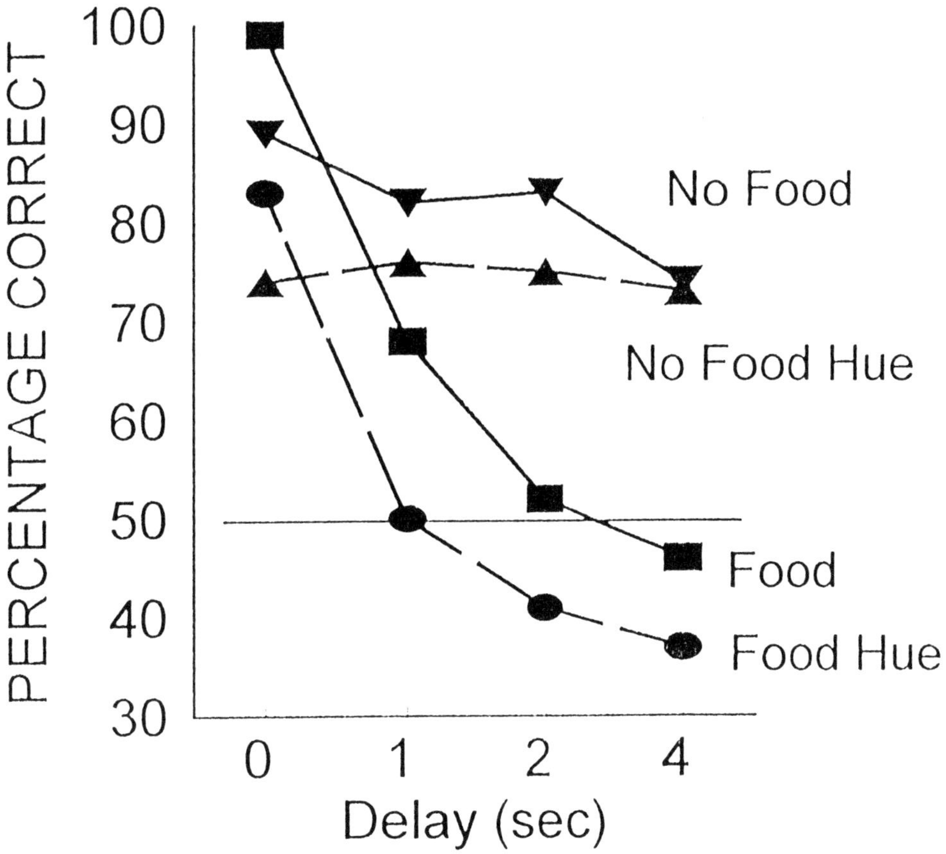

Figure 2. Retention data for pigeons trained with many-to-one matching-to-sample involving food and no-food samples (that typically yield asymmetrical retention functions) and red and green samples (that typically yield symmetrical retention functions). Food Hue trials are those involving the hue sample associated with the same comparison as the Food sample. No Food Hue trials are those involving the hue sample associated with the same comparison as the No Food sample.

The results of these matching-to-sample experiments involving the many-to-one mapping of samples onto comparisons provide extensive convergent support for the notion that animals (and specifically pigeons) have the capacity to form arbitrary classes of stimuli through the common association of those stimuli with another event. Furthermore, under certain circumstances, not only can we ascertain that those stimuli are represented by a common code, but we may also be able specify what the nature of that code might be. In the case of many-to-one matching of hue and food/no-food samples, for example, there is suggestive evidence that pigeons encode the red sample as if it were a food event and the green sample ass if it were a no-food event. When humans engage in analogous stimulus-class-formation behavior, as they do when acquiring and using language, we refer to it as learning the *meaning* of words. Just as the color red, under certain conditions, can come to represent a food event for the pigeon, so too does the word food, under certain conditions, come to represent a food event for humans. A more complete discussion of the relation between stimulus class formation and language will be presented later.

Equivalence Classes

To this point, I have used the terms, *stimulus class* and *common code*, but have avoided the term, *equivalence class* because that term has been used in a much more narrow sense than the others. According to Sidman (1990), the term equivalence is defined in its formal sense to include evidence of the *emergence* of three different kinds of relations between stimuli. In order to satisfy the definition of a set of equivalent stimuli first, one must be able to show that an organism can respond appropriately to reflexive relations. Reflexivity, also called generalized matching, suggests that the organism has learned to respond to the sameness relation between two stimuli, without explicit training with those stimuli. In other words, if an organism has been trained to select comparison A when the sample is A, and select comparison B when the sample is B, will it then select the novel comparison C when the sample is C, and comparison D when the sample is D.

Second, an organism must be able to demonstrate symmetry. Symmetry, or bidirectional learning, is indicated, for example, when in the context of a matching task, after training an animal to select comparison B when the sample is A, and to select D when the sample is D, the animals now selects comparison A when the sample is B and selects comparison C when the sample is D.

Finally, an organism must be able to show transitivity. Transitivity involves an emergent relation that develops between two stimuli that has been mediated by a third. For example, if an organism has learned to respond to comparison B when the sample is A and to respond to comparison C when the sample is B, will it then, without further training, respond to comparison C when the sample is A?

The three components of formal, stimulus equivalence have been examined in animals independent of Sidman's focus on stimulus equivalence because generalized identity, bidirectional or backward-associative learning. and mediated learning have implications for the role of cognitive processes in animal learning.

Reflexivity. The ability of animals to show generalized identity learning has been viewed as an indication of their cognitive capacity see Premack, 1976). The simplest was to test for generalized identity is to train an animal on an identity matching task (e.g., to respond to comparison A when the sample is A and to respond to comparison B when the sample is B) and then to look for evidence of transfer to novel stimuli. Examination of this literature indicates that positive transfer occurs under a variety of condition (Urcuioli & Nevin, 1975; Zentall & Hogan, 1974, 1975, 1976, 1978). There may be a problem, however, with use of a transfer test involving novel stimuli when evidence for positive transfer is weak, because pigeons have a tendency to be neophobic (i.e., they tend to avoid novel stimuli). Any tendency to avoid the novel test stimuli could result in an underestimation of identity transfer. Although significant positive transfer has often been found on early trials of the transfer test (Zentall & Hogan, 1974), to allow for the possibility of a temporary disruption in performance, the rate of acquisition of the transfer task, relative to an appropriate negative transfer control group. Such a control group generally consists of subjects originally trained on a mismatching (or oddity) task with the training stimuli and then transferred to the same matching task as subjects in the experimental group (see Table 4 for the complete design of this experiment; after Zentall & Hogan, 1976). The advantage of this design is that it controls for prior conditional discrimination experience not only with the stimuli used in training but also with the specific stimulus configurations. Furthermore, it controls for whatever difference there might be in the rates of acquisition between matching and oddity tasks. An example of transfer data obtained with such a design is presented in Figure 3 (the data are from Zentall & Hogan, 1978, Figure 2).

TABLE 4
Test of reflexivity: matching and oddity transfer

Condition	Group	Phase 1	Test
Nonshifted	M - M	C --> C P --> P	R --> R G --> G
	O - O	C --> P P --> C	R --> G G --> R
Shifted	M - O	C --> C P --> P	R --> G G --> R
	O - M	C --> P P --> C	R --> R G --> G

Note. M = matching, O = oddity, R = red, G = green, C = a white line-drawn circle, P = a white plus sign. The shapes were projected on a black background. For each trial type the first letter represents the sample and the second the correct comparison.

An alternative approach to the study of identity transfer in animals avoids the problem associated with the use of novel transfer stimuli by training the animals with either identity or oddity and then testing with novel combinations of the familiar stimuli (Zentall, Edwards, Moore, and Hogan, 1981). Pigeons were first trained on a four-hue version of the matching task (either identity or oddity). For each sample hue trained, however, the mismatching comparison was one of only two of the remaining three hues (e.g., if the sample was red the mismatching comparison could have been green or blue, but not yellow). The third, mismatching, hue was reserved for test trials. Test trials were of two types: Either the remaining hue replaced the matching comparison (i.e., in the example presented involving red samples, the yellow comparison replaced the red comparison and on some test trials the comparisons following a red sample were yellow and green, while on other test trials the comparisons were yellow and blue), or the remaining hue replaced the mismatching comparison (i.e., in the example presented, the yellow comparison replaced the green comparison on some trials and replaced the blue comparison on other trials, such that the comparisons were always red and yellow). The complete design of this experiment appears in Table 5.

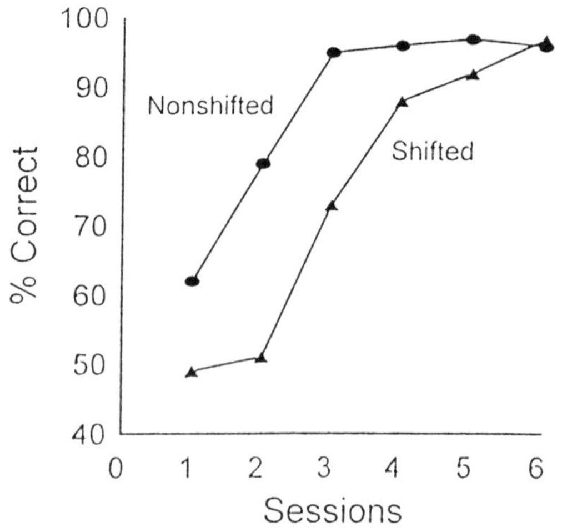

Figure 3. Acquisition of hue matching and oddity following training with shape matching and oddity. Pigeons that were nonshifted were trained with shape matching and were transferred to hue matching or were trained with shape oddity and were transferred to hue oddity. Pigeons that were shifted were trained with shape matching and were transferred to hue oddity or were trained with shape oddity and were transferred to hue matching (after Zentall & Hogan, 1978). The design of this experiment is presented in Table 4.

Two questions were asked: First, would it matter which of the two comparisons (the matching or the mismatching comparison) was replaced, and second would the baseline task (identity or oddity) make a difference. In the case of the matching task, replacing the matching comparison led to a sharp decrease in matching accuracy (pigeons failed to consistently chose the replaced stimulus), whereas replacing the mismatching comparison had little effect on matching accuracy (see Figure 4). This finding was quite compatible with Skinner's (1950) suggestion that the basis of reinforced responding on any matching task was the relation established between the sample and correct comparison, regardless of whether there was a physical match or not (i.e., whether the task was identity, oddity, or an arbitrary relation).

TABLE 5
Design of experiment to determine the basis of matching and oddity learning by pigeons

	Samples			
	Red	Green	Yellow	Blue
Training Trial Types	RRG	GGB	YYR	BBY
	RRB	GGY	YYG	BBR
Test Trial Types	RRY	GGR	YYB	BBG
	YRG	RGB	BYR	GBY
	YRB	RGY	BYG	GBR

Note. R = red, G = green, Y = yellow, B = blue. The middle letter in each triad indicates the sample hue. For pigeons in the matching group, the left letter indicates the hue of the correct comparison and the right letter, the hue of the incorrect comparison. For pigeons in the oddity group, right is correct and left is incorrect. For the test trial types, the top line represents new correct comparisons for pigeons in the matching group and new incorrect comparisons for pigeons in the oddity group. The bottom two lines of test trial types represent new incorrect comparisons for pigeons in the matching group and new correct comparisons for pigeons in the oddity group.

The results from the group trained with oddity, however, were not at all in keeping with Skinner's (1950) predictions. Once again, replacing the comparison that matched the sample led to disrupted matching performance. In this case, however, the stimulus that matched the sample was the incorrect alternative. Thus, in spite of the fact that the correct alternative was present, accuracy was disrupted. Results equally incompatible with Skinner's predictions occurred on trials in which the correct alternative was replaced but the comparison the matched the sample remained. On these trials there was little disruption of accuracy (the oddity-trained pigeons continued to avoid the matching comparison, see Figure 4).

The results of this experiment indicate that pigeons trained on identity and oddity tasks use the matching comparison as the basis of their choice. In the case of identity, they presumably look for the matching comparison and respond to it. In the case of oddity, they presumably look for the matching comparison and respond to the alternative comparison. The fact that the correct comparison controls comparison choice in the identity task while the incorrect comparison controls comparison choice in the oddity task indicates that, contrary to popular belief among animal learning researchers, identity does play an important role for pigeons.

In spite of data such as those described above, the role of stimulus identity been questioned (Cumming & Berryman, 1961; Premack, 1976; Wilson, Mackintosh, & Boakes, 1985). Furthermore, in general, transfer to novel stimuli has been accompanied by a drop in overall level of matching accuracy. Thus, although pigeons often respond to the novel stimuli at a level significantly above chance (or significantly better than the matching accuracy of a control group), they typically do not respond at the high level they had reached prior to transfer test. These deficits may be due to some neophobia (i.e., a tendency to avoid responding to novel stimuli) on the part of pigeons. The stronger effects seen when tests involve all familiar stimuli is consistent with this neophobia hypothesis (Zentall et al., 1981).

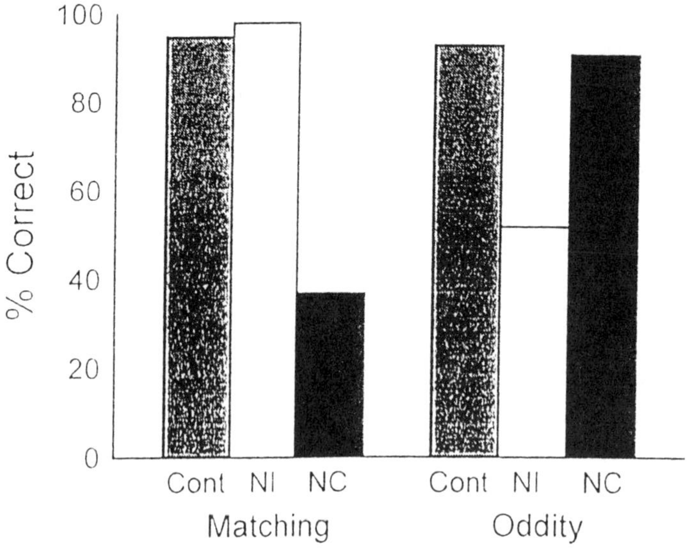

Figure 4. Performance on trial types involving replacement of the correct or the incorrect comparison with a new (but not novel stimulus) for pigeons originally trained on either a matching or an oddity task (after Zentall, Edwards, Moore, & Hogan, 1981). The design of this experiment is presented in Table 5.

Others have proposed that the transfer effects are real enough but they can be explained in terms of a relative familiarity account (i.e., development of a rule based on the comparison stimulus that relatively more familiar, in the case of identity and relatively more novel, in the case of oddity). But the only way that relative familiarity can be judged is by recognizing the similarity between the sample and the matching comparison. It is not sufficient for a pigeon to learn to "ask itself" whether it had just seen the circle or the plus stimulus most recently. If it is going to be able to perform a hue identity task without further training it must also acquire a general relative-familiarity rule. I suggest that this is merely another way of describing generalized identity.

Finally, Premack (1976) has proposed that there is a fundamental conceptual difference between the matching-to-sample procedures used with pigeons and same/difference procedures used with chimpanzees. In the case of same/different judgements, two stimuli are presented to the subject and if they are the same, a left response is reinforced and if they are different, a right response is reinforced. Whether Premack is correct or not the matching-to-sample transfer data reflect the development of a generalized identity class. Furthermore, there is evidence that pigeons too can acquire and show positive transfer to novel stimuli when trained on the same/different procedure (Edwards, Jagielo, & Zentall, 1983).

A third approach to the search for reflexivity in animals is to examine their ability to match the response of another animal. Although response matching, generally referred to as imitation or observational learning, is not typically thought of as an example of reflexivity, the ability to match behavior certainly warrants such characterization. In the case of response matching one looks for the facilitation of acquisition of a response through observation of that response performed by another.

The literature on animal imitation is difficult to interpret because a clear definition of imitation has been difficult to find. In fact, imitation has generally been defined by what it is not rather than by what it is (Galef, 1988; Whiten & Ham, 1992; Zentall, 1988; Zentall, 1996). For example, facilitated acquisition following observation could result from a number of motivational or attentional factors. Some, like the facilitation of eating behavior resulting from the presence of others (sometimes called contagion, Thorpe, 1963), are relatively easy to rule out by selection of an arbitrary behavior. Others, like the increase in activity produced by the mere presence of another animal (sometimes called social facilitation, Zajonc, 1965) may more subtly increase (or in some cases decrease) the probability of the chance occurrence of the behavior being assessed. In the case of observation of a bar press, for example, the observers attention may be drawn to the location of the demonstrator (local enhancement, Thorpe, 1963), or to the movement of the bar itself (stimulus enhancement, Galef, 1988).

A procedure that adequately controls for all of these confounding factors is the *two-action* method. In this procedure, first described by Dawson and Foss (1965). Animals observe a response (e.g., depression of a treadle for food reward, Zentall, Sutton, & Sherburne, in press) being made in one of two different ways (i.e., by a demonstrator either pecking at the treadle or by stepping on the treadle). The dependent measure is the correlation between observer's response topography and that of its demonstrator. That is, does the observer make the treadle response in the same way as it observed it being made. Using variations on the two-action method, evidence has been found for imitative learning in rats (Heyes & Dawson, 1990), pigeons (Zentall et al., in press), Japanese quail (Akins & Zentall,1996) and budgerigars (Dawson & Foss, 1965).

Evidence for imitative learning suggests not only that an animal can identify the similarity between two behaviors, but more remarkably, that the animal can identify the similarity between the behavior performed by itself and the behavior performed by the other animal. From the perspective of the imitator the stimulus complex produced by its own behavior must be perceptually quite different from the stimulus complex produced by the demonstrator. The question is, how does the observer detect their similarity. One possibility suggested by Piaget (1928) is that the observer is able to "take the perspective" of the demonstrator. Although such a process could account for imitative learning in adult humans (and perhaps even in children), it is unlikely that rats and pigeons are capable of such extrapolations (see Zentall, 1988). Whatever the mechanism underlying imitative learning in animals, the finding suggest that identity matching is a phenomenon that applies not only to stimuli, but to responses as well.

Symmetry. Although there has been some research on symmetry relations in animals using simple instrumental conditioning procedures (Hearst, 1989), most of the research has been carried out in the context of Pavlovian backward associative learning (see Spetch, Wilkie, & Pinel, 1981). Some attempts have been made to find evidence for bidirectional associations in animals using the matching-to-sample procedure (Gray, 1966; Hogan & Zentall, 1977; Richards, 1988; Rodewald, 1974). In this research animals have been trained on a symbolic or arbitrary matching task in which the comparisons are different from the samples. Following acquisition, the samples are replaced by the comparisons and the comparisons are replaced by the samples. Evidence of the establishment of bidirectional associations is indicated if presentation of one of the former comparisons (as a sample) results in choice of the appropriate sample (as one of the comparisons). In most cases, little evidence of symmetry has been found. The problem with this approach to the assessment of symmetry in pigeons is that the exchange of samples and comparisons results in the presentation of those stimuli in unexpected locations (but see Lipkins, Kop, Matthijs, 1988; Sidman, Rauzin, Lazar, Cunningham, Tailby, & Carrigan, 1982). Thus, Hogan and Zentall found that on test trials after pecking the sample to produce the comparison stimuli, pigeons often looked at the "correct" comparison and then returned to the sample and resumed pecking it (i.e., they attempted to perform the appropriate forward association).

Good evidence for the establishment of bidirectional associations have been found in a matching task when differential outcomes were used as the vehicle (Zentall, Sherburne, & Steirn, 1992). In the differential outcomes procedure, each sample-correct-comparison association is reinforced with a different outcome (see Peterson, 1984). On test trials, Zentall et al. presented the outcomes as samples and asked if the comparisons would be selected differentially, based on the outcome that preceded it. The advantages of this procedure are first, that the locations of the outcomes and the comparisons can remain as they are in training and second, as in the case of Pavlovian conditioning, one of the stimulus events in each association was biologically meaningful (i.e., an outcome). The design of this experiment is presented in Table 6.

Unfortunately, because differential outcomes were used during training, it is possible differential responding to samples associated with a food outcome versus those associated with a no-food outcome might have acted as an overt mediator. In other words, when pigeons learn that the outcome following one sample is food, they tend to peck at that sample. When they learn that the outcome following the other sample is no-food, however, they peck very little at the sample. On test trials, when food and no food replace the samples, one could argue that they peck at (eat) one of the samples and fail to peck at the other. To the extent that his overt differential responding gained control over comparison choice, differential transfer effects might be found. To control for this possibility, Zentall et al. (1992, Experiment 3) trained pigeons to peck the samples *nondifferentially* by presenting response-independent food following some of the sample associated no-food outcomes. Thus, some trials with that sample were followed with comparison choice and a no-food outcome for correct responding but on other trials presentation of that sample was followed immediately with food (i.e., no comparison stimuli were presented). In spite of this manipulation, significant differential transfer effects (consistent vs. inconsistent groups) were found.

Additional evidence, although less direct, for the establishment of bidirectional associations using a matching design has been reported by Sherburne and Zentall (1993). Pigeons were trained on matching-to-sample using food samples associated with one comparison and the-absence-of-food samples associated with the other comparison and they were tested with delays inserted between the offset of the sample and the onset of the comparisons. Sherburne and Zentall noted that when the delays were short, the pigeons showed an overall preference for the comparison stimulus that was associated with a food sample. Furthermore, a similar presence-sample short-delay preference was not found when the samples were nonhedonic events (i.e., the presence vs. the absence of a yellow hue).

TABLE 6
Design of experiment to test for symmetry (backward associations)

Training	Test	
	Consistent	Inconsistent
Red --> Red ==> Peas	Peas --> Red	Peas --> Green
Green --> Green ==> No Peas	No Peas --> Green	No Peas --> Red

Note. In training, the sample is indicated by the first hue, the correct comparison by the second, and the outcome follows. In test, peas and no peas are presented as samples, the hue indicates the correct comparison, and the outcomes for correct responding are always mixed grain.

Sherburne and Zentall reasoned that the comparison preference resulted from the backward association between comparison and food sample. In a follow-up study Sherburne and Zentall (1995) argued that evidence for the backward association reported by Sherburne and Zentall (1993) may have been weakened by the presence of the strong forward associations established between the comparisons and food outcomes. To test this hypothesis, Sherburne and Zentall (1995) attempted to manipulate the strength of the forward comparison-outcome associations by reinforcing correct comparison responses on only half of the trials. Consistent with the hypothesis that the food-sample short-delay comparison preference resulted from a backward association between the comparison stimulus and food sample that preceded it, when tested with delays, these pigeons showed a significantly stronger preference for the food-sample-associated comparison than control pigeons that received reinforcement for all of their correct comparison responses.

To summarize the findings from these symmetry experiments, when the location of the stimulus events does not change and only the temporal order is varied, and in addition, if one of the associated stimuli is an important hedonic event, such as food, good evidence of bidirectional associations (symmetry) can be found.

Transitivity. Pavlovian conditioning experiments are also the source of much of the evidence for the development of transitive relations in animals. Two kinds of transitive relations have been demonstrated using Pavlovian conditioning procedures. In the first, Pavlovian second order conditioning, Phase 1 of training involves the pairing of a conditioned stimulus (CS1) with an unconditioned stimulus (US, e.g., food), resulting in the establishment of a conditioned response (CR) to CS1. In Phase 2, pairing a novel conditioned stimulus (CS2) with the original CS1 is sufficient to result in the establishment of the CR to CS2 (see Rescorla, 1980, for a review).

In the second kind of transitive relation that has been demonstrated using Pavlovian procedures, sensory preconditioning, two conditioned stimuli, CS1 and CS2 are paired in Phase 1. In Phase 2, one of the conditioned stimuli, CS1 is paired with the US, resulting in a CR to CS1. In Phase 3, presentation of CS2 by itself results in the CR (see Seidel, 1959, for a review).

Recently, evidence for emergent transitive relations in pigeons has been reported in an instrumental context (Steirn, Jackson-Smith, & Zentall, 1991). Pigeons were first trained in a simple successive discrimination to associate a red stimulus, for example, with food, and a green stimulus with the absence of food. In Phase 2, the pigeons were trained on a symbolic matching task involving samples of food and the absence of food, and line-orientation comparisons. Finally, the pigeons were tested in a positive- versus negative-transfer test, with hue samples and line-orientation comparisons. The design of this experiment appears in Table 7.

On test trials, although there was a strong bias on the part of all the pigeons to respond to the comparison that in training had been associated with the absence-of-food sample (in fact, during the test phase, none of the samples were food), there was also a significant difference in matching accuracy between the positive and negative transfer groups. Thus, although no association between the hue sample and line-orientation comparisons was explicitly trained, an emergent relation apparently developed between them.

The research presented in this section suggests that pigeons are capable of demonstrating emergent relations involving each of the three components of stimulus equivalence. Although the three components of stimulus equivalence were demonstrated in separate experiments, there is no reason to believe that all three could not, in principle, be demonstrated in the same experiment. In fact, Sidman and Tailby (1982) have suggested that both symmetry and transitivity are demonstrated in the many-to-one transfer design (refer back to Table 1) used by Urcuioli et al. (1995). Transitivity is indicated because the relation between samples must be mediated by the comparison stimulus that is common to both. Symmetry is indicated because the path from one sample to the other must involve a backward association between the common comparison and the second sample.

According to Sidman (1990), one can incorporate a test for both symmetry and reflexivity into the transitivity design by simply reversing the roles played by the samples and comparisons in Phase 3. Thus, Phase 1 consists of training with A samples and B comparisons, Phase 2 consists of training with B samples and C comparisons, and the test involves C samples and A comparisons. Reversing the roles played by A and C on test trials tests for symmetry because it represents a relation that is in the backward direction from that which is hypothesized to occur in training. According to Sidman, this design also tests for reflexivity because in order for positive transfer to occur, an emergent relation between the A sample and A comparison (as well as between the C sample and C comparison) is implied. Emergent equivalence relations have been found in some species of animals when this design has been used (e.g., in sea lions, Schusterman & Kastak, 1993, and in chimpanzees, Yamamoto & Asano, 1995). It is of particular interest that in both studies the animals had initial difficulty in demonstrating emergent symmetry relations (i.e., given AB training, initially, emergent BA relations were not reliably found). Once trained, however, the animals not only generalized the emergent symmetry relation to other stimulus pairs, but also demonstrated both emergent transitive relations (i.e., given AB and CD training, they showed good transfer to AC) and emergent equivalence relations (i.e., good transfer to CA).

TABLE 7
Design of experiment to test for transitivity (mediation)

Phase 1 Train	Phase 2 Train	Test	
		Consistent	Inconsistent
Red --> Fd Green --> No Fd	Fd --> Vert No Fd --> Horiz	Red --> Vert Green --> Horiz	Red --> Horiz Green --> Vert

Note. Fd = Peas, Vert = vertical, Horiz = horizontal. Phase 1 training involved Pavlovian pairings of hue with outcome. Phase 2 training involved matching-to-sample with peas or no peas as samples and line-orientation comparisons. On matching-to-sample test trials, hue samples were presented with line-orientation comparisons.

The implications of these findings are important not only because they demonstrate emergent equivalence relations in (nonverbal) animals, but also because they suggest that when the spatial location of stimuli change from training to test (as they often do when testing for symmetry) animals may have to be trained not to use spatial location as a relevant cue. This methodological problem may account for many of the past failures to find evidence for emergent equivalence relations in animals.

Humans Versus Other Animals: The Great Divide

Recently, there has been some criticism of animal experiments on symmetry, transitivity, and on evidence for stimulus equivalence in general (Horne & Lowe, 1996; Saunders & Williams, in press; Saunders, Williams, & Spradlin, 1996). The bulk of this criticism has focused on the possibility that these emergent relations may be mediated by a common response. For example, as mentioned earlier, in the case of the symmetry experiment reported by Zentall et al., (1992), pigeons trained with differential outcomes (see Table 6) will often respond differentially to the samples. Then on test trials when the differential outcomes replace the samples they will respond differentially to the new samples. To the extent that (1) the differential responding rather than the stimuli themselves control comparison choice and (2) those differential responses to the sample stimuli in training are similar to the differential responses to the outcomes that replaced the samples on test trials, there may be no need to posit bidirectional associations. However, when Zentall et al. (Experiment 3) controlled for differential sample responding, evidence for bidirectional associations was still found.

Furthermore, response mediation cannot account for the results reported by Sherburne and Zentall (1993, 1995) in which a preference emerged for the comparison associated with a food rather than a no-food sample. Nor can it account for the results reported by Hearst (1989), in which a stimulus that was preceded by food in training was more easily converted into a positive stimulus than a stimulus that was preceded by the absence of food.

A similar criticism has been made of the transitivity experiment reported by Steirn et al. (1991). According to this alternative account, differential responding is likely to occur to the two stimuli in Phase 1, associated with food and no-food respectively (see Table 7). Differential responding most certainly also occurs to the food and no-food samples presented in Phase 2. Again, to the extent that (1) differential responding to the samples in Phase 2 controlled comparison choice and (2) the differential responding to food and no food samples in Phase 2 was similar to the differential responding to the stimuli associated with food and no-food in Phase 1, then the differential responding rather than transitivity may have mediated the emergent relations between the initial stimuli in Phase 1 and comparisons in Phase 2.

Steirn et al. (1992), indicated that food and no-food were selected as mediators because they were salient hedonic stimuli that should readily enter into association with both the initial stimuli and the comparison stimuli. On the other hand, the case against response mediation would have been stronger if similar results could be shown using more hedonically neutral stimuli as mediators.

A related argument has been proposed to account for the fact that emergent relations typically found following the many-to-one mapping of samples onto comparisons in matching-to-sample by pigeons are weaker (Zentall,Sherburne, Steirn, Randall, Roper, & Urcuioli, 1992) or are almost absent (Urcuioli & Zentall, 1993; Urcuioli, Zentall, & DeMarse, 1995) when training involves a one-to-many mapping (originally used successfully by Sidman, 1971). Urcuioli et al. (1995), have proposed that a form of response mediation may account for these differences in finding. They propose that Hull's (1939) mediation model based on secondary generalization could be applied to these procedures.

According to Hull, when a sample, A, is associated with a comparison, B, A will acquire the capacity to produce, implicitly, a mediator "b." Similarly, if sample C, is also associated with comparison B (as it is in a many-to-one design) it too will acquire the capacity to produce mediator "b." During Phase 2, when A is then associated with a new comparison, D, the mediator "b," produced by A, will now be associated with D. Finally, on test trials, when sample C, which also produces mediator "b" is presented, the previously established association between "b" and D allows the relation between C and D to emerge (see model as it is applied to many-to-one, matching-to-sample, presented in Table 8).

The same mediational model cannot be applied successfully to the one-to-many matching procedure. Although sample A will acquire the capacity to produce, implicitly, mediator "b,"it will also produce mediator "c" (see Table 8). In Phase 2, sample D, which initially produces no mediator, will, after training come to produce mediator "b." On test trials involving sample D and comparison C, D will produce mediator "b" but there will be no link between "b" and C, so transfer will not take place.

This analysis has been taken by some to be evidence for the absence of true emergent relations in animals (Saunders et al., in press). But as is clear from Urcuioli et al's (1995) presentation of Hull's model, no overt response mediator is implied by this analysis. Furthermore, to propose that pigeons would make a distinctively different overt response to one shape or color sample than another is wildly speculative and without foundation. Furthermore, the implicit response-mediation model cannot account for evidence of symmetry described earlier. Thus, it must be an incomplete account of emergent stimulus relations in animals.

TABLE 8
A mediated generalization account of common coding in many-to-one matching

Phase 1	Phase 2	Test
P --> V	P -- "v"--> R	
S --> H	S -- "h"--> G	
B --> V		B --"v"--> R
W--> H		W--"h"--> G

Note. P = a white line-drawn plus, S = a white line-drawn square, V = three vertical black lines, H = three horizontal black lines, R = red, G = green. "v" = presumed representation of anticipated vertical comparison. "h" = presumed representation of anticipated horizontal comparison. Lines were projected on a white background and shapes were projected on a white background. For each trial type the first letter represents the sample and the second the correct comparison. In test, samples from Phase 1 that were not presented in Phase 2 were presented with the comparisons from Phase 2.

It is ironic that criticism of the evidence for emergent stimulus relation in animals based on overt response mediation should come from researchers who study emergent stimulus relations in human populations because it is precisely in those human populations that overt response mediation (in the form of verbal naming responses) has been hypothesized to occur (Horne & Lowe, 1996). Furthermore, Horne and Lowe have gone so far as to propose that it may not be possible to find emergent stimulus relations of the type required to support stimulus equivalence in humans who do not have language. Thus, while unobservable response mediation has been proposed to explain away findings of emergent stimulus relations relevant to questions of stimulus equivalence in animals, it has also been proposed to be the only source of true stimulus equivalence in humans. Clearly, the overt response mediation shown by humans who are demonstrating stimulus equivalence should not explain away this remarkable capacity. Similarly, covert response mediation should not be used to explain away the evidence for emergent stimulus relations necessary to establish equivalence relations in animals.

The Question of Stimulus Equivalence and Human Language

The ability of an organism to form stimulus classes that have the property of equivalence may have important implications for behavior thought to be primarily human. Sidman, for example, has proposed that stimulus equivalence is a linguistic prerequisite (Sidman, 1986, p.226) and it has been suggested that equivalence relations allow us to use *words* that have *meaning* to *name* objects (Hayes, & Hayes, 1989, p.182).

Horne and Lowe (1996) have taken an even stronger position with regard to the relation between equivalence and language. They have suggested that language *allows* humans to form equivalences. This occurs because of the unique correspondence between the stimulus (word) heard by the listener and the response (word) spoken by the speaker. When the listener produces a word, the stimulus properties of that spoken word can be matched to the stimulus properties of the word spoken by others. Thus, there is a natural symmetry that occurs when a listener repeats a word that has been heard.

This position has important implications for the likelihood of finding equivalence in nonverbal organisms. Any organism that does not have adequate language skills should not be capable of developing equivalences. Horne and Lowe argue, in fact, that stimulus equivalence has not been demonstrated in any nonverbal species. However, they also offer the curious disclaimer that even if a particular nonhuman animal did succeed in showing stimulus equivalence, it would not mean that the functional determinants of the behavior would be the same as those which enable verbally competent humans to succeed on these tasks. Thus, according to Horne and Lowe, humans need language to form equivalences but, perhaps recognizing that animals too may be shown to form equivalences, they suggest it is possible that animals may do so by other means. As Pilgrim (1996) has noted, however, Horne and Lowe's position may not be falsifiable and thus, may be of limited value. Clearly, language plays an important role in the use of equivalences by humans. Language probably accounts for the ability of humans to form equivalences involving stimuli that bare little formal resemblance to one another, but rather are based on complex abstract properties (e.g., forms of transportation). More important, however, language allows a single stimulus to be a member of multiple equivalences (e.g., a car is not only a member of the class "form of transportation," but it can also be a member of equivalence classes such as "objects that are red," "made of metal," "box-shaped," etc.). Furthermore, the stimulus (class member) "car" is itself a class made up of a large number of members, as well as being made up of a large number of sometimes-overlapping subclasses (e.g., sedans, Fords, black cars, cars registered in Kentucky, etc.).

Horne and Lowe (1996) stress the unique property of verbal behavior in which feedback from the speaker's response also serves as a stimulus that can be matched with the stimulus that was originally only heard (but see Palmer, 1996, for exceptions in nonverbal language and may examples of the absence of reproductive *parity* in children's language acquisition). But what makes language truly remarkable is that members of a category designated by a word or name are equivalent only in the context of that category and are clearly not equivalent in other contexts.

In that regard, categories used by animals may share some of those attributes. As mentioned earlier, although pigeons show no evidence of "confusing" red and yellow hues (i.e., treating them generally as the same), when they are presented as samples in independent matching and oddity tasks, they appear to use the same code to represent red and yellow samples. Furthermore, even hues, such as red and green, that are not coded in the same way when presented in the context of other hues, appear to be coded similarly by pigeons when they appear in the context of achromatic shapes. Thus, pigeons appear to have some capacity for the hierarchical categorization characteristic of human language use.

Finally, the great gulf between research with humans and nonhumans has its basis in religious and humanistic philosophies that set humans apart from other animals in ways that are convenient for our purposes. Certainly, important differences exist that should not be underestimated. But, perhaps in an effort to justify our superiority over other life forms, we have tended to ignore or even deny many important similarities. While this denial is self serving, it is not becoming of the image we have of ourselves as objective scientist. A better appreciation of the similarities between us and other animals can only benefit our attempt to understand our own capacities and limitations.

Author Notes

The research described in this chapter was supported by National Institute of Mental Health Grants MH 35376 and MH 45979, and by National Science Foundation Grants BNS 8418275, RII 8902792, BNS 9019080, and IBN 9414589. Important contributions to the research were provided by Peter J. Urcuioli, David E. Hogan, Charles A. Edwards, Joyce A. Jagielo, Pamela Jackson-Smith, Janice N. Steirn, Lou M. Sherburne, Karen L. Roper, and Daren H. Kaiser.

References

Akins, C., & Zentall, T. R. (in press). Imitative learning in Japanese quail: *(Coturnix japonica*)using the two-action method. *Journal of Comparative Psychology.*

Astley, S. L., & Wasserman, E. A. (1996). Mediating associations, essentialism, and nonsimilarity-based categorization. In T. R. Zentall & P. M. Smeets (Eds.), *Stimulus class formation in humans and animals* (pp. 111-136). Amsterdam: Elsevier.

Berlin, B., & Kay, P. (1969). *Basic color terms: Their universality and evolution.* Berkeley, CA: University of California Press.

Blough, D. S. (1959). Delayed matching in the pigeon. *Journal of the Experimental Analysis of Behavior, 2,* 151-160.

Cerella, J. (1979). Visual classes and natural categories in the pigeon. *Human Perception and Performance, 5*, 68-77.

Cumming, W. W., & Berryman, R. (1961). Some data on matching behavior in the pigeon. *Journal of the Experimental Analysis of Behavior, 4*, 281-284.

Dawson, B. V. & Foss, B. M. (1965). Observational learning in budgerigars. *Animal Behaviour, 13*, 470-474.

Edwards, C. A., Jagielo, J. A., & Zentall, T. R. (1983). Same/different symbol use by pigeons. *Animal Learning and Behavior, 11*, 349-355.

Estes, W. K. (1950). Toward a statistical theory of learning. *Psychological Review, 57*, 94-107.

Farthing, G. W., Wagner, J. W. Gilmour, S., & Waxman, H. M. (1977). Short-term memory and information processing in pigeons. *Learning and Motivation, 8*, 520-532.

Galef, B. J., Jr. (1988). Imitation in animals: History, definition, and interpretation of data from the psychological laboratory. In T. R. Zentall & B. G. Galef, Jr. (Eds.), *Social learning: Psychological and biological perspectives* (pp. 3-28). Hillsdale, NJ: Erlbaum.

Grant, D. S., & Spetch, M. L. (1993). Analogical and nonanalogical coding of samples differing in duration in a choice-matching task in pigeons. *Journal of Experimental Psychology: Animal Behavior Processes, 19,* 15-25.

Gray, L. (1966). Backward association in pigeons. *Psychonomic Science, 4,* 333-334.

Greene, S. L. (1983). Feature memorization in pigeon concept learning. In M. L. Commons, R. J. Herrnstein, & A. R. Wagner (Eds.), *Quantitative Analyses of Behavior: Vol. 4. Discrimination Processes* (pp. 209-229). New York: Ballinger.

Hall, G. (1991). *Perceptual and associative learning.* New York: Oxford University Press.

Hayes, S. C., & Hayes, L. J. (1989). The verbal action of the listener as a basis for rule-governance. In S. C. Hayes (Ed.). *Rule-governed behavior: Cognition, contingencies, and instructional control* (pp. 153-190). New York: Plenum.

Hearst, E. (1989). Backward associations: Differential learning about stimuli that follow the presence versus the absence of food in pigeons. *Animal Learning and Behavior, 19,* 280-290.

Heider, E. R. (1972). Universals in color naming and memory. *Journal of Experimental Psychology, 93,* 10-20.

Herrnstein, R. J., & deVilliers, P. A. (1980). Fish as a natural category for people and pigeons. *The Psychology of Learning and Motivation, 14,* 59-95.

Herrnstein, R. J., & Loveland, D. H. (1964). Complex visual concept in the pigeon. *Science, 146,* 549-551.

Herrnstein, R. J., Loveland, D. H., & Cable, C. (1976). Natural concepts in pigeons. *Journal of Experimental Psychology: Animal Behavior Processes, 2,* 285-301.

Heyes, C. M., & Dawson, G. R. (1990). A demonstration of observational learning in rats using a bidirectional control. *Quarterly Journal of Experimental Psychology, 42B,* 59-71.

Honig, W. K., & Urcuioli, P. J. (1981). The legacy of Guttman and Kalish (1956): 25 years of research on stimulus generalization. *Journal of the Experimental Analysis of Behavior, 36,* 405-445.

Hogan, D. E. & Zentall, T. R. (1977). Backward associative learning in the pigeon. *American Journal of Psychology, 90,* 3-15.

Horne, P. J., &Lowe, C. F. (1996). On the origins of naming and other symbolic behavior. *Journal of the Experimental Analysis of Behavior, 65,* 185-241.

Hull, C. L. (1939). The problem of stimulus equivalence in behavior theory. *Psychological Review, 46,* 9-30.

Hull, C. L. (1943). *Principles of behavior.* New York: Appleton-Century

Kaiser, D. H., Sherburne, L. M., Steirn, J. N., & Zentall, T. R. (1995). *Common coding in many-to-one matching by pigeons as evidenced by reduced discriminability of commonly coded stimuli.* Unpublished manuscript.

Lashley, K. S., & Wade, M. (1946). The Pavlovian theory of generalization. *Psychological Review, 53,* 72-87.

Lawrence, D. H. (1949). Acquired distinctiveness of cues: I. Transfer between discriminations on the basis of familiarity with the stimulus. *Journal of Experimental Psychology, 39,* 770-784.

Lawrence, D. H. (1963). The nature of a stimulus: Some relationships between learning and perception. In S. Koch (Ed.), *Psychology: A study of a science* Vol. 5 (pp. 179-212). New York: McGraw-Hill.

Lipkins, R., Kop, P. F., & Matthijs, W. (1986(. A test of symmetry and transitivity in the conditional discrimination performance of pigeons. *Journal of the Experimental Analysis of Behavior, 49*, 395-410.

Mostofsky, D. I. (Ed.). (1965). *Stimulus generalization*, Stanford, CA: Stanford University Press.

Nakagawa, E. (1986). Overtraining, extinction, and shift learning in a concurrent discrimination in rats. *Quarterly Journal of Experimental Psychology, 38*, 313-326.

Palmer, D. C. (1996). Achieving parity: The role of automatic reinforcement. *Journal of the Experimental Analysis of Behavior, 65*, 289-290.

Peterson, N. (1962). Effect of monochromatic rearing on the control of responding by wavelength. *Science, 136*, 774-775.

Peterson, G. B. (1984). How expectancies guide behavior. In Roitblat, H. L, Bever, T. G. & Terrace, H. S. (Eds.) *Animal cognition* (pp. 135-147). Hillsdale, NJ: Erlbaum.

Piaget, J. (1928). *Judgement and reasoning in the child*. New York: Harcourt.

Pilgrim, C. (1996). Can the naming hypothesis be falsified? *Journal of the Experimental Analysis of Behavior, 65*, 284-286

Poole, J., & Lander, D. G. (1971). The pigeon's concept of pigeon. *Psychonomic Science, 25*, 157-158.

Premack, D. (1976). *Intelligence in ape and man*. Hillsdale, NJ: Erlbaum.

Rescorla, R. A. (1980). *Pavlovian second-order conditioning*. Hillsdale, NJ: Erlbaum.

Richards, R. W. (1988). The question of bidirectional associations in pigeons' learning of conditional discrimination tasks. *Bulletin of the Psychonomic Society, 26*, 577-579.

Riley, D. A., & Leuin, T. C. (1971). Stimulus generalization gradients in chickens reared in monochromatic light and tested with a single wavelength value. *Journal of Comparative and Physiological Psychology, 75*, 389-402.

Roberts, W. A. (1972). Short-term memory in the pigeon: Effects of repetition and spacing. *Journal of Experimental Psychology: Animal Behavior Processes, 6,* 217-237.

Rodewald, H. K. (1974). Symbolic matching-to-sample by pigeons. *Psychological Reports, 34*, 987-990.

Saunders, K. J., & Williams, D. C. (in press). A study of bidirectional transfer in parakeets: An alternative interpretation with suggested control procedures. *Journal of the Experimental Analysis of Behavior.*

Saunders, K. J., Williams, D. C., & Spradlin, J. E. (1996). Derived stimulus control: Are there differences among procedures and processes? In T. R. Zentall & P. M. Smeets (Eds.), *Stimulus class formation in humans and animals* (pp. 93-110). Amsterdam: Elsevier.

Schusterman, R. J., & Kastak, D. (1993). A California sea lion (*Zalophus californianus*) is capable of forming equivalence relations. *Psychological Record, 43*, 823-839.

Seidel, R. J. (1959). A review of sensory preconditioning. *Psychological Bulletin, 56*, 58-73.

Sherburne, L. M., & Zentall, T. R. (1993). Coding of feature and no-feature events by pigeons in a delayed conditional discrimination. *Animal Learning and Behavior, 21*, 92-100.

Sherburne, L. M., & Zentall, T. R. (1995). Delayed matching in pigeons with food and no-food samples: Further examination of backward associations. *Animal Learning and Behavior, 23*, 177-181.

Sidman, M. (1971). Reading and auditory-visual equivalences. *Journal of Speech and Hearing Research, 14*, 5-13.

Sidman, M. (1986). Functional analysis of emergent verbal classes. In T. Thompson & M. D. Zeiler (Eds.). *Analysis and integration of behavioral units* (pp 213-145). Hillsdale, NJ: Erlbaum.

Sidman, M. (1990). Equivalence relations: Where do they come from? In H. Lejeune & D. Blackman (Eds.). *Behavior analysis in theory and practice: Contributions and controversies* (pp. 93-114). Hillsdale, NJ: Erlbaum.

Sidman, M., Rauzin, R., Lazar, R., Cunningham, S., Tailby, W., & Carrigan, P. (1982). A search for symmetry in the conditional discrimination of rhesus monkeys, Baboons, and children. *Journal of the Experimental Analysis of Behavior, 37*, 23-44.

Sidman, M., & Tailby, W. (1982). Conditional discrimination vs. matching-to-sample: An expansion of the testing program. *Journal of the Experimental Analysis of Behavior, 37*, 5-22.

Skinner, B. F. (1950). Are theories of learning necessary? *Psychological Review, 57*, 193-216.

Spence, K. W. (1937). The differential response in animals to stimuli varying within a single dimension. *Psychological Review, 44*, 430-444.

Spetch, M. L., Wilkie, D. M., & Pinel, J. P. J. (1981). Backward conditioning: A reevaluation of the empirical evidence. *Psychological Bulletin, 89*, 163-175.

Steirn, J. N., Jackson-Smith, P., & Zentall, T. R. (1991). Mediational use of internal representations of food and no-food events by pigeons. *Learning and Motivation, 22*, 353-365.

Thorpe, W. H. (1963). *Learning and instinct in animals* (2nd ed.). Cambridge, MA: Harvard University Press.

Urcuioli, P. J., DeMarse, T., & Zentall, T. R. (1994). Some properties of many-to-one matching with hue, response, and food samples: Retention and mediated transfer. *Learning and Motivation, 25*, 175-200.

Urcuioli, P. J., & Nevin, J. A. (1975). Transfer of hue matching in pigeons. *Journal of the Experimental Analysis of Behavior, 24*, 149-155.

Urcuioli, P. J., & Zentall, T. R. (1986). Retrospective memory in pigeons' delayed matching-to-sample. *Journal of Experimental Psychology: Animal Behavior Processes, 12*, 69-77.

Urcuioli, P. J., & Zentall, T. R. (1993). A test of comparison-stimulus substitutability following one-to-many matching by pigeons. *Psychological Record*, 43, 725-744.

Urcuioli, P. J., Zentall, T. R., & DeMarse, T. (1995). Transfer to derived sample-comparison relations by pigeons following many-to-one versus one-to-many matching with identical training relations. *Quarterly Journal of Experimental Psychology, 48B*, 158-178.

Urcuioli, P. J., Zentall, T. R., Jackson-Smith, P., & Steirn, J. N. (1989). Evidence for common coding in many-to-one matching: Retention, intertrial interference, and transfer. *Journal of Experimental Psychology: Animal Behavior Processes, 15*, 264-273.

Vaughan, W., Jr. (1988). Formation of equivalence sets in pigeons. *Journal of Experimental Psychology: Animal Behavior Processes, 14*, 36-42.

Wasserman, E. A., DeVolder, C. L., & Coppage, D. J. (1992). Non-similarity based conceptualization in pigeons via secondary or mediated generalization. *Psychological Science, 6*, 374-379.

Whiten, A., & Ham, R. (1992). On the nature and evolution of imitation in the animal kingdom: Reappraisal of a century of research. In P. J. B. Slater, J. S. Rosenblatt, C. Beer, and M. Milinski (Eds.), *Advances in the Study of Behavior* (Vol. 21, pp. 239-283). New York: Academic Press.

Wilson, B., Mackintosh, N. J., & Boakes, R. A. (1985). Matching and oddity learning in the pigeon: Transfer effects and the absence of relational learning. *Quarterly Journal of Experimental Psychology, 37B*, 295-311.

Wright, A. A., & Cumming, W. W. (1971). Color-naming functions for the pigeon. *Journal of the Experimental Analysis of Behavior, 15*, 7-17.

Yamamoto, J., & Asano, T. (1995). Stimulus equivalence in a chimpanzee (*Pan troglodytes*). *Psychological Record, 45*, 3-21.

Zajonc, R. B. (1965). Social facilitation. *Science, 149*, 269-274.

Zentall, T. R. (1973). Memory in the pigeon: Retroactive inhibition in a delayed matching task. *Bulletin of the Psychonomic Society, 1*, 126-128.

Zentall, T. R. (1988). Experimentally manipulated imitative behavior in rats and pigeons. In T. R. Zentall & B. G. Galef, Jr. (Eds.), *Social learning: Psychological and biological perspectives* (pp. 191-206). Hillsdale, NJ: Erlbaum.

Zentall, T. R. (1996). An analysis of imitative learning in animals. In B. G. Galef, Jr., & C. M. Heyes (Eds.), *Social learning and tradition in animals* (pp. 221-243). Hillsdale, NJ: Erlbaum.

Zentall, T. R., & Edwards, C. A. (1984). Categorical color coding by pigeons. *Animal Learning and Behavior, 12*, 249-255.

Zentall, T. R., Edwards, C. A., Moore, B. S., & Hogan, D. E. (1981). Identity: The basis for both matching and oddity learning in pigeons. *Journal of Experimental Psychology: Animal Behavior Processes, 7*, 70-86.

Zentall, T. R., & Hogan, D. E. (1974). Abstract concept learning in the pigeon. *Journal of Experimental Psychology, 102*, 393-398.

Zentall, T. R., & Hogan, D. E. (1975). Concept learning in the pigeon: Transfer of matching and nonmatching to new stimuli. *American Journal of Psychology, 88*, 233-244.

Zentall, T. R., & Hogan, D. E. (1976). Pigeons can learning identity, difference, or both. *Science, 191*, 408-409.

Zentall, T. R., & Hogan, D. E. (1978). Same/different concept learning in the pigeon: The effect of negative instances and prior adaptation to the transfer stimuli. *Journal of the Experimental Analysis of Behavior, 30*, 177-186.

Zentall, T. R., Hogan, D. E., Howard, M. M., & Moore, B. S. (1978). Delayed matching in the pigeon: Effect on performance of sample-specific observing responses and differential delay behavior. *Learning and Motivation, 9*, 202-218.

Zentall, T. R., Jackson-Smith, P., & Jagielo, J. A. (1990). Categorical color and shape coding by pigeons. In M. L. Commons, R. J. Herrnstein, & S. Kosslyn (Eds.), *The quantitative analyses of behavior: Vol. 8. Pattern recognition and concepts in animals, people, and machines* (pp. 3-21). Hillsdale, NJ: Erlbaum.

Zentall, T. R., Jackson-Smith, P., Jagielo, J. A., & Nallan, G. B. (1986). Categorical shape and color coding by pigeons. *Journal of Experimental Psychology: Animal Behavior Processes, 12*, 153-159.

Zentall, T. R., Sherburne, L. M., & Steirn, J. N. (1992). Development of excitatory backward associations during the establishment of forward associations in a delayed conditional discrimination by pigeons. *Animal Learning and Behavior, 20*, 199-206.

Zentall, T. R., Sherburne, L. M., Steirn, J. N., Randall, C. K., Roper, K. L., & Urcuioli, P. J. (1992). Common coding in pigeons: Partial versus total reversals of one-to-many conditional discriminations. *Animal Learning and Behavior, 20*, 373-381.

Zentall, T. R., Sherburne, L. M., & Urcuioli, P. J. (1993). Common coding in a many-to-one delayed matching task as evidenced by facilitation and interference effects. *Animal Learning and Behavior 21*, 233-237.

Zentall, T. R., Sherburne, L. M., & Urcuioli, P. J. (1995). Coding of hedonic and nonhedonic samples by pigeons in many-to-one delayed matching. *Animal Learning and Behavior, 23,* 189-196.

Zentall, T. R., Steirn, J. N., Sherburne, L. M., & Urcuioli, P. J. (1991). Common coding in pigeons assessed through partial versus total reversals of many-to-one conditional discriminations. *Journal of Experimental Psychology: Animal Behavior Processes, 17*, 194-201.

Zentall, T. R., Sutton, J. E., & Sherburne, L. M. (in press). True imitative learning in pigeons. *Psychological Science*.

Zentall, T. R., Urcuioli, P. J., Jagielo, J. A., & Jackson-Smith, P. (1989). Interaction of sample dimension and sample-comparison mapping on pigeons' performance of delayed conditional discriminations. *Animal Learning and Behavior, 17*, 172-178.

5

Virtual Associative Networks: A Framework for Cognitive Modeling

Yan M Yufik, Ph D
Institute of Medical Cybernetics, Inc.
Potomac MD 20854

Introduction

"What little understanding we have so far gained about the workings of the brain has come only after we had already understood a problem in the world that the brain has evolved to solve." R. N. Shepard, 1995, p 51.

The cognitive modeling proposal in this chapter is founded on a premise that the human brain solves the problem of satisfying some inborn values in a manner maximally efficient under the constraints changing in the course of the life cycle. Modeling such value satisfaction has many aspects, some of which are underscored in the following questions

1. Neuronal processes in the brain underlie intelligent performance. Since people outperform cats, are human neurons more intelligent than feline neuron?

2. What allows a chess player, capable of a few operations per second, to compete with a computer carrying over 108 operations per second (e.g., Kasparov vs. Deep Blue)?

3 As I am starting to write this article, I know precisely what points I am going to make but do not know how my argument will come out. Presently, there is no text in my memory that I can read from. In that case, what is it that I know, in what form does this knowledge exist, and how am I going to use it?

This introductory section will preview answers offered by the Virtual Associative Network (VAN) model, starting with the last question. Presently, I have a set of entities in my memory, such as words, concepts and images, that I hope to explicate and interrelate in the article. Some of the arguments have been structured before, others will be shaped for the first time and combined with earlier ones, depending on how the article unfolds.

My knowledge is a repertoire of resources allocated dynamically to satisfy my objectives. If my arguing points and/or their priorities change, the flow and the structure of the exposition will change too. If my computer breaks down and a few sentences are lost, I will not be able to reproduce them word-by-word in the new version. Fortunately, I will not have to if the resources in my memory remain accessible and my allocational ability does not fail. Resource Allocation (RA) is the central formalism in the VAN model, applied to define the organization of memory structures and operations that underlie performance. This leads to the second question.

Performance is negotiated between Short Term Memory (STM) and Long Term Memory (LTM). At this very moment, I am aware of only a few words that I am typing, and of the overall intention of this paragraph. How these words are being unearthed from LTM and brought to the surface, is completely unknown to me. Somehow, out of all the conceivable alternatives, my LTM zeroes on a few words, and delivers them to STM for attentive evaluation. Efficiency of this process is determined, apparently, by at least two factors: first, the resource set is globally integrated so that, for example, I can draw on my entire vocabulary in writing this article and, second, the retrieval mechanism, without denying the access to the entire set, automatically applies my earlier experiences to narrow it to some smaller subsets appropriate for my current objectives.

Accordingly, the VAN model hypothesizes a self-organizing LTM subject to self-integration and self-partitioning, and proves that the architecture combining a small capacity STM and a large capacity self-organizing LTM indeed enables highly efficient performance, as compared to the conventional (i.e., computer-like) memory. In the former, accumulation of experiences in small-scale tasks accelerates solution of new large-scale tasks, and drastically reduces the amount of processing involved. In the latter, nothing of the sort takes place. A chess master outperforms fast algorithms because self-organizing biological memory dynamically collapses large sets of alternatives to just a few alternatives. By contrast, a static computer memory is stuck with an unabridged set.

Answers to the third and second questions render the first one incongruous: What differentiates humans from other species is not what occurs in the neurons but mainly how the neurons are put together in the system (Alkon, 1989). This account disagrees with equating neurons to computing devices. According to the VAN framework, computation is a form of abstractive thinking attainable in a highly organized nervous system, such as that of a grown human, but not in any of its components. Delegating computation to neurons circumvents the problem of uncovering material organizations underlying behavior, and replaces it with designing algorithms for behavior simulation.

Presumably, computational models would answer the first question in the affirmative. If, for example, "artificial neurons" can learn to produce spoken English from written English text (Sejnowski et al, 1987), then either cats should be able to read or their neurons must have inferior computing skills. This conclusion, along with the observation that a "neural-network computer containing a million synapses has about one ten-thousandth the mental capacity of a housefly" (Restak, 1995, p. 28) indicates that something is amiss between the computational models and biology. This chapter avoids the dilemma by removing computation from the nervous system and shifting the inquiry to other concerns: 1) What organization makes thinking, including computation, possible, 2) What conditions cause its success or failure, and 3) What problems does it help to solve?

The VAN model insists that memory integration is the key determinant of human cognitive advantages, but acknowledges that the arguments will remain speculative for as long as neurophysiology, according to a Nobel laureate in that discipline, continues to "have no idea how information is integrated in the nervous system" (Sakmann, 1994, p.157).

The chapter has four parts. Part I presents the main concepts and poses questions the model is designed to solve. Part II outlines the model. Part III reviews some applications, and Part IV discusses how the model fits in with other theories.

Part I. Self-organization.

"Order always seeps away toward chaos, but the ways of entropy can be passing subtle." H.C. von Bayer, 1997, p. 15.

1. Things, animals, and humans.

Self-organization takes place under two conditions: the system resides sufficiently far from equilibrium, and can exchange entropy with its surroundings (Nicolis & Prigogine, 1989; Prigogine, 1980; Goldfield, 1995; Kelso, 1995). System components constitute organizational "resources" that either improve the overall order in the system by joining some coherent ensembles, or degrade that order by disengagement (Swenson, 1992).

For example, in experiments dating back to the beginning of this century viscous fluid is heated between a heater and heat sink. Initially, molecular movements are completely disorganized. However, increasing the heat supply beyond a certain threshold causes emergence of macroscopic "cells" containing large ensembles of coherently moving molecules. Every such "cell" acts to amplify itself by "pulling in" new resources, in "competition" with other "cells." The emergent organization reduces entropy within the system and accelerates entropy production outside the system, allowing a steep increase in the energy transport from the heater to the sink. The system affects "resource grouping" in a way that minimizes internal entropy and maximizes coherence in the liquid volume, given the constraints (e.g., the rate of energy supply) (Swenson, 1992, 1989).

Natural evolution can be understood "as a change from an incoherent homogeneity into a coherent heterogeneity, accompanying the dissipation of motion and integration of matter" (Spencer, 1937). On that basis, things, animals and humans belong to the spectrum of self-organizing systems but differ in the efficiency of entropy reduction and the degree of integration they can reach, contingent upon energy availability.

Things have no control over the energy supply and, therefore, the lowest efficiency. Organisms improve their efficiency by seeking energy sources, within a scope. Finally, the human system reaches the highest efficiency by expanding the scope indefinitely and by controlling energy distribution inside the system. This chapter will argue that the latter two factors are reciprocal, and determine the main differences between the human condition and that of an animal. Most distinctly the differences are manifested in values.

Seeking energy in the environment gives different values to different external conditions. Control over the internal energy distribution gives values to the internal states. In the animal, the internal states are externalized (e.g., food has high value to a hungry animal and low value to a satiated one). By contrast, in humans the external conditions are internalized. Here lies the divide between conscious and unconscious, automatic and deliberate: "deliberation" emerges as "weighing in mind" (Webster's Dictionary), that is, self-motivated choice inside the system between its own valued structures. Values coordinate with the degree of autonomy: internal values make the system self-directed, the external ones allow limited separation, while things are inseparable from the environment.

Research reported in this paper is centered on the mechanisms allowing a system with a small capacity STM to respond on line to a number of variables far exceeding that capacity, with the limits for off-line processing yet to be discovered. How the human cognitive system accomplishes this feat is a profoundly puzzling problem.

In modeling animal behavior this problem hardly arises since animals are rigidly restricted in the scope of variables they can process and the interconnections they can make. For example, rats can learn to press a bar for food, but not to avoid shock. Conversely, they learn to jump to avoid a shock, but not to receive food. Similarly, pigeons peck at a spot to obtain food but will not learn to hop on a treadle for food reward; they learn to avoid shock by hopping but not by pecking, etc. (Gould et al, 1987).

Humans appear to be unrestricted in the number of variables they can process and the intervariable associations they can make, including creation of novel links that have never been experienced. Animals connect stimuli based almost exclusively on their contiguity in time and space, while for humans this factor is not at all decisive, as evidenced, for example, by the references in this paper.

Unrestricted interconnectivity subserves "Integrative Tendency" (Koestler, 1967), entailing a unified global organization in the human Central Nervous System (CNS). Figure 1 chronicles its development during the individual's life time.

Figure 1 combines the stages of psychological maturation (after Fisher & Pipp, 1984) with those of cerebral maturation (Hudspeth & Pribram, 1992). Classification in Fisher & Pipp, 1984 has been extended here to include synthesis of diverse principles into laws, and synergistic application of laws from diverse knowledge areas. Figure 1 is in close correspondence with the stages of cognitive development identified by Piaget, 1954, 1950.

How such global organization builds up in the human brain is not known. According to psychophysiological measurements, each stage engages different brain areas to a varying degree, and each starts at a rapid pace, followed by deceleration, and ending in a plateau. In general, cerebral maturation proceeds from the posterior to the frontal regions, with the period between 17 and 21 years of age dedicated mostly to the development of frontal "executive" functions (Hudspeth & Pribram, 1992; Grafman, 1995).

The system is holonomic, that is, operates as an indivisible functional whole subject to "holonomic constraints" (McFarland, 1971; Pribram et al, 1975). Growing organization reduces the number of the degrees of freedom, i.e., "generalized coordinates required to describe the state of the system" (McFarland, 1971. p. 217), as in Figure 1.

In functional terms, Figure 1 depicts a goal-directed system (Granit, 1977). A taxonomy by Mayr, 1974 will help to clarify the notion. Teleomatic goal-directed systems change their state spontaneously by virtue of some physical processes (self-organizing liquid is an example). Teleonomic systems are driven to the goal state genetically and as a result of both external and internal changes, although they have no conscious awareness of either the goals or the changes. Finally, teleological systems are conscious of their goals.

Reserving the latter characteristic for humans until compelled otherwise, the next section proposes yet another and more formal definition that combines all three taxonomic classes, and leads to a concept of consciousness adopted in the rest of the paper.

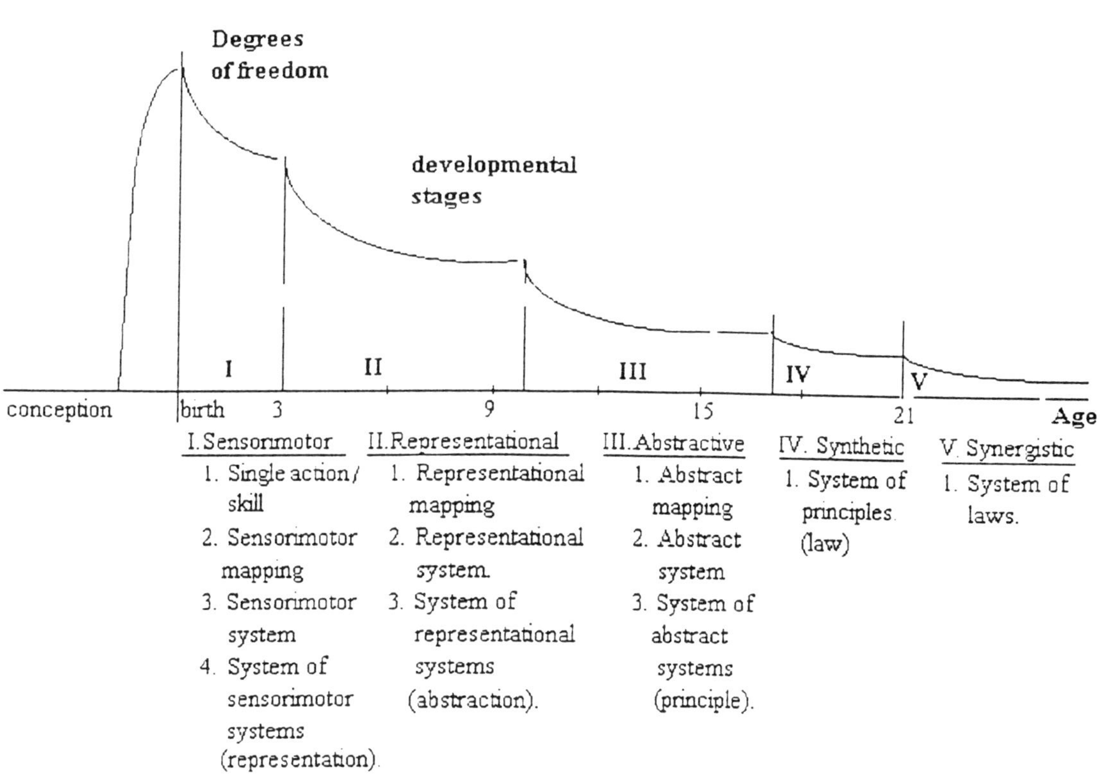

Figure 1. Neuronal population in the nervous system grows at the rate of 250,000 per minute during the period from egg fertilization to birth. Orderly migration of neurons along specific pathways in the emerging structure (Tessier-Lavinge et al, 1996) results in a system with potentially more than 10^{11} degrees of freedom (approximate number of neurons). Development after birth is characterized by a growing functional organization. As a result, behavior changes from largely erratic responses to stimulation in the neonatal period, to highly coordinated proactive activities in adults. The number of degrees of freedom drops due to the growing interconnectivity and improved capability to constrain present responses by wealth of earlier experiences.

2. Virtual systems.

Virtual systems have a limited time span and a limited envelope of possible developmental trajectories all of which emanate from the same origin and converge in the same target area in the system's phase space. Interaction with the environment selects a trajectory, and changes the system's organization in such a way that entropy of the trajectory is reduced, and the system's advance to the target area is streamlined (Yufik, 1980).

The definition stems from two beliefs by this author: (a) the human nervous system progresses throughout the life time toward some genetically defined functional organization, and (b) the system's behavior is driven by the need for external stimulations facilitating its development towards the target "synergistic" condition.

Subjectively, striving towards the target organization correlates with the experience of self and self-awareness transcending all the developmental stages. Objectively, the organizational goal is the source of the "downward causation" (Sperry, 1972; 1995), or a global utility function operating throughout the life time and participating in the value assignments in each episode of system-environment interaction so that stimulation that "helps toward attainment of the goal becomes good and that which obstructs becomes bad. Similarly everything that helps to attain all the subsidiary things, which in turn help to reach the main goal, becomes valued accordingly" (Sperry, 1972, p. 122).

This premise does not concur with the current emphasis on parallel and distributed processes in the nervous system that, purportedly, behave as if controlled by a unifying objective while in reality there is none. On the present account, this view eliminates the two upper classes in the taxonomy of goal-directed systems, and the subject of psychology with them. Postponing criticism, the chapter will proceed in a constructive fashion, by formulating a model of global memory and demonstrating its performance.

Entropy of the nervous system's trajectory cannot be measured and remains a theoretical abstraction. However, the chapter will show that this abstraction provides a basis for the model, and entails verifiable conclusions. Entropy measures are defined as follows. Let ε denote the minimal distance in the developing system's phase space corresponding to some minimal appreciable changes in the organization. The topological entropy measure h determines the degree of local trajectory divergence computed as

$$h = \lim_{\varepsilon \to 0} \lim_{t \to \infty} \ln(N(\varepsilon, t)) / t,$$

where $N(\varepsilon, t)$ is the number of alternative trajectories emanating from the current point X_i in the phase space and intersecting in a cube with the side size ε drawn around $\mathbf{X}_i$.

If $\mathbf{X}_i$ and $\mathbf{X}_g$ are the original and target points in the system's phase space, the metric entropy measure K_h determines the degree of remote trajectory divergence (proximity between the minimal trajectory and the actual one), computed as

$$K_h = \lim_{\varepsilon \to 0} \lim_{t \to \infty} \lim_{N \to \infty} \{-1/N \sum_{i_1 \ldots i_N} p(i_1, \ldots\ldots, i_N\} \ln(i_1, \ldots\ldots, i_N)\},$$

where $p(i_1, \ldots i_N)$ is the probability of retaining the actual trajectory in the ε vicinity of the minimal trajectory during the consecutive time periods dt_1, $dt_2 \ldots$, from the start until during the final interval dt_N the trajectory reaches the ε neighborhood of the target $\mathbf{X}_g$.

Roughly, topological entropy reflects the system's ability to constrain its current responses based on past history, while metric entropy represents its ability to prepare for the future.

The nervous system undergoes two self-organization periods during its life cycle: morphological organization (neuron generation and axonal growth prior to birth), and functional organization based on the already established morphology. In that, postembryonic plasticity of the neuronal organization is an extension of developmental plasticity in the embryonic period (Roush, 1996). Functional organization includes synaptic pruning, branching, changes in the synaptic geometry (Rosenzweig, 1985) and efficacy (Markram et al, 1997), and formation of global dendritic networks (Pribram, 1991) (more about such networks in the following sections).

The first self-organizing period is characterized by a rapid expansion of the system's volume in the phase space, succeeded by a protracted collapse in the second period due to the growing functional organization (as shown in Figure 1). During both periods the initial and target organizations are genetically determined, while the connecting trajectories vary depending on the environment.

The following notion is central to the present model: facing new conditions, the human nervous system can choose behavior that takes advantage of all its past experiences but is not bound to replicate any of them. Weighing the alternatives and overcoming the inertia of past history is the function of consciousness, contributing to the reduction of entropy.

The concept of consciousness was articulated by a French philosopher Henri Bergson in a few phrases capturing both the essence of the phenomenon and the underlying functional mechanism. Consciousness was defined as the

> "power of choice.. coextensive with the fringe of possible actions that surrounds the real action." By contrast, the animal "is shut up in the habits of the species; by pulling on its chain, it succeeds only in stretching it. With man, consciousness breaks the chain. Doubtless man owes it to the superiority of his brain which enables him to build an unlimited number of motor mechanisms, to oppose new habits to the old ones unceasingly, and by dividing automatism against itself, to rule it" (Bergson, 1955, p.138).

This chapter argues that Bergson's aphorism is a technically precise description of organizational changes in the human CNS underlying learning and performance in all domains, from motor control to problem solving. The remainder of this part will list the main observations, setting the stage for the model in part II.

3. Motor control.

Even the simplest repetitive motor activities, such as hammering nails, reveal wide variations in each execution cycle. The only invariable parameter in the activity is the goal, while all the intermediate steps are subject to variation. In general, the more the activity is practiced, the wider range of condition changes the performer can tolerate (Bernstein, 1967). How are these goal-directed activities controlled, and in what way does the control mechanism attain such a high degree of robustness?

Figure 2 shows relationships between the efferent stimulus I, relative muscle contraction dL, and skeleton tension P. Observe that efferentation can select one of the inclined lines but not a point on the line. As a result, there can exist no one-to-one correspondence between the muscle's tension and movement. In principle, the relationship is that of uncertainty. Two consequences follow from the "uncertainty principle": first, goal-directed movements cannot be controlled by fixed memory "programs" and, second, such control cannot address individual muscles but must engage interdependent muscle groups.

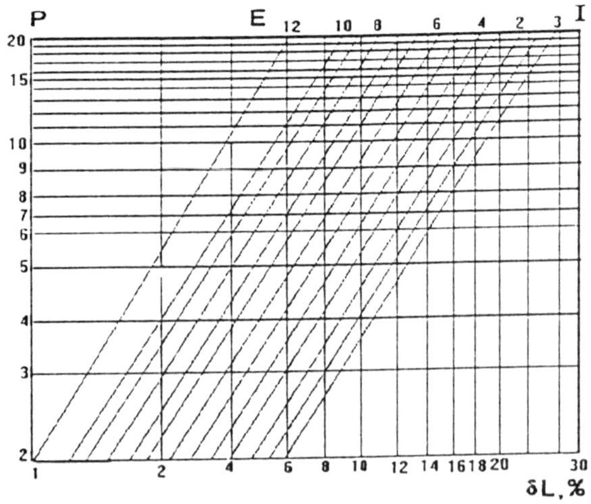

Figure 2. Relationship between the efferent stimulus I, relative muscle contraction δL, and muscle tension P. The inclined lines correspond to efferentation increments in the range from the lowest level I to the maximum level E (Bernstein, 1947). Uncertainty of the relationship underscores the impossibility of movement coordination via fixed "motor programs." This concept is pervasive but untenable.

The mechanism of motor coordination was formulated first by Feldman (1981, 1980, 1974, 1966a, 1966b). Briefly, in a muscle-load system the parameters under control are the stretch reflex threshold: the lower the threshold, the smaller the muscle length at which the load is equilibrated. Load transfers by a joint are obtained via reciprocal threshold shifts in the agonist and antagonist muscles, as follows.

Based on earlier experiences, the nervous system selects, in advance of the movement, threshold values for both muscles that correspond to the desired load position. Substitution of the initial threshold value for the target threshold makes the present position unstable, and causes load transfer to the target position where the stability is restored. Since only the initial and the target positions are fixed, all the transfer trajectories allowed by the uncertainty envelope in Figure 2 converge around the same destination area. After the threshold's substitution, the law of the least action takes over, ensuring an optimal trajectory under the constraints.

The classical muscle-load model by Feldman has been generalized to include coordination of multiple anatomically independent muscles groups, or synergies. Coordination results from "recruiting" motor neurons into cohesive although temporary ensembles that constrain the corresponding muscles to act together as functional units. Goal-directed movements are produced by matching the goal with one of the equilibrium states represented in the ensemble. This allows the goal to be reached economically from any starting position, even under disturbances (Kugler et al, 1980; Ma & Feldman, 1995).

Complex movements are produced by multiple equilibrium shifts. Their trajectories are composed of near-optimal segments connected at the points of feedback interference. Experience improves selection of the intermediate points, allowing them to be maximally spaced and creating the sensation of execution automaticity (e.g., Flanagan et al, 1993).

In all cases, while the goals are attentively selected, robustness and optimality in the resulting action are attained by yielding control and coordination to the lower level processes. "Whenever one throws a ball and gives attention to the throwing rather than to the target one is quite sure to miss the target" (Koffka, 1925, p. 258). The next section gives a closer look to control and coordination in skilled performance.

4. From motor control to problem solving.

Variability is not restricted to motor control but is the key feature of any goal-directed activity. In problem solving, variability is manifested in an individual solving related problems, solving a single problem, and even within a single trial (Siegler, 1994). The present approach accounts for this fundamental characteristic of skilled performance in a unified fashion. Return first to the issue of motor coordination.

I type slowly and use only two fingers. Skilled typists use all ten fingers, and type much faster. What is the mechanism of such acceleration? Regardless of how many fingers one uses, only one key can be punched at a time. If the fingers were controlled independently, I would be able to accelerate my typing only by moving faster, in which case the number of fingers involved will not make a big difference. In reality, acceleration results from cross-coordination and prepositioning: frequent letter sequences produce flexible muscle-joint configurations where the wrist and finger positions are adjusted early to accommodate the forthcoming letters (Grudin et al, 1982). Consequently, it is not the motion speed that is increased, rather the amount of motion is reduced. Skilled behavior is more economic. How are those savings obtained?

As I learn to type the letter **A**, my finger, wrist, and elbow are placed in a certain position with respect to the keyboard. An ensemble of motor neurons executing this configuration constitutes a "virtual finger A" (the term from Arbib, 1985), so that typing **B** produces "virtual finger **B**," etc. Figure 3a portrays two such "virtual fingers," and Figure 3b indicates an overlap as a result, for example, of the wrist and elbow positions shared by both configurations. In Figure 3c the overlapping component is separated out, and a network of three motor neuronal sets is formed, corresponding to a "virtual finger **AB**."

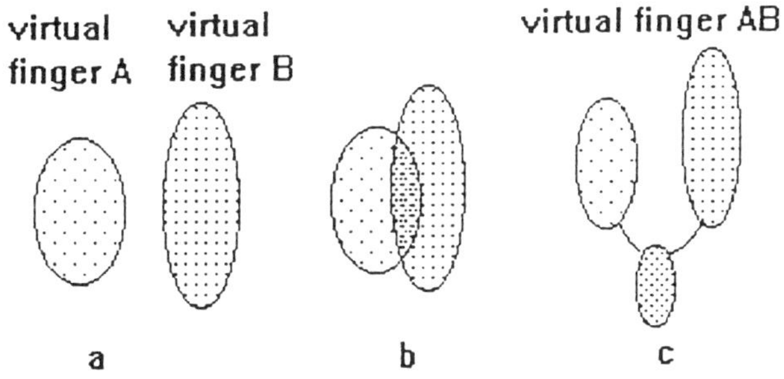

Figure 3. Acquisition of complex movement sequences proceeds by extracting invariants from the participating configurations. As a result, the amount of movement is minimized, and so is the amount of memory (in comparison with the "fingers" stored independently).

The notion of "virtual fingers" is an abstraction but relations expressed in Figure 3 are real. First, there must exist bounded configurations of neuronal activities underlying distinct movement patterns and, second, there must be a way to take advantage of the existing configurations in the acquisition of new movements. A mechanism that extracts the overlapping subsets in the configurations will do the job (Yufik, 1996b; 1990b). In typing AB, the mechanism will adjust the wrist and elbow configuration first, so that, for example, the A and B keys can be subsequently depressed without any wrist transfer.

If learning continues, the network grows as new overlaps emerge and extend the 3c structure in all directions. Figure 4 shows the stages in the network growth, with the emerging groups of nodes, and groups of groups, etc., corresponding to the typical letter combinations, words, phrases, etc. In such a structure, each elemental component appears only once, but can enter into an unlimited number of configurations, in the same way as a letter participates in an unlimited number of words.

The concept of movement construction underlies a theory of motor control in (Gelfand et al, 1971) (building on the seminal research in Bernstein, 1947). In the theory, elemental muscular activities constitute "letters" combined synergistically into "words," "phrases," and larger structures of "motor kinetic melodies."

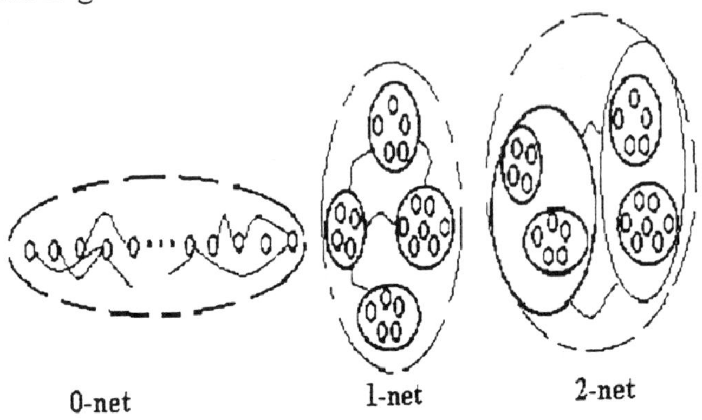

Figure 4. Emergence of cascaded integrated networks. The 0-net comprises a disorganized set of elements (e.g., letters). The 1-net partitions the set into ensembles (typical letter combinations), the 2-net forms ensembles of ensembles (e.g., frequent words and expressions).

Connections between the groups establish a variable flexible order for their execution so that if the initiated path turns out to be obstructed the system can divert to one of the collaterals without interruption, making the cascade organization inherently robust. Note that "opposing new habits to the old ones unceasingly" and "automatism divided against itself" (Bergson, 1955) are befitting descriptions for Figures 3 and 4. To develop a better feel for the mechanism, I will look at the Figures from a different perspective.

Let the cascade structure in Figure 4 represent memory organization of a competent chess player, with the innermost elements standing for the chess pieces, and their groupings standing for the typical combinations a player has learned (such groupings hold no pieces but only pointers to them; more on that later.). Responding to the current position on the board amounts to matching it with a few candidate groupings, followed by evaluating the alternatives within the groups' confines. As a result, the amount of processing is limited to what must be a tiny subset of the exploding combinatorial space (Yufik, 1997; 1996a). Combinatorial collapse in the integrative memory explains the observation that

> "a master rarely looks ahead any further than a novice does - and moreover, a master usually examines only a handful of possible moves! Master-level players have built up higher levels of organization in the way they see the board: consequently, to them, bad moves are as unlikely to come to mind as illegal moves are, to most people. This might be called *implicit pruning* of the giant branching tree of possibilities. By contrast, *explicit pruning* would involve thinking of a move, and after superficial examination, deciding not to pursue examining it any further" (Hofstadter, 1980, p. 286) (this conclusion summarizes results in the study of chess psychology in (deGroot, 1966)).

The "giant tree" corresponds to the 0-net in Figure 4, while the n-net narrows its exploration to a few branches. Problem solving is expedited due to the problem space partitioning. Chess algorithms need 10^8 operations/sec. to climb the tree, while the human player obtains a comparable efficiency with incomparably slower processing rates. The mammoth task of *implicit pruning* is carried by the self-organizing LTM, leaving the STM with a much smaller task of *explicit pruning*. Other examples follow.

5. Concept formation.

Lions and tigers look similar and both are recognized as cats. But how similar are they? One of the best-validated theories in cognitive psychology determines the similarity of two concepts as a function F of the number of the shared attributes minus the distinctive attributes, as follows:

$$S(A, B) = \theta F(a \cap b) - \alpha F(a - b) - \beta F(b - a),$$

where S(A, B) is the perceived degree of similarity between concepts A and B, a and b are the lists of their attributes, correspondingly, $(a \cap b)$ is the overlap, (a - b) and (b - a) are the features of A and B named as distinctive when comparing A to B and B to A, correspondingly, and θ, α, β are parameters reflecting feature salience and their relative importance (Tversky, 1977; Tversky et al, 1978). Figure 3 is compatible with that result.

Similarity judgments have been found to be asymmetrical, with the similarity of A to B not necessarily equal to the similarity of B to A. Furthermore, in expressing a judgment the more important or salient concept serves as a referent for the less salient concept. For example, a child having a cat at home and visiting a zoo for the first time, is more likely to compare the tiger to the well familiar pet than the other way round.

This asymmetry reveals involvement of subjective values in the concept formation and alternative resolution: the choice between the alternatives is biased by values, thus avoiding a potential impasse. Changing the values can re-direct the bias, affecting all the levels in the cascade in the manner of "downward causation."

Figure 3 suggests that in the stimuli categorization the invariants are processed first, and the outcome can be left undecided if the process aborts prematurely. For example, a briefly displayed image can be recognized as cat like, with a more fine differentiation contingent upon the available time. Symbols in a visual display are categorized as numbers or letters first, and identified precisely next (Shiffrin et al, 1977).

In often cited experiments, chess players briefly exposed to a board image recognize the overall situation (e.g., the left flank is stronger), and reproduce it in a way that vary the positions of individual pieces while the game is left strategically the same (Hofstadter, 1980). Certain forms of brain damage impair recognition of unique entities, while leaving recognition at the categorical level intact (Damasio et al, 1994) (more on that in part III).

6. Holonomic memory.

Figure 4 shows a system of non-overlapping but interrelated groups, while Figure 1 expands this view to a "system-of-systems." That is, a representational system comprises sensorimotor systems and participates as a member in an abstractive system, and so on.

Duality inherent in the notion of a system was best expressed in Koestler, 1967: Each unit in the system has some degree of internal *cohesion,* as well as boundaries separating it from other units. "Cohesion" implies that members of the unit "function together, cohere, interact, much more with each other... than with the members of other units." Such functional units are called "holons." Despite their relative autonomy and separation, jointly they still constitute a functional whole, in that changes in any holon can influence the way all other holons function and interact, while the system has properties attributable to the organization of holons but to none of them individually.

The key feature of holonomy is cross-coordination among units in the entire the system. For example, learning produces highly synchronized and similar event-related potential waveshapes in the brain regions distant from one another, "suggesting some kind of resonant oscillation" (John, 1990, p. 33). Electroencephalographic (EEG) recordings from the surface of the scalp reveal that responses to stimuli and other behavioral measures depend systematically on the momentary layout of the EEG landscapes (Lehmann, 1990).

In the holonomic memory, units form, change and dissolve, as elements switch membership between units in a flexible dynamic manner, adapting to the conditions to which the system is exposed. If the typist cuts one finger, all of the "virtual fingers" will be eventually rearranged to accommodate the loss. Similarly, typing in a different language will cause changes in the virtual organization. Speed gains, memory savings, the capability to adapt and expand are all rooted in efficient cross-coordination.

7. Fragmented memory.

If one remembers several chess positions, how many chessmen must one's memory hold? (The issue of cognitive economy was raised in Conrad, 1972; Collins et al, 1969.)

Return to 1- net in Figure 4, sever the ties between the groups, and let each group represent a position held in an isolated "box." Each such "memory box" must contain all the constituent pieces. Furthermore, since thinking about the position can engage any piece, the entire set and the board must be held in each box. This organization will not only consume much "real estate" but will pay a next to nothing return on the investment.

For example, how can this organization be updated? If you come up with a new move in the first position, can all the other boxes benefit from that finding if the there is no mechanism for cross-coordination? All the boxes must be attended to in turn, each time deciding what adjustments to make in light of the new discovery.

By the same token, choosing each move would involve comparing the present position with all those stored. Going piece by piece or relying on some statistical measures appears hopeless since even a slight deviation in the position can change the situation entirely.

Memory composed of "boxes" allows only *explicit pruning*, providing no mechanism for *implicit pruning*. As a result, learning is confined to isolated performance pockets, with no possibility for transfer and generalization. Consider the examples in Figure 5.

Note first that both constructs in Figure 5 are "memory boxes." After the weights are fixed, the boxes are closed. To obtain a different mapping, the weights must be recomputed (discarding the old ones and the previous mapping with them), or new boxes must be added (this critical shortcoming is called "catastrophic interference" in neural nets (McCloskey et al, 1989) and "brittleness" in expert systems (Lenat et al, 1986)).

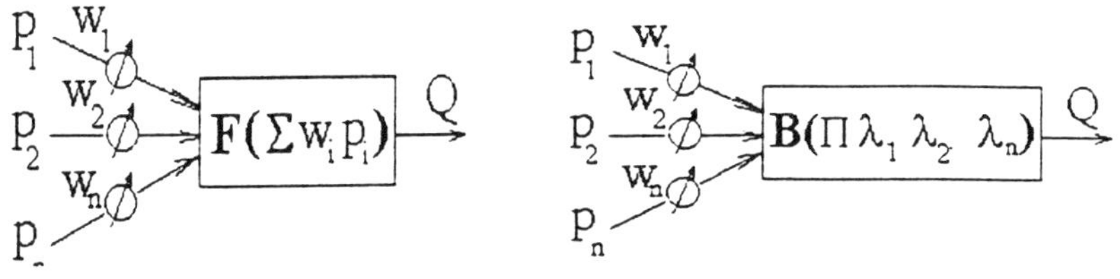

5a 5b

Figure 5. "Neural network" in 5a outputs the sum of its inputs subject to a non-linear (e.g., threshold) function F. Any input-output mapping can be obtained on such a summator, given a large enough set of weights, computer, and enough time to compute weight adjustments (Judd, 1990). Figure 5b transforms the "neural net" into an "inference net" by replacing summation with Bayesian multiplication. p_i stands for evidence, Q for hypothesis, and w_i for the "strength of belief" that Q is true given evidence p_i, $\lambda_i = 1 - w_i$.

Another observation is even more discouraging, and concerns what Figure 5 does not show: an intelligent agent external to the networks and responsible for weight computation. A neuron like that shown in Figure 5a can produce nothing but the sum of its inputs. Some agency other than neurons is therefore needed to specify the mapping and to figure out the weights. The problem is similar to figuring out a combination lock, and becomes intractable quickly as the size of the network grows (Judd, 1990).

Models in Figure 5 do not scale up and allow no cross-coordination. They demonstrate, in fact, that fragmented memory can operate only if all the "weights" are genetically preset except for some narrow variations, as is the case with the teleonomic systems.

8. Memory units.

Where and how are the integrative memory structures formed, and what do the "letters" look like? There are no definitive answers to these questions. A hypothetical one adopted in this chapter is due to (Pribram, 1991).

The site of global integration resides in the interpenetrating axodendritic and dendo-dendritic structures. Organizations are formed of clusters, or "ensembles of mutually interacting pre- and postsynaptic events distributed across limited extents of the dendritic network." Neurons interact with the networks as "thresholding devices that spatially and temporally segment the results of the dendritic microprocesses into discrete packets for communication and control of other levels of processing. These packets are more resistant to degradation and interference than the graded microprocesses. They constitute the channels of communication and not the processing element" (Pribram, 1991, pp. 7, 16). Figure 6 illustrates the dynamics of cluster formation (after Dalva & Katz, 1994).

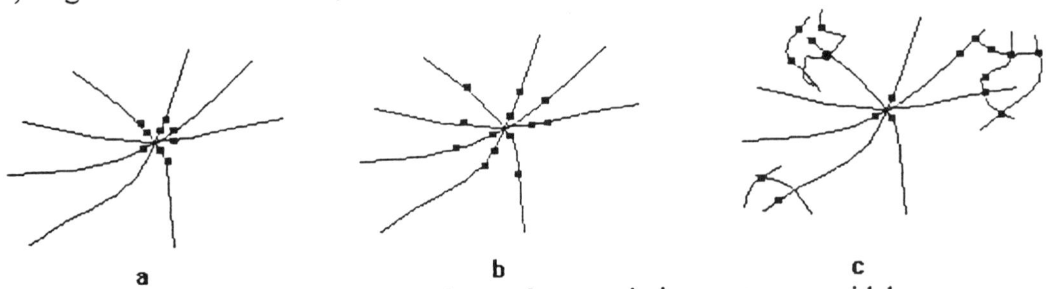

Figure 6. Changes in the topology of synaptic inputs to pyramidal cells in the cat's visual cortex before eye opening (a), immediately prior and after opening (b), and during adulthood (c) (Dalva & Katz, 1994; Katz & Shatz, 1996; Gilbert, 1992).

Patterns of synaptic connectivity in the visual cortex were found to assume three distinct forms, corresponding to three development periods. In the postnatal period (a), although the long extensively arborized axons are already present, all functional synapses congregate in the immediate vicinity of neuronal cells. As the animal matures, local synaptic connections radiate outward and increase in number until the animal's eyes first open (b), after which time the number steadily declines. In the adult period (c), well-formed clusters appear at remote locations (Dalva & Katz, 1994; Katz & Shatz, 1996).

The hypothetical global memory organization consists of horizontal dendritic nets and vertical columns extending across the layers. Axonal grids provide functional scaffolding for the dendritic networks. Discrete metastable network formations such as packets and clusters have proved to be polyfunctional in their responses to stimuli, and amenable to regulation by the forebrain (more evidence later). These formations can constitute the "letters," or the basic functional units in the VAN system.

9. What is memory for?

In the ascending line of self-organization, the human CNS is distinct from other systems in its efficiency of energy processing and entropy reduction. Due to integrated memory, the system can extract invariants and cross-variable regularities in large variable sets. Extraction of the invariants (laws, principles) causes increased organization in the system, which enables further increase in the scope of interaction and in the efficiency of entropy reduction. Expansion and self-organization efficiency appear to be the problems the human brain has evolved to solve, individually and jointly with other individuals.

As a gross simplification, I will suggest the following sketch. Picture a vessel with self-organizing matter (which represents memory) carrying its own energy production unit and a distribution unit capable of directing the energy flow inside the vessel. Let the organization in the material develop. This organization then influences both the rate of energy supply and the energy distribution decisions, which makes the three units a system. Furthermore, it all works in a way so that the organization grows. This section will substantiate the sketch following (Luria, 1973), and the rest of the chapter will explain it.

The *energy generation* module coincides largely with the hindbrain, which is the oldest part of the brain, and includes the Reticular Activating System (RAS) located at the center of the brainstem. The earliest function of the RAS was regulating sleep and waking, and maintaining sufficient non-specific activation during the waking state.

In the human brain the RAS becomes richly interconnected with the prefrontal cortices in the *executive, or energy distribution* module, allowing energy delivery to specific locations and providing a "mechanism by means of which the functional patterns of excitation arising in the cortex recruit the systems of the reticular formation of the 'old' brain and receive from them their charge of energy" (Luria, 1973, p. 45).

Stimulation of the RAS components evokes arousal and sharpens sensitivity to stimulation by lowering both the absolute and differential sensitivity thresholds. By contrast, lowering the arousal dulls sensitivity, and arrests the mobility of the nervous processes.

The *associative module* includes the primary and higher-order sensory cortices, and the parietal-temporal-occipital association cortex which is intercalated between the associative areas and integrates the outputs from different sensory modalities (Kupfermann, 1996). The function of this module "is connected with spatial organization of discrete impulses of excitation entering the various regions and with the conversion of *successive stimuli into simultaneously processed groups*" (Luria, 1973, p. 73). Such groups are processed in succession under the control of the *executive module* (Grafman, 1995).

The *executive module* includes prefrontal and orbitofrontal cortices in conjunction with the premotor and supplementary motor cortex and limbic association cortex. A prominent feature of this module is a rich system of interconnections with other brain structures, allowing synthesis of afferent signals arriving from them and organization of efferent impulses for their control (Mishkin et al, 1987). Summarily, the cortical architecture can be divided into primary, unimodal, heteromodal and supramodal (executive) association cortices, with the latter cortex occupying frontal and prefrontal brain areas (Restak, 1995).

It must be reiterated that the *executive module* has "a particularly powerful modulating influence on the reticular formation, giving its activating impulses their differential character, and making them conform to the dynamic schemes of behavior which are directly formed with the frontal cortex." In turn, regions in the *executive module* receive "powerful streams of impulses from the first unit by which they are 'charged' to the appropriate energy tone" (Luria, 1973, p.86). The *executive - energy generation modules* implement a regulatory loop (attention) unique in its precision (more on that in part II).

The executive functions include Priorities, Proprieties, and Practicalities (Pribram, 1991), which correspond roughly to goal-setting, value assignment (via the limbic system), and selection and ordering of actions to satisfy the goals. The underlying CNS architecture manifests in those functions, giving their execution a uniquely human character: humans can take advantage of earlier experiences but are not limited by them.

More precisely, the *executive module* selects responses among alternatives that include possible action combinations as well as those that have already taken place. The effort of overcoming the inertia of previous selections and evaluating new alternatives correlates with consciousness, and is subserved by language and logic.

Associative groupings in memory are shaped by invariants, that is, repetitive combinations of input features and response components. The number of such combinations far exceeds the number of units in them. To accommodate the combinatorial possibilities, the number of neurons in the association layers in LTM grows inward from the sensory periphery (inverted cone). The tip of the cone is narrow so that only a few variables at a time can be attended to in the STM. The entire internal machinery in the system is geared, in a sense, towards balancing the cone on its tip under two reciprocal tendencies: to maximize the number of variables processed through the tip, and to minimize the number of cross-variable groupings accumulating at the bottom of the cone.

These processes will be accounted for in part 2 in a model addressing four mechanisms: deliberate resource allocation in the STM, automatic resource grouping in the LTM, group stabilization, and group manipulation via controlled energy distribution (attention). Most of the concepts introduced in part 1 will be revisited, although within a better formalized context. Of the four mechanisms, only the first two have been developed in detail. Their models have been implemented, and the results will be reviewed in part 3.

To summarize the main points in part I:

1. Development in the human CNS is a process of entropy reduction.

2. A global memory organization is likely to emerge in the dendrititic networks in the cortex, in the form of circumscribed relatively stable symatodendritic clusters.

3. Clusters are influenced by the intrinsic values in the system, and are the units of LTM/STM exchange.

4. The clusters change flexibly, depending on the history of stimulation and response. Functional cortical "elements" are "recruited" into the clusters, and "demobilized" from them dynamically, as required for the input processing and response generation.

5. Clusters caused by the stimulation overlap self-partition in such a way that the overlapping components are extracted from the clusters, without the structure loosing its connectivity. These extracted components capture stimulation invariants.

6. As a result of the subsequent overlaps, a network of clusters can grow indefinitely.

7. Cross-coordination among multiple activities is due to cluster interconnections allowing dynamic transfer of "elements" between them.

8. When recognizing inputs and forming responses, attentive STM processing is limited to the contents of a few clusters. As a result, the amount of processing is drastically reduced, in comparison with what it would have been if the entire LTM were involved.

Part II. Virtual Associative Networks.

"In the brain the entropy reduction process involves a match between an input pattern and a pattern inherent in the synaptodendritic network by virtue of genetic or learning experience. ... The match is probabilistic and is expressed in changes in the probability amplitude weighing functions of synaptodendritic polarizations" (Pribram, 1991, p. 41).

Ontogenetically (Figure 1), a match is sought between the input organization and the target CNS organization. I will suggest some general characteristics of that target at the conclusion of this part. The match is obtained in a piecemeal fashion, by reconciling segments of the input stream with the synaptodendritic structure. Each cycle *assimilates* a new segment into the structure, and *accommodates* the structure to the segment (Piaget, 1954; 1950). The STM/LTM interface is the site of such reconciliation.

1. The STM.

In STM components of the incoming stimuli are matched with the contents of STM via serial-exhaustive scanning and item-by-item comparison (Sternberg, 1969, 1966).

For example, visual "percepts" are matched against the features of the stimuli (e.g., Sowa, 1987). The nature of such "percepts" is not known except that they are likely to be polyfunctional units matching different stimuli to a different degree. Pribram (1991) demonstrates that cell formations in the visual cortices join multiple features in their receptive fields. "Precepts" can be composed of such formations.

Successful matches produce constructs where features are "glued" together as a result of attention focusing (Treisman, 1988). As summarized in Kosslyn (1988, p. 1621), "although objects in visual mental images may seem to appear all of a piece, when the time to form images is measured this introspection is revealed to be incorrect; objects in images are constructed a part at a time." The stages of the construction process become apparent when the parts fail to be integrated seamlessly in the final product.

When the process breaks down, its stages are revealed in the distorted outcome. For example, a patient drawing a watch can draw all the numbers crowded into one half of the circle. In severe cases, the image cannot be constructed at all; patients are unable to follow clues and recognize familiar objects (Ornstein at al, 1984).

Drawings by young children indicate fragile constructs and inability to reassemble them from the parts at will. For example, a child asked to draw a square produced a beetle-like object, with a sphere in the center and lines radiating from it. Prompted to explain, she identified two lines as standing for the vertical sides, two others for the horizontal ones and, separately, two more lines standing for the corners (Michie at al, 1984).

In tactile recognition, the process is particularly evident. If an object is placed into one's open palm, its footprint, although available all at once, is difficult to recognize, which is done easily by following the shape with the tip of one's finger (Vekker, 1981).

To appreciate the role of STM, consider a primitive organism living in a liquid world of two variables (turbulence and light intensity). The organism can either move towards the light or away from it. Compare this predicament with a situation where an organism is sensitive to multiple environmental variables and several response options are available.

For example, think of returning a ball that somebody has thrown to you: you want to send it back (goal G_1) without hurting yourself (goal G_2). You can hit it (response **a**) or kick it (response **b**). If the former, the ball might not go far enough; if the latter, you might hurt your strained ankle. Which way is it going to be? The choice will be biased by the values (Duncan, 1990) (e.g., whether you are playing competitively or for exercise), and you can change your mind as the ball is approaching.

The STM process can be conceptualized as a stream of microdecisions in selecting and combining resources available to the organism in order to satisfy a few immediately demanding goals, which includes deciding on what those conditions are (combining the "percepts") and what actions to take (combining the responses). The process is that of combinatorial microplanning, or dynamic on-line Resource Allocation. Figure 7 suggests two generic allocation (matching) strategies the STM can employ.

The choice of a strategy can depend, e.g., on whether a single occurrence of a unique feature (Strategy I), or a conjunction of features (Strategy II) are sought. Inhibition of the intermediate choices, as in 7(3), makes the process effortful, requiring an influx of energy to support it. In the absence of adequate attention, the features can be combined randomly into "illusory conjunctions" (Treisman, 1991, 1988; Treisman et al, 1980).

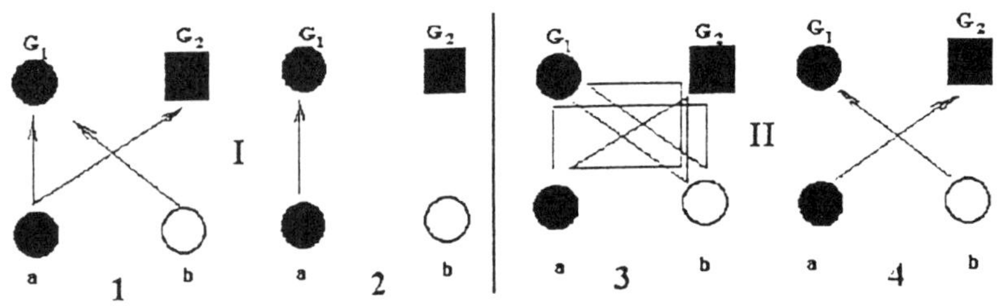

Figure 7. Matching strategies. Unit **a** matches items G_1 and G_2, to an extent, while unit **b** matches only G_2. Strategy I accepts for each unit the very first match found, resulting in a poor overall match in 7(2). Strategy II inhibits decisions until all the combinations have been evaluated, as in 7(3), producing the best match in 7 (4). The strategies can vary, depending on the task (Shiffrin et al, 1977).

Consider the possibilities for increasing the number of variables the system can handle. One way would be to keep the STM's capacity growing. Such a solution would be unproductive: the STM cycle will grow in duration and intensity, with the organism frozen until each cycle's completion and with all the effort going to waste as soon as the next cycle starts ("memory box"). Another solution would be to keep the STM small and the processing cumulative so that each cycle links uninterruptedly with the next one.

The fact that viewing objects requires continuous foveation mediated by saccadic movements, with the object disappearing from view as soon as the movement is arrested (MacKay, 1972; Vossius, 1972), indicates that the latter solution has been favored.

2. The LTM.

a. Assimilation. Figure 8 makes the first step towards the system/environment reconciliation. The idea is that a satisfactory match in the STM causes associative binding of the participating resources in the LTM, following the classical theory that "any two cells or system of cells that are repeatedly active at the same time will tend to become "associated," so that activity in one facilitates activity in the other" (Hebb, 1949, p.70).

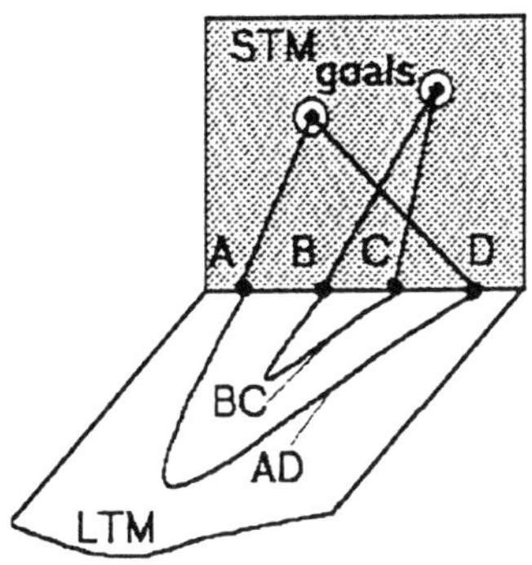

Figure 8. STM resources allocated to the same goal become connected by associative links (AD and BC). The strength of such links depends on the relative goal significance, and other factors. Repetitive co-allocation of the resource pair to the same or other goals strengthens the link between them. The phenomenon of synthesizing individual items in the STM into larger units is known as "chunking" (Miller, 1956). Sufficiently strong re-occuring links register in the LTM permanently. Otherwise, they are subject to decay.

Neuronal implementation of such associative links is not precisely known, although some critical evidence has accumulated, to be referenced in conjunction with the present model.

Consider input synchronicity and neuronal responses that must take place for the association to occur. Figures 7 and 8 suggest that in the STM nothing occurs precisely "at the same time." Consequently, for a link to emerge between two co-allocated resources, there must be some latitude in the simultaneity requirement, at least within the STM cycle. Recent data suggests that such latitude is indeed feasible, allowing associations between asynchronous inputs even at the cellular level (Fuster, 1995, p. 31).

Importantly, the role of simultaneity diminishes in human learning, to allow associations across arbitrary time intervals. By contrast, in animal learning proximity of the stimuli remain the decisive factor, with no or minimal tolerance for the intervening events.

The other concern is as follow. Action potentials propagating back from the axon to the dendrites can cause Hebbian modification of coactive synaptic terminals (Magee et al, 1997). At the same time, action potentials propagating forward can lead ultimately to an overt action. If associative bindings were predicated by the neuron firing, internal activities in the CNS and overt actions would be difficult to disentangle. Fortunately, that is not the case: associations can be formed on neurons prevented from firing (Carew et al, 1984; Fuster, 1995). Again, in the human CNS the phenomenon is amplified. allowing contemplative internal activities decoupled from the overt actions, while in the animal the distance between the internal and its expression remains short.

To summarize, inputs are assimilated into memory as associations of resources (e.g., "percepts") selected to act on them (e.g., match them). Associations yield two performance advantages. First, the next time one of the resources gets selected, it brings about the other associated resources automatically, thus cutting down on the selection time and effort. Second, by connecting the resources and causing them to behave in a coherent fashion, associations improve the overall degree of coherence in the system.

Associations have a dual nature: that of a physical object when following them produces an action, and that of a topological object when no such action follows. Responding by action R to stimulus S is, in essence, a physical event, while acknowledging the S - R connection without acting on it is a mental event. Thinking is an internal process operating on topological relationships (e.g., Lewin, 1936). This thesis will be elaborated shortly.

b. Conflict. Figure 9 depicts two STM cycles processing two different inputs G_1 and G_2. However, the same resource A was selected in both cycles.

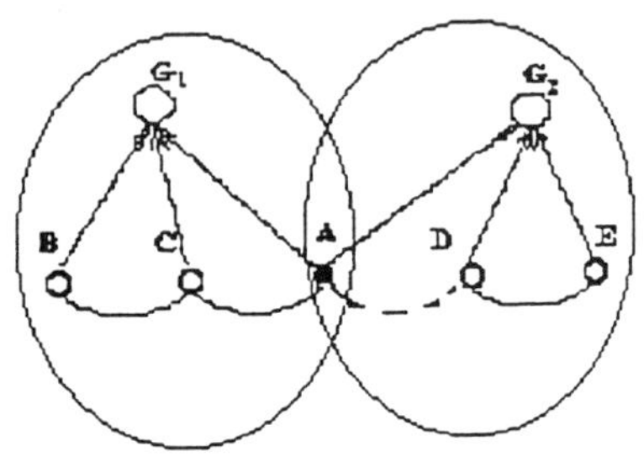

Figure 9. Two resource allocation episodes stitch together two overlapping associative structures. This creates a conflict, as follows: next time A is selected, which one of the structures will be pulled along? If both, the mechanism will be soon pulling in the entire resource set. This will render the associations useless.

The dilemma is as follows. If no overlap is permitted, there is no experience transfer between the episodes, so that each allocation is done independently. This strategy will severely restrict the number of variables the organism can deal with. On the other hand, a similar problem is faced if the overlapping sets are merged indiscriminately. A mechanism is needed for integrating the overlapping experiences without destroying them.

For example, lions and tigers are recognized as cats but remain fairly distinct: one does not form a chimera cat image mixing their features together. What mechanism allows the system to balance fusion and separation in the overlapping sets?

A comprehensive account of the process dynamics was obtained in observations of infants exposed to overlapping visual tracking tasks. Since the skill is one of the first people learn, such experiments offered a view of the process unobstructed by other experiences.

Infants were performing two tasks: tracking a moving object, and fixing their eyes on a motionless one. In the overlapping version, the object would stop and resume motion, intermittently. At one week of age, infants failed in both tasks, that is, they kept their eyes moving after the object stopped, and stared at the empty spot after the object was gone.

At sixteen weeks, the infants performed well in the overlapping task. In the interim period, however, they could perform the tasks separately, but failed at the overlapping version, randomly choosing one of the two tracking strategies.

Importantly, the babies did not improve gradually but abruptly acquired new capabilities after which they performed almost flawlessly (Bower, 1979).

Similar although less pronounced effects are observed in the adult visual tracking. For example, in compensatory tracking of several simultaneous signals, the operators' response time as a function of training duration shown distinctly nonmonotonic character for each signal, with periods of performance improvement interspersed with plateaus and even brief periods of performance decline (Venda, 1986).

On a different scale, abrupt changes are demonstrated in the acquisition of composite skills requiring cross-coordination among multiple activities. This is clear when people learn to ride a bike or to swim. People's skill normally advance slowly at first but then there is a sharp increase (grasp) in the learning pace and performance quality (Bernstein, 1967). After such grasp, the mastery of the skill is, to a large extent, permanent.

Similarly, in the acquisition of problem solving skills a "grasp of a concept," once attained, is hardly ever lost. For example, after grasping the idea of some tactical device a chess player is unlikely to forget it. Before such grasp the player does not reliably execute or recognize the tactics, which is liable to disappear from memory.

Piaget analyzed results in (Bower, 1979), placing conflict at the center of learning and development. According to Piaget, there is no progress without conflict, which builds up for some time before getting resolved in an abrupt memory reorganization (Piaget, 1954, Bower, 1979). Adding a new variable to the skill base entails its reconstruction. Performance can decline in the process and improve steeply after its completion.

Mainly in overconstrained repetitive tasks (e.g., adding two numbers) with no expansion in the skill base, performance demonstrates steady uneventful improvement with practice (and decline without it) (Anderson, 1980). Figure 10 underscores these differences.

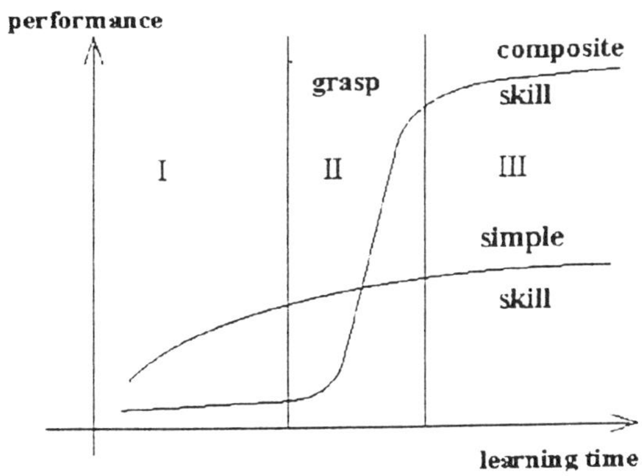

Figure 10. Composite underconstrained learning undergoes steep acceleration (grasp) (II), following initial slow advancement (I). By contrast, simple overconstrained learning exhibits steady improvement. (after Bernstein, 1967), (Bower, 1979), and (Anderson, 1980)).

Summarily, two factors must be accounted for in a model of memory dynamics: functional overlap that creates a conflict, and an abrupt restructuring that resolves the conflict and causes a qualitative change in performance. A new experience that partly coincides with an earlier one and partly contradicts it, destabilizes memory and creates tension. As the experiences accumulate and the tension reaches a threshold, the system snaps into a new organization that restores the stability temporarily, until the experience base is further expanded. The process can continue indefinitely, adding new skills and integrating them into the base. Where can such memory build up and integration occur?

The synaptodendritic networks in the upper cortices as the likely site of integration. Such networks proved responsive to stimulations, with the topological changes in them occurring within minutes (Gilbert et al, 1995). Interpenetrating synaptodendritic structures provide flexible and geographically unbounded medium for the global organization.

Changes in the networks accompanying conflicts and their resolution are articulated next. Crucially, those changes are conceived to be both automatic and near-optimal, allowing performance to take maximum advantage of the earlier experiences, without any supervisory agency being responsible for such optimality.

c. Accommodation: automatism divided against itself.

Figure 11 shows how an integrated structure builds in the LTM due to the emergence of new connections.

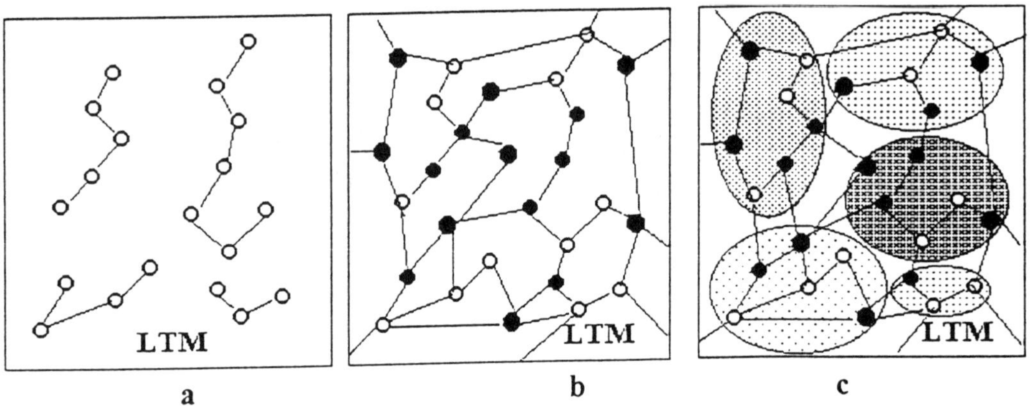

Figure 11. Development of a global associative network. Links are modified by experiences, and can be produced by spawning new synaptic connections, and/or by pruning (disabling) the existing unexercised ones. In this way, the functional connections (shown) are embedded in a thicket of dysfunctional ones (not shown).

Figure 12 highlights one of the clusters in Figure 11, and explains its formation.

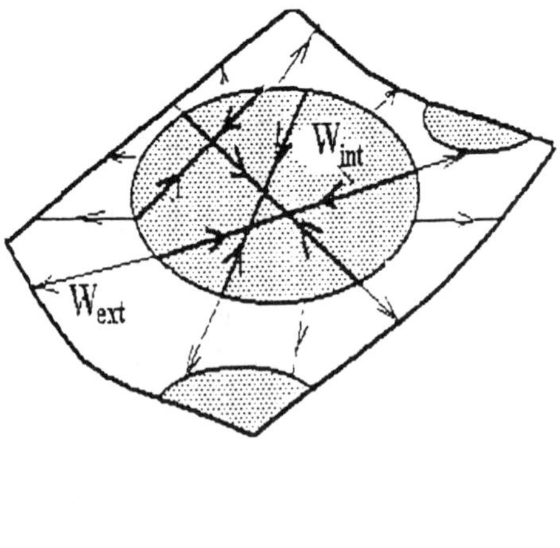

Figure 12. Stimulation causes neuronal structures to organize into groups such that interaction within the groups is stronger than interaction between them (Edelman, 1978). Accordingly, in the present model associative networks are assumed to self-partition into clusters such that in each cluster the cumulative strength of internal associations (Σw^n_{int}) exceeds the cumulative strength of the external ones (Σw^m_{ext}).

Each cluster remains intact for as long as the internal associative forces or, more generally, entropic forces (the term is due to von Bayer, 1997)) that keep it together prevail over the external ones that pull it apart, so that the stability factor σ exceeds unity

$$\sigma = \Sigma w^{n}_{int} / \Sigma w^{m}_{ext} > 1.$$

Clusters loose stability when the strength of the external connections start to increase due to their more frequent exercise, and/or the strength of the internal connection diminishes due to lack of exercise. When the value of the stability factor falls below unity $\sigma < 1$, the cluster disintegrates and its components are absorbed into the surrounding clusters, which in this way expand both territorially and functionally.

Networks amenable to self-partitioning as defined in Figure 12 ($\sigma > 1$) are called Virtual Associative Networks. Clusters produced by such self-partitioning are called packets whose role is to "spatially and temporally segment the results of the dendritic microprocesses..." (Pribram, 1991, p.7). The concept of packets in the VAN model generalizes that of "neuronal pools" and/or "neuronal ensembles" invoked in other models in explaining motor control and other functions (e.g., Feldman, 1981, 1980, 1974).

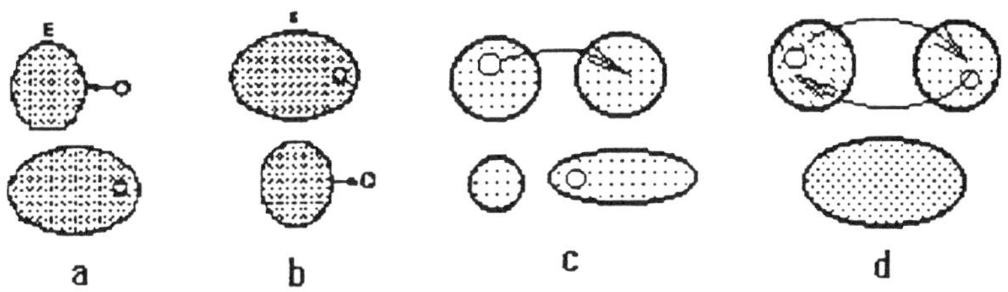

Figure 13. Dynamics of "neuronal ensembles" corresponding to the associative processes in Figure 12: **a** - unit is "recruited" into the ensemble E, **b** - unit is released from the ensemble E, **c** - unit switches membership from E1 to E2; **d** - ensembles E1 and E2 merge due to the increased unit traffic across their boundaries.

To appreciate the model, think of the lion-tiger puzzle and apply Figure 12 to resolve it. According to the relationship in Figure 11, conceptual structures self-partition to obtain maximally cohesive attribute packets, with a minimal external coupling between them. The structure is maintained automatically, for as long as the stability conditions are preserved. Adding new features (consider adding leopard, cheetah ...animals, living creatures, etc.) extends the structure and causes its periodic spontaneous reorganization as "the old habits are being opposed to the new ones." In this way, an unlimited number of overlapping concepts can be absorbed into and balanced in the memory network.

Changes in any location in the network trigger a wave of changes through the entire net and leaving structural alterations in its wake commensurate with the stability conditions the wave encounters. That is, the solidified packets will not be vulnerable to spurious influences, while the less-tested ones might succumb to them (thus resolving the stability/plasticity dilemma posed in Grossberg, 1987). The self-dividing network remains, however, a unified whole at all times.

Figure 14 accounts for the fact that packets are both topological and physical entities. As physical entities, they are subject to the laws of thermodynamics. Correspondingly, an energy barrier is hypothesized at the packet boundary determining the packet's stability. The height of the barrier U is derived from the Gibbs-Helmholtz law as

$$U = \sigma - T\,(d\,\sigma/dT),$$

where σ is stability factor and T is equated to the level of arousal (or activation). In the nervous system, the arousal control circuits center on the amygdala and define the phasic physiological responses to input, while the activation circuits center on the basal ganglia and forebrain and define the readiness to respond (Pribram et al, 1975).

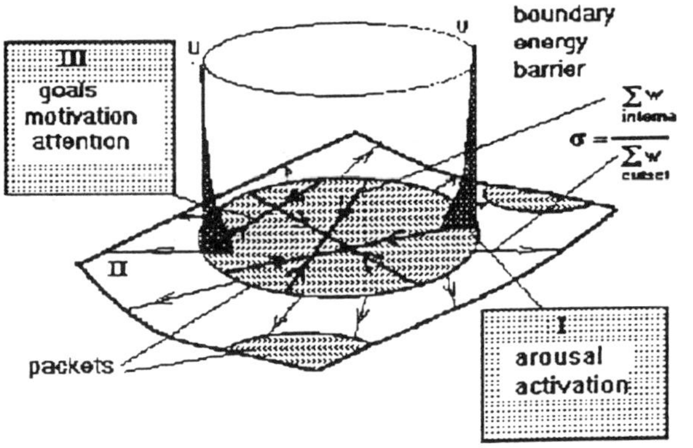

Figure 14. Interaction between the nervous system modules determines the dynamics of LTM structures. Priorities controlled by the *executive* (III), and the arousal/activation level contingent on the energy supply (I) determine stability in the memory system.

The higher the relative frequency of associations within the packet and the higher their relative subjective value, the fewer reasons the system has to deviate from them. Correspondingly, increases in the σ ratio cause the boundary barrier U to grow. On the other hand, when conditions and/or values change and the packet becomes unsuccessful or irrelevant, the system responds by increasing T and lowering the barrier, which eases access to the connections outside the packet. The more frequent use of such external connections further decrements stability σ, precipitating packet dissolution.

The neuronal mechanism underlying self-partitioning is likely to employ lateral, or associative inhibition (the term is due to Kline, 1921). To experience such inhibition, consider learning some stimulus - response association A → B, and then replacing it with an A→ C. If the response B has settled in, it will inhibit the replacement so that performing the new response will take an extra effort until the memory consolidates. Such inhibition allows associations to perpetuate themselves and resist dissolution. Perhaps, the same inhibitory mechanism is responsible for establishing stability thresholds. Attentive effort must be applied to overcome the threshold before the packet dissolves.

Associative inhibition can be implicated in producing "repetition blindness" in subjects viewing a rapid series of overlapping stimuli (words or pictures). In one experiment subjects had difficulty reporting stimuli components that appeared twice in the series. For example, in their report of two sentences they would omit occurrence of a word in the second sentence if it was present in the first one, even if the resulting phrase becomes meaningless. For example, stimulus "It was work time so work had to get done" is reported as "It was work time so had to get done" (Park & Kanwisher, 1994).

In the assimilation/accommodation iterations, associative inhibition adds "cognitive inertia," or resistance to restructuring, thus compelling new skills to take maximal advantage of the already existing skill base while causing its minimal disruption. Such inertia is crucial in making new responses consistent with the prior experiences.

A study of memory organization as a function of stimuli overlap has produced circumstantial but compelling evidence that the areas of overlap in the cortex indeed obtain special properties and respond differently to subsequent stimulation, in comparison with the responses from the surrounding structures (Vlasova, 1989).

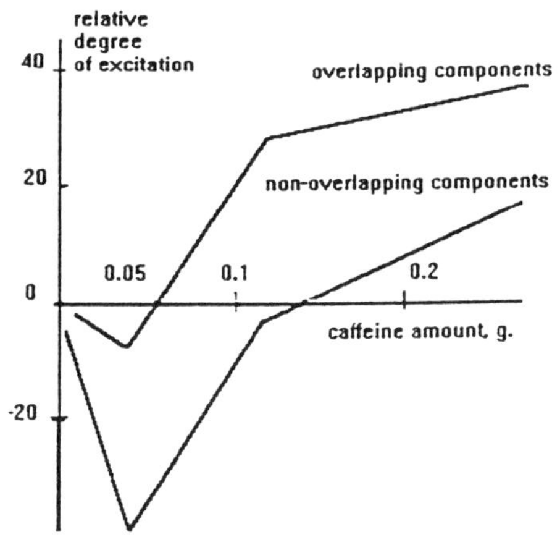

Figure 15. Measurements in (Vlasova, 1989) reveal different dynamic properties of memory structures ("relative degree of excitation") produced by overlapping and non-overlapping stimuli. The differences manifest in the way the structures respond to the subsequent stimulation and caffeine induced arousal. These results are consistent with the packet model in Figure 11.

Subjects had to learn interconnections between unmarked buttons and indicators variously located in control panels. In the learning phase the subjects were shown independent and overlapping control sequences. During testing the task was to recollect the sequences and to change the state of a few select indicators in the shortest time and in a minimum number of trials. While performing the task, subjects had to respond to probe signals emitted from areas in the panel occupied by the overlapping and non-overlapping sequences. Response delays and other characteristics were used as measures of the relative excitation in the developing memory structures. The results demonstrated significantly higher excitation levels in the areas of overlap, and distinct overlap boundaries responsive to the level of subject's arousal (e.g., induced by caffeine).

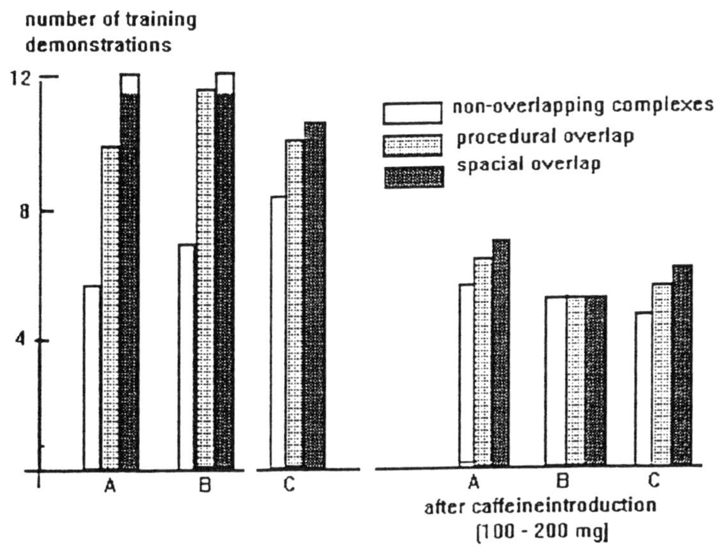

Figure 16. Subjects memorize instrument groupings and control procedures in a series of brief demonstrations. Learning is affected by the degree of group overlap (spatial and procedural) and the arousal level (after Vlasova, 1989).

3. The LTM/STM interaction.

What does the LTM/STM mechanism buy in performance efficiency? According to the present model, self-organization in the LTM is the single most important contributor to such efficiency, allowing the CNS to assimilate exceedingly large numbers of variables and respond to changing conditions in real time in a manner optimally constrained by previous history. Figure 17 illustrates, symbolically, how the process works.

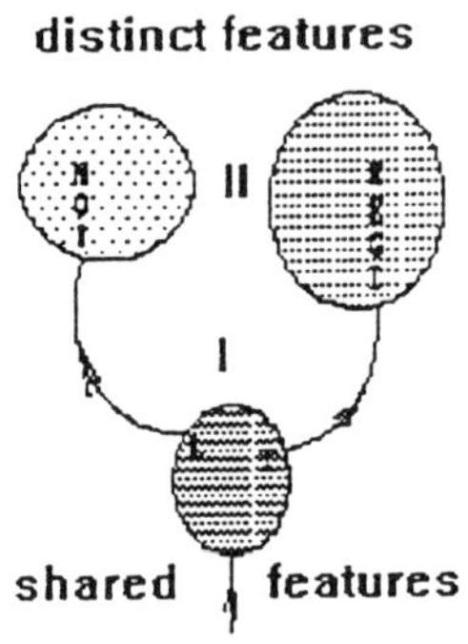

Figure 17. Stages of input processing in disambiguating two concepts. The first stage identifies features shared by both concepts, followed by confirming the distinctive features in order allowing their quickest and most reliable differentiation. Letter chains in the packets indicate that packets are not structureless "pools" but sequences of perceptual steps, along with their collaterals to ensure performance robustness.

The LTM organization in Figure 11 yields an economic scanning strategy, determining what to identify first and where to go next at each subsequent step, depending on the previous matching results. In short, the scanning mechanism strives towards disambiguating inputs in a minimal number of steps, as afforded by the history of earlier experiences with other inputs both similar and dissimilar with the current one. The mechanism biases the order of input processing towards the "heaviest" packets comprising prototypical attributes of some class. After the class qualifications are met, the subclasses are determined (or a few iterations follow, as indicated in the next section).

Experimental evidence in support of this contention is presented in (Vlasova, 1989). Subjects were trained to categorize visual stimuli in two ways: by identifying the overlapping features first and the discriminating ones next, and in the reverse order, that is, by identifying the discriminating features first. Figure 18 shows that the former strategy reduces processing time and error incidence, in comparison with the latter one.

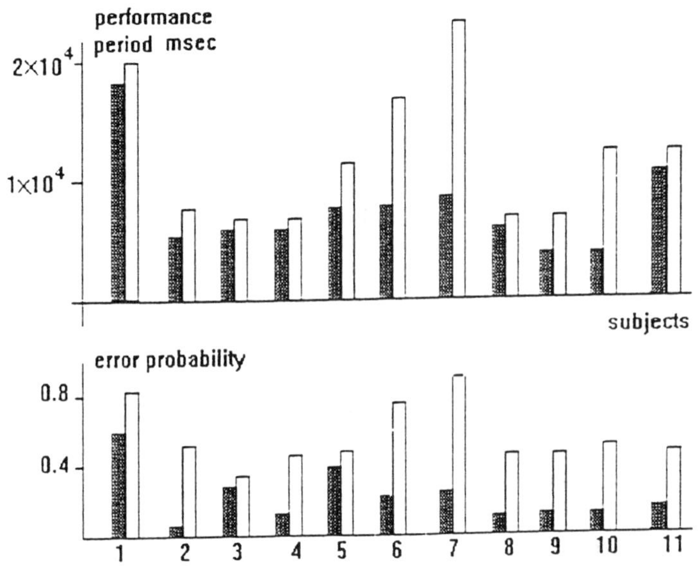

Figure 18. Processing strategy shown in Figure 17 speeds up performance and reduces error rate in stimuli categorization tasks (dark bars), in comparison with a reversed strategy (light bars) (after Vlasova, 1989).

Explanation of these and similar results in the present model is sharply different from the conventional notion of pattern recognition. In the chess example, a conventional view would correspond to an archive of position snapshots showing specific pieces in specific locations. In the present model, there can be no such fixed archives. For example, a pin prevents the movement of a piece from the line lest it exposes a more valuable piece. This basic maneuver can be realized in a literally infinite number of ways. Consequently, there is neither a fixed configuration of pieces, nor a set of such configurations, nor a statistical tendency of any sort that can be labeled as a pin and stored in an archive.

More generally, the model claims that the notion of "pattern recognition" as the foundation of competent performance is inadequate. There are no fixed patterns in the environment except those imposed by the performer. The situation is precisely as follows:

> "We come often to this conclusion: Processes write their history chronologically, so that, as page is laid on page, each page gives a description of a moment, and the sequence of pages tells us of past time from far back up to the present. This metaphor is, however, not altogether adequate: There are no separate pages. Rather, in the dimension of past time the record is a continuum just as it is in its representation of space. And, instead of pages, which are indifferent to the content, the record exhibits words, sentences and chapters which are set off from each other precisely according to this content. ...The traces are by no means inert objects. Pervaded by forces, and sometimes by currents, the record will tend to revise its own text spontaneously. If we consider that it is also under the permanent dissolving influence of metabolism, we can only expect the final edition to be a sketchy abstract of a distorted story" (Kohler, 1938, p. 244).

4. The LTM hierarchy.

Figure 19 completes the basic VAN formalism by defining a hierarchy of virtual networks.

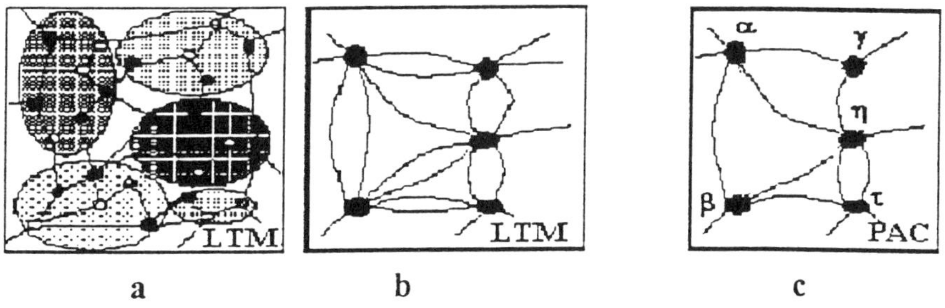

Figure 19. Then LTM is a hierarchy of virtual networks where each node in the upper level net (**b**) represents a packet in the level below (**a**). Network in the top layer (**c**) has symbols associated with the nodes. Presumably, the top layers reside in the Prefrontal Associative Cortices (PAC, or supramodal cortex (Restak, 1995)).

Figure 20 highlights a vertical component in the hierarchy. The AB and BC connections have been exercised relatively more often than the AC. Correspondingly, connections in the upper packet reflect the preferred paths for navigating the lower level region.

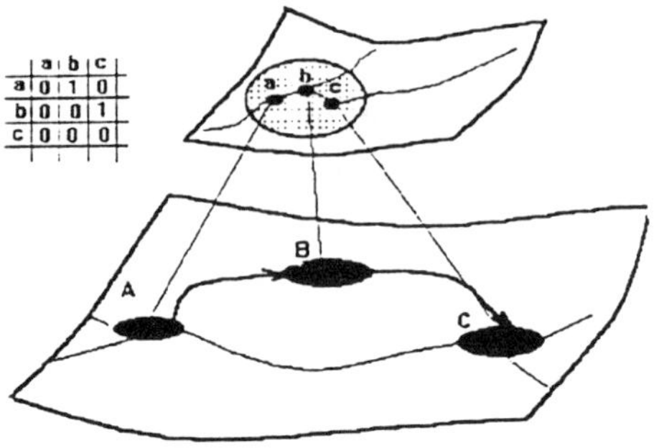

Figure 20. The upper level packet is a "map" of significant interrelationships within some region in the lower level network. Such maps can form after the lower nets become sufficiently stable. Since both the map and the network region it describes belong to the same system, map formation is a form of self-description.

To experience "associative maps," think of some movie you saw years ago. You can tell almost instantaneously whether you remember it quite well, or that you only remember a few scenes. To make such a judgment, you do not have to retrieve all the episodes and verify that they are remembered. Instead, only the "map" of the memory structure has to be accessed to determine whether it is available and its connectivity preserved. Such "maps" are topological entities, similar, perhaps, to topological projections of the external space in the motor control system which

> "must be congruent with external space but only topologically and in no sense metrically. ..the topological properties of the projection of space in the CNS may prove to be very strange and unexpected; we must not expect to find in the cortex some sort of photographic space, even an extremely deformed one" (Bernstein, 1967, p.109).

Animals can form spatial maps (Tsien et al, 1996), but are unable to form other topological constructs to represent higher order regularities. For example, apes can learn as many as a few hundred signs and apply each of them properly, but cannot learn rules for connecting such signs into regular structures, however simple (Sebeok et al, 1980).

5. Accommodation in the LTM hierarchy.

In Figure 17 features in two overlapping sets have been re-grouped, with two results. First, an economic representation was obtained for both sets by separating the common features out, as opposed to replicating them. Second, the overlap has been resolved and interdependencies between the residual sets minimized. In the "map" in the next level up the representation structure will be further reduced by omitting the features altogether, and preserving the intergroup relationships only. Figure 19 generalizes this process for a feature set of an arbitrary size. One can see that accommodation in the LTM hierarchy is a process by which the expanding feature set is being dynamically partitioned into non-overlapping and minimally interdependent units.

6. *Ruling the automatism.*

This section considers interactions between the associative memory and the *executive module* that allow the CNS to deviate from experiences, and profit from such deviations. In this way the system exercises its "power of choice.. coextensive with the fringe of possible actions that surrounds the real action" (Bergson, 1955). Without introducing new assumptions into the model, the section will suggest how such fringes of possibilities can be formed in the LTM. The notion of construction remains central, defining the CNS as a system "adapted to the reception of stimuli traveling to the brain from the peripheral receptors, to their analysis into a very large number of component elements, and to their "synthesis into whole functional systems" (Luria, 1973, p.68). As it ascends in the neocortex hierarchy, the process transforms from direct stimuli synthesis to symbol manipulation.

a. Cognitive effort and effort regulation. Following strong associations within the packets is relatively effortless, while transiting from one packet to another takes an effort due to barrier crossing. The barriers are, in fact, the key means of performance optimization: given the goals, the LTM supplies packets of resources expected to satisfy them; additional resources outside the packets can be reached but are less accessible due to the required effort. In this way, the system suppresses spurious responses, making the current ones consistent with past history (Yufik et al, 1995a; 1995b).

The system can adjust the packets and regulate network navigation by changing the activation or arousal level (Figure 14), by adjusting stability thresholds in the packets, and by selectively modifying the strength of associative links in packets and their surroundings. A subjective correlate of such selective modifications is attention focusing.

b. Attention. Current theories describe attention as a "processing resource" that can be varied from uniform distribution to highly focused concentration (Ericsen & Yeh, 1985; Theeuwes, 1993; Crick, 1984). According to the present model, the resource is energy. Traversals inside the packets are inattentive or low attention activities, interspersed with attentive, that is, energy demanding boundary crossings.

Another function of attention focusing is amplification of associative links. By moving the attention "searchlight" (Crick, 1984) inside the packet and in its vicinity, one can hit on a link whose amplification upsets the local stability balance and triggers restructuring. As a result, one can retrieve items previously defying recall, and/or become aware of some overlooked relationships. The familiar experience of having a word "at the tip of one's tongue" can be explained by following a link that goes through a barrier. This gives one the sensation of knowing what one is looking for, without being able to recollect it until a sufficient effort is mounted for getting over the barrier.

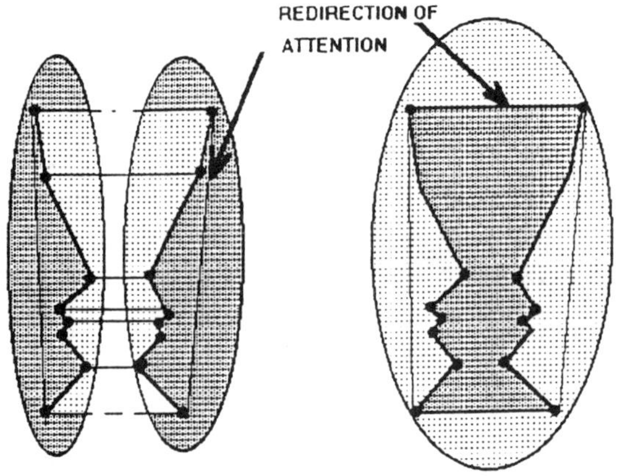

Figure 21. Packet restructuring as a result of attention shift. Some configurations are unstable (under the $\sigma > 1$ stability condition) so that attention focusing on some links causes the network to break in two parts, while shifting attention to other links causes the parts to merge, as in the "faces-vase" and other ambiguous figures.

Diffuse energy distribution sets the "temperature " T in a network area, or the local "cortical tone" (Luria, 1973; Benson, 1994), causing variation in the height U of energy barriers at the packet boundaries (Figure 14). Roughly, based on the approximation offered by the Gibbs-Helmholtz equation, lowering the "tone" causes the barriers to grow and recede inward, which makes it more difficult to deviate from the kernel of overlearned connections inside the packet. By contrast, increasing the tone lowers the barriers and shifts them outward, allowing easier access to the less exercised connections. (This approximation can be fine tuned in many ways).

However, under these forms of attentive regulation performance can neither deviate from the existing paths, nor account for future eventualities. In other words, performance can only "stretch the chain" of habits. To break the chain, other forms of regulation are needed, more elaborate than tone adjustment and random wielding of the "spotlight."

c. Action planning. In planning, one exercises the "power of choice.. coextensive with the fringe of possible actions that surrounds the real action." The exercise has four ingredients: first, the habitual action must be suppressed, second, the fringe of possible actions must be generated, third, the alternatives must be evaluated without committing to any of them, and forth, one alternative must be selected and followed through.

Let some input A be associated with the condition B, and absence of A^* with the absence of B (condition B^*). The fringe of possibilities includes four combinations $a = A \times B$, $b = A \times B^*$, $c = A^* \times B$, and $d = A \times B^*$, further combined into a lattice of sixteen alternatives ( ), (a, b, c, d), ... ((a + b), (a + c) ....) .... (a + b + c +d).

In the course of normal maturation, memory structures of the lattice type become available to the child after the sensorimotor stage (Flavell, 1963; Piaget, 1950).

By what mechanism can those structures be produced in memory if they have not occurred in actuality? Figure 22 demonstrates that attentive packet manipulation can generate lattices, by deliberate shifting of the packet's boundaries.

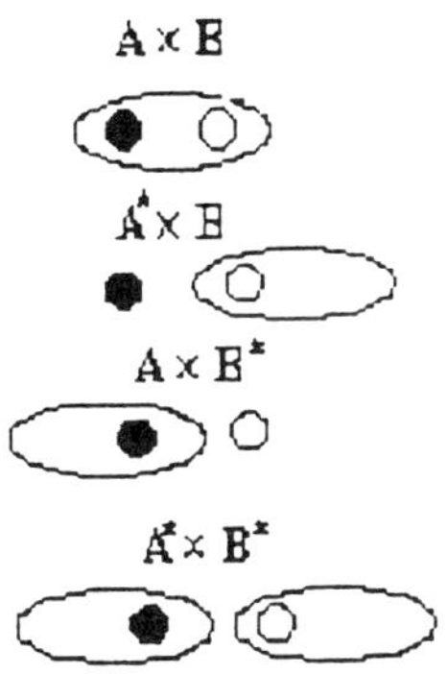

Figure 22. Generation of hypothetical action alternatives (lattice) via packet manipulations. The mechanism is the same as in the motor control (see Figure 16), but addresses packet stability rather than muscle group stability, and manipulates packet stability thresholds σ instead of the muscle activation thresholds. Positioning of A and B inside the packet establishes an association, while their separation by the boundary inhibits it. A precisely focused attention is needed for such manipulations. Alternative selection includes their prioritization and item-by-item comparison in the STM, as in Figure 3.

Upward projection and downward regulation between the layers in Figures 18 and 19 correspond, perhaps, to the notion of "reentrant mapping" advanced in (Edelman, 1992), and that of prefrontal "executive" in Grafman (1995).

d. Planning mechanism and attention focusing. Data in Pribram (1991) offers neurobiological substantiation of the model. Measurements of the neuronal receptive fields in the animal's lateral geniculate nucleus and the primary visual cortex demonstrated their susceptibility to prefrontal regulation. In that, the inhibitory surrounds and the organization of inhibitory flanks can be augmented or diminished by electrical excitation applied to different forebrain locations. For example, stimulation of the far frontal cortex diminishes the surround, while stimulations in the posterior association cortex cause their enlargement. Accordingly, expansion of the adjacent fields leads to their eventual merger, while shrinking entails decoupling.

A regulative mechanism centered in the thalamus is proposed in (Crick, 1984). According to the searchlight hypothesis, bursts of activities in thalamic neurons produce transient alterations in the synaptic connections in the cortex where the neurons project. Prefrontal regulation can be mediated by the basal ganglia. "The output from the basal ganglia is inhibitory, and therefore it can also *inhibit inhibition*. This either excites them or prepares them for excitatory input, a state important for attention" (Edelman, 1992, p.142). The role of the basal ganglia has been confirmed by the recent positron emission tomography measurements indicating their involvement in two attentional systems. The posterior system moves the attention spotlight to the target cortex area, and amplifies the target. The anterior system is engaged only in deliberate alternative selection (Posner et al, 1994).

The mechanism hypothesized in Figure 22 agrees with these findings. The model suggests that the regulative system, physiologically available in higher animals, have been perfected by evolution to become functionally available in normally maturing humans.

e. Abstraction. The chain of automatism is broken when packet boundaries become amenable to deliberate adjustment. No other mechanism but manipulation of the packet stability thresholds needs to be involved in the process. Extending the scope of the mechanism underlies abstraction, that is, deliberate re-combination of the existing structures in the formation of new ones. A diagram below illustrates progression from the association of contiguous stimuli, to lattice, to abstractive re-combination. In the latter, association between B and C enables combinations of A and C although no temporal or spatial proximity between them has been experienced.

$$
A \times B \Rightarrow \begin{matrix} A \times B \\ A^* \times B \\ A \times B^* \\ A^* \times B^* \end{matrix} \Rightarrow (B \times C) \rightarrow \begin{matrix} A \times C \quad A \times C \times B^* \\ A^* \times C \quad \ldots\ldots\ldots\ldots \\ A \times C^* \quad \ldots\ldots\ldots\ldots \\ A^* \times B^* \quad \ldots\ldots\ldots\ldots \\ A \times B^* \times C^* \end{matrix} \rightarrow (C \times D) \rightarrow \ldots\ldots
$$

Abstractive re-combination, in a sense, transports lattice across the entire associative structure, making it possible to reach towards past experiences, however remote, and relate them to the present condition, however indirectly. Abstraction benefits performance since reaching further into the past enlarges the "fringe of possibilities" and extrapolates it further into the future. This ability is due to memory integration which compresses all past experiences into a unified structure available simultaneously in the present.

Only a contiguous relation ($A \times B$) is available to an animal. As a result, animals live in the present while for the humans present is a fleeting overlap between the past and the future.

Selfrestrain and suppression of automatisms remain the key performance ingredients in abstraction, as underscored in the definition: "Abstraction consists in asserting that a given sign is applicable instead of merely applying it" (Roberts, 1973, p. 64).

f. Signs.

Packet formation accounts for gradual separation of objects from the background of sensory flux in the CNS of a maturing human. Ascendance in the associative hierarchy coordinates with the extraction of sensory invariants, which culminates in associating symbols with the stable components of upper level networks (Yufik, 1994; 1987). This aspect of the model draws on the central premise of gestalt psychology which holds that

> "sensory units have acquired names, have become richly symbolic, and are now known to have certain practical uses while nevertheless they have existed as units before any of these further facts were added. Gestalt psychology claims that it is precisely the original segregation of circumscribed wholes which makes it possible for the sensory world to appear so utterly imbued with meaning to the adult; for in its gradual entrance into the sensory field, meaning follows the lines drawn by natural organization; it usually enters the segregated wholes" (Kohler, 1947, p. 82).

The VAN model holds that operations on the stabilized networks exercising "the power of choice" and serving the purpose of determining the "fringe of possible actions that surrounds the real action" (Bergson, 1955) give rise to language and grammar.

g. Language.

"My words fly up, my thoughts remain below;

Words without thoughts never to heaven go."

W. Shakespeare, Hamlet, III, 3.

Highly stable packets are processed as consolidated wholes. They can be segregated from the rest of the structure and connected to other packets in a manner determined only by the links. Such a packet along with the N links emanating from it constitutes a symbol with N attaching points. Combinations of such symbols are called plexes (Feder, 1971).

Abstractive re-combination is the same as plex generation. Grammars determining plex interconnections define combinations of phonemes in the spoken languages, combinations of letters in written languages, and combination of "percepts" in sensory processing (Feder, 1971; Sowa, 1984). In this way, consolidated packets available for abstractive re-combination constitute units of a "language of the brain" of which English is a subset.English letters are symbols with (-t-) (-w-) (-o-) attaching points. As a result, they can only be connected into one- dimensional strings, as in this text. Visual patterns are combinations of two-dimensional "percepts" (Fu, 1977), as in the Figures below.

Unrestricted plex grammars allow plex constructions where any subconstruct can be cut out and replaced by any other. Context-sensitive grammars make such replacement contingent on the structure into which the replacing construct is embedded. Context- free grammars allow replacement of certain plexes regardless of the surrounding structure. Grammars are generative in that they define a generation of composite structures from a finite set of terminal units (e.g., percepts) specifying the choice of units and the order in which they are to be connected, as in Figure 23.

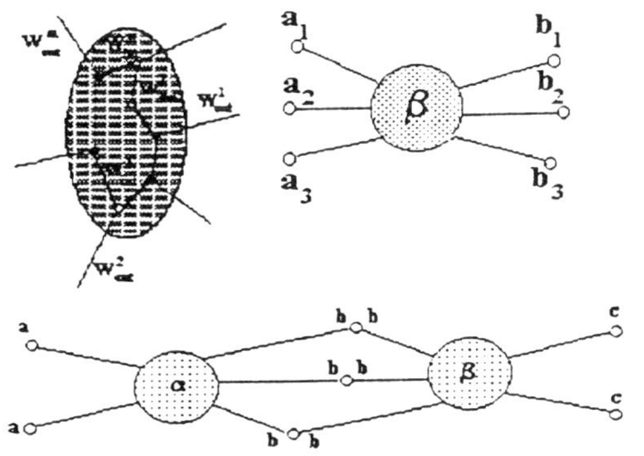

Figure 23. Transition from a consolidated packet to plex. Symbols become associated with stabilized packets in the upper cortices. Rules for re-combination of plexes can underlie grammars, such as English grammar.

For example, a context-free plex grammar below defines construction of letters **A** and **H** using a single terminal unit [SL] (Straight Line) which denotes a line without regard to orientation, with attaching points 1 and 2 at each end (specifications of how the attaching points connect, which is a part of the grammar, are not included for simplicity) :

[LETTER] →[A] | [H] Construct [LETTER] is assembled of [A] or [H].

[A] →[SIDE] [SIDE] [SL] Construct [A] is assembled of two [SIDE]s and [SL].

[H] →[SIDE] [SIDE] [SL] Construct [H] is assembled of two [SIDES]s and [SL].

[SIDE] →[SL] [SL] Construct [SIDE] is assembled of two [SL]s.

In this way, plex grammars define generation of cascade structures, as in Figure 4.

In the brain, spoken languages operate in conjunction with non-spoken ones, such as visual and motor-kinesthetic. This is possible if the languages share the basic machinery for assembling and disassembling their constructs. In this interaction, symbols seem to form only a thin crust on the top of other structures, and only after they have sufficiently organized. The "crust" does not get separated from the underlying modality-specific processes, as has been demonstrated by direct measurements. For example, PET studies have shown involvement of the left prefrontal/premotor regions in the retrieval of words denoting actions, and the temporal pole regions in the retrieval of words denoting persons. These results indicate that those regions may hold reconstruction maps for the phonemic structures of the corresponding words (Damasio et al, 1996). In view of the underlying machinery, acquiring "words withouht thoughts," as in (Sejnowski et al, 1987), appears to be a mechanistic exercise revealing hardly anything about language.

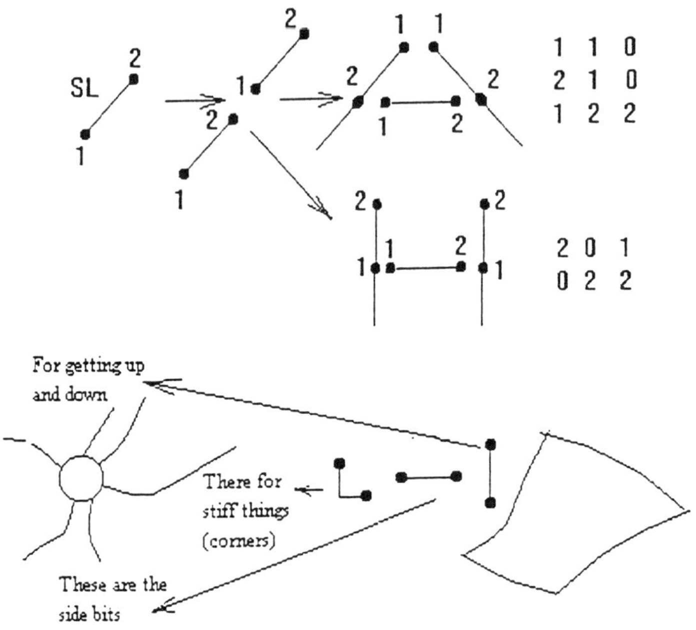

Figure 24. Construction of letter shapes from a single element Straight Line [SL] with two attaching points. Construction "maps" are matrices determining how the attaching points are joined. The A- map shows that the point 1 of the first [SIDE] connects to point 1 of the second [SIDE] (first row), etc.

The 3.5 years old girl could not quite follow her "map" of a square although already having one, as demonstrated by her drawing a "proper" square (to the right) (after Michie, 1984).

Perhaps, at this point the testimony by the little girl (cited in Michie, 1984) can be corroborated by more general observations, to assert the notion that a system that squeezes its inputs and outputs through a narrow STM aperture has no choice but to be generative and constructive:

> "A conjunction of representations is not to be found in objects themselves, nor can it be borrowed from them and taken up into the understanding by perception, but it is on the contrary an operation of the understanding itself, which is nothing more than the faculty of conjoining a priori, and of bringing the variety of given representations under the unity of apperception. This principle is the highest in all human cognition" (Kant, 1990, p. 78).

Finally, Figure 25 comments on a theory of grammar acquisition (Chomsky, 1965).

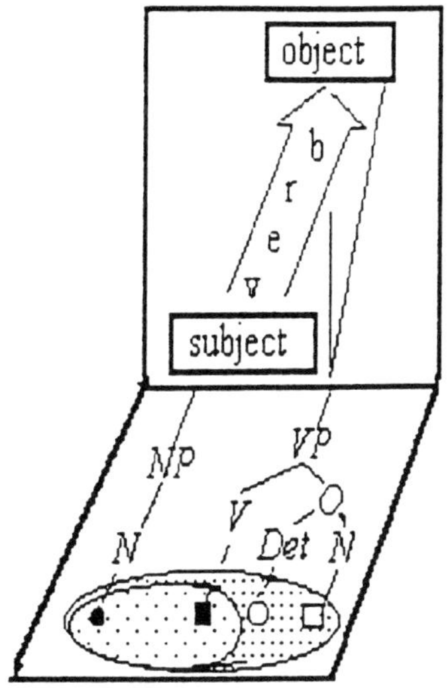

Figure 25. A universal generative grammar includes generative base S →NP ^ Aux ^VP (John kicked the ball) and rewriting (assembly) rules. The suggestion is that the deep structure reflects the basic STM cycle, with a hierarchy of maps in the LTM implementing the "rewriting rules," down to the level of elementary "lexical entries" that combine specific syntactic and phonetic features drawn from a set of such items. (Chomsky, 1965). The theory of linguistic competence asserts a small number of fixed universal features and combination procedures tacitly available to the child and instantiated to a specific language in the course of development.

h. Logic

"Rejection is the basis of logical thinking. Logical thinking is the management of NO."
E. de Bono, 1970, p. 197.

Logical concepts can be expressed in terms of topological relationships (e.g., Sowa, 1987). Logical calculus by Pierce makes the coupling between symbol manipulation and topological operations particularly explicit. This calculus, called existential graphs, is consistent with the virtual network model. The relationship will be suggested in this section, and developed elsewhere. The point of the suggestion in the present discussion is that the same core machinery can be engaged at all cognitive levels, from sensory processing to formal logic, and, if so, can be accounted for within a unified framework. Such synthesis is accomplished in (Sowa, 1987) where composition of perceptual, syntactic, and semantic structures is accounted for uniformly in graph-theoretic concepts.

According to Pierce (cf. Roberts, 1973), the basic logical relationship of the proposition "If P then Q" is expressed as a graph consisting of two closed lines, one inside the other, with the antecedent and consequent in separate compartments. Each of the lines determines a "cut" (or a "sep," from saepes, a fence). "The cut may be imagined to extend down to one or another depth into the paper, so that the overturning of the piece cut out may expose one stratum or another, these being distinguished by their tints; the different tints represent different kinds of possibilities" (p. 88).

On this definition, cut corresponds to a packet surrounded by a barrier. A consolidated packet is manipulated as a whole. Accordingly, "cut precisely denies its contents" (p. 78) expressing the inaccessibility of the packet's inside from outside.

Connecting an element inside a cut to one in the outside area expresses the relationship of "excluded possibility." For example, a cut that has inside the expression "is a dog" connected to the expression outside "is an animal" asserts that some animal is not a dog.

Finally, cuts cannot overlap but can be made within other cuts, which entails cascading. Iterations in such cascades are like "a march to a band of music, where every other step only is regulated by the beat, while the alternate steps go on of themselves" (p. 43). (i.e., path traversals, with alternating packet crossings and interpacket transitions). These and other logical operators in existential graphs are in close correspondence with the basic packet manipulations shown in Figure 22.

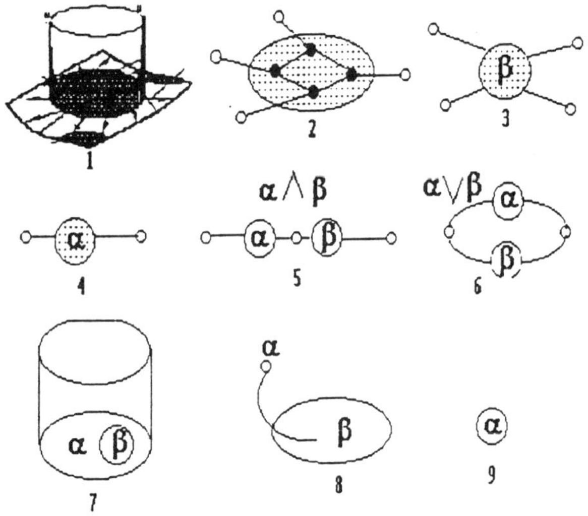

Figure 26. Progression from topological to symbolic manipulations. 1 - energy barrier at the packet boundary; 2 - "consolidated," i.e. stable and strongly associated packets; 3 - two-dimensional symbol with 4 attaching points; 4 - symbol with two attaching points; 5 - conjunction as serial concatenation; 6- disjunction as parallel concatenation; 7 - cuts "extending down in the paper" and expressing "If α then β" relation; 8 - α is "other than" β (left outside the packet); 9 - α negated (no access).

7. The developing organization.

This section pulls together the elements of the model, and defines a mathematical formalism used in its implementation in part III. The unifying concepts are those of value and complexity of the developing organization in the LTM.

In the hierarchical LTM, each component is of a dual nature: that of an object, and that of a description, or a "map" of other objects. In this sense, the overall memory system not only contains its own description but also acts upon it in a manner of downward causation. The predicament is similar to that of DNA which is an object and at the same time a description of itself and the way it will develop.

This section suggests that the CNS acts in the direction of optimizing the complexity of its own description, which leads to minimizing the entropy of its trajectory. Since physical laws, genetic determination, and conscious choice are all entangled in this activity, the tendency unifies all three forms of goal-directed behavior.

Let system X have a binary description of length D (like a binary matrix in Figure 19). Such a system is said to have D-diameter complexity. Partition X into N non-overlapping parts x_i of diameter $d_i < D$, and consider all possible partitions. K(X), or Kolmogorov's complexity measure, is determined by the length of the system's description, including descriptions of all the parts plus the description of the assembly procedure, minimized over the set M of possible partitions

$$K(X) = \min_{M} [k(P) + \sum_{N} k(x_i)],$$

where P is the length of the description of the assembly procedure and $k(x_i)$ is the length of the description of the part x_i (Kolmogorov, 1987; Chaitin, 1979).

The suggestion is that the human CNS progresses towards some optimal K(X) value.

Each such optimization step is entirely pragmatic. Since the response to every input must be assembled, the better the parts are organized the less effort and time the next such assembly will take. By the same token, the more thorough the accommodation in the present cycle, the less restructuring the next assimilation/accommodation cycle will incur.

Summarily, this amounts to a system seeking to assemble itself from the simplest parts by the simplest procedures. Since there is a trade-off between the two, the optimal complexity is sought by way of dynamic reconciliation between the internal structures and external conditions. Such optimum preserves a degree of instability in the system, to allow easier reorganization in the case of abrupt and irregular external changes.

Interplay of all those factors results in a system that, stated colloquially, strives throughout its life time towards a comprehensive world view that would unify the description of the world with that of the self in a minimum number of non-overlapping concepts.

Internal actions are those of attentive re-partitioning and re-arrangement of the parts. It is worthwhile to reiterate that a) these activities are based on the division of labor between the functional modules so that damage to the prefrontal cortex entails inability to shift from a learned concept to a novel one (Grafman et al, 1992), that is, to manipulate packet boundaries and b) attention shifting is a physical process demanding energy and time. Shifting from one object to another requires at least 400 msec, and can take longer after frontal, parietal or basal ganglia strokes (Husain et al, 1997).

Progression towards optimal complexity requires "optimal brain neurodynamics" (Luria, 1973) manifested as the ability to obtain all possible partitions in the maps, and all possible part recombinations, without restriction. Figure 27 illustrates this conclusion.

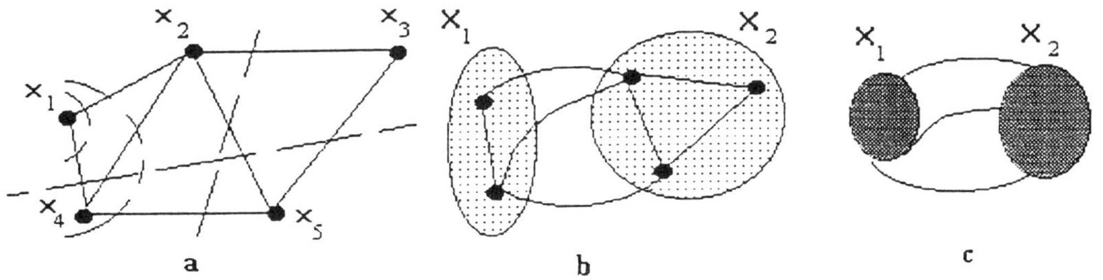

Figure 27. Cognitive complexity of associative structures. **a** - Kolmogorov's complexity of a "map" is determined by its partitions obtained by cutting links in the map. **b** - formation of packets establishes a privileged partition. **c** - packet stabilization produces a minimal description of the privileged partition in the form of its "map" in the next layer above.

Figure 27a shows that partitions break the network into two separate parts, and are obtained by cutting the links in the corresponding cutsets (a set of links whose cutting or removal from the net leaves it disconnected is called a cutset). The set of partitions includes individual nodes, pairs of nodes, and all other possible node groupings.

Figure 27b shows a special partition resulting in two maximally independent parts, or packets. Figure 27c shows a map of that partition in the next level up the hierarchy.

Automatic integration and self-partitioning into packets in LTM are minimizing topological entropy of the system's trajectory by limiting trajectory divergence in the immediate vicinity of the present state. Conscious re-partitioning and re-combination reduces metric entropy by limiting the trajectory's divergence in the future.

Operationally, consciousness resides in the split between the physical and topological in the association, and within the time window open for interfering with the activity after the onset of neuronal processes underlying the activity. For example, EEG and other measurements reveal a protracted interval from the moment one starts willing the action and the moment the action actually begins (Deecke et al, 1976). Recent measurements have indicated several schedules followed by different regions in the brain: the pre-supplementary motor areas show activity ahead of the movement, while the supplementary areas are locked with the movement (Hikosaka et al, 1996).

The automatic and the conscious complement each other in optimizing the Kolmogorov's complexity of the overall developing organization (Yufik et al, 1995a; 1989).

The following summary of the main points concludes Part II:

1. Self-organization in the LTM is modeled as integration of link-weighted associative networks that self-partition into cohesive and weakly coupled subnets, or packets.

2. Self-partitioning is near-optimal and does not involve any supervisory agency. The mechanism is such that the entire network responds adaptively to any local change.

3. Packet stability depends on the balance of the external and internal associative forces in the packet, and on the "cortical tone" in the cortex area where the packet resides. The system controls stability by shifting stability thresholds at the packets boundaries, in the same way it shifts muscle activation thresholds in motor control.

4. Attention focusing can change the local "cortical tone" in the packet's region, and amplify selectively some of the associative links, causing the packet's destabilization.

5. Thinking consists of shifting packet boundaries in the networks of interconnected packets. This manipulation is sufficient for implementing logic, language, hypothetical and abstractive reasoning.

6. A hierarchy of self-organizing networks becomes self-descriptive.

7. The system strives towards an optimally complex self-description comprising a maximum number of significant variables in a minimum number of concepts.

8. Attaining such self-description is a uniquely human form of goal-directed behavior.

Part III. Modeling in the virtual associative networks.

"The purpose of thinking is not to be right but to be effective. Being right means being right all the time. Being effective means being right only at the end." E. de Bono, 1970, p. 95.

An aircraft is an immensely complex system of complex systems. However, the principle of flying is simple, and boils down to a special shape of the wing creating lifting force. Most everything in the aircraft is dedicated to overcoming gravity and increasing the speed and the distance of flying, while minimizing the cost. This article takes the position that most everything in the human cognitive system is dedicated to overcoming combinatorial explosion and increasing the number of variables and the processing speed, while minimizing the energy expenditure. Like with a flying wing, the principle is simple.

The first two sections in this part will demonstrate that a) memory organization as postulated in the model allows on-line near-optimal solving of large-scale RA problems with small memory platforms, and b) this postulated organization corresponds to the characteristics of the human cognitive system established in the developmental psychology. The subsequent sections will review other evidence of the model's validity.

1. Speed up in resource allocation.

Return to the problem of finding a "match between the input pattern and a pattern inherent in the synaptodendritic network by virtue of genetic or learning experience. ... The match is probabilistic and is expressed in changes in the probability amplitude weighing functions of synaptodendritic polarizations" (Pribram, 1991, p.41).

Let the matching be controlled by value distribution over the feature set, seeking the overall matching quality F in excess of some subjective threshold, $F = \sum_k A(S_k)\varpi_{kj} > F_{\min}$. where $A(S_k)$ stands for the relative value of the feature S_k, and ω_{ij} stands for the degree of match between the S_k and the jth "percept." How does the CNS obtain satisfactory solutions of this problem in real time on large feature sets?

According to the VAN model, the solution principle is based on dynamic memory organization. To validate this assumption, the process was replicated computationally in conjunction with a standard method of combinatorial optimization (Papadimitriou, 1982).

Experiments were designed to allow precise measurement of the model's performance. For that purpose, the same set of problems was solved by the same standard method, with and without dynamic memory organization. The results were compared to measure the relative solution acceleration and the relative error caused by not using dynamic memory organization (Yufik, 1995). In the experiments, the matching problem was generalized by assuming the possibility of composite features matched by several combined percepts.

In that problem, an allocation matrix $\|\delta_{ij}\|$ had to be obtained to maximize the quality function F on a set of S prioritized composite features and N percepts,

$$F = F(\delta) = \sum_{i=1}^{S} A_i \left(1 - \prod_{j=1}^{N} \varepsilon_{ji}(\delta_{ji})\right); \quad \begin{array}{ll} \delta_{ji} \in \{1;0\}, & \sum_{i=1}^{S} \delta_{ji} = 1, j = 1,\dots,N, \\ 1 \geq (\varepsilon_{ji} = 1 - \omega_{ji}) \geq 0, & j = 1,\dots,N, \\ A_i > 0, & i = 1,\dots,S. \end{array}$$

The solution takes the form of a binary matrix $\|\delta_{ji}\|$ where $\delta_{ji} = 1$ if percept N_i is allocated to the feature S_j, and $\delta_{ji} = 0$ if otherwise. Experiments involved sets of 20 composite features and 300 percepts, with randomly generated probabilistic matching matrices $\|\omega_{ij}\|$. Each experiment included an initialization stage and a performance stage.

In the initialization stage, solutions were obtained for randomly selected subproblems, and percepts allocated to the same feature were connected by minimally weighted links. Each time a link was exercised, its weight was incremented. The process continued until the percept set was connected into a network, subsequently partitioned into packets.

In the performance stage, new full size problems were generated, and solved in two ways: disregarding the packets (control allocation), and processing the packets (test allocation). The ratio of control time/test time served as a measure of acceleration, while the difference in the utility value in the test allocation and control allocation determined the relative error. Experiments were conducted in batches of seventy , varying the relative packet size and other parameters. The results of one such batch are shown in Figure 28.

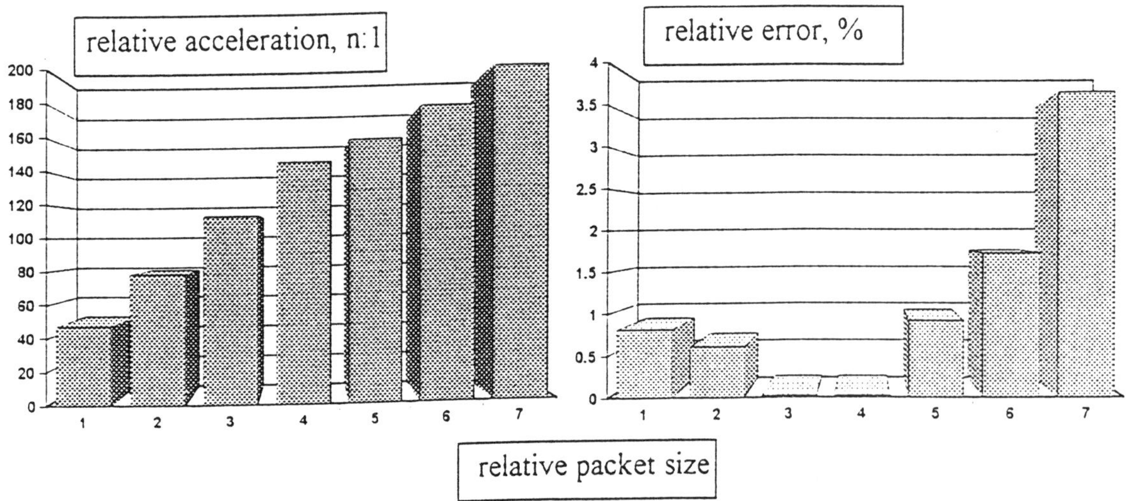

Figure 28. The relative acceleration factor depends on the packet size (limited by the STM capacity). Within some size ranges, there is substantial acceleration with no tangible accuracy loss (< 0.1%). Acceleration is due to the diminishing amount of processing.

The lifting power in a wing is due to the difference in the speed of the air above and below the wing. In the biological system, the "lifting power" is due to the difference in the amount of processing in allocating resources from large sets and from the properly formed packets. Excluding the interpacket combinations causes the combinatorics to collapse.

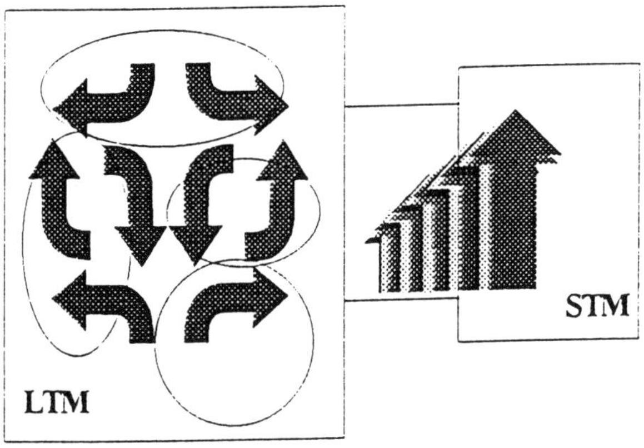

Figure 29. Large allocation problems are solved near-optimally as quick successions of small ones, due to the integrative memory taking full advantage of the earlier experiences (Yufik, 1996; Yufik et al, 1996).

Theoretically, the system can add variables indefinitely, with no changes in the architecture and with the performance improving after each assimilation/adaptation cycle. The "cone" organization, subject to the "law of diminishing modal specificity," is compatible with the operational principle formulated in this section. The lower zones at the tip of the cone have maximal specificity, which decreases in the secondary zones, and becomes minimal in the tertiary overlap zones. "The law of diminishing specificity... is thus responsible for the transition from discrete reflection of particular modal-specific cues to the integrated reflection of general and abstract schemes of the perceived world" (Luria, 1973, p. 76).

Resource allocation in static memory faces diminishing return, that is, it slows as the size of the problem grows. A self-organizing LTM yields progressive return, that is, the relative acceleration grows with the size of the problem (as compared to static solution).

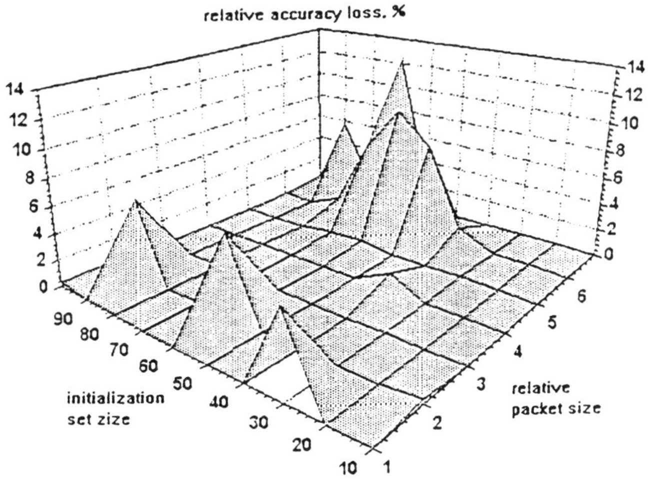

Figure 30. Interdependency between the initialization set size, packet size, and relative accuracy loss in the 300:1 acceleration range. Connectivity is reached when the initialization set size equals ten. Peak losses appear sporadically within some packet size regions. Cascades with feedback yield multiplicative speed gains, bringing acceleration to the 10^5 - 10^6 range.

As a rule, experimental networks stabilize shortly after reaching connectivity, so that adding new samples causes no significant performance changes.

In a multi level LTM, input processing starts by propagating the feature set upward in the hierarchy until convergence is reached in one of the upper level "maps." Based on the upper "map," the lower level packets are selected for STM processing (see Figure 20). If a satisfactory match is not obtained, alternatives are tried via a few iterations between the lower and intermediate levels in the hierarchy. Failure to match calls for re-organization.

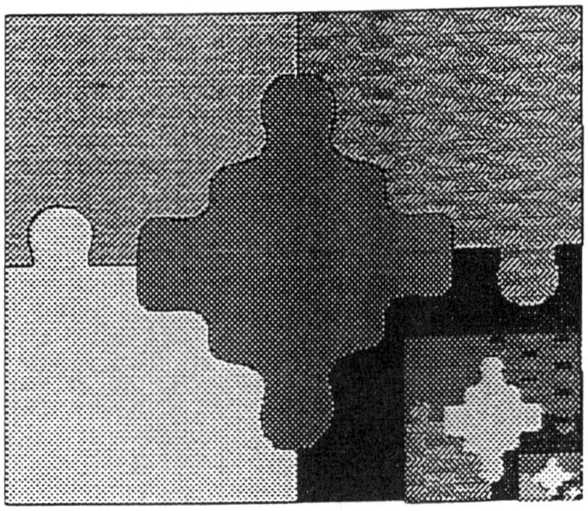

Figure 31. Input processing as solving a multi stage jigsaw puzzle in LTM. Each piece in the puzzle corresponds to a node in the network, and combination of pieces to a packet. The pieces "fit" if they belong to the same packet. Feature convergence in an upper level packet identifies the constituent packets below, which in turn point to the packets in the next lower level, and so on until the bottom level is reached containing the feature set. Recognition proceeds as in Figure 17, from the common to distinct feature combinations

2. The machinery of thinking.

Four claims were made in the previous sections:
first, that mature thinking consists of deliberate repartitioning and recombination of circumscribed regions in dendritic networks formed as a result of stimulation overlaps,
second, that these operations are conducted by shifting stability thresholds (inhibition/disinhibition) in those regions,
third, that dendritic networks can be modeled by weighted connected networks, and
fourth, that stability shifts can be modeled by cutset manipulations in such networks.

There is no way to validate those claims directly. However, if some significant defining characteristics of observable behavior were proved consistent with the significant intrinsic properties of the networks and cutsets, the claims will be supported.

The theory of cognitive development by Piaget asserts that in normally maturing adults operations of disciplined reasoning acquire the properties of algebraic groups. It must be shown, therefore, that cutset operations in the virtual nets have the same property.

Each cut divides the network in two halves (see Figure 27a), so that a series of all possible cuts generates a set of networks. There are three basic operations for re-combining those networks. The *union* of two nets produces a network that includes all the nodes and all the links of both of them. A subset of the *union*, called *link-disjoint union*, excludes the shared links. The *intersection* keeps only those nodes and links in the combination that are shared by the constituents. Finally, the *ring sum* produces a network that includes all the nodes of the constituents, and keeps only the distinct links. The *ring sum* and the *link-disjoint union* completely satisfy the notion of novel combinations of the earlier formed associations. Therefore, for the model to be consistent with Piaget's findings, the set of all possible cutsets and their *link-disjoint unions* in the connected network must form a group under the *ring sum* operation.

Theorem. The set of cutsets G_1, G_2 ....G_n and their link-disjoint unions in the associative network G must satisfy the following criteria ($\oplus$ denotes ring sum):

1. Closure: if G_i and G_j are in G, then $G_i \oplus G_j$ is also in G.
2. Association: if G_i, G_j, G_k are in G, then $(G_i \oplus G_j) \oplus G_k = G_i \oplus (G_j \oplus G_k)$.
3. Identity: there exists a unique G_0 in G such that for any G_i, $G_i \oplus G_0 = G_0 \oplus G_i = G_i$.
4. Inverse: for every G_i in G there exists a unique G_i' such that $G_i \oplus G_i' = G_i' \oplus G_i = G_0$.
5. Commutation: if G_i and G_j are in G, then $G_i \oplus G_j = G_j \oplus G_i$.

Proof of the Theorem is well known (e.g., Deo, 1974), and will be referenced shortly, after the meaning of the criteria is reviewed, following (Flavell, 1963). (Note that in cutsets under the *ring sum* the identity element G_0 is an empty set, $G_0 = \varnothing$.)

a) Under Identity, the operation causes no changes. One can observe (follow) some alternative A without altering it (or anything at all, which would include releasing some action as a result of observing A). One can also return to A again, and find it "equal to itself," i.e., unaltered.

b) Under Inverse, the operation reverses the previously taken step. One can follow A, and then return to exactly the same point where one started, with nothing changed.

c) Under Commutation and Association, one can change the order of the steps one is taking, and arrive at the same point in several alternative ways.

d) Under Closure, all the alternatives remain within the confines of the original associative structure (network G).

Since all the alternatives remain within the bounds of the network, are all reversible, and all have the same initial and final points, they represent the invariant properties of the network, and, consequently, input invariants, although of a higher order than those captured in the packets of which the network is formed. Summarily, satisfaction of the criteria one through five corresponds to the "power of choice.. coextensive with the fringe of possible actions that surrounds the real action" (Bergson, 1955).

Proving the Theorem does not guarantee that abstractive thinking involves partitioning and recombination operations on dendritic networks but it does support the theory. That is, these operations do have group-algebraic properties manifested in adult reasoning.

An example close at hand is the proof of the Theorem, which consists of nothing else but partitioning and re-combinations, and observing of what those two bring about.

Proof. **Start.** Let G_1 and G_2 be cutsets in G. Let N_1 and N_2 be a disjoint partitioning of the node set in G corresponding to G_1 (partition), and N_3 and N_4 be those corresponding to G_2 (partition). The subset $N_5 = (N_1 \cap N_4) \cup (N_2 \cup N_3)$ (combine) equates to $N_1 \oplus N_3$ (observe). Similarly, $N_6 = (N_1 \cup N_3) \cap (N_2 \cup N_4)$ (re-combine) equates to $N_2 \oplus N_4$ (observe). Consequently, the ring sum $G_1 \oplus G_2$ (combine) partitions N in G into N_5 and N_6 such that $N_5 \cup N_6 = N$ and $N_5 \cap N_6 = \varnothing$ (observe). $G_1 \oplus G_2$ is either a cutset in G if N_5 and N_6 each remain connected after $G_1 \oplus G_2$ is removed from G, or $G_1 \oplus G_2$ is a link-disjoint union in G (observe). Hence, Closure. Commutativity and Associativity of G_1 and G_2 are apparent (observe). That empty set is the identity element (observe). A cutset or an edge-disjoint union of cutsets is its own Inverse. Hence, Theorem. **Target.**

Note that only the starting and the target points are fixed in advance while the line of proof can vary. For a less abstract example of partitions and re-combinations, consider the problem of taking a wolf, a goat, and a cabbage across the river in a boat (so that one part of your cargo does not consume the other part), or think of a chimpanzee discovering the use of a stick for reaching a banana outside its cage (Kohler, 1925). In all cases, the process is that of establishing a set of variables followed by variable grouping.

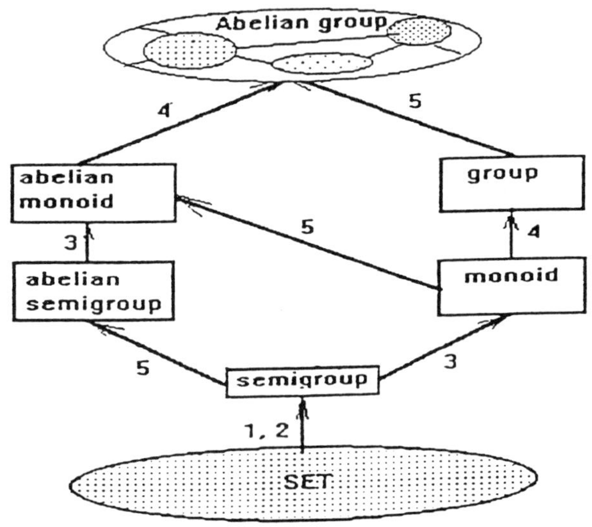

Figure 32. A hierarchy of system organizations. Numbers indicate group-algebraic criteria the system satisfies (counted along the paths). Cutsets under the *ring sum* satisfy all the criteria, and form a group corresponding to the highest degree of organization. Ascendance from the set to the Abelian group in the hierarchy is that of "a change from an incoherent homogeneity into a coherent heterogeneity" (Spencer, 1937).

A set corresponds to the lowest organization in memory in the neonatal period. At that time, no or minimal functional connections are established. As a result, infants demonstrate erratic responses to stimulation. In the Abelian organization, permanent objects have emerged from the stimulation flux, and the invariant relationships between them have been explicated.

The characteristics of the target CNS organization and, consequently, the experiences sought by the system, are genetically determined. Transition from the initial to the target organization during the life cycle is influenced by the environment. Intelligence manifests in maintaining a near-minimal, (e.g., Fernandez et al, 1989) low entropy transition trajectory under the changing constraints. One can define human behavior as fulfillment of an assignment: "All of the information we learn and remember, all of the patterns we process, classify, abstract and compare, are used to complete our assignments" (Alkon, 1992, p.240). Intelligence is the measure of our ability to streamline the process.

3. Learning dynamics.

The learning process is often divided into three phases. In the first phase (instructional or cognitive) the learner is instructed in the task or studies it himself. In the second phase (associative), first, errors in the initial understanding are eliminated, and second, the connections among the various elements required for successful performance are strengthened. The outcome of this phase is a successful performance procedure. The third phase (autonomous) is an extension of the second one, with the procedure becoming more automated and rapid (Anderson, 1980, p. 236).

We shall look closer at what it takes to obtain an initial understanding of the task, which can be a problem when the learner starts with the conditions and the goal, and neither the task elements nor the connections between them are laid out up front. For an account on what leads to understanding, we will consult the "Rules for the Direction of the Mind" (Descartes, 1952). The method consists of two steps: "deduction" (assembly/disassembly) and "intuition" (observation). It is the latter that is of particular interest now.

Step 1. "Rule IV. There is need of a method for finding out the truth.
Rule V. Method consists entirely in the order and disposition of the objects towards which our mental vision must be directed ...[to] reduce involved and obscure propositions step by step to those that are simpler, and then starting with the intuitive apprehension of all those that are absolutely simple, attempt to ascend to the knowledge of all others by precisely similar steps" (p.7).

Step 2. " Rule VII... If I first found out by separate mental operations what the relation is between the magnitudes A and B, then what between B and C, between C and D, and finally between D and E this does not entail my seeing what the relation is between A and E... To remedy this I would run them from time to time, keeping the imagination moving continuously in such a way that while it is intuitively perceiving each fact it simultaneously passes on to the next; and this I would do until I had learned to pass from the first to the last so quickly that no stage in the process was left to the care of the memory, but I seemed to have the whole in intuition before me at the same time.. (p. 10).

The two steps may result in "real knowledge" such that those who have it

> "discern the truth with equal facility whether they evolve it from matter that is simple or that is obscure, they grasp each fact by an act of thought that is similar, single and distinct, after they have once arrived at the point in question." (p.15).

In the second step, the elements of the structure, without being altered, are attentively "glued" together (Treisman, 1988) into an integrated simultaneous whole. In terms of the present model, both the steps and the outcome remain fundamentally the same, whether one is learning bicycle riding or theorem proving. Integration of the disjointed elements leads to a connected structure, as captured in Figure 11. In problem solving, relations that were potentially present become functionally available, so that one can discern, for example, that $N_5 = (N_1 \cap N_4) \cup (N_2 \cup N_3)$ equals $N_1 \oplus N_3$, leading to the proof of the Theorem. A similar mechanism makes bicycle riding robust and adaptive. In all cases, the result is a connected structure amenable to adaptive reorganization.

Before the pieces come together, one can neither hold the bicycle steady nor follow a theorem, while after the integration the performance sharply improves. If that is so, the mechanism responsible for such a sharp upturn can be associated with the intrinsic characteristics of network connectivity growth studied in (Kauffman, 1993). In his computational experiments, random connections were added to a set of initially disjointed nodes. At each step, the degree of connectivity in the growing structure was determined by the relative size of the largest connected cluster. The connectivity grew steeply after a certain threshold is reached, as illustrated in Figure 33.

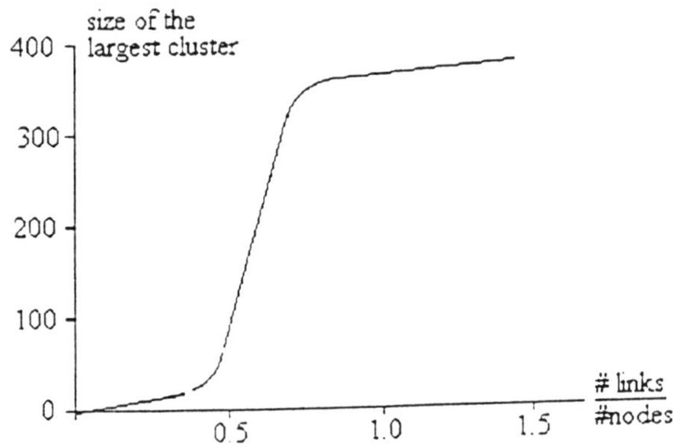

Figure 33. Connectivity dynamics in randomly connected networks. A null graph of 400 nodes is connected by adding links arbitrarily. The degree of connectivity is measured by the size of the largest emerging cluster. As the number of links approaches one half of the number of nodes, the structure "crystallizes" quickly into a unified network (Kauffman, 1993).

If the present account is plausible, the curve in Figure 33 depicts memory transformation between the stages in Figure 11a and 11b. In this way, abrupt changes in performance must accompany two such transformations: initial integration, and packet restructuring, with the former preceding the latter and making it possible (Yufik et al, 1996).

In these terms, the instructional and associative learning phases involve iterations between two activities: expanding the number of variables first, and cross-correlating them next. In the first activity, the tendency is to vary the conditions and to populate memory with a sufficient number of fragments by trying out new combinations (Figure 11a). In the second one, the fragments are integrated (Figure 11b and c), allowing new variations..

If the task is tightly constrained and involves one or a few variables, the variations are suppressed and transformations become less pronounced or impossible. Data in (Anderson, 1980) demonstrating steady performance improvement with practice, and results in (Kauffman, 1993) showing sharp connectivity increase in the networks, are both compatible with the model and reveal changes in learning under different conditions.

4. Learning improvement.

Learning can be improved by facilitating memory integration. Experiments in (Yufik et al, 1995) were designed to assess how procedural learning is influenced by the extent to which instruction and stimuli organization are responsive to integration requirements. The latter were derived from the model, with the expectation that compliance with the requirements will entail appreciable performance improvements. These expectations were confirmed by the following data (Yufik et al, 1995a, 1995c).

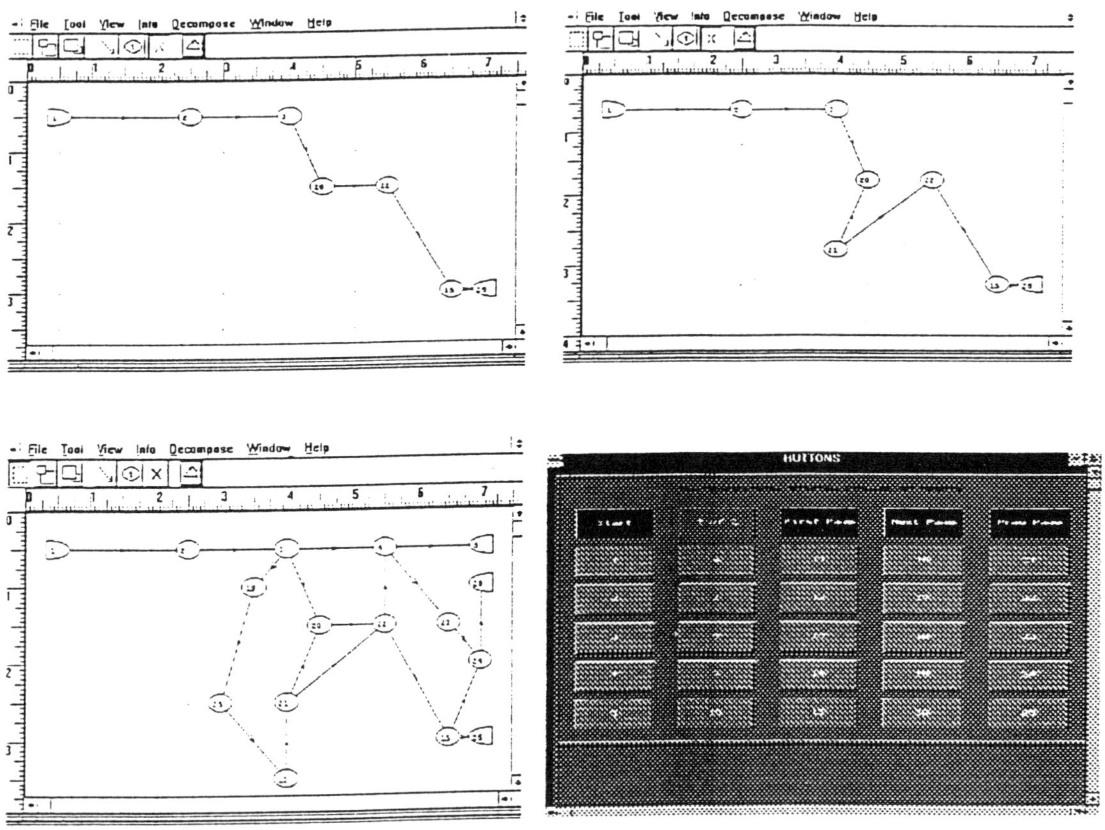

Figure 34. Experimental design to assess the impact of display organization and instruction on procedural learning. Procedures were obtained as paths in the network, and displayed as sequences of button depressions. The underlying network was never revealed.

The subjects were 29 undergraduate psychology students at The Johns Hopkins University. The task was to learn partially overlapping procedures obtained as paths in a network (sample procedures and the network are shown in Figure 29). Procedures were implemented as sequences of button depressions in multi function displays, with the network nodes represented as buttons. Neither the paths nor the underlying network were shown to the subjects. During learning and practice procedures were identified by automatic highlighting of the appropriate button sequences. During testing, the highlighting was turned off, and subjects performed the procedures by depressing the buttons. The timing and correctness of the depressions were automatically recorded.

There were two sessions, lasting a total of approximately 3 hr., with one week between the sessions. The first session included 11 exercise sets for different procedures, with each set including 7 learning trials, 3 practice trials, and one test trial. In the second session each exercise included 3 learning trials, 5 practice trials, and one test trial. The subjects were asked to go as fast as they could (although the speed was discounted in the analysis). Prior to the experiments, links in the network were weighted based on the number of their traversals in the exercises, resulting in the network partitioning into packets.

The subjects were divided between tree conditions: 1) conventional display layout. (10 subjects), 2) optimized display layout (10 subjects), and 3) optimized layout and instruction (9 subjects). In the first condition the buttons were grouped arbitrarily, in the second one the grouping was consistent with the packets, and in the third one the packets were pointed out to the subjects at the beginning of each exercise. Data in both sessions were combined. Figure 35 summarizes the results.

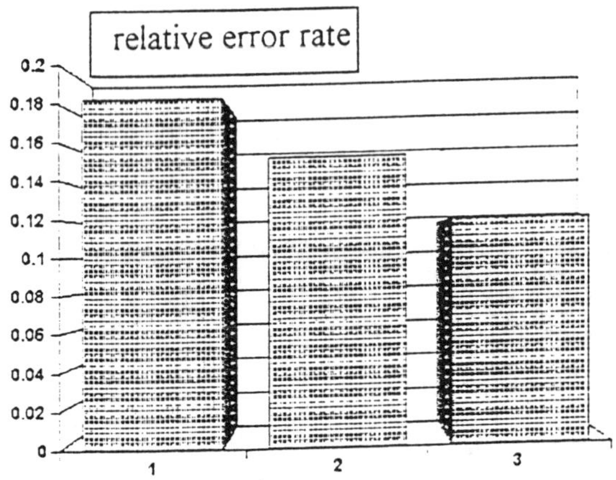

Figure 35. Error rate reduction in procedure retrieval as a result of layout optimization and instruction, based on the model. 1 - conventional layout, 2 - optimized layout, 3 - optimized layout and instruction. Optimized layout facilitated integration of procedures into a unified network in the subject's memory. Instruction directed subject's attention to the overlapping procedural components.

The study produced a practical technique (called OneCanSeeWhatOneCanDo, or OCSWOCD) for optimizing display layout by grouping instruments based on their functional proximity. This minimizes interpage traffic and lowers the operator's workload (for more details, please see Yufik et al, 1997a, 1995; Yufik, 1997; Corker, 1996).

5. Compensatory adjustments.

The cortex has been observed to reorganize in response to persistent changes in stimulation. When stimulation is removed, the silenced region becomes responsive to stimulation from the region's periphery. The boundaries recede inward, until the entire region is filled (Gilbert et al., 1995). Such compensatory reorganization is pervasive, taking place in somatosensory and motor cortices. The rate of reorganization varies widely (Kaas, 1995; Wang et al, 1995). Compensotary adjustment and memory distortions entailing systematic error in recall (Schacter, 1996) are manifestations of spontaneous reorganization in the memory structures.

Reorganization is predicted by the core mechanism of the model, as articulated in Figure 11 and Figure 12. Loss of stimulation weakens connections in the region, causing dissolution of packets and their absorption into the surrounding structure.

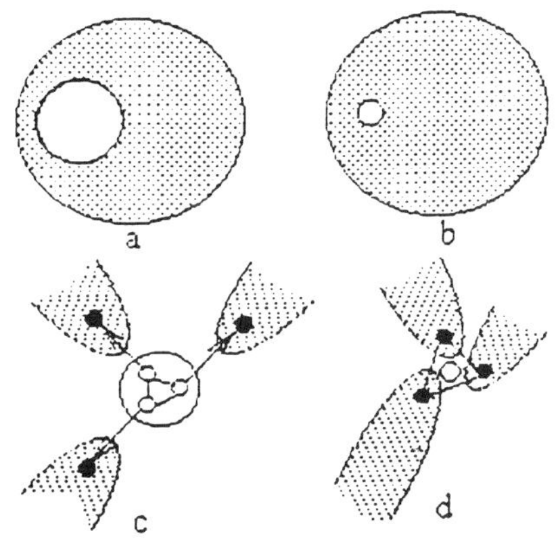

Figure 36. Cortex restructuring after stimulation redistrtibution. A region deprived of stimulation (a) starts responding to stimulation from the margins, with the margins gradually receding inward into the region (b). According to the VAN model, decay of the associative connections in the deprived packets (c) causes packet dissolution, with the constituent units being absorbed into the surrounding packets that are gradually encroaching on the region (d).

Network partitioning into packets predicts "fair occupancy" in the cortex, that is, distribution of the cortex areas among all the stimuli as necessary and sufficient for their processing. This suggests that acquisition of new activities will cause initial expansion of the interested region, followed by its contraction after performance stabilization.

Such cortical dynamics has been consistently observed in (Bechtereva, 1988), where "minimization of cortical areas to an extent necessary and sufficient for supporting functions not undergoing further development" (p. 59) is identified as one of the key factors discriminating normal and pathological conditions in the brain. In pathology, cortical flexibility is compromised, preventing adaptive and compensatory reorganizations.

The "fair occupancy" principle inherent in the model is congruent with the notion that cortices have finite surface areas contested by different neuronal ensembles (Edelman, 1987; 1994; Squire, 1987). A persistent sensory input projected into the cortex due to recurring experiences and rehearsal (learning) favors the recipient ensembles, entailing their expansion into and "colonization" of the adjacent areas. Interplay between the neuronal specification and expansion produces cortical zones with flexible and dynamically adjusted boundaries (Kilackey, 1990; Rakic, 1988).

The model is also consistent with the stability/plasticity principle in (Grossberg, 1978), requiring an optimal balance in the memory organization so that persistent environmental changes cause restructuring, while temporary deviations make no lasting impact.

6. Performance efficiency and cognitive effort.

The Gibbs-Helmholtz thermodynamic law underlies the Yerkes-Dodson performance law correlating optimal performance with the optimal level of arousal or anxiety (Yerkes et al, 1908). The VAN model relates performance to the "cortical tone," as shown in Figure 35.

Prefrontal regulation of the "cortical tone" participates in performance adjustments based on a non-specific and/or postponed feedback. Success or failure in multi step tasks often cannot be attributed to any particular choice in the course of the performance. In that case, the entire sequence of packets along the performance paths has to be strengthened or destabilized, depending on the outcome. A non-specific regulation of the "cortical tone" can subserve this adaptive function.

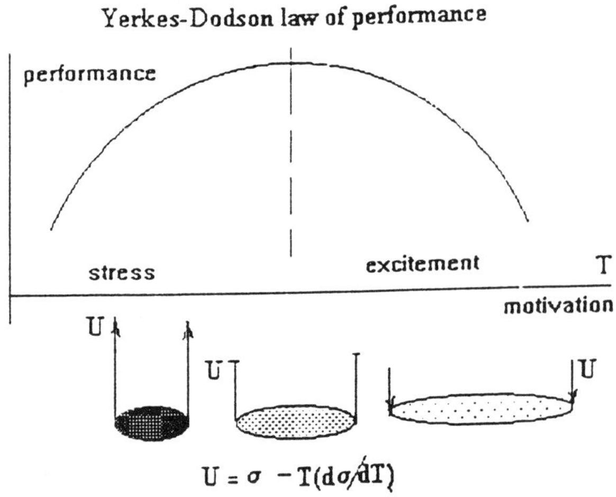

Figure 37. Interrelationship between the "cortical tone" T and memory dynamics. Increase of the "cortical tone" lowers energy barriers at the packet boundaries and causes them to extend outward from the core of the packet. By contrast, lowering the tone causes the barriers to grow and recede inward inside the packet. Pathological enhancement of these adaptive adjustments can interfere with goal-directed activities.

The fundamental adaptive role of the mechanism can be associated with two basic strategies in responding to unfamiliar conditions. A non-threatening condition can stimulate exploratory deviations from the overlearned responses and venturing outside the packet, while a threat can limit responses to the packet's kernel.

A non-specific regulatory mechanism in Figure 37 is in line with the view that "the controls on arousal and activation ..operate on the constraints operating within and between channels rather than on the information being processed. These constraints involve the neuronal model and may be conceived of as operating on memory rather than on input information" (Pribram et al, 1975, p.136).

The "tone" regulation can be pathologically amplified, causing memory disturbances. Pathologically low packet barriers manifest in an incessant flood of associations and abnormal distractibility by irrelevant stimuli. For example, asked to repeat a phrase, the patient repeats the first word, which associates with whatever comes in mind in conjunction with that word, which pulls in another irrelevant associations, etc. An opposite dysfunction manifests as rigidity, fixation on a stimuli, and inability to change the response pattern. For example, when given two sentences to repeat, such patients would keep repeating the first one, unable to make the transition (Luria, 1973).

Clinically, these dysfunctions accompany lesions in the visceroautonomic system (the amygdala circuits) and prefrontal lobes. In the absence of pathology, similar effects are produced by extremal conditions. For example, agitation can cause distractability but also flexibility, capability to respond in a non-standard fashion, and "openness" to different views. By contrast, stress and fatigue can entail "tunnel vision" and vacillation.

Adverse conditions can increase the height of the energy barriers, but at the same time mobilize additional energy to overcome them. Interdependencies between the "processing resources" and cognitive effort has been addressed in a number of performance theories (Kahneman, 1973; Ellis at al, 1988), and expressed clearly in the notions of performance effectiveness and processing efficiency in (Eysenck et al, 1992). The former notion refers to the quality of performance as a function of "processing resources" allocated to the task, while the latter one reflects the amount of effort invested in the performance.

Figure 37 suggests that performance can be maintained under adverse conditions at the expense of greater effort. This suggestion coincides with the central prediction of the processing efficiency theory, supported by experimental findings (Eysenck et al, 1992).

According to the present model, cognitive effort is invested in navigating memory structures beyond the packet boundaries, and in the deliberate shifting of those boundaries in the course of reasoning and problem solving. The mechanism of boundary shifting is presumed to be essentially the same as that responsible for shifting the muscle activation thresholds in motor control.

This view is consistent with the explanation of effort by Pribram and McGuinness (1975) suggesting that "the neuromuscular processes demanding effort are to a large extent initially peripheral in their origin and manifestation. But as problem solving progresses and skill develops, the effort becomes more and more a concomitant of the brain processes involved - apart from, or only reflected in, the peripheral manifestations - and thus become truly cognitive" (Pribram et al, 1975, p. 141).

The model in Part III can be briefly summarized as follows:

1. Human CNS is a purposeful, synergistic and integrative system.

2. An interplay between the tendency to expand and resistance to change determines the system's dynamics.

3. Efforts to overcome the resistance and decouple the stimuli from the habitual responses underlie consciousness as a specifically human form of goal-directed behavior.

4. Progressive internalization of system activities and unification of experiences give rise to the higher cognitive functions as operations on self-organizing associative networks.

5. Stable networks and operations on them acquire the properties of Abelian groups.

6. Formation of stable networks accelerates input matching and response generation.

7. Input matching is a multi stage iterative process where input components are matched with the elements of previously memorized structures. The process is two-directional: packets are sought in LTM that satisfy feature combinations in the present input, and feature combinations are sought in the input that confirm the selected packets.

8. Learning can be improved by organizing the material in a manner commensurate with the learner's memory organization and its changes in the course of learning.

Discussion

This part will review correspondence and the lack of such between the VAN model and other theories, focusing on their defining features.

1. Convergence.

The notions of interlevel mapping in the LTM hierarchy, with the higher level maps guiding the assembly of the lower level entities, concur precisely with the Convergence Zone Framework (CZF) in (Damasio et al, 1994; Damasio, 1989). The two critical components in the CZF architecture are

> " the fragment record of feature- based sensory or motor activity, and the convergence zone, an amodal record of the combinatorial arrangements that bound the fragment records as they occurred in experience. There are convergence zones of different orders; for example, those that bind features into entities, and those that bind entities into events or sets of events, but all register combinations of components in terms of coincidence or sequence, in space and time.. The records of those perceptuomotor interactions, both at fragment level and at combinatorial level, are inscribed in superimposed and overlapped fashion; yet, because of the different conditions according to which they are grouped, they become committed to separate neural regions " (Damasio, 1989, p.26).

The function of the convergence zones is to recreate geographically separate sets of neural activities that "coincided during the time window necessary to us to attend to it and be conscious of it, which means hundreds of milliseconds." Such recreation does not entail simultaneous activation of multiple perceptual and motor fragments but rather "an extremely fast sequence of activations that would make separate neural regions come on-line in some order imperceptible to consciousness" (Damasio, 1994, p. 72). Formation of fragments and zones is conceived as "recruiting a neuron population for a limited range of cortical inputs... Within that population, however, different synaptic patterns define individual features, or entities, or events. In simple terms one might say that generally similar material stacks up together within the same regions and systems" (Damasio 1989, p. 43). Fragments can vary in size, with the relatively large ones representing unique feature constellations such as human faces and body parts, and small ones that can be shared by numerous constructs and use interchangeably in different recreation attempts.

Convergence zones are conceived as microscopic neuronal ensembles located within macroscopic convergence regions. The latter are cytoarchitectonically defined and number about one hundred, while the number of zones is envisioned on the order of thousands. Both convergence regions and zones are genetically defined and epigenetically modified.

The overall CZF architecture is conceptualized as comprising three sets of systems: those in the early sensory and motor cortices containing the primary "explicit" records of the interactive events, those in the higher cortices that orchestrate reanactment of the "explicit" representation in the lower cortices, and those in the forebrain that provide attentional enhancement required for the concerted operation of the other two systems.

The CZF architecture ensures perceptuomotor interaction with the surroundings, learning of those interactions at the representational level, internal activation of experience-replicative representations in a recall, and higher functions such as problem solving, planning and creativity. The VAN model supports and substantiates this claim.

2. Holonomy.

Functional units in the VAN model can be viewed either as discrete packets that re-combine into larger spatio-temporal structures, or as Fourier analyzers (Pribram, 1991). The two views are complementary, with the choice between them depending, in part, on mathematical expediency. Sinusoidal components in the Fourier series correspond to unbounded waves extending to the infinity. Processes having specific localization in the dendritic networks must be represented by "wave packets," that is, superpositions of sinusoidal waves. The advantage of the latter perspective is an exceedingly simple mathematical formalism (Resource Allocation), and a shortcut in performance modeling. Since such modeling is central to the present inquiry, the "packet" view is preferred. Another more fundamental motivation for the RA formalism will be identified shortly.

The centerpiece of the VAN model is a mechanism enabling automatic coordination of changes throughout the resource organization in a wave-like process of change propagation. In that sense, an aspect of Fourier representation is retained in the model.

The model is holonomic in its fundamental premise that all the functional components in the CNS, and the "brain and body are inseparably yoked" (Pribram at al, 1975, p. 140). Accordingly, the system is thought to develop as a functional whole, subject to the Law of the Least Action in its development and interactions with the environment. The Law is expressed as entropy minimization in the developmental trajectory. Values experienced by the system reflect discrepancies between the actual and minimal trajectories, which makes the system purposeful, synergistic, and integrative. The system is purposeful because the genetic determinants of its behavior translate into conscious motivations. It is synergistic due to global coordination across the components (Basar et al, 1983; Haken, 1990). It is integrative due to representation of the past "series of events" (Grafman, 1994) and, ultimately, the entire past history in a unified memory structure available in the present.

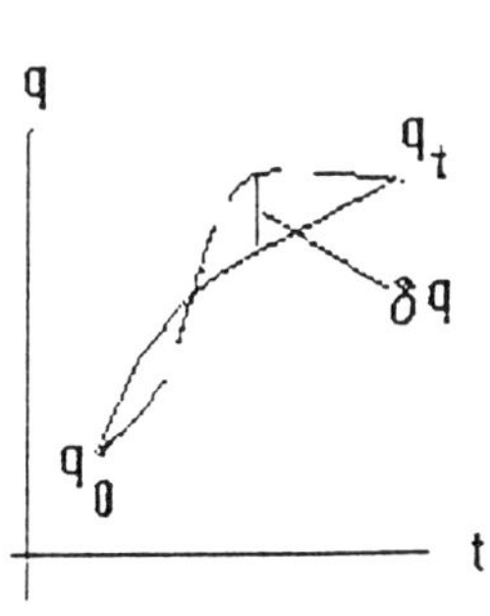

Figure 38. In the transformation of physical systems, multiple potential trajectories connecting the initial q_0 and the end q_t points in the system's phase space are all variations δq of some minimal trajectory, to be followed in actuality. In the CNS life cycle, among multiple potential trajectories connecting the initial and the target organization, the actual trajectory lies in the vicinity of the minimal one. Values in the system depend on its current condition q and discrepancy δq between the trajectories.

3. Computation.

Deriving cognition from calculating neurons advances a circular argument about both. In the argument, calculation that approximates behavioral patterns replaces analysis of the underlying mechanism, by virtue of the assumption that calculation is the mechanism.

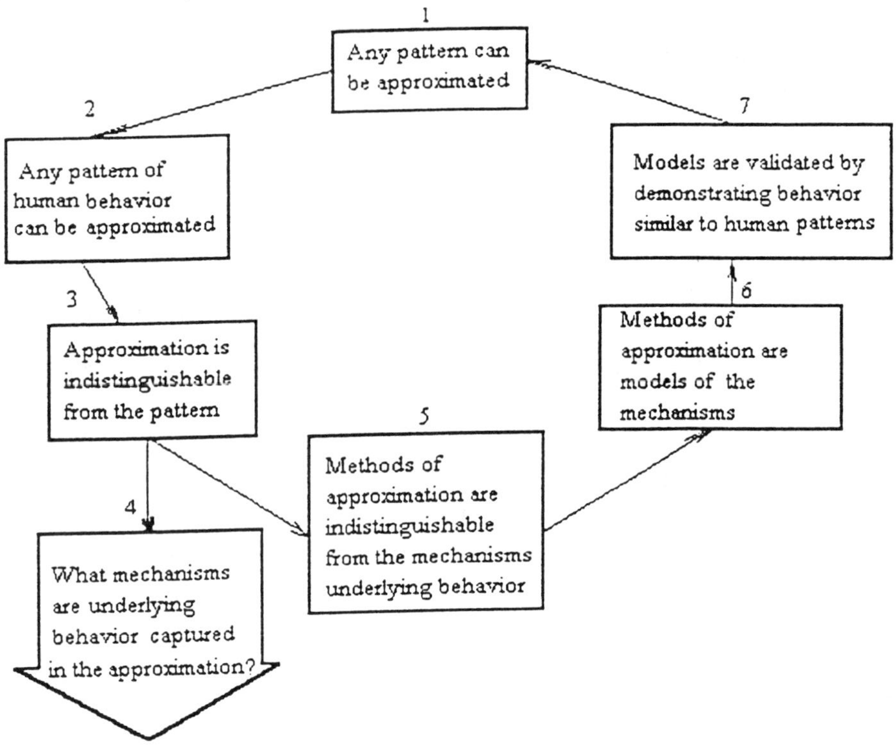

Figure 39. Steps 1, 2 and 3 establish the basis for performance analysis. Step 5 executes a U-turn, replacing performance analysis with performance simulation and implanting a mathematically savvy homunculus into the nervous system. The way out of the circle goes downward to the analysis of the underlying mechanism rather than upward to the design of more elaborate approximation techniques.

Integrative development in the human CNC proceeds from combining sensory experiences into cohesive separable units, to recognizing that such units can maintain their identity over time, to attaching symbols like "apple" and "orange" to different units, to realizing that one apple and one orange make two objects, and, finally, to realizing that one and one equals two whichever objects are counted. Set, or manifold is "a collection into a whole of definite distinct objects of our intuition or our thought" (Cantor, 1972, p. 21). Collecting the "objects of thought" into a whole and breaking this whole into its constituents, as in Figure 22, are the basic abstractive operations. Without them, there is no calculation. Neurons that compute are neurons that think.

Thinking of apples can bring to mind different apples but adding two numbers must always bring the same result, making this form of thinking maximally autonomous from the input variations and impervious to entropy inside the system. In that sense, computation is an ultimately regularized zero entropy process. Conditions that undermine regularity in the brain would cause a breakdown in the counting ability. For example, parieto-occipital lesions that impair voluntary switching from one activity to another would also replace a regular series of counting operations by inert stereotypes. As a result, when counting backward, such patients keep repeating the same final digit (100 -7 = 93, 83, 73, 63 .....etc.), and fail other counting tasks (Luria, 1973, p. 217).

Biological self-organization proceeds from extracting a few environmental invariants in primitive organisms (Alkon, 1989, 1988), to the extraction of cross- regularities in an expanding variable set in the human nervous system. Outside that system and prior to a certain degree of its maturity, the notion of computation has no meaning. For example, an infant is not disturbed when seeing multiple mirror images of its mother, and addresses them at random. At a later stage, after the identity of the mother-construct has been established, the same situation causes distress (Bower, 1979). One cannot surmise that before that stage the infant had the experience of multiple mothers and found it agreeable; rather, its nervous system did not allow differentiation between many and one. To the system in this condition the notion of computation does not apply. Under no conditions does it apply to the neurons of which the system is composed.

Computation is often understood as "what a computer does" (e.g., Penrose, 1997, p. 105). I disagree; computation is what people do using their fingers, a bundle of sticks, or a CRAY to help them. My computer is a big help (and an irritation) in writing this paper but not a co-author; its function is that of a "glorified pencil" (Popper, 1994, p.23). This "pencil" amplifies my skills by adding the skills of its creators but has none its own.

To illustrate how the circle in Figure 39 works, I will follow it once by creating a "neuron" and teaching it to recognize, for example, cats, dogs, and wolves, and then causing this recognition to fail. Here comes the neuron with three synapses, as in Figure 5a:

$$p_{11}w_1 + p_{12}w_2 + p_{13}w_3 = q_1$$

$$p_{21}w_1 + p_{22}w_2 + p_{23}w_3 = q_2$$

$$p_{31}w_1 + p_{32}w_2 + p_{33}w_3 = q_3$$

Let there be, accordingly, three features to distinguish the animals, valued between 10 and -2. I should want the neuron to state its answers clearly so that I can interpret the signal and announce the result. To make sure of that, I want the signal $q_1 = 6$ for a cat, $q_2 = 7$ for a dog, and $q_3 = 8$ for a wolf. The desired mapping is shown below.

	climbs trees	wags the tail	lives in the wild
cat	10	-2	-2
dog	- 2	10	- 1
wolf	-1	-1	10

$$10w_1 - 2w_2 - 2w_3 = 6$$

$$-2w_1 + 10w_2 - w_3 = 7$$

$$- w_1 - w_2 + 10w_3 = 8$$

Now I will "teach" the equations to do the mapping, by solving them via iterative error reduction (the technique dates back to Gauss). I will sensitize the system to the errors R_i (by dividing the first row by $p_{11} = 10$, the second row by $p_{22} = 10$, and the third row by $p_{33} = 10$), and will administer the first lesson where I will set all the weights to zero.

Prepare to compute the errors:	Take arbitrary weights:	Compute the errors:
$R_1 = 0.6 - w_1 + 0.2\,w_2 + 0.2\,w_3$	$w_1 = w_2 = w_3 = 0$	$R_1 = 0.6$
$R_2 = 0.7 + 0.2w_1 - w_2 + 0.2w_3$		$R_2 = 0.7$
$R_3 = 0.8 + 0.1w_1 + 0.1w_2 - w_3$		$R_3 = 0.8$

The largest error turned up in the third row, so I will eliminate it by adjusting the corresponding weight $w_3 = 0.8$ (error backpropagation). Now $R_3 = 0.0$.

In the second lesson I plug $w_3 = 0.8$ into the first two rows and compute the new errors as

$R_1 = 0.6 + 0.2 \times 0.8 = 0.76$, $R_2 = 0.7 + 0.2 \times 0.8 = 0.86$, and $R_3 = 0.0$ as before.

Now R_2 is out of order, and w_2 will have to take the correction, $w_2 = 0.86$.

	w_1	w_2	w_3
Start	0	0	0
lesson 1			0.80
lesson 2		0.86	
lesson 3	0.93		
lesson 4			0.18
lesson 5		0.13	
lesson 6	0.07		
lesson 7			0.02
lesson 8		0.01	
Result	1.00	1.00	1.00

Altogether, eight lessons will be needed before the "neuron" shapes up and performs the mapping flawlessly so that $R_1 = R_2 = R_3 = 0.00$.

A trained neuron will have $w_1 = w_2 = w_1 = 1.00$. and will not only recognize the original animals but will identify other creatures as one of the three, based on the preponderance of evidence.

Having the neuron fully trained completes step 3 in Figure 39, setting the stage for step 5.

For example, brain damage is modeled by damaging the equations (Hinton et al, 1993).

$10w_1 - 2w_2 -$ [redacted] $= 8$

$-2w_1 + 10w_2$ [redacted] $= 8$

$-w_1 - w_2 +$ [redacted] $= -2$

This neuron has a lesion in the third synapse. As a result, it will announce "wolf" when seeing cats and dogs, while the sight of a wolf can leave it speechless (if cut off values are defined for q).

I will suggest now how to model learning disabilities. The reason for a small number of lessons in my example lies in a prudent choice of values. Iterations converge quickly if the diagonal coefficients are greater than the non-diagonal ones. Otherwise the process might take longer, or not converge at all. By modifying the coefficients, one can postpone convergence. In keeping with the computational view, this will represent the disorder. Using more elaborate methods (Rumelhart et al, 1986) but in a similar manner one can simulate acquisition of any skill and deterioration of it (Hornik et al, 1989).

The question is this: what do such simulations have to do with the causes of the behavior they replicate? The computational view maintains that organization of the algebraic system represents organization of knowledge in the CNS. Accordingly, by observing how coefficients change during iterations one finds out how knowledge builds up, and by corrupting the procedure one sheds light on the cause of malfunctions in the brain. If the assertion of a circular approach in Figure 39 is valid, this diagnostic method, in a sense, takes temperature of the mirror instead of the patient. What is missing?

4. What is missing?.

Descartes placed human soul in the pineal gland. Contemporary neuroanatomy underlying this surprising choice is no longer relevant but the motivation behind it is still valid. The gland was considered the only unitary organ in the brain while all the other parts came in pairs. Consequently, as Descartes concluded, the gland was the only conceivable place where "the two images coming through the two eyes, or the two other impressions coming from a single object through the double organs of other senses, can coalesce into one before reaching the soul so that they do not represent two objects to it instead of one" (Descartes, 1989, p.37). Recent realization that there is no single convergence point in the brain made the problem of integration even more urgent and bewildering. Presently, it has no solution. Perhaps, within the computational framework there is no solution.

Computational approach revolves around the Turing's argument: If a computer performs the same function as a human does, what basis is there to say that one performer thinks and the other does not, and what is there to thinking other then computing? The argument blurs the distinction between material processes and their simulations and, more generally, between material things and mathematical ideas. Note that "neuron" in the example is a system of equations. Network in Figure 5a is a diagram of intervariable relations in the system which plays no role in how the equations are solved except that the term "network" connotes a material thing. As a result, features of an algebraic construct are projected on the neuronal process making the latter parallel because the equations are parallel and distributed because the solution is a matrix. As argued elsewhere, computational models give biological interpretations to mathematical facts, presumably, under a license from the Turing's argument (Yufik et al, 1996). Is the license valid?

The Turing's machine and other abstractions of universal computability (e.g., Markov's machine) are based on the premise of infinite time and infinite resources. The essence of biological conditions is limited time and limited resources. Conceptualizing biological systems in terms of universal computability misses the mark by no small measure which is the distance between the ideal and corporeal. The conception is sterile since it allows no contact with the core determinants of biological phenomena except behavior simulation. As a result, the problems of experience unification, development, and constructive responses, central to the efficiency of resource utilization, are left entirely out of focus

Cognitive theory has rejected analogies between brain and serial computer: "It has always bothered me that models of psychological processing were thought to be inspired by our understanding of the computer. The statement has always been false" (Norman, 1988, p. 534). Instead, it has adopted the view of neurons as computational elements (Koch, has adopted the view of neurons as computational elements (Koch, 1997). Consequently, the paradigm shift is from a single large computer to a confederation of small ones. What concept of memory does the paradigm support?

The common feature of models 5a and 5b is fragmentation and absence of development. Think of the "neural nets" and "if-then" rules as "knowledge atoms" (Smolensky, 1986) or pieces of sand in a box: leave the box alone and the pieces will not move; shake the box or change its shape, and the pieces will come to rest in a new configuration. There is no structure, no development, and no continuity in the subsequent configurations.

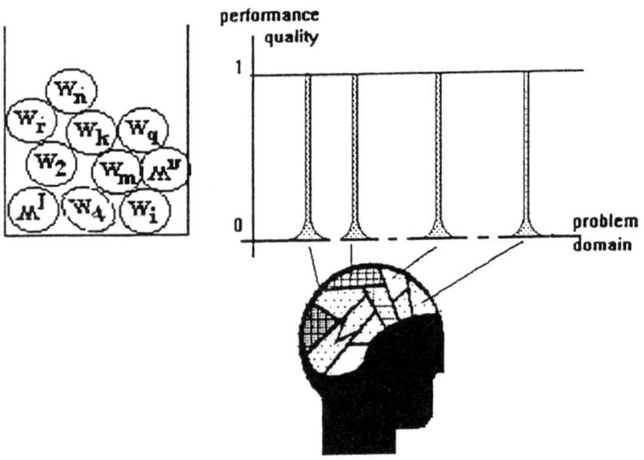

Figure 40. Computation-based memory is a mosaic of specialized fragments. Performance of each fragment is tailored to a narrow domain and declines steeply outside it. The notion that biological memory comprises multiple elements concurs with the computational paradigm. It is the question of what binds all these elements together that defies it.

There seems to be a consensus that the "key feature of biological nervous systems that radically distinguishes them from present-day computers is their ability to learn from previous experience, presumably by adjusting synaptic weights" (Koch, 1977, p. 207). This chapter has argued that this distinction can hardly be elucidated let alone reduced if the CNS is viewed as a computer(s) and neuronal processes as a form of computation. The suggestion is that cognitive theory does not have to pass through the gate of computational paradoxes. It appears that the "gate," although locked, is not attached to any fence so that instead of forcing the gate one can just walk around it, and keep going.

This chapter has portrayed human memory as a holonomic system where flexible components are bound together by what can be broadly defined as entropic forces. It suggested that cognitive mechanisms based on such forces can be described in precise although non-computational terms as, for example, the fall of a stone can be described without making the stone compute its trajectory on the way down. The other purpose of the chapter was to explore how holonomic features of biological memory can be reverse engineered to obtain a general-purpose information fusion model of high utility.

Conclusion

It is informative to compare the goals and dynamics of individual development with those of the scientific process. The following two quotations suggest the parallels.

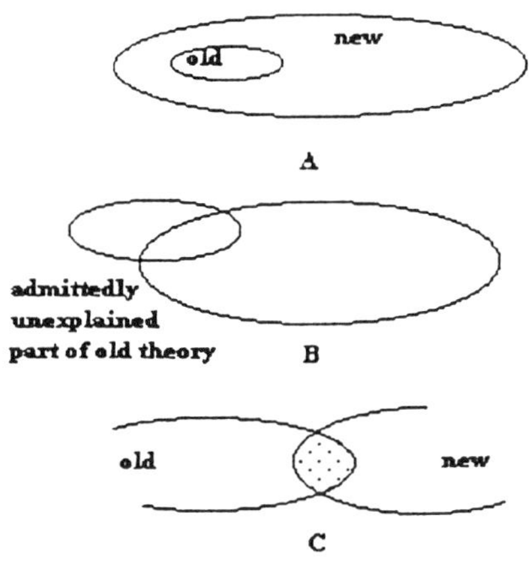

"Not only is the description of every single fact dependent on some theory but there exist facts which cannot be unearthed except with the help of alternatives to the theory.. This suggests that the methodological unit .. of test and empirical content is constituted by a *whole set of partially overlapping, factually adequate, but mutually inconsistent theories.* ...

Comparing the old and the new is thus appears that the relation of empirical contents is like this (A), or perhaps, like this (B) while in actual fact it is much more like this (C)" Feyerabend, 1996, pp. 27, 156).

Finally, Albert Einstein has defined the purpose of science as follows:

> "Although it is true that it is the goal of science to discover rules which permit the association and foretelling of facts, this is not its only aim. It also seeks to reduce the connections discovered to the smallest possible number of mutually independent conceptual elements. It is this striving after the unification of the manifold that it encounters its greatest successes, although it is precisely this attempt which causes it to run the greatest risk of falling prey to illusion" (Einstein, 1994, p.52).

Acknowledgments

This research was supported in part by NASA grants NAS2-1370 and NAS-13283, NSF grant DMI-9460057, and U.S. Army Strategic Defense Command grant DASG6097C0019. Tom Sheridan of the MIT was consultant and collaborator in the research. Steven Yantis of the Johns Hopkins University helped with the experimental design, and made other contributions. Danielle Burke conducted the experiments. I am grateful to Jordan Grafman of the NIH for his thoughtful comments on an early draft of this paper, and particularly to Karl Pribram for advice and discussions that helped me to correct some of my misconceptions. All the remaining ones are my responsibility. My special thanks are to Leo Vekker and other colleagues who contributed in the development of views that partially overlap but are inconsistent with theirs.

References

Alkon, D.L. 1992. Memory's Voice. Harper Perennial.

Alkon, D.L. 1989. Memory storage and neural systems. Scientific American, 42-50.

Alkon, D.L. 1988. Memory Traces in the Brain. Cambridge University Press.

Anderson, J.R. 1980. Cognitive Psychology and Its Implications. W.H. Freeman.

Arbib, J.H. 1985. Schemas for the temporal organization of behavior. Human Neurobiology, 4, 63-72.

Basar, E., Flohr, H., haken, H., Mandell, A.J. 1983. Synergetics of the Brain. Springer-Verlag.

Von Bayer, H.C. 1997. Disorderly conduct. The Sciences, May/June 1997, 15- 17.

Bernstein, N. 1947. The coordination and Regulation of Movement. Moscow. In Russian. Translation: 1967. The Coordination and Regulation of Movements. Oxford, Pergamon.

Bekhtereva, N.P. 1988. Healthy and Sick Human Brain. Nauka. (In Russian).

Benson, D.F., 1994. The Neurology of Thinking. Oxford University Press.

Bergson, H. 1955. Animal and Human Consciousness. In: Runes D.D. Treasury of Philosophy. Philosophical Library. New York.

de Bono, E. 1970. Lateral Thinking. Penguin Books.

Bower, T.G.R. 1979. Development in Infancy. W.H. Freeman

Brown, J.W., Kiriss, E.W., Keenan, C.L. 1990. Hebbian synapses: biophysical mechanisms and algorithms. Annual Rev. Neuroscience, 13, 475-511.

Cantor, G. 1972. The Encyclopedia of Philosophy. I. Macmillan, Inc. .

Carew, T. J., Hawkins, R.D., Abrams, T.W., Kandel, E. 1984. A test of Hebb's postulate at identified synapses which mediate conditioning in Aplysia. J. Neurosc., 4, 1217-1224.

Chaitin, G.J. Toward a mathematical definition of life. In: Levine, R.D., Tribus, M. (Eds.) The Maximum Entropy Formalism. The MIT Press.

Chomsky, N. 1965. Aspects of the Theory of Syntax. The MIT Press.

Crick, F.H. C. 1984. The function of the thalamic reticular complex: The searchlight hypothesis. Proc. National Academy of Sciences, 81, 4586-4590.

Collins, A.M., Quillian, M.R., 1969. retrieval time from semantic memory. J. Verbal Learning and Verbal Behavior, 9, 432-438.

Conrad, C. 1972. Cognitive economy in semantic memory. J. Exp. Psy., 92, 148-154.

Corker, K. 1996. MIDAS: Modeling and simulation in complex piloting tasks. Proc. 1996 Intern.Conf on Intelligent Systems: A Semiotic Perspective. Gaithersburg, MD. 330 -333.

Dalva, M.B., Katz, L.C. 1994. Rearrangement of synaptic connections in visual cortex revealed by laser photostimulation. Science, 265, 255- 258.

Damasio, H., Grabowski, T.J., Tranel, D., Hichwa, R.D., Damasio, A.R. 1996. A neural basis for lexical retrieval. Nature, 380, 499- 505.

Damasio, R.D., Damasio, H. 1994. Cortical systems fore retrieval of concrete knowledge: The Converegence Zone Framework. In: Koch, C., Davis, J.L. (Eds.) Large-scale neuronal theories of the brain. The MIT Press. 61-74.

Damasio, A.R. 1989. Time-locked multiregional retroactivation: A systems-level proposal for the neural substrates of recall and recognition. Cognition, 33, 25-62.

Deeke, L., Grotzinger, B., Kornhuber, H.H., 1976. Voluntary finger movements in man: cerebral potentials and theory. Biological Cybernetics, 23, 99 -117.

Deo, N. 1974. Graph Theory with Applications. Prentice-Hall.

Descartes, R. 1989 (1641). The Passions of the Soul. Hackett Publishing Company.

Descartes, R. 1952 (1639) Rules for the Direction of the Mind. University of Chicago.

Duncan, J. 1990. Goal weighting and the choice of behavior in a complex world. Ergonomics, 33, 1265-1279.

Edelman, G. M. 1992. Bright Air, Brilliant Fire. Basic Books.

Edelman, G. M. 1978. Group selection and phasic reentrant signaling: a theory of higher brain function. In: Edelman, G.M., Mountcastle, V.B. (Eds.) The Mindful Brain: Cortical Organization and the Group-Selective Theory of Higher Brain Function. MIT, 55-100.

Einstein, A. 1994. Ideas and Opinions. The Modern Library. N.Y.

Eriksen, C.W., Yeh, Y.Y. 1985. Allocation of attention in the visual field. J.Exp. Psy.: Human Perception and Performance, 11, 583-597.

Eysenck, M.W., Calvo, M.G. 1992. Anxiety and performance: The processing efficiency theory. Cognition and Emotion, 6, 409-434.

Feder, J. 1971. Plex languages. Information Sciences, 3, 225-241.

Feldman, A. G. 1981. Composition of central programs subserving horizontal eye movements in man. Biological Cybernetics, 42, 107-116.

Feldman, A. G. 1980. Superposition of motor programs. . Neuroscience, 5, 81-95.

Feldman, A. G. 1974. Control of the length of the muscle. Biofizika, 19, 4, 749-753.

Feldman, A. G. 1966a. Functional tuning of the nervous system Controllable parameters of the muscle. Biophysics, 11, 565-578.

Feldman, A.G. 1966b. Functional tuning of the nervous system. Mechanographic analysis of the execution by man of the simplest motor tasks. Biophysics, 11, 766-775.

Fernandez, L., Levinson, H., Seglin, M. 1989. A discrete mathematical model for learning and intelligence. In: Capobianco, M.F., Guan, M., Hsu, F., Tian, F. (Eds.) Graph Theory and Its Applications: East and West. Annals N.Y. Academy of Sciences, 576, 170-177.

Feyerband, P. 1993. Against Method. Verso. London, New York.

Fisher, K.W., Pipp, S.L. 1984. Development of the Structure of Unconscious Thought. In: Bowers, K.S., Meichenbaum, D. (Eds.). The Unconscious Reconsidered. Wiley.

Flanagan, J.R., Ostry, D.J., Feldman, A.G. 1993. Control of trajectory modifications in target-directed reaching. J. Motor Behavior, 25, 140-152.

Flash, T. The organization of human arm trajectory control. In: Winters, J.M., Woo, S.L.-Y.(Eds.) Multiple Muscle Systems, Biomechanics and Movement Organization. Springer-Verlag, N.Y.

Flavell, J.H. 1963. The Developmental Psychology of Jean Piaget. D.Van Nostrand.

Fu, K.S. 1977. Syntactic Methods in Pattern Recognition. Academic Press.

Fuster, J.M. 1995. Memory in the Cerebral Cortex. The MIT Press.

Gelfand, I.M., Gurfinkel, V.S., Tsetlin, M.L., Shik, M.L. 1971. Some problems in the analysis of movements. In: Gelfand, I.M., Gurfinkel, V.S., Fomin, S.V. Models of the Structural-Functional Organization of Biological Systems. The MIT Press. 329-345.

Gilbert, C.D., Darian-Smith, C. 1995. The dynamic nature of adult visual cortex. In: McGaugh, J.L., Weinberger, N.M. , Lynch, G. (Eds.) Brain and Memory: Modulation and Mediation of Neuroplasticity. Oxford University Press., 193- 205

Gilbert, C.D. 1992. Horizontal integration and cortical dynamics. Neuron, 9, 1-20.

Goldfield, E.C. 1995. Emergent Forms. Origins and Early Development of Human Actions and Perception. Oxford University Press.

Gould, J. L. , Mahler, P. 1987. Learning by Instinct. Scientific American, 256, 74-86.

Gould, J. H. III, Cusick, C.G., Pons, T.P., Kaas, J.H. 1986. The relationship of corpus callosum connections to electrical stimulation maps of motor, supplementary motor and frontal eye fields in owl monkeys. J. Comparative Neurology, 247, 297- 325.

Grafman, J. 1995. Similarities and distinctions among current models of prefrontal cortical functions. In: Grafman, J., Holyoak, K.J., Boller, F. (Eds.) Structure and Functions of the Human Prefrontal Cortex. Annals N. Y. Acad. Sciences, 769, 337- 368.

Grafman, J. 1994. Neuropsychology of the Prefrontal Cortex. In: Neuropsychology. Academic Press. 159- 181.

Grafman, J., Litvan, I., Massaquoi, S., Stewart, M., Sirigu, A., Hallett, M. 1992. Cognitive planning deficit in patients with cerebellar atrophy. Neurology, 42, 1493- 1496.

Granit, R. 1977. The Purposive Brain. The MIT Press.

Grossberg, S. 1987. Competitive learning: From interactive activation to adaptive resonance. Cognitive Science, 11, 23-63.

de Groot, A. 1966. Thought and Choice in Chess. The Hague: Mouton.

Grudin, J.T., Larochelle, S. 1982. Digraph frequency effects in skilled typing. Center for Cognitive Science, University of California, San Diego.

Haken, H. Synergetic Computers and Cognition. Springer Verlag.

Hebb, D.O. 1955. The Organization of Behavior: A Neuropsychological Theory. Wiley.

Hikosaka, O., Sakai, K., Miauchi, S., Takino, R., Sasaki, Y., Putz, B. 1996. Activation of human presupplementary motor area in learning of sequential procedures: A functional MRI study. J. Neuropsychology. 76,1, 617- 621.

Hinton, G.E., Plaut, D.C., Shallice, T. 1993. Simulating Brain Damage. Scientific American, 269, 76-83.

Hofstadter, D.R. 1980. Godel, Esher, Bach: An Eternal Golden Braid. Vintage Books.

Hornik, K., Stinchcombe, M., White, H. 1989. Multilayer feedforward networks as universal approximators. Neural Networks, 2, 359- 371.

Hudspeth, W. J., Pribram, K. H. 1992. Psychophysiological indices of cerebral maturation. Int. J. Psychophysiology, 12, 19-29.

Husain, M., Shapiro, K., Martin, J., Kennard, C. 1997. Abnormal temporal dynamics of visual attention in spatial neglect patients. Nature, 385, 154- 156.

John, E.R. 1990. Representation of information in the brain. In: John, E.R. (Ed.). Machinery of the Mind. Birkhauser, Boston.

Judd, S.J. 1990. Neural Network Design and the Complexity of Learning. The MIT Press.

Kaas, J. H. 1995. The plasticity of sensory representations in adult primates. In:

Kant, I. 1990. (1781) Critique of Pure Reason. Prometheus Books.

Katz, L.C., Shatz, C.J. 1996. Synaptic activity and the construction of cortical circuits. Science, 274, 1133 - 1138.

Kauffman, S. 1995. At Home in the Universe. Oxford University Press.

Kelso, J. A.S., 1995. Dynamic Patterns. The Self-Organization of Brain and Behavior. The MIT Press.

Kilhstrom, J.F. 1984. Consciuos, subconscious, unconscious: A cognitive perspective. In: Bowers, K. S. Meichenbaum, D. (Eds.) The Unconscious Reconsidered. Wiley, 149- 211.

Killackey, H.P., 1995. Neocortical expansion: An attempt toward relating phylogeny to ontogeny. J. Cognitive Neuroscience, 2, 1, 1-17.

Kline, L.M. 1921. Experimental study of associative inhibition. J.Exp. Psyc.4, 270-299.

Koch, C. 1997. Computation and the single neuron. Nature, 385, 207- 210.

Koestler, A. 1967. The Ghost in the Machine. Arkana.

Koffka, K. 1925. The Growth of the Mind. Harcourt, Brace, & Company, Inc.

Kohler, W. 1947. Gestalt Psychology. Mentor Book.

Kohler, W. 1938. The Place of Value in a World of Facts. Liveright, N.Y.

Kohler, W. 1925. The Mentality of Apes. Harcourt Brace Jovanovich. N.Y.

Kolmogorov, A.N. 1987. Theory of Information and Algorithms. Moscow. In Russian.

Kosslyn, S. M., 1988. Aspects of a cognitive neuroscience of mental imagery. Science, 240, 1621-1626.

Kugler, P.N., Kelso, J.A.S., Turvey, M.T. 1980. On the concept of coordinative structures. In: Stelmach, G.E., Requin, J. (Eds.) Tutorials in Motor Behavior. North-Holland, Amsterdam.

Lakoff, G., Johnson, M. 1980. Metaphors We Live By. Chicago University Press.

Lehmann, D. 1990. Basic Electric Microstates and Cognition: The Atoms of Thought. In: John, E. R. (Ed.). Machinery of the Mind. Birkhauser, Boston.

Lenat, D., Prakash, M., Shapard, M. 1986. CYC: using common sense knowledge to overcome brittleness and knowledge acquisition bottlenecks. AI Magazine, 6(4), 65-85.

Lewin, L. 1936. Principles of Topological Psychology. McGraw- Hill, N.Y.

Lumsden, A., Krumlauf, R. 1996. Patterning the veterbrate neuraxis. Science, 274, 1109-1115.

Luria, A.R. 1973. The Working Brain: An Introduction to Neuropsychology. Basic Books Inc.

Ma, S., Feldman, A.G. 1995. Two functionally different synergies during arm reaching movements involving the trunk. J. Neurophysiology, 73, 5, 2120-2122.

Magee, J.C., Johnston, D. 1997. A synaptically controlled, associative signal for Hebbian plasticity in hippocampal neurons. Science, 275, 209- 212.

MacKay, D. M. 1972. Voluntary eye movements as questions. In: Dihgans, J., Freiburg, B., Bizzi, E. (Eds.) Cerebral Control of Eye Movements and Motion Perception. Basel.

Markram, H., Lubke, J., Frotsher, M., Sakman, B. 1997. Regulation of synaptic efficacy by coincidence of postsynaptic APs and EPSPs. Science, 275, 213- 215.

Mayr, E. 1974. Teleological and teleonomic: A new analysis. Boston Studies in the Philosophy of Science. 14, 91-117.

McCloskey, M., Cohen, J.N. 1989. Catastrophic interference in connectionist networks: The sequential learning problem. The Psy. of Learning and Motivation, 34, 109 - 165.

McFarland, D.J. 1971. Feedback Mechanisms in Animal Behavior. Academic Press.

Michie, D., Johston, R. 1984. The Creative Computer. Machine Intelligence and Human Knowledge. Viking.

Miller, G.A. 1956. The magic number seven plus or minus two: Some limits on our capacity for processing information. Psychological Review, 63, 81-97.

Mishkin, M., Appenzeller, T. 1987. The anatomy of memory. Sci. Am., 256, 16, 80-90.

Nicolis, G., Prigogine, I. 1989. Exploring Complexity. W.H. Freeman and Company.

Norman, D.A. 1986. Reflections on cognition and parallel distributed processing. In: Rumelhart, D.E., McClelland and the PDP Research Group. Parallel Distributed Processing. The MIT Press, 2, 531-546.

Norman, D. A. 1968. Toward a theory of memory and attention. Psy. Rev., 75, 522-536

Ornstein, R., Thompson, R.F. 1984. The Amazing Brain. Houghton Mufflin, Boston.

Palm, G. 1982. Neural Assemblies. Springer, Heidelberg.

Papadimitriou, C.H., Steiglitz, K. 1982. Combinatorial Optimization. Algorithms and Complexity. Prentice-Hall, Inc.

Park, J., Kanwisher, N. 1994. Determinants of repetition blindness. J.Exp. Psyc: Human Perception and Performance, 20, 3, 500-519.

Penrose, R. 1997. The Large, the Small and the Human Mind. Cambridge.

Penrose, R. 1994. Shadows of the Mind. A Search for the Missing Science of Consciousness. Oxford University Press.

Piaget, J. 1954. The Construction of Reality in the Child. Basic Books.N.Y.

Piaget, J. 1952. The Origins of Intelligence in Children. Basic Books. N.Y.

Piaget, J. 1950. The Psychology of Intelligence. Basic Books. N.Y.

Popper, K. R. 1994. Knowledge and the Body-Mind Problem. Routledge.

Posner, M.I., Raichle, M.E., 1994. Images of Mind. Scientific American Library. N.Y.

Pribram, K.H. 1997. The Deep and Surface Structure of memory and Consciuos Learning: Toward a 21st Century Model. In: Solso, R.L. (Ed.) The Science of the Mind: The 21st Century. To be published.

Pribram, K. H. 1995. Neural circuitry and behavior: Synaptodendritic microprocesses and conscious experience: data and theory. Proc. INNS Meeting, Washington, D.C.

Pribram, K.H. 1991. Brain and Perception. Holonomy and Structure in Figural Processing. Lawrence Erlbaum Associates, Hillsdale, New Jersey.

Pribram, K.H 1990. The holographic hypothesis of brain functions. In: Grof, S. (Ed.) Ancient Wisdom and Modern Science. State University of New York Press, 167- 179.

Pribram, K.H., McGuinness, D. 1975. Arousal, Activation, and Effort in the Control of Attention. Psychological Review. 82, 2, 116-149.

Prigogine, I. 1980. From Being to Becoming. Freeman, San Francisco.

Rakic, P. 1988. Specification of cerebral cortical areas. Science, 241, 170-176.

Restak, R. M. 1995. Brainscapes. Hyperion, N.Y.

Roberts, D.D. 1973. The Existential Graphs of Charles S. Pierce. The Hague: Mouton.

Roush, W. 1996. The supple synapse: An affair that remembers. Science, 274, 1102-1103.

Rosenzweig, M. R. 1985. Neuronal plasticity related to cognition. In: Klix, F., Naatanen, R, Zimmer, K. (Eds.) Psychophysiological Approaches to Human Information Processing. Elsevier Science Publishers, 31- 45.

Rumelhart, D.E., Hinton, G.E., Williams, R.J. 1986. Learning internal representations by error propagation. In: Rumelhart, D.E., McClelland and the PDP Research Group. Parallel Distributed Processing. The MIT Press, 1, 318-362.

Sakmann, B. 1994. In: Bass, T.A. Reinventing the Future. Conversations with the World's Leading Scientists. Addison- Wesley.

Schacter, D. L 1996. Searching for Memory. The Brain, the Mind and the Past. Basic Books.

Sebeok, A.T., Umiker-Sebeok, J. (Eds.) 1980: Speaking of Apes. N.Y, Plenum Press.

Sejnowki, T.J., Rosenberg, C. R. 1987. Parallel networks that learn to pronounce complex English text. Complex Systems, 1, 145-168.

Squire, L. 1987. Memory and Brain. Oxford University Press.

Shepard, R.N. 1995. Mental Universals: Toward a Twenty First Century Science of Mind. In: Solso, R.R., Massaro, D.W. (Eds.). The Science of the Mind: 2001 and Beyond. Oxford University Press. 50- 62.

Shiffrin, R.M., Schneider, W. 1977. Controlled and automatic human information processing: II. Perceptual learning, automatic attending, and a general theory. Psychological Review, 84, 127-190.

Siegler, R.S. 1994. Cognitive variability: A key to understanding cognitive development. Current Directions In Psychological Science, 3, 1, 1-5

Singer, W. 1993. Neuronal representations, assemblies and temporal coherence. In: Hicks, T.P., Molotchenkoff, S., Ono, T. (Eds.). Progress in Brain Research. 95, Elsevier.

Sowa, J.F. Conceptual Structures: Information Processing in Mind and Machine.Addison-Wesley, Reading, MA.

Spencer, H. 1937 (1862). First Principles. Harting, London.

Sperry, R.W. 1995. The Impact and Promise of the Cognitive Revolution. In: Solso, R.R., Massaro, D.W. (Eds.). The Science of the Mind: 2001 and Beyond. Oxford University Press. 35-49.

Sperry, R.W. 1972. Science and the Problem of Values. Perspectives in Biology and Medicine. 115-130.

Smolensky, P. 1986. Information processing in dynamical systems: Foundations of harmony theory. In: Rumelhart, D.E., McClelland and the PDP Research Group. Parallel Distributed Processing. The MIT Press, 1, 194-281.

Sternberg, S. 1966. High-speed scanning in human memory. Science, 153, 652-654.

Sternberg, S. 1969. Memory-scanning: Mental processes revealed by reaction time experiments. American Scientist, 57, 421- 457.

Swenson, R. 1992. Order, evolution and natural law: Fundamental relations in complex system theory. In: Negoita, K. (Ed.) Cybernetics and Applied Systems. Marcel Dekker, Inc.125- 148.

Swenson, R. 1989. Emergent attractors and the law of maximum entropy production: Foundation to a theory of general evolution. Sys. Res. 6, 187-197.

Tessier-Lavinge, M., Goodman, C.S. 1996. The molecular biology of axon guidance. Science, 274, 1123- 1132.

Theeuwes, J. 1993. Visual selective attention: A theoretical analysis.Acta Psychologica, 83, 93-154.

Treisman, A.M. 1991. Search, similarity, and integration of features between and within dimensions. J. Exp. Psychology: Human Perception and Performance, 17, 652-676.

Treismman, A.M. 1988. Features and objects. Quarterly Journal of Experimental Psychology. 40A, 201-237.

Treisman, A.M., Gelade, G. 1980. A feature integration theory of attention. Cognitive Psychology, 12, 97- 136.

Tsien, J.Z., Huerta, P.T., Tonegawa, S. 1996. The essential role of hippocampal NMDA recepror-dependent synaptic plasticity in spatial memory. Cell, 87, 1327-1338.

Turing, A.M. 1950. Computing machinery and intelligence. Mind, 59, 236.

Tversky, A. 1977. Features of similarity. Psychological Review. 84, 327- 352.

Tversky, A., Gati, I. 1978. Studies of similarity. In: Rosch, E., Lloyd, B.B. (Eds.) Cognition and Categorization. LEA, Hillsdale, N.J.

Vekker, L. M. 1981. Psychological Processes. v. 1, 2, 3. Leningrad University. In Russian.

Vekker, L.M. 1964. Perception: Introduction to Modeling. Leningrad. In Russian.

Venda, V. 1986. On transformation learning theory. Behavioral Science, 31, 1-11.

Vlasova, M. M. 1989. Mechanisms for Recognizing the Unknown. Space Biology. Moscow. In Russian.

Vossius, G. 1972. Adaptive control of saccadic eye movements. In: Dihgans, J., Freiburg, B., (Eds.) Cerebral Control of Eye Movements and Motion Perception. Karger, Basel.

Wang, X., Merzenich, M.M., Scmeshima, K., Jenkins, W. M. 1995. Remodeling of hand representation in cortex determined by timing of tactile stimulation. Nature, 378, 71- 75.

White, N. M. 1995. Emotional memory: Conceptual and Methodological Issues. In: McGaugh, J.L., Weinberger, N.M. , Lynch, G. (Eds.) Brain and Memory: Modulation and Mediation of Neuroplasticity. Oxford University Press, 93- 102.

Yerkes, R.M., Dodson, J.D. 1908. The relation of strength of stimulus to rapidity of habit-formation. J. Comparative Neurological Psychology, 18, 459-482.

Yufik, Y.M. 1997. Operator cognitive modeling in complex systems. Presentation to National Academy of Sciences Committee on Cognitive Modeling.

Yufik, Y.M., Sheridan, T.B. 1997a. Virtual Associative Networks: New formalism for operator cognitive modeling. Annual Review in Control, 20, 179-192.

Yufik, Y.M. 1996a. Resource Allocation System with Self-Adaptive Capabilities. U.S . Patent 5586219.

Yufik, Y.M ?96b. Memory, complexity, and control in biological and artificial systems. Proc. Int. C :. on Intelligent Systems, NIST. 311-317.

Yufik, Y.M Voloshin, M., Zhukarev, V.A. 1996c. Memory integration in complex learning: A el network modeling approach. Soc. Neuroscience, 549.6.

Yufik, Y.M heridan, T.B. 1995a. Assessment of Cognitive Complexity. NASA Report, NASA Am Research Center, CA.

Yufik, Y. Sheridan, T.B. 1995b. Virtual Network for Supervisory Control. NSF Report.

Yufik, Y. 1994. Virtual network formalism for modeling cognitive functions. IEEE Internatio Conf. on Systems, Man and Cybernetics. San Antonio, 952-958.

Yufik, Y. .f., Sheridan, T.B., Venda, V.F. 1992. Knowledge measurement and cybernetic mutual man- machine adaptation. In: Negoita, C.S. (Ed.). Applied Systems and Cyber ics. Marcel Dekker. New York. 187-238.

Yufik, Y. 1990a. Application of cognitive modeling and knowledge measurement in diagnosis training of complex skills. Human-Computer Interaction - INTERACT' 90, D. Diaper al. (Eds.) Elsevier Science Publishers. North-Holland; 887-892.

Yufik, Y. 1990b. Computational and cognitive models of configurative synthesis. Proc. IEEE Int tional. Conf. on Intelligent Control. Philadelphia, PA. 89-96.

Yufik, Y . Sheridan, T.B. 1986. Hybrid knowledge-based decision aid for operators of large scale systems. ge Scale Systems, North Holland, 133-146.

Yufik, Y , Hartzell, E.J. 1989. Assessment of Cognitive Complexity in Man-Machine Systems. : Salvendy, G., Smith, J. (Eds.) Designing Human-Computer Interfaces. Elsevier,)-167.

Yufik, Y . 1980. Simulation of cognitive strategies and systems with virtual structure. CSI Pap Cognitive Science Institute. University of California, San Diego.

6

Self-Organization of Cortical Information Processing

Thomas P. Vogl[1]
Kim T. Blackwell,
*Daniel L. Alkon**
Environmental Research Institute of Michigan
1101 Wilson Blvd, Arlington VA 22209
and
George Mason University
Fairfax, VA 22030
**Laboratory of Adaptive Systems, NINDS, NIH, Bethesda, MD 20892*

Self-Organization of Cortical Information Processing

Summary

Over the past decade, significant progress has been made in the understanding of cortical organization within the context of extracting information from images and comprehending their content. In those areas of striate cortex in which retinotopic relationships and spatial frequency information is preserved, the organization of receptive fields and relationships of structure and function are beginning to be revealed. At the other end of the chain of visual cortical information processing, inferior temporal (IT) cortex cells have been shown to respond selectively to highly complex stimuli, including specific faces. The cortical information processing paradigm presented here suggests how relatively local features derived from the visual scene by the striate cortex could be transformed into composite features of great specificity that are identifiable in IT cortex. Biophysical properties of dendrites and spines, in particular mechanisms of synaptic modifiability and memory storage in conjunction with often ignored but critically important statistical properties of sensory inputs, facilitates a self-organization of neuronal connectivity for information processing that allow for increasing specificity in successive processing layers.

Key words: Information processing, visual cortex, Inferiotemporal cortex, dendrites, spines, synapses, models, clusters, neurophysiology.

[1] Corresponding Author. Fax: 703-524-3527; e-mail: tvogl@wo.erim.org

Introduction

Recent neurophysiologic, neuroanatomic, psychophysical, and computational modeling studies have begun to clarify the cortical organization and neuronal pathways through which an image on the retina is transformed into recognition of a face. Although many questions remain, the mechanisms involved in the transformation of the retinotopic image by the retina (Miller, 1976 a,b, c), LGN (Hubel, 1961), and striate cortex (Hubel 1959, 1962) which have received the longest and most intense scrutiny, are beginning to be understood. Experimental evidence and models of the organization, development, structure, and function of simple cells, complex cells, and the orientation columns, are rapidly accumulating (Rose, 1985; Obermeyer, 1993; Miller, 1994; Gilbert, 1996a,b). Early visual processing (up to and including the striate cortex) extracts a variety of features (referred to here as microfeatures) from the retinal image (Van Essen, 1992), including edges and ends of contrast boundaries (Von der Heydt, 1989), local color (Zeki, 1983), texture, and energy in the spatial frequency bands (De Valois, 1982; Movshon, 1978).

Further along the processing pathways (Baizer, 1991; McIntosh, 1994; Mishkin, 1982), cells in the inferior temporal (IT) cortex have been shown to respond selectively to highly complex stimuli (Baylis, 1987; Desimone, 1984; Tanaka, 1993, 1996; Sary, 1995; Ito, 1995) including facial expressions and faces (Baylis, 1985; Desimone, 1991; Rolls, 1994). However, the information processing paradigm in the intermediate stages between striate and IT cortex are largely unknown and it is the paradigm and its neurobiological plausibility, rather than the details of the cellular and subcellular mechanisms that support it, that is the primary subject of this paper.

Accumulated observations, surveyed below, suggest that learning extends through multiple neuronal layers linking striate to IT cortex and can produce the observed responses of IT cortical cells. Learning rules followed within these layers can be derived from neuroanatomic and biophysical observations that implicate clusters of dendritic spines as fundamental architectural loci of information processing and memory storage. These loci are distributed throughout large arrays of neurons comprising multiple layers of brain structure. Taken together, these results suggest the formulation of three principles used to construct a unifying conceptual framework of information processing between striate and IT cortex:

(1) Each axon makes multiple synaptic connections on a neuron (Deuchars, 1994, Kaneko, 1994), providing a one-to-many connectivity between axons and synaptic spines that are located along a well-elaborated dendritic tree (Larkman, 1991;Chicurel, 1992; Trommald, 1995),suggesting a rich spectrum of afferent combinations that enervate the dendritic spine clusters (DSCs).

(2) A cluster of interacting synaptic spines, co-located on dendritic branches responds maximally to specific patterns of inputs (Alkon, 1987; Harris, 1995; Liu, 1995; Eilers, 1995) and form a basic information processing unit of the CNS.

(3) The signals corresponding to the micro-features generated by the striate cortex are not all independent; subsets of these signals are subject to functional constraints imposed by the properties of the external world. Membership in these subsets can be determined from the signals (Saffran, 1996; Becker, 1996; Ghose, 1997).

These three biologically suggested and statistically plausible assumptions permit the conceptualization of a neuronal network that is self-organizing in all its aspects, and that has the capacity to generalize as well as recognize highly indiosyncratic characteristics. It also allows a relatively simple description of how complex pattern recognition can naturally derive from from aggregates of object-specific microfeatures. The straightforward interaction between the assumptions, the framework, and the derived behavior enhances the plausibility of both the asumptions and the framework.

Before embarking on a description of the framework and the concepts that can be derived therefrom, the three principles are examined.

The Loci of Associative Memory Storage

Associative learning and memory require sites for convergence of multiple signals that become associated. Synaptic spines that cluster on dendritic branches receive the preponderance of synaptic inputs from pre-synaptic afferent terminals (Larkman, 1991; Chicurel, 1992; Trommald, 1995). For any given integrative neuron there are fewer afferents than synapses (Friedlander, 1991; Douglas, 1995) with each of these afferents making multiple connections to the 10,000 to 100,000 synaptic spines (Larkman, 1991; Chicurel, 1992; Trommald, 1995). The one-to-many connectivity (one afferent to many synapses on the same dendritic tree) provides for a spectrum of signal combinations from the afferents. Furthermore, many of the afferents to a neuron convey closely related information (Gilbert, 1989; Douglas, 1991 and 1995; Fujita, 1996). This rich, one-to-many connectivity (one afferent to multiple synapses), creates sufficient juxtaposition of synapses to allow flexibility in the formation of a variety of combinations of information from the available afferents. Thus, multiple signals converge and interact through post-synaptic dendritic spines clustering on common branches.

Detailed studies of dendritic morphology, from the camera lucida drawings of Golgi stained neurons to the more recent serial slice reconstructions of Dehay (1991) and Claiborne (1986 and 1990) reveal a complicated dendritic branching pattern. Interactions among voltage and current signals propagating down these branches are highly non-linear, especially at branch points (Zador, 1992; Agmon-Snir, 1993). Non-linear information processing within the dendritic tree implies that the output of a neuron cannot be a simple linear function of the sum of its inputs. Rather, neuronal output is highly sensitive to specific combinations of inputs. This has been shown in modeling studies (Mel, 1992; Softky, 1993 and 1994), as well as in biophysical studies (Optican, 1987;Richmond, 1990), which further show that significant information is carried by the pattern of output spikes.

Experiments over the last two decades have provided a substantial and diverse body of evidence implicating dendritic spine clusters in memory storage. Studies that correlated actual associative learning and memory of living animals with long-term changes of specific cellular parameters provided critical insights. These studies began with Pavlovian conditioning of the mollusc *Hermissenda* for which a precise point-to-point synaptic network could be determined. Changes of particular voltage-dependent K^+ currents within post-synaptic membranes were causally implicated in the learning and memory of the Pavlovian conditioned response (Alkon 1979, 1984, 1985; Collin, 1988; Werness, 1992 and 1993). Persistent reductions of such currents as I_A and I_{CA} 2+$_-$ $_K$+ was shown to result from a cascade of second messenger (Ca^{2+}, DAG, AA) activation of protein kinase C (PKC) and its subsequent phosphorylation of signal transduction G-proteins, such as the G-protein cp20, that occurred only when the animal learned an association (Alkon, 1988; Nelson, 1990; McPhie, 1993).

These same channels and their regulatory biochemical cascades were found to change with Pavlovian conditioning paradigms in the rabbit. In mammals, the evidence indicates that these long-lasting changes are located on dendritic branches rather than the cell bodies. With rabbit nictitating membrane conditioning, for example, I_C changes occurred in the hippocampal pyramidal cells in the absence of any changes of the current-voltage relations or membrane excitability of the cell somata (Disterhoft, 1986; Sanches-Andres, 1991). Summation of EPSP's (but not individual EPSP's) elicited by electrical stimulation of presynaptic Schaeffer collaterals, however, are greatly enhanced for the conditioned rabbits, correlated with post-synaptic K^+ current reductions (Coulter, 1989).

Similarly, the excitability of intracellularly recorded Purkinje cell dendritic branches are enhanced for conditioned rabbits in the absence of changes in the dendritic membrane potential or input resistance (Schreurs, 1990 and 1996). This enhancement of postsynaptic dendritic excitability can be simulated by perfusion with 4-aminopyridine which blocked I_A. These changes of Purkinje cell dendrites within the H6 cerebellar cortex (implicated by lesion and pathway studies as a critical locus for nictitating membrane conditioning), were not accompanied by the synaptic changes studied as in vitro cellular models of learning, such as LTD or LTP (Schreurs, 199, 1993, 1996). Similarly, no evidence of LTP or LTD was apparent in hippocampal recordings from conditioned rabbits (LoTurco, 1988).

The above electrophysiologic evidence demonstrating post-synaptic dendritic loci for associative memory, is corroborated by a number of biochemical measurements made during associative learning and memory (Olds, 1989). Twenty-four hours after classical conditioning, calcium mediated increases in membrane-associated PKC were observed in both the somata and dendrite regions of the hippocampal array of CA1 pyramidal cells. Seventy-two hours after learning the distribution markedly shifts into the CA1 dendritic branches. This result was confirmed and extended when monoclonal antibodies to PKC were found to increase (under similar conditions) inside the dendritic branches themselves (Van der Zee, 1992). These observations have since been generalized to a variety of other associative learning paradigms including spatial maze learning of the rat (Olds, 1990) and olfactory discrimination learning (Olds, 1994). Calcium dependent long term changes in synaptic efficiency have also been measured in neocortex (Artola, 1996; Chen, 1996).

Thus, the neuroanatomic design for afferent convergence, the electrophysiologic changes observed in close correlation with associative learning and memory, and the imaged biochemical correlates, collectively implicate post-synaptic loci on dendritic branches in memory storage.

Dendritic Spine Clusters (DSCs) as Information Processors

The essential hypothesis relative to the proposed information processing paradigm is that clusters of interacting synapses (DSCs), rather than individual synapses, are the basic unit of information storage. Simulations in computer models (Blackwell, 1994 and 1995) examined the long-term transformation of a proto-DSC into a DSC, and evaluated the pattern matching behavior of such DSCs. In the computer model, training a proto-DSC with repeated presentation of a particular synaptic input causes a decrease in potassium channel conductance, and a decreased sensitivity of the calcium-dependent potassium channel to increased calcium concentration. After training, DSCs generate a greater depolarization whenever all of their synaptic inputs are similar to (match) the inputs that were responsible for the training. Conversely, inputs that vary from those in the training set produce smaller depolarization, irrespective of whether that variation is composed of stronger or weaker inputs than the training set. This formulation of non-linear synaptic integration, i.e., that DSCs are sensitive to the difference between the pattern of synaptic input and a previously learned pattern, is supported both by biophysical studies (Eilers, 1995; Denk, 1995; Teunissen, 1996) that show that dendritic spines can accumulate calcium independently of each other and that several spines on a branchlet are activated as a group, and by modeling studies (Mel, 1992 and 1994; Hopfield, 1995; Zador, 1992).

A general purpose associative learning algorithm, called Dystal, is based on dendritic spine clusters illustrated in figure 1, and utilizes a similarity measure to relate groups of incoming signals to previously stored patterns (Blackwell, 1992; Alkon, 1994). That algorithm appears to validate the assumptions regarding the importance of clusters for pattern recognition by demonstrated efficacy in the recognition of handwritten Zip codes and Kanji characters (Vogl, 1991; Blackwell, 1992), the identification of complex organic mixtures from spectroscopic data (Dettmar, 1996); and face recognition (Vogl, 1992; Alkon, 1994; Blackwell, 1997).

Information processing in DSCs is necessary, but by itself insufficient, to account for the transformation from microfeatures extracted by the visual system and striate cortex to the individual-specific face recognition demonstrated in IT cortex. That this transformation could in fact be performed by a self-organizing network of neurons, requires the third, and last, assumption. Specifically, it results from the interaction of DSCs and of the constraints on the microfeatures that derive from the external world.

Functional Constraints (FCs) on Micro-features

When considered over the range of all possible external stimuli, the array of micro-features produced by the striate cortex can be grouped into a multiplicity of subsets. Each subset is composed of micro-features that are functionally constrained (FC). By functional constraint on a subset of microfeatures is meant a relationship among the subset members which can be either an equation or an inequality. The simplest constraint is linear, i.e., $\sum a_i x_i = \text{constant}$, where x_i is the ith component of the array and a_i is a constant coefficient (or weight). In this example, when some components of the subset are high in value, others must be low in order to keep their sum constant. In general, and in the case of the cortex in particular, there is no requirement that the FCs be linear. The concept of FC is important because the likelihood of components of differing patterns incident on a DSC being consistent over diverse inputs is higher if those components include only members of a single FC subset. The repeated pattern presentation required for DSC formation selects for DSCs whose afferents all belong to the same FC subset (Ghose, 1997; Gilbert, 1996a,b). Note that Fcs are a high order statistic (c.f. correlation -- a second order statistic).

In the context of developing the framework, careful distinction must be made between two situations that differ in time scale.

(a) Events occuring within a fractions of a second allow the extraction of *correlations* among the components of one image and among temporally sequential images of the same scene. (The world usually does not change much in a few millisecond.) These <u>correlations control dendritic spine cluster (DSC or 'patch') formation</u>, since DSCs will form in response to consistent pattern repetition;

and

(b) Event occuring on time scales of seconds and longer will tend to be uncorrelated but allow for the extraction of *functional constraints (FC)* among multiple *disperate images* that can only be learned from the presentation of multiple diverse images that encompass samples of all possible inputs. Note that FC does not require correlation but the value of a component within an FC subset can be predicted from the values of all other components of the subset. <u>FCs control receptive field (RF) selection and neuron recruitment</u>.

This distinction is displayed in figure 2, where each column of the matrix is one of the micro-features provided by striate cortex. Each row contains the collection of micro-features that derive from the single input pattern that is incident at any one time. Thus correlations among rows imply repeated presentation of the same or similar patterns of activation across sets of afferents, e.g., rows 1-4, 5-7, and 8 & 9. FCs among columns are among micro-features and do not require that the activation pattern (over time) of one afferent be similar to that of another; e.g., the first four rows of the leftmost four columns in figure 2 are correlated (the row repeats); the first four columns are not correlated but are functionally constrained, because when some are high, others must be low by virtue of the physical properties that constrain the inputs that in this hypothetical case, define 'color'. Despite the FC within subsets of columns, the relationship between a subset (and any of its components) and any other subset or any individual micro-feature outside the subset is random (unconstrained), e.g., in figure 2, any of the four leftmost columns (colors) are unconstrained relative to any of the columns numbered five through twelve (line direction and texture). Furthermore, there is no restriction on the number of FC subsets in which a particular micro-feature can participate, as shown in Figure 3 which derived from Table 1.

Table 1. Pairwise measure of co-occurence between neurons depicted in Figure 3.

	1	2	3	4	5	6	7	8
1	-	**0.9**	0.2	0.4	0.3	0.1	0.5	0.4
2		-	**0.85**	**0.8**	0.55	0.5	0.3	0.1
3			-	**0.95**	**0.7**	**0.8**	0.4	0.2
4				-	**0.9**	**0.85**	0.45	0.6
5					-	**0.95**	0.55	0.45
6						-	**0.85**	**0.8**
7							-	**0.9**
8								-

Table 1 tabulates a measure, 0 to 1, of the pairwise co-occurence among afferent neurons. A measure of 0.7 or greater is assumed to constitute a significant level of co-occurence. Figure 3 maps the resulting FC groupings. Neurons 2, 3, 4, and 6 participate in more than one FC group. The learning rule based upon the interaction of DSC and FC subsets can be considered a generalization of the learning rule that is based on the pairwise correlation of inputs to independently responding individual synapses (see, e.g., Linsker (1986) or Kossel (1995)). While the biological mechanism that allows the formation of FC subsets, namely DSC formation, may be computer modeled directly, we note that it may be computationally more efficient to determine FC subsets by tabulating co-occurrences of afferent signals (Bickel, 1977).

Note that multiple exemplars of the micro-feature sets are required to determine which components are FC and therefore comprise a subset; that determination cannot be accomplished from a single exemplar, nor from multiple exemplars drawn from the same scene / image. Ten to 30 input samples are sufficient for the long-term transformation of potassium channel properties (Schreurs, 1996; Alkon, 1987 and 1992; Collin, 1995), which suggests that this number of inputs may also be sufficient for the transformation of proto-DCSs into DSCs. That this number of input samples is available in a reasonable amount of time follows from psychophysical experiments (Purves, 1996) and from the flicker-fusion rate of the retina, about 50 milliseconds, thereby providing a reasonable basis for assuming that afferent visual information is refreshed at approximately the same rate, i.e., 20 times a second.

DSC-based Information Processing

Three principles, explained in the foregoing sections, are necessary and sufficient components of the proposed framework. To recapitulate, these are:

(1) One-to-many connections between axons and synaptic spines, together with a well elaborated dendritic arborization.

(2) Dendritic spine clusters (DSCs).

(3) Subsets of micro-features that have functional constraints which may be estimated from the input signals.

How do these assumed properties interact to produce a content-specific neuronal response? The answer is based on the learning and response properties of individual DSCs that form only if the pattern of inputs to the spines is sufficiently consistent over time to allow long term retention of associative learning. Consider a pyramidal or other principal cell with a well elaborated dendritic arborization with several tens of thousands of dendritic spines (Larkman, 46; Chicurel, 1992; Trommald, 1995); the axons of a far fewer number (a few tens to many hundreds (Friedlander, 1991)) of neurons provide the presynaptic input to these spines. This architecture allows for a rich diversity of proto-DSCs, with varying combinations and numbers of presynaptic inputs. To facilitate description, but without loss of generality, assume that the afferents to this particular neuron consist largely of "color" specific signals with an admixture of signals of other micro-features, e.g. related to "shape" and "texture" (Gilbert, 1989, 1996a,b; Douglas, 1995; Ghose, 1997). A time series of images on the retina will produce a time series of inputs to the neuron. As the visual input shifts or changes with retinal, head, or in-scene movement, some of the inputs will remain the same and others will vary. DSCs will begin to form from proto-DSCs if a group of clustered spines repetitively receives the same pattern of inputs. As discussed above, repetitive patterns are more likely within FC subsets. That is, since the learning and response of each DSC is highly dependent on consistency of signal across all of its inputs, only that group of spines whose ensemble of inputs are highly consistent will form a DSC that, with further consistent stimulation, will be trained to the extent necessary for long term memory changes to take hold. I.e., only ensembles of inputs that are functionally constrained will have sufficiently consistent values over time. The dendritic tree of the hypothetical neuron being considered will contain primarily "color" related DSCs because it has primarily "color" related afferents; note that those DSCs that contain some inputs that are uncorrelated with the rest will not be fully formed or trained. For example, referring to our hypothetical neuron, DSCs relating to a mixture of a specific "color" and some specific other attribute, e.g.,"glossy texture", may be created if those attributes happen to co-occur with sufficient frequency. However, as the visual input ranges over diverse scenes, those DSCs will not be fully formed and will revert to proto-DSCs.

The end result of this process is a neuron (most probably an ensemble of population coded neurons (Douglas, 1991)) whose individual DSCs respond to specific patterns of "color" information afferents (e.g., DSCs that respond primarily to "sky blue", or to "fuchsia", etc., while the neuron as a whole can be thought of as a "color" specific neuron, i.e., having a "color"-specific receptive field), while filtering out the "noise" consequent to other inputs that are not FC with "color", e.g., "texture" or "shape". Since DSC creation process is self-selecting for optimal specificity of input patterns, one neuron's noise becomes another neuron's signal. Note that enhancement (or depression) of the response of *individual* spines (as suggested by an LTP model of learning (Bliss, 1993; Malinow, 1994)) will not provide the needed functionality, because associative interactions among the multiplicity of spines within a proto-DSC are required to produce a DSC with a pattern-specific response.

The framework suggests the occurrence of the following chain of events:

1. An image is incident on the retina and an array of micro-features and their image-specific values are extracted by striate cortex.

2. The micro-feature outputs stimulates a large number of spine (proto-clusters) that could, potentially, store memory.

3. A second image is incident on the retina and is processed by striate cortex.

4. The second array of features is incident on the same synapses as the first. If, and only if, this synaptic input is sufficiently similar to the previous one, will associative learning begin to occur and DSCs begin to form.

5. Subsequent presentations will not only strengthen the learning, but the response of the DSCs to the new stimuli will be critically dependent on whether *all* the synaptic spines within the

cluster receive the stimuli they received at the previous time intervals. If all do, the output of the DSC will be large; if even one input significantly differs from the training value, the output will be weaker. Thus the DSCs are self-selecting for FC components of the input array and the development of DSCs from proto-DSCs that contain unconstrained components will tend to be suppressed.

By creating a series of layers in which the output of one layer is the input to the next, the specificity implicit in face recognition can be created. The result is a system whose input is the large array of micro-features that is the output of striate cortex and whose outputs are associations of increasing specificity.

An ensemble of neurons in a layer of cortex, transmits to the next layer a content-specific signal characteristic of the pattern of inputs that has been processed. The signaling characteristics of the content-specific signal have not been elucidated experimentally, but at least four different, but not mutually exclusive, hypotheses have been put forward. Hopfield (1995) suggests that input specific information is coded into the explicit timing of action potentials by virtue of differences in delay times in different paths through the dendritic tree. Sophisticated spike train analysis, (e.g., Mainen and Sejnowski (1995); Richmond (1990)) suggest that the information is encoded into the temporal patterns of spike trains. Lukashin ((1994) and Georgopoulos (1993) propose that neuro-muscular motion information is transmitted through population coding of ensembles of neurons each with relatively broad tuning that varies from cell to cell, a mechanism that Sary's (1995) results suggest have broad implications for CNS information tuning. Non-linear interactions among signals originating at various sites along the dendritic tree as they propagate toward the soma facilitates sophisticated information processing of a complex of incoming signals. In this context, basket cell inhibition (Collin, 1995) may serve to delay neuron firing. In turn, this delay retards the initiation of the retrograde pulse propagating into the dendritic tree that quenches further PSPs (Spruston, 1995), allowing longer integration time among the signals in the dendritic branches. Alternatively, those interneurons whose axonal arborization extends to the distal branches of the principal neuron's dendritic tree may serve to assist or circumscribe DSC formation. Which of these mechanisms, or which combination of these mechanisms, is active in the inter-neuronal signaling process remains to be determined.

However, independent of the details of the mechanism, just as for the first layer inputs, the range of second layer inputs are also limited by functional constraints, constraints that are different, and in general more complicated, from those on the first layer (e.g., an image of grass is functionally constrained to a limited range of textures and shades of green and brown). Thus the DSCs formed on the dendrites of this layer will be new subsets comprised of the combinations of subsets developed by the previous layer. In the linguistic convention, these DSCs will be correlated combinations of "color" and "texture", or possibly "color" and "texture" and "gross shape" depending on the content of the original retinal image (see Figure 2, layer 2). This is the beginning of object specificity (Ghose, 1997).

This chain of events can be concatenated to produce a multi-layer system in which DSCs instantiate ever more complex combination of combinations of FC subsets that respond to scene content of ever increasing specificity (and complexity). Multicellular recordings (Baylis, 1987; Desimone, 1984;Tanaka, 1993) demonstrate the increase in specificity of layers in IT, as well as the groupings of FC subsets into columns that "required moderately complex features for their activation" (figure 4). The critical features are more complex than orientation, size, color, and simple texture, which are known to be extracted and represented in V1 (Gilbert, 1996a,b). Some of the features are shapes that are moderately complex, whereas others are combinations of such shapes with colors and textures (Tanaka, 1996). Sary (1995) investigated the response of IT cells to gratings defined by motion, luminance, and texture and found that only a small portion of the 330 single cells recorded were orientation sensitive to more than one type of cue (see figure 5) whereas the tuning width of all the responses was roughly the same. They conclude that "the existence of a class of high-level visual neurons with cue-invariant selectivities is in agreement with psychophysical data and our daily perceptual experience." The commonality of tuning width shown in figure 5 also suggests that population coding for fine tuning may be common in the CNS.

Note that in this progression, the size (number of afferents) of individual DSCs does not necessarily increase. Rather, the input to each spine represents a richer combination of subsets, so that an ever more content-specific representation is created in each DSC of subsequent layers. As the subsets processed by each subsequent layer become increasingly diverse in content, this ingathering of FC information (increases in specificity) is probably roughly balanced by the increase in the number of different specific objects within the purview of the subsets. "Recognition" occurs when this process has been concatenated through a sufficient number of layers to ensure that the content-specific representation has become unique, thereby allowing an unambiguous association with a semantic label or other input-specific "UCS" response.
That is, the construction of further layers is self-limiting in that no additional, higher-level FC subsets can be created.

Discussion

This paper presents a novel framework for conceptualizing cortical information processing, a framework that is parsimonious with current data on the architecture and function of the neocortex and provides a working hypothesis for mechanisms that transform large sets of features provided by striate cortex into a representation with recognition-specific components. Associative learning by constellations of spines on DSCs of dendritic membrane that cluster an array of inputs into subsets as a consequence of functional constraints within each subset, but not among subsets, is necessary and sufficient to construct a model of information processing that is self-organizing at successive levels of information complexity. In distinction from previously proposed models of neural information processing (such as Back-propagation, self-organizing maps, the Neocognitron, adaptive resonance, etc.), our paradigm does not require the *a priori* specification of the number of neuronal elements in each layer. Rather, neuronal elements are instantiated (self-selected) based on the functional constraints (FCs) among inputs, while the DSC's on the dendritic tree of these neuronal elements are self-selected for specific combinations of constrained inputs. Development of this computer model has begun as an extension of the Dystal associative learning algorithm (Alkon, 1994).

A number of relevant, interdependent, issues that affect future directions for the framework and, eventually, the computer model, have not been explicitly addressed. They include:

(1) The signal processing that occurs when signals from DSCs distributed along dendritic trees traverse the tree structure, a process that can further refine selectivity and responsivity. These processes are only beginning to be explored experimentally (Magee, 1995; LoTurco, 1988;

Spruston, 1995) as well as modeled (Zador, 1992; Mel, 1992).

(2) The role and circuitry of inhibitory interneurons. The accumulating evidence for differences in connectivity and biophysical properties of inhibitory interneurons suggests that each inhibitory interneuron plays a different role in neuronal information processing (Debanne, 1995; Buhl, 1994). As mentioned above, a hypothesized role of basket cells is to delay neuronal firing in order to allow increased interactions among afferent signals within the dendritic tree (Collin, 1995; Sik, 1994). The multiple roles of inhibitory interneurons suggests that they may also be the mechanism that allows the accumulation of information over time, thereby serving as the mechanism that determines (or assists in the determination) of membership in FC subsets. Possible mechanism to this end may include processes similar to the role of stratum oriens-alveus interneurons in switching emphasis between monosynaptic and trisynaptic circuits in hippocampus (Maccafferi (1995) and the differing roles of inhibitory interneurons that synapse on distal dendritic trees compared to those that synapse more proximally (Soltesz, 1995) These and other ongoing investigations of the functions and roles of inhibitory interneurons will undoubtedly clarify many of these unresolved issues and influence the structure of the resulting models.

(3) The active roles of intention and attention and the neuromodulatory feedback mechanisms through which they are presumed to function (Gilbert, 1996a,b).

(4) In addition to “direct” mono-synaptic inputs from striate cortex and modulation of neuronal response by chemical messengers, there are “indirect” multi-synaptic inputs from other areas of the brain. A logical extension of the framework suggests that the function of some of these inputs is to provide ancillary information with which the signals arising from striate cortex become associated (Pribram, 1991; Tanenhaus, 1993). This ancillary information can, for example, consist of signals from the language centers that have previously become associated with the striate cortex input, e.g., the word "green" or “cat”; it may consist of temporally contiguous inputs from other sensory, motor, or emotional sources that become associated with the striate cortex input, e.g., the tactile response to a texture; or it may be a reinforcing feedback signal that further stimulates the more global process of "recognition," based on the stimulation of an association from a higher layer of processing (Pribram, 1991). In the higher mammals, this can be considered a generalization of Pavlovian conditioning, that allows an association to develop between arbitrary stimuli (not just with a reflexive UCS as in Pavlovian conditioning (Schreurs, 1990 and 1995)), providing only that the two stimuli occur within a suitable inter-stimulus interval.

To test the validity of the framework, experimental verification of the formation and function of DSCs is crucial. An extension of Liu & Tsien’s method (1995) of stimulating individual CA1 synapses to stimulate identifiable groups of synapses might allow for a series of experiments that not only test for the existence of DSCs, but also explore DSC formation and modification. For example, it is not known whether multiple DSCs can share common synapses, whether a proto-DSC can “recruit” additional synapses depending on the size of the FC subset and to what extent such recruitment may be limited by physical placement of synapses on the dendritic branch.

On a more functional level, the methods of Tanaka (1993, 1996), Ito (1995), and Sary (1995), when exten-ded to allow simultaneous recordings in multiple areas of IT cortex in the fashion of Nicolelis (1995), would elucidate the convergence of the responses to increasingly complex "features" to the final recognition of faces observed by Desimone (1991) and Rolls (1994).

The development of a computer simulation of a single layer of the proposed structure which could then be concatenated into a multi-layer system would provide an interesting tool to test of the underlying concepts. For example, the ability of such a multi-layer system to generate composite features that result in the specificity suggested by face recognition would provide additional support for the underlying concepts. We conclude by noting that the "Constructionist Manifesto" of Quartz and Sejnowski (1996) requires a neural system that is, to all intents and purposes, completely self-organizing and that

> "does not involve a search through a pre-defined hypothesis space, and is not one of selective induction. Instead, the constructivist learner builds this hypothesis space as it learns. This has shifted the problem from one of parameter space estimation to a nonparametric regime. We must accordingly ask, what is the effect of allowing a system to add structure -- to build representations -- as it learns."

These conceptualization presented herefulfills these requirements and is currently being explored as an extension of the Dystal associative learning algorithm (Alkon, 1994).

Acknowledgments

The partial support of this research effort by NIH contract N01NS32304 and ONR contract N00014-92-C-0018 is gratefully acknowledged.

References

Agmon-Snir, H. And Segev, I. Signal Delay and Input Synchronization in Passive Dendritic Structures. *J. Neurophysiol.* 70 (1993) 2066-2085.

Alkon, D.L. *Memory Traces in the Brain.* Cambridge University Press, Cambridge, U.K. (1987).

Alkon, D.L. Voltage-dependent calcium and potassium ion conductances: A contin- gency mechanism for an associative learning model. *Science* 205 (1979) 810-816.

Alkon, D.L. Calcium-mediated reduction of ionic currents: A biophysical memory trace. *Science* 226 (1984) 1037-1045

Alkon, D.L., Sakakibara, M., Forman, R., Harrigan, J., Lederhendler, I. and Farley, J. Reduction of two voltage-dependent K^+ currents mediates retention of a learned association. *Behavioral and Neural Biology* 44 (1985) 278-300.

Alkon, D.L., Naito, S., Kubota, M., Chen, C., Bank, B., Smallwood, J., Gallant, P. and Rasmussen, H. *J. Neurochemistry* 51 (1988) 903-917.

Alkon, D.L., Sanches-Andres, J.V., Ito, E., Oka, K., Yoshoka, T. and Collin, C. Long-term Transformation of an Inhibitory into an Excitatory GABA-ergic Synapse Response. *Proc. Nat. Acad. Sci. USA*, 89 (1992) 11862-11866.

Alkon, D.L., Blackwell, K.T., Barbour, G.S., Werness, S.A., and Vogl, T.P., Biological plausibility of synaptic associative memory models, *Neural Networks* 7 (1994) 1005 - 1017.

Artola, A., Hensch, T., and Singer, W. Calcium-Induced Long-Term Depression in the Visual Cortex of the Rat In Vivo. *J. Neurophysiol.* 76 (1996) 984-994.

Baizer, J.S., Ungerleider, L.G. and Desimone, R. Organization of Visual Inputs to the Inferior Temporal and Posterior Parietal Cortex in Macaques. *J. Neurosci.* 11 (1991) 168-190.

Baylis, G.C., Rolls, E.T. and Leonard, C.M. Selectivity between faces in the response of a population of neurons in the cortex in the superior temporal sulcus of the monkey. *Brain Research* 342 (1985) 91-102.

Baylis, G.C., Rolls, E.T. and Leonard, C.M. Functional Subdivisions of the Temporal Lobe Neocortex. *J. Neurosci.* 7 (1987) 330-342.

Becker, S. Mutual information maximization: models of cortical self-organization. *Ntework: Computation in Neural Sytems* 7: 7-31 (1996).

Bickel, B.J. and Doksum, K.A. *Mathematical Statistics*, Chapter 8. Holden-Day, Inc., Oakland, CA (1977).

Blackwell, K.T., Vogl, T.P., Hyman, S.D., Barbour, G.S. and Alkon, D.L., A new approach to the classification of hand written characters, *Pattern Recognition* 25 (1992) 655-666.

Blackwell, K.T., Vogl, T.P. and Alkon, D.L., Pattern discrimination in an impedance matching model of dendrites, *J. Neurosci. (Supl.)* 20 (1994) 1209.

Blackwell, K.T., Vogl, T.P. and Alkon, D.L., Pattern recognition by a circuit model of dendritic spines, *J. Neurosci. (Supl.)* 21 (1995) 1225.

Blackwell, K.T., Vogl, T.P., Dettmar, H.P., Brown, M.A., Barbour, G.S., Alkon, D.L. Identification of Faces Obscured by Noise: A Neural Network that Outperforms Human Observers. In press, *Journal of Experimental and Theoretical Artificial Intelligence* (1997).

Bliss, T.V.P. and Collingridge, G.L. A Synaptic Model of Memory: Long-term Potentiation in the Hippocampus. *Nature* 361 (1993) 31 - 39.

Buhl, E.H., Halazy, K. and Somogyi, P. Diverse Sources of Hippocampal Unitary Inhibitory Postsynaptic Potentials and the Number of Synaptic Release Sites. *Nature* 368 (1994) 823-828.

Chen, W.R., Sunghoon, L., Kato, K., Spencer, D.D., Shepherd, G.M., and Williamson, A. Long-term modifications of synaptic efficacy in the human inferior and middle temporal cortex. *Proc. Natl. Acad. Sci. (USA)* 93 (1996) 8011-8015.

Chicurel, M.A. and Harris, K.M., The Three-Dimensional Analyusis of the Structure and Composition of CA3 Branched Dendritic Spines and Their Synaptic Relationship with Mossy Fiber Boutons in the Rat Hippocampus. *J. Comp. Neurol.* 325 (1992)169-182.

Claiborne, B.J., Amaral, D.G. and Cowan, W.M. A Light and Electron Microscopic Analysis of the Mossy Fibers of the Rat Dentate Gyrus. *J. Comp. Neurol.* 246 (1986) 435-458.

Claiborne, B.J., Amaral, D.G. and Cowan, W.M. Quantitative, Three-Dimensional Analysis of Granule Cell Dendrites in the Rat Dentate Gyrus. *J. Comp. Neurol.* 302 (1990) 206-219.

Collin, C.H. , Ikeno, J.F., Harrigan, I. and Lederhendler, I.I. Sequential modification of membrane currents with classical conditioning. *Biophys. J.* 54 (1988) 955-960.

Collin, C.H., Devane, W.A., Dahl, D., Lee, C.J., Axelrod, J. and Alkon, D.L. Long-term synaptic transformation of hippocampal CA1 γ-aminobutiric acid synapses and the effect of anamidine. *Proc. Nat. Acad. Sci. (USA)* 92 (1995) 10167-10171.

Coulter, D.A., Lo Turco, J.J., Kubota, M., Disterhoft, J.F., Moore, J.W. and Alkon, D.L. Classical Conditioning Reduces Amplitude and Duration of Calcium-dependent Afterhyperpolarization in Rabbit Hippocampal Pyramidal Cells. *J. Neurophysiol.* 61 (1989) 971-981.

De Valois, R.L., Albrecht, D.G. and Thorell, L.G. Spatial Frequency Selectivity of Cells in Nacaque Visual Cortex. *Vision Research* 22 (1982) 545-559.

Debanne, D., Guerinaeu, N.C., Gahwiler, B.H. and Thompson, S.M., Physiology and Parmacology of Unitary Synaptic Connections between Pairs of Cells in Areas CA3 and CA1 of Rat Hippocampal Slice Cultures. *J. Neurophys.* 73 (1995) 1282-1294.

Dehay, C., Douglas, R.J., Martin, K.A.C. and Nelson, C. Excitation by geniculocortical synapses is not vetoed at the level of dendritic spines in cat visual cortex. *J Physiology* 440 (1991) 723-734.

Denk, W., Sugimori, M. and Llinas, R. Two Types of Calcium Response Limited to Single Spines in Cerebellar Purkinje Cells. *Proc. Nat. Acad. Sci. (USA)* 92 (1995) 8279-8282.

Desimone, R., Albright, T.D., Gross, C.G. and Bruce, C. Stimulus-selective Properties of Inferior Temporal Neurons in the Macaque. *J. Neurosci.* 4 (1984) 2051-2062.

Desimone, R. Face-Selective Cells in the Temporal Cortex of Monkeys. *J. Cogn. Neurosci.* 3 (1991) 1-8.

Dettmar, H.P., Barbours, G.S., Blackwell, K.T., Vogl, T.P., Alkon, D.L., Fry, F.S., Totah, J.E. and

Chambers, T.L., Orange juice classification with a biologically based neural network, In Press *Computers and Chemistry* (June,1996).

Deuchars, J., West, D.C., Thomson, A.M., Relationship between morphology and and physiology of pyramid-pyramid single axon connections in rat neocortex in vitro. *J. of Physiol. (Cambridge)* 478 (1994) 423-435.

Disterhoft, J.F., Coulter, D.A. and Alkon, D.L., Conditioning-specific membrane changes of rabbit hippocampal neurons measured in vitro. *Proc. Nat. Ac. Sci. USA* 83 (1986) 2733-2737.

Douglas, R.J., Martin, K.A.C. and Whitteridge, D. An Intracellular Analysis of the Visual Responses on Neurones in Cat Visual Cortex. *J. Physiol.* 440 (1991) 659-696.

Douglas, R.J., Koch, C., Mahowald, M., Martin, K.A.C. and Suarez, H.H. Recurrent Excitation in Neocortical Circuits. *Science* 269 (1995) 981-985.

Eilers, J., Augustine, G.J. and Konnerth, A. Subthreshold synaptic Ca^{2+} signalling in fine dendrites and spines of cerebellar Purkinje neurons. *Nature* 373 (1995) 155-158.

Friedlander, M.J., Martin, K.A.C. and Wassenhove-McCarthy, D. Effects of Monocular Visual Deprivation on Geniculocortical Innervation of Area 18 in Cat. *J. Neurosci.* 11 (1991) 3268-3288.

Fujita, I., Fujita, T. Intrinsic connections in the Macaque Inferior Temporal Cortex. *J. Compar. Neurol.* 368 (1996) 467-486.

Georgopoulos, A.P., Taira, M. and Lukashin, A.V. Cognitive Neurophysiology of the Motor Cortex. *Science* 260 (1993) 47-52.

Gilbert, C.D. and Weisel, T.N., Columnar specificity of intrinsic horizontal and corticocortical connections in cat visual cortex, *J. Neurosci.* 9 (1989) 2432-2442.

Gilbert, C.D., Das, A., Ito, M., Kapadia, M., Westheimer, G. Spatial integration and cortical dynamics. *Proc. Natl. Acan. Sci. USA* 93: 615-622 (1996a).

Gilbert, C.D. Plasticity in visual perception and physiology. *Current Opinions in Neurobiology* 6:269-274 (1996b).

Ghose, G.M., T'so, D.Y. Form processing modules in primate area V4. In press, *J. Neurophysiol.* (1997).

Harris, K.M., How Multiple-Synapse Boutons Could Preserve Input Specificity During an Interneuronal Spread of LTP. *TINS* 18 (1995) 365-369.

Hopfield, J.J. Pattern Recognition Computation Using Action Potential Timing for Stimulus Representation. *Nature* 376 (1995) 33-36.

Hubel, D.H. and Wiesel, T.N. Receptive Fields of Single Neurones in the Cat's Striate Cortex. *J. Physiol. (London)* 148 (1959) 574-591.

Hubel, D.H. and Wiesel, T.N. Integrative Action in the Cat's Lateral Geniculate Body. *J. Physiol. (London)* 155 (1961) 385-398.

Hubel, D.H. and Wiesel, T.N. Receptive Fields, Binocular Interaction and Functional Architecture in the Cat's Visual Cortex. *J. Physiol. (London)* 160 (1962) 106-154.

Ito, M., Tamura, H., Fujita, I. and Tanaka, K. Size and position invariance of neuronal responses in monkey inferotemporal cortex. *J. Neurophsiol.* 73 (1995) 218- 226.

Kaneko, T., Caria, M.A., Asanuma, H., Information processing within the motor cortex: II. Intracortical connections between neurons receiving somatosensory cortical input and motor output neruons in the cortex. *J. Compar. Neurol.* 345 (1994) 172-184.

Kossel, A., Lowel, S. and Bolz, J. Relationships between dendritic fields and functional architecture in striate cortex of normal and visually deprived cats. *J. Neurosci.* 15 (1995) 3913 - 3926.

Kovacs, G., Vogels, R. and Orban, G.A. Selectivity of macaque inferior temporal neurons for partially occluded shapes. *J. Neurosci.* 15 (1995) 1984 - 1997.

Larkman, A.U., Dendritic Morphology of Pyramidal Neurones of the Visual Cortex of the Rat: III. Spine Distribution. *J. Comp. Neurol.* 306 (1991) 332-343.

Linsker, R. From basic network principles to neural architecture: Emergence of orientation columns. *Proc. Nat. Acad. Sci. (USA)* 83 (1986) 8779-8783.

Liu, G. and Tsien, R.W., Properties of Synaptic Transmission at Single Hippocampal Synaptic Boutons. *Nature* 375 (1995) 404 -408.

LoTurco, J.J., Coulter, D.A. and Alkon, D.L. Enhancement of Synaptic Potentials in Rabbit CA1 Pyramidal Neurons Following Classical Conditioning. *Proc. Nat. Acad. Sci. (USA)* 85 (1988) 1672-1676.

Lukashin, A.V. and Georgopoulos, A.P. Directional Operations in the Motor Cortex Modeled by a Neural Network of Spiking Neurons. *Biological Cybernetics* 71 (1994) 79-85.

Maccaferri, G. and McBain, C.J. Passive Propagation of LTD to Stratum Oriens-Alveus Inhibitory Interneurons Modulates the Temporoammonic Input to the Hippocampal CA1 Region. *Neuron* 15 (1995) 137-145.

Magee, J.C. and Johnston, D. Synaptic Activation of Voltage-Gated Channels in the Dendrites of Hippocampal Pyramidal Neurons. *Science* 268 (1995) 301-304.

Mainen, Z.F. and Sejnowski, T.J. Reliability of Spike Timing in Neocortical Neurons. *Science* 268 (1995) 1503-1506.

Malinow, R., LTP: Desperately Seeking Resolution. *Science* 266 (1994) 1195-1196.

McIntosh, A.R., Grady, C.L., Ungerleider, L.G., Haxby, J.V., Rapoport, S.I. and Horwitz, B. Network Analysis of Cortical Visual Pathways Mapped with PET. *J. Neurosci.* 14 (1994) 655-666.

McPhie, D.L., Matzel, D.L., Olds, J.L. and Alkon, D.L. Learning specific changes in the distribution of protein kinase C within the type B photoreceptor of *Hermissenda crassicornis*. *J. Neurochem.*, 60 (1993) 646-651.

Mel, B.W. NMDA-Based pattern discrimination in a modeled cortical neuron. *Neural Computation* 4 (1992) 502-517.

Mel, B.W. Information processing in dendritic trees. *Neural Computation* 6 (1994) 1031 - 1085.

Miller, R.F. and Dacheux, R.F. Synaptic Organization and Ionic Basis of On and Off Channels in Mudpuppy Retina. I. Intracellular Analysis of Chloride-Sensitive Electrogenic Properties of Receptors, Horizontal Cells, Bipolar Cells, and Amacrine Cells. *J. General Physiol.* 67 (1976) 639-659.

Miller, R.F. and Dacheux, R.F. Synaptic Organization and Ionic Basis of On and Off Channels in Mudpuppy Retina. II. Chloride-Dependent Ganglion Cell Mechanisms. *J. General Physiol.* 67 (1976) 661-678.

Miller, R.F. and Dacheux, R.F. Synaptic Organization and Ionic Basis of On and Off Channels in Mudpuppy Retina. III. A Model of Ganglion Cell Receptive Field Organization Based on Chloride-Free Experiments. *J. General Physiol.* 67 (1976) 679-690.

Miller, K.D. A Model for the Development of Simple Cell Receptive Fields and the Ordered Arrangement of Orientation Columns through Activity-Dependent Competition between ON- and OFF-Center Inputs. *J. Neurosci.* 14 (1994) 409-441.

Mishkin, M. and Ungerleiter, L.G. Contribution of Striate Inputs to the Visiospatial Functions of Pieto-preoccipital Cortex in Monkeys. *Behav. Brain. Res.* 6 (1982) 57-77.

Movshon, J.A., Thompson, I.D. and Tolhurst, D.J. Spatial and Temporal Contrast Sensitivity of Neurons in Areas 17 and 18 of Cat's Visual Cortex. *J. Physiol.* 283 (1978) 101-120.

Nelson, T.J., Collin, C. and Alkon, D.L. Isolation of a G protein that is modified by learning and reduces potassium currents in *Hermissenda*. *Science* 247 (1990) 1479-1483.

Nicolelis, M.A., Baccala, L.A., Lin, R.C.S. and Chapin, J.K. Sensorimotor encoding by synchronous ensemble activity at multiple levels of the somatosensory system. *Science* 268 (1995) 1353 - 1358.

Obermayer, K. and Blasdel, G.G. Geometry of Orientation and Ocular Dominance Columns in Monkey Striate Cortex. *J. Neurosci.* 13 (1993) 4114-4129.

Olds, J.L., Anderson, M.L., McPhie, D.L., Staten, L.D. and Alkon, D.L., Imaging of memory-specific changes in the distribution of protein kinase C in the hippocampus, *Science*, 245 (1989) 866-869.

Olds, J.L., Golski, S., McPhie, D.L., Olton, D., Mishkin, M. and Alkon, D.L., Learning-specific changes in rat hippocampal protein kinase C distribution for two water maze discrimination tasks, *J. Neurosci.*, 10 (1990) 3707 -3713.

Olds, J.L., Bhalla, U.S., McPhie, D.L., Bower, J. and Alkon, D.L. Lateralization of Membrane-Associated PKC Distribution in Rat Olfactory Cortex: Specific to learning in the Olfactory Modality. *Behavioral Brain Research* 61 (1994) 37 - 46.

Optican, L.M. and Rishmond, B.J. Temporal Coding of Two-dimensional Patterns by Single Units in Primary Visual Cortex: III Information Analysis. *J. Neurophysiol.* 57 (1987) 162-178.

Purves, D., Paydarfar, J.A., Andrews, T.J. The wagon whell illusion in movies and reality. *Proc. Nat. Acad. Sci. USA* 93: 3693-3697 (1996).

Pribram, K.H. *Brain and Perception: Holonomy and Structure in Figural Processing.* Lawrence Erlbaum Assoc., Hillsdale, NJ (1991)

Quartz, S.R. and Sejnowski, T.J. The Neural Basis of Congnitive Development: A Constructivist Manifesto. In press, *Behavioral & Brain Sciences* (target article).

Richmond, B.J. and Optical, L.M. Temporal Encoding of Two-dimensional Patterns by Single Units in Primary Visual Cortex II. Information Transmission. *J. Neurophysiol.* 64 (1990) 370 - 380.

Rolls, E.T., Tovee, M.J., Purcell, D.G., Stewart, A.L. and Azzopardi, P. The Responses of Neurons in the Temporal Cortex of Primates, and Face Indentification and Detection. *Exp. Brain Res.* 101 (1994) 473-484.

Rose, D. and Dobson, V.G., Eds. Models of the Visual Cortex. John Wiley and Sons, Inc, Chichester, UK (1985).

Saffran, J., Aslin, R.N., Newport, E.L. Statistical learning by 8-month-old infants. *Science* 274: 1926-1928 (13 December 1996).

Sary, G., Vogels, R., Kovacs, G. and Orban, G.A. Response of Monkey Inferior Temporal Neurons to Luminance-, Motion-, and Texture-Defined Gratings. *J. Neurophysiol.* 73 (1995) 1341 - 1354.

Schreurs, B.G. and Alkon, D.L. US-US conditioning of the Rabbit's Nictitating Membrane

Response: Emergence of a conditioned Response without Alpha Conditioning. *Psychobiology* 18: 312-320 (1990)

Schreurs, B.G., Sanches-Andres, J.V. and Alkon, D.L., Learning-Specific Differences in Pukinje-Cell Dendrites of Lobule HVI (Lobus Simplex): Intracellular Recording in a Rabbit Cerebellar Slice. *Brain Research* 564 (1991) 18-22.

Schreurs, B.G. and Alkon, D.L., Rabbit Cerebellar Slice Analysis of Long-Term Depression and Its Role in Classical Conditioning. *Brain Research 631 (1993) 235-240.*

Schreurs, B.G., Oh, M.M., Hirashima, C. and Alkon, D.L. Conditioning-Specific Modification of the Rabbitt's Unconditioned Nictitating Membrane Response. *Behavioral Neuroscience* 109: 24-33 (1995)

Schreurs, B.G. (1996) Personal Communication. (Note to Editor: A manuscript is in preparation and we expect to provide a journal reference in proof.)

Sik, A., Ylinen, A., Penttonen, M. and Buzaki, G. Inhibitory CA1-CA3-Hilar Region Feedback in the Hippocampus. *Science* 265 (1994) 1722-1724.

Singer, W., Development and Plasticity of Cortical Processing Architectures. *Science* 270 (1995) 758 - 764.

Softky, W.R. and Koch, C. The Highly Irregular Firing of Cortical Cells is Inconsistent with Temporal Integration of Random ESPSs. *J. Neurosci.* 13 (1993) 334-350.

Softky, W. Sub-millisecond Coincidence Detection in Active Dendritic Trees. *Neuroscience* 58 (1994) 13 - 41.

Soltesz, I., Smetters, D.K. and Mody, I. Tonic Inhibition Originates from Synapses Close to the Soma. *Neuron* 14 (1995) 1273-1283.

Spruston, N., Schiller, Y., Stuart, G. and Sakmann, B. Activity-Dependent Action Potential Invasion and Calcium Influx into Hippocampal CA1 Dendrites. *Science* 268 (1995) 297-300.

Tanaka, K. Neuronal Mechanisms of Object Recognition. *Science* 262 (1993) 685-688.

Tanaka, K. Representation of visual features of objects in the inferotemporal cortex. *Neural Networks* 9: 1459 - 1475 (1996).

Tanenhaus, M.K., Spivey-Knowlton, M.J., Eberhard, K.M. and Sedivy, J.C. *Science* 268 (1995) 1632 - 1634.

Theunissen, F., Roddey, J.C., Stufflebeam, S., Clague, H., Miller, J.P. Information theoretic analysis of dynamical encoding by four identified primary sensory interneurons in the cricket cercal system.

J. Neurophysiol. 75: 1345-1364 (1996). See also J. E. Levine and J. P. Miller, Broadband encoding in the cricket cercal sensory system enhanced by stochastic resonance. *Nature* 380: 165-168 (14 March 1996).

Trommald, M., Jensen, V, and Andersen, A. Analysis of Dendritic Spines in Rat CA1 Pyramidal Cells Intracellularly Filled with a Fluorescent Dye. *J. Comp. Neurol.* 353 (1995) 260-274.

Van der Zee, E.A., Compaan, J.C., de Boer, M. and Luiten, P.G.M. Changes in PKCγ immunoreactivty in mouse hippocampus induced by spatial discrimination learning. *J. Neuroscience* 12 (1992) 4808 - 4815.

Van Essen, D.C., Anderson, C.H. and Felleman, D.J. Information Processing in the Primate Visual Syste: An Integrated Systems Perspective. *Science* 255 (1992) 419-423.

Vogl, T.P., Blackwell, K.T., Hyman, S.D., Barbour, G.S. and Alkon, D.L., Classification of hand-written digits and Japanese kanji, *Proc. IJCNN'91*, 1 (1991) 97-102.

Vogl, T.P., Blackwell, K.T., Barbour, G.S. and Alkon, D.L., Dynamically Stable Associative Learning (DYSTAL): A neurobiologically based ANN and its applications. *SPIE Proceedings* 1710 (1992) 165-176.

Von der Heydt, R. and Peterhans, E. Mechanisms of Contour Perception in Monkey Visual Cortex. I. Lines of Pattern Discontinuity. *J. Neurosci.* 9 (1989) 1731-1748.

Werness, S.A., Fay, D.S., Blackwell, K.T., Vogl, T.P. and Alkon, D.L., Associative learning in a network model of *Hermissenda crassicornis* I. Theory, *Biol. Cybern.*, 68 (1992) 125 - 133.

Werness, S.A., Fay, D.S., Blackwell, K.T., Vogl, T.P. and Alkon, D.L., Associative learning in a network model of *Hermissenda crassicornis* II. Experiments, *Biol. Cybern.*, 69 (1993) 19 - 28.

Zador, A.M., Claiborne, B.J. and Brown, T.H., Nonlinear Pattern Separation in Single Hippocampal Neurons with Active Dendritic Membranes. In *Advances in Neural Information Processing Systems 4*, pp 51-58. J. E. Moody, S.J. Hanson and R.P. Lippman, Eds. Morgan Kaufmann, Publ., San Mateo, CA (1992)

Zeki, S. Color Coding in the Cerebral Cortex: The Reaction of Cells in Monkey Visual Cortex to Wavelengths and Colours. *Neuroscience* 9 (1983) 741-765.

Figure 1. A pyramidal cell and its afferents at various scales illustrating DSC formation.

1a. A tracing of a pyramidal cell reconstructed from multiple sections obtained from a laser-scanning confocal microscope. (Image from M. Trommald (1995), used with permission).

1b. A schematic representation of a small number of the branching dendrites shown in figure 1a and their afferents. The presynaptic afferents are represented by the six horizontal lines. The four afferents represented by heavy lines (1 - 4) are FC inputs representing the four color inputs in figure 2. The two lighter lines are “texture” inputs that are independent of “color” inputs. Regions a - f represent areas of dendritic membrane where DSCs might form. In regions a and f, DSCs will not form because the constrained inputs are not fully represented, resulting in incoherent signals that will not lead to DSC formation. Area c will not form a DSC under the assumption that there are many more alternative “texture” inputs. Area d is a proto-DSC whose afferents include the covariant “color” afferents together with a subset of “texture” afferents. Because of the presence of the two texture afferents that are unconstrained with respect to the “color” afferents, this proto-DSC will not form a DSC or will form only a partially formed DSC. However, areas b and e, whose afferents are all of the constrained afferents within one FC and no others, will form a fully formed active DSC representative of particular patterns within that FC. If the afferents shown are consistent across the dendritic tree of this neuron, the neuron will become specialized to the FC of the micro-feature subset described by afferents 1 - 4.

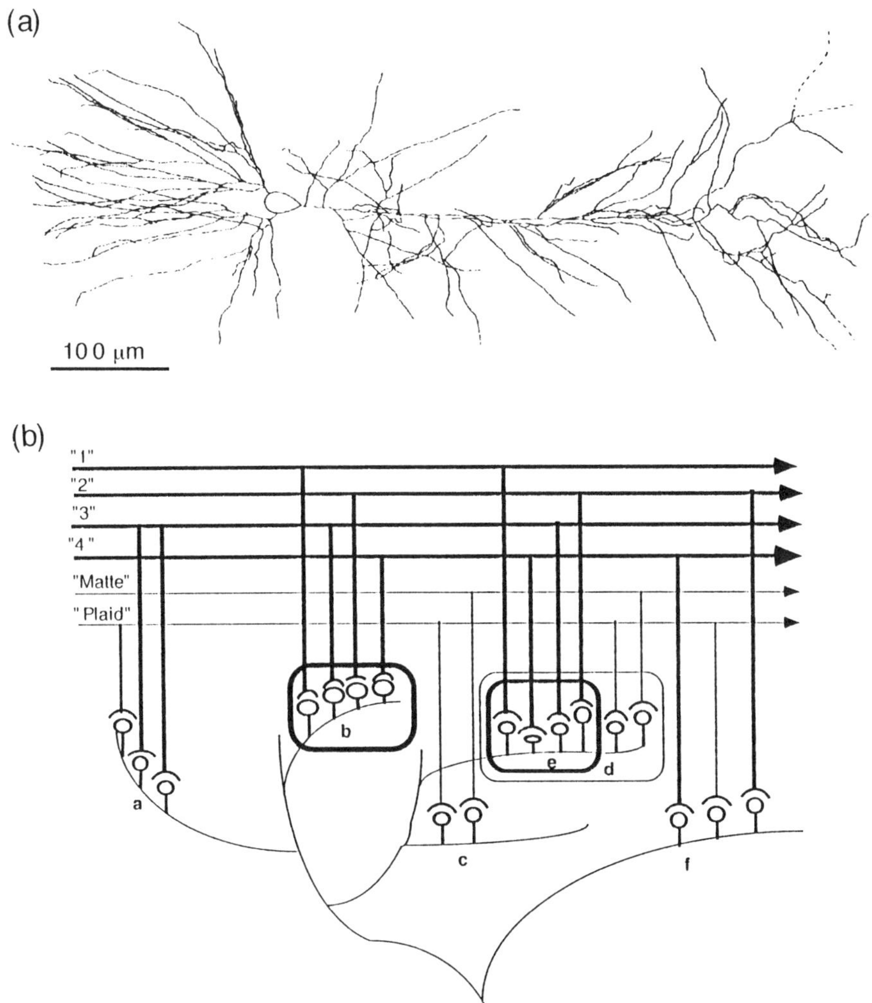

1c. An image of a dendritic segment reconstructed from 57 sections, 0.1 microns apart, projected onto a single plane. Stars and arrow indicate spines. (Image from M. Trommald (1995), used with permission).
1d. A schematic drawing of a DSC derived from 1c.

(c)

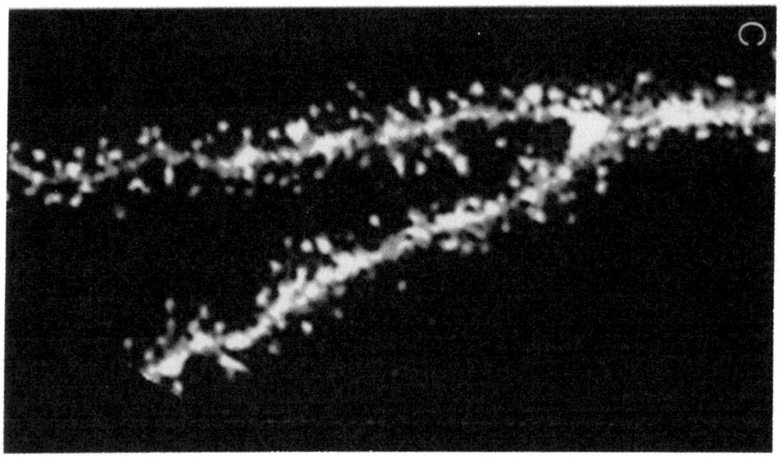

(d)

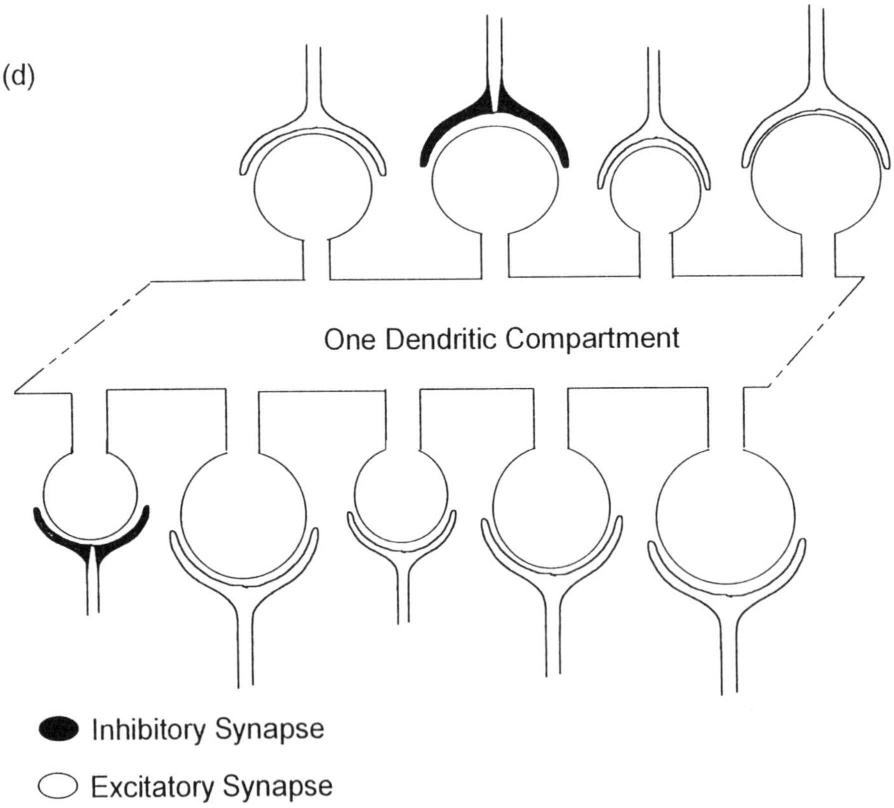

Figure 2. A representation of the signals on the presynaptic afferents. The size of the circles or lines is indicative of the strength of the signals. Columns represent the different axonal outputs that provide signals for each micro-feature. R, B, Y, G, P, and T represent Red, Blue, Yellow, Green, Purple and Turquise, respectively. Rows are the signals from the constellation of micro-features that are consequent to the presentation of an individual image (an exemplar). Noise, although present, is not depicted. Rows are time sequential. Note that all the inputs in columns 1 - 4 cannot be simultaneously high or low, thus forming an FC subset. In general, no constraints exist among any of columns 1 - 4 and any of the other columns. However, columns 5 - 8 and 9 - 12 form independent FC subsets. The output of the first layer is input to the second layer, shown in figure 2b. Note that neuron ensemble 1.1 has taken the color specific afferents and organized them into color-specific DSCs; the DSC-specific outputs (individual colors) become the inputs to the next layer, which also forms FC subsets, albeit different ones, based upon its input stream. The output of layer 2 is shown in figure 2c. Note that inputs 1 and 6 do not produce colored line outputs because the associations between that color and a line in that direction are insufficiently repeated to cause a DSC to form on neuron ensemble 2.1.

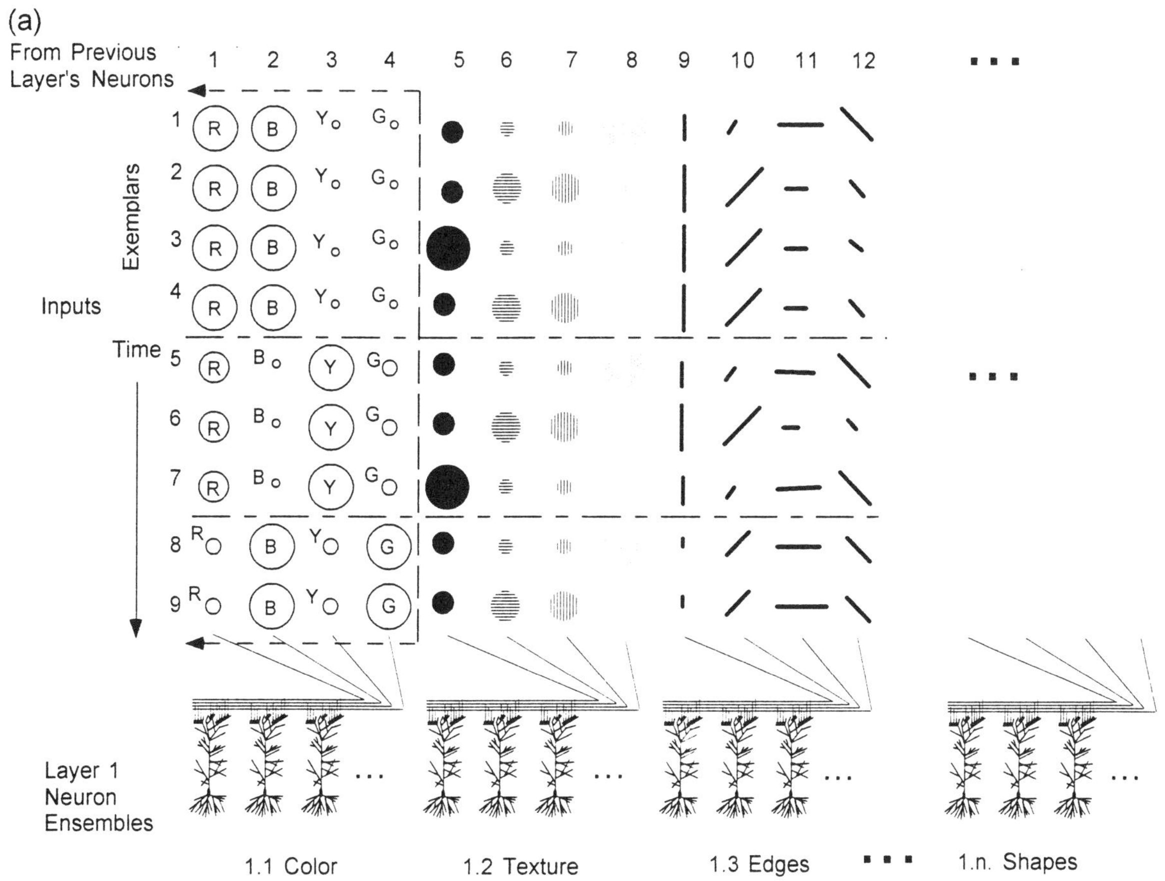

(b)

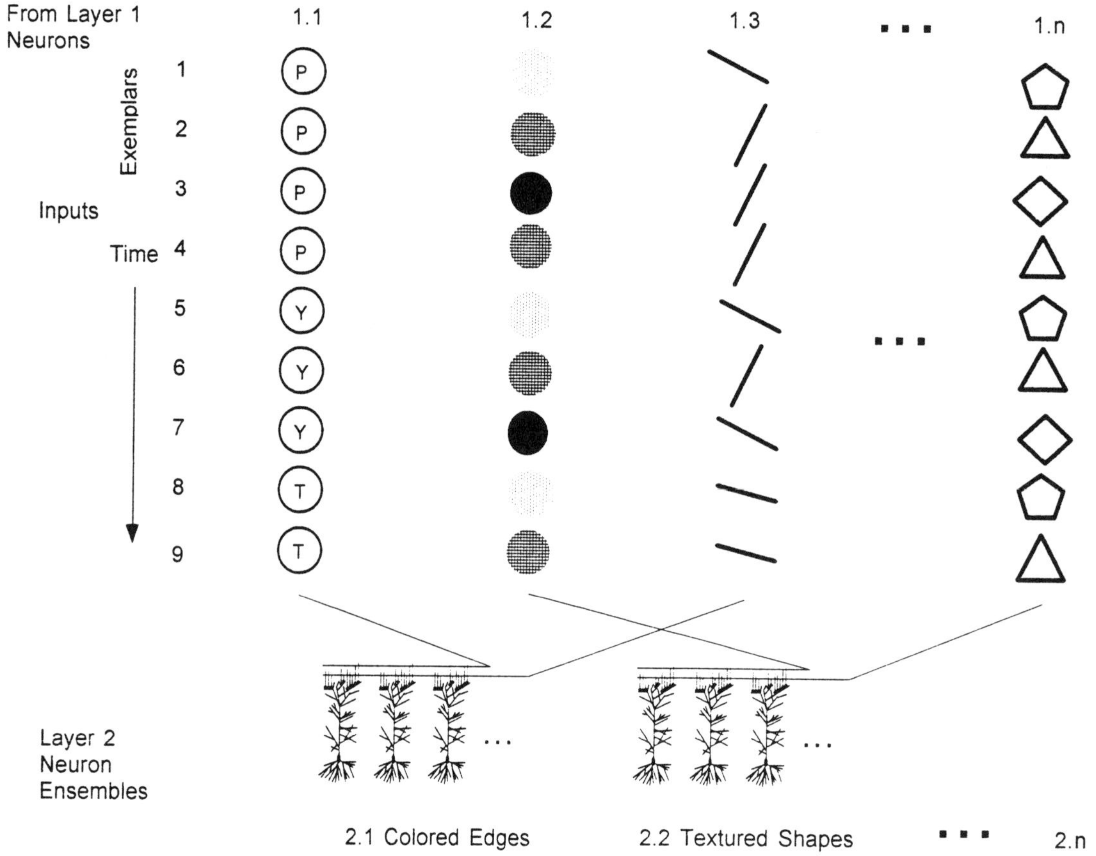
From Layer 1
Neurons
1.1
1.2
1.3
1.n
Exemplars
Inputs
Time
1
2
3
4
5
6
7
8
9
P
P
P
P
Y
Y
Y
T
T
Layer 2
Neuron
Ensembles
2.1 Colored Edges
2.2 Textured Shapes
2.n

(c)

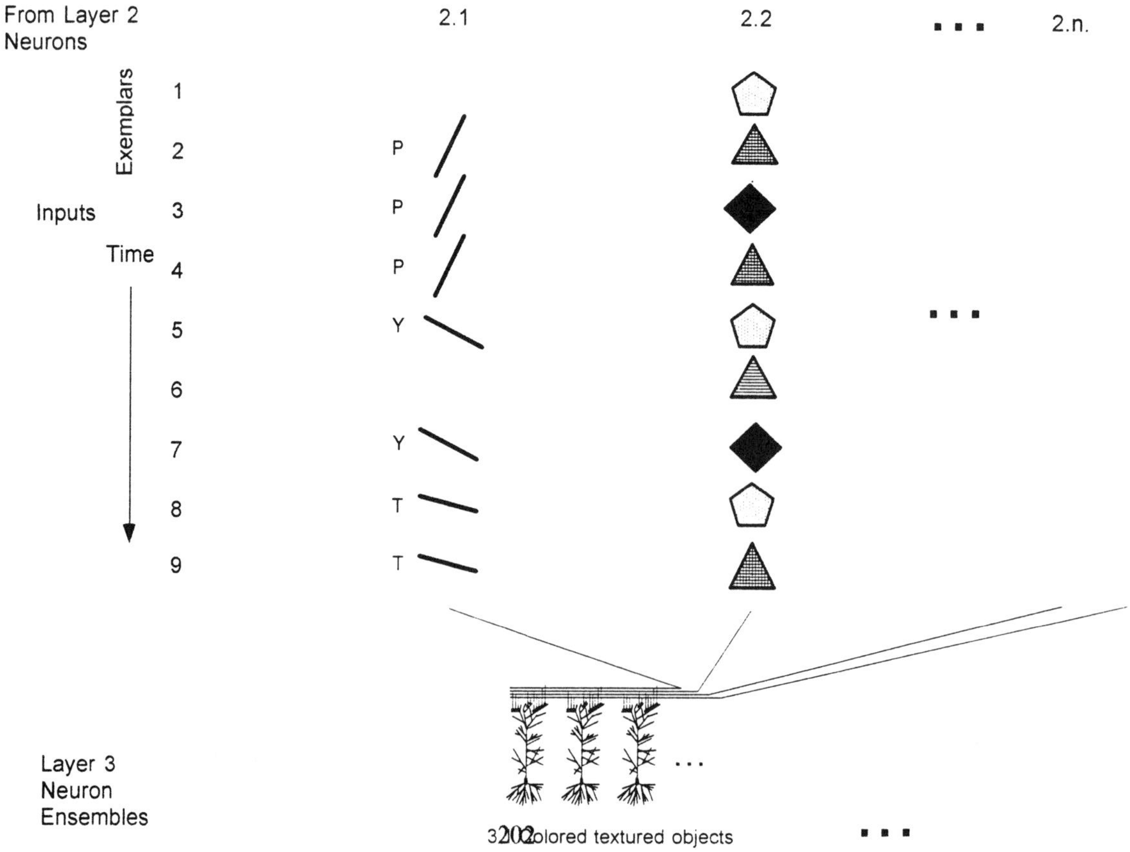
From Layer 2
Neurons
2.1
2.2
2.n.
Exemplars
Inputs
Time
1
2
3
4
5
6
7
8
9
P
P
P
Y
Y
T
T
Layer 3
Neuron
Ensembles
colored textured objects

Figure 3. A map of the FC groupings resulting from the co-occurence matrix in Table 1. The threshold for significance is assumed to be 0.7. Only a representative subset of co-occurence measures is shown explicitly. Note that neurons numbered 2, 3, 4, and 6 are members of more than one subset.

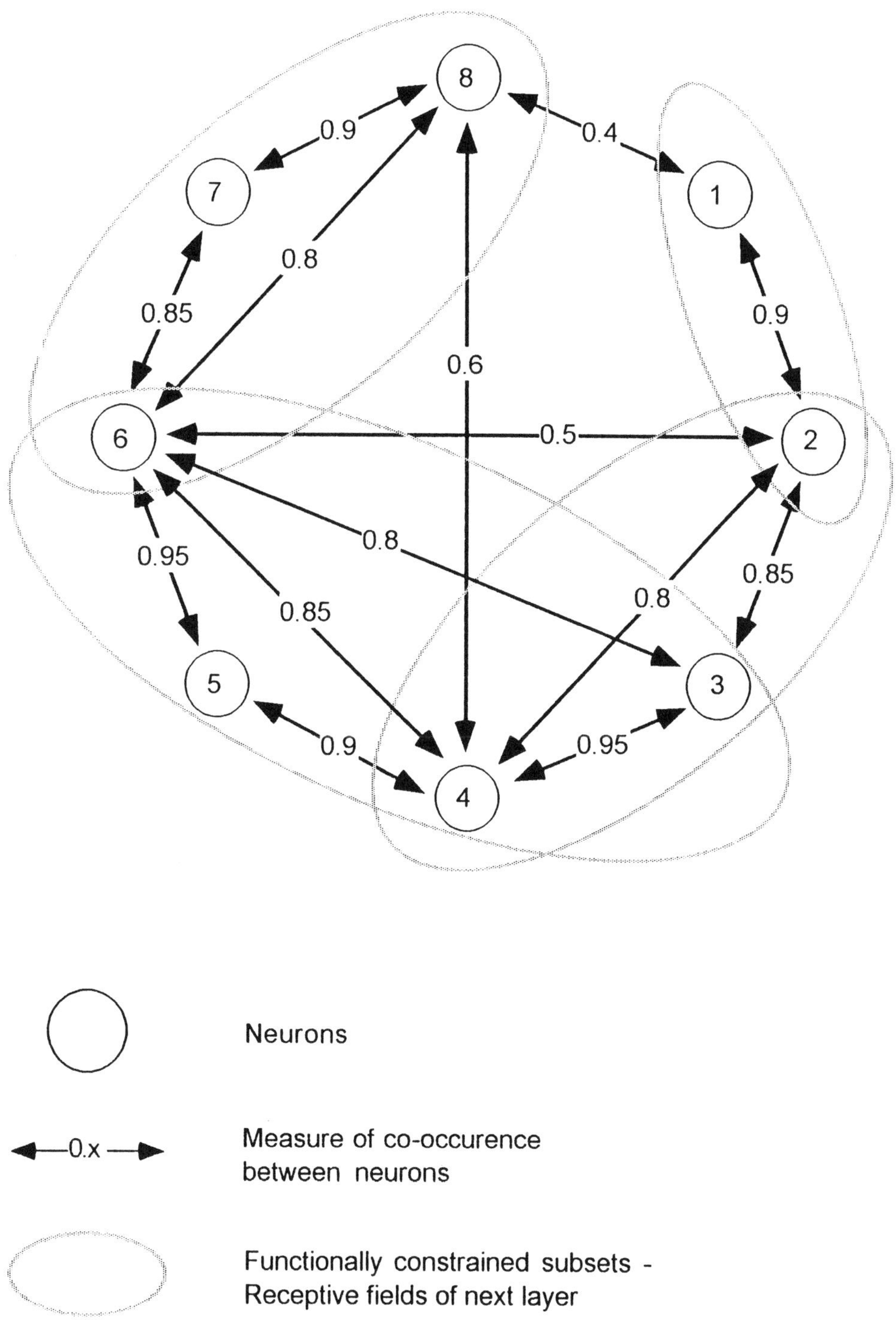

Figure 4. A schematic representation of columns in TE cortex, illustrating that cells with similar but slightly different selectivity cluster into columns. From Tanaka (1993), used with permission.

Figure 5. Shape selectivity of an inferior temporal single unit. Peri-stimulus time histograms of the spikes for each of the shape outlines shown above the histograms. NO and MV refer to the absence or presence, respectively, of partial occlusion. From Kovacs (1995), used with permission.

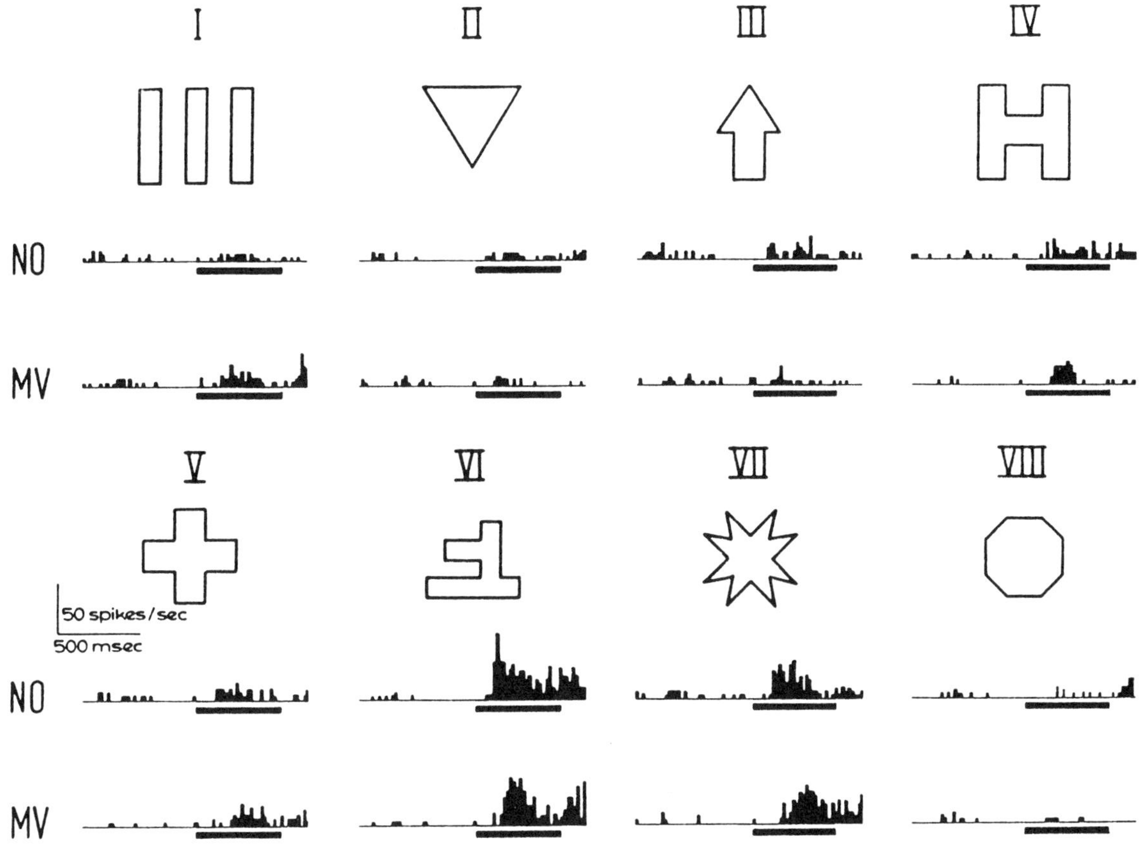

7

The Self-Organizing Map, a Possible Model of Brain Maps

Teuvo Kohonen
Helsinki University of Technology,
Neural Networks Research Centre
Rakentajanaukio 2 C
FIN-02150 Espoo
Finland

Abstract

It is stipulated that all the following three categories of dynamic phenomena must occur in any realistically behaving neural-network model: (i) Activation effects. (ii) Adaptation effects. (iii) Plasticity control effects.

In most neural models one only has activation and adaptation present. The Self-Organizing Map (SOM) algorithm devised by this author (Kohonen, 1982, 1995a) is the only neural-network model that involves all the three phenomena. Its modeling laws include the following partial functions:
1.~Some parallel computing mechanism for the specification of a cell in a piece of cell mass whose parametric representation matches with or which responds to the afferent input best. This cell is called the ``winner." 2.~Control of some learning factor at the cells in the neighborhood of the ``winner" so that only this neighborhood will be adapted to the present input. By virtue of the ``neighborhood learning," the SOM will form spatially ordered maps of sensory experiences, very much resembling the brain maps.

The newest version of the SOM is the ASSOM (Adaptive-Subspace SOM). The adaptive processing units of it are able to represent signal subspaces, not just templates for original patterns. A signal subspace is an invariance group: therefore the processing units of the ASSOM are able to respond invariantly, e.g., to moving and transforming patterns, in a somewhat similar way as the complex cells of the cortex do.

Introduction

One of the the most salient features in brain physiology is the spatial order of the neural responses, reflected in various kinds of ``maps." In many cases the topological order of a receptive surface is preserved in its projection on a brain area, and an explanation could simply be the ordered growth of axons. However, one may also find examples, such as the color map in area $V4$ of the primary visual cortex (Zeki, 1980), or the sonar-feature map in the auditory cortex of the mustache bat (Suga & O'Neill, 1979), whereby no ordered receptive surfaces for the corresponding sensory qualities exist. One can find many kinds of similar spatially ordered maps of sensory experiences at different levels of abstraction in the nervous systems.

As most neural functions and their general organization are determined prenatally, whereby most sensory stimuli are not yet present, the rough order of the projections must be determined genetically. However, this order will be refined and eventually changed by some prenatal and all the postnatal experiences. Therefore there must also exist some kind of self-organizing function in physiology, closely associated with the learning mechanisms, by virtue of which the order can be forced to and maintained in the neural maps.

As the synaptic connections and their plasticities possess a vast capacity for forming various neural structures and functions, the prevailing view in the neural-network theory is that all learning results are supposed to be explainable by models, the main action of which is mutual modifiable transsynaptic control of the neurons. This tradition in theoretical thinking overlooks the many diffuse control actions, chemical and electrotonic, through which the neurons are known to control each other. In this work a more general view is held, and some diffuse, eventually chemical control factors are introduced to the model. Although the variable elements are still identified with synapses in the analysis, in principle even this assumption is not necessary: similar adaptive and self-organizing processes may occur on the molecular level, too. However, we shall only exemplify these general processes by neural circuits augmented by simple chemical effects, because the latter are easiest to concretize.

The central assertion made in this paper is that for effective self-organization one must have *three* different kinds of control action between the neurons (or other adaptable units): 1. Excitatory and inhibitory *control of the activity* of a neuron by the activities of other neurons. 2. *Adaptation* of the coupling strengths as some function of the activities at both sides of the coupling. This is a *parametric* control. 3. A mechanism for the *control of plasticity*, in other words, the proportionality factors that determine the rates of change of the coupling strengths. This control may not take part in the direct activation of a neuron by another one; it is a *parametric control of the parameters.*

There exists only one neural-network model, viz. the Self-Organizing Map (SOM) (Kohonen, 1995a) in which all three of the above types of control have been taken into account, and by which the spatial order of neural responses can be achieved. Plenty of models based on the SOM system equations have already been presented, and the following kinds of simulation results have been obtained: 1. Emergence of many kinds of ordered sensory maps. 2. Emergence of invariant feature analyzers and decoder functions for signals at various levels of abstraction. 3. Learning of nontrivial robot actions and control functions for complex nonlinear processes. 4. Development of effective nonlinear statistical methods for "data mining."

This paper describes the basic SOM process and tries to motivate the role of some neurophysiological components and phenomena necessary for this kind of self-organization.

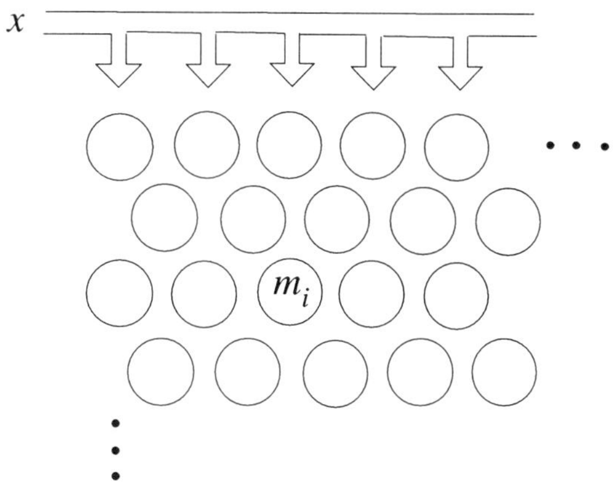

Figure 1: The SOM array.

One of the most central objectives in neural modeling indeed ought to be to devise how the organizational order, manifesting itself in the various brain maps, emerges. It is surprising that most neural models do not pay the least attention to it. Only the Self-Organizing Map (SOM) has effectively demonstrated unsupervised emergence and self-ordering of many brain functions.

The SOM principle, however, has even more far-reaching consequences. We shall show that if its processing units are defined in a particular way, they will develop into analyzers of signal patterns and sequences such that each analyzer will become tuned to a particular *class of transformations* of input representation. The architecture called the *ASSOM (Adaptive-Subspace SOM)* is a new concept in neural-network theory, and is believed to describe the formation of various *invariant-feature maps* in the brain.

2 Preliminary description of the SOM process

There already exist many different versions of the basic SOM, but all of them share the following two operations:

1. Spatial clustering of activity into a center of high activity, the location of which is a function of the input pattern. This location is called the *winner.*
2. Adaptive modification of the input connections in the local area around the winner such that all neural functions in this area become tuned better to the prevailing input activity and will respond stronger to it.

It may be helpful to discuss the SOM in the context of a very simple flat array of neurons (Fig. 1), to every neuron of which the same set of afferent lines, carrying the signal vector $x \in \Re^n$, is connected. If this array were referred to any brain area, it might only have a diameter of a few millimeters. The neural systems are then thought to be composed of many arrays of this kind, each forming its own "map."

The input (synaptic) weights are denoted by the scalar numbers μ_{ij}, where i identifies the neuron and j its synapse. The "synaptic weight vector" of neuron i in matrix notation reads $m_i = [\mu_{i1}, \ldots, \mu_{in}]^{\mathrm{T}}$. Without essentially losing any generality in this discussion, we assume that every neuron has exactly n synapses. If the neurons had different numbers of synapses, the mathematical analysis would become more tedious, while the results would be essentially the same.

2.1 Winner-take-all

The first operation we need for effective self-organization is the *winner-take-all (WTA)* function. It is indispensable for all so-called competitive-learning models to which the SOM also belongs. The most plausible neural implementation of WTA is lateral feedback interconnectivity, mono- or polysynaptic, between the neurons in the array such that at longer distances the neurons inhibit each other, whereas at short distances they excite each other. In Sec. 3.1 we shall discuss a relatively simple but still effective WTA model. For a discussion of more realistic WTA circuits of this type, cf. (Kaski & Kohonen, 1994).

The essential operation of the WTA is that if input x is held steady for a while, the activity of one neuron called the "winner" is enhanced, whereas the activities of the rest of the neurons will be suppressed to zero. This, of course, will only happen within a relatively small neural assembly describable by the simple model where all the neurons of the assembly are assumed to cooperate mutually. The winner in this subset of neurons will be the neuron that was activated strongest by x. In the simplest physiologically motivated model assumption it means that the dot product $m_i^{\mathrm{T}} x$ is highest for some neuron indexed by $i = c$, or

$$m_c^{\mathrm{T}} x = \max_i \{m_i^{\mathrm{T}} x\} \; ; \tag{1}$$

however, much more complex input functions than the dot product can also be used, as we will see in Sect. 2.4. It will be useful to remember that index c always denotes the "winner" in the following discussions.

2.2 Adaptation of input connectivities

Most contemporary neural-network models assume *Hebb-type synaptic modifications* (Hebb, 1949) that depend on input and output activities of the *same* neuron. A much more powerful learning principle, however, ensues from the assumption that *nearby neurons modify each other's synapses.* We shall first ignore the details of plasticity control; let the combination of all possible effects be described in a general way by the so-called *neighborhood function* h_{ci}, which depends on the lateral distance of neurons c (winner) and i, and also time. Let us recall that the *plasticity control is not supposed to affect the activities of the neurons directly.*

Users of the simple Hebb law also frequently overlook the fact that the synaptic changes defined by it are one-directional, whereby the memory traces would saturate sooner or later. The changes must be made reversible, and the simplest modification is to introduce a *conditional (active) forgetting effect* such that the synaptic efficacies tend to be equalized or partially *redistributed* within a neuron; this *active* redistribution effect is supposed to depend on the amount all the synapses activate this neuron. The *activation* of neuron i by its inputs may be approximated by the entity $a_i = \sum_j \mu_{ij}\xi_j$. Active learning also means that both the formation of new memory traces due to input ξ_j, where ξ_j is the afferent activity at synapse j, or the jth component of input vector x, and active fading of old memory traces must be proportional to the neighborhood function h_{ci}. Moreover it will be natural to assume that the amount of change of synaptic strength is directly proportional to the synaptic strength μ_{ij} itself.

When we take the active forgetting and neighborhood effects into account in a general way, we obtain the following type of adaptation equation for the synapse j of neuron i:

$$d\mu_{ij}/dt = h_{ci}(\xi_j - a_i\mu_{ij}) \; . \tag{2}$$

Here the effect of the output activity of Hebb's law has been replaced by a function of the weighted output activities of nearby neurons, represented by the neighborhood function h_{ci}. The presynaptic effect of ξ_j is modified by the active forgetting term $a_i\mu_{ij}$, where $a_i = \sum_j \mu_{ij}\xi_j$.

Without lengthy discussion it may now be mentioned that (2), together with the WTA function (1), has been proven to define one kind of SOM process (Kohonen, 1995a). This algorithm, written in vector form and as discrete-time steps for numerical simulation reads

$$\text{WTA:} \qquad m_c^{\mathrm{T}}(t)x(t) = \max_i \{m_i^{\mathrm{T}}(t)x(t)\} \; , \tag{3}$$

$$\text{Adaptation:} \qquad m_i(t+1) = m_i(t) + h_{ci}(t)[x(t) - m_i(t)m_i^{\mathrm{T}}(t)x(t)] \; . \tag{4}$$

As a matter of fact, (4) will automatically normalize the synaptic weight vectors to a constant length.

Yet another approximately equivalent form of (2) or (4) is

$$m_i(t+1) = \frac{m_i(t) + h_{ci}(t)\cdot x(t)}{\|m_i(t) + h_{ci}(t)\cdot x(t)\|} \; . \tag{5}$$

It can be shown that (4) is the first-order term in the Taylor expansion of (5) in terms of $h_{ci}(t)$ when the $m_i(t)$ and $x(t)$ are already normalized.

Eqs. (3) and (4), or (3) and (5) define the so-called "dot-product SOM." On the other hand, another almost equivalent mapping, characterizable as the "Euclidean-distance SOM," is

$$\text{WTA:} \qquad \|x(t) - m_c(t)\| = \min_i \{\|x(t) - m_i(t)\|\} \; , \tag{6}$$

$$\text{Adaptation:} \qquad m_i(t+1) = m_i(t) + h_{ci}(t)[x(t) - m_i(t)] \; . \tag{7}$$

The last formulation is very effective for practical applications, but one can hardly find any physiological justification for this particular formulation. However, (3) and (5) on one hand, and (6) and (7) on the other are *mathematically homological*, and thus their solutions behave in the same way. The functional similarity of the different pairs of WTA and adaptation equations may become more clear from Sec. 2.4.

When the process defined by (3) and (5), or (6) and (7) has converged asymptotically, the weight vectors m_i will have acquired values, each of which approximates a different domain of the input signal space *in an orderly fashion*: since the different m_i respond to different input, the SOM array acts like a set of decoders, each neuron being sensitized to a different domain of vectorial input values. The spatial locations of the different neural responses will be ordered along some feature dimensions like in the various sensory maps in the brain. The mapping of the domains onto the different neurons is also determined by the density function of input, whereby more frequently occurring input (receptor) signals occupy a larger area of the SOM array, corresponding to a larger *magnification factor* like in the brain maps. In this sense the SOM can also be regarded as an optimal, nonlinear, dimensionality-reducing *projection* of the high-dimensional input space onto the two-dimensional array of neurons. We shall exemplify a couple of artificially generated maps in Sec. 2.3.

2.3 Two examples of ordered maps formed by the SOM algorithm

The SOM is a "black box" that learns its parameters from the input data, in an unsupervised computing process that resembles the *clustering methods*. In that capacity it can be used to *visualize* sets of complex data such as states of processes and machines. An advantage of the SOM over the other neural-network models is its ability to represent the high-dimensional states as points on an *ordered map*, a kind of "monitor panel." An example of these is the representation of various faults of an anaesthesia system on the SOM display (Fig. 2) (Vapola et al., 1994).

The SOM algorithm is able to organize any information, the samples of which can be compared metrically. This kind of information is abundantly available in medical tests and biophysical measurements. For instance, an application in which we as well as several other laboratories have obtained promising results is the classification of brain states on the basis of EEG curves (Fig. 3) (Joutsiniemi et al., 1995). Also MEG, EMG, and ECG analyses have been implemented by the SOM algorithm (for references, see Kohonen, 1995a).

There exist over 2300 articles written on the SOM (Neural Networks Research Centre, 1995). We must abandon further application examples in this paper that was meant for a physiological modeling approach.

2.4 What is the learning result in the SOM?

The ordering result can also be defined mathematically. We shall approach this problem in several steps. To start with we discuss the simplest competitive-learning model, in which

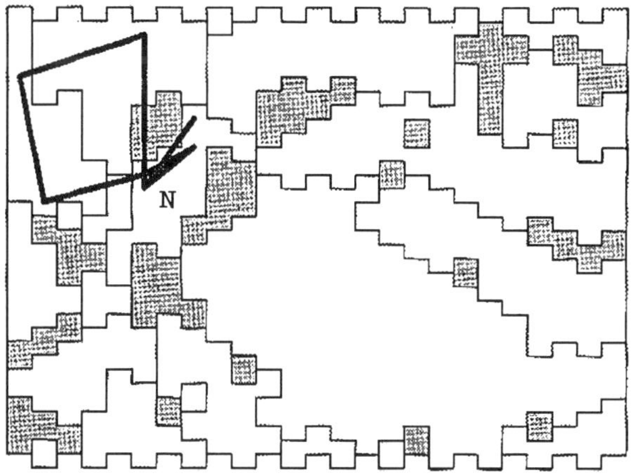

Figure 2: Fault identification of an anaesthesia system tested in a true situation. Fifteen measurements from the system constituted the input vector x; this display is the ordered clustering diagram produced by the SOM where differently faulty states of the system correspond to different zones of the "map." N = normal state. The position of the patient was changed and the intubation tube was obstructed for a short period of time. The trajectory of the response (winner) was moving from the area corresponding to the normal situation to the area that correspond to an obstruction in the specific part of the system.

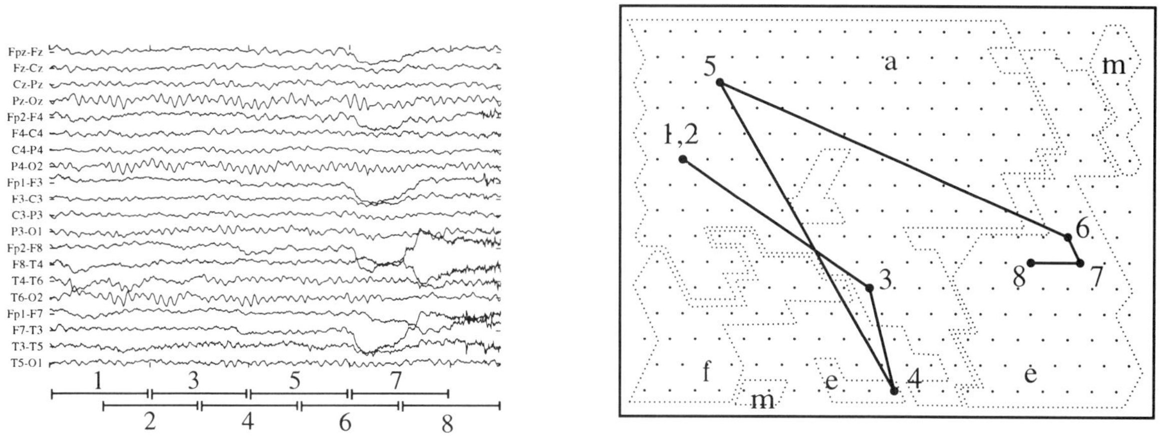

Figure 3: Mapping of different EEG segments into points on the SOM. The input vector to the SOM consisted of 140 components, describing short-time frequency contents of the EEG in multiple bands and multiple locations on the scalp. The letters in the SOM on the right refer to the alpha rhythm (a), alpha attenuation (f), muscular artifacts (m), and eye movements (e), respectively.

the neurons do not yet modify each other; only the winner will be updated at each learning step. Let c denote the "winner" neuron for a particular domain of x values, defined by (6). The so-called (optimal) *vector quantization* of the input signal space is first defined by the minimum of the following objective function that represents the average expected (squared) *quantization error*:

$$E = \int \|x - m_c\|^2 p(x) dx \, . \tag{8}$$

Here $p(x)$ is the probability density function of x, and in the integration over the whole x space, m_c is always the weight vector of that neuron that approximates the value of x best, cf. (6). The norm is henceforth assumed as Euclidean. Notice that index c is a function of x and all the m_i, and changes abruptly at the borders of the so-called Voronoi tessellation in the x space where the "winner" is changed to another. Also dx is a shorthand notation meaning a "volume differential" in the n-dimensional x space.

It can be shown that the following stepwise "delta rule" in the discrete-time formalism ($t = 0, 1, 2, \ldots$) defines the optimal values asymptotically. However, its strict derivation is not quite straightforward (Kohonen, 1991). Let $m_c = m_c(t)$ be the closest codebook vector to $x = x(t)$ in the Euclidean metric. The steepest-descent gradient-step optimization of E in the m_i space yields the sequence

$$\begin{aligned} m_c(t+1) &= m_c(t) + \alpha(t)[x(t) - m_c(t)] \, , \\ m_i(t+1) &= m_i(t) \quad \text{for} \quad i \neq c \, , \\ \text{where} \qquad c &= \arg \min_i \{\|x(t) - m_i(t)\|\} \, , \end{aligned} \tag{9}$$

and $\alpha(t)$ is a suitable, monotonically decreasing sequence of scalar-valued gain coefficients, $0 < \alpha(t) < 1$ that determines the step size in optimization. This is then the simplest analytical description of competitive learning, and it already resembles the SOM algorithm to some extent, especially if we write it as

$$m_i(t+1) = m_i(t) + \alpha(t)\delta_{ci}[x(t) - m_i(t)] \, , \tag{10}$$

where δ_{ci} is the Kronecker delta ($= 1$ for $i = c$, $= 0$ for $i \neq c$).

When the neighborhood effects described by function h_{ci} are then taken into account, we may try to minimize another objective function closely related to (8):

$$E_1 = \int \sum_i h_{ci} \|x - m_i\|^2 p(x) dx \, . \tag{11}$$

Here the squared quantization error $\|x - m_c\|^2$ of (8) has been replaced by the *spatially weighted* squared quantization error $\sum_i h_{ci} \|x - m_i\|^2$, where h_{ci} acts like a *smoothing kernel* in statistical calculations.

However, attempts to minimize E_1 in the general topology of the SOM array by standard optimization methods have not yet been successful. The reasons for it have been discussed in (Kohonen, 1991); let it be mentioned that the essential difficulty is that index c in the integrand is not constant but depends on x and all the m_i through (6). It also makes a sudden jump when x passes the border of the division of the Voronoi tessellation. In the derivation of (9) or (10) this problem could be circumvented by a mathematical trick explained in (Kohonen, 1991); this trick is no longer possible here. As the same problem occurs with other comparison metrics, too, we shall embed the problem into a more general setting. The values of the m_i that minimize E_1 can be shown to ensue as the asymptotic values of a process similar to that defined by (6) and (7).

If the comparison of the input vector with the m_i is made in an arbitrary metric where $d(x, m_i)$ denotes the distance between x and m_i in the signal space, in the general case one has to find those values of the m_i that minimize the objective function

$$E_2 = \int \sum_i h_{ci} d^2(x, m_i) p(x) dx \ , \tag{12}$$

whereby index c is defined by

$$d(x, m_c) = \min_i \{d(x, m_i)\} \ . \tag{13}$$

An approximate minimization solution of (12) can generally be found by the traditional *Robbins-Monro stochastic approximation method* (Robbins & Monro, 1951). The class of these solutions has been studied with mathematical rigor and especially the convergence properties are thoroughly known. In applying this method we first define the *sample function* $E_2^{(s)}(t)$:

$$E_2^{(s)}(t) = \sum_i h_{ci}(t) d(x(t), m_i(t)) \ . \tag{14}$$

An approximate solution to (12) is obtained by the gradient-descent techniques, whereby the gradient is approximated by the gradient of the stochastic sample function (14), as the last term of a sufficiently long sequence defined by

$$m_i(t+1) = m_i(t) - \lambda(t) \mathrm{grad}_{m_i(t)} E_2^{(s)}(t) \ , \tag{15}$$

where $\lambda(t)$ is called the learning-rate factor, and defines the step size. The sequence of the $\lambda(t)$ must always fulfil the following conditions:

$$\sum_{t=0}^{\infty} \lambda(t) = \infty \, , \quad \sum_{t=0}^{\infty} \lambda^2(t) < \infty \ . \tag{16}$$

For instance, if $d(x, m_i) = ||x - m_i||$, and we denote $\alpha(t) = 2\lambda(t)$, we obtain the "Euclidean-distance SOM" algorithm (7).

It can be shown that if x and the m_i are kept *normalized* and we define the squared quantization error as $d^2(x, m_i) = 1 - \cos^2 \theta$, where θ is the angle between vectors x and m_i, then also (4) and (5) can be derived by optimization of (12). We shall not do it here, however, because the problem can be embedded into a still wider class of projection problems to be discussed in Sec. 4.

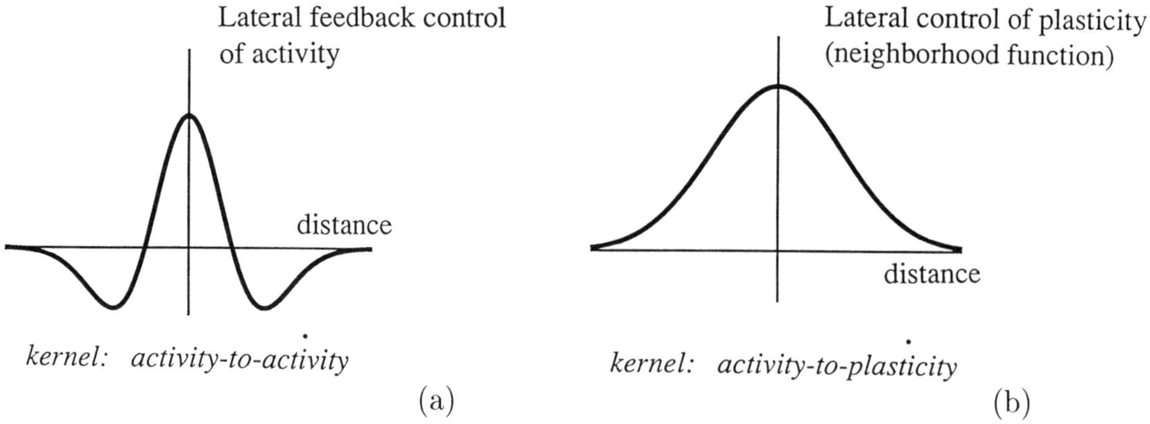

Figure 4: The two types of lateral interaction: (a) activity-control kernel, (b) plasticity-control kernel.

In general, due to the spatially smoothing effect of kernel h_{ci}, the condition for the m_i values to minimize the objective function is that they must be distributed into the signal space in a *topologically ordered* fashion. This is perhaps the simplest way to point out why the SOM tends to become ordered.

3 More detailed physiological modeling of the SOM

It is possible to ask just how "neural" or "physiological" the SOM algorithms are. The same question could be posed about the traditional feedforward and feedback "neural networks", too, the analytical descriptions of which have been developed further in the computational than in the biological direction. Therefore it may be of considerable interest to find that the SOM can be related to a biological neural network in many essential details, as shown below.

A comprehensive discussion of this physiological SOM model can be found in (Kohonen, 1993, 1995a) and (Kaski & Kohonen, 1994); let only its main features be reviewed here.

The most central feature in my attempt to model the self-organizing process in terms of physiological components is *separation of the following two lateral feedback actions* (Fig. 4): 1. The control of *activity* of a neuron by the activity of another laterally situated neuron, whereby this control depends on lateral distance as a "Mexican-hat function" (Fig. 4a); this function is essential to the WTA network discussed in the next section. 2. The control of *plasticity*, or the learning-rate factor of the synapse of a neuron by the activity of another laterally situated neuron, whereby this function may have the form of a Gaussian function Fig. 4b); this is then the "neighborhood function" h_{ci} that occurs in the learning law, but does not take part in the WTA function and does not directly control the activity of a neuron.

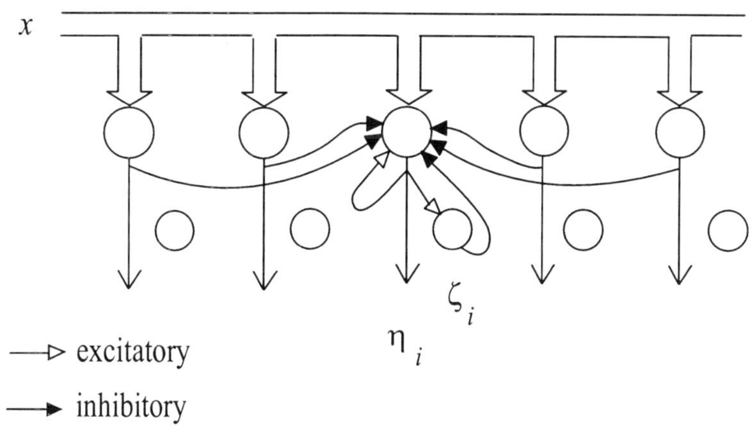

Figure 5: Simplified model of a distributed neural network (cross section of a two-dimensional array). Each location consists of an excitatory principal input neuron and an inhibitory interneuron that feeds back locally. The lateral corrections between the principal input neurons may or may not be made via interneurons.

3.1 The WTA Network

Consider a planar neural network, of which Fig. 5 represents a one-dimensional cross section; each cell receives afferent input from some external sources, and the cells are further interconnected by abundant lateral feedbacks. In the simplest case the same set of afferent input signals is connected to all cells in this piece of network.

Assume that the output activity η_i (spiking frequency) of the input neuron i in the network is described (Kohonen, 1993, 1995a) by

$$d\eta_i/dt = I_i - \gamma(\eta_i) \; , \tag{17}$$

where I_i is the combined effect of all inputs, i.e. afferent inputs as well as lateral feedbacks, on cell i embedded in the layered network. Let $\gamma(\eta_i)$ describe the net effect of all loss or leakage effects that oppose to I_i. This is an abbreviated way of writing: since $\eta_i \geq 0$, (17) only holds when $\eta_i > 0$, or when $\eta_i = 0$ and $I_i - \gamma(\eta_i) \geq 0$, whereas otherwise $d\eta_i/dt = 0$.

If we write $I_i = I_i^a + I_i^f$, where the superscript a means afferent input and f the lateral feedback, respectively, then in the simplest model we have

$$I_i^a = m_i^{\mathrm{T}} x = \sum_j \mu_{ij} \xi_j \; , \tag{18}$$

$$I_i^f = \sum_k g_{ik} \eta_k \; . \tag{19}$$

Here $x = (\xi_1, \xi_2, \ldots, \xi_n)^{\mathrm{T}} \in \Re^n$ again means the input data vector, or the vector of signal activities on a set of axons that is assumed to be connected in parallel to all cells of this network, and $m_i = [\mu_{i1}, \mu_{i2}, \ldots, \mu_{in}]^{\mathrm{T}} \in \Re^n$ is redefined to be the corresponding vector of synaptic strengths of cell i. The $g_{ik} \in \Re$ describe effective lateral coupling strengths of cells in the planar array. For simplicity, but without significant loss of generality, it is assumed that g_{ii} is independent of i, and the g_{ik}, $k \neq i$ are mutually equal.

In the above system of equations, the only and rather general further restrictions shall be $\mu_{ij} > 0$, $\xi_j > 0$ (i.e., the external inputs are mainly excitatory), $g_{ii} > 0$, $g_{ik} < 0$ for $k \neq i$, $|g_{ik}| > |g_{ii}|$, $\gamma > 0$, and $d^2\gamma/d\eta^2 > 0$ (i.e., the loss effects are progressively larger with increasing activity η).

Starting with arbitrary (nonnegative) different m_i and $\eta_i(0) = 0$, the output η_c of that cell for which $m_i^{\mathrm{T}} x$ is maximum ("winner") will converge to an asymptotic high value, whereas the other η_i, $i \neq c$ tend to zero. This convergence property and the uniqueness of the winner can be proved mathematically (Kohonen, 1993, 1995a).

As this network must respond to new inputs continually, the output state $\{\eta_i\}$ must be *reset* before a new input can select a new winner. In this model resetting is done automatically and asynchronously with respect to the input signals by providing each cell with an extra local inhibitory integrating feedback loop, eventually through an interneuron with feedback variable ζ_i (Fig.5) described by the system equation

$$d\zeta_i/dt = b\eta_i - \theta \; , \tag{20}$$

where b and θ are scalar parameters. Also here we must state that (20) only holds if $\zeta_i > 0$, or if $\zeta_i = 0$ and $b\eta_i - \theta \geq 0$, whereas otherwise $d\zeta_i/dt = 0$. The complete equation corresponding to (17) reads

$$d\eta_i/dt = I_i - a\zeta_i - \gamma(\eta_i) \; , \tag{21}$$

where a is another scalar parameter.

This WTA circuit operates in *cycles* like a multivibrator, and each cycle can be thought to correspond to one discrete-time phase in (2). Normally the input is also thought to change at each new cycle; however, if the input is steady for a longer time, the next cycle selects the "runner-up", after which the "winner" is selected again, etc.

The continuous operation of this WTA circuit is illustrated in Fig. 6.

3.2 Local diffusion of the learning factor

In Sec. 2.2 we pointed out that the Hebb-type law of synaptic plasticity must be modified in order to make the changes *reversible*. Another severe problem is that the neural cells are usually ramified, and according to Hebb's hypothesis, the postsynaptic effect ought to reach even the most distal branches and synapses *unattenuated* . It is hardly believable that such an effect could be propagated along the cell membrane or intracellularly.

If it is essential, as it seems, that the plasticity (modifiability) of a synapse is a function of the cell's output, then the quickest and most uniform retrograde messenger for this information would be a chemical agent that is produced in proportion to the postsynaptic activity and spreads back to the synaptic sites *extracellularly*. As a matter of fact, some

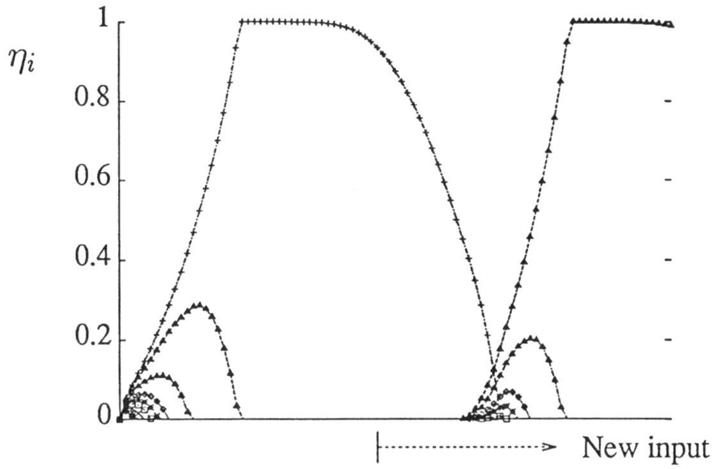

Figure 6: Demonstration of the WTA function provided with automatic reset. The first inputs were applied at time zero. New inputs were applied as indicated by the dotted arrow. The network consisted of 20 cells, and the inputs $I_i^e = m_i^T x$ were selected as random numbers from the interval (0, 1). The g_{ii} were equal to 0.5 and the $g_{ij}, i \neq j$, were -2.0, respectively. The loss function had the form $\gamma(\eta) = 0.1\ln\frac{1+\eta}{1-\eta}$; other simpler laws can also be used. The feedback parameters were $a = b = 1, \theta = 0.5$. The network operates as follows: The first "winner" is the cell that receives the largest input; its response will first stabilize to a high value, while the other outputs tend to zero. When the activity of the "winner" is temporarily depressed by the dynamic feedback, the other cells continue competing. The solution was obtained by the classical Runge-Kutta numerical integration method, using a step size of 10^{-5}.

experimental evidence for such agents, of which the most primary one is nitric oxide (NO), has accumulated in a few recent years (cf., e.g., Fazeli, 1992 and the thematic issue of *J. Chemical Neuroanatomy*, vol. 10, Nos. 3,4, 1996). Among the numerous chemicals that are present at the cells, there probably also exist other similar factors. Such chemicals could spread evenly over the cells' surfaces, even to their most distal branches.

An even more important aspect of the above chemical interaction is *lateral diffusion* of the "learning factor." The postsynaptic chemical messenger can then also affect *nearby* cells, not only the active cell itself. This property would be central in the SOM process where the "winner", or the cell with the highest activation, is supposed to control adaptation in its neighboring cells, too. Attempts to explain such an interaction through collateral synaptic control have not been particularly successful, and one must also note that the anatomy of the biological neural networks is not isotropic enough for to define a symmetric neighborhood function $h_{ci}(t)$ using nonspecific transsynaptic control only. If, on the other hand, the lateral interaction during learning were predominantly implemented by diffuse chemical substances, its range would be rather isotropic, independent of the microanatomy of the network, and thus very beneficial for the SOM process.

However, in order to forestall a potential misinterpretation of the WTA circuit of Fig. 5 it will be necessary to emphasize that each formal neuron i of the simplified model may in reality stand for a group of, say, hundreds of tightly coupled biological neurons. Therefore it would be more accurate to talk of *site i*. It is not necessary to specify whether such a site is a microanatomical aggregate, or whether the activity of an active neuron simply triggers the activity in nearby cells. The most important condition for the chemical-interaction assumption is that a sufficient amount of the "learning factor" is released from the "winner site," the total activity of which we simply denote by η_i in simulations.

Let us assume that the active site constitutes a thin cylindrical local source, perpendicular to the cortex, and the control agent for plasticity is diffused from this source proportional to its activity η_j. Consider now the dynamics of cylindrical diffusion from a line source of strength $\alpha(t)\eta_j(t)$; this model need not be quite accurate. The slowly decreasing coefficient $\alpha(t)$ describes temporal variations in the source strength. In a homogeneous medium, the density of the agent at time t and at cylindrical distance r from the source is

$$D_j(t,r) = \int_0^t \alpha(\tau)\eta_j(\tau)\frac{1}{4\pi(t-\tau)}e^{-r^2/4(t-\tau)}d\tau. \tag{22}$$

For better readability, the time constants and scaling coefficients needed in the actual simulations have been omitted above. In simulations we describe the neighborhood function by

$$h_{ij}(t) = D_j(t, r_{ij}) \ , \tag{23}$$

where r_{ij} denotes the lateral distance between sites i and j.

3.3 Simulation of the physiologically motivated SOM

The most important message of the above theoretical discussion was to motivate the "idealized" SOM theory biologically. The new features of this discussion were: 1. A natural, robust way for the implementation of the WTA function. 2. A realistic interpretation of the Hebb-type law of synaptic plasticity. 3. A natural explanation of the lateral interaction (neighborhood function) between neurons in learning.

A more complete discussion of the physiological SOM model can be found in (Kohonen, 1993, 1995a).

The next step is to combine the WTA function, the modified Hebb-type learning law, and the diffusion effect into the same simulation model. The neurons form a regular, hexagonal array as in Fig. 1.

The output activities are first made to oscillate as demonstrated in Fig. 6. A lateral control effect of plasticity, in proportion to $\eta_j(t)$, is made to spread from each neuron j according to the diffusion model described by (22) and (23).

In the simulations the control factor describing the plasticity (23) was complemented with two additional details: first, the possible mechanisms responsible for the *inactivation* of the diffusing factor after some time were modeled simply by using a finite integration length in the discrete-time approximation of (22). Second, the plasticity was required to exceed a very small threshold before taking effect.

For faster simulation, (22) can also be approximated by the expression $D_j(t,r) \approx \alpha(t)h(r)\eta_j(t)$, where h is a coefficient that depends neither on η_j nor t, if the activity $\eta_j(t)$ changes slowly compared to the time rate of the diffusion.

In principle it is possible to demonstrate the formation of many kinds of "sensory feature maps" by the "physiological" SOM model. In practice it has to be taken into account that this model is several million times slower computationally than the algorithmically simplified models described by (3) and (4), or (6) and (7). The slowdown is mainly caused by the modeling of the diffusion effect that has to be simulated using plenty of small steps; but the "multivibrator" type WTA function needs an almost similar computing time. Therefore we must rest content with a standard test of the SOM algorithm (Kohonen, 1995a). In it, the input consists of two abstract signals ξ_1 and ξ_2 that make up the vector $x = [\xi_1, \xi_2]^{\mathrm{T}} \in \Re^2$, and the array is two-dimensional, say, hexagonal. The probability density function $p(x)$ has a very simple (artificial) form in this test: it is constant within a square in the ξ_1, ξ_2 coordinate system, and zero outside the square. In Figs. 7 and 8 the square has been demarcated by the framed area.

As the weight vectors $m_i = [\mu_{i1}, \mu_{i2}]^{\mathrm{T}}$ tend to approximate $p(x)$ in an *orderly* fashion, they can be plotted as points in the same diagram where x is presented. To show which points belong to spatially neighboring "neurons" in the SOM array, the closest neighbors are connected with auxiliary lines. In the beginning of the process the m_i can be selected as random vectors, whereby the network of the auxiliary lines is still messed up. When topological ordering of the m_i takes place, the network becomes ordered and tends to approximate $p(x)$.

A simulation of the system behavior was first performed using the more accurate diffusion model, and its solution was illustrated by an animation film. Like in many earlier simulations, the input $x(t)$ was defined as a series of successive samples picked up at random from the framed rectangular area in Fig. 7. Now $x(t)$ has to be approximated by steady values during the WTA cycles. As the hexagonal topology was used in the neuron array, the auxiliary lines drawn to Fig. 7 also mean that the distance r_{ij} between the neighboring nodes, used in (23), is evaluated at the discrete nodes of the hexagonal lattice only.

Three frames from this film are shown in Fig. 7. It took about three weeks to run this simulation on a modern workstation, and even so the final asymptotic state was not yet achieved; we had to leave the process halfway, whereby the m_i were ordered but did not yet approximate $p(x)$ particularly well. However, the self-organizing tendency is already clearly visible. Most of the time was spent for the computation of (22) and also for the Runge-Kutta computation of the time functions $\eta_i(t)$. A corresponding similar simulation based on the simplest SOM model, e.g. (3) and (5), would only have taken less than a second on the same computer. This simulation, however, already proves that the general behavior of both the physiological SOM model, say, and the "Dot-product SOM" is qualitatively similar, and therefore *the "Dot-product SOM" can henceforth be regarded as a fast phenomenological SOM model that has a physiological motivation.*

Just for a reference, we also included here another simulation that is otherwise similar to the above one except that the diffusion effect has been significantly simplified, making

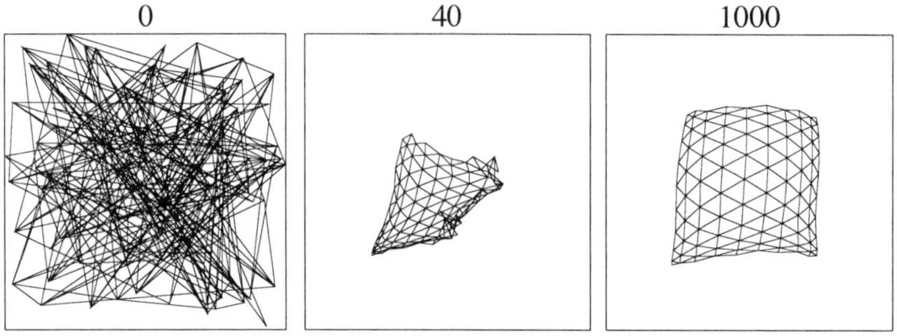

Figure 7: Three frames from an animation film (learning steps 0, 40, and 1000, respectively) that illustrates the operation of a complete physiological SOM system, after which the simulation was stopped.

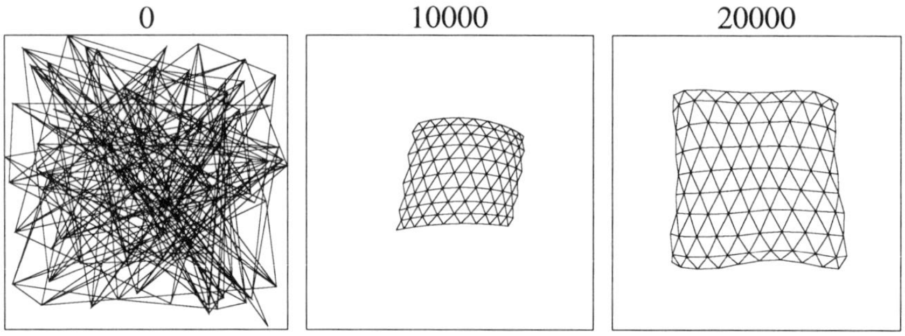

Figure 8: Three frames from an animation film based on the precomputed diffusion solution (24).

the simulation much faster. Now the spread of the "learning factor" was assumed to be of the form:

$$D_j(t,r) = \alpha(t)\eta_j(t)e^{-r^2/2[\sigma(t)]^2} \ , \tag{24}$$

where $\alpha(t)$ and $\sigma(t)$ are (linearly) decreasing parameters like in the "Dot-product SOM." Then again, (23) can be used to define $h_{ij}(t)$ as a sum of contributions from the different neurons. A simulation based on (24) was also performed. Again, an animation film was made, and three frames from it are shown in Fig. 8.

4 Emergence of invariant-feature analyzers in the adaptive-subspace SOM (ASSOM)

A long-standing problem in the theory of perception has been *invariance* of observations to moving and transforming objects. The prevailing view is that if the representations of the objects are decomposed into elementary *features* such as locally defined stripes of various width, and each of the latter can be detected by a special filter, the response of which is invariant to a modest amount of some transformation of the feature, also the whole object is recognizable under the same transformations.

In this section we shall discuss an adaptive process in which such filters emerge in response to input data that contain the corresponding transformations. The discussion is rather terse; for a more extensive explanation, see, e.g., (Kohonen, 1995a).

A rather little known approach in pattern recognition techniques is to model the data in pattern classes directly as analytically definable *manifolds*. For instance, in practice, many important transformation groups are automatically taken into account if the classes are put to correspond to *linear subspaces*. Thereby the class-affiliation of an unknown input pattern is expressed in terms of the *distance of its representation from such a subspace*. The subspace itself is defined by its *basis vectors*, for which one can select, for instance, representative sample vectors of the class or principal components, and the input pattern is classified according to its similarity with the *general linear combination of the basis vectors*. This makes a big difference with respect to direct comparison of unknown vectors with weight vectors, usual in many neural-network approaches.

Let us recall that in traditional neural-network models, each node of a signal processing network is identified with a single neuron. However, in the sensory cortices such as the primary visual cortex the neurons are organized as modules or tightly cooperating groups; it is not necessary to identify these modules with the columns. We argue that, for instance, several simple cells may cooperate with one complex cell as such a module. Such an organization is now shown to be favorable to the formation of invariant-feature analyzers.

4.1 The architecture

Consider the network delineated in Fig. 9. Each dotted line separates a module, a processing unit in the special SOM array. The first-layer neurons are linear and they output the sums of dot products of x with the various synaptic weight vectors. The second-layer neurons (Q) form some quadratic functions of the first-layer neuron outputs. If the weight vectors of the linear layer were identified with the orthonormalized basis vectors of the corresponding linear subspace, the output layer only has to form the sums of squares of its inputs.

The classifier represented by Fig. 9 can be shown to compare input patterns x with certain linear subspaces. If the basis vectors of the latter can be defined in a way in which their linear combinations represent some transformation groups, approximately at least, then

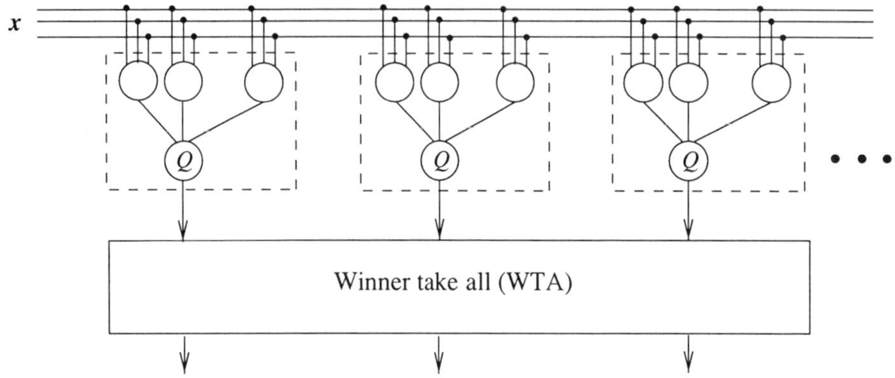

Figure 9: The ASSOM architecture.

matching becomes invariant with respect to such groups.

The outputs from the modules of Fig. 9 shall further be compared by a WTA function.

The WTA function specifies the "winner" module indexed by c; module c and its neighboring modules in the array will be updated in proportion to the neighborhood function h_{ci} as discussed below. The WTA function may be integrated with the modules, for instance by their lateral interaction.

4.2 The subspace formalism

There exist "neural" processes by which the input weight vectors of each module in Fig. 9 can be *orthonormalized* automatically in the adaptive process (Oja, 1982). As their physiological motivation is still unclear, we shall not attempt any more sophisticated biophysical modeling here. Notwithstanding it may be interesting to inspect what really happens if the input weight vectors of each module can be kept orthonormal by some learning mechanism, and this mechanism further *tends to minimize a certain error function.* From such a general objective function one can then derive a new, equivalent optimization algorithm to demonstrate these very intriguing learning results by computer simulation.

Let us denote the input weight vectors of module i by $b_h^{(i)}$, where h is the index of neuron h in the module. As the $b_h^{(i)}$ are now assumed orthonormal, they will be regarded as the *orthonormal basis vectors of some linear subspace* $\mathcal{L}^{(i)}$. The latter is defined as a *manifold*, or a set of vectors x, where every x can be expressed as a general linear combination of the $b_h^{(i)}$.

For an arbitrary x that need not belong to $\mathcal{L}^{(i)}$ one can define its *distance* d from $\mathcal{L}^{(i)}$, defined by

$$d^2 = d^2(x, \mathcal{L}^{(i)}) = \|x\|^2 - \|\hat{x}^{(i)}\|^2 \ , \tag{25}$$

where $\hat{x}^{(i)}$ is the *orthogonal projection* of x on $\mathcal{L}^{(i)}$, or

$$\hat{x}^{(i)} = \sum_h b_h^{(i)^{\mathrm{T}}} x \ . \tag{26}$$

The *vectorial projection error* is the residual

$$\tilde{x}^{(i)} = x - \hat{x}^{(i)} \ , \tag{27}$$

and $\hat{x}^{(i)}$ and $\tilde{x}^{(i)}$ can easily be shown to be orthogonal.

For an arbitrary x, its *minimum projection error* can be defined as the *distance of x from the closest subspace* $\mathcal{L}^{(i)}$, and the "winner subspace" with index c is again defined by

$$\|\tilde{x}^{(c)}\| = \min_i\{\|\tilde{x}^{(i)}\|\} \ , \text{ or} \tag{28}$$

$$\|\hat{x}^{(c)}\| = \max_i\{\|\hat{x}^{(i)}\|\} \ . \tag{29}$$

We are now ready to introduce the objective function that defines the *average expected normalized projection error*

$$E_3 = \int \sum_i h_{ci} \frac{\|\tilde{x}^{(i)}\|^2}{\|x\|^2} p(x) dx \ , \tag{30}$$

where h_{ci} is the neighborhood function that defines the interaction of modules c and i, and c is again the index of the winner subspace $\mathcal{L}^{(i)} = \mathcal{L}^{(c)}$. *Notice again that c is a function of x and all the basis vectors $b_h^{(i)}$.* We use also here the shorthand notation dx for a volume differential in the x space.

Minimization of (30), i.e., selection of the basis vectors $b_h^{(i)}$ for all subspaces $\mathcal{L}^{(i)}$ such that the average expected distance of x from the closest subspace is minimized, is another complicated process (Kohonen 1995a,b). Some extra problems are caused by the stability of the recursion by which E_3 is minimized. Without quoting all the details it may be mentioned that if the Robbins-Monro stochastical approximation process is used, the optimal values of the $b_h^{(i)}$ are obtained in the recursion

$$b_h^{(i)}(t+1) = b_h^{(i)}(t) + \lambda(t) h_{ci}(t) \frac{x(t) x^{\mathrm{T}}(t)}{\|x(t)\|^2} b_h^{(i)}(t) \ . \tag{31}$$

4.3 The ASSOM

Our ultimate goal, however, was to show how certain *invariant feature analyzers* emerge in the architecture of Fig. 9 that is described by the subspace formalism.

The key insight into the invariance problem is that if a set of successive input patterns must be recognizable invariantly when they are changing continuously, they all must belong to the same subspace, approximately at least. These signal subspaces will be learned by the architecture of Fig. 9.

Consider an "episode" $\mathcal{S}$ that consist of a finite set of successive sampling times t_p; denote $\mathcal{S} = \{t_p\}$. The set of samples $X = \{x(t_p)|t_p \in \mathcal{S}\}$ has now to be recognized as one class, such that any member of X and even an arbitrary linear combination of the $x(t_p), t_p \in \mathcal{S}$ shall be decoded by the same module of Fig. 9 (subspace $\mathcal{L}^{(i)}$). In learning, the vector set X, which is a Cartesian product of the $x(t_p), t_p \in \mathcal{S}$, must be taken as one batch, instead of optimizing the error using single patterns $x(t_p)$ one at a time. The error minimization problem will now be modified by defining the new objective function in terms of the average squared projection error over the episode:

$$E_4 = \int \sum_{t_p \in \mathcal{S}} \sum_i h_{ci} \frac{\|\tilde{x}^{(i)}(t_p)\|^2}{\|x(t_p)\|^2} p(X) dX \ . \tag{32}$$

Here $p(X)$ is the joint probability density for the samples $x(t_p), t_p \in \mathcal{S}$ that make up the Cartesian product set X to occur, and dX is a shorthand notation meaning a volume differential in the Cartesian product space of the $x(t_p)$.

The basis vectors $b_h^{(i)}$ that define the subspaces $\mathcal{L}^{(i)}$ which optimally distinguish between typical episodes will constitute the solution to the optimization problem. Incidentally, for natural, dynamic signal data, these $b_h^{(i)}$ will define a set of analyzers that are *optimally invariant to the typical transformations that occur in the input signal patterns.* For instance, if the $x(t_p)$ represent two-dimensional moving images or their parts, the $b_h^{(i)}$ that emerge resemble the *Gabor filters* (Gabor, 1946), (Daugmann, 1983). If the pattern transformations are different (rotation, scaling, perspective tranformations, etc.), the filters that emerge are different, but always optimized with respect to the true transformations and tending to recognize the episodes as invariantly as possible.

The Robbins-Monro stochastic approximation is applicable to the minimization of E_4, too, when the gradient step is made to consist of the whole episode $\mathcal{S}$. The learning phase is then desribed by the following equation:

$$b_h^{(i)}(t+1) = b_h^{(i)}(t) + \lambda(t) h_{cr}^{(i)} \sum_{t_p \in \mathcal{S}(t)} \frac{x(t_p) x^{\mathrm{T}}(t_p)}{\|s(t_p)\|^2} b_h^{(i)}(t) \ . \tag{33}$$

Earlier (Kohonen, 1995a) we have described the "episode learning" by a product of elementary projection operators, each one corresponding to one pattern $x(t_p), t_p \in \mathcal{S}$:

$$b_h^{\prime(i)} = \prod_{t_p \in \mathcal{S}} \left[I + \alpha(t_p) h_{cr}^{(i)} \frac{x(t_p) x^{\mathrm{T}}}{\|\hat{x}^{(i)}(t_p)\| \, \|x(t_p)\|} \right] b_h^{(i)}(t_p) \ . \tag{34}$$

The special choice of the learning-rate factor in (34) as a function of the projection of x on subspace $\mathcal{L}^{(i)}$ has been found necessary for stability reasons (Kohonen, 1995a).

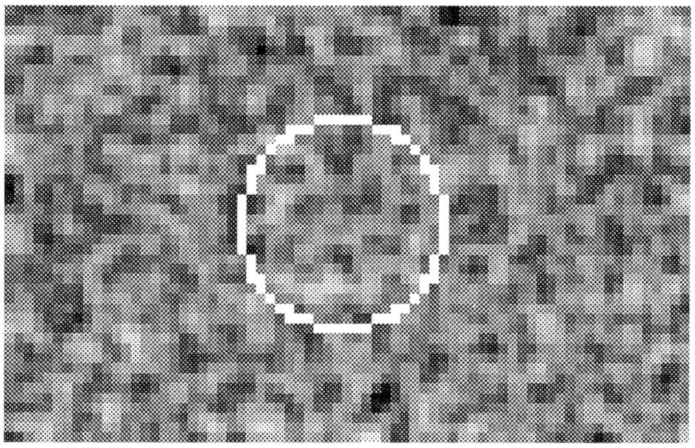

Figure 10: Colored noise (second-order Butterworth-filtered white noise with cut-off frequency of 0.6 times the Nyquist frequency of the lattice) used as input data. The receptive field is demarcated by the white circle.

It can easily be seen that when the learning-rate factor is small, and we select

$$\lambda(t_p) = \alpha(t_p)\frac{||x(t_p)||}{||\hat{x}^{(i)}(t_p)||} , \tag{35}$$

the corrections ensuing from (33) and (34) can be found equivalent, and (33) may be regarded as a batch version of (34), where the whole episode is used as one operation.

4.4 Computer experiments

We have produced various ASSOM filters for very different input data (Kohonen, 1995a,b). Here a demonstration, illustrating the basic idea, is shown in the first three subsections. Two-dimensional filters, reflecting different kinds of invariances, were formed although the same algorithm was used, and the basic pattern type was the same. What was essentially different in the first three experiments reported below was the transformation group of the basic pattern over the episode. The following kinds of filters thereby emerged: translation-invariant filters, rotation-invariant filters, and scale-invariant filters. The receptive field of the input layer consisted of a circular sampling lattice (equivalent to 316-dimensional pattern vectors), shown in Fig. 10. Over the input field we generated patterns consisting of colored noise (white noise, low-pass filtered by a second-order Butterworth filter with cut-off frequency of 0.6 times the Nyquist frequency of the sampling lattice). The input episodes for learning were formed by taking samples from this data field. The mean of the samples was always subtracted from the pattern vector.

We have also produced ASSOM filters for real speech signals (Kohonen, 1995a,b, 1996).

Translation-invariant filters. In the *translation-invariant filter* experiment, the ASSOM was a hexagonal, 9 by 10 unit array, with two basis vectors at each unit. The episodes were

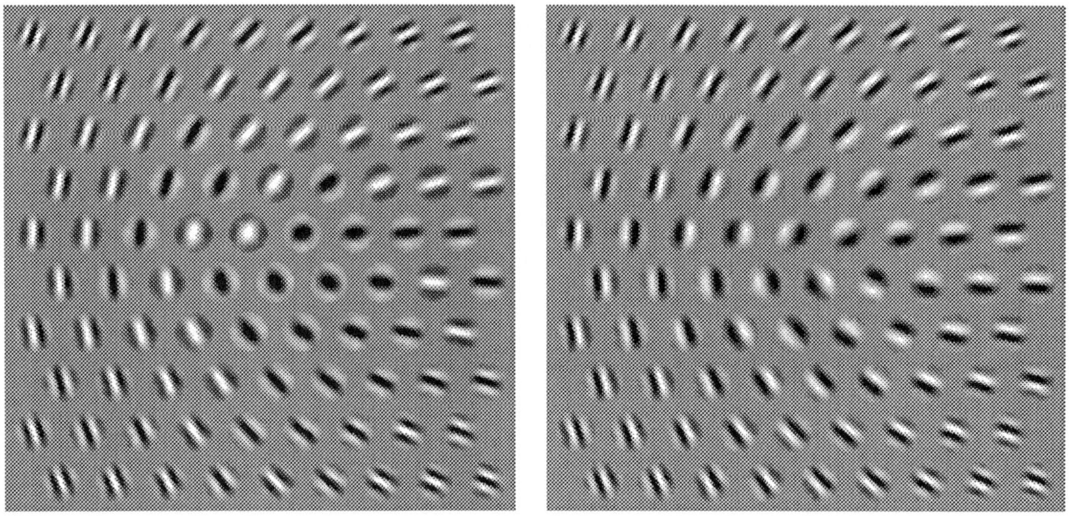

(a) (b)

Figure 11: The ASSOM that has formed Gabor-type filters: (a) The b_{i1}, (b) The b_{i2}.

formed by shifting the receptive field randomly into five nearby locations, the average shift thereby being ± 2 pixels in both dimensions. The next episodes were taken from completely different places. In order to stabilize and symmetrize the filters with respect to the center of the receptive field (because the filters are easily formed as eccentric) we weighted the input samples by a Gaussian function that was symmetrically placed over the receptive field, and had a full width at half maximum that varied linearly during the learning process from one to 15 sampling lattice spacings. The radius of the circular neighborhood function $N^{(c_r)}$ decreased linearly from five to one ASSOM array spacings during the process, and the learning-rate factor was of the form $\alpha(t) = T/(T + 99t)$, where t was the index of the learning episode, and T the total number of episodes, respectively; in the present experiment T was 30000. The initial values of the basis vectors were selected as random but normalized vectors. The basis vectors were orthonormalized at each learning step.

Fig. 11 shows the basis vectors b_{i1} and b_{i2} graphically in a gray scale at each array point. One should notice that the spatial frequencies of the basis vectors are the same, but the b_{i1} and b_{i2} are mutually 90 degrees out of phase. (The absolute phase of b_{i1} can be zero or 180 degrees, though.)

Rotation filters. The ASSOM lattice in this experiment was linear, consisting of 24 units, whereas the receptive fields were the same as in the first experiment. The episodes were formed by *rotating* the input field at random five times in the range of zero to 60 degrees, the rotation center coinciding with the center of the receptive field. Each new episode was taken from a completely different place of the input field. In this experiment, no Gaussian weighting of the samples was necessary. Fig. 12 shows the rotation filters thereby formed at the ASSOM units; clearly they are sensitive to azimuthal optic flow.

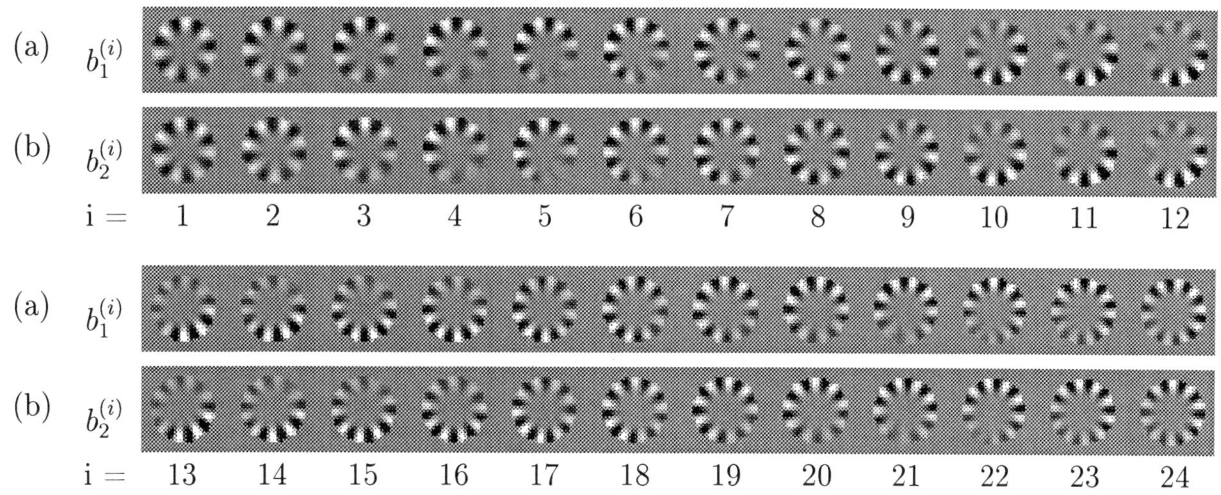

Figure 12: One-dimensional rotation-invariant ASSOM. (a) Cosine-type "azimuthal wavelets" (b_{i1}), (b) Sine-type "azimuthal wavelets" (b_{i2}). Notice that the linear array has been shown in two parts.

Scale-invariant filters. This time the episodes were formed by *zooming* the input pattern field, with the center of the receptive field coinciding with the zooming center. No Gaussian weighting was used. The zooming range was 1 : 2, from which five patterns were picked up to an episode; the next episode was formed from a quite different place of the input field. The filters thereby formed, shown in Fig. 13, have clearly become sensitive to radial optic flow, corresponding to approaching or withdrawing objects.

4.5 The ASSOM as a learning preprocessing stage in classification

A practical pattern recognition or classification task is usually organized in two stages, as delineated in Fig. 14: 1. Preprocessing, whereby a number of *invariant features* is extracted from the sensory signals. 2. Statistical decision on the basis of the extracted features, whereby plenty of alternatives exist for the classification operation.

As our intention was to study how the classification accuracy is improved when the feature analyzers are adapting to data, we selected for the classifier the most fundamental algorithm, the k-nearest-neighbor classifier (for a textbook account of different classifiers, cf., e.g., Kohonen, 1995a). In other words, no "neural" models were tested in the latter stage; the results would have been qualitatively similar with them.

For the practical example we selected classification of *textures*; the case example was extremely simple and only intended for a demonstration. The objective was to show that even when the initial parameter values of the feature analyzers are selected as random, the analyzers will learn to describe specific sensory experiences, whereby the classification accuracy will be increased along a *"learning curve."*

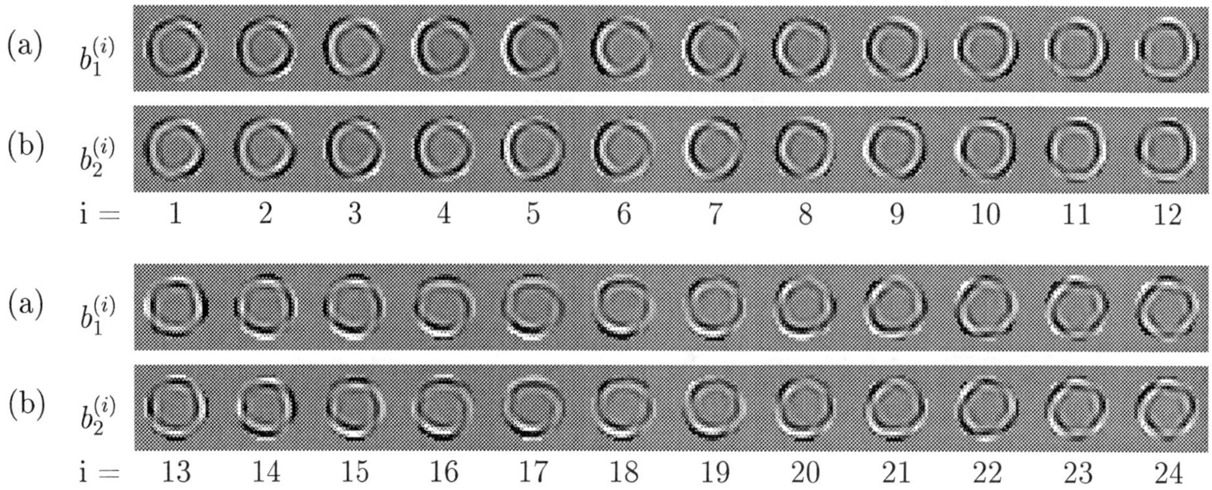

Figure 13: One-dimensional zoom-invariant ASSOM. (a) Cosine-type "radial wavelets" (b_{i1}), (b) Sine-type "radial wavelets" (b_{i2}). Notice that the linear array has been shown in two parts.

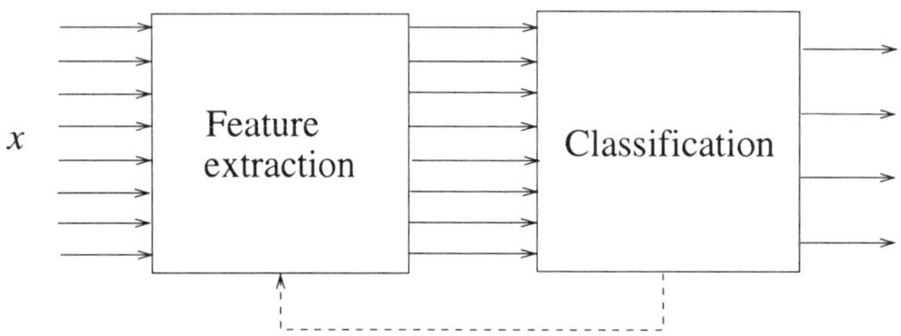

Figure 14: Organization of a pattern recognition system.

The input patterns were derived from standard texture samples (Brodatz, 1968). Fig. 15 shows in the left column the samples used for training; the textures in the right column are statistically independent of the training samples, and they were used for testing of the classifier accuracy. In one of the series of experiments the inputs were taken from "receptive fields," over which the patterns were *translated* during learning, whereafter the classifier was supposed to be able to classify the textures independent of their arbitrary translation. In another series of experiments, the patterns were *rotated* during learning, whereafter the classifier was supposed to classify these textures independent of their arbitrary rotation. In the two series of experiments the learning curves were qualitatively similar, but the analyzers that emerged were totally different, resulting from different kinds of sensory experiences. For exemplification, the wavelets corresponding to rotation of the textures in a two-dimensional ASSOM after 60 000 training cycles are shown in Fig. 16.

It may suffice to present here two exemplary learning curves, one (Fig. 17a) for translation, and another (Fig. 17b) for rotation. As it might have been expected, the ASSOM learns the translational invariance much better (with over 90 per cent accuracy), whereas

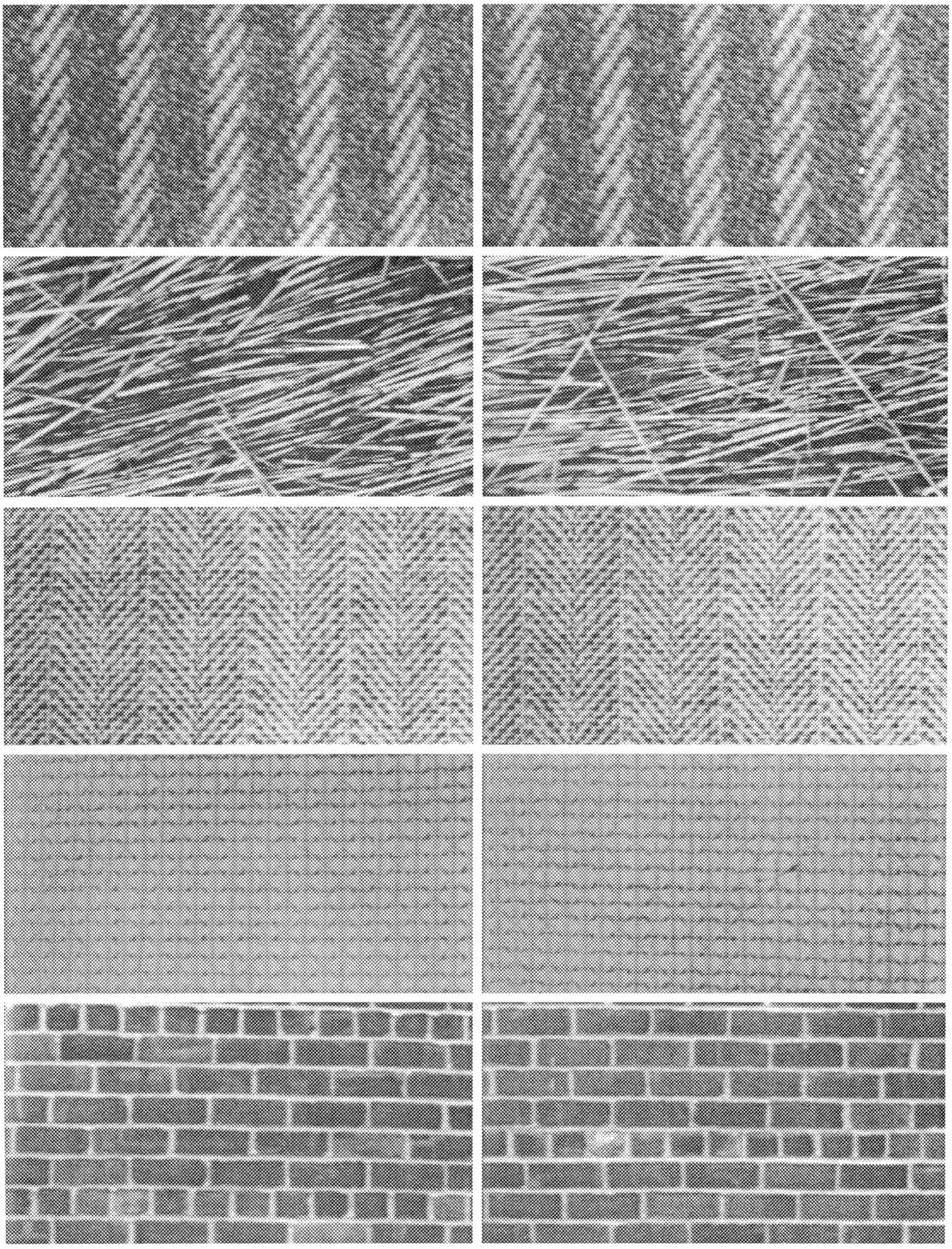

Figure 15: Examples of textures used in the experiment.

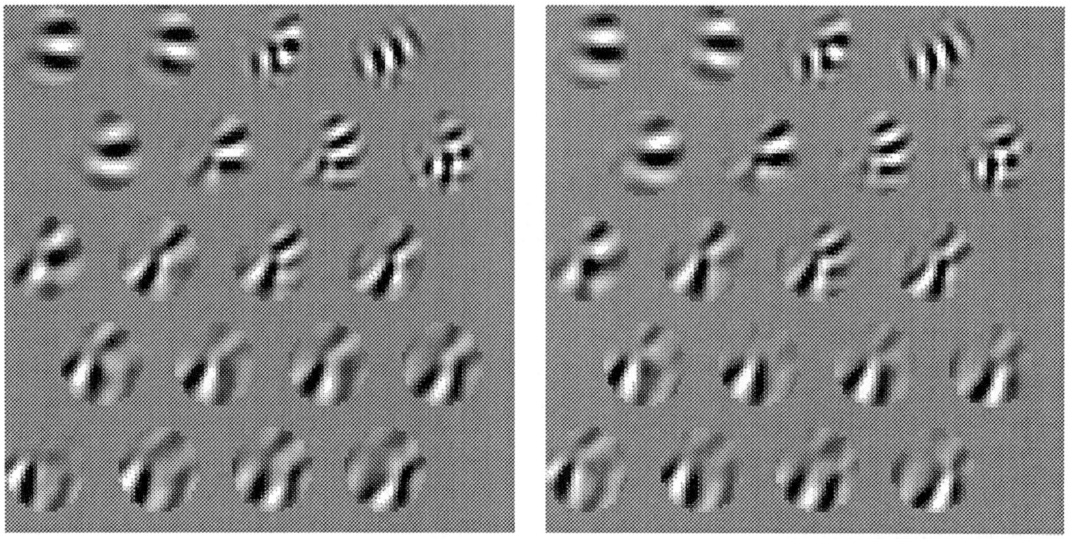

(a) (b)

Figure 16: The wavelets corresponding to rotation of the given textures, in a two-dimensional ASSOM, after 60 000 training cycles. (a) The b_{i1}, (b) The b_{i1}.

the achieved rotational invariance was worse (typically less than 80 per cent accuracy). One explanation may be that in the experiment, only a few highly structured artificial textures, containing plenty of long straight lines were used, which the translational filters then learned better. The "rotation filters" were not as clear as those depicted in Fig. 12. The results might have been somewhat different for more uniform, granular and statistically defined textures.

At any rate, the purpose of this demonstration was by no means to aim at the highest possible classification percentage, but to illustrate the learning curves resulting from adaptive preprocessing only.

5 Discussion

The following two questions can be posed to assess the various artificial neural network models: 1. Can the models generalize their responses over variations of input data presented at different levels of abstraction of the input information? 2. Can the models generate feature analyzers that operate invariantly with respect to some basic transformations of the input patterns?

Most of the extensive literature on the Self-Organizing Map (cf. Kohonen, 1995a, Neural Networks Centre, 1996) deals with the previous question, and the general notion seems to be that the SOM can be used to find many kinds of abstraction of the primary data. This article, after a general SOM discussion, emphasized the second aspect. Emergence of certain specific feature analyzer functions was thereby demonstrated. To answer the

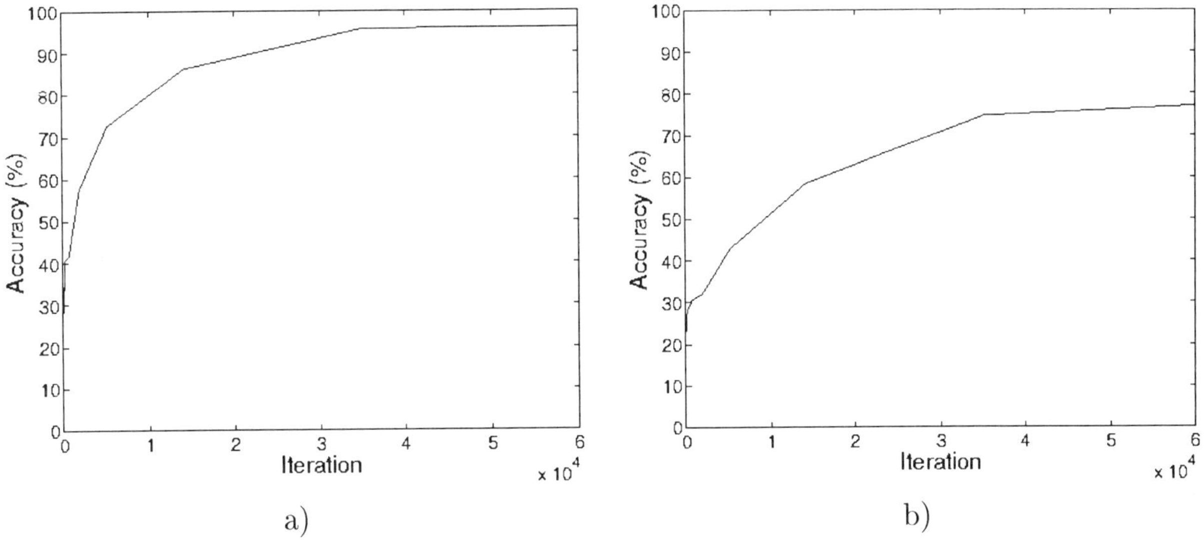

Figure 17: Learning curves in the texture classification experiment. a) The patterns were translated during learning. b) The patterns were rotated during learning.

question whether these models are also models of the physiological functions of the brain, it will be necessary to point out that many models used in science are *phenomenological models*, whereby they demonstrate the mathematical processes that are at work in producing the observable phenomena, without relating them to every detail of the concrete structures underlying them. In this sense the Self-Organizing Map and ASSOM experiments may tell some essential facts about the processes taking place in the brain.

What is the most significant difference between the Self-Organizing Map and other contemporary neural-model approaches? Most of the latter strongly emphasize the aspect of distributed processing, and only consider spatial organization of the processing units as a secondary aspect. The map principle, on the other hand, is in some ways complementary to this idea. The intrinsic potential of this particular self-organizing process for creating a localized, structured arrangement of representations in the basic network module is emphasized.

Actually we should not talk of the localization of a "function": it is only the *response* that is localized. I am thus not opposed to the view that neural networks are distributed systems. The massive interconnects that underlie all neural processing are certainly spread over the network; their effects, on the other hand, may be "focused" on local sites.

It seems inevitable, however, that any complex processing task requires organization of information into separate parts. Distributed processing models in general underrate this issue. Consequently, many models that process features of input data without structuring exhibit slow convergence and poor generalization ability, usually ensuing from ignorance of the localization of the adaptive processes.

On the lower perceptual levels, localization of responses in topographically organized maps has already been demonstrated long time ago, and it is known that such maps need not be prespecified in detail, but can instead organize themselves on the basis of the statistics of the incoming signals. Such maps have already been applied with success to many complex pattern recognition and robot control tasks.

On the higher levels of representation, relationships between items seem to be based on more subtle roles in their occurrence, and are less apparent from their immediate intrinsic properties. Nonetheless it has also been shown recently that even with a simple modeling assumption of semantic roles, topological self-organization of *semantic data* will take place (Ritter & Kohonen, 1989; Honkela et al., 1995; Kohonen et al., 1996). To describe the semantic role of an item, it will be sufficient that the input data are presented together with a sufficient amount of context. This then controls the adaptation process.

Although the basic one-level map has already been shown to be capable of creating hierarchical (ultrametric) representations of structured data distributions, it might be expected that the real potential of the map lies in a genuine hierarchical or otherwise structured system that consists of several interconnected map modules. In a more natural system, such modules might also correspond to contiguous areas in a single large sheet, where each area receives a different kind of external input, as in the different areas in the cortex. In that case, the borders between the modules might be diffuse. The problem of hierachical maps, however, has turned out to be very difficult. One of the particular difficulties arises if the inputs to a cell come from very different sources; it then seems inevitable that for the comparison of input patterns, an asymmetrical distance function, in which the signal components are provided with adaptive tensorial weights, must be applied (Kangas et al., 1990). Another aspect concerns the interfaces of modules in a hierarchical map system: the signals merging from different modules may have to be combined nonlinearly (Ritter, 1989).

It was also demonstrated in this paper that wavelet-type filters emerge in the ASSOM architecture if the primary patterns that define the successive input sample vectors in an episode are translated. If the patterns are zoomed or rotated, different kinds of feature analyzers will emerge, although the algorithm is the same. Still other types of filters developed for natural speech.

Emergence of invariant filters means that the "neural" subspaces represented by these filters become adapted to corresponding signal sets that are also describable as linear subspaces. Thus the network is selectively sensitized to such components in the signal mixtures that satisfy certain linear constraints, and desensitized to the rest of them.

Sometimes it may be beneficial to introduce three or more basis vectors for each subspace, and the ASSOM array might be, say, three-dimensional. The ASSOM architecture may be used as a genuine, adaptive *feature-extraction layer* in a pattern recognition system (Kohonen, 1995a). It is also possible to use decision-directed training of the feature-extraction layer (see the so called FASSOM architecture discussed in Kohonen, 1995a).

It is an intriguing question what possible biological counterparts the various neuron types in this architecture might have. In a series of interesting studies, Daugman (1980, 1985), Marcelja (1980), and Jones and Palmer (1987) advanced the hypothesis that the receptive fields of the *simple cells* in the mammalian visual cortex are describable by Gabor-type functions. What transpired in the present study is that such functions in fact emerge in the *input layer* of the ASSOM structure, whereby the cells of this layer act in the similar way as the simple cells. If the output of the quadratic neuron were recorded, this neuron might have been seen to operate like a *complex cell* of the primary visual cortex; notice, however, that this function needs the integration of all cells in the whole module. It is not necessary that there be exactly *two* input neurons in each module; the same ASSOM operation is obtained with three or more input neurons. It also seems that the assumption of exactly quadratic operation can be alleviated to some extent, as long as the frequency-to-frequency transfer function of the output neuron is convex and smooth. If the transfer function is sigmoidal, invariance properties are mainly obtained at low-intensity signals, whereas at high degree of stimulation the module becomes more sensitive to the location of the stimulus.

References

Brodatz, P. (1968). *Textures – A Photographic Album for Artists and Designers.* New York: Reinhold.

Fazeli, M. (1992). Synaptic plasticity: On the trail of the retrograde messenger. *Trends in Neuroscience*, 15, 115-117.

Daugmann, J. (1980). Two-dimensional spectral analysis of cortical receptive field profiles. *Vision Research*, 20, 847-856.

Daugmann, J. (1983). Six formal properties of two-dimensional anisotropic visual filters: Structural principles and frequency/orientation selectivity. *IEEE Trans. Syst., Man, Cybern.*, 13, 882-887.

Daugmann, J. (1985). Uncertainty relation for resolution in space, spatial frequency, and orientation optimized by two-dimensional visual cortical filters. *J. Opt. Soc. Am.*, 2, 1160-1169.

Gabor, D. (1946). Theory of communication. *J. IEE*, 93, 429-457.

Hebb, D. (1949). *Organization of Behaviour.* New York: Wiley.

Honkela, T., Pulkki, V., and Kohonen, T. (1995). Contextual relations of words in Grimm tales, analyzed by self-organizing map. *Proc. ICANN'95, Int. Conf. on Artificial Neural Networks*, vol. 2, pp. 3-7. Paris: EC2 & Cie.

Jones, J. and Palmer, L. (1987). An evaluation of the two-dimensional Gabor filter model of simple receptive fields in cat striate cortex. *J. Neurophysiol.*, 58, 1233-1258.

Joutsiniemi, S.-L., Kaski, S., and Larsen, T.A. (1995.) Self-organizing map in recognition of topographic patterns of EEG spectra. *IEEE Trans. Biomedical Engineering*, 42, 1062-1068.

Kangas, J., Kohonen, T., and Laaksonen, J. (1990). Variant of self-organizing maps. *IEEE Trans. Neural Networks*, 1, 93-99.

Kaski, S. and Kohonen, T. (1994). Winner-take-all networks for physiological models of competitive learning. *Neural Networks*, 7, 973-984.

Kohonen, T. (1982). Self-organized formation of topologically correct feature maps. *Biological Cybernetics*, 43, 59-69.

Kohonen, T. (1991). Self-organizing maps: Optimization approaches. In Kohonen, T. et al. (eds.), *Artificial Neural Networks*, vol. 2, pp. 981-990. Amsterdam: North Holland.

Kohonen, T. (1993). Physiological interpretation of the self-organizing map algorithm. *Neural Networks*, 6, 895-905.

Kohonen, T. (1995a). *Self-Organizing Maps.* Springer Series in Information Sciences, vol. 30. Heidelberg: Springer.

Kohonen, T. (1995b). Emergence of invariant-feature detectors in self-organization. In Palaniswami, M. et al. (eds.), *Computational Intelligence, A Dynamic System Perspective*, pp. 17-31. New York: IEEE Press.

Kohonen, T., Kaski, S., Lagus, K., and Honkela, T. (1996). Very large two-level SOM for the browsing of newsgroups. In v.d. Malsburg, W. et al. (eds.), *Lecture Notes in Computer Sciences*, Vol. 1112, pp. 269-274. Berlin, Heidelberg: Springer.

Marcelja, S. (1980). Mathematical description of the responses of simple cortical cells. *J. Opt. Soc. Am.*, 70, 1297-1300.

Neural Networks Research Centre (1996). 2358 studies of the self-organizing map (SOM) and learning vector quantization (LVQ): A bibliography. Obtainable via Internet at http://nucleus.hut.fi/nnrc/refs/references.ps.

Oja, E. (1982). A simplified neuron model as a principal component analyzer. *J. Math. Biol.*, 15, 267-273.

Ritter, H. (1989). Combining self-organizing maps. *Proc. Int. Joint Conf. on Neural Networks*, pp. II-499-II-502. Piscataway: IEEE Service Center.

Ritter, H. and Kohonen, T. (1989). Self-organizing semantic maps. *Biological Cybernetics*, 61, 241-254.

Suga, N. and O'Neill, W.E. (1979). Neural axis representing target range in the auditory cortex of the mustache bat. *Science*, 206, 351-353.

Vapola, M., Simula, O., Kohonen, T., and Meriläinen, P. (1994). Representation and identification of fault conditions of an anaesthesia system by means of the self-organizing map. In Marinaro, M. and Morasso, P.G. (eds.), *Proc. ICANN'94, Int. Conf. on Artificial Neural Networks*, vol. 1, pp. 350-353. London: Springer.

Zeki, S. (1980). The representation of colours in the cerebral cortex. *Nature (London)*, 284, 412-418.

8

Pragmatic Approach to Consciousness

Henry P. Stapp
[1]Lawrence Berkeley National Laboratory
University of California
Berkeley, California 94720

Abstract

Physical scientists were driven during the late twenties to abandon a fundamental idea that had reigned since the time of Issac Newton. To obtain a rationally coherent and practically useful theory of all physical phenomena they turned to a pragmatic approach. The core idea was that the basic physical theory was no longer directly about a physical world that was conceived to exists apart from anyone's knowledge of it. Rather the theory was regarded as being directly about certain of our knowings. This switch appears to be exactly what is needed to establish a rationally coherent theoretical foundation for the science of consciousness. For it converts the immediate objects of psychological and physical theories into things of the same kind, namely human experiencings, rather than things of disparate kinds separated by an unbridgable conceptual gap. Within this pragmatic quantum approach certain particular aspects of human brain structure entail the existence of macroscopic quantum effects that are linked to our conscious experiences. Moreover, our conscious thoughts have causal effects that can both enhance our prospects of survival, and work effectively against thermal noise in the creation of the brain states that guide our behaviour.

---

[1]This work was supported by the Director, Office of Energy Research, Office of High Energy and Nuclear Physics, Division of High Energy Physics of the U.S. Department of Energy under Contract

1998. In N. Osaka (Ed.), Neural Correlates of Consciousness. To be published.

Disclaimer

Introduction

In modeling the connection between a person's conscious experiences and the activities of his body and brain it is often assumed that a conception of the human body/brain concordant with the principles of classical physics will be adequate. This assumption would seem injudicious in view of the well known failure of the classical conception of matter at the fundamental level. But it is often argued that quantum effects pertain to atomic-scale processes rather than the large-scale brain activities that govern our conscious experiences, and that this scale difference renders classical description adequate in practice even though it is wrong in principle.

It turns out, however, in the case of human brains, that simple general considerations show that the failure of classical concepts at the atomic scale necessarily has, according to the pragmatic quantum precepts, large macroscopic effects that are linked to conscious experiences. In particular, the irreducible Heisenberg uncertainties in the locations of presynaptic calcium ions entail that the observed macroscopic behaviour of the body/brain will be determined by reductions of quantum wave packets that, according to the pragmatic quantum principles, are representations of conscious knowings. These sudden quantum reductions have no counterpart in classical mechanics. Consequently, within the pragmatic framework, the irreducible Heisenberg uncertainties inject our conscious thoughts into the dynamics of the body/brain in a way that is neither allowable nor imaginable in classical physics.

This result, which will be explained in detail below, resolves a fundamental philosophical problem of long standing: the classical-physics conception of a person's body/brain renders his conscious experiencings causally inert, and hence superfluous with respect to the course of physical events. But many scientists and philosophers find it unnatural that something so different from classically conceived matter as our conscious experiences would become tacked onto the material universe if they play no causal role in the unfolding of physical events.

This problem is nicely laid out by the philosopher Daniel Dennett. Some of Dennett's earlier books were interpreted by critics as evading this difficulty with consciousness by denying, unreasonably, the existence of consciousness. He certainly did deny any sort of ontological mind-matter duality in nature. But in his new book ``Kinds of Minds-Toward an Understanding of Consciousness"

(Dennett, 1996) he starts right off using the word ``mind" to denote, in a perfectly normal way, a personal realm of conscious experience. He quickly distinguishes questions about what exists (ontology) from questions what we know (epistemology), and says that the goal of his book is to show that these two kinds of questions have to be answered together. That thesis, developed in another way, constitutes the essence of what I shall be saying here.

Dennett then continues: ``I will argue that we already know enough about minds to know that one of the things about them that makes them different from everything else in the universe is the way we know about them. For instance, you know you have a mind and you know you have a brain, but these are different kinds of knowledge. ....each of us knows exactly one mind from the inside... No other thing is known in that way."

Having thus specified some epistemological aspects of mind Dennett gets, near the end of the book (p. 155), to the ontological side:

``Mental contents become conscious... by winning the competition against other mental contents for domination in the control of behavior, ... A common reaction to this suggestion about human consciousness is frank bewilderment, expressed more or less as follows: `Suppose all these strange competitive processes are going on in my brain, and suppose, as you say, that the conscious processes are simply those that win the competition. How does that make them conscious?' "

This question appears to reflect the questioner's tacit assumption that the physical brain conforms to the classical-physics conception of it. The classical-physics conception of the physical world is that this world consists of a huge collection of tiny bits of structureless matter. This collection is separated into swarms that constitute stars, planets, apples, tornadoes, locomotives, computers, human bodies, and all of the other objects that we can see, and also the many others that, for one reason or another (e.g., too small, too distant), we cannot see.

According to the classical-physics conceptualization, these things that are seen from afar are to be conceptualized as being, in essence, like collections of tiny moving ``billiard balls", where each ball is essentially structureless, but is connected by some forces to its nearest neighbors.

The essential idea of the classical-physics conceptualization of the physical world is that this reality can be conceived of by *stripping away from what we might actually see*, which can be colored by all of our personally appended ornamentations and significances and meanings, every property but the spacetime trajectories of the structureless bits: these trajectories alone constitute the basic classical-physical reality. (Even the fields can be conceived of as consisting of such trajectories.) Nothing beyond this stripped-down ontology is needed to make the principles of classical physics work.

The motions of the ``billiard balls" that constitute the activities of the classically described physical brain can, in the mind of someone who knows that brain from the outside, be analyzed in many ways. However, no matter how many layers of modules monitoring modules are present, and no matter how many physical memories tracks are being laid down, and no matter how many feed-back structures are operating, if that brain is built out of nothing but the particles and field of classical physics, and these entities have no properties or qualities not specifically assigned to them by the principles of classical physics, then within that classical conceptual framework, there is no logical necessity for there to exist, by virtue of the existence of that brain and its activities, any conscious knowings beyond the `knowings from the outside' that the classical framework is built upon. There is, within the conceptual framework of classical physics, no logical requirement for there to be, in association with your brain, any special conscious knowing that can be identified with your conscious knowings. For your brain itself, and all its activities, are, according to the classical-physics conceptualization of it, exactly the set of spacetime trajectories of its component bits: nothing more or less. Within the classical-physics framework all aspects of your brain beyond what is represented by the idea of these trajectories themselves, gathered into various and sundry swarms, are ornamentations attached by knowings from the outside.

This fact that the existence of your brain and its activities does not, within the classical-physics framework, logically require the existence of any `knowings from the inside' (of your brain by your brain) does not mean that these knowings do not exist. It only means that classical mechanics does not require them to exist. Yet if the existence of these conscious knowings is, within classical-physics framework, not required by the existence of the physical body/brain and its activities, then these knowings could be left out of the classically conceived reality without upsetting the classical-physics causal description of the behaviour of the body/brain. Since that description is causally complete, these knowings must, within that framework, be causally inert. However, a conceptual framework that includes your conscious knowings but makes your behaviour causally nondependent upon your knowings lacks full logical coherence.

This logical deficiency motivates turning to a quantum-mechanical treatment in which mind and body are causally interpenetrating in a way that is both mathematically describable and pragmatically useful. I begin by documenting the subjective/experiential character of the orthodox Copenhagen interpretation of quantum mechanics.

The Subjective Character of the Orthodox Interpretation of Quantum Mechanics

In the introduction to his book ``Quantum theory and reality" the philosopher of science Mario Bunge (1967) said: ``The physicist of the latest generation is operationalist all right, but usually he does not know, and refuses to believe, that the original Copenhagen interpretation --- which he thinks he supports --- was squarely subjectivist, i.e., nonphysical."

Let there be no doubt about this.

Heisenberg (1958a): ``The conception of objective reality of the elementary particles has thus evaporated not into the cloud of some obscure new reality concept but into the transparent clarity of a mathematics that represents no longer the behavior of particles but rather our knowledge of this behaviour."

Heisenberg (1958b): ``...the act of registration of the result in the mind of the observer. The discontinuous change in the probablitity function... takes place with the act of registration, because it is the discontinuous change in our knowledge in the instant of registration that has its image in the discontinuous change of the probability function."

Heisenberg (1958b:) ``When old adage `Natura non facit saltus' is used as a basis of a criticism of quantum theory, we can reply that certainly our knowledge can change suddenly, and that this fact justifies the use of the term `quantum jump'. "

Wigner (1961): ``the laws of quantum mechanics cannot be formulated...without recourse to the concept of consciousness."

Bohr (1934): ``In our description of nature the purpose is not to disclose the real essence of phenomena but only to track down as far as possible relations between the multifold aspects of our experience."

In his book ``The creation of quantum mechanics and the Bohr-Pauli dialogue" (Hendry, 1984) the historian John Hendry gives a detailed account of the fierce struggles by such eminent thinkers as Hilbert, Jordan, Weyl, von Neumann, Born, Einstein, Sommerfeld, Pauli, Heisenberg, Schroedinger, Dirac, Bohr and others, to come up with a rational way of comprehending the data from atomic experiments. Each man had his own bias and intuitions, but in spite of intense effort no rational comprehension was forthcoming. Finally, at the 1927 Solvay conference a group including Bohr, Heisenberg, Pauli, Dirac, and Born come into concordance on a solution that came to be called ``The Copenhagen Interpretation". Hendry says: ``Dirac, in discussion, insisted on the restriction of the theory's application to our knowledge of a system, and on its lack of ontological content." Hendry summarized the concordance by saying: ``On this interpretation it was agreed that, as Dirac explained, the wave function represented our knowledge of the system, and the reduced wave packets our more precise knowledge after measurement."

Certainly this profound shift in physicists' conception of the basic nature of their endeavor, and the meanings of their formulas, was not a frivolous move: it was a last resort. The very idea that in order to comprehend atomic phenomena one must abandon ontology, and construe the mathematical formulas to be directly about the knowledge of human observers, rather than about the external real events themselves, is so seemingly preposterous that no group of eminent and renowned scientists would ever embrace it except as an extreme last resort. Consequently, it would be frivolous of us simply to ignore a conclusion so hard won and profound, and of such apparent direct bearing on our effort to understand the connection of our knowledge to our bodies.

Einstein never accepted the Copenhagen interpretation. He said: ``What does not satisfy me, from the standpoint of principle, is its attitude toward what seems to me to be the programmatic aim of all physics: the complete description of any (individual) real situation (as it supposedly exists irrespective of any act of observation of substantiation)." (Einstein, 1951, p.667) and ``What I dislike in this kind of argumentation is the basic positivistic attitude, which from my view is untenable, and which seems to me to come to the same thing as Berkeley's principle, {\it esse est percipi}. (Einstein, 1951, p. 669). Einstein struggled until the end of his life to get the observer's knowledge back out of physics. But he did not succeed! Rather he admitted that: ``It is my opinion that the contemporary quantum theory...constitutes an optimum formulation of the [statistical] connections." (ibid. p. 87). He referred to: ``the most successful physical theory of our period, viz., the statistical quantum theory which, about twenty-five years ago took on a logically consistent form. ... This is the only theory at present which permits a unitary grasp of experiences concerning the quantum character of micro-mechanical events." (ibid p. 81).

One can adopt the cavalier attitude that these profound difficulties with the classical conception of nature are just some temporary retrograde aberration in the forward march of science. Or one can imagine that there is simply some strange confusion that has confounded our best minds for seven decades, and that their absurd findings should be ignored because they do fit our intuitions. Or one can try to say that these problems concern only atoms and molecules, and not things built out of them. In this connection Einstein said: ``But the `macroscopic' and `microscopic' are so inter-related that it appears impracticable to give up this program [of basing physics on the `real'] in the `microscopic' alone." (ibid, p.674).

The present paper is based on my judgement that our probings into nature have revealed a profound failure of the classical-physics conceptualization of nature precisely at the point under discussion here, namely the connection of epistemology to ontology.

Epistemology and Ontology

The Copenhagen interpretation of quantum theory is epistemological: the theory is construed to be about our knowledge, and the quantum jumps---the famous "reductions of the wave packets" --- are considered to be the kind of sudden changes that naturally accompany changes in someone's knowledge. Yet nature must surely contain some reality besides our knowledge: for what happens to my knowledge when I am sleeping?

From a mathematical point of view the simplest way to represent a reality compatible with quantum phenomena is to let the quantum state represent both our knowledge of reality, in the way claimed by the Copenhagen interpretation, and also reality itself.

This move is philosophically attractive, because it moves us in the direction of an increased comprehension of the underlying unity of nature, a unity revealed by the amazing way that aspects of nature that appear at first to be completely different eventually turn out to be the same thing in different guises. The course of science has been essentially the process of uncovering this underlying unity: celestial and terrestial dynamics were unified by Newton; electro-magnetism and light were unified by Maxwell; space and time were unified by Einstein, as were spacetime and gravity; heat and molecular motion were unified by Boltzmann and others: these are just a few highlights of the unrelenting scientific process of revealing underlying oneness behind apparent diversity.

In view of these precedents it is reasonable examine the consolidation of mind and matter brought about by positing that the quantum state represents not only our knowledge, in the way specified by orthodox quantum theory, but also aspects of the reality that surrounds it.

This apparently obvious move seemed untenable to the founders of quantum mechanics because of the notorious `quantum jumps'. These jumps seemed natural for a representation of `knowledge of reality', but unacceptable as a behavior of reality itself.

Physicists were reluctant to accept the physical realness of these jumps because behaviour of this kind contradicted a strongly held prejudice stemming from the theory of relativity. But overcoming irrational prejudice is a virtue, particularly if it allows science to describe reality.

The irrational prejudice was the conviction that a free choice made by someone in one region could not affect a faraway real physical situation at a speed faster than light. This idea, that no effect of a free choice can travel faster than light, originated in the context of a strictly deterministic classical theory. But it is an open the question whether it must continue to hold in a context where determinism itself is question.

It turns out, in fact, that this no-faster-than-light demand is logically incompatible with the *predictions* of quantum theory themselves, quite apart from all matters of ontological interpretation (Stapp, 1997). So abandoning this prejudice not only removes a logical contradiction: it also makes possible the simplest mathematical representation of physical reality consistent with quantum phenomena.

The theory to be described here is based on this identification of the quantum state with both reality itself and also our knowledge of reality in the way specified by the Copenhagen interpretation. This move does indeed carry the overtones of Berkeley's *esse est percipi* that Einstein objected to, but was unable to circumvent in three decades of trying. In the present context our core problem is, as emphasized by Dennett, precisely the breakdown of the normal idea of the separation between epistemology and ontology. Hence it makes sense to examine the alternative conception of this relationship that physics is pressing so hard upon us.

Quantum Effect of Presynaptic Calcium Ion Diffusion

Let me assume here, in order to focus attention on a particular easily analyzable source of an important quantum effect, that the propagation of the action potential along nerve fibers is well represented by the classical Hodgson-Huxley equation, and that indeed all of brain dynamics is well represented by the classical approximation apart from one aspect, namely the motions of the pre-synaptic calcium ions from the exit of the micro-channels (through which they have entered the nerve terminal) to their target sites. The capture of the ion at the target site releases a vesicle of neurotransmitter into the synaptic cleft.

The purpose of the brain activity is to process clues about the outside world coming from the sensors, within the context of a current internal state representing the individual's state of readiness, in order to produce an appropriate ``template for action", which can then direct the ensuing action (Stapp, 1993). Let it be supposed that the classically described evolution of the brain, governed by the complex nonlinear equations of neurodynamics, will cause the brain state move into the vicinity of one member of a set of attractors. The various attractors represent the various possible templates for action: starting from this vicinity, the state of the classically described body/brain will evolve through a sequence of states that represent the macroscopic course of action specified by that template for action.

Within this classically described setting there are nerve terminals containing the presynaptic calcium ions. The centers of mass of these ions must be treated as quantum mechanical variables. To first approximation this means that each of these individual calcium ions is represented as if it were a statistical ensemble of classically conceived calcium ions: each individual (quantum) calcium ion is represented as a cloud or swarm of virtual classical calcium ions all existing together, superposed. This cloud of superposed virtual copies is called the wave packet. Our immediate interest is in the motion of this wave packet as it moves from the exit of a microchannel of diameter 1 nanometer to a target trigger site for the release of a vesicle of neurotransmitter into the synaptic cleft.

The irreducible Heisenberg uncertainty in the velocity of the ion as it exits the microchannel is about 1.5 m/sec, which is smaller than its thermal velocity by a factor of about 4×10^{-3}. The distance to the target trigger site is about 50 nanometers. Hence the spreading of the wave packet is of the order of 0.2 nanometers, which is of the order of the size of the ion itself, and of the target trigger site. Thus the decision as to whether the vesicle is released or not, in an individual instance, will have a large uncertainty due to the large Heisenberg quantum uncertainty in the position of the calcium ion relative to the trigger site: the ion may hit the trigger site and release the vesicle, or it may miss it the trigger site and fail to release the vesicle. These two possibilities, yes or no, for the release of this vesicle by this ion continue to exist, in a superposed state, until a ``reduction of the wave packet" occurs.

If there is a situation in which a certain particular *set of vesicles* is released, due to the relevant calcium ions having been captured at the appropriate sites, then there will be other nearby parts of the (multi-particle) wave function of the brain in which some or all of the relevant captures do not take place---simply because, for those nearby parts of the wave function, the pertinent calcium ions miss their targets---and hence the corresponding vesicles are not released.

More generally, this means, in a situation that corresponds to a very large number N of synaptic firings, that, until a reduction occurs, all of the 2^N possible combinations of firings and no firings will be represented with comparable statistical weight in the wave function of the brain/body and its environment. Different combinations of these firings and no firings can lead to different attractors, and thence to very different macroscopic behaviours of the body that is being controlled by this brain.

The important thing, here, is that there is, *on top of the nonlinear classically described neurodynamics*, a quantum mechanical {\it statistical effect} arising from the spreading out of the wave functions of the centers of mass of the various presynaptic calcium ions relative to their target trigger sites.The spreading out of the wave packet is unavoidable, because it is a consequence of the Heisenberg uncertainty principle. This spreading is extremely important, because it entails that every vesicle release will be accompanied by a superposed alternative situation of comparable statistical weight in which that vesicle is not released. This means that wave function of the entire brain must, as a direct consequence of the Heisenberg uncertainty principle, disperse into a shower of superposed possibilities arising from all the different possible combinations of vesicle releases or non-releases. Each possibility can be expected to evolve into the neighborhood of some one of the many different attractors. These different attactors will be brain states that will evolve, in turn, if no reduction occurs, into different possible macroscopic behaviors of the brain and body.

Thus the effect of the spreadings of the wave functions of the centers of the presynaptic calcium ions is enormous: it will cause the wave function of the person's body in its environment to disperse, if no reduction occurs, into a profusion of branches that represent all of the possible actions that the person is at all likely to take in the circumstance at hand. The eventual reduction of the wave packet becomes, then, the decisive controlling factor: in any given individual situation the reduction selects---from among all of the possible macroscopically different large-scale bodily actions generated by the nonlinear (and, we have supposed, classically describable) neurodynamics---the single action that actually occurs.

In this discussion I have generated the superposed macroscopically different possibilities by considering only the spreading out of the wave packets of the centers-of-mass of the pertinent presynaptic calcium ions relative to the target trigger sites, imagining the rest of the brain neurodynamics to be adequately approximated by the nonlinear classically describable neurodynamics of the brain. Improving upon this approximation would tend only to increase the quantum effect I have described.

It should be emphasized that this effect is generated simply by the Heisenberg uncertainty principle, and hence cannot be simply dismissed or ignored within a rational scientific approach. The effect is in no way dependent upon macroscopic quantum coherence, and is neither wiped out nor diminished by thermal noise. The shower of different macroscopic possibilities created by this effect can be reduced to the single actual macroscopic reality that we observe only by a reduction of the wave packet.

Brain and Consciousness

The classical-physics conception of the physical world was described above: the basic point was that the physical world is conceived to consist of ``knowables". But in the pragmatic/quantum approach we take ``knowings" to be the basic reality, and think of the objective reality as, basically, a representation of a set of possible knowings. Then when we learn something we project our new knowing onto the pragmatic/quantum objective reality by restricting the prior set of possible knowings to a new more restricted set. Specifically, the quantum reality is conceptualized as a set of knowings compatible with a wave packet that is represented by a statistical ensemble of visualizable classical worlds. When a new knowing occurs, this event is represented by reducing the prior wave packet to the part of it that is compatible with the new knowing. Thus ontology and epistemology become reconcilable because the ontology is built basically out of idealized epistemological stuff.

By adopting a quantum-theoretical approach we open the way, of course, to a quantum treatment of various chemical processes that are important to the functioning of the brain. But that is NOT the point here. Those atomic processes can surely be treated to sufficient accuracy by a quasi-classical model that merely adjusts atomic-scale properties that have little to do directly with our consciousness. The point of going to the pragmatic/quantum framework is to accommodate the huge macroscopic quantum effects that are directly forced upon us by the Heisenberg uncertainty principle, and that make the reduction of the wave packet of decisive importance in the determination the large-scale behaviour of the body/brain. These reductions of the wave packet are, within the pragmatic/quantum framework, projections of our knowings onto our mathematical representation of physical reality .

It is sufficient to consider a model of the brain that is mainly classical. To a good first approximation the introduction of quantum theory merely involves introducing on top of the normal classical statistical ensemble arising from our incomplete knowledge a further statistical ensembles of classical motions arising from the irreducible quantum uncertainties.

At first sight this just seems to overlay the classical statistical ensemble of brain states by another layer of statistical uncertainty that adds nothing perceptible to the uncertainties already present.

But there is a basic difference. In any single empirical instance only one member of the classical component of the statistical ensemble is actually present, but all of the members of the quantum superposition that are forced to be present by the uncertainty principle are necessarily all present simultaneously, until a reduction occurs. This presence in principle of the various superposed possibilities is the essence of quantum theory: it is entailed by the fact that the different superposed members of the quantum ensemble can interact with each other.

The presence of these superposed possibilities means that in any given empirical instance, no matter which classical element of the ensemble is actually present, the quantum ensemble spreads over all of the various possible attractors. Consequently, this quantum reduction exercises an overriding control over the choice of attractor: this choice *could* be the same for each of the alternative possible members of the classical ensembles, and hence independent of which of the alternative classical states (generated, for example, by the thermodynamical mixture of possibilities) is present. For the quantum principles are absolutely mute on this sort of unphysical question: *What would the choice have been if the occurring situation had been other than what it actually is?*

Thus the quantum choice could in principle be independent of which member of the classical statistical ensemble is present. But in that case the quantum choice would wipe out the classical uncertainties introduced by the thermal noise.

This point is raised merely to emphasize that the quantum choice is the decisive control element in cases---such as the human brain---where the irreducible quantum uncertainties are so great that essentially all of the alternative macroscopic possibilities are included within the range spanned by the quantum uncertainties.

The quantum choices, in this pragmatic/quantum framework, are projections of our knowings onto our representation of physical reality. If the knowing is the `knowing of my decision to raise my arm' then the projection of that knowing reduces the wave packet of my brain to one in which a certain attractor is actualized, namely an attractor that embodies the template for action that initiates and directs the raising of my arm. The reduction restricts the wave packet to one that is compatible with the new knowing, in complete parallel to the way that a `knowing of the location of a pointer on some measuring device' reduces the wave packet of the pointer, and hence also the wave packet of the entire world, to a form that is compatible with that knowing.

There is no problem here with the fact that a mere ``knowing'' can do something. For `knowings' are recognized to be real elements of the ontological structure, supplementing the `material' things that, although formerly thought to be independent basic realities, appear now as dynamically entwined parts of a greater whole that includes our minds.

This way of constructing the theory of consciousness and its connection to experiment is practical, and coherent, and it meshes nicely with our basic physical theory, quantum theory, without adding any significant mathematical burden to the theory of consciouness beyond what is needed in classical mechanics: for most purposes it will be adequate to regard the brain as merely a statistical ensemble of classical brains. But the theory nevertheless puts our knowings, which are certainly real parts of nature, into the ontology in a natural way, and allows them to play a key dynamical role, freed from the rigid control of local purely mechanical laws.

Bringing consciousness in the physical theory in this natural and efficacious manner points the way to natural resolutions of the basic philosophical problems that have beset the philosophy of mind and psychology for most of this century. It resolves the puzzlement about how something like a ``feeling'' can arise out of the motions of a set of ``billiard balls'' by simply asserting what quantum theory has already shown, namely that the world cannot be understood as being built out of such classically conceived matter: it is better understood as being built out of an entwined complex consisting of two kinds of stuff: potentialities that obey laws analogous to those that govern classical matter, and the experiential actualities that those potentialities are potentialities for. Adopting to this quantum mechanical conception of nature does not lead to any loss of the practical computational power provided by classical physics. Quite the opposite: this more modern conception of nature was essentially forced on physical scientists by the demand for a logical framework that provides an accurate, comprehesive, rational, and practically useful account of our experiences in what had formerly been thought to be the domain of the purely physical sciences. And in the study of mind/brain phenomena it will *usually* be adequate to regard the brain as simply a statistical ensemble of classically described brains, with this ensemble being always compatible with everything we know. So we get the philosophical advantages essentially for free in cases where the classical and quantum predictions agree.

I have focused here on human science, and hence on human body/minds. But reductions of the wave packet associated with other kinds of systems should also occur, and be connected to other kinds of `knowings''. However, the effort to extend the theory to nonhuman``knowers'' should, I believe, be built upon the development and testing of the theory in the human case, where the empirical evidence is more direct.

Quantum Enhancement of Psycho-states

An important question for science is this: Does the quantum dynamical involvement of our knowings, and the reductions that represent them, alter the predictions of this theory relative to those of the parallel classical statistical account. The answer is a qualified ``yes''. One possible cause of difference is now described.

Let a ``psycho-state'' be a brain state, or patterns of brain/body activity, that corresponds to a conscious experience: according to the pragmatic theory, a psycho-state is a (functional) representation of an actual knowing.

For reasons to be described now, the quantum dynamics should increase, relative to classical mechanics, the statistical likelihood of the occurrence of psycho-states relative to states having no phenomenal/epistemological content.

Within the classical approximation to quantum mechanics the predictions of the classical and quantum theories are the same. This approximation is obtained, in our simplified example where the quantum effects were confined to the motions of presynaptic calcium, by allowing the wave packets of the presynaptic calcium ions to be just statistical ensembles of classical trajectories. Then the whole model becomes classical. There can still be reductions of wave packets, but these are classical reductions that can be understood in the normal classical way as just picking out from a ensemble of imagined copies of the system those members that are compatible with a new knowing. That reduction is mathematically essentially the same in the classical and quantum theories. It is rather the dynamical evolution that is different. The quantum wave packet evolves (when no reduction is occurring) according to the Schroedinger equation, and this evolution is not identical to the evolution of a statistical ensemble of classical copies.

The difference can be dramatic. In the evolution of a classical statistical ensemble the individual trajectories evolve independently: they are together only in our imagination. But in the quantum case neighbors act on each other. Indeed, the Schroedinger equation is essentially an equation for hydrodynamical flow (Feynman, 1965).

Imagine scooping water out of a bucket with a repeated motion that takes the water always from the same place: one can nearly empty the bucket, even though the water is always taken from one place. For water rushes in to fill a void, due to the action of the different molecules upon their neighbors.

Similarly, a repeated taking away of the same substate out of the state of the brain can be expected to take away more when the dynamics is controlled by quantum theory, in contrast to classical statistical mechanics, because in the quantum case the `diffraction effect' works to fill up a void created by removing superposed neighbors, but there is no analogous effect in the corresponding classical statistical theory.

We do not yet know what physical criteria distinguishes psycho-states from the others, apart from the capacity of these states to represent ``knowings": that, indeed, may be all that is needed. But the hydrodynamical effect described above entails that it would be advantageous for an organism to be organized so that templates for actions that increase its chances of survival are psycho-states: the hydrodynamical effect will cause such states to acquire more statistical weight in a quantum system than in the corrsesponding classical system because a repeated taking away of the state corresponding to a certain knowing will cause probability to flow into that state to fill the void, and thus contribute more to a repeated action. Thus organisms that are conscious of the actions that enhance their chances of survival could become more likely to survive. This effect could follow just from the pragmatic quantum statistical rules themselves, without any assumption that mind acts to bais the probabilities specified by the basic principles of quantum theory.

Another possibly important effect of the collapses could arise if the workable templates for action are so finely honed that the thermal noise makes it difficult for them to be formed: it might be like trying to get a key into a lock with a palsied hand. In such a situation the quantum hydrodynamic effect described above would automatically act to `suck' the key into the lock, and thus to counteract the uncontrollable thermal agitation.

The most important thing, however, is that although these quantum enhancement effects might be difficult to observe directly with the technologies available today, their possible existence in principle shows that the pragmatic account is not, in principle, empirically identical to its classical approximation. Hence it is reasonable to hope that the philosophical coherence of the pragmatic quantum approach reflects a movement toward a truer picture of nature that eventually will be supported by empirical ramifications.

Acknowledgement

I thank Aaron Sloman for helpful criticism of an earlier draft.

References

Bohr, N (1934), *Atomic Theory and the Description of Nature* (Cambridge: Cambridge University Press).

Bunge, M. (1967), *Quantum Theory and Reality* (Berlin: Springer).

Einstein, A. (1951) *Albert Einstein: Philosopher-Scientist* ed. P.A. Schilpp (New York: Tudor).

Fogelson, A.L. \& Zucker, R.S. (1985),`Presynaptic calcium diffusion from various arrays of single channels: Implications for transmitter release and synaptic facilitation', *Biophys. J.*, *48*, pp. 1003-1017.

Feynman, R., Leighton, R., and Sands, M., (1965) *The Feynman Lectures in Physics.* (Vol. III, Chapter 21). New York: Addison-Wesley.

Heisenberg, W. (1958a) `The representation of nature in contemporary physics', *Deadalus 87*, 95-108.

Heisenberg, W. (1958b) *Physics and Philosophy* (New York: Harper and Row).

Hendry, J. (1984) *The Creation of Quantum Theory and the Bohr-Pauli Dialogue* (Dordrecht: Reidel).

Stapp, H.P. (1993) *Mind, Matter, and Quantum Mechanics* (Berlin: Springer), Chapter 6.

Stapp, H.P. (1997) `Nonlocal character of quantum theory', *American Journal of Physics, 65*, 300-304.

Wigner, E. (1961) `The probability of the existence of a self-reproducing unit', in *The Logic of Personal Knowledge* ed. M. Polyani (London: Routledge \& Paul) pp. 231-238.

Zucker, R.S. \& Fogelson, A.L. (1986), `Relationship between transmitter release and presynaptic calcium influx when calcium enters through disrete channels', *Proc. Nat. Acad. Sci. USA*, *83*, pp. 3032-3036.

9

Do All Dynamical Systems Have Memory? Implications of the Systemic Memory Hypothesis for Science and Society[1]

Gary E. Schwartz, Ph.D. & Linda G. Russek, Ph.D.
Department of Psychology
The University of Arizona
P.O. Box 210068
Tucson, AZ 85721-0068
Telephone: 520-621-5497 / Fax: 520-621-9306
E-Mail: gschwart@ccit.arizona.edu

Abstract

Memory is one of the most fundamental and far reaching phenomena in nature. Memory technology is currently evolving at a rapid rate, improving our ability to store not only information but energy as well. However, is the process of memory unique and selective, limited to nervous systems, immune systems, and storage systems of modern technology? Or, is the storage of information and energy an essential and ubiquitous process in nature, inexorably tied to the very process of the existence of objects as dynamical systems? When a dynamical energy systems approach is used to examine the process of interaction (especially recurrent feedback interaction) in systems, the logic leads to the systemic memory hypothesis which posits that all dynamical systems store information and sustain memory. The chapter outlines the fundamentals of systems theory, information and energy, and illustrates the profound implications that follow when these concepts are integrated. Using resonance between and within two tuning forks as a model system, the logic of how recurrent feedback interaction generates dynamic memory is explained. Recurrent feedback interaction was anticipated by William James (1890), was implicit in the writings of Warren McCulloch (1951), is implicit in the holonomic brain theory of Karl Pribram (1991), and is implicit in the neural network concept of the recurrent feedback loop. Recurrent feedback interaction not only occurs between neurons in neural networks, but occurs within all cells and molecules. Heretofore unexplained observations reported in homeopathy, kinesiology, organ transplants, and parapsychology can be understood in terms of the systemic memory hypothesis. One of sciences greatest challenges is to develop mathematical models for the retrieval of holistic information inherently stored in dynamical systems. Future research can confirm or disconfirm the wide ranging predictions of the systemic memory hypothesis for science and society.

[1]In Pribram KH and King JS, eds (1997). *Brain and Values: Behavioral Neurodynamics V*. Hillsdale, NJ: Lawrence Erlbaum Associates, in press (January 23, 1997).

The real voyage of discovery consists not in seeking new lands but in seeing with new eyes.

(Marcel Proust, 1871-1922)

In this regard I would caution the reader to adhere to a maxim once issued by Warren McCulloch: "Do not bite my finger; look where I am pointing."

(Pribram, in Laszlo, 1995)

Introduction and Overview

Memory is one of the most fundamental and far reaching phenomena in nature. The human brain has the extraordinarily advanced ability to store highly diverse and complex forms of information. It also has the remarkably creative capacity to invent new electronic, chemical and mechanical systems for storing ever increasing amounts of information. Memory technology is currently evolving at a rapid rate, improving the capacity to store energy as well as information.

In lay usage, the term memory refers to the conscious *retrieval* of information as well as the *storage* of information. Historically, the term memory came from the Latin *memoria*, which meant "the faculty of remembering" (Webster, 1977). However, this use of the word memory (which in psychology is termed explicit memory) is only one form of memory. Substantial research on memory without awareness, termed implicit memory, documents that information can be stored in the absence of conscious awareness, and evidence for memory storage can be found in behavior in the absence of explicit recall (Schacter, 1996). Implicit memory refers to the demonstration that information has been stored per se, even if the information can not be retrieved consciously. The modern use of the term memory in computer science (e.g. hard disk memory) and electrical engineering (e.g. battery memory) also refers to the storage of information per se. In this chapter, we use the term memory in the generic sense to refer to the storage of information per se, regardless of whether the information can be retrieved consciously.

Is the process of memory unique and selective, limited to the nervous system, the immune system, and storage systems invented through technology? The purpose of this chapter is to outline the logic that leads to the far reaching hypothesis that all dynamical systems generate and sustain memory, and to consider some of the implications of this hypothesis, termed the *systemic memory hypothesis*, for science and society. As will become clear, implicit (unconscious) storage of information and energy may be the rule, rather than the exception, in all levels of nature --- explicit memory, the conscious retrieval of information and energy, may be a special case of implicit memory.

The systemic memory hypothesis occurred to the first author while he was a professor at Yale University in the early 1980's working on applications of systems theory to behavioral medicine and modern physics (e.g. Schwartz, 1984, 1987). In the process of attempting to explain to a seminar of students the basic mathematics of how feedback interactions worked, he unexpectedly came to the realization that feedback interactions, especially cyclic, repeated feedback interactions (termed here *recurrent feedback interactions)*, inherently involved the dynamic storage of information. Because he considered the systemic memory hypothesis too controversial to communicate at that time, he did not publish the hypothesis.

However, inspired by the second author's understanding and extension of the hypothesis to the storage of energy in energy medicine (Russek and Schwartz, 1996a;b), the decision was reached to publish the logic and implications of the hypothesis with the hope that this will stimulate thoughtful debate and research designs to test the hypothesis.

The systemic memory hypothesis addresses a persistent mystery in modern science, the origin of holism and emergent properties in dynamical systems. The simple dictum "the whole is greater than the sum of its parts, and its corollary, "the whole depends upon the interaction of its parts for its unique properties" together express the foundation of general and living systems theory (von Bertalanffy, 1968; Miller, 1978) and modern complexity and chaos theory (Kauffman, 1993) as they are employed in the physical, biological, behavioral, social and environmental sciences. The problem of the origin of holism in dynamical systems is potentially resolved when circular feedback is viewed as recurrent feedback interactions that inexorably lead to the storage of information and the creation of dynamical systemic memories in all dynamical systems at all levels. The complex linear and non-linear interactions that naturally accumulate through the circulation of information and energy turn out to be holistic memories that reflect the evolving identity of systems as emerging wholes.

As will become clear below, the logic used to generate the systemic memory hypothesis is relatively simple and straightforward. The logic has been evolving over the past 100 years, and inexorably follows from modern systems theory. We will describe how the systemic memory hypothesis was anticipated by William James in 1890, how it was implicitly recognized by Warren McCulloch in 1951, how it is implicit in the holonomic brain theory of Karl Pribram (1991), and how it is the explanation for the creation of memory in neural networks. The concept of the recurrent feedback loop in neural networks (e.g. Lisberger and Sejnowski, 1992) turns out to be a special case of the general principle of recurrent feedback interaction.

In addition to explaining the origin of dynamic memory and the expression of holism and emergent properties in systems, the logic of recurrent feedback interaction also provides an explanation for self-organization, self-awareness, and evolution in systems. As will become clear, some of the predictions that follow from the concept of recurrent feedback interaction prove to be novel, and some prove to be very controversial. This is unavoidable. *Conceptual integrity and logical completeness requires that the deep and challenging implications be entertained openly and addressed seriously.*

Our current ideas about memory explicitly shape science and society, especially concerning our openness to phenomena that challenge our current beliefs because they appear inexplicable. We logically dismiss controversial claims for the clinical effectiveness of homeopathy and well-controlled double-blind studies documenting significant effects of selective homeopathic preparations (reviewed in Bellavite and Signorini, 1995), because our theories lead us to hypothesize that while neurons and immune cells can have memory for molecules, water per se cannot. We logically dismiss historical claims of native populations and well documented psychics (e.g. Smith, 1964) that detailed information about persons and situations can sometimes be retrieved from ancient artifacts and personal effects, and explain these claims as due to coincidence or superstition, because our theories lead us to hypothesize that while neurons can have detailed information about persons and situations, material artifacts cannot. We logically dismiss emerging reports from certain organ transplant patients that their memories and values have changed in surprising ways since receiving their new organs (e.g. Siegel, 1995), and explain their post surgical symptoms as side effects of immunosuppressant drugs or stress, because our theories lead us to hypothesize that the donor's personal history is only stored in his or her central nervous system, and not in his or her donated organs (for example, the heart or the lungs).

However, how would science and society change if a theory, already well accepted within modern science, turned out to predict that the dynamic storage of information and energy was an essential and ubiquitous process in nature, inexorably tied to the very process of the existence of objects as dynamical systems? How would the practice of basic science change if such a theory predicted that electromagnetic energy and information, expressed for example as organized patterns of visible light, was continuously stored in all objects to varying degrees, especially in objects that were black in color? How would the practice of medicine change if such a theory predicted that sensory and perceptual information was continuously stored in all biological tissues, including the heart and lungs? How would the practice of our religious institutions and rituals change if such a theory predicted that personal information historically termed "spiritual" was stored in the personal effects of our loved ones?

This chapter begins with a brief overview of systems theory and the concepts of information and energy. The logic of dynamical recurrent feedback interaction is then analyzed, and the emergence of the systemic memory hypothesis is explained. Various implications of the hypothesis for science and society are then outlined. The relationship of dynamic memory to valence is discussed in this context. Though neurons are clearly especially gifted in storing sensory and psychological information (because neurons are so highly interconnected, creating profoundly complex recurrent feedback interactive networks), it may be time to evolve our intellectual heritage and re-envision the brain as being a marvelously special case of a ubiquitous dynamic systemic (holistic) memory process in nature.

What is Systems Theory? The Key is Dynamic Interaction

Historically, systems theory, including general systems theory (Von Bertalanffy,1968) and living systems theory (Miller, 1978), was developed as a conceptual tool to organize and integrate knowledge within and across disciplines, from the physical and biological sciences to the behavioral and social sciences (Schwartz, 1982; 1984; 1987; 1989). Some recent advances stemming from systems theory include information theory (Abu-Mostafa, 1957), self-regulation theory (Mithaug, 1993), complexity theory (Flood, 1988), self-organizing theory (Yates, 1987; Kauffman, 1993), and chaos theory (Cambel, 1993). Chaos theory has been applied to both classical and quantum physics (NATO, 1985), an example of its integrative capacity. The further use of systems theory as a conceptual framework for integrating classical and quantum physics in the psychobiology of personality and health is illustrated in Schwartz (1984; 1987). The integration of systems theory with modern concepts of energy reflects the continued evolution of these ideas (Russek and Schwartz, 1996a,b).

As described by Miller (1978);

> "General systems theory is a set of related definitions, assumptions, and propositions which deal with reality as an integrated hierarchy of organizations of matter and energy. General living systems theory is concerned with a special subset of all systems, the living ones." (page 9).

The meaning of "system" is often confused and is used differently in different contexts. A "system" of measurement units (e.g. the measurement of energy, see below) simply means "structure" (static information), whereas a "system" of cells comprising an organ means much more - "interactions", "relationships", "non-linear summations" (dynamic information). Hence, whereas the "structural" meaning of system is static and closed, the "interactions" meaning of system is dynamic and open. Note that the term interaction is used broadly here --- the concept of interaction in statistics (e.g. analysis of variance) reflects a special (limited) case of the concept of interaction in systems.

The root meaning of the word "system.", which derives from the Greek *synhisanai* ("to place together") is the concept of an *integrated whole whose essential properties arise (emerge) from the relationships between its parts.*

A classic example of holism in a system is how atoms of hydrogen and oxygen, each with their own unique set of holistic properties, can combine to form H_20, a molecule whose unique (holistic) properties are qualitatively different from the individual properties of hydrogen and oxygen, and only emerge when hydrogen and oxygen interact recurrently. The same logic applies to the atoms of hydrogen and oxygen themselves. Electrons, protons, and neutrons each have their own unique set of properties. They can combine to form hydrogen, an atom whose unique (holistic) properties are qualitatively different from the individual properties of the subatomic particles, and only emerge when the particular combination of the subatomic particles interact recurrently and create hydrogen.

In systems theory, a system is defined as follows (Miller, 1978):

> "A system is a *set* of *interacting units* with *relationships* among them. The word 'set' implies that the units have some *common* properties. These common properties are essential if the units are to interact or have relationships. *The state of each unit is constrained by, conditioned by, or dependent on the state of the other units.* The units are *coupled.* Moreover, there is at least one measure of the sum of its units which is *larger than the sum* of that measure of its units." (page 16, italics added).

These ideas, so to speak, are the heart of systems theory. Table 1 illustrates

Table 1: Comparison of Non-Systemic and Systemic Concepts

Non-Systemic Concepts	Systemic Concepts
Independence	Interdependence
Separation	Relationship
Static	Dynamic
Closed	Open
Active	Interactive
Disconnected	Connected
Linear summation	Non-linear as well as linear summation
Linear prediction	Emergent (non-linear) as well as linear prediction
Linear action	Circular (recurrent) interaction
Randomness	Complex order
"Either - Or" logic	"And" logic as well as "either-or" logic
State independent	State dependent
Parts	Wholes
Fixed	Flexible, creative

some core concept differences between a non-systemic and systemic approach to nature.

Living Systems theory (Miller, 1978; Capra, 1996) adds functional units which generate and process information, including memory, decoder, and reproducer subsystems, that operate at multiple levels (from cells to societies). Paraphrasing Miller (1978), we have defined *living* systems as *dynamic organizations of intelligent information expressed in energy and matter* (Russek and Schwartz, 1996a). Because systems theory is deliberately general (i.e. it is not unique to any discipline or sub-discipline) it might more appropriately be called a meta-theory or meta-framework. In any case, we have taken these general ideas, applied them to the concept of energy, and reframed them as dynamical energy systems hypotheses. However, we must first review what is meant by the terms information and energy before we can explain how systems theory revises our vision of memory in nature.

What are Information and Energy? Energy Does the "Work" of Information

Information is one of the most mundane yet complex concepts in modern psychology. In one sense, everyone knows what information is. We use everyday language such as the "Information Highway" and the "Information Age." According to Miller (1978), information in systems refers to the abstract concept of patterns (literally, to give form). Information may be simply descriptive (structure and mathematical order), it may reflect complex knowledge, or it may convey deep wisdom. Information is the heart, so to speak, of the concept of mind, and is the foundation of consciousness and intention (Chalmers, 1996).

Everyday language reminds us that there is much more to information than just the "surface" pattern, form or order. For example, the pattern of letters "bon jour" is explicitly different than the pattern of letters "hello" even though both patterns of letters implicitly mean the same thing (e.g. warm greeting). Patterns can have "depth" or layers of meaning --- which is, of course, the essence of language. Language involves the translation of intended meaning into sequences of words (encoding by the sender) and pattern recognition and interpretation of the sequences of words (decoding by the receiver). Systems theory emphasizes that components "process" information --- information is "interpreted", and interpretation influences "memory." Since the behavior (functioning) of systems is influenced by both the physical and interpreted aspects of information, the systemic (holistic) memories that are stored will reflect a complex integration of pattern and meaning.

Energy is one of the most mundane yet mysterious concepts in modern physics. In one sense, everyone knows what energy is. We use everyday language such as "the sun generates energy." Electricity, magnetism, sound, pressure, temperature, these common terms all involve energy in some way. The reader not familiar with the measurement of energy, force and fields, especially as they are applied to biological systems, will find a clear introduction in *The Body Electric*(Becker and Selden, 1988) and *Cross Currents* (Becker, 1990). The sophisticated reader will find a detailed review in *Bioelectromagnetism* (Malmivuo and Plonsey, 1995). Research on the effects of electromagnetic energies on cellular and organ function are regularly published in the journal *Bioelectromagnetics.*

The concept of energy addresses the fundamental question of how things move in space. Classically, energy is defined as the "ability to do work." (Miller, 1978, page 11). Physics speaks of two general types of energy: (1) kinetic energy, the energy that is expressed when an object accelerates, and (2) potential energy, the energy that is potentially available. In classical physics, quantities such as force and energy are usually measured either in meters, kilograms, and seconds (termed the "MKS system") or in centimeters, grams, and seconds (termed the "CGS system") (Orear, 1962). Note that the term "system" was used here by a physicist (Orear, 1992) simply to mean "structure" -- in a static sense -- not to mean "interactive units" -- in a dynamic sense -- as is used by systems theorists.

Though the measurement of energy may be defined precisely, the interpretation of energy is abstract and is difficult to comprehend (even by seasoned physicists). The inherent difficulty in understanding energy can not be understated. The abstract nature of energy may be one reason why the concept of energy, though appreciated by certain early scholars of psychology and medicine (e.g. Freud and Jung), has yet to make its way into the mainstream of modern psychology, neuroscience, and medicine (Russek and Schwartz, 1996a,b).

In this chapter, we use the term energy in the most general sense to reflect any process that has the "ability to do work" and therefore can influence the motion, and hence functioning, of objects (or systems). Examples of objects or systems include atoms, molecules, cellular processes, physiological functioning, consciousness, and overt behavior at the level of the organism. We will speak of biological systems as being energy "generating" systems, and we will speak of electrostatics (stationary charge), electricity (the flow of electrons), magnetism, sound, temperature, and pressure as examples of "energies" since they all have the ability to do work.

However, systems theory reminds us that what we measure is the *functioning of systems*. Energy and force are *concepts we invent* to make sense of the *observation* that iron particles do jump through space to a permanent magnet, and we use this observation to explain how, for example, the heart can be observed to contract when an electromagnet is placed near the chest (Ragan, Wang, and Eisenberg, 1995).

In summary, we can think of information as the form or structure of a system - and energy as the force or function that moves the system and enables it to emerge and evolve. In living systems, information and energy typically go hand in hand --- energy contains information, and information is expressed through energy. Not only does energy convey information, in a deep sense energy does the work of information. It follows that organized energy both expresses and actualizes information (Schwartz, 1997; Russek and Schwartz, 1996a). As will become clear below, information and energy are both stored in systems as memory.

Conservation of Information and Energy

Classical physics reminds us of a profoundly important yet little appreciated fact --- *information contained in energy, once produced, does not spontaneously vanish* (Orear, 1962). For example, the discipline of astrophysics is based on the assumption that the energy and information from stars, once released, travel at the speed of light indefinitely. This assumption explains the well known fact that the visual experience of the stars at night reflects the electromagnetic history of stars at different ages depending upon their distance from the earth.

It logically follows that the visual pattern created by light reflected from a person's face, though obviously very weak compared to the energy of stars, also travels into space and continue indefinitely. Similarly, the electromagnetic pattern generated by a person's organs (e.g. the electrocardiogram, generated by the heart, which is the largest electromagnetic signal generated by a single organ in the body), also travels into space and continue indefinitely. The idea that electromagnetic energy and information have a kind of immortality is not, per se, controversial. However, what becomes controversial is the realization of what happens when energy and information interact with dynamical systems.

Dynamical Energy Systems --- Resonance as Interaction

When the essence of the mechanics of dynamic interactions is understood, the emergence of the systemic memory hypothesis becomes self-evident. We illustrate the logic using a simple two element system --- the interaction of two tuning forks.

Consider the phenomenon of resonance between two tuning forks (A and B) that vibrate at similar rates. If one tuning fork (A) is struck, a second tuning fork (B) some distance from A will begin to "resonate" (make a sound), especially if B is identical in shape, size and substance to A.

How is this phenomenon explained? The classical, non-systemic interpretation is to say that tuning fork A generates sound "waves" which reach tuning fork B, and if B naturally vibrates at a frequency similar to the frequency generated by A, B will begin to move in synchrony with A. B therefore responds to A. A acts on B, hence B reacts to A.

It is worth noting that radar, a common measurement device, requires resonance. Waves are emitted (by A), travel to B and resonate with it, waves from B return to C (the receiver), and the waves that return are compared with the waves that left. However, *radar is designed to be minimally interactive* in the sense that the purpose here is to *hold A relatively constant, regardless of changes caused by B that might in turn cause changes in A*. The tuning fork example is more complex because *tuning forks have the potential to interact ---- A can react to B just as B can react to A.*

A dynamical energy systems interpretation requires that we reinterpret the relationship between tuning forks A and B as interactive, dynamically coupled, and cyclically connected. A does not simply *act on* B, A *interacts with* B. A and B are connected through the air (they are also connected electromagnetically, but we will ignore this for the time being). Since A and B are connected (coupled) when the energy from A begins to move B (cause vibration in B), B begins to generate a sound (energy), which returns to A. Therefore, *B begins to cause vibration in A shortly after A begins to cause vibration in B.* This implies that the functioning (behavior) of A (e.g. measured by the sound A emits) will be modulated in some complex ways by the functioning (behavior) of B (e.g. measured by the sound B emits).

To observe these predicted interactive effects, it becomes necessary to measure the sound coming from both A and B simultaneously. Using highly directional microphones, one microphone can be focused on A - pointed away from B - the other microphone can be focused on B ---pointed away from A. The sounds must be monitored simultaneously and displayed over time. Data need to be collected striking A in the absence of B and striking A in the presence of B. *The dynamical energy systems prediction is that we should observe that A's behavior is quantitatively and qualitatively different (e.g. more complex) when struck in the presence of B as compared to when A is struck in the absence of B.* Conversely, we should observe that B's behavior is quantitatively and qualitatively different (e.g. more complex) when struck in the presence of A as compared to when B is struck in the absence of A.

This seemingly trivial interaction is not so simple to mathematically model over time. Complex phase interactions (dynamic interference patterns) need to be calculated over time and distance, and as will become clear below, the interaction will naturally evolve over time due to the circulation and sharing of energy and information between A and B.

Tuning forks A and B can be shown to be interdependent and interactive, processing information generated by each of them as they interact. From a dynamical energy systems perspective, since all systems are open to varying degrees at various times, it follows that all systems should interact energetically in complex and sometimes powerful ways.

What Happens *Between* A and B Also Happens *Within* A and B

Tuning forks A and B are obviously material objects, they are each material systems. As illustrated above, when tuning forks A and B resonate, they become a two tuning fork system. Each tuning fork functions as a subsystem in the two tuning fork system (or we can say each tuning fork is a system in the two tuning fork suprasystem - the principle is the same).

Tuning forks A and B each contain molecules. Molecules are subsystems within each tuning fork (or we can say each molecule is a system within a single tuning fork suprasystem - the principle is the same). Hence, it follows that resonance not only can occur *between* tuning forks A and B, resonance can occur *within* tuning forks A and B. In fact, tuning forks A and B cannot vibrate as individual tuning forks unless their molecules can vibrate (resonate) as a whole system. In other words, the logic of what happens between tuning forks A and B applies to what happens within tuning forks A and B as well.

Modern quantum physics tells us that even at the temperature of absolute zero, matter vibrates, and hence, resonates. As described by Forward (1996) in his tutorial on quantum mechanics:

> "Matter is quantized. A block of matter, although seemingly a continuously divisible divisible substance, is ultimately found to be made of 'quanta 'called atoms. An atom consists of a small massive nucleus surrounded by a large cloud of electrons. The electron cloud acts as a "spring" suspension for the mass of the nucleus, and suspends it in its place in the block of matter. This mass-spring system can vibrate....These vibrational quanta have been named 'phonons'..... Now here comes the interesting part. When the equations of quantum mechanics are used to determine the 'average energy' $\langle E \rangle$ of the vibrations of the atoms, the answer is $\langle E \rangle=[n(T)+1/2]hf$, where the number of phonons $n(T)$ is a function of temperature such that when T=0 K, $n(T)$=0. Thus, even at zero temperature, quantum mechanics predicts that each of the atoms will have an average residual energy of $\langle E \rangle=hf/2$. This residual energy is an average. It is not that the energy of each atom is a 'half a phonon' but that roughly half the atoms have one (perhaps more) vibrational quanta or phonons, while the others have no phonon. *The phonon distributions rapidly change with time as the phonons are passed back and forth between the many atoms.* This residual energy at zero absolute temperature predicted by the equations of quantum mechanics is the so-called 'Quantum Mechanical Zero-Temperature Vibrational Fluctuations of Matter." (p. 328-329).
>
> Forward (1996) goes on to say:
>
> "This quantum mechanical fluctuation energy of the atoms in matter has been measured by measuring the vibrations in a crystal as the temperature of the crystal is lowered. The experimental data agree with the predictions of the equations of quantum mechanics, so the quantum mechanical zero-temperature vibrational functions of atoms in matter is real. It is this residual quantum mechanical vibrational energy that keeps liquid helium from freezing even when it is cooled to within microdegrees of absolute zero temperature." (page 329).

When matter is viewed from a dynamical energy systems perspective, logic dictates that interactive resonance not only occurs *between* material systems (e.g. between tuning forks A and B), but also occurs *within* material systems (e.g. within tuning fork A and within tuning fork B) as well. For this reason, the logic that leads to the hypothesis outlined below, that recurrent feedback interaction creates dynamic memory *between* tuning forks A and B, also requires that we entertain the hypothesis that recurrent feedback interaction simultaneously creates dynamic memory *within* tuning fork A and within tuning fork B as well, and this intra-tuning fork memory is sustained, even after tuning forks A and B have been separated. In psychological terms, the "short term" memory of the interaction between A and B may be sustained as a "long term" memory by the interactions within A and B.

From Interaction to Dynamic Memory - Recurrent Feedback Interaction and the Systemic Memory Hypothesis

The interaction between tuning forks A and B (or between any A's and B's ---- e.g. photons, subatomic particles, atoms, molecules, cells, organs, organisms, groups of organisms, etc.) contains a profound implication. As A and B interact with each other, they literally create a dynamic memory of their interaction over time through the circulation of their information and energy. *In fact, this dynamic memory is part of the expression of their interactions as a whole, and is a natural requirement for them to interact.* The dynamic memory is the relationship. The relationship information is systemic information, and systemic information is an expression of the whole. Moreover, it follows that the information of the whole of a system is stored dynamically within each of the parts comprising the system to varying degrees.

This idea, surprisingly, is not well appreciated, even by seasoned systems theorists. However, the logic is relatively simple and straightforward, and inevitably follows when feedback interact is analyzed systemically. If A and B do not interact (the classic non-systemic approach), dynamic memory between A and B is not an issue. However, if A and B do interact, and interact recurrently, *logic dictates that dynamic memory processes must unfold over time naturally.* We illustrate this point by considering memory from the point of view of tuning fork A. The logic is displayed in Figure 1.

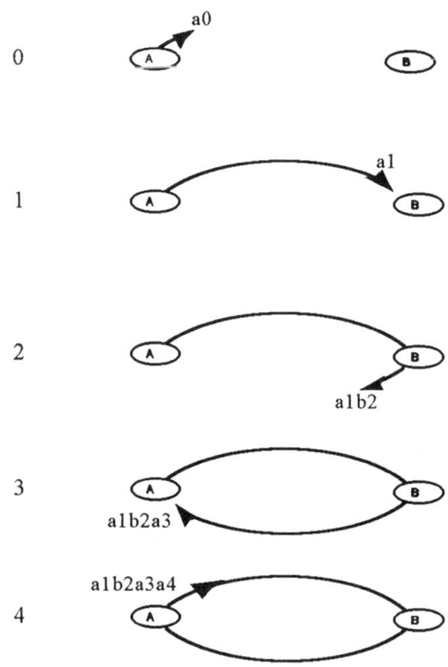

Figure 1: Graphic description of the logic of dynamic (systemic) memory.

At time 0, before A is struck, A is vibrating dynamically as predicted by quantum dynamics (e.g. its electrons, protons and neutrons are vibrating and moving in complex, interacting, resonating ways as mentioned above). When A is struck at time 1, it vibrates with a frequency "a1." Now, frequency "a1" moves (at the speed of sound) to tuning fork B. B begins to vibrate at time 2, which is some complex product of "a1" (we will, for the sake of simplicity, assume for the moment that "a1" has not changed much as it travels to B) and B's state at time 2. We will call this complex product "a1b2".

As B vibrates, the sound now returns to A at time 3. The sound that returns to A at time 3 is a complex product of "a1b2" plus whatever interference occurs with the continued sound generated by A at time 3 (a3). This gives us the even more complex product "a1b2a3" that returns to A at time 3.

Now, let us hypothesize that A is influenced by this returned sound at time 3 to some degree, and starts to interact with A's vibration at time 4. Not only will A's vibration have changed spontaneously by time 4 (e.g. it might be decreasing - the simplest case), *but it will have further changed by the complex "a1b2a3" interaction reaching it*. The resulting sound will be a4 modulated by "a1b2a3", or "a1b2a3a4."

In other words, in one complete "cycle" of A - B interaction (A influences B, B responds in part as feedback from A), the sound generated by A at time 4 contains the *complex history of the A-B interaction (their a1b2a3 relationship) over the cycle*. We see an "image" of a1, modulated by b2, interacting back with a3, returning to A, so that the vibrational pattern of A at time 4 includes the *history of the first A-B interaction in its next interaction*. Meanwhile, from the perspective of B, a somewhat parallel set of interactions is also simultaneously occurring.

Each cycle includes the previous information, hence the history (dynamic memory) continues to grow. *As long as the units are connected, the memory trace will be retained, albeit modified as time goes on.* All things being equal, the dynamic memory trace will "grow" with time.

Now, if we take a dynamical energy systems approach, we realize that *systems are always interconnected to various degrees in various ways*. Hence, energy (and the information riding with the energy) is continually being exchanged to various degrees in various ways. As a result, the dynamic memories continue naturally.

The logic and simple mathematics of the two tuning fork example are obviously grossly oversimplified. We have chosen only four time points, and have described the complex interaction only from the perspective of tuning fork A. However, the essence of the idea should be self-evident. Any time two (A + B) or more (A + B + n) things interact, information concerning their history accumulates in some form over time. A and B can be a an electron and a proton, a thermostat and a furnace, the two strands of DNA, the heart and the brain, a mother and a fetus, even the planets and the sun. From a dynamical energy systems perspective, *at whatever levels the systems are interacting, the interactive history of the energy and information should be contained in a complex way.* Simply stated, all memory is dynamic (systemic) memory.

As described in the previous section, every system is, by definition, composed of multiple subsystems, and these subsystems normally are in constant interaction with each other. *Therefore, it logically follows that after tuning forks A and B have interacted, tuning fork A and tuning fork B will have each stored information about their A-B interactions.* This information will continue to circulate within each tuning fork, even if the tuning forks are separated from each other. In fact, if quantum mechanics is correct, this information will continue to circulate within each tuning fork, even if the tuning forks were cooled to absolute zero.

Of course, outlining the logic that dynamic memory is intrinsically created and stored in systems does not imply that this information, once stored, can be accessed and retrieved (at least in human consciousness). *The deep question of retrieving dynamic memory, once stored, will be briefly addressed later.*

Some Implications for Basic Science - Recurrent Feedback Interactions and Dynamic Memory in Neurons, Neural Networks, Cells, DNA and Water.

The concept of recurrent feedback interaction has been emerging for at least the past 100 years. The concept of recurrent feedback interaction was anticipated at the neural level by William James (1890). In chapter 16 of James *Psychology (Briefer Course)*, James stated:

> "When two elementary brain-processes have been active together or in immediate succession, one of them, on re-occurring, tends to propagate its excitement into the other."

When the word "subsystems" is substituted for "brain-processes", the systemic memory hypothesis is anticipated.

> "When two subsystems have been active together or in immediate succession, one of them, on re-occurring, tends to propagate its excitement into the other.

McCulloch, in his seminal paper "Why the Mind is in the Head", not only proposed the idea of "reverberatory memory", but pointed out that:

> "The reverberating activity patterned after something that happened retains the form of the happening but loses track of when it happened. Thus it shows that there was some time at which such and such occurred. The 'such and such' is the idea wrenched out of time."

He went on to state:

> "It is an eternal idea in a transitory memory wherein the form exists only so long as the reverberation endures. When that ceases, the form is no longer anywhere."

McCulloch did not extend his logic to physical and biological systems other than neurons. Had he done so, he would have discovered that "reverberation" (circulating interaction) in systems is the rule, not the exception, and that according to modern quantum dynamics, reverberation persists, even at absolute zero temperature. In fact, recurrent feedback resonance would be predicted to occur in its most "effortless" (frictionless) form in "super-conducting" states of matter. Only in the case of presumed complete entropy would it be predicted that dynamic memory should be eliminated completely (note that entropy is a concept derived from classical physics ---systemic physics would predict that recurrent feedback interactions would likely continue in the entropic state). Note also that the psychological distinction between "short" term and "long" term memory blurs when a dynamical energy systems perspective is taken (e.g. "short" term memory may differ from "long" term memory in terms of the difficulty of retrieving the "short" term memory rather than the presumed lack of the storage of the short term memory per se).

The concept of recurrent feedback interaction is implicit in the writings of Karl Pribram (e.g. Pribram, 1991) in his holonomic brain theory of perception and language. His writings are replete with descriptions such as "local circuit interactions", "ensembles" with "iterations", "spatial interactions", "back propogations", and "cooperative interactions." For example, Pribram writes:

> "A microprocess is conceived in terms of *ensembles* of *mutually interacting* pre- and postsynaptic events distributed across limited extents of the dendritic network. The limits of *reciprocal interaction* vary as a function of input (sensory and central) to the network --- limits are not restricted to the dendritic tree of a single neuron. In fact, reciprocal interaction among pre and postsynaptic events often occurs at a distance from one another." (Pribram, 1991, pg. 16).

It is this capacity for "reciprocal interactions" to occur between highly interconnected, distributed, "mutually interacting" dendritic processes that allows for the emergence of holographic-like information to be stored throughout complex dendritic (network) systems. At the end of the Appendix (Pribram, 1991), Yasue, Jibu and Pribram note that information "comes to be stored in the new neural channel *between the units A and B*. The inference process thus has a procedure for enlarging the scope of inference." Though not emphasized by Pribram, the logic he has used to explain dynamic memory formation in an interconnected, mutually interacting neural network system can be applied to dynamaic memory formation in *any highly interconnectted, mutually interacting dynamical network system*.

It is well known that recurrent feedback loops in neural networks foster dynamic memory and learning in neural networks (e.g. Lisberger and Sejnowski, 1992). Neural networks are mathematical models that reflect our attempt to envision how recurrent neural interactions allow for memory to occur in the brain. It has not been appreciated that the logic used to explain recurrent feedback loops in neural networks can be equally applied to recurrent feedback networks in all dynamical systems at all levels. However, the reasoning is fundamentally the same. Moreover, the insight that circulating recurrent feedback interactions provide an explanation for holism in all dynamical systems has not been previously explicated.

Hence, information should be stored dynamically not only in neurons, but in all biophysical systems, from the very small (e.g. electron - proton interactions in the atom) to the very large (e.g. atmosphere-ground interactions in the planet). The systemic memory hypothesis clearly stretches and challenges our capacity to envision the dynamical interactive nature of systems (e.g. do recurrent feedback interactions occur between the moon and the earth? Do recurrent feedback interactions occur between the earth and the sun?). By definition, the mathematics of dynamic recurrent feedback interactions is not limited to a subset of scientists in a subset of disciplines studying a subset of systems. For this reason the hypothesis is termed the systemic memory hypothesis and the memory is term dynamic memory.

Certain molecules may be particularly good at storing complex information. DNA, the profoundly complex double-helix, may be especially designed to store informational and energetic interactions between its two strands. However, all cells, theoretically, should store information --- not just neural cells and immune cells, but heart cells and lung cells as well. Because individual neurons are so extensively interconnected with other neurons, and because there are literally billions of extensively interconnected neurons, neurons are exquisitely designed to generate profoundly rich and complex parallel recurrent feedback interactions, which may be why they are especially successful in the storage of rich and complex information.

Of course, each cell can only store the information it receives, and it will process this information in its own way. Systems can potentially store only what they are capable of responding to (and hence processing). Technology reminds us, for example, that photographic plates store visible light patterns much better than they do audible sound patterns.). It logically follows that the nature of information stored between subatomic particles, for example, will be of a different order than the nature of information stored between neurons. Also, the more reliable and flexible the components of the system (and hence, the more complex the system), the more reliable and flexible should be the storage of information.

It logically follows that the more rapidly recurrent feedback interactions occur in a given system, the more rapidly a stabile holistic history should emerge. For example, atoms vibrate billions of times a second. Therefore, although it would be predicted that it should take a finite amount of time for a molecular holistic history to form when hydrogen and oxygen come together as H_20, the time it actually takes may be a few nanoseconds.

Of course, the fact that information is theoretically stored does not necessarily mean that the information can be retrieved, at least consciously. Everyday experience and substantial empirical research (Schacter, 1996) reminds us that our ability to recognize information is typically far greater than our ability to recall information. For example, the senior author's recall of basic chemistry is minimal--- yet his recognition of chemistry, particularly in the textbooks he studied as an undergraduate minor in chemistry --- is substantial.

Failure to show recall does not necessarily imply that dynamic memory has not occurred. Forgetting, therefore, does not necessarily imply that dynamic memory has been erased --- the process may involve an alteration in retrieval. This point is emphasized by Pribram (1991):

> "By using primes and probes, Fergus Craik in an elegant program of experiments (Craik, 1988), has shown that disturbances of remembering are almost always due to interference with the process and not with a lost of stored items. Neuropsychological evidence (e.g. Pribram, 1986; Weiskrantz, 1986) has also repeadedly demonstrated that "engrams" are not "lost" as such as a result of brain damage. Rather engrams are reconstructions that can appear as intrusion errors when amnesics are examined in a systematic fashion; that is, during recall, reconstruction of an engram occurs but in an inappropriate context." (page xxvii).

The systemic memory hypothesis predicts that information, once received, is retained in some form forever, so long as the system remains intact and recurrent feedback interactions (cycling) continue. Not only will the information continue, but it potentially will evolve over time. In fact, in a deep sense, it may be impossible to ever erase information completely in an intact system. Theoretically, if the experiment is sensitive enough, evidence for savings (see Schacter, 1996) or other subtle measures of change in functioning (behavior) should be demonstrable in all systems as a function of the evolution of the hypothesized dynamic memory process.

It is generally assumed that psychosocial information is not stored in the genes, and therefore it is not transferred biologically from parents to fetus. This assumption is consistent with the empirical observation that rats, for example, do not show obvious transfer of maze learning across generations, though not all scholars interpret the data in this way (Sheldrake, 1981). Failure to show transfer of learning does not necessarily imply that information has *not* been transferred --- *it could equally be the case that the information had been transferred in some form but was not retrieved.* This conceptual distinction is very important, because the systemic memory hypothesis requires that we seriously entertain the hypothesis that recurrent feedback interaction involves dynamic memory to some degree. Data that are inconsistent with the hypothesis that the information is stored (dynamic memory), therefore, may require an alternative interpretation (failure to retrieve). As will be seen, understanding how information is retrieved, especially the complex kind of information accumulated through the process of recurrent feedback interaction, is one of science's greatest challenges.

As mentioned earlier, the concept of recurrent feedback interaction provides us with a new approach to understanding the phenomenon of emergent properties (Scott, 1995). The classic example is how two molecules, hydrogen and oxygen, invisible gases at room temperature, combine to create a unique molecule, water, which is clearly visible at room temperature. No other combination of atoms is known to form a molecule with the identical properties of water. Water is one of the few molecules that increase in size when it solidifies. Also, water is one of the few molecules that in the process of freezing creates a seemingly infinite number of beautifully complex structures termed snow flakes. The capacity for water to form such unique combinations of order is challenging for science to understand.

Life as we understand it requires water. Although all molecules have unique, emergent properties --- water may be especially unique. From a systems perspective, water as a liquid may function as a complex dynamical system (e.g. the whole ocean can be viewed as a gigantic dynamical system - the circulation of blood in organisms can also be viewed as a large dynamical system) and it may have a special capacity to store information and energy.

When a dynamical energy systems approach is applied to quantum mechanics, and these concepts are applied to water, questions can be raised about the storage of information not only in water as a system, but in the individual molecules comprising the water system (e.g. the recurrent dynamic interactions between hydrogen and oxygen - see Jibu, Pribram and Yasue, 1996). Moreover, the logic requires that we entertain the possibility that if hydrogen and oxygen are subsequently separated after having existed in the water molecule, the hydrogen and oxygen will each retain some information about their history (due to recurrent feedback interactions between the electrons and the nuclei) that may be expressed in terms of increased dimensional complexity. Though clearly controversial, the systemic memory hypothesis requires that we posit the possibility of recurrent feedback resonance and dynamic memory between and within atoms.

Emergent properties (i.e. novel properties of the whole) may reflect a recurrent feedback interaction process --- through dynamical recurrent feedback interaction, unique, creative, holistic memories can emerge and be expressed.

It is worth noting that the systemic memory hypothesis has implications for understanding normal and pathological aging. It is conceivable that systems can only accumulate so much information / complexity. It is possible that certain systems (e.g. living systems) ultimately "collapse under the weight", so to speak, of too much information (e.g. complex vibrations that destroy the integrity of the system to function as a whole). Moreover, if a system is unable to sustain dynamic circulation, it will not be able to sustain it's holistic memories, and therefore, theoretically, it should "die."

Implications for Predicting and Explaining Certain Heretofore Unexplained Observations and Claims - Homeopathy, Kinesiology, Transplant Surgery and Parapsychology

The concept of recurrent feedback interaction, and the systemic memory hypothesis, provide a possible explanation for a host of heretofore unexplained observations and claims. In the emerging field of alternative medicine, two areas that challenge our traditional models are homeopathy and kinesiology. Homeopathy involves treatments using dilution's of substances in water that are so weak that virtually no molecules of the substance dissolved in the water can be detected. However, double blind studies (reviewed in Bellavite and Signorini, 1995) show that certain homeopathic dilution's can have statistically and clinically significant effect in reducing certain symptoms and treating certain diseases. Proponents of homeopathy claim that water contains a memory for the molecules. The systemic memory hypothesis provides a rationale for predicting and explaining these observations.

Kinesiology (e.g. Diamond, 1976) involves diagnoses and treatments that presume that the muscles and organ tissues store information and energy, for example, about traumas. Since traditional models presume that memories are stored only in the central nervous system, the observations and interpretations of kinesiologists are interpreted as being in error. However, the systemic memory hypothesis suggests that all cells may indeed store information, and therefore this information can be potentially retrieved from muscles and organ tissues. Having a patient "tune into memories stored in specific bodily regions" may be more than just a metaphor --- organs may store patterns of information that are unique and can be added to information stored in the brain.

A particularly remarkable implication of the systemic memory hypothesis concerns predictions about organ transplants, especially heart transplants. As described in Russek and Schwartz (1996a,b), the heart can be viewed as a dynamical energy generating system. Of all the organs within the body, the heart is preeminent in terms of the centrality of its location, the richness of its connections to all the cells within the body, and particularly relevant here, the intensity of its energy transmission. This energy aspect of the heart does not receive much attention. But just as the heart not only pumps patterns of biochemical nutrients to every cell within the body through the circulation, it also "pumps" patterns of energy and information to every cell within the body through the circulation as well. For example, it is well known that the electrical potential generated by the heart, identified by the electrocardiogram, can be recorded from *any site on the body* because of volume conduction, a mechanism that is well known in physics and biology and is not, in and of itself, controversial (Malmivuo and Plonsey 1995).

A common phrase is that we learn things "by heart", and that we know things "by heart." Is this simply a misplaced metaphor, or is there some deep significance to this particular choice of words? If all cells store information, if the heart is especially involved because of the centrality of its location and connections, and if the heart is especially involved in emotion, then memories (in particular, implicit emotional memories), may literally involve the heart in addition to the brain (Russek and Schwartz, 1996a,b).

If the heart of one person is transplanted into another (something that does not happen naturally in nature), some aspects of the history of the donor (as processed and stored by the heart) will be potentially accessible (implicitly if not explicitly) to the recipient. Since the recipient's body typically treats the donor's heart as foreign matter to be rejected, drugs are required to suppress this natural reaction. Is it possible that the rejection response is not simply the rejection of the matter comprising the heart, but the rejection of stored energy and information contained in the heart as well?

It would take a special recipient, indeed, to possibly retrieve the hypothesized information stored in the donor's heart. However, sporadic reports have surfaced suggesting that strange changes in recipient's perceptions and preferences sometimes occurs (e.g. Siegel, 1995). The critical question is, do the recipient's new perceptions and preferences correspond to the donor's known preferences?

We recently learned of the experiences of a former dancer, Claire Sylvia, who received a heart/lung transplant in Boston. Her's was the first successful heart/lung transplant in New England. According to an article by Leonore Fleisher (Subrights Letter, March 15, 1996, page 7), describing a book currently being co-authored with William Novak, and tentatively entitled *A Change of Heart*,

Sylvia (and other unique transplant recipients) report that their lives sometimes change in strange ways after their transplants. For example, it was reported in the article that Sylvia remembered that six weeks after her transplant, when she was allowed to drive again she had driven "straight to the nearest Kentucky Fried Chicken, a place she'd never been before, and this former dancer and fit, thin person had ordered chicken nuggets." She later learned that the 18 year old person whose heart and lungs now lived inside her had a fondness for fried chicken nuggets. Moreover, at the time of the young man's death, uneaten chicken nuggets were found stuffed inside the pocket of his leather jacket.

Claims such as these are typically and understandably treated as nonsense --- that is, they do not make any sense --- from a traditional non-systemic perspective. They are explained as coincidences or misperceptions, side-effects of the immunosuppressant drugs, or the expression of preexisting psychopathology interacting with the stress of surgery. However, when systems are viewed not only as material systems, but as informational and energy systems as well, claims such as these from select transplant patients begin to make sense.

As William James put it, "If you wish to upset the law that all crows are black....it is enough if you prove one single crow to be white." It is possible that Claire Sylvia may be the "white crow." Claire Sylvia may be a special case because she is artistic, a skilled dancer, in touch with her body and emotions, open-minded and spiritual. Maybe her experiences do not simply reflect her creative consciousness and / or effects of the drugs, but reflect actual information retrieval (explicit or implicit) from the essential organs she has inherited. Whether this information is truly episodic and cognitive in nature, or is more stylistic and emotional / motivational in nature, is a completely open question that can and should be addressed in future research.

Probably the most controversial heretofore unexplained observations involving memory and memory retrieval are reported in parapsychology. It is commonly claimed in parapsychological folklore as well as some well documented case studies (e.g. Smith, 1964) that "sensitives" and "psychics" can "read" information about a person from personal effects, including photographs. These claims truly challenge our traditional theories of memory. However, quantum physics, viewed from a dynamical energy systems perspective, provides a possible systemic explanation for these effects.

Consider what happens when an object (such as a person) spends time in front of a surface such as a black wall, or the black pupils of someone else's eyes. Physics tells us that objects appear black because they "absorb" the visible spectra of light (Orear, 1962). Complex patterns that comprise the organization of visible light (the visual portion of the electromagnetic spectrum) are "absorbed" by the wall, the pupil (or any black surface). Complex patterns comprising the organization of invisible light (e.g. infrared that is related to heat) are also "absorbed."

Once these energy patterns are "absorbed", where do they go? Do the dynamic patterns of information simply disappear? This is the assumption of classical, non-systemic physics. But a dynamical energy systems approach, accepting recurrent feedback interaction and resonance as common and natural, and accepting the storage of information in "A - B" situations as similarly common and natural, would hypothesize that the patterns become part of the on-going systemic energetic (recurrent) interactions occurring within the wall or eye system, and therefore will be "stored" (i.e. become part of the wall's or pupil's systemic history).

A black wall, for example, being a less complex system than the pupil of the eye, should "store" this information in a less complex (and potentially physically more "pure") manner. Theoretically, however, people interact with walls in a manner that is similar to the way people interact with eyes (though obviously less complexly). People may vary in their ability (or desire) to "resonate" with walls or pupils, and, therefore, resonate differently with the stored information in the walls or pupils. *Since the storage of information (the "membering" part of re-membering) is not the same as the retrieval of information (the remembering), the task of retrieving all of this hypothetically stored information remains a major challenge for future science.*

It is challenging to raise the question about whether electromagnetic information, at least in the visible spectrum, is "stored" in a black object. It would be predicted, for example, that black objects should store more patterned light information than white objects. It is even more challenging to raise the question about whether such information, once stored, is "living." When systems are viewed as energetic and dynamic, each comprised of continually interacting subsystems (e.g. recall that water is comprised of billions of continually interacting water molecules) creating recurrent information (a water system), the simple distinction between non-life and living becomes blurred. Though beyond the scope of the present paper, it is clear that the criteria we use to determine whether information and energy are "living" or not (Miller, 1978; Capra, 1996) needs to be specified in order to address this question.

A logical extension of these hypotheses applies to the ancient idea that spaces and settings store information and energy, even over centuries. For example, some people claim that certain places of worship (e.g. St. Patrick's and St. John the Divine Cathedrals in New York City) accumulate very different information and energy (e.g. love and compassion), compared with certain places of transportation (e.g. tension and frustration in the Port Authority terminal and Grand Central Station in New York City) (Brennan, 1993).

These hypotheses may also help explain Jung's controversial concept of archetypes and the so-called "collective unconscious" stored in the so-called "quantum sea" (see Laughlin, 1996). The systemic memory hypothesis encourages us to be open to the possibility that some aspects of parapsychological folklore may, in part, reflect a genuine dynamic memory phenomenon.

Some Implications for Society - The Information Bomb and Survival of Consciousness Hypothesis

If future research were to establish that some version of the systemic memory hypothesis were true, the implications for society would be far reaching. One profound implication is the possibility that the past really exists in the present, and that this information is potentially accessible. The amount of information potentially stored is extraordinarily challenging to envision. However, what would happen if someone figured out a procedure for retrieving this information?

To the extent that knowledge is power, access to this information could not only be potentially very useful, it could also be potentially abused. In the same way that the discovery of $e=mc^2$ led to the development of atomic power and atomic destruction, the discovery of a parallel formula for information and memory retrieval could have both positive and negative effects.

For example, most of us are not flooded with all the implicit information we have stored in our lifetime. Is it possible that systems evolve to keep most of this implicit information in check? Is it critical for human consciousness that we not be flooded with past information? Are some disorders of mental illness the breakdown of inhibitor mechanisms that keep implicit information out of awareness? What would happen if someone discovered a drug that removed the capacity to keep information in check? Could this become a new method of chemical warfare, leading solders and citizens to be swamped literally by their own information? What is science fiction today sometimes becomes the science of tomorrow. The hypothesis that information is continually stored and is potentially retrievable has many potential science fiction implications, some of which may turn out to be tomorrow's science.

Probably the most controversial implication of the systemic memory hypothesis concerns the survival of consciousness after death. Throughout recorded history, people have entertained the hypothesis, and the wish, that some aspects of consciousness continue after physical death. Interest in the bridge between science and spirituality continues to grow, and scholarly books addressing the psychology of consciousness at all levels in systems (Chalmers, 1996) and the physics of immortality (Tipler, 1994) are appearing. If it were the case that the history of people's lives, their information and energy, is not only stored in their brains and bodies, but is to some degree dynamically stored in all systems (especially in their living loved ones and objects with which they have spent significant time), and if consciousness is a fundamental aspect of information (Chalmers, 1996), then the survival of consciousness hypothesis after physical death becomes a greater possibility. The systemic memory hypothesis provides an unusually challenging and intimate bridge between science and spirituality.

The Power of Logic and Its Limitations: The Answer Lies in Future Research

Who would have guessed (except with hindsight) that in the process of attempting to understand the logic of circular feedback (recurrent interactions) from the perspective of information, that the logic would inevitably lead to the systemic memory hypothesis? As we attempt to explicate in this chapter, when the implications of dynamical recurrent feedback interaction are understood, some surprising and far reaching predictions follow that can be confirmed or disconfirmed in future research.

In a review of a recent paper (Russek and Schwartz, 1996a) outlining some dynamic energy systems implications for cardiology, mind-body medicine, and alternative medicine, Puthoff (1996), a physicist, noted that "On this trail we are led inexorably to consider not only hypothesized deep connections between brain and heart, consciousness and cardiovascular activity, but between the whole organism and its environment, including others, by means of patterned energetic mechanisms." Puthoff's selection of the term "inexorably" seems well chosen. When a systems approach to energy and information is taken, and the process of interaction is analyzed, a series of novel and controversial predictions do follow inexorably.

One could argue that the logic of systems theory is circular, and that the reasoning of recurrent feedback interactions is circular. Interestingly, this turns out to be literally and unavoidably true. In fact, the essence of the meaning of the word "circular" (and the concept of a circle) is "ending in itself" (the 5th definition of the word "circular" in Webster's Unabridged Second Edition Dictionary).

Capra (1996), in his book on living systems, emphased that feedback is in fact *circular feedback*, that relationships are *circular relationships*, and that organization is *circular organization*. Though Capra (a seasoned systems theorist) emphasized that circular organization is involved in self-organization, self-awareness, and evolution, and he implicitly recognized that self-organization, self-awareness, and evolution require the storage of information and energy, he apparently did not take the logic far enough to discover that circular feedback involves recurrent (circular) interaction which naturally creates dynamic memory --- the systemic memory hypothesis (see Schwartz and Russek, 1997, for further discussion of circular causality in science and medicine).

Recurrent feedback interaction means cyclic, repeated interaction (the word recurrent comes from the word "recur", which means to "return"), and this implies "circulation." It turns that the hypothesis of the storage of information and energy as dynamic memory (e.g. the preserving of order) was historically understood in the concept of circulation. According to Webster, the obsolete definition of the word circulation is *"a series in which the same order is preserved, and things return to the same state."* When the concept of dynamical interaction is added to Webster's historic description of the concept of circulation, logic leads to the hypothesis of dynamical (evolving) order (storage of evolving information and energy) in systems --- the systemic memory hypothesis.

Circular logic also suggests that the emergence and expression of holism in systems may be a natural result of dynamical interactive circulation of accumulating relationship information and energy. Dynamic systemic memory, then, is holistic memory.

Of course, this does not mean that the logic is correct, or that the predictions are valid. Though the authors have been unable to discover an obvious flaw in the logic, the logic may still be misplaced or mistaken. Even if the logic turns out to be correct, nature may not work the way logic predicts, suggesting that the basic assumptions are in error. For example, though general systems theory makes the assumption that systems in nature are always open to various degrees, and therefore interact recurrently with other systems to various degrees, the openness assumption could be wrong, and therefore systemic memories should not form.

However, it is possible that the logic has not been taken far enough. Is it possible that energy interacts with energy in the absence of matter? *For example, do organized patterns of light / electromagnetic waves interact in a vacuum? Do these " packets" or "quanta" of energy in a vacuum combine and create pure energy systems? If pure energy systems do in fact exist, is information stored in these systems too?*

New quantum mechanical predictions about electromagnetic fluctuations of the vacuum suggest that some version of this hypothesis is tenable. According to Forward (1996):

> "Now comes the real problem, and the major reason why we need to carry out experiments to verify that the quantum mechanical electromagnetic fluctuations of the vacuum behave as the equations of quantum mechanics predict. The block of matter has a large, but finite, number of atoms and therefore a finite total quantum mechanical vibrational fluctuation energy. *The region of vacuum, however, can support an infinity of electromagnetic vibrations....*" (page 330).

It is conceptually challenging to ponder that between every "particle" of "matter" is a "space" (a "region of vacuum") that is potentially teaming with energy and information. If the region of the vacuum can support an "infinity of electromagnetic vibrations" (the quantum mechanics hypothesis) and electromagnetic vibrations can function as pure energy systems, then an "infinity of recurrent feedback interactions" can occur as well (the systemic memory hypothesis). Since the vacuum (V) theoretically exists between components A and B in the simplist, two component system, V may serve as a dynamic mechanism allowing stored information and energy to leave the particular A-B system selectively and enter the larger V suprasystem (that includes A and B). It follows that all subsystems, systems, and supra systems, are ultimately interconnected by V.

Using the concept of "quantum vacuum interaction" to explain the formation of memory, Laszlo (1995) provides a new explanation for the wave / particle paradox observed in the well known single slit / double slit quantum physics experiment.

> "It appears, then, that photons successively emitted by a given light source pass

through only one of the slits and then interfere with each other. This suggests that not self-interference, but *memory* is a work. The previously emitted photons are no longer physically 'there', but their *traces* can be: the most reasonable interpretation is that the successively emitted photons interfere with the traces of the photons that were emitted previously. But how are these traces registered and conveyed? For a realistic interpretation a physical medium carrying the traces is required: a medium that transcends the limitations of relativisitic spacetime....the indicated medium in the guise of the scalar-mediated spectrum of the quantum vacuum." (page 54, italics by Laszlo, 1995).

It is possible that the systemic memory hypothesis, integrated with modern quantum dynamics of the vacuum (Milonni, 1994; Laszlo, 1995), may help explain the profoundly challenging hypothesis about memory in the universe proposed by Sheldrake (1981). Sheldrake has proposed that information in the universe is continuously stored in so-called morphogenetic (i.e. form) fields, that systems emerge through morphogenetic resonance, and therefore, that the physical laws of nature are better understood to be evolving "habits" (Sheldrake, 1981) or memories of a dynamic universe rather than fixtures of a static universe. *If modern quantum mechanics is correct, the vacuum may turn out to be the ultimate storage device for recurrent feedback energy system interactions and the "habits" of nature.*

The logic that leads to the conclusion that the vacuum may be the ultimate storage device for recurrent feedback energy system interactions is implicated in Bohm's recurrent process explanation of Sheldrake's morphogenetic resonance hypothesis (Sheldrake and Bohm, 1982). After Bohm published his seminal book on wholeness and the implicate order(Bohm, 1980), Bohm went on to employ the circular concepts of "re-injection" , "re-projection", and "*recurrent* actuality", to explain the origin of memory and wholeness in nature. Simply stated, Bohm hypothesized that implicate memories emerge through "repeated cycles of re-injections and re-projections." Repeated cycles of re-injections and re-projections described at the quantum level can be viewed as being a special case of the process of recurrent feedback interactions that occur in all systems at all levels - the implicit logic is the same. According to Bohm, our everyday concept of "actuality" (what we term "reality") should be reconceptualized as being a dynamical process, a "*recurrent* actuality."

However, the deep challenge still remains, how can implicate memories stored in dynamical systems become explicate? It is difficult enough to attempt to explain, conceptually and mathematically, how information, once stored as memory in neural systems, can be retrieved. It is even more difficult to attempt to explain, conceptually and mathematically, how information, possibly stored as memory in other organ systems, could be retrieved. It is extraordinarily difficult to attempt to explain, conceptually and mathematically, how information, potentially stored as memory in a vacuum (pure energy systems), could be retrieved. Though this chapter has not addressed the profound challenge of the retrieval of systemic memory, it seems possible that the solution to the retrieval of systemic memory will involve an even deeper understanding of recurrent feedback resonance in systems, integrating harmonic resonance and stochastic resonance (e.g. Astumian, Weaver, and Adair, 1995) from a recurrent feedback interaction perspective. To "re-member" may be to "re-resonate" in some deep systemic way.

Pribram (personal communication) has suggested that valence (the foundation of value) is a prerequisite for recurrent feedback interaction and hence the creation of dynamic memory. His suggestion is consistent with dynamical energy systems theory. The concept of a system requires that the components be connected energetically,. Energetic connection implicitly involves the concept of valence. In humans, emotion (in particular love) functions as valence, and emotion (in particular love and loving relationships) fosters memories Valence in turn fosters the creation and maintenance of relationships. The systemic memory hypothesis predicts that the kind of memory that is stored is dynamic relationship memory. It is conceivable that the capacity to "re-resonate" with dynamic memory may involve patterns of valence as well.

The purpose of this chapter has been to raise fundamental questions about dynamical recurrent feedback interaction, to encourage open discussion about the storage of information and energy and the creation of dynamic memory, and stimulate empirical research to determine whether dynamic memory exists in all systems at all levels. As mentioned in the beginning, though neurons are especially gifted in storing sensory and perceptual information (because neurons are so highly interconnected, creating profoundly complex recurrent interactive networks), it may be time to evolve our intellectual heritage and re-envision the brain as being a marvelous special case of a ubiquitous dynamic (systemic) memory process in nature. It is our belief that in the process of addressing the question of recurrent feedback interaction, science will gain a deeper understanding of the essence of memory, emergent properties and holism, and self-awareness and evolution, and in the process, make better sense of heretofore unexplained observations in medicine such as homeopathy (Bellavite and Signorini, 1995), and in psychology such as parapsychology, (Bem and Hornoton, 1994) that strain our current theories of how memory works.

Acknowledgments:

Our sincerest thanks to John Allen, John Kline, Lynn Nadel, and Lonnie Nelson, from the University of Arizona, for their help in challenging and clarifying the thesis of this chapter, and to Karl Pribram, for his inspiration and insights in preparing this chapter. The chapter is dedicated to the memory of Heinz Pagels, the distinguished physicist, who encouraged the first author at Yale University to risk integrating concepts from modern physics and systems theory with psychology and neuroscience to foster both basic and applied science, and to the memory of Henry I. Russek, M.D., the distinguished cardiologist, who encouraged the second author during their twenty year research and clinical collaboration to risk integrating biological, psychosocial, and spiritual concepts to foster both the art and science of clinical medicine.

References

Abu-Mostafa, Y.S. (1957). *Complexity in Information Theory.* New York, NY: Springer-Verlag.

Astumian, R.D., Weave,r J.C., and Adair, R.K. (1995). Rectification and signal averaging of weak elecric fields by biological cells. *Proceedings of the National Academy of Sciences.* 92:3740-3743.

Becker, R.O (1990). *Cross Currents.* New York, NY: Jeremy P. Tarcher / Perigee.

Becker, R.O. and Selden, G. (1988). *The Body Electric.* New York, NY: Quill, William Morrow.

Bellavite, P. and Signorini, A. (1995). *Homeopathy: A Frontier in Medical Science.* Berkeley, CA: North Atlantic Books.

Bem, D.G., and Honorton, C. (1994). Does psi exist? Replicable evidence for an anomalous process of information transfer. *Psychological Bulletin,* 115, 4-18.
Bohm, D. (1980). *Wholeness and the Implicate Order.* London: Routledge and Kegan Paul.
Brennan, B.A. (1993). *Light Emerging.* New York, NY: Bantum Books.
Cambel A.B. (1993). *Applied Chaos Theory: A Paradigm for Complexity.* Boston, MA: Academic Press.
Capra, F. (1996). *The Web of Life: A New Scientific Understanding of Living Systems.* New York, NY: Anchor Books.
Chalmers, D.J. (1996). *The Conscious Mind: In Search of a Fundamental Theory.* New York, NY: Oxford University Press.
Craik, F.I.M. (1988). On the making of episodes. In Roediger, H.L. III and Craik, F.I.M. (eds.). *Varieties of memory and Consciousness: Essays in Honour of Endel Tulving.* Hillsdale, NJ: Lawrence Erlbaum Associates.
Diamond, J. (1979). *Behavioral Kinesiology.* New York, NY: Harper and Row.
Flood, R.L. 1988. *Dealing with Complexity: An Introduction to the Theory and Application of Systems Science.* New York, NY: Plenum.
Forward, R. (1996). Mass modification definition study. *Journal of Scientific Exploration.* 10:325-254.
James, W. (1890). *Psychology (Briefer Course).* New York, NY: Holt.
Jibu, M, Pribram, K.H. and Yasue, K. (1996). From conscious experience to memory storage and retrieval: The role of quantum brain dynamics and boson condensation of evanescent photons. *International Journal of Modern Physics B*, 10:1735-1754.
Kauffman, S.A. (1993). *The Origins of Order.* New York, NY: Oxford University Press.
Laszlo, E. (1995). *The Interconnected Universe: Conceptual Foundations of Transdisciplinary Unified Theory.* Singapore: World Scientific Publishing.
Laughlin, C. D. (1996). Archetypes, neurognosis, and the quantum sea. *Journal of Scientific Exploration.* 10: 375-400.
Lisberger, S.G. and Sejnowski, T.J. (1992). Motor learning in a recurrent network model based on the vistibulo-ocular reflex. *Nature*, 360: 159-161.
Malmivuo, J. and Plonsey, R. (1995). *Bioelectromagnetism.* New York: Oxford University Press.
McCulloch, W.S. (1951). Why the mind is in the head. In Jeffress, L.A. (ed.). *Cerebral Mechanisms in Behavior.* New York, NY: John Wiley.
Miller, J.G. (1978). *Living Systems.* New York, NY: McGraw-Hill.
Milonni. P.W. (1994). The Quantum Vacuum: An Introduction to Quantum Electronics. New York, NY: Academic Press.
Mithaug, D.E. (1993). *Self-Regulation Theory: How Optimal Adjustment Maximizes Gain.* Westport, CT: Praeger.
NATO, 1985. *Chaotic Behavior in Quantum Systems: Theory and Applications.* New York, NY: Plenum.
Orear J, 1962. *Fundamental Physics.* New York, NY: John Wiley & Sons.

Pribram, K. H. (1991). *Brain and Perception: Holonomy and Structure in Figural Processing.* Hillsdale, NJ: Lawrence Erlbaum Associates.
Pribram, K. H. (1986). The cognitive revolution and mind/brain issues. *American Psychologist.* 41: 507-520.
Puthoff, H.E. (1996, in press). Technological problems, bold possibilities. *Advances: The Journal of Mind-Body Health.*
Ragan, P.A., Wang, W., and Eisenberg, S.R. (1995). Magnetically induced currents in the canine heart: A finite Element Study. *IEEE Transactions on Biomedical Engineering.* 42: 1110-1115.
Russek, L.G. and Schwartz, G.E. (1996a). Energy Cardiology: A dynamical energy systems approach for integrating conventional and alternative medicine. *Advances: The Journal of Mind-Body Health.* 12(4): 4-24.
Russek, L.G. and Schwartz, G.E. (1996b). The heart, dynamic energy, and integrated medicine. *Advances: The Journal of Mind-Body Health.* 12(4):36-45.
Schacter, D.L. (1996). *Searching for Memory.* New York, NY: Basic Books.
Schwartz G.E. (1982). Cardiovascular Psychophysiology: A Systems Perspective. In Cacioppo, J.T. and Petty, R.E. (eds.). *Perspectives in Cardiovascular Psychophysiology.* New York, NY: Guilford Press.
Schwartz,G.E. (1984). Psychobiology of Health: A New Synthesis. . In. Hammonds, B.L. and Scheirer, C.J. (eds.). *Psychology and Health: The Master Lecture Series. Volume 3* Washington, D.C.: American Psychological Association.
Schwartz, G..E. (1987). Personality and the unification of Psychology and Modern Physics: A systems approach. In Aronoff, J., Robin, A.I. and Zucker, R.A. (Eds.), *The Emergence of Personality.* New York: Springer, 1987.
Schwartz, G.E. (1989). Disregulation Theory and Psychosomatic Disease: A Systems Approach. In Cheren, S. (ed.). *Psychosomatic Medicine: Theory, Research and Practice.* New York, NY: International University Press.
Schwartz, G. E. (1997). Information and Energy: The soul and spirit of mind-body medicine. *Advances: The Journal of Mind-Body Health,* in press.
Schwartz, G. E. and Russek, L. G. (1997). The challenge of one medicine: Theories of health and eight world hypotheses. *Advances: The Journal of Mind-Body Health,* in press.
Scott, A. (1995). *Stairway to the Mind: The Controversial New Science of Consciousness.* New York, NY: Springer-Verlag.
Sheldrake, R. (1981). *A New Science of Life.* Los Angelos, CA: Tarcher.
Sheldrake, R. and Bohm, D. (1982). Morphogenetic fields and the implicate order. *ReVision.* 5:41-48.
Siegel, B.S. (1995). Exploring what can't be explained. *Advances: The Journal of Mind-Body Health.* 11:2-3.
Smith, S. (1964). *The Mediumship of Mrs. Leonard.* New Hyde Park, NY: University Books.
Tipler, F. (1994). *The Physics of Immortality.* New York, NY: Macmillan.
von Bertalanffy, L. (1968). *General System Theory.* New York, NY:Braziller.
Webster, N. (1977). *Webster's New Twentieth Century Dictionary of the English Language. Unabridged, 2nd Edition.* (New York, NY: Collins World.

Weiskrantz, L. (1986). *Blindsight: A Case Study and Implications.* Oxford: Clarendon Press.

Yates, F.E. (ed.) (1987). *Self-Organizing Systems: The Emergence of Order.* New York, NY: Plenum Press.

10

Preserved Vocabulary and Reading Acquisition in an Amnesic Child

Charles A. Ahern, Frank B. Wood, Christine M. McBrien
Section of Neuropsychology
Bowman Gray School of Medicine
Medical Center Blvd.
Winston-Salem, NC 27157
Telephone: 415-206-5270 / Fax: 415-206-4722
E-Mail: ChuckA2651@aol.com

Abstract

This chapter presents the case of a nine year old boy who is amnesic as a result of congenital brain damage. Despite severe memory impairment for the events that constitute his experience, reading and vocabulary acquisition are shown to be within normal limits at age 9. This is interpreted as evidence of a dissociation between semantic and episodic memory systems. Observational data supporting this conclusion are also presented. The wider theoretical implications of this view are discussed. Methodologically, this study suggests there is value in attending to psychological variables in studies which aim to measure cognitive capabilities of brain-injured individuals.

This chapter will present findings concerning a remarkable nine year old boy, T.J., who is amnesic due to congenital brain damage. Specifically, we will focus on the implications of areas of preserved memory performance for a rather long-standing disagreement in the literature concerning theories which seek to describe the separate systems which constitute human memory.

While memory impairment in amnesia is diffuse, it is not global. Thirty years ago, Milner (1966) and Corkin (1968) focused renewed attention on the fact that some memory functions appear to be preserved in human amnesia. Case studies of two patients in particular, H.M. and N.A., demonstrated that amnesia involves selective memory impairment. These data converged with data from animal studies which indicated that the memory deficits which resulted from specific lesions were circumscribed and definable (Pribram 1966,1969).

These works engendered intensified interest in the nature of amnesia that has continued to the present day. When understood as a selective impairment, the nature of amnesia becomes subtle and fascinating. Since the Milner and Corkin studies, hundreds of papers have explored empirically which memory functions are spared in amnesia. The findings have served as the basis for theoretical accounts of distinct memory systems in amnesia. These accounts have also been considered relevant to understanding the structure of normal memory; that is, to identifying and defining distinct systems operating in normal memory (Schacter, 1987).

Evidence for a distinct memory system usually features claims that the system can be behaviorally dissociated from other memory systems. Dissociation is demonstrated when performance on memory tasks assumed to represent one system is shown to be at a different level than performance on tasks assumed to represent a second system. Amnesic studies have constituted a particularly helpful subset of this type of data because they have demonstrated dissociations which are not readily apparent in normal subjects. It is assumed that the same separate systems exist in normals but, because memory functions are typically overlapping and redundant in non-impaired individuals, dissociations cannot be as readily demonstrated. Dissociations in amnesics have therefore been interpreted as evidence for the existence of corresponding distinct systems in normals as well as amnesics.

There is consensus at present that an obvious and consistent impairment in amnesia is for detailed recall of specific events or episodes (Tulving, 1991). For example, the amnesic will show impairment on memory tasks such as recalling what he had for breakfast, the fact that he was recently taught some new facts by a researcher, or the details of a story which he was told an hour before. Memory of this type is referred to as episodic memory (Tulving, 1972) and it is identical with what was earlier called context memory (Pribram, 1969). Episodic memory is bound to a temporal and spatial context (Tulving, 1972). It is the autobiographical memory, rich in detail, for events that occurred at a specific place and time. It carries with it the ability to re-experience, to recall the feelings and state of mind, of the episode in question. In Pribram's terms, it is the memory which allows one to monitor and recall personal events; these memories constitute the narrative self. Impairment in this sphere is commonly considered the hallmark of the amnesic syndrome (Wood et al., 1989).

Controversy exists concerning whether another type of memory, semantic memory, can be preserved in amnesia. This issue is basic to an understanding of the nature of the amnesic syndrome and, moreover, to theory of normal memory structure. Semantic memory involves recall of information which is factual and non-context bound. Tulving refers to semantic memory as "the memory necessary for the use of language, organized knowledge about words (and) their meanings and referents." Semantic memory is essentially what Pribram earlier called reference memory. One calls on reference memory as one might call upon reference books, to recall facts about the world.[1]

A variety of studies have provided evidence of preserved reference or semantic memory in amnesic individuals. Wood et al. (1989) reported the case of a young woman who suffered onset of amnesia at age 12, but showed normal academic progress through high school. Glisky and Schacter (1987, 1989) have shown that amnesics can learn and use computer commands, when taught using a technique based on priming. Tulving (1991) tutored the amnesic K.C., teaching him short factual sentences over a three month period which were retained on nine month follow-up. Kinsbourne and Wood (1982) have shown that when given an association task, amnesics performed in the average range for producing semantic memories (while they were impaired for producing episodic memories.) Amnesics have been shown to learn fictional statements of fact, although they were impaired in recalling the recent episode in which they learned the information. (Schacter et al., 1984).

However, the conclusion that these studies demonstrate preserved semantic memory has been challenged on various grounds. It has been argued that in some instances, e.g. preserved academic progress after onset of amnesia, subjects might be drawing on premorbid memory and therefore are not clearly demonstrating preserved capability for new semantic learning (Osteergard 1987). Critics have also argued that semantic memory may appear preserved in comparison to episodic memory

---

[1] Episodic memory is recall of the singing of the national anthem before a sporting event, remembering where one was, when it was, what one saw and how one felt. Semantic memory is the knowledge of the content of the anthem and the fact that it is typically sung before sporting events in order to honor the country.

because subjects may receive numerous exposures to semantic knowledge, while, by definition, they may receive only one exposure to any given episode (Squire et al., 1993).

The question of whether there is preserved semantic memory in amnesia is also blurred by different responses to the fact that amnesics have generally learned semantic information more slowly than normal subjects. Some commentators have asserted that fully normal performance is the only acceptable criterion for preserved performance (Ostergaard and Squire, 1990; Squire 1992). It follows from this assumption that evidence of semantic memory in amnesia which is not within normal limits is irrelevant to the question of preserved memory. Others assume that evidence of preserved semantic memory relative to episodic memory is significant, and to be accounted for theoretically, whether it is preserved within normal limits or not (Wood et al., 1989).

The most prominent current theoretical account of the nature of amnesia is organized around the distinction between declarative and nondeclarative (previously called procedural) memories. This distinction was first proposed in 1980 by Squire and Cohen. Drawing upon computer terminology, they stated that there were two types of memory apparent in the amnesic syndrome: procedural memory for "knowing how" and declarative memory for "knowing that." Procedural (or nondeclarative) memory can be expressed unconsciously, as in improved performance on a skill. Declarative memory can be brought to conscious awareness and verbally expressed or declared. This theoretical view rests upon the premise that the major distinction in memory performance in amnesia is between preserved nondeclarative memory and impaired declarative memory. In fact, this distinction is said to fully describe the pattern of preserved and impaired memory in amnesia (Squire 1992b). In this view, semantic and episodic memories are both manifestations of the declarative memory system.

A prior theoretical view anticipated many of the main concepts in the declarative/nondeclarative hypothesis but also contrasts with it in significant ways. Pribram (1966, 1969) proposed a scheme featuring three main types of memory: reference (called semantic by Tulving); context (now episodic); and skill (later known as procedural). Several points separate this theoretical view from the declarative/nondeclarative view. First, semantic memory and episodic memory are seen as separate memory systems, not manifestations of a single declarative system. Second, semantic memory is assumed to be frequently preserved in amnesia, relative to episodic memory. Finally, the memory impairment seen as best characterizing the amnesic syndrome is the lack of the narrative self. It is this aspect of memory, related to the monitoring of personal experience, which is most characteristically impaired in amnesia in this view.

The issue of whether semantic memory is preserved in amnesia is central to choosing between these two theoretical positions. If semantic memory is preserved relative to episodic memory, this supports the conclusion that they are separate memory systems. Moreover, if semantic memory is preserved, it cannot be true that only memory out of conscious awareness is preserved in amnesia.

If the persistent criticisms of the evidence for preserved semantic functioning in amnesia were answered in even a single case, it would seem to cast the accumulated data in a different light. To do this, a study would have to meet two main criteria. First it would have to show unambiguously that semantic learning occurred after the onset of amnesia, to relieve concerns of a confound with premorbid learning. Second, semantic learning would need to be preserved at normal limits or, at least, significantly greater than expected given the subject's level of episodic memory (Squire and Ostergaard, 1990).

The present study was formalized in response to clinical and historical data which suggested the performance of the subject, T.J., might well meet these criteria. T. J. suffered congenital brain damage, and underwent two consequent surgeries before the age of six months. As we shall describe below, he is profoundly amnesic in that he has never demonstrated reliable episodic memory. Despite his life-long amnesia, T.J. appeared capable of learning semantic information. For example, we found that he had learned verbal language to age-appropriate levels when tested at age 7. However, at the end of first grade, T.J. had shown no progress in learning to read, despite long-standing intensive efforts to prepare him for reading. Reading acquisition presented itself as the most important learning task before him, particularly because continued failure to read would drastically limit all future learning. Moreover, most compensations for his memory deficit would involve reading. Therefore, a year-long reading intervention was undertaken with T.J. which will be described below (see Ahern et al., under editorial review, for detailed description of reading acquisition).

Glisky and Schacter (1986,1988) pioneered neuropsychological studies in which the conditions include serial attempts, over several months, to improve the subjects' performance through instruction. The present study followed this general approach of extended intervention and, and in certain respects, modified it. The modification can be succinctly described as a determination to be as responsive as possible to moment by moment behavioral feedback on both pedagogical and psychological levels. This meant, for example, continuously adapting such variables as rate of presentation and difficulty of trials, so that T.J. was always working as closely as possible to his capacity, while failed trials were minimized. Psychologically, it meant attending to issues such as motivation, anxiety and fear as they appeared to become salient.

Method

Overview

The focus of this study is a description of memory and learning in T.J., a nine year old amnesic boy. Data were obtained at three test administrations. Baseline performance on tests of memory, attention, vocabulary, and reading skills were measured prior to reading intervention. Testing was repeated during and after tutorial intervention.

Subject

The following summarizes T.J.'s development to the time the first phase of intervention concluded at age 9 years, no months (9-0).

Medical History

Immediately after birth, T.J. was diagnosed with a midline brain cyst, with agenesis of the Corpus Callosum, clinidactycalty of the right hand and retinal hemorrhages. Surgery was performed twice in the first six months to excise a multicompartmental cyst and a shunt was implanted for drainage. At birth, the cystic material occupied the space where the right hemisphere should have been, thus preventing normal development. Serial CT and MRI studies document some rebound of right hemisphere tissue after the cyst was surgically removed. A recent MRI images idiosyncratic structural anatomy, including loss of right hemisphere mass, thalamic involvement, and diminished frontal lobes.

Following a post-operative seizure after his second operation, T.J. was placed on Tegretol which he continues to take (700 mg per day). He has continued to have occasional seizure activity. At age 5, he was placed on Dexedrine for attentional problems, but was discontinued shortly because it caused headaches. By observation, attentional problems are characterized by impulsivity, overactivity, and visual and auditory distractability.

Prior Cognitive Assessments

At nine months, T.J. was noted by his physician to have achieved "acceptable intellectual milestones." Bayley scales of infant development administered at age 2-0 yielded low average performance although the examiner felt this was an underestimate of T.J.'s capabilities. Significantly, the Bayley does not test any function analogous to episodic memory. A neurodevelopmental assessment at age 5 years, ten months (5 -10) noted impaired memory and diminished attention with relative sparing of language functioning. School evaluations consistently noted memory difficulties. At age 5-7 it was noted that T.J. "remembers nothing at all from playtime." At age 5-11 it was noted that he could not remember what activities he had participated in and could not remember if he had had a turn while playing. In a neuropsychological evaluation at age 7-0, memory performance was very impaired on immediate and delayed trials of both memory tests administered: memory for the Narrative Prose Memory Test and the Rey Auditory Verbal Learning Test.

Parental Report on Memory Functioning

T.J.'s parents report that memory capabilities for specific instances have never been demonstrated by T.J. They first noticed memory problems at around age 2 when they began to give him simple instructions that he would forget after a delay. They also found that he very easily lost his toys and did not have ideas about where he last played with them. They spontaneously noticed a marked difference between T.J.'s performance on these tasks and that of his older siblings at the same stage. They also report that as a toddler T.J. did not remember if he had eaten. They report the following memory failures as typical throughout childhood and to the present day. He forgets what he has been told he is going to do, even after having been told many times. He forgets if he has been to school, and where he has been during the day and the week.

Areas of preserved memory functioning were also noted by T.J.'s parents. They believe that he has always been able to learn social rules at a normal pace and they note that he has been able to learn about the people, places, and things which are consistently a part of his experience. They also note that as a toddler he would get very agitated whenever a routine changed, such as when his mother took an alternate route to a regular Occupational Therapy appointment.

His parents have never noticed flattening of affect. To the contrary, they report that he has always displayed a wide range of affect. They have not been aware of any deficit in his emotional attachment to them and to his siblings, relatives, and friends.

Clinical Observation

At age 8-0, the following memory failures were observed during an assessment at BGMS. He was unable to recall what he had eaten for breakfast a few hours earlier. He was unable to correctly identify an examiner that he had worked with that morning from among a group of four people. He was unable to say what he had eaten for lunch after returning from a restaurant. As part of a separate study, his forehead was smudged with red lipstick. When he saw himself in the mirror he laughed. Forty minutes later T.J. had no recollection of this episode.

T.J. was able to give his name, age, birthday, and names of family members. He reported the same favorite game, television program, and favorite color across trials. Expressive language capabilities appeared age appropriate and there were no obvious weaknesses in grammar.

Motor Development

TJ was delayed 1 to 4 months in reaching milestones for turning, crawling and walking. Through age 2, he generally lagged an average of three months on reaching developmental milestones. At age 3-6, an evaluation showed gross and fine motor functioning within normal limits. At the time of this evaluation (age 9-0), he exhibited mild coordination deficits such that his gait was mildly awkward at times. Running, jumping, and climbing were obviously awkward and difficult, as were fine motor coordination functions.

Speech and Language

At eight months T.J. was assessed as four months behind in speech and language development. He continued to lag approximately four months until age 2, at which point verbal development appeared grossly within normal limits. Clinical observation at age 9-0 revealed no evidence of speech or language deficit

Social History

T. J. lives with his parents and two siblings. He is the last of three children. He has received a variety of special educational services since age 4. T.J. was enrolled in speech and language therapy as well as occupational therapy while in a preschool handicapped program. Since entering kindergarten he has attended regular classrooms, with special assistance provided.

Socially, T.J. is often timid with children, but is able to make friends. He is well-liked by his classmates, many of whom appear to relate to him often in a mode of support and assistance. T.J. is more outgoing with adults, and often strikes up conversations with strangers. He is generally perceived as very charming and engaging. He is well-mannered and socially adept. He asks intelligent questions about the activities and preferences of others.

Research Design and Procedure

Intervention

Intervention took the form of reading tutorials. There were two phases. In the first, T.J. was followed (by C.A.) from ages 8-4 to 9-0. One-on-one instruction was planned for one hour, 3 times a week, for thirty-two weeks. However, a total of 25 sessions were lost due to vacation, illness, and other scheduling conflicts. Therefore, 71 tutoring sessions were conducted over thirty-two weeks. In the second phase, T.J. was followed (by C.M.) from age 9-2 to age 9-5. One-on-one instruction was planned for one hour, two times a week. 24 sessions were conducted over a twelve week period. Tutoring sessions were held at a private room in a public library which contained several tables with chairs, and a blackboard.

Reading intervention was based on a phonemic decoding approach. It was conducted in three phases. The principle of this method is to assist the student in mastering the complex relationship between letters and sounds (Lieberman 1989; Felton and Brown, 1991). The first phase was phonemic awareness training. In the second phase, phonemic awareness training was applied to reading texts. In the third phase, explicit instruction regarding the phonemes represented by certain letter patterns was added.

Flexibility was employed in the rate and manner of presentation of these exercises. The difficulty of learning trials and the type of tasks presented were adjusted on an ongoing basis in response to T.J.'s performance. For example, the number of repetitions of an exercise and the difficulty of the target words were varied according to the subject's success, motivation, and concentration. The goal was to allow T.J. to practice at the highest level at which he could perform successfully and to maximize successful trials. Cues of various sorts were provided to help him recall phonemic principles, but cues which would support guessing at a word were never provided.

Steps were taken at times to increase interest and motivation. For instance, based on his interest in cars, the names of cars in a car magazine were sometimes used as a basis for phonemic drills. Also, apparent anxiety and concerns were discussed as they arose. Successful practice trials were maximized not only to build proper habits, but also because this appeared to increase motivation and quiet anxiety. Occasionally, when topics were introduced by T.J., anxiety was dealt with more directly. At one juncture he wondered aloud whether his mother would still read to him if he learned how to read. This inquiry appeared to be an expression of concern about his relationship with his mother as he became more independent. After some discussion, we agreed that they could take turns reading to each other and he was visibly relieved. A small percentage of tutorial time was spent on such issues.

Data

Baseline testing data were obtained at age 8-0 (Time 1)[2]. Post-intervention tests were administered at the close of the first phase of intervention (Time 2) and the second phase of intervention (Time 3). Standardized tests measuring memory, attention, and basic reading skills, were administered.

Instruments:

Memory:

Reye Auditory Verbal Learning Test (RAYVLT). This is a serial learning task which consists of five consecutive presentations of a fifteen word list with immediate free recall after each trial. Trial 6 consists of presentation of a distracter list with immediate free recall. This is followed by a free recall trial of the original list. After a thirty minute delay, there is a forced-choice, delayed recognition trial.

Narrative Prose Memory. This test consists of an orally presented narrative with immediate and delayed (30 minutes) free recall. The subject listens to a short story and is asked to repeat it immediately. After a thirty minute delay, the subject is again asked to recount the story.

These two verbal memory tests allow for ready evaluation of loss of registered material after delay. These tests provide standardized measurement by which to evaluate the extent of T.J.'s amnesia.

Attention

Digit Span Forward. This measures the subject's ability to repeat number strings which are presented orally. This test provides further information regarding T.J.'s ability for initial registration of information.

---

[2] Two of the tests included at Time 1 were administered at age 8-3

Vocabulary
Peabody Picture Vocabulary Test-Revised (PPVT-R). This is a test of vocabulary in which the patient is asked to select one of four drawings which best represent an orally presented word.

Wechsler Intelligence Scale for Children (WISC-R and WISC III), vocabulary subtest. These are expressive tests in which the subject is required to define words.

The Boston Naming Test (BNT). This is a test of confrontational naming which taps vocabulary knowledge. It requires precise naming of line drawings of objects and animals.

These three tests were used to track vocabulary knowledge. They were used in the initial clinical assessments of T.J. and were also appropriate for ongoing research.

Basic Reading Skills
Woodcock -Johnson Revised Psycho-Educational Battery (WJ-R) : reading cluster subtests:
• Letter-Word Identification. This subtest requires identifying letters and phonetically regular and irregular words within a 5 second time limit.
•Passage Comprehension. This subtest requires reading sentences which are missing a word and supplying a word which makes sense in the sentence or passage.
•Word Attack. This subtest requires the reading of a list of words defined by Woodcock as non-words or extremely low-frequency words. Since these are not words that would have been encountered previously, this is a test of decoding ability.

Decoding Skills Test (DST). Part Two, used here, requires untimed reading of a graded list of phonetically regular words selected to represent common orthographic patterns, and nonwords that are analogous to the real words. Raw scores are generated for four types of words: monosyllabic real words; monosyllabic nonsense words; polysyllabic real words; polysyllabic nonsense words. Comparison of nonsense word performance to real word performance allows for inferences concerning the learners' application of phonemic principles to nonsense words.

All of these measures are commonly used in reading research, and each correlates highly with reading level. In the context of our questions concerning memory, these tests are a measure of T.J.'s ability to learn reading, and , therefore, a measure of semantic memory.

Baseline Testing
Memory and Attention

Table 1 summarizes baseline testing data. Performance on memory for Narrative Prose documents the psychometric hallmark of amnesia: after a thirty minute delay, those details of the story that T.J. had recalled are completely forgotten. Performance was in the borderline range for immediate recall (raw score 4 ½), but he recalls no details--and denies having heard a narrative--after delay.

RAVLT results reveal impaired learning of a rote word list on repeated trials. T.J.'s performance did not improve consistently across repetitions of the word list. During the course of the learning trials, T.J.'s best immediate recall is 3/15, which is in the impaired range. Delayed recognition was also impaired, near chance level. Taken together, the Narrative Prose and RAVLT memory tests reflect inefficient initial learning and clearly document loss of learned information over time. T.J.

scores broadly within normal limits on Forward Digit Span. Thus, consistent with a typical amnesic profile, initial registration appears relatively spared.

TABLE 1. Baseline Testing: Memory and Attention

		AGE: 8.0 & 8.3	
		Standard Score	Raw Score
MEMORY			
	Narrative Prose:		
	- Immediate Recall	81	4.5
	- Delayed Recall	<69	0
	RAVLT:		
	- Best Trial	70	3/15
	- Recall After Distractor	<69	2/15
	- Delayed Recognition	<69	9FA/2*
ATTENTION			
	Digit Span	85	7

NA = Not Available
NT = Not Tested
* at or near chance level

Vocabulary

Table 2 summarizes baseline vocabulary performance generally within normal limits. Performance on the WISC-III test of expressive vocabulary was within low normal limits. Normal performance was also found on confrontational naming, as measured by the Boston Naming Test. In contrast, performance on the PPVT-R test of receptive vocabulary was impaired. This poor performance may be attributable to difficulties in visual analysis of relatively complicated line drawings. In any case, intact WISC and BNT performance demonstrates there was no true impairment for vocabulary.

Basic Reading Skills

Table 2 shows impaired performance on four measures of basic reading and pre-reading skills. T.J. was able to identify letters but not words on the Word/Letter ID subtest of the WJ-R. He received a raw score of 0 on subtests tapping Passage Comprehension and Word Attack. These results describe an individual at the pre-reading stage. Tests results, corroborated by teacher reports and parental reports, indicate that T.J. was essentially a non-reader.

TABLE 2. Baseline Testing: Vocabulary and Reading

		AGE: 8.0 & 8.3 GRADE: 1.9 & 2.0 Standard Score	 Raw Score
VOCABULARY			
	PPVT-R	64	62
	Boston Naming	85	33
	WISC-III Vocabulary	85	14
	WISC-R Vocabulary	NT	NT
UNDERLYING READING SKILLS			
	TAAS	<65	3
	RAN (seconds)		
	- Colors/Objects	<69	: 254
	- Numbers/Letters	<69	: 231
	Digit Span	85	7
READING SKILLS			
	Woodcock Johnson:		
	- Word/Letter ID	57	13
	- Word Attack	<55	0
	- Passage Comp	<55	0
	Decoding Skills		
	- Real Words	NT	NT
	- Nonsense Words	NT	NT

NA = Not Available
NT = Not Tested

Comment

In summary, baseline testing indicates that T.J. was a non-reader with severely impaired memory for recently presented material. Results confirm preserved vocabulary acquisition.

Post-Intervention Results

Memory and Attention

Post intervention results are summarized in Table 3 . Tests were administered at the close of the first (Time 2) and second phases (Time 3) of intervention. Memory testing results at Times 2 and 3 both reflect a similar profile to baseline testing. Improvement is noted on immediate but not delayed raw scores. Immediate Narrative Prose recall rose to raw score of 6 on one occasion. In both cases, the severity of T.J.'s amnesia is documented by complete loss of information after a 30 minute delay. Rote learning of a word list (RAVALT) continued to be in the impaired range for both immediate recall and delayed recognition. Analysis of raw scores reveals material learned over time was completely lost after delay. T.J.'s score reached 5/15 on the learning trials, and he retained 3 items after a brief distracter, but delayed recognition was at chance level.

As in baseline testing, results show that material which is recalled on immediate trials is forgotten completely after delay, and that learning is inefficient, particularly on a rote learning task.

TABLE 3. Post Intervention Testing: Memory and Attention

		TIME 2		TIME 3	
		AGE: 8.11 & 9.0 **GRADE: 2.8 & 2.9**		**AGE: 9.5** **GRADE: 3.4**	
		Standard Score	*Raw Score*	*Standard Score*	*Raw Score*
MEMORY					
	Narrative Prose:				
	- Immediate Recall	84	6	78	4
	-Delayed Recall	<69	0	<69	0
	RAVLT:				
	- Best Trial	NT	NT	70	5/15
	- Recall After Distractor	NT	NT	81	3/15
	- Delayed Recognition	NT	NT	<69	14FA/O*
ATTENTION					
	Digit Span	108	10	NT	NT

NA = Not Available
NT = Not Tested
* at or near chance level

Forward Digit Span performance was in the above average range, documenting intact initial registration.

Vocabulary

WISC-R Expressive vocabulary and Boston Naming performance continued to be within normal limits (Table 4). Performance on the PPVT-R was again impaired. These scores indicate that T.J.'s vocabulary continued to progress at a normal rate.

Basic Reading Skills

Results at Time 2

Results are summarized in Table 4. After 8 months of tutoring, standard scores were in the low average range for Word ID and Word Attack, and in the impaired range for Passage Comprehension. Performance on the Decoding Skills Test (DST) was borderline impaired for real words, but was within normal limits for nonsense words. This indicated he was relying heavily on phonological skills rather than recognition when encountering new words.

TABLE 4. Post Intervention Testing: Vocabulary & Reading

	TIME 2 AGE: 8.11 & 9.0 GRADE: 2.8 & 2.9			TIME 3 AGE: 9.5 GRADE: 3.4		
	Standard Score	*Raw Score*	*GE*	*Standard Score*	*Raw Score*	*GE*
VOCABULARY						
PPVT-R	NT	NT	NA	68	76	NA
Boston Naming	96	36	NA	90	35	NA
WISC-III Vocabulary	NT	NT	NA	90	21	NA
WISC-R Vocabulary	85	19	NA	NT	NT	NA
UNDERLYING READING SKILLS						
TAAS	87	9	1.7	100	13	3.3
RAN (seconds)						
-Colors/Objects	<69	:200	NA	<69	:312	NA
-Numbers/Letters	<69	:117	NA	<69	:132	NA
Digit Span	108	10	NA	NT	NT	NA
READING SKILLS						
Woodcock Johnson:						
- Word/Letter ID	82	25	1.8	88	31	2.6
- Word Attack	89	7	2	96	14	3
- Passage Comp	66	6	1.3	87	13	2.4
Decoding Skills Test						
- Real Words	72	14	NA	86	34	NA
- Nonsense Words	84	13	NA	81	20	NA

NA = Not Available
NT = Not Tested

Results at Time 3

Normal performance was sustained on Word ID and Word attack, and DST nonsense words. Scores now rose to within normal limits on Passage Comprehension, and DST real words. Thus, performance on all basic reading tests was within normal limits (Table 4).

Comment

Post-intervention results on the Narrative Prose and RAVLT are very similar to baseline. They document a failure to recall registered information after delay.

In contrast, results at Time 2 and Time 3 indicate sustained normal performance on measures of expressive vocabulary and confrontational naming. At Time 3, all basic reading skills were also broadly within normal limits (Table 5). Results show that reading was acquired at an average, and in some cases, accelerated rate (Table 6). Grade equivalency improvement averaged 2.3 academic years in an 18 month period.

TABLE 5. Preserved Semantic Memory Performance

		AGE: 9.5 GRADE: 3.4 Standard Score
READING SKILLS		
	Woodcock Johnson:	
	- Word ID	88
	- Word Attack	96
	- Passage Comp	87
	Decoding Skills	
	- Real Words	86
	- Nonsense Words	81
VOCABULARY		
	Boston Naming	90
	WISC-III Vocabulary	90

TABLE 6. Reading Acquisition Gains

Expressed in Grade Levels

	Time 1 Baseline	Time 2 Post	Time 3 Post	Grade Level Gains Time 1 to Time 3*
READING SKILLS				
Woodcock Johnson:				
- Word ID	1.0	1.8	2.6	+1.6
- Word Attack	<K	2.0	3.0	+3.0
- Passage Comp	<K	1.3	2.4	+2.4

* 18 month period

Observational Data

Behavioral Observations During First Phase of Tutorial Intervention

T.J. was typically unable to correctly answer questions about what he had done earlier in the day (e.g. gone shopping with his mother). He could not reliably state whether he had been to school on a given day. When asked if he had been to school earlier, he usually answered yes, once reasoning aloud that he only did tutoring on school days. When asked on a school holiday, he replied that he had been to school. He was generally not able to recall details concerning the specific activities he had done in a current or previous tutoring session. He often repeated himself several times in a session regarding a current topic of interest, with no apparent awareness that he was repeating himself. He was unable to demonstrate even an approximate sense of how many tutoring sessions he had completed, or how long we had been working at a given session. T.J. was unable to say whether he had seen any of the four regular librarians before, even after dozens of interactions.

T.J. at first needed assistance in finding the tutoring room. Over time, with prompting, he was able to master a procedure in which he would ask the librarian for the key, lead us past a door through a room, unlock the next door, then return the key and return to the room. He learned this through trial and error. He was able to verbalize the steps of this procedure. He was also able to learn the location of a store that we visited on breaks. He was proud the first time he correctly used the key to open the tutoring room, and mentioned that it was the first time he had done this on his own. He proudly said this again the next 2 times he opened the door. Similarly, when questioned, he stated for three consecutive sessions that it was the first tutoring session since vacation.

He clearly grew more familiar with me over time. He referred to themes we had discussed and regular activities spontaneously, and interacted in certain habitual ways. His manner of relating interpersonally revealed no sign of memory impairment. T.J. demonstrated consistently that he was readily familiar with the names and appearances of family members, teachers, and friends, and could recall their names immediately whether they were present or not. He frequently recalled the name of a tutor he had worked with 3 years earlier. T.J. learned my first name, appearance, and my role after two meetings. As I approached the library for the third meeting, T.J. saw me at approximately 150 feet, and identified me by name to his mother. Her view was obscured and she stated that I was not approaching. He replied adamantly that I was.

T.J. occasionally spoke of a very limited set of important specific events from his past. When questioned, however, he could provide no details or context. For instance, he recalled several times that his father had cut his finger in the past. When asked about details or context he could only say, "he cut his finger and it bled and bled". There appeared to be a strong emotional concomitant to all of these memories.

He could readily state the usual features of his life, e.g. going to school, watching cartoons, helping at his father's job. He was able to name the activities that characterized our meetings (e.g. working on reading, drawing, going on a break) and those that generally characterized his other usual activities as well. After several sessions spent on an exercise, he was able to say that he had done it previously but he could not say when.

Perhaps his most impressive memory performance was in noticing variances in the usual environment. He mentioned spontaneously that a banner in the library was new and that posters in the tutoring room were new. He also learned quickly that a back door had an alarm and was not to be used and he rehearsed this fact often, perhaps cued by the bright red sign. He did not notice when the arrangement of chairs and tables in the tutoring room had been changed.

Observations during tutoring sessions were consistent with parental observations that T.J.'s affect is not blunted or constricted. His emotional reactions to both the instructor and the material were varied and cumulative.

Comment

These observations are consistent with test data and previous structured and unstructured observation. They show that T.J. was unable to recall specific events yet was able to learn effectively about the general features of his environment and to retain this information.

Discussion

On the empirical level, this study demonstrates two significant forms of memory functioning which have not previously been documented in amnesia. The present data indicate that reading acquisition was accomplished at a normal rate despite profound amnesia, and that vocabulary acquisition was also in the normal range through age 9-5. These findings converge with the previously reported case of normal academic progress in a case of acquired amnesia at age 12 (Wood et al., 1989). The present study extends these findings because T.J. did not have well-established basic language and cognitive skills prior to the onset of amnesia as the subject in the Wood et al. study did.

The pivotal theoretical question posed in this chapter is whether semantic memory can be shown to be preserved in amnesia. Put another way, the question is whether semantic memory can be preserved relative to episodic memory. This case convincingly demonstrates preserved semantic memory in contrast to profoundly impaired episodic memory. The demonstration is especially valuable because the data successfully address the major criticisms leveled at previous studies showing preserved semantic memory in amnesia.

Critics have questioned whether some demonstrations of preserved semantic memory have in fact depended on knowledge learned premorbidly. Because of the congenital nature of T.J.'s condition, he had no opportunity to establish knowledge premorbidly. In the virtual absence of episodic memory, he not only improved academic skills, but acquired them. He had no premorbid knowledge to serve as a basis for this learning

Critics of studies supporting preserved semantic memory have also wondered whether semantic memory might only "appear" to be preserved relative to episodic memory because of a difference in exposure to relevant information. This critique culminated in a proposed criterion that semantic memory must be preserved beyond what would be expected given the level of episodic memory (Ostergaard and Squire, 1990). This criterion, even in its most stringent possible interpretation, is addressed satisfactorily by the present data. Psychometrically, reading and vocabulary performance (Table 5), which rely on semantic memory, are far beyond what might be expected given devastated recall performance on the Narrative Prose memory task, a standard psychometric measure of episodic memory.

The psychometric data are corroborated by behavioral observation which shows extensive preserved semantic knowledge about, for example, people that T.J. knows, including their mannerisms and appearance, places he is familiar with, including activities he typically engages in there and geographical and structural features, and routines such as preparing for school. All this is in dramatic contrast to virtually absent moment by moment episodic memory so that T.J. typically cannot recall, for example, where he has been, want he has done and what he has eaten.

The present data can also be considered in light of a recent definition of declarative memory which states that it is "fast and flexible... cross-modal " (Squire et al, 1993). In learning to read, T.J. learned symbol-sound relationships, and orthographic patterns which varied these symbol-sound relationships. In the act of reading, he simultaneously appreciated the relationship between the letters and the phonemes, the relationship between the phonemes themselves, and the meaning created by those phonemes when properly grouped. He was able to do this in response to novel combinations, and was able to use words in a sentence or paraphrase a sentence. Training in phonemic awareness was generalized to learning how to pronounce blends (e.g. *bl*) without explicit instruction. Reading training was also generalized without explicit instruction such that T.J. was able to pronounce and comprehend a similar set of words when letters are presented orally rather than presented visually.

Thus, there is ample evidence that T.J.'s reading performance met recent criteria[3] for declarative memory.

As further evidence of preserved semantic memory, consider that T.J. was socially adept in a variety of situations. His speech was consistent with the general rules of social interaction and with the tone and content of past shared conversation and discussion. He proved capable of being polite, of asking appropriate questions, and of angling for something he wants (e.g. a snack or a break) without directly asking for it. He was also capable of describing (at times spontaneously) his social behavior and the principles he was following. Such social adeptness requires subtle application of well learned principles.

Skillful social behavior has often been found in amnesic individuals, but has seemed insignificant theoretically because it was assumed this trait simply represented the application of knowledge over learned premorbidly. This case clearly suggests consideration of the issue of whether complex social skills can be gained by amnesics, and reminds us that they retain the ability to express social skills in response to new information in new environments. The ability to converse appropriately, for example, often requires that detailed knowledge gained from previous interactions be applied in a flexible and creative manner in response to rapidly shifting contingencies.

Tulving (1984) has speculated that semantic and episodic memory functioning, while separate, are closely interdependent such that if episodic memory is impaired, semantic memory is likely to suffer some decrement as well. He stated that if it was indeed true that the two systems were separate and dissociable, he expected that "sooner or later", a case would make it apparent. Measures of semantic memory here are clearly within normal limits: reading acquisition took place at a normal rate and proficiency on tests of basic reading skills and vocabulary acquisition are in the normal range. These data give credence to the notion that previous studies presenting evidence of semantic memory which is preserved, but not within normal limits, are documenting a true dissociation between semantic and episodic memories. Therefore, the present study not only adds to the body of existing evidence for preserved semantic memory in amnesia, it strengthens already existing evidence.

The conclusion that semantic memory is preserved in amnesia has further theoretical implications. These results are consistent with the memory classification in which semantic and episodic memories are distinct. In turn, this conclusion is inconsistent with several facets of the nondeclarative/declarative account of memory systems in amnesia. First, these data counter the claim that episodic and semantic memories are expressions of a single, declarative, memory system. The fact that semantic memory can be preserved relative to episodic memory argues strongly for the position that they represent separate systems of memory. Second, these data are problematic for the claim that only nondeclarative memory is preserved in amnesia. They show semantic memory, assumed to be one of two main expressions of declarative memory, to be preserved.

In the light of accumulated evidence for preserved semantic memory in amnesia, it is difficult to see what is to be gained by retaining declarative and nondeclarative as overarching categories. It

---

[3] The definition of declarative memory has grown more restrictive over the past 16 years, as the definition of procedural memory has expanded.

seems clearer and more consistent with the full range of available data to return to the classification suggested by Pribram and followed by Tulving and others in which there are three main memory systems: semantic (reference) ,episodic (context), and procedural (skill).

The original inspiration for the declarative/nondeclarative hypothesis was the distinction between "knowing that" and "knowing how." If reading, vocabulary, and other academic and factual learning is possible in amnesia, this defies the notion that the ability to bring knowledge to consciousness, to "know that", is the defining impairment in amnesia. It appears that amnesics can "know that" when the knowing involves general knowledge that is not temporally bound.

To the extent that such generalizations can be made, the defining impairment in amnesia appears to be for episodic memory, a subset of "knowing that " which depends upon, or perhaps makes possible, self-monitoring capacity and the accumulation of a detailed personal narrative. The central deficit in amnesia is not the inability to recall and declare conscious knowledge, but the inability to recall richly detailed and temporally coherent *self*-conscious knowledge--to possess, in Pribram's term, the "narrative self."

The present data also reflect on a well-documented phenomenon known as source amnesia. Source amnesia (Schacter et al, 1984) refers to a dissociation between preserved ability to acquire new factual information and impaired ability to recall how or where the new facts were acquired. In the prototypical research paradigm, a subject answers a question correctly based on information recently learned from the researcher but claims not to know how he learned the information. Typically, the learning episode and the test all occur within the same day. In the present case, data was collected over an eight month tutorial intervention, providing a different perspective than short-term laboratory designs. This allows us to look at the same issue in a different time frame.

Although this was tested only informally, it appears that T.J. would demonstrate source amnesia in the typical research design, as described above. This conclusion is drawn because his ability to recall or identify learning episodes from earlier in a given session was shown again and again to be impaired. However, in the full course of this study, he did not demonstrate source amnesia. He knew, for instance, that he learned reading at tutoring sessions, that he did other types of learning at school, that he learned about the pirate Blackbeard from his sister, and that he learned about soccer from his uncle.

The present data show that, longitudinally, T.J. did not exhibit source amnesia, but rather impaired episodic memory. Source amnesia experiments ask first about a decontextualized fact and then about an episode. The current data suggest that these studies may, in fact, demonstrate preserved semantic memory versus impaired episodic memory—rather than any characteristic specific to the memory of sources.

We have argued that the fact that T.J. has been amnesic since birth makes the present data in many ways uniquely informative. The unique nature of his neurological condition also, of course, suggests due caution in the interpretation of these data. T.J.'s neurological status was influenced by cystic intrusion, surgical intervention, and, thus, the forces of plasticity. This case differs from cases in which a brain that has developed normally receives a focal and identifiable insult, surgically or otherwise. T.J.'s neurological development has been idiosyncratic since before his birth. Although we have documented ways in which his performance appears normal, there is little doubt that this performance is the result of a central nervous system that underwent a unique developmental process. It is significant, then, that the data are not being used to support new or highly unusual theoretical claims, but rather theoretical positions which are over twenty years old (Pribram 1969; Kinsbourne and Wood, 1975; Tulving, 1972). The long-established conceptual framework of preserved semantic and impaired episodic memory in amnesia readily brought coherence to the data. This makes it reasonable to entertain that we are observing the brain's functioning through a unique window, rather than seeing brain functioning which is too unique to be informative.

This study followed Glisky and Schacter (1986a, 1986b, 1987, 1989) and Tulving (1991) in building intensive intervention into the research design. One virtue of such an approach in this case is clear. It seems likely that if there had been no focused and flexible intervention, T.J. would not have acquired reading skills. Serial testing might have continued to document, over the years, T.J.'s failure to acquire reading skills and we would be grappling with a very different set of data, perhaps centered on the question of why he was able to learn spoken but not written language.

Our impression is that the stance of responsiveness to psychological as well as cognitive factors facilitated T.J.'s learning to a significant extent. The process of reading acquisition appeared to be emotionally difficult for T.J.. At the beginning, difficulties appeared most often to be related to fear of failure. As his proficiency grew, anxiety arose because reading represented to him a level of independence about which he was ambivalent.

Although T.J. at first presented himself as quite motivated, it soon became clear that he was anxious, frightened and expectant of failure. In responding to this, we found successful trials tended to assuage his anxiety and fear. He was readily demoralized by unsuccessful trials. He responded well to encouragement, to reassurance, and to allowing him to structure most of the nonessential elements of the session (e.g. when and where the study break was taken). Particularly because learning to read was an emotionally charged process for T.J., it seemed to make a difference that he felt free to discuss or act out his emotional concerns from time to time.[4] In general, then, it seemed that the quality of the interpersonal relationship was relevant to learning in that it ameliorated certain anxieties. It also seemed to serve as a motivating factor. T.J.'s mother reported many times on her observation that the relationship forged in the midst of tutoring was important to T.J. In the sessions, it was clear that he was motivated at times by showing that he could do well.

The nature of the learning task may also have been important to the outcome. Research with amnesic subjects has not typically asked them to engage in learning tasks which are relevant to day to day living. People are apt to approach relevant tasks with greater motivation, attention, and affective valence. These variables might be particularly important in the performance of individuals with memory impairment. The surprising success in Glisky and Schachter's studies and in the present study may be due in part to the practical importance which the learning material held for the subjects.

These observations suggest that thorough consideration of psychological issues may be advisable prior to drawing any final conclusions concerning memory impairment in any individual or group study. Other subjects may, as T.J. did, appear eager and motivated out of compliance when their performance is actually hindered by their psychological state. The quality of the relationship between subject and investigator may also be considered as a possible influence on performance. Psychological variables cannot be assumed to be normally distributed in amnesics, because their experiences—particularly their experiences in learning situations—have not been typical. At whatever age amnesia is acquired, it is likely to result in experiences which are frustrating, puzzling, and otherwise difficult.

---

[4] This was something I did not invite, but I did not discourage him when it occurred.

Further direct study of interplay between amnesia and psychological factors might well be profitable. Psychological factors may have definable effects on memory performance in amnesia, and the amnesic syndrome may have definable psychological sequelae. These issues can be pursued with T.J. using data already collected, and with other subjects in the future. Such a pursuit may thoroughly integrate psychological and neuropsychological factors in seeking to understand the complex nature of human memory.

References

Ahern CA, Wood FB, Hart LA, Meyer M (1997) Reading Intervention with an Amneisc Child. Under editorial review

Brooks DN, Baddely AD (1976) What can amnesic patients learn? Neuropsychologia 14,111-122

Caramazza A (1986) The Case For Single Patient Studies. Brain and Cognition 5, 41-46

Cermak LS, Bleich RP, Blackford SP (1988) Deficits in implicit retention of new associates by alcoholic Korsakoff's patients. Brain and Cognition, 1, 145-156.

Cohen NJ, Eichenbaum H, Deaced BS, Corkin S (1985) Different memory systems underlying acquisition of procedural and declarative knowledge. Annals of New York Academy of Sciences 444: 54-71

Cohen NJ and Squire LR (1980) Preserved learning and retention of pattern- analyzing skill in amnesia: dissociation of "knowing how" and "knowing that". Science 210, 207-209.

Corkin,S (1968) Acquisition of motor skill after medial temporal lobe excision. Neuropsychologia 6:215-234.

Daum I, Channon S, Canavar A (1989) Classical conditioning in patients with severe memory problems. In: J. Neurol. Neurosurg. Psychiatry 52, 47-51

Glisky EL and Schacter DL (1988). Long-term retention of computer learning by patients with memory disorders. J. Learn. Disabil. 21, 333-39.

Glisky EL and Schacter DL (1987a). Acquisition of domain-specific knowledge in organic amnesia: training for computer-related work. Neuropsychologia 25, 893-906.

Glisky EL and Schacter DL (1987b) Extending the limits of complex learning in organic amnesia: computer training in a vocational domain. Neuropsychologia 27, 107-120

Glisky EL, Schacter DL, Tulving E (1986a) Computer learning by memory-impaired patients: acquisition and retention of complex knowledge. Neuropsychologia 24, 313-328.

Gliskey EL, Schacter DL, Tulving E (1986b) Learning and retention of computer-related vocabulary in memory-impaired patients: method of vanishing cues. J. Clin. Exp. Neuropsychol.

Heilman K, Valenstein E (1993) Clinical Neuropsycholgy. Oxford University Press, New York.

Kinsbourne M, Wood FB (1982) Theoretical considerations regarding the episodic-semantic memory distinction. In: Cermak LS (ed) Human Memory and Amnesia. Erlbaum, Hillsdale/NJ,

Kinsbourne M and Wood FB (1975) Short term memory and the amnesic syndrome. In: Short-Term Memory,

D Deutsch and JA Deutsch (editors). Academic Press, New York. pp 195-217

Knowlton BJ, Squire LR (1994) The information acquired during artificial grammar learning. Journal of Experimental Psychology: Learning, Memory, and Cognition 20,79-91
Knowlton BJ, Ramus SJ, Squire LR (1992) Intact artificial grammar learning in amnesia. Psychological Science 3, 172-177
McCauley, RN (1984) Inference and temporal coding in episodic memory. The Behavioral and Brain Sciences 7, 246-247
Milner B (1970) memory and the medial temporal region of the brain. In: Pribram KH and Broadbent D (eds) Biology of Memory. New York. Academic Press.
Milner, B (1969). Residual intellectual and memory deficits after head injury. In Walker A, Caveness W, and Critchley, McD (eds), The Late Effects of Head Injury. Springfield, Il: Charles Thomas, pp 84-97
Milner B, Corkin S, Teuber HL (1968) Further analysis of the hippocampal amnesic syndrome: 14 year follow-up study of H.M. Neuropsychologia 6:215-234.
Milner B (1966) Amnesia following operation on the temporal lobes. In Whitty, CWM and Zangwill, OL (eds), Amnesia. London: Butterworths, pp 109-133.C.
Moscovitch M (1986) Multiple Dissociations of function in amnesia. In: Cermak LS (ed), Human Memory and Amnesia. Hillsdale NJ: Erlbaum
Musen G, Shimanura AP, Squire LR, (1990) Intact text-specific reading skills in amnesia. J. Exp Psychol. Learn. Mem. Cogn 6, 1068-1076
Ostergaard A, Squire, L (1990) Childhood amnesia and distinctions between forms of memory. Brain and Cognition 14:127-133
Ostergaard A (1987) Episodic, semantic and procedural memory in a case of amnesia at an early age. Neuropsychologia 25, 341-357
Pribram KP (1991) Brain and Perception --Holonomy and Structure in Figural Processing. Hillsdale, NJ. Lawrence Erlbaum Associates
Pribram KH (1971) Languages of the Brain: Experimental Paradoxes and Principles in Neuropsychology. Englewood Cliffs, NJ: Prentice-Hall
Pribram KH (1969) The amnestic syndrome: disturbances in coding? In Talland, GA and Waugh, N.C. (eds.) Psychopathology of Memory. New York: Academic: pp 127-157
Pribram KH (1966) Some dimensions of remembering: steps toward a neuropsychological model of memory. In Gaito, J. (ed.) Macromolecules and Behavior. New York: Academic: pp 165-187
Schacter DL, Cooper LA, Tharan M, Rubens AB (1991) Preserved priming of novel objects in patients with memory disorders. Journal of Cogn. Neurosci. 3:118-31
Schacter DL(1987) Implicit expression of memory in organic amnesia: Learning of new facts and associations. Human Neurobiology 6:107-118.
Schacter, DL (1985) Priming of old and new knowledge in amnesic patients and normal subjects. Ann. N. Y. Acad. Sci. 444, 41-55
Schacter DL, Harbluk JL, McLachlan DR (1984) Retrieval without recollection: an experimental analysis of source amnesia. Journal of Verbal Learning, Verbal Behavior 23:593-611
Squire LR, Knowlton B and Musen, G (1993) The structure and organization of memory. Annual Review of Psychology 44, 453-495.

Squire LR, McKee R (1992) Influence of prior events on cognitive judgments in amnesia. J. Exp. Psychol. Learn. Mem. Cogn. 18:106-15.

Squire LR (1992) Memory and the hippocampus. Psychological Review 99 (2): 195-231

Squire LR (1992b) Declarative and nondeclarative memory. Journal of Cognitive Neuroscience 4(3): 232-242

Tulving E, Hayman CAG, MAcDonald CA (1991) Long-lasting perceptual priming and semantic learning in amnesia: a case experiment. J. Exp. Psychol. Learn. Mem. Cogn. 17:595-617

Tulving E (1987) Multiple memory systems and consciousness, Human Neurobiology 6:67-80.

Tulving E (1984) Precis of Elements of Episodic Memory. The Behavioral and Brain Sciences 7:223-268

Tulving E (1983) Elements of Episodic Memory. Oxford: Claredon Press,

Tulving E (1972) Episodic and semantic memory. In: Tulving E and Donaldson W (eds) Organization of Memory. New York: Academic Press.

Weiskrantz I, Warrington EK (1979). Conditioning in amnesic patients. Neuropsychologia 17: 187-194.

Wood FB, Brown I, Felton R (1989) Long-tern Follow-up of a Childhood Amnesic Syndrome. Brain and Cognition 10:76-86

Wood FB (1987) Focal and diffuse memory activation assessed by localized indicators of CNS metabolism: the semantic-episodic distinction. Human Neurobiology 6: 141-151

Wood FB, Ebert V, Kinsbourne M (1982) The episodic-semantic distinction in memory and amnesia: clinical and experimental observations. In: Cermak L.S. (Ed.) Human Memory and Amnesia. Hillsdale, NJ. Erlbaum Associates.

Wood FB, Kinsbourne M (1982) Theoretical considerations regarding the episodic-semantic memory distinction. In: Cermak L.S. (Ed.) Human Memory and Amnesia. Hillsdale, NJ. Erlbaum Associates.

11

The Role of Memory in Brain, Values, and Choice

Bennet B. Murdock
Department of Psychology
University of Toronto
Toronto, ON M5S 3G3
Canada
Telephone: (416) 978-3175/FAX: (416) 978-4811
E-mail: murdock@psych.toronto.edu

Abstract

The paper discusses a memory model TODAM that presents a particular point of view on the relation between data and brain formalisms. TODAM, a theory of distributed, associative memory is based on the holographic approach of Pribram (1971) and the convolution-correlation formalism of Borsellino and Poggio (1973). It attempts to explain the storage and retrieval of item, associative, and serial-order information, and is tested by data from psychological studies of short-term human episodic memory. A brief account of the application of the model to these three areas and illustrative experimental data will be presented.

Introduction

The general theme of the Appalachian conferences is the relation between data and brain formalisms. The theme of Appalachian V is brain, values, and choice. This chapter will deal with the role of memory in brain, values, and choice. I will focus my remarks on the understanding of memory per se, and leave the broader questions to others.

Memory is not a single entity; rather, there are different kinds of memory. The basic partitioning is in turns of the types of information stored and retrieved. At least for simple episodic memory there are three types of information: item information, associative or relational information, and serial-order or positional information. Item information allows us to recognize objects and events--names, faces, sounds, the myriad objects and events that make up the world around us. It underlies the feeling of familiarity when we see or hear them. Associative information allows us to remember the relation between two objects or events: lightning and thunder, the taste and smell of a particular food, or the names and faces of our friends. Serial-order information allows us to remember temporal order: the days of the week, the letters of the alphabet, or how to spell words. Experimental evidence for these distinctions may be found in Murdock (1974).

Until recently the standard model for memory was a node-and-link model where the items were stored in the terminal nodes of a semantic network and the associations were the links between the nodes. Serial order could be represented as the order in which the links were traversed, and search was a universal metaphor for retrieval. The nodes were organized on the basis of semantic similarity, and spreading activation accounted for retrieval (Collins & Loftus, 1975).

One problem for this model was the lexical decision time. You would have to search the whole network to decide that "tible" was not a word but in fact experimental studies showed you could make this decision about as quickly as you could to decide that "table" was a word. More generally, the notion that you had to search a large network to retrieve any information at all seemed quite unlikely; our memory is very large but reaction times are generally quite short.

About 20 years ago a radically different approach to memory appeared on the scene. Thanks to the pioneering work of Pribram (1971), Kohonen (1977), and James Anderson (1970), holographic or distributed-memory models were proposed as an alternative. These were direct-access rather than search models that used distributed rather than localized storage. Information was stored in common memory system, and search was not necessary. The advantages of this type of model have been summarized by Anderson (1973).

Since then work has proceeded in two directions. One has been the remarkable development of connectionist models which try to implement these ideas in so-called neural networks. The other has been to try to forge links with the behavioral data from psychological laboratories, and this is where my interests lie. The neural-network work is well documented (e.g., Arbib, 1995) so I shall focus on the relation between distributed-memory models and behavioral data.

More particularly, I will discuss TODAM (a theory of distributed associative memory) and how it explains some of the experimental data on the storage and retrieval of item, associative, and serial-order information. TODAM uses the convolution-correlation formalism of Borsellino and Poggio (1973) which is an extension of earlier ideas by Pribram (1971) and others. (see Hinton & Anderson, 1981, for other important contributions). The reason I prefer the convolution-correlation approach rather than outer-product (or matrix approach) generally used in neural-network models is that it is easier to provide a unified account of item, associative, and serial-order information (Murdock, 1993).

TODAM

Assumptions

TODAM has five basic assumptions:

1. Items are represented as N-dimensional random vectors. A random vector is a vector composed of random variables, and each item vector has N such random variables. You can think of them as computational units or as abstract features, according to your preference. This approach was proposed by Anderson (1973) and is common in both connectionist and distributed-memory models.
2. Associations are represented as convolutions of random vectors, and convolution is denoted by the star sign (*). Connectionist models generally use the outer- (or matrix-) product for associations.
3. Items, associations, and chunks (for serial organization) are stored in a common memory vector, and the memory vector is symbolized by **M**. This assumption is common to the other global-matching models such as the SAM model of Shiffrin and of Raaijmakers (1972), the MINERVA model of Hintzman (1986; 1988), the matrix model of Humphreys (Humphreys, Bain, & Pike, 1989), and the CHARM model of Metcalfe-Eich (1982; 1985). Global-matching models are memory models which assume that all information enters into the comparison process in recognition; for a review see Clark and Grunlund (1996).
4. Retrieval involves the correlation of the probe item with the memory vector **M**, and correlation is denoted by the number sign (#). If convolution is the operation that binds two items into an association, then correlation is the appropriate retrieval operation.

5. These operations (convolution, correlation, and summation) go on in Working Memory, and Working Memory is part of the system architecture. The system architecture is shown in Fig. 1 where the P- (or perceptual) system is for encoding, the R- (or response) system is for response and decision, and the Q- (for query) system is where item, associative and serial-order information is stored. We will discuss Working Memory further in the section on serial-order information.

Definitions

The convolution of random vectors is the discrete analogue of the convolution of continuous functions or waveforms. At the component level, the x-th element of the convolution of the random vectors **f** and **g** is defined as

$$(f*g)(x) = \Sigma_i f(i)g(x-i).$$

Similarly, at the component level, the x-th element of the correlation of the random vectors **f** and **g** is defined as

$$(f\#g)(x) = \Sigma_i f(i)g(x+i).$$

For convolutions, the sum of the subscripts is x whereas, for correlation, the difference (second subscript minus the first) is x.

Processes

The two standard methods of testing memory (or responding to queries) are recognition and recall. If I ask you whether you have seen a particular face or heard a particular piece of music, that is recognition. If I ask you what is the capital of Australia, that is recall. With recall you have to produce or generate the answer, whereas with recognition all you have to do is make a binary (or n-ary) response.

Recognition involves two processes, comparison and decision. You must compare the probe item (the particular face or piece of music) with memory (i.e., the memory vector **M**) and then you must make a decision. With different probes the output from the comparison process will vary continuously over a wide range from very familiar to very unfamiliar, but you must map this into a discrete binary (yes or no) or n-ary (which of many) response.

Recall also involves two processes, but they are not quite the same. Recall involves generation or retrieval and deblurring. Given the probe (the question) you must correlate it with **M** to generate a response, but this response is only an approximation to (a noisy version of) the target item (in this case, Canberra). You must clean up or deblur this noisy approximation, but that it not modeled in TODAM. It is a function of the R system, and TODAM deals with the Q system.[1]

The comparison process for recognition is the inner (or dot product) where the dot product of two item vectors **f** and **g** is $\mathbf{f}\cdot\mathbf{g} = \Sigma_i f(i)g(i)$. The dot product is actually a poor man's correlation; that is, the correlation at the zero index. If items **f** and **g** are associated by convolution, correlation of the probe item with convolution generates an approximation (**g**') to the target item (**g**) because

$$\mathbf{f}\#(\mathbf{f}*\mathbf{g})=(\mathbf{f}\#\mathbf{f})*\mathbf{g}=\delta'*\mathbf{g}=\mathbf{g}'$$

where δ is the delta vector where $\delta = (...0,0,1,0,0,...)$. You can also use **g** to retrieve **f** because

$$\mathbf{g}\#(\mathbf{f}*\mathbf{g})=\mathbf{g}\#(\mathbf{g}*\mathbf{f})=(\mathbf{g}\#\mathbf{g})*\mathbf{f}=\delta'*\mathbf{f}=\mathbf{f}'.$$

Thus there is associative symmetry; recall works equally well in either direction. Associative symmetry seems to characterize short-term associative memory but not short-term serial-order memory (Kahana, submitted).

The recall recognition identity relates recall and recognition at a conceptual level. This identity is

$$\mathbf{g}\cdot\mathbf{g}'=\mathbf{g}\cdot\{\mathbf{f}\#(\mathbf{f}*\mathbf{g})\}=(\mathbf{f}*\mathbf{g})\cdot(\mathbf{f}*\mathbf{g})$$

and a proof can be found in Weber (1988). This identity applies to the comparison process in recognition and the retrieval process in recall. It does not take into account the deblurring process in recall or the decision process in recognition.

Storage Equation

Since TODAM assumes that item and associative information are stored in a common memory vector, the simplest storage equation would be

$$\mathbf{M}_j=\mathbf{M}_{j-1}+\mathbf{f}_j+\mathbf{g}_j+\mathbf{f}_j*\mathbf{g}_j \tag{1a}$$

where the subscript j indexes the j-th pair in an on-going stream of events (e.g., introductions at a meeting where **f** would be the name of the person and **g** would be the person).

However, memory would soon be overloaded so we need a forgetting process. We introduce a forgetting parameters α where $0\leq\alpha\leq1.0$. Then,

$$\mathbf{M}_j=\alpha\mathbf{M}_{j-1}+\mathbf{f}_j+\mathbf{g}_j+\mathbf{f}_j*\mathbf{g}_j \tag{1b}$$

so the addition of new information would be balanced by the forgetting of old information. But one can pay differential attention to item and associative information so we introduce an attentional parameter γ where $0\leq\gamma\leq0.5$. Then

$$\mathbf{M}_j=\alpha\mathbf{M}_{j-1}+\gamma\mathbf{f}_j+\gamma\mathbf{g}_j+(1-2\gamma)(\mathbf{f}_j*\mathbf{g}_j). \tag{1c}$$

Attention is limited so the more attention we pay to one type of information the less attention we pay to the other type of information.

On first encounter not all features are encoded (or computational units activated) so we assume probabilistic encoding. If p is the probability that any particular feature of an item vector is encoded (added to memory) then

$$\mathbf{M}_j=\alpha\mathbf{M}_{j-1}+\gamma(p\mathbf{f}_j)+\gamma(p\mathbf{g}_j)+(1-2\gamma)(p\mathbf{f}_j*p\mathbf{g}_j). \tag{1d}$$

This is the basic TODAM storage equation for simple item and associative episodic memory.

Item and Associative Information

Item Information

The matched-filter model of Anderson (1973) is a special case of Eq. 1d when only single items are presented and $\gamma=p=1.0$. (Anderson did not explicitly include a forgetting parameter but it could easily be added.) This model shows some of the basic features of distributed-memory models; there is direct access not search (the probe item is compared directly with the memory vector), all items are stored in a common memory (**M**) so there are no separate nodes for separate items, and if one assumes the individual elements (or computational elements) are independent then memory loss will be gradual.

Forgetting curves for experimental studies of item information generally show extensive recency and little or no primacy. Memory for the most recent items is best, and accuracy decreases monotonically until the beginning. These characteristics are shown in Fig. 2 which presents the hit rate (proportion of yes responses) for old items as a function of lag (number of items intervening between the study and test of an item) and the false-alarm rate (proportion of yes responses to new items) as a function of test position. Although these data are averaged over four subjects, each subject individually showed this pattern though the slopes varied somewhat from subject to subject.

The model predicts this pattern because, if L is list length and t is test position, the expected value of the dot product for the k-th item in the list is

$$E[\mathbf{f}_{kt}\cdot\mathbf{M}] = E[\mathbf{f}_{kt}\cdot(\alpha_{kt}\mathbf{f}_k + \sum_{l\neq k} \alpha_{lt}\mathbf{f}_l)]$$

$$= E[\mathbf{f}_{kt}\cdot\alpha_{kt}\mathbf{f}_k] + E[\mathbf{f}_{kt}\cdot \sum_{l\neq k} \alpha_{lt}\mathbf{f}_l]$$

$$= \alpha_{kt}\cdot E[\mathbf{f}\cdot\mathbf{f}] = \alpha^{L-k+t-1}.$$

Since lag=L-k+t-1 the greater the lag the less the familiarity of an old item as measured by the dot product. New items have no lag, so the fact that they increase with test position is due to changes in the decision criteria used by the subject.

Associative Information

The issue of associative information immediately brings up the question of learning. Why does performance improve with practice? When we start to learn a foreign language our performance is slow and unreliable, but with practice we may become fluent. Learning a foreign language is usually a complex cognitive process, but perhaps TODAM can throw some light on simple associative learning.

The standard paradigm used by experimental psychologists is paired-associative learning. In a study-test procedure, pairs of words matched at random are presented to the subject, and each study trial is followed by a test trial where on each test one of the words (usually the first word of the pair) and the subject tries to recall the second word. The pairs are symbolized as A-B, so in the test trials the A items would be presented as cues or probes for recall of the B items.

An extension of this paradigm is the so-called A-B, A-C retroactive-inhibition paradigm used in tests of interference theory. The original (A-B) learning is continued until subjects reach some level of mastery (generally one perfect trial), and then the interpolated (A-C) learning follows in which subjects learn new associations to old stimuli. According to interference theory, the unlearning of the B responses during the learning of the C response accounts for the interference (reduced recall of the B responses) that generally occurs.

In a classic study Barnes and Underwood (1959) tracked the unlearning of the B responses by testing for recall of both the B and C responses after various numbers of A-C interpolated-learning trials. Their results are shown on Fig. 3 where the left-hand panel shows the A-B learning and the right-hand panel shows the recall of B and C items after 1, 5, 10, and 20 trials of A-C learning. The recall of B items clearly decreases as A-C learning proceeds.

This interference effect has been troublesome for connectionist models because they seem to show "catastrophic" interference (McCloskey & Cohen, 1989). Simulations of these models show that the A-B forgetting drops off much too rapidly (Fig. 4), and this is a major problem for these models if they are proposed as a model for human learning. However, it turns out that it is possible to modify these models so catastrophic interference does not occur; for an example, see Sloman & Rumelhart (1992).

Does TODAM predict catastrophic interference? The results of a simulation are shown in Figure 5, and the answer is clearly no. If anything it does not show enough interference. The parameter values in this simulation were N=499, α=.99, p=.25, and γ=.25 and these are the mean results over 100 replications. Recall was scored as correct if the retrieved item ($\mathbf{g}'_k$) resembled the target item ($\mathbf{g}_k$) more closely than its competitors ($\mathbf{g}_l$, l in L) where the competitors were list specific;

see Murdock (1989) for further details.

Item and Associative Information

There is not time to consider the other empirical effects in studies of item and associative information; instead, we will discuss briefly two recent studies of the interaction of item and associative interaction which are more difficult for TODAM to explain. The first is the differential-forgetting effect reported by Hockley (1991). If you present pairs of items but test recognition memory for both individual items and pairs of items, the retention of items drops off over lag ranges where retention of the pairs is constant. This differential forgetting of item and associative information is shown in Fig. 6.[2]

This is a problem for TODAM because, as shown in Eq. 1b, items and associations are stored in a common memory vector and the forgetting parameter (α) operates on M, not selectively on **f** and **g** or on **f*g**. Before discussing the solution let us discuss the other problem, the emphasis results. As shown in Eq. 1b, the attention parameter (γ) predicts a tradeoff between item and associative information. The more attention you pay to one the less you pay to the other. Consequently, the better you encode the item information, the less you should remember the associative information, and vice versa.

Hockley and Cristi (1996) presented many short lists of pairs and told subjects in one condition to concentrate on the items and on the pairs in the other condition. After each list they were tested on what they had been told to remember. The surprise came when on a final test the subjects in both groups were tested on both items and pairs. The results are shown in Fig. 7 and, as can be seen, the item group did poorly on the pairs but the pair group did about as well on the items as the item group.

Obviously there is something wrong with TODAM, and these two sets of findings led to TODAM2. In the revision, the general idea was to embed items in context which would be used in tests of item information but not in tests of associative information. The rationale was that context is essential for these episodic tests because the only thing that differentiates old and new items is list membership. Pair recognition tests compared intact and rearranged pairs and, since the items in the rearranged pairs had all been members of the lists, context would not be useful.

For the emphasis results, it seems necessary to include item information in associative information. We borrowed an idea from Cooper (1973) and added the **f** and **g** item vectors before doing the convolution. Since

$$(\mathbf{f}+\mathbf{g})*(\mathbf{f}+\mathbf{g})=\mathbf{f}*\mathbf{f}+\mathbf{g}*\mathbf{g}+2(\mathbf{f}*\mathbf{g})$$

the associative information contains both item information (**f*f** and **g*g**) and the associative information (**f*g**).

In a distributed-memory model it is not enough simply to add items embedded in context to the memory vector; if you just did that the items would be free floating and not attached to context. We borrowed an idea from the CHARM model (Metcalfe-Eich, 1985) and used autoassociation to bind items to context. If we use "*2" to denote autoassociation (i.e., $\mathbf{f}^{*2}=\mathbf{f}*\mathbf{f}$) and capital letters to represent items in context, we can write the revised storage equation as

$$\mathbf{M}_j=\alpha\mathbf{M}_{j-1}+\gamma(p\mathbf{F}_j)^{2}+\gamma(p\mathbf{G}_j)^{*2}+(1-2\gamma)(p\mathbf{F}_j+p\mathbf{G}_j)^{*2} \quad (2)$$

where probabilistic encoding is included as in Eq. 1d.

It is also necessary to add mediators to improve the absolute level of performance for pair recognition so the curves cross. Now TODAM2 is able to account for these troublesome findings. The differential forgetting results predicted by the model are shown in Fig. 8 and the emphasis results predicted by the model are shown in Fig. 9. More parameters were needed in these fits, but the parameter values were the same for both sets of data. For a more detailed account see Murdock (in press).

Serial-Order Information

Item information involves single items, associative information involves two items, and serial-order information involves three or more items. Just as item information is included in associative information so item information is included in serial-order information. If I show you a short list of items, not only can you recall them in order but also you can tell whether or not a single probe item had been included in the list.

How do we remember the serial order of a short list of items, or how do we learn much longer strings such as the months of the year or the letters of the alphabet? This has been a seminal issue in the study of memory ever since the classic paper of Lashley (1951) who raised these issues in the context of language. The answers must involve the operations and processes involved in the storage and retrieval of serial-order information.

Chunking

The key notion in TODAM is chunking where, according to Miller (1956), a chunk is a "glob" of information. As I am using the term, a chunk is a higher-order unit that can function as single item but can also be unpacked into its constituents if necessary. Phonemes are composed of distinctive features, letters are composed of phonemes, words are composed of letters, and sentences are composed of words, but generally we remember the gist of a sentence. However, we can also recognise the words in a sentence, the letters in a word, etc. Thus, chunks at one level may be components at a higher level.

Chunks provide the basis for organization in memory. Organisation was a key concept in semantic-memory models, and is just as important in distributed-memory models as well. What are chunks, how are they formed, how are they retrieved, and how do they allow us to remember the temporal order of events? In TODAM the key notion are multiple convolutions, n-grams, and n-chunks, and we discuss these in turn.

Multiple Convolutions

If **a**, **b**, **c**, and **d** are random vectors then **a*b** is a two-way convolution, **a*b*c** is a three-way convolution, **a*b*c*d** is a four-way convolution, etc. Just as **a*b** is a vector (although its dimensionality is approximately twice that of its constituents **a** and **b**), so **a*b*c** and **a*b*c*d** are also vectors although their dimensionality is approximately three and four times that of their constituents.[3]

We use X (the capital Greek letter chi) for multiple convolutions, and it is analogous to Σ for summations. Just as

$$\sum_{i=1}^{n} \mathbf{f}_i = \mathbf{f}_1+\mathbf{f}_2+\ldots+\mathbf{f}_n$$

so

$$\mathop{X}_{i=1}^{n} \mathbf{f}_i = \mathbf{f}_1*\mathbf{f}_2*\ldots*\mathbf{f}_n.$$

We can also have multiple autoassociations; e.g., a*a (n=2), **a*a*a** (n=3), etc. This is like squaring or cubing a number or, more generally, raising it to the n-th power. To symbolize multiple autoassociations we define any vector "star n" as

$$\mathrm{f}^{*n} = \mathbf{f}*\mathbf{f}*\ldots*\mathbf{f} \qquad \text{(n times)}$$

or

$$\mathbf{f}^{*n} = \mathbf{f}*\mathbf{f}^{*(n-1)},\ \mathbf{f}^{*0} = \delta.$$

n-grams

An n-gram is the n-way autoassociation of the sum of n item vectors. We symbolize this by G(n) where for item vectors **a**, **b**, and **c**:

$$\mathbf{G}(1) = \mathrm{a}^{*1} = \mathbf{a},$$
$$\mathbf{G}(2) = (\mathbf{a}+\mathbf{b})^{*2} = (\mathbf{a}+\mathbf{b})*(\mathbf{a}+\mathbf{b}) = \mathbf{a}*\mathbf{a}+\mathbf{b}*\mathbf{b}+2(\mathbf{a}*\mathbf{b}),$$
$$\mathbf{G}(3) = (\mathbf{a}+\mathbf{b}+\mathbf{c})^{*3} = \mathbf{a}*\mathbf{a}*\mathbf{a}+\mathbf{b}*\mathbf{b}*\mathbf{b}+\mathbf{c}*\mathbf{c}*\mathbf{c}$$
$$+\ 3(\mathbf{a}*\mathbf{a}*\mathbf{b}+\mathbf{a}*\mathbf{a}*\mathbf{c}+\ldots+\ \mathbf{b}*\mathbf{b}*\mathbf{c})+6(\mathbf{a}*\mathbf{b}*\mathbf{c})$$

or, in general,

$$\mathbf{G}(n) = \Big(\sum_{i=1}^{n} \mathbf{f}_i\Big)^{*n}.$$

We used G(2) for associative information in Eq. 2 but we did not call it an n-gram. Expanding an n-gram is just like raising a series to a power; compare the terms of G(2) with $(x+y)^2$ or G(3) with $(x+y+z)^3$.

Like multiple convolutions and multiple autoconvolutions an n-gram is still a vector. It is also a higher-order unit. It can function as a unit but with the right retrieval cues, it can also be unpacked into its constituents. This will be explained shortly in more detail.

n-chunks

We can now define an n-chunk as the sum of n n-grams, j=1 to n. We symbolise an n-chunk by **C**(n) where

$$\mathbf{C}(n) = \sum_{j=1}^{n} \mathbf{G}(j) = \sum_{j=1}^{n} \Big(\sum_{i=1}^{j} \mathbf{f}_i\Big)^{*j}. \tag{3}$$

An n-chunk is still a vector because it is a sum of vectors, and it serves as the basic unit in the storage and retrieval of serial-order information.[4]

We form n-chunks in Working Memory, and we retrieve single items from chunks in Working Memory if it is necessary to unpack a chunk (or recall a short string of items in serial order). We assume Working Memory has five slots or storage registers which can be used for convolution and correlation. These registers are specialized for vector processing, and could easily be implemented in a neural network. It takes three registers for convolution and correlation because these functionals (or functions) have two arguments and an outcome. It takes a fourth register for the item information and a fifth register to accumulate the n-grams; for further details see Murdock (1992).

Retrieval

Given that we have formed an n-chunk, how do we unpack it into its constituents? We use an idea from CADAM (Liepa, 1977). As he suggested, you can use a multiple convolution of the items recalled so far as the retrieval cue or probe for the next item. Starting with the δ vector and using → to indicate deblurring,

$$\delta\#\mathbf{C}(n) = \mathbf{a}' \rightarrow \mathbf{a},$$
$$\mathbf{a}\#\mathbf{C}(n) = \mathbf{a}'+2\mathbf{b}' \rightarrow \mathbf{b},$$
$$(\mathbf{a}*\mathbf{b})\#\mathbf{C}(n) = 3\mathbf{a}'+3\mathbf{b}'+6\mathbf{c}' \rightarrow \mathbf{c},$$

or in general to recall item j

$$\mathop{X}_{i=1}^{j-1} \mathbf{f}_j\#\mathbf{C}(n) = \mathbf{f}_j'.$$

The multiple convolution of the first j-1 items correlated with the n-chunk retrieves the j item because that is next in line. It also retrieves all the other items recalled to date, but the target item (i.e., the next item) is always twice as strong as its competitors. Recall is a sequential process, and this shows how it could happen.

Since convolution is commutative, ACB or CBA should be just as effective retrieval cues for D as ABC if these cues were presented by the experimenter rather than generated by the subject. However, if you tried BDC, BCD, ACD, or even ABD for the recall of E none of them should be as effective as ABCD because you need the multiple convolution of all j-1 prior items to recall the j-th item.

Transpositions occur occasionally in serial recall; a subject might recall a 5-item list as ABDCE rather than as ABCDE. ABD would be as good a retrieval cue for C as ABC would be for D, but what is unlikely that AB would retrieve D. However, given that it happened, recall could continue, and this feature of the chunking model could probably obviate one of the major criticisms of the chaining model of Lewandowsky and Murdock (1989) by both Mewhort, Popham, and James (1994) and Nairne and Neath (1994).

Recall Data

If you present a short list of items to a subject and ask for serial recall, you get marked primacy and little or no recency. The first item is the item recalled most frequently, and the proportion of correct recalls drops off monotonically with serial position. This is exactly opposite to item recognition where recency predominates (see Fig. 2). Typical serial-position curves for recall are shown in Fig. 10, and the results for auditory and visual presentation are quite similar. Note that, at each serial position, the proportion of correct recalls is inversely proportional to list length.

Simulations

I have run a number of simulations to see if the chunking model can mimic these data, and it certainly can. Typical results are shown in Fig. 11. With one exception (Serial Position 6) the proportion of correct recalls is again proportional to list length.

For each list length we formed an n-chunk according to Eq. 3. We then unpacked the chunk item by item using the multiple-convolution algorithm. At each step in the process we scored the best match as the item recalled. If all matches were negative we scored on omission. What is retrieved at each step must enter into the retrieval cue for the next step. If the recall was successful we used the target item; otherwise we used whatever item had been recalled. For further details see Murdock (1995).

While the serial-position curves from these simulations were quite good, further analyses did reveal a problem. The error gradients either showed primacy or were approximately flat. In experimental data error gradients are usually sharply peaked around the target serial position (Lee & Estes, 1977; Nairne, 1991). One possible reason for this discrepancy is that experimental subjects group the longer lists into two chunks and the intrusions generally come from the appropriate chunks. In the simulations we used a single chunk even for the longest lists.

Summary

In this chapter I have presented a general account of TODAM, a theory of distributed associative memory that deals with the storage and retrieval of item, associative, and serial-order information. It attempts to explain behavioral data from psychological studies of short-term episodic memory. It is reasonably successful in explaining many patterns of data, and hopefully can be useful in understanding the role of memory in some of the broader issues that are the theme of this conference.

References

Anderson, J.A. (1970). Two models for memory organization using interacting traces. Mathematical Biosciences, 8, 137-160.

Anderson, J.A. (1973). A theory for the recognition of items from short memorized lists. Psychological Review, 80, 417-438.

Arbib, M.A. (Ed.). (1995). The handbook of brain theory and neural networks. Cambridge, MA: MIT Press.

Barnes, J.M., & Underwood, B.J. (1959). "Fate" of first-list associations in transfer theory. Journal of Experimental Psychology, 58, 97-105.

Borsellino, A., & Poggio, T. (1973). Convolution and correlation algebras. Kybernetik, 122, 113-122.

Clarke, S.E., & Grunlund, S.D. (1996). Global matching models of recognition memory: How the models match the data. Psychonomic Bulletin & Review, 3, 37-60.

Collins, A.M., & Loftus, E.F. (1975). A spreading-activation theory of semantic processing. Psychological Review, 82, 407-428.

Cooper, L.N. (1973). A possible organization of animal memory and learning. In B. Lundquist & S. Lundquist (Eds.), Proceedings of the Nobel Symposium on Collective Properties of Physical Systems (pp. 252-264). New York: Academic Press.

Hertz, J., Krogh, A., & Palmer, R.G. (1991). Introduction to the theory of neural computations. Redwood City, CA: Addison-Wesley.

Hinton, G.E., & Anderson, J.A. (Eds). (1981). Parallel models of associative memory. Hillsdale, NJ: Erlbaum.

Hintzman, D.L. (1986). "Schema abstraction" in a multiple-trace memory model. Psychological Review, 93, 411-428.

Hintzman, D.L. (1988). Judgments of frequency and recognition memory in a multiple-trace memory model. Psychological Review, 95, 528-551.

Hockley, W.E. (1991). Recognition memory for item and associative memory: A comparison of forgetting rates. In W.E. Hockley & S. Lewandowsky (Eds.), Relating theory and data: Essay on human memory in honor of Bennet B. Murdock (pp. 227-248). Hillsdale, NJ: Erlbaum.

Hockley, W.E., & Crisi, C. (1996). Tests of encoding tradeoffs between item and associative information. Memory & Cognition, 24, 202-216.

Humphreys, M.S., Bain, J.D., & Pike, R. (1989). Different ways to cue a coherent memory system: A theory for episodic, semantic, and procedural tasks. Psychological Review, 96, 208-223.

Kahana, M.J. (submitted). Associative symmetry and the serial order problem in human memory.

Kohonen, T. (1977). Associative memory: A system-theoretical approach. Berlin: Springer-Verlag.

Lashley, K.S. (1951). Cerebral mechanisms in behavior. In L.A. Jeffries (Ed.), The problem of serial order in behavior (pp. 112-136). New York: Wiley.

Lee, C.L., & Estes, W.K. (1977). Order and position in Primary Memory for letter strings. Journal of Verbal Learning and Verbal Behavior, 16, 395-418.

Lewandowsky, S., & Li, S-C. (1994). Memory for serial order revisited. Psychological Review, 101, 534-538.

Lewandowsky, S., & Murdock, B.B. (1989). Memory for serial order. Psychological Review, 96, 25-57.

Liepa, P. (1977). Models of content addressable distributed associative memory (CADAM). Unpublished manuscript. University of Toronto.

McCloskey, M., & Cohen, N.J. (1989). Catastrophic interference in connectionist networks: The sequential learning problem. The Psychology of Learning and Motivation: Advances in Research and Theory, 24, 109-165.

Metcalfe-Eich, J.M. (1982). A composite holographic associative recall model. Psychological Review, 89, 627-661.

Metcalfe-Eich, J.M. (1985). Levels of processing, encoding specificity, elaboration, and CHARM. Psychological Review, 92, 1-38.

Mewhort, D.J.K., Popham, D., & James, G. (1994). On serial recall: A critique of chaining in TODAM. Psychological Review, 101, 534-538.

Miller, G.A. (1956). The magical number seven, plus or minus two: Some limits on our capacity for processing information. Psychological Review, 63, 81-96.

Murdock, B.B. (1974). Human memory: Theory and data. Potomac, Md: Lawrence Erlbaum Associates.

Murdock, B.B. (1989). Learning in a distributed memory model. In C. Izawa (Ed), Current issues in cognitive processes: The Tulane Floweree symposium on cognition (pp. 69-106). Hillsdale, NJ: Erlbaum.

Murdock, B.B. (1992). Serial organization in a distributed memory model. In A.F. Healy, S.M. Kosslyn, & R.M. Shiffrin (Eds.), From learning theory to connectionist theory: Essays in honor of William K. Estes, Vol 1 (pp. 201-225). Hillsdale, NJ: Lawrence Erlbaum Assoc.

Murdock, B.B. (1993). TODAM2. A model for the storage and retrieval of item, associative, and serial-order information. Psychological Review, 100, 183-203.

Murdock, B.B. (1995). Primacy and recency in the chunking model. In C. Weaver, S. Mannes, & C. Fletcher, (Eds.), Discourse comprehension: Essays in honor of Walter Kintsch (pp. 49-63). Hillsdale, NJ: Erlbaum.

Murdock, B.B. (1996). Item, associative, and serial-order information in TODAM. In S. Gathercole (Ed.), Models of short-term memory (pp. 239-266). Hove, Sussex, England: Erlbaum.

Murdock, B.B. (in press). Context and mediators in TODAM2. Psychological Review.

Nairne, J.S. (1991). Positional uncertainty in long-term theory. Memory & Cognition, 19, 332-340.

Nairne, J.S., & Neath, I. (1994). A critique of the retrieval/deblurring assumptions of TODAM. Psychological Review, 101, 528-533.

Pribram, K.H. (1971). Language of the brain. Englewood Cliffs, NJ: Prentice-Hall.

Schonëmann, P.H. (1987). Some algebraic relations between involutions, convolutions, and correlations with applications to holograph memories. Biological Cybernetiks, 56, 367-374.

Shiffrin, R.M., & Raaijmakers, J. (1992). The SAM retrieval model: A retrospective and prospective. In A.F. Healy, S.M. Kosslyn, & R.M. Shiffrin (Eds.), From learning theory to connectionist theory: Essays in honor of William K. Estes, Vol. 1 (pp. 69-86). Hillsdale, NJ. Erlbaum.

Sloman, S.A., & Rumelhart, D.E. (1992). Reducing interference in distributed memories through episodic gating. In A.F. Healy, S.M. Kosslyn, & R.M. Shiffrin (Eds.), From learning theory to connectionist theory: Essays in honor of William K. Estes, Vol. 1 (pp. 227-248). Hillsdale, NJ: Erlbaum.

Weber, E.U. (1988). Expectation and variance of item resemblance distributions in a convolution-correlation model of distributed memory. Journal of Mathematical Psychology, 32, 1-43.

Footnotes

[1]One possibility for deblurring is a Hopfield net (see Hertz, Krogh & Palmer, 1991). This was used by Lewandowsky and Li (1994) to model deblurring in a serial-order task.

[2]The pair retention curves seem to go up in Exps. 3 and 4, but this is not a statistically significant effect.

[3]More specifically, the dimensionality is 2N-1, 3N-2, and 4N-3 for two-, three-, and four-way convolutions where N is the dimensionality of the single item vectors **a**, **b**, **c**, and **d**.

[4]To add vectors of different dimensionality, we assume they are always center-justified and padded with zeros at both ends; see Schonëmann (1987).

Fig. 1. System architecture showing the interactions of the Perceptual-, Query-, and Response-Systems and Working Memory. Fig. 10-4 from Murdock (1996).

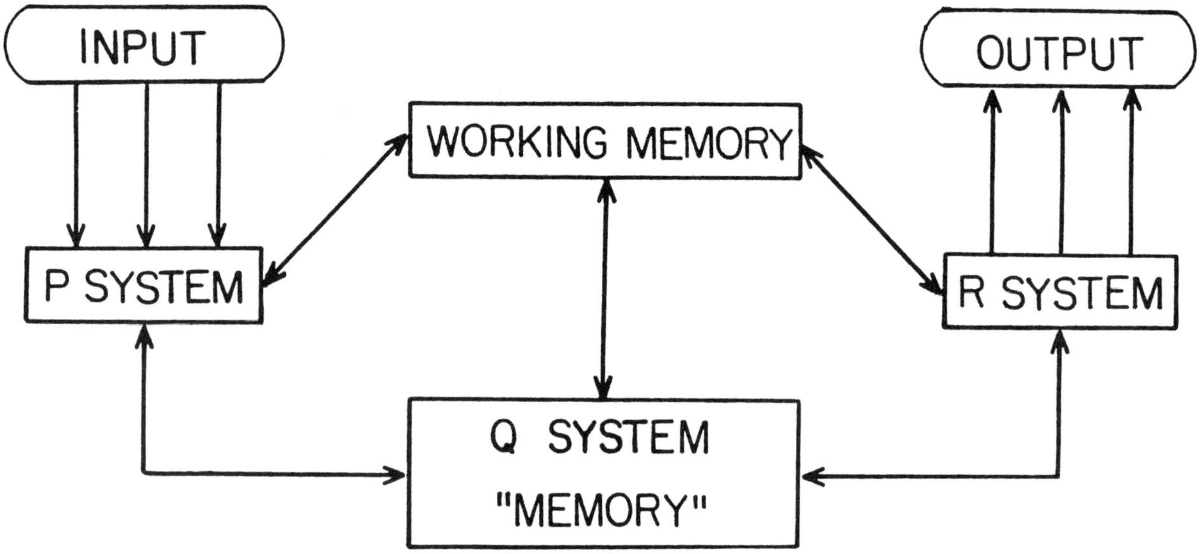

Fig. 2. Hit rates for old items as a function of lag and False Alarm rates for new items as a function of test position. Fig. 10-2 from Murdock (1996).

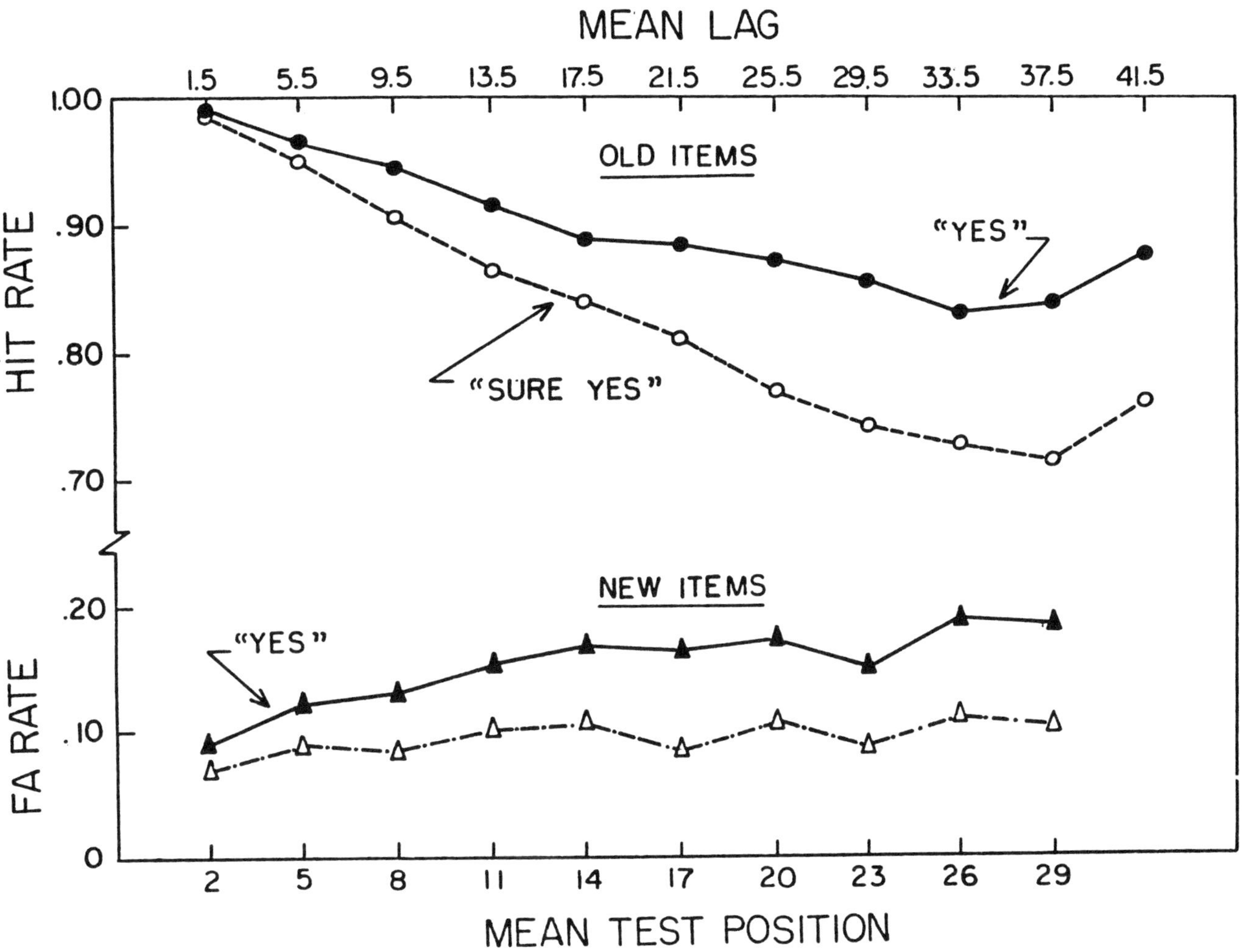

Fig. 3. Number recalled from the A-B and A-C lists as a function of the number of A-C learning trials. Data from Barnes & Underwood (1959).

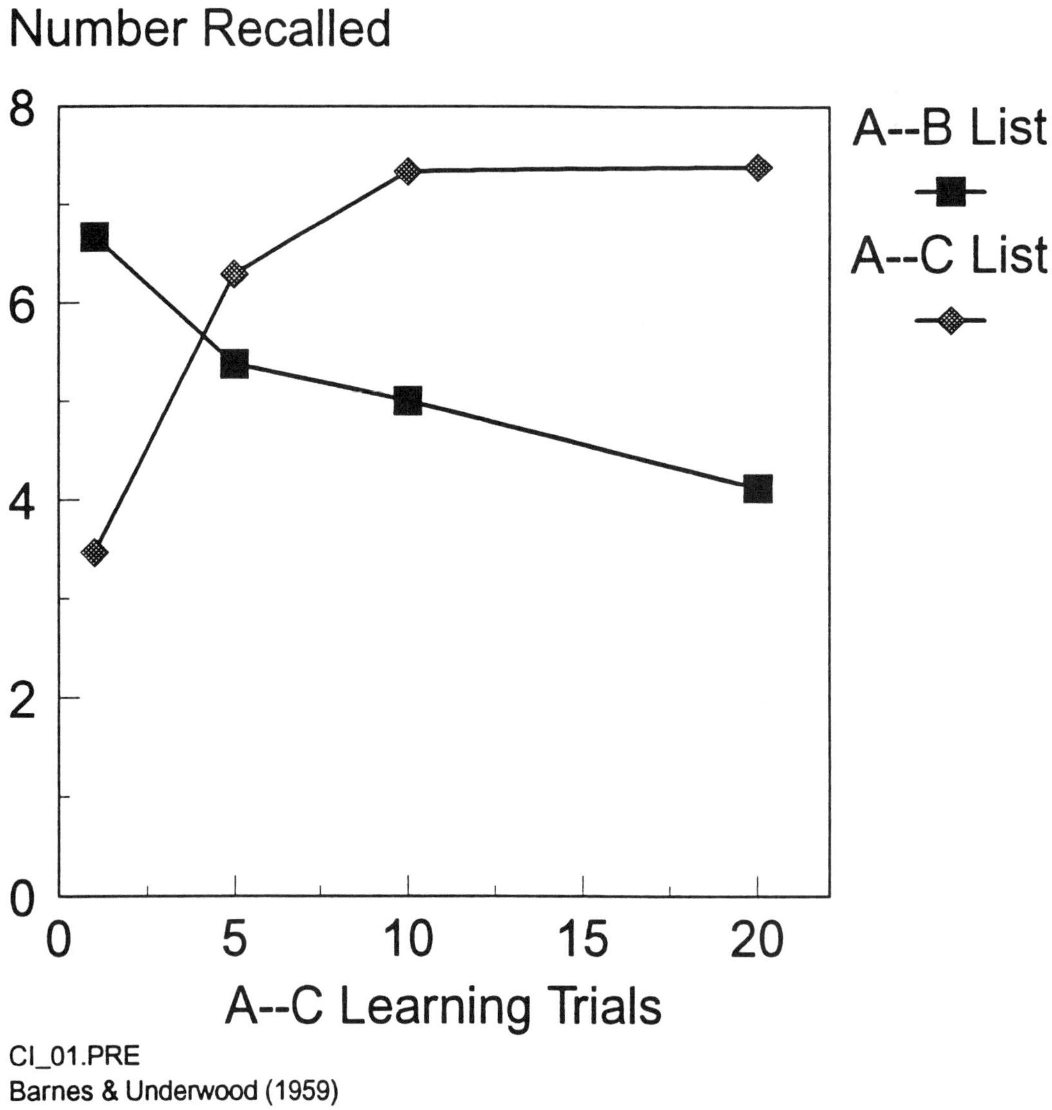

Fig. 4. Number recalled from the A-B and A-C lists as a function of the number of A-C learning trials. Data from McCloskey & Cohen (1989).

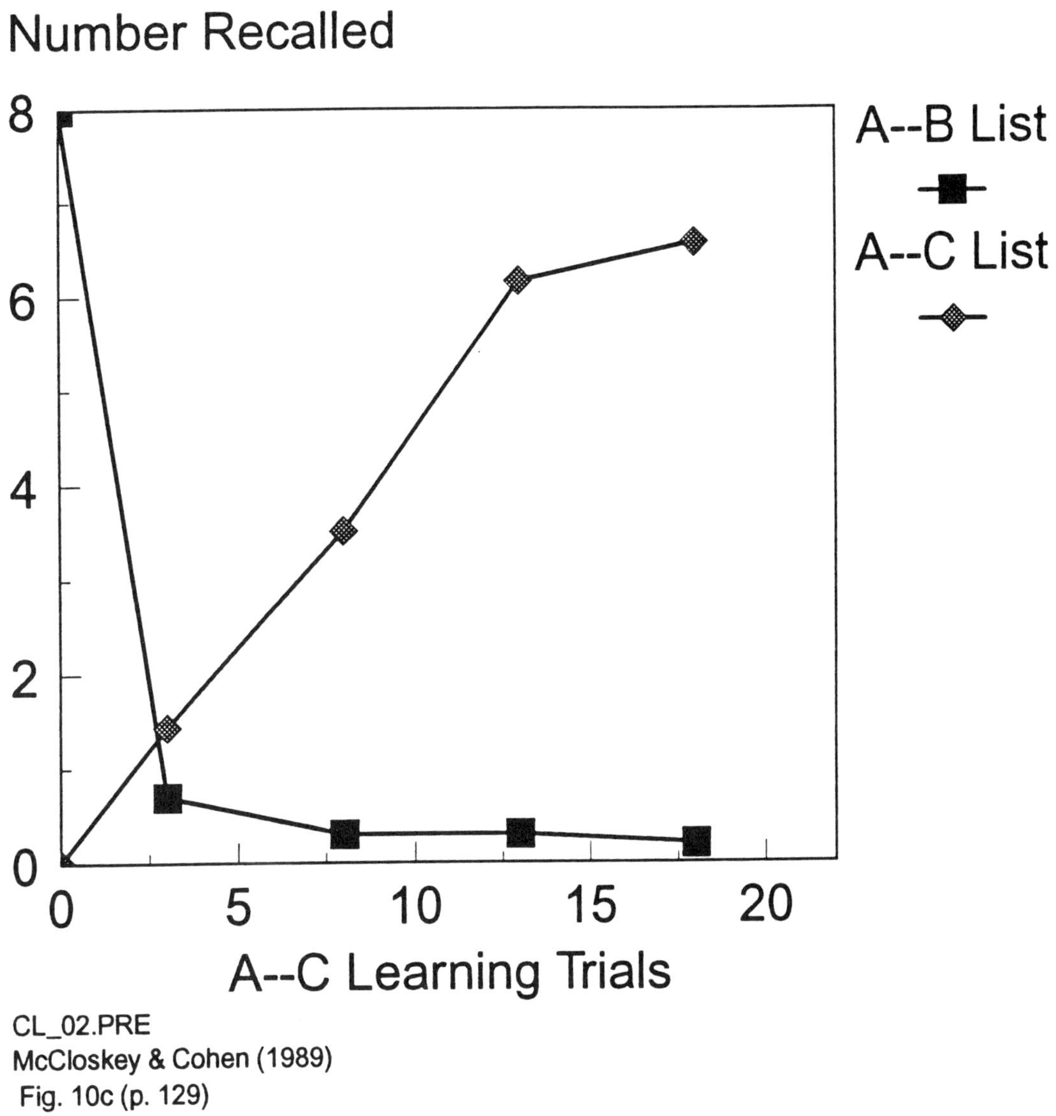

Fig. 5. Number recalled from the A-B and A-C lists as a function of the number of A-C learning trials. Data from a TODAM simulation.

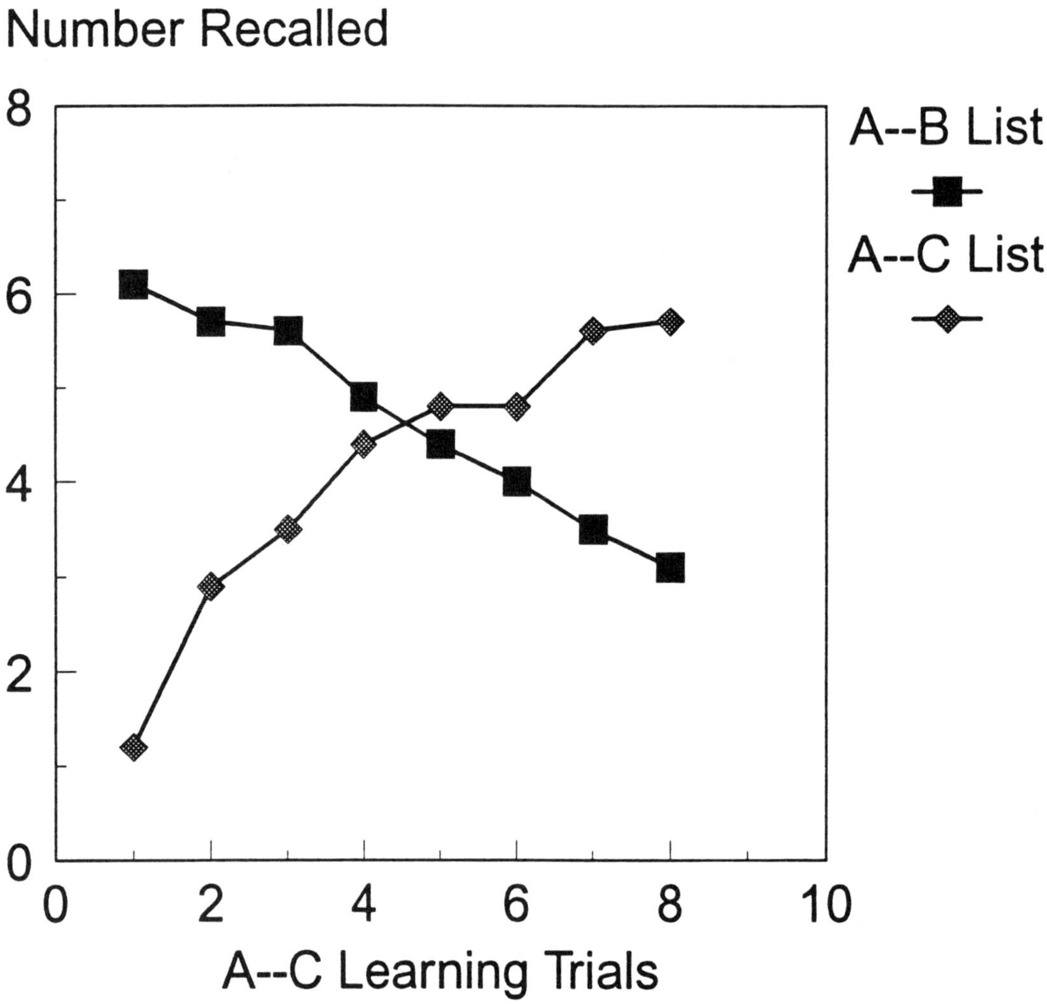

CI_03.PRE
CDM209A.FOR
16/7/96

Fig. 6. d' values for item recognition (IR) and associative recognition (AR) as a function of lag. Data from Hockley (1991).

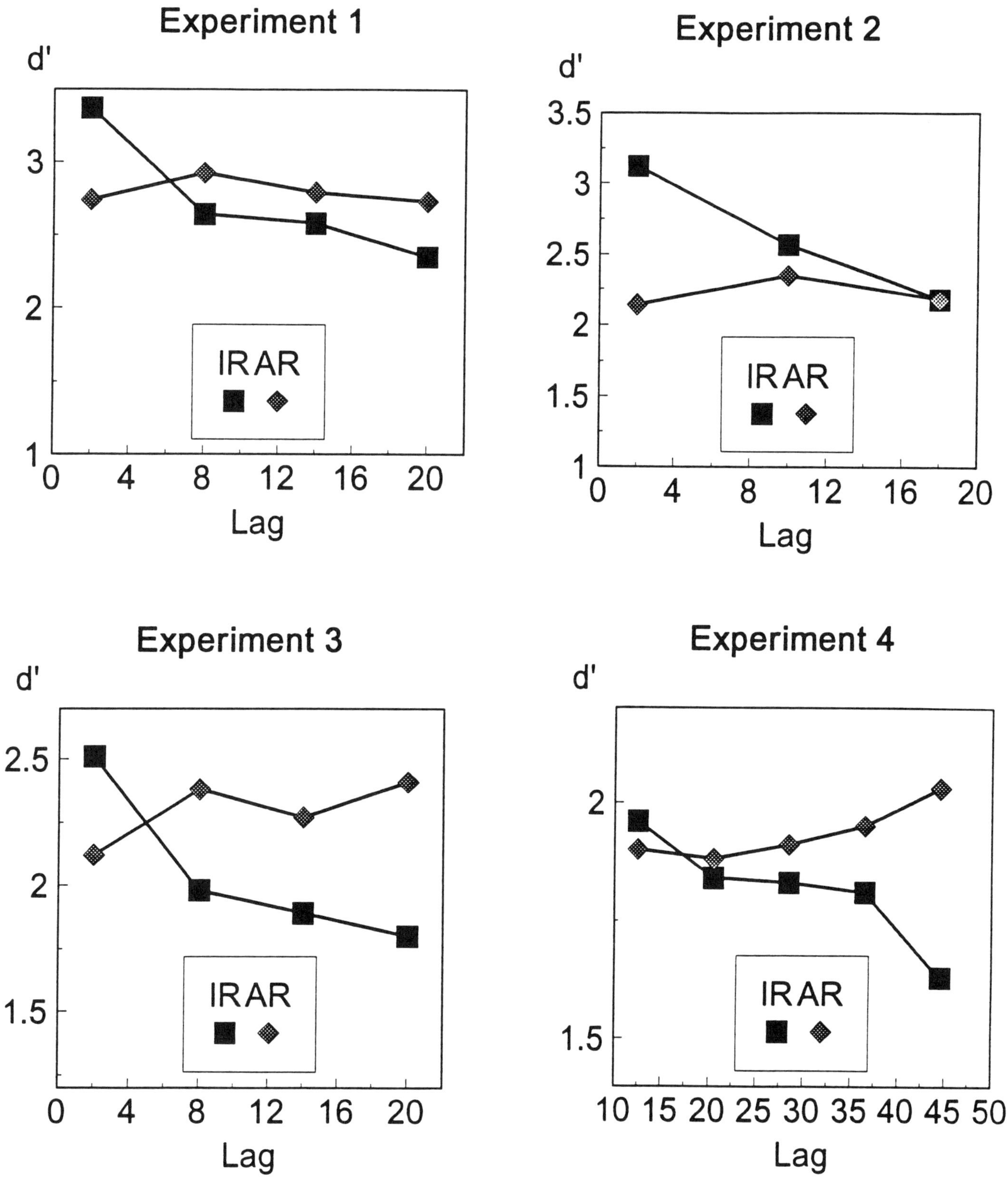

Fig. 7. Percent correct for associative information (AI) and item information (II) for four experiments. For each experiment, the left-hand bar for each information type is for the pair-emphasis condition, the right-hand bar is for the item-emphasis condition. Data from Hockley & Cristi (1996).

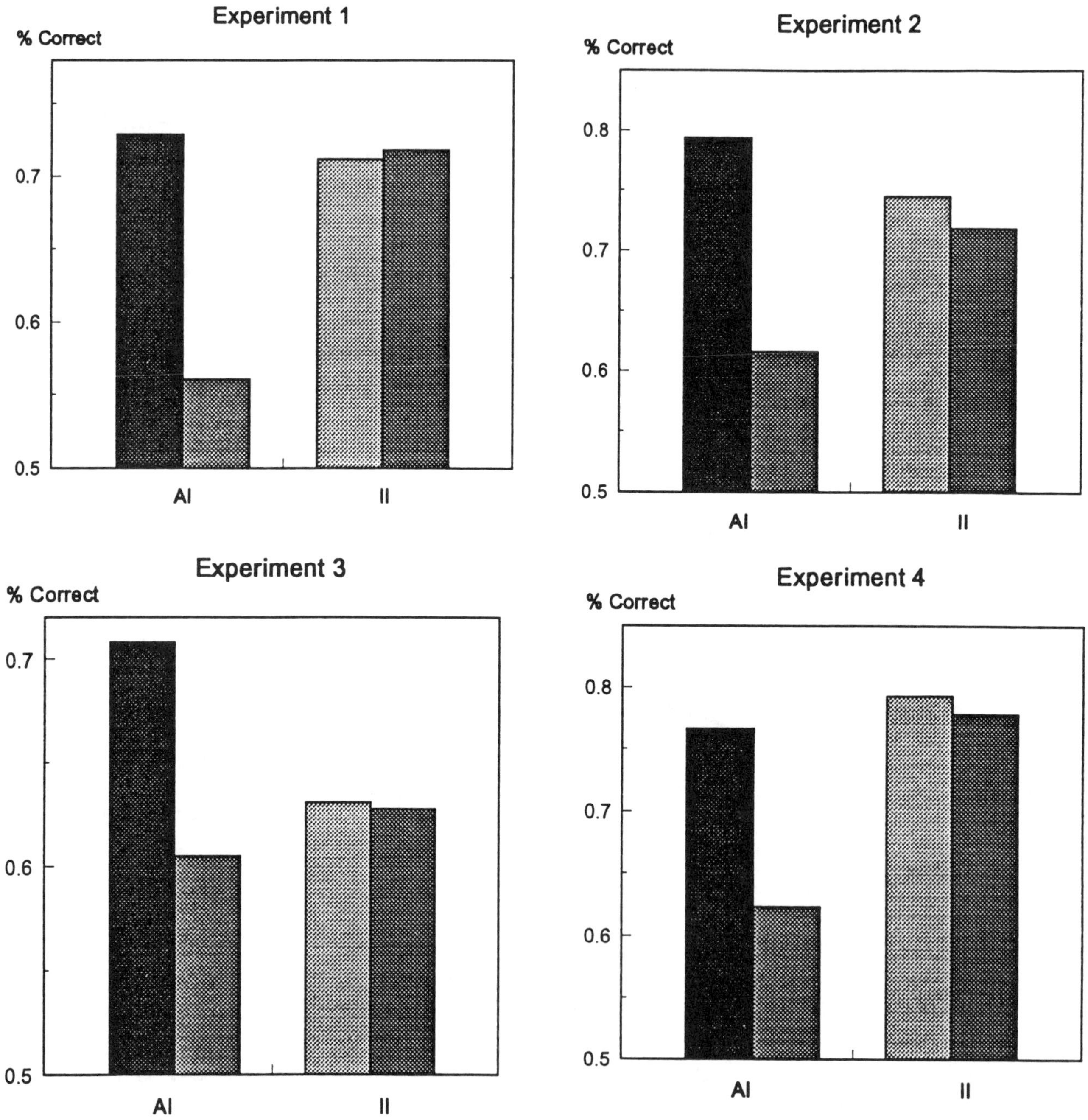

Fig. 8. d' as a function of test position for item information (II) and associative information (AI) from a simulation of TODAM2. Fig. 10 from Murdock (in press).

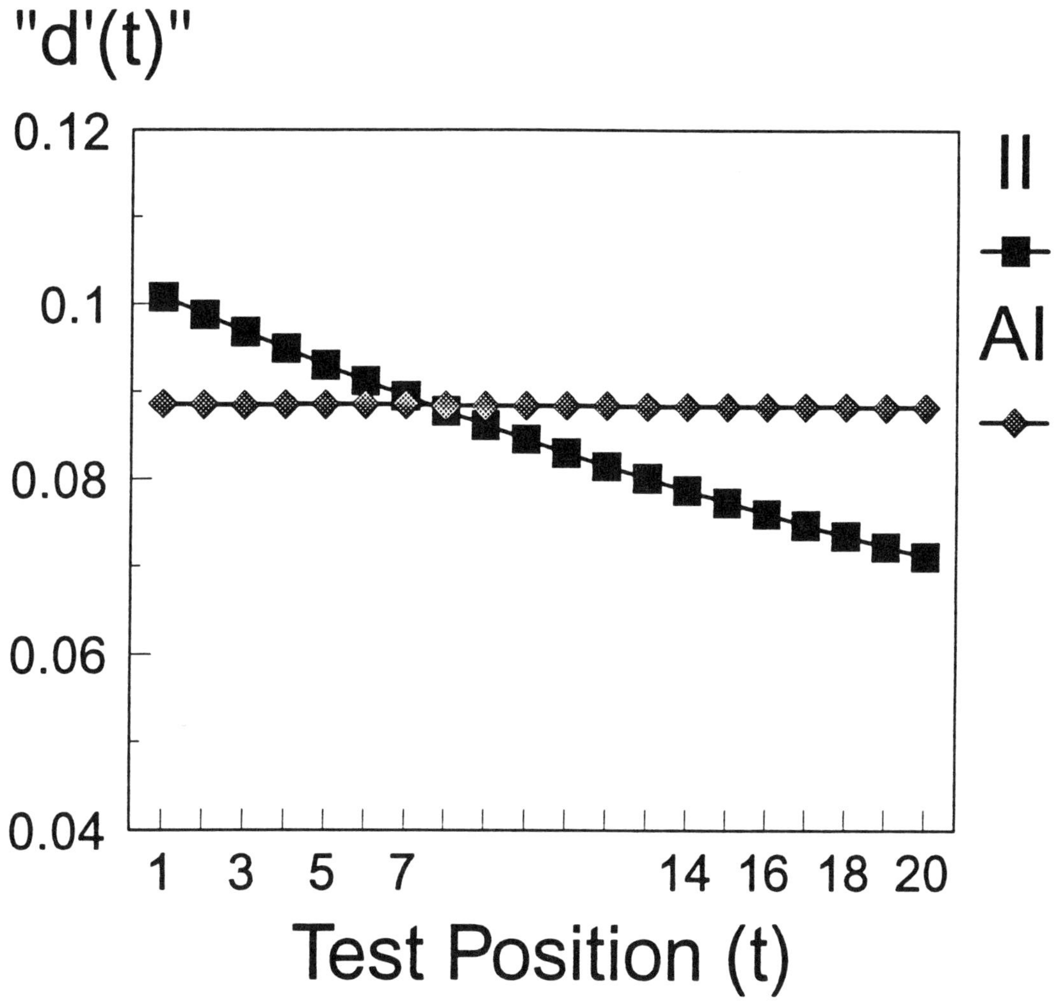

PRIB_03.PRE
(CAM04X.PRE

Fig. 9. Percent correct for associative information (AI) and item information II) for four experiments from a simulation of TODAM2. For each experiment, the left-hand bar for each information type is for the pair-emphasis condition, the right-hand bar is for the item-emphasis condition. Fig. 11 from Murdock (in press).

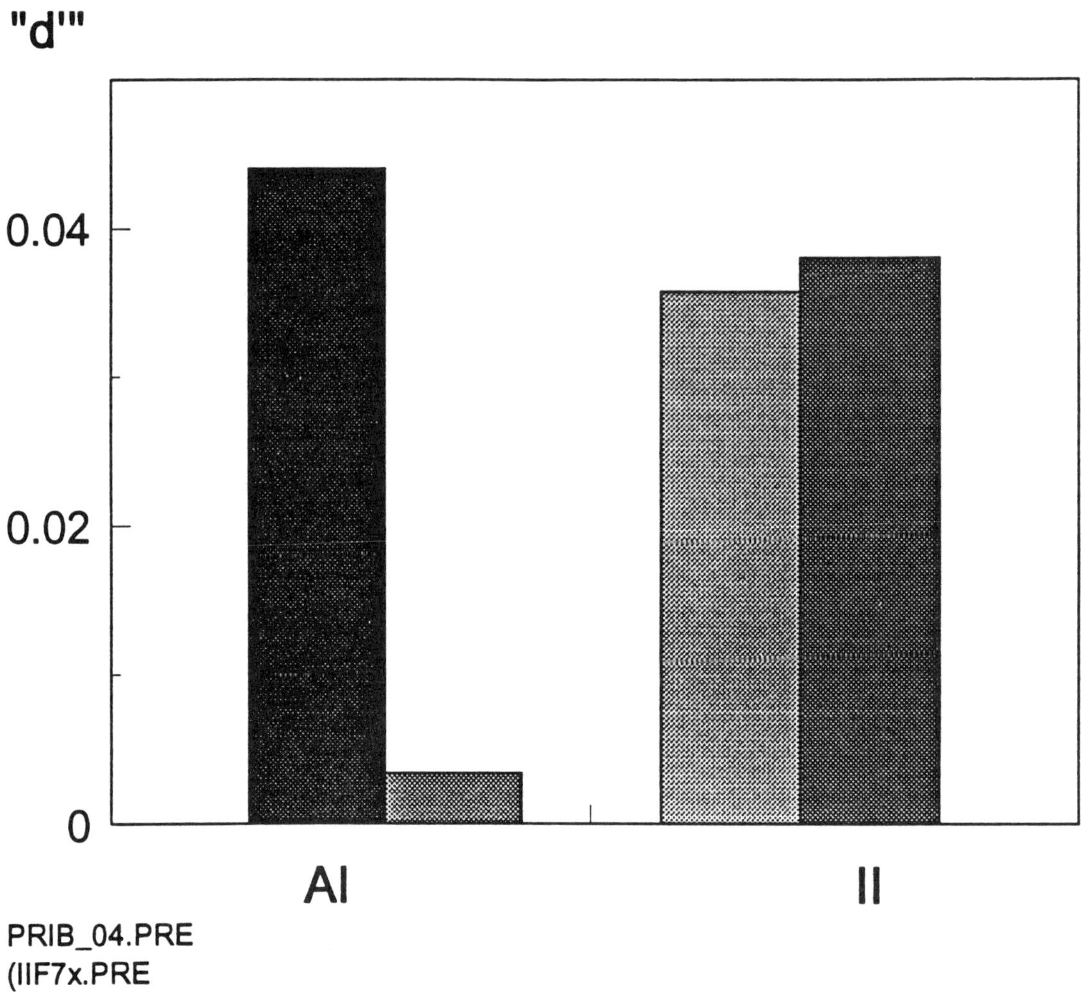

Fig. 10. Proportion of correct recalls as a function of serial position for lists of 3, 4, 5, 6, or 7 items following auditory or visual presentation. Fig. 4.2 from Murdock (1995).

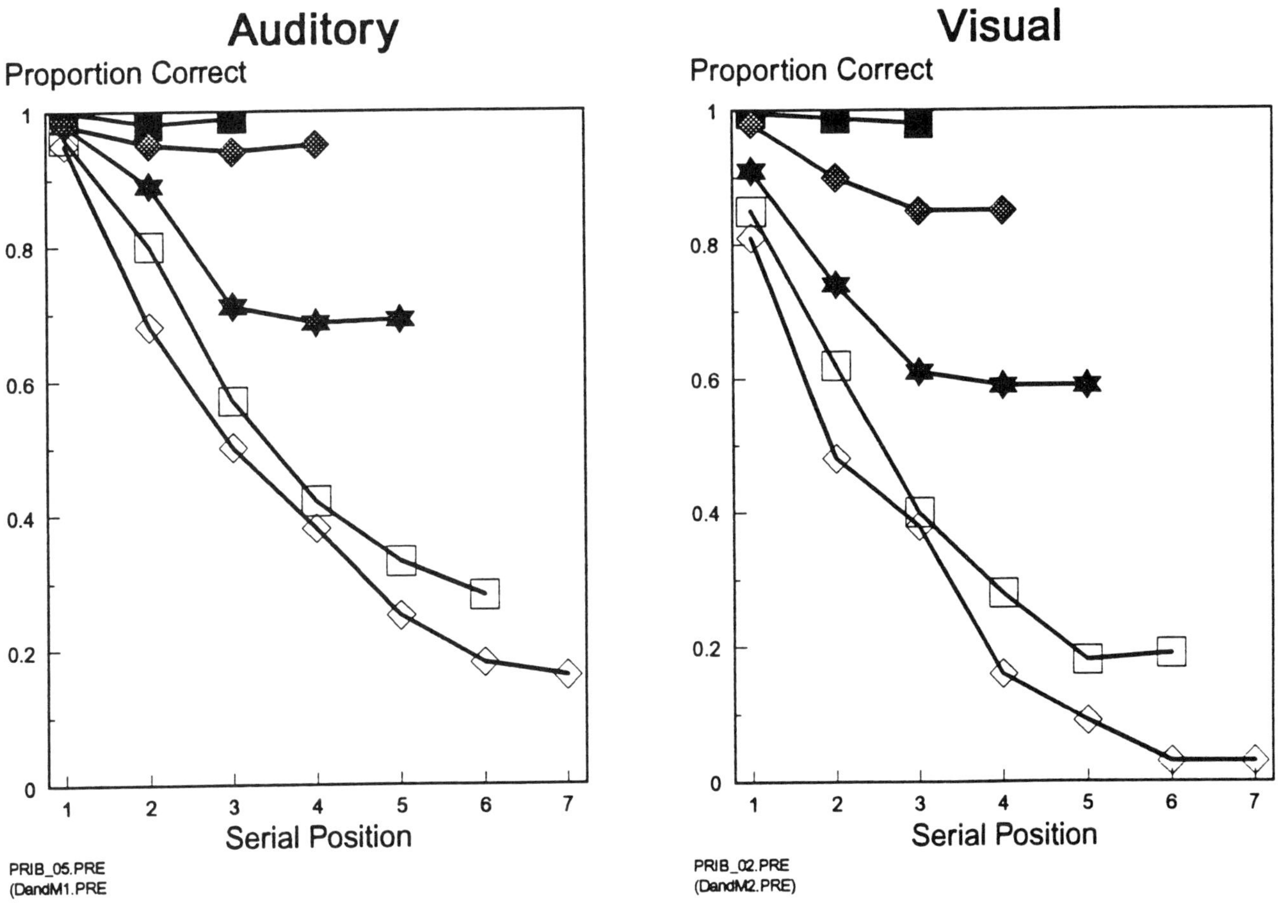

Fig. 11. Recall probability as a function of serial position for lists of 3, 4, 5, 6, or 7 items following auditory or visual presentation from a simulation of the chunking model. Fig. 4.8 from Murdock (1995).

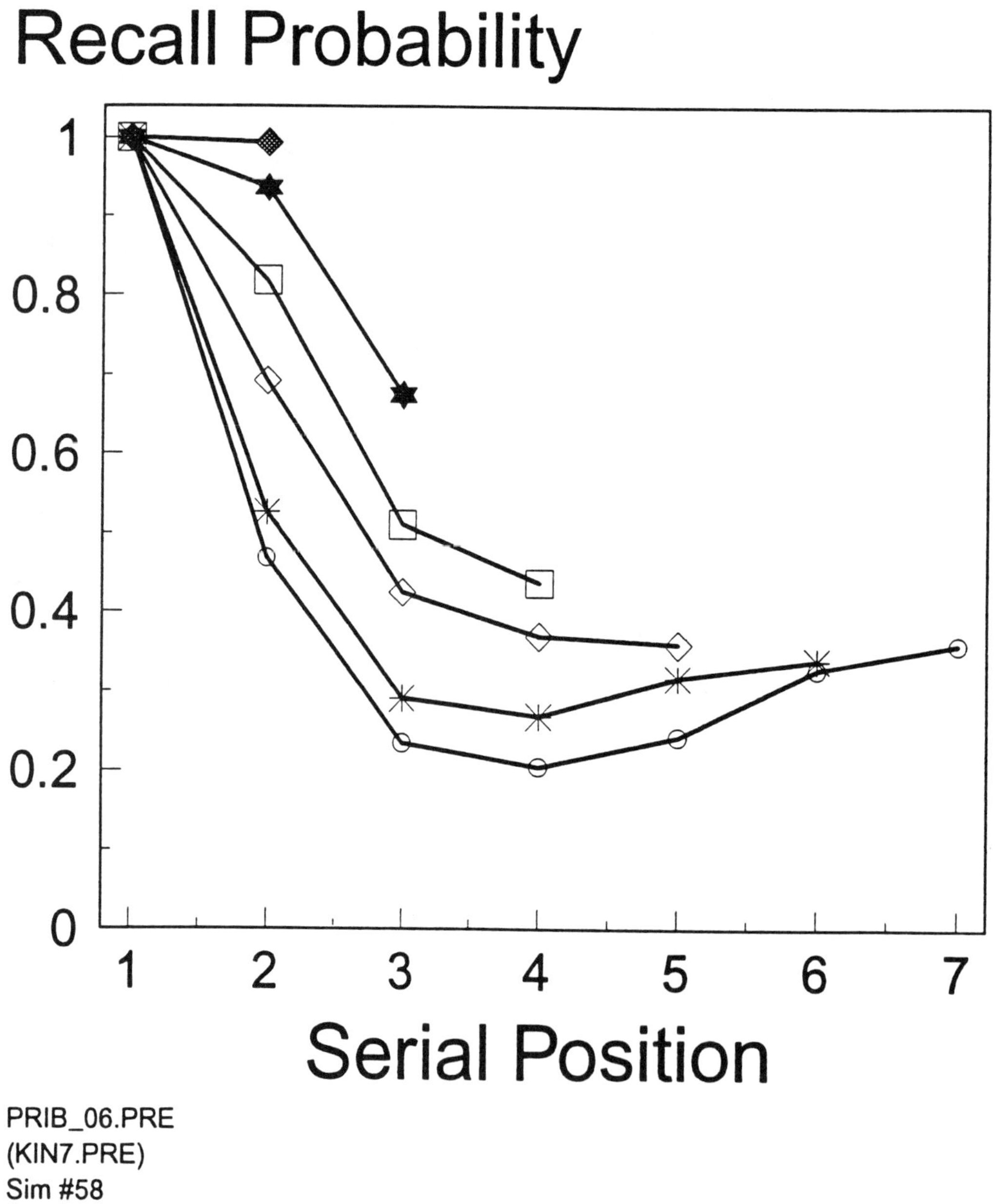

12

The Transfer of Value in Simultaneous Discriminations: Implications for Cognitive and Social Processes

Thomas R. Zentall and Lou M. Sherburne
Department of Psychology
University of Kentucky
Lexington, KY 40506-0044
Telephone: 606-257-4076/Fax: 606-323-1979
E-Mail: Zentall@pop.uky.edu

Abstract

The transitive inference task was developed by Piaget (1928) as a tool for the assessment of cognitive development in children. A child is said to make a transitive inference when, based on a set of relational propositions, s/he infers a relation that has not been provided explicitly. Such relations are said to have emergent properties because they were not trained. An example of such a transitive inference is, if a child is told that an object A is larger than B, and that B is larger than C, s/he may infer than A is larger than C. If such an inference is made, the presumption has been that B serves to mediate the emergent relation between A and C. Although there is some disagreement about the age and level of development at which transitive inference appears in children (see Bryant & Trabasso, 1971), the transitive inference task remains an important index of relative cognitive development in humans.

Interest in the cognitive capacity of animals has lead to the design of procedures that have attempted to translate tests of cognitive development into a form that can be used with nonverbal organisms. For example, Doré (1986) has examined object permanence in cats and Woodruff, Kennel, and Premack (1978) have looked at conservation of various materials in a chimpanzee..

The strength of the transitive inference task, when it used with humans, is that it is presented in verbal form. Because the proposition, A is larger than B, is presented verbally, without regard to perceivable objects, the stimuli need have no absolute properties. Typically, they have only relational properties (i.e., the communication says nothing about the absolute size of A or B). However, its strength with humans is a disadvantage with (nonverbal) animals. If one were to present the propositions of a transitive inference task to an animal (e.g., A is better than B, and B is better than C), A would have some inherent value, as would B and C. On the test trial, one could compare directly, the inherent value of A to the inherent value of C. Thus, one could select A over C, without having to make an inference involving B. Thus, such a nonverbal form of the transitive inference task would defeat the purpose of the procedure.

McGonigle and Chalmers (1977) were the first to propose, as a solution to this problem, the five-term series with equal *absolute value* given to the two stimuli (i.e., the better-than stimulus and the worse-than stimulus) in each proposition. They suggested that five stimuli could be arranged in four propositions (simultaneous discriminations) that could maintain the appropriate logical relation among the stimuli, while avoiding differential absolute value for the stimuli in the test pair. For example, if responses to A are reinforced in the presence of B (A+B-), responses to B are reinforced in the presence of C (B+C-), responses to C are reinforced in the presence of D (C+D-), and responses to D are reinforced in the presence of E (D+E-), such training may establish the relational series A>B>C>D>E (see Table 1). Although responses to A are always reinforced (so A should be more preferred than all of the others) and responses to E are never reinforced (so E should be less preferred than all the others), responses to B, C, and D are all sometimes reinforced and sometimes not. Furthermore, of these three stimuli, B and D were never experienced together in training and thus, they can serve as an appropriate test pair. If, as a result of the above training, an organism selects B over D, it has been proposed that relational learning has occurred (Davis, 1992). (For descriptive purposes, such a result will be viewed as an empirical finding of transitive inference even when relational learning is not implied.) Using this and similar procedures, evidence for transitive inference (in the form of test pair transfer performance) has been found in a number of nonverbal species, including chimpanzees (Gillan, 1981), monkeys (McGonigle & Chalmers, 1977), rats (Davis, 1992), and pigeons (Fersen, Wynne, Delius, & Staddon, 1991; Steirn, Weaver, & Zentall, 1995; Weaver, Steirn, & Zentall, in press; Wynne, 1995).

TABLE 1
Nonverbal analog of the verbal transitive inference task with nonendpoint test stimuli.

Training:	A+B-
	B+C-
	C+D-
	D+E-
Test:	B vs. D

Note - The letters A, B, C, D, and E are arbitrary representations of unrelated stimuli.

Much of the research with animals has been done with a mixed-pair training (e.g. Fersen et al., 1991; McGonigle & Chalmers, 1977) in which all of the discriminations are presented over trials, within a session. Actually, training generally starts with one pair (e.g., A+B-) and then a second pair is added (e.g., A+B-, B+C-), and so on, until all four pairs are presented on randomly alternating trials during a single session.

Surprisingly, recent evidence suggests that blocked training (i.e., each session involves training with a single discrimination) results in very similar findings of transitive inference (Davis, 1992; Steirn et al., 1995; Weaver et al., in press). Furthermore, the order in which the pairs are trained (A+B- first or D+E- first) does not appear to affect the magnitude of the transitive inference effect (Steirn et al., 1995).

Theoretical Accounts of Transitive Inference

Cognitive Theories

Two cognitive theories have been proposed to account for transitive inference effects. The first, suggests that subjects have the acquired discriminations (or premise pairs) stored in memory as verbal propositions that can be logically linked together when needed (i.e., at the time of test). To answer the question, "which is greater, A or C?" a subject must wend his or her way through the propositions "A is greater than B" and "B is greater than C." Alternatively, subjects establish an ordered (linear) array of the stimuli as the propositions are being learned (A<B<C<D<E) and then they consult the linear array to answer questions about untrained pairs (Bryant & Trabasso, 1971).

Noncognitive Theories

Reinforcement History. Although findings of transitive inference have been taken by some as evidence of relational learning (Davis, 1992; Weaver et al., in press) others have interpreted such results in terms of simpler mechanisms (Couvillon & Bitterman, 1992; Fersen et al., 1991; Steirn et al. 1995; Wynne, 1995). The simplest alternative theory attributes the transitive inference effect to the difference in reinforcement history associated with the two test stimuli (Couvillon & Bitterman, 1992; Wynne, 1995).

Although the history of reinforcement associated with the test stimuli should be nominally the same (because each of the test stimuli has served first as an S+ and then as an S-), the actual histories of reinforcement may have been different. For example, if the A stimulus is positive, the A+B- discrimination might be performed at a high level because A is the one positive stimulus that is always positive. If the A+B- discrimination is performed at a high level, it means that few responses are made to B-. If few responses are made to B-, relative to D-, less inhibition may accrue to B than to D during training, and B may be preferred over D on test trials. On the other hand, the D+E- discrimination should be relatively easy as well, because E is the one negative stimulus that is always negative. Thus, one might expect more reinforced responses to D in the D+E- discrimination than to B in the B+C- discrimination. If this is the case, however, one might expect D to be preferred over B. Thus, given that these two factors have their effects in opposite directions, it is not clear, a priori, from the above analysis which of the B and D test stimuli should be preferred on test trials. Nevertheless, differential histories of reinforcement associated with the test stimuli could be responsible for the transitive inference effect.

Steirn et al. (1995) proposed two ways to empirically test the hypothesis that history of reinforcement could account for the transitive inference effect. First, they asked if pigeons trained on such a task experienced either more reinforced responses to B than to D or more nonreinforced responses to D than to B. In both cases the answer was no. Second, they asked whether pigeons that had *relatively* more reinforced choices of B than D or relatively more nonreinforced choices of D than B (compared with other pigeons) showed a larger transitive inference effect. Again, the answer was no (see also Zentall & Sherburne, 1994).

The appeal of reinforcement-history models is that once the parameter values of the model are selected, the range of outcomes is relatively constrained. Unfortunately, it is often very difficult to specify, prior to data collection, what those parameter values should be. Thus, although various reinforcement history models have been proposed to account for the transitive inference effect (see Wynne, 1995), all of them are much better at accounting for transitive inference effects once the model and parameter values are estimated from the collected data (i.e., post hoc).

Value Transfer Theory. An alternative interpretation of the transitive inference finding, value transfer theory, has been proposed by Fersen et al. (1991). According to value transfer theory, when an organism acquires a simple simultaneous discrimination, some of the positive value acquired by the S+ transfers to the S- (and some of the negative value acquired by the S- may also transfer to the S+).

At first thought, the idea that some of the value of an S+ can transfer to the S- with which it is presented may seem quite compatible with traditional views of stimulus generalization (see e.g., Spence, 1937). Spence proposed, for example, that the value of (or excitation associated with) a stimulus acquired through training will generalize to other stimuli, including the S-, in proportion to the similarity of the test stimuli to the training stimulus. Spence also proposed that inhibition associated with the S- would generalize to other stimuli, including the S+. When one combines these two principles, this relatively simple theory of stimulus generalization has considerable explanatory power. For example, following training with a simple simultaneous discrimination involving training to respond to an S+ that is brighter than the S-, it can account for an organism's choice of a novel stimulus (still brighter than the S+) over the S+ experienced during training (i.e., transposition).

But value transfer theory does not rely on stimulus similarity. According to value transfer theory, value should transfer from an S+ to an S- independent of the similarity between them. Thus, stimulus generalization theory should play no direct role in the effect. Of course, stimulus similarity could augment value transfer effects but the effects of stimulus generalization should average out if the training stimuli are counterbalanced.

According to value transfer theory, the fact that the S+ and the S- are presented simultaneously should be sufficient for some of the value of the S+ to transfer to the S-. Furthermore, the greater value of the S+, the more value the S+ has to transfer to the S-. In the case of the transitive inference task, B and D should receive comparable direct value from their presence in the B+C- and D+E- discriminations, respectively. In addition, B should receive indirect value from A because of its participation in the A+B- discrimination, and stimulus D should receive indirect value from C because of its participation in the C+D- discrimination. However, stimulus A, which never appears as an S-, should have more value to give than does stimulus C, which sometimes appears as an S+ but at other times appears as an S-. Thus, according to the value transfer account of the transitive inference effect, it is this differential transfer of value from the S+ to the S- that explains the preference for B over D on test trials.

Although value transfer theory appears to be able to account for findings of transitive inference, at the time it was proposed, it had no empirical support. Furthermore, the assumption on which it is based may be thought of as rather counter intuitive. Given the history of research with contrast effects (i.e., incentive contrast, e.g., Crespi, 1941, and behavioral contrast, e.g., Reynolds, 1961), one might predict that in the transitive inference task, B would be *less* preferred than D because B was presented in the context of (or in contrast to) the more positive A, whereas D was presented in the context of the less positive C.

Empirical Support for Value Transfer. One way to evaluate value transfer theory is to create a five term series in which relational learning (i.e., a transitive inference) should not be possible, but in which the stimuli paired with the test stimuli during training were themselves differentially associated with reinforcement, as they are in the standard nonverbal transitive inference design (Steirn et al. 1995). Such a series is presented in Table 2. Pigeons were trained such that A>B and that A>E>C>D. Thus, no logical or direct links should exist between B and D as a result of such training. The stimuli consisted of the hues red, green, yellow, blue, and white and the assignment of hue to letter (A-E) was counterbalanced over pigeons. The pigeons were successively trained on the four simultaneous discriminations, and following training they were tested with the BD pair. During the test session responses to B and D were reinforced nondifferentially. Given this training, although no obvious inference, relating B to D, should be possible, the direct values associated with B and D should be the same and the value that could transfer to B from the always-positive A presumably would be greater than the value that could transfer to D from the sometimes-positive-sometimes-negative C. Thus, according to value transfer theory, B should be preferred over D.

TABLE 2
"Pseudo" transitive inference task in which no transitive inference should be possible but, according to value transfer theory, differential value should transfer (after Steirn, Weaver, & Zentall, 1995).

Training Phase:	1	2	3	4	Test
Group A+:	A+B- :	C- E+ :	C+D- :	A+E- :	B vs.D
Group A- :	A- B+:	C+E- :	C- D+ :	A- E+ :	B vs.D

Note - The letters A, B, C, D, and E are arbitrary representations of unrelated hues (red, green yellow, blue, and white) counterbalanced over subjects within groups.

Alternatively, if contrast plays a role in simultaneous discrimination learning, although B and D should have equal direct value, the greater contrast between A and B, as compared to that between C and D, should result in a preference for D over B. In support of value transfer theory, Steirn et al. (1995) found that pigeons showed a strong preference for B over D.

More direct support for value transfer theory comes from experiments in which training involves only two simple simultaneous discriminations, but the reinforcement schedule associated with the untested stimulus in each pair is allowed to vary in a way analogous to the way it varies in the transitive inference task. Instead of using five stimuli, however, only four are used, and there are no stimuli common to the two discriminations. Thus, there should be no mediational links capable of providing a true transitive inference. Zentall and Sherburne (1994) trained pigeons on a hue discrimination (e.g., red-green) in which responses to one stimulus were always reinforced and responses to the other stimulus were never reinforced (A+B-). On other, interspersed trials they trained the pigeons on a second hue discrimination (e.g., yellow-blue) in which responses to one stimulus were reinforced on only half of the trials and responses to the other stimulus were never reinforced (C±D-, see Table 3).

TABLE 3
Design of value transfer experiment (after Zentall & Sherburne, 1994).

Train:	A+B- and C±D-
Test:	B vs.D

Note - The letters A, B, C, and D are arbitrary representations of unrelated hues (red, green yellow, and blue) counterbalanced over subjects.

According to value transfer theory, more value should transfer to B from the always positive

A, than to D from the sometimes negative C. As in the case of Steirn et al. (1995), however, there should be no basis for relational learning between the stimulus pairs, so it should not be possible for the pigeons to draw an inference about that relation. Following training on this task, to a criterion of two consecutive sessions (of 96 trials) with discrimination accuracy at 90% correct or better, pigeons were given a choice between the two S-s (B and D) with nondifferential reinforcement provided for responding.

Again, in support of value transfer theory, Zentall and Sherburne (1994) reported a strong preference for B over D. Thus, it appears that in a simultaneous discrimination, the value of the S- is directly dependent on the value of the S+ with which it is presented. Once again, there was little evidence for a differential-reinforcement-history account. First, there were not significantly more nonreinforced responses to D- than there were to B-. Second, the relative number of nonreinforced responses to D- was not correlated with the magnitude of the preference for B over D.

Theoretical Accounts of Value Transfer. Value transfer theory provides a prediction of the relation between S+ and S- in a simultaneous discrimination, but it does not provide an explanation. How does this transfer of value between an S+ and its S- take place? As noted earlier, stimulus generalization, based on the similarity between S+ and S- may occur but the phenomenon does not appear to depend on it (i.e., the amount of value transfer does not appear to depend on the similarity between the S+ and the S-).

The fact that responding to the S- is not reinforced suggests that the process by which value transfers might be Pavlovian in nature rather than instrumental. Two Pavlovian mechanisms might be involved. The first is direct trace conditioning. Specifically, as the animal learns to avoid selecting the S-, whenever it sees the S-, food follows shortly (i.e., after selecting the S+). According to this view, selecting the S+ merely fills the interval between observing the S- and food. Thus, both the S+ and the S- are associated with the food outcome but the S+-food association is stronger because there is no delay between observation of the S+ and food (as there is with the S-), and there is a reinforced instrumental response to the S+ (but not to the S-, see Table 4).

TABLE 4
Two possible Pavlovian mechanisms that could account for value transfer from the S+ to the S- in a simultaneous discrimination.

Hypothesis	Mechanism
(1) Direct Conditioning:	CS_2(B-) ---------------->US(reinforcement) CS_1(A+) -->US(reinforcement)
(2) Higher Order Conditioning:	CS_2(B-) -->CS_1(A+) -->US(reinforcement)

Note. The length of the arrow corresponds to the duration of the interstimulus interval.

Alternatively, as the animal learns to avoid selecting the S-, observing the S- comes to serve as a signal for the presence of the S+. Thus, the S- may acquire its value indirectly, as in Pavlovian second order conditioning, through its association with the S+. On some trials the animal sees the S+ and responds to it. On other trials the animal sees the S- and then the S+. This inadvertent pairing of S- with S+ is strengthened by the explicitly trained association between the S+ and food (see Table 4). Although the training procedure is a bit different from that used to assess the development of second order conditioning, the underlying mechanism may be similar.

One way to distinguish between these two theories is to ask what would happen if, after discrimination training, the value of the S+ was reduced (e.g., responding to the S+ was extinguished). If the S- acquired its value through direct association with food, devaluation of the S+should have relatively little effect on the value of the S-. If, on the other hand, the S- acquired its value as a signal for the presence of the S+, devaluation of the S+ should reduce the value of the S- as well.

To test these hypotheses, Zentall, Sherburne, Roper, and Kraemer (1996, Exp. 1) trained pigeons on two simultaneous discriminations, A+B- and C±D-. For half of the pigeons, responding to A was then extinguished. For the remaining pigeons (which served as a control group) responding to C was extinguished. The design of this experiment is presented in Table 5. If, in a simultaneous discrimination, the value acquired by the S- is direct, then both groups should retain their strong preference for B over D. If, however, the S- acquires its value through its ability to signal the presence of the S+, then the devaluation of A should result in a reduced preference for B over D on test trials, whereas the devaluation of C, if it has any effect, should augment the preference for B over D. Results reported by Zentall et al. (1996) indicate that, although pigeons in the control group continued to show a strong preference for B (92.7%), those in the experimental group did not (46.9%). Thus, it appears that, in training, the S- served as a signal or higher order conditioned stimulus for the presence of S+.

TABLE 5
Design of experiment to evaluate the direct and second order theories of value transfer (after Zentall, Sherburne, Roper, & Kraemer, 1996, Exp. 1).

Group	Training	Test 1	Devaluation	Test 2
A-	A+B- and C±D-	B vs. D	A-	B vs. D
C-	A+B- and C±D-	B vs. D	C-	B vs. D

Note: The hues, red, green, yellow, and blue served equally often as A, B, C, and D.

In a follow-up experiment, Zentall et al. (1996, Exp. 3) initially trained pigeons on two simultaneous discriminations, both involving a positive stimulus to which responses were partially reinforced (S±), and a negative stimulus to which responses were never reinforced (S-), A±B- and C±D-. Thus, both training pairs were treated similarly. Following acquisition, the pigeons were trained on a simultaneous discrimination involving the two original positive stimuli, A+C- such that the probability of reinforcement associated with one of the S+s increased from 50% to 100%, while the probability of reinforcement associated with the other S+decreased from 50% to 0%. The design of this experiment is presented in Table 6. According to the signaling hypothesis, although the values of B and D should have be comparable at the end of original training, both the effect of increasing the value of A and the effect of decreasing the value of C should result in a preference for

B over D. A significant preference for B was, indeed, found. Thus, it appears that the preference for B over D that has been reported in other research resulted from the differential signaling value of each of the stimuli. In original training, B signaled the presence of A, and D signaled the presence of C. When A was increased in valued and C was decreased in value, however, B now signaled the presence of an even better S+, whereas D now signaled the presence of an S-.

The results of this experiment also argue against the reinforcement history account of transitive inference. According to the reinforcement history account, altering the value of A or C should have had no effect on the relative values of B and D, because only reinforcement associated with the test stimuli themselves should affect their value. Given that original training (A±B-, C±D-) should not produce a differential preference for B over D, and none should be produced by A+C- training, none should have appeared on test trials.

TABLE 6
Design of alternative experiment to evaluate the direct- and second-order-conditioning theories of value transfer (after Zentall, Sherburne, Roper, & Kraemer, 1996, Exp. 2).

Phase 1: Train:	A±B- : C±D-
Phase 2: Train:	A+C-
Test:	B vs.D

Note: The hues, red, green, yellow, and blue served equally often as A, B, C, and D.

Negative value transfer. According to value transfer theory, not only should positive value transfer from the S+ to the S-, but negative value should transfer from the S- to the S+. In principle, the design of the experiment to test for negative value transfer follows directly from the design used by Zentall and Sherburne (1994). Pigeons should be trained on two simultaneous discriminations with the value of the two S+s equated and the values of the two S-s allowed to vary (A+B-, C+D±). On test trials, a preference for C over A would suggest that negative value had transferred from B to A (or perhaps that some positive value had transferred from the less positive D to the more positive C).

A pilot study that was conducted using this design, included training with a single positive stimulus (E+) to provide a baseline against which to assess both negative value transfer from B to A and positive value transfer from D to C. Results indicated that the pigeons showed indifference between A and C, between A and E, and between C and E. We concluded that the pigeons may have had insufficient experience with B- and with D± to allow them to affect A and C, respectively. Furthermore, it may be that the 100% schedule of reinforcement associated with responding to A, C, and D may have produced a ceiling effect (i.e., with such a schedule, on test trials, the pigeons may have responded nondifferentially to the first S+ that they saw).

In the next experiment, to allow for sufficient experience with B and D, these stimuli were presented on their own, as well as with their respective S+s (A and C). Furthermore, the reinforcement schedule associated with responding to A and C was reduced from 100% to 75%. Finally, to ensure that the original discriminations were not too difficult, the schedule of reinforcement associated with D was reduced from 50% to 25%. The design of this experiment is presented in Table 7.

TABLE 7
Design of negative value transfer experiment.

Train:	A(75%)B(0%) : C(75%)D(25%) : B(0%) : D(25%)
Test:	A vs.C

Note - Probability of reinforcement associated with each stimulus is indicated in parentheses. The letters A, B, C, and D are arbitrary representations of unrelated hues (red, green yellow, and blue) counterbalanced over subjects.

The results of this experiment were quite surprising. On test trials the pigeons consistently selected A over C. Thus, it appears that not only does B not transfer negative value to A but also, D does not transfer positive value to C. Instead, in a simultaneous discrimination, it appears that the effect of a negative stimulus on its associated positive stimulus is to *enhance* its value. This result is suggestive of a positive behavioral contrast effect that has typically been reported when, following training with two successively presented positive stimuli, the schedule associated with one of them is made substantially leaner (e.g., it is changed to extinction). Not only does responding decline to the stimulus associated with the extinction schedule, but responding typically increases to the stimulus associated with the unchanged schedule (see Reynolds, 1961). In the present case, A is preferred over C because, having been trained in the context of the negative B, A is judged to be more positive than C (which was trained in the context of the less negative or more positive D).

The implications of these findings (that positive value transfer can enhance the value of a negative stimulus with which it appeared and positive behavioral contrast can enhance the value of a positive stimulus with which it appeared) are that stimuli presented in the context of a simultaneous discrimination should be preferred over stimuli with a similar reinforcement history, but presented alone. This hypothesis has not yet been tested.

Findings of positive behavioral contrast, rather than negative value transfer, also have implications for the value-transfer account of the transitive inference effect. According to the value transfer account, B acquires added value from the always-positive A, and D loses some of its value to the always-negative E. But if the always-negative E actually enhances the value of D, this effect should tend to compensate for the positive effects of A on B. Thus, one would expect that transitive inference effects would be small, relative to the two opposing effects (positive value transfer form A to B and positive behavioral contrast from E to D) that give rise to them. Apparently, this is not the case.

Value Transfer in Humans. If Pavlovian processes are responsible for value transfer effects, then similar effects may be found in other species that can learn through Pavlovian conditioning. We have recently asked if value transfer can be demonstrated in humans. The rationale for this series of experiments was not simply to test the generality of value transfer and its underlying mechanisms (although the research is perhaps justifiable on those grounds alone) but also to establish a possible link between these basic learning processes and more complex social psychological phenomena (to be discussed later) that might be controlled by similar processes.

Our first attempt to replicate with humans, the value transfer effect we had found in pigeons, was a failure. When we used the A+B-, C±D- design reported earlier by Zentall and Sherburne (1994), in which the probability of reinforcement associated with the S+ was varied while neither of the test stimuli was associated with reinforcement, our human participants showed no systematic preference for either B or D. Fortunately, post session interviews suggested to us that there might be a problem. Many of the participants reported that when they responded to C and they got no "reward," they thought that had they responded to D, they would have been rewarded. In other words, human participants may assume that on each trial there is a correct response that will earn them a reward. Thus, some participants reported that D had value because it was correct *whenever C was not correct*. Interestingly, humans appear to make inferences about the consequences of their behavior that go beyond the feedback that they receive. This inference may result from their extensive prior experience discriminations in which one of the stimuli was always rewarded.

In our next attempt (Lonon & Zentall, 1996, Exp. 1), we removed this apparent "ambiguity" by assigning a number of points to each stimulus, and that number did not vary from trial to trial [A(1000)B(0) and C(100)D(0)]. When this modification of the original procedure was used, a significant proportion of participants selected B over D on test trials.

Once we had replicated the basic value transfer effect, we asked whether the mechanism underlying value transfer in humans might be similar to that responsible for value transfer in pigeons. To answer this question humans were trained with the nondifferential procedure used by Zentall et al. (1996, Exp. 3) but consistent points replaced the probablistic reward used with pigeons [i.e., A(100)B(0) and C(100)D(0)]. We then modified the values of the positive stimuli from training [A(1000)C(0)] in the absence of the negative stimuli and then tested with B versus D (see Table 8). Again we found a significant preference for B over D, suggesting that in humans too, in original training, B must have signaled the presence of A, and D signaled the presence of C. When we later increased the value of A and decreased the value of C, the signals for those two stimuli now reflected their new relative values. Thus, it appears that for humans too, not only does value transfer from S+ to S- in a simultaneous discrimination, but also the mechanism responsible for the transfer of value appears to be the signaling value of the S-.

TABLE 8
Design of experiment to evaluate the direct and second order theories of value transfer

Phase 1: Train:	A(100)B(0) C(100)D(0)
Phase 2: Train:	A(1000)C(0)
Test:	B vs.D

Note. Numbers in parentheses refer to the number of points awarded for a response to that stimulus. Hues and shapes served equally often as A, B, C, and D.

Value Transfer Applied to Complex Social Behavior. As suggested above, value transfer may be applicable to more than simultaneous discriminations involving simple stimuli. Although the role played by value transfer in more complex social behavior is somewhat speculative at this time, it is presented here to suggest possible directions for future research. For example, attractive or famous models are frequently used in product advertising. It could be argued that advertisers use attractive models, because it is expected that some of the model's value (e.g. beauty, desirability, respect) will transfer to the product. Such transfer of value (rather than a more cognitive attribution) may be most plausible when there is little evidence that the model should have particular knowledge or expertise regarding the product. Smith and Engle (1968), for example, reported that participants who saw a new car advertisement that included a young woman model, rated the car as better designed, faster, more appealing, and more expensive looking than those who saw the same advertisement without the model. More surprisingly, the participants did not believe that the presence of the model altered their perception of the car (Cialdini, 1988).

Gorn (1982) reported similar results based on an experiment in which participants were shown video advertisements for pens that were presented with musical backgrounds. The musical background were selected based on ratings that the participants had made previously. The results indicated that significantly more participants selected pens that were paired with “liked” music than with “disliked” music. Thus, some of the value of the liked music appears to have transferred to the pen with which it was paired.

Another social psychological phenomenon in which value transfer may be involved has been referred to as, "basking in the reflected glory of others" (or BIRGing; see Cialdini, et al., 1976). BIRGing occurs when attempts are made to enhance one's self-presentation by promoting one's association with famous or highly valued individuals. For example, people might take pride in the fact that they come from the same town as a famous athlete. Presumably, they believe that they attain value from the common association they and the athlete have with the town (see also Tesser's, 1988, reflection effect).

The "mere ownership effect" (Beggan, 1992) may also be related to value transfer, although in this case it is the self that operates as the more positive event. The mere-ownership effect occurs when objects are rated as more desirable after they are given to (or owned by) the participant than when the participant is merely asked to rate the same objects. Ownership can be viewed as an acquired association between an object and a person (its owner). Because of this association, Beggan proposed that evaluation of an owned object is linked with self perception. Thus, the association between the object and the person may allow value to transfer from the valued self to the previously relatively neutral object.

Whether these complex social psychological phenomena are controlled by the same mechanism as the simpler effects we have referred to as value transfer will require further study. The present results suggest, however, that animal models of human behavior may be useful in attempting to integrate what may appear to be unrelated complex human behavior.

Transitive Inference Revisited

Although value transfer provides an adequate account of findings of transitive inference in animals, it may be that other processes are at work as well. Research described here suggests that the presence of differential value associated with the A and C stimuli in a transitive inference task is sufficient to account for findings of transitive inference (i.e., a preference for B over D) It had been assumed that it was sufficient to provide a reasonable, simpler alternative to transitive inference to discount the likelihood of relational learning. But that may not be enough to rule out an inferential process. An stronger case against transitive inference could be made if one could show that in the absence of differential value transfer, findings of transitive inference will *fail* to occur. Weaver et al. (in press), have argued that if the transitive inference effect were multiply determined (i.e., if it were produced by both differential value transfer *and* by relational learning) demonstrations of value transfer, in the absence of possible transitive inference through mediated associations, would not be informative about the capacity of the organism to form a transitive inference.

We attempted to rule out differential value transfer in a number of ways (Weaver et al., in press). In the most convincing variation of the procedure, responses to all of the stimuli not involved in the B versus D test were treated alike. Thus, the A, C, and E stimuli in each of the discriminations were consistently partially reinforced, such that the four discriminations in training were, A±B-, B+C±, C±D-, D+E±. In this way, the values of A, C, and E all should have been comparable and thus, the transfer of value from A to B and from C to D should have been comparable. In addition, should it be the case that in a simultaneous discrimination, value can transfer from the stimulus associated with the lower probability of reinforcement (the S-) to the stimulus associated with the higher probability of reinforcement (the S+), this design would also control for the possible transfer of value from C to B and from E to D.

One could argue, perhaps, that this design, might bias the animals against demonstrating transitive inferences. Although with this design it is still possible for relational learning to occur (i.e., it is still true that A>B, B>C, C>D, and D>E), the degree of difference in value between stimuli within a discrimination should be somewhat smaller (i.e., in this design, the relatively easy S+S- discriminations typical of the standard transitive inference task, have been replaced by harder S±S- and S+S± discriminations). In spite of this reduction in discriminability of the stimulus pairs and the elimination of differential value transfer, a significant transitive inference effect was found that was as strong as that typically found with the more standard (A+B-, B+C-, C+D-, D+E-) design (Weaver et al., in press, Exp.2). These results suggest either that the transitive inference effect *is* multiply determined (i.e., both differential value transfer and relational learning are present and either can control choice on test trials) or that, although differential value transfer is clearly a reliable phenomenon, it does not occur in the context of a transitive inference task (e.g., perhaps the prior value of a stimulus, C-, does not affect its current value, C+), and C+ and A+ actually have comparable value to give to B and D, respectively (but see Steirn et al., 1995, Exp. 3). In either case, the present results suggest that some form of relational learning can occur when the nonverbal form of the transitive inference procedure is used with pigeons.

Conclusions

The present line of research began with an attempt to test various alternative accounts of findings of transitive inference in animals. Differential reinforcement history associated with the test stimuli did not appear to provide a consistent account of transitive inference findings. On the other hand, differential transfer of value from the positive to the negative stimulus in a simultaneous discrimination appears to provide a more parsimonious account of the findings. Value transfer appears to be a reliable finding that can be attributed to Pavlovian secondary conditioning. Furthermore, value transfer effects can be found in humans and may contribute to a number of what have been presumed to be more complex social psychological phenomena. A potential problem for the value transfer account of the transitive inference effect is the fact that negative stimuli appear to enhance the value of the positive stimuli with which they are presented. Thus such an effect would be expected to reduce or eliminate the difference in value between the test stimuli in the transitive inference task. Furthermore, recent evidence suggests that the transitive inference effect may be multiply determined. When differential value transfer is eliminated, transitive inference effects remain. Thus, the present findings suggest that relational learning does, in fact, occur when animals acquire the discriminations that define the transitive inference task.

Authors Notes

The research described in this chapter was supported by National Institute of Mental Health Grants MH 35376 and MH 45979, and by National Science Foundation Grants BNS 8418275, RII 8902792, BNS 9019080, and IBN 9414589. Lou Sherburne is now at Wabash College, Crawfordsville, IN 47933. Important contributions to the research were provided by, Janice N. Steirn, Janice Weaver, Karen L. Roper, and Philipp Kraemer.

References

Beggan, James K. (1992). On the social nature of nonsocial perception: The mere ownership effect. *Journal of Personality and Social Psychology, 62,* 229-237.

Bryant, P. E., & Trabasso, T. (1971). Transitive inferences and memory in young children. *Nature, 232,* 456-458.

Cialdini, R. B., Borden, R. J., Thorne, A., Walker, M. R. Freeman, S. & Sloan, L. R.(1976). Basking in reflected glory: Three (football) field studies. *Journal of Personality and Social Psychology, 34,* 463-476.

Cialdini, R. B. (1988). *Influence: Science and Practice, 2nd Edition.* United States of America: Harper Collins Publishers.

Couvillon, P.A., & Bitterman, M.E. (1992). A conventional conditioning analysis of "transitive inference" in pigeons. *Journal of Experimental Psychology: Animal Behavior Processes, 18,* 308-310.

Crespi, L. P. (1944). Amount of reinforcement and level of performance. *Psychological Review, 51,* 341-357.

Davis, H. (1992). Logical transitivity in animals. In W. K. Honig & J. G. Fetterman (Eds.), *Cognitive aspects of stimulus control* (pp. 405-429). Hillsdale, NJ: Erlbaum.

Doré, F. Y. (1986). Object permanence in adult cats (*Felis catus*). *Journal of Comparative Psychology, 100,* 340-347.

Fersen, L. von, Wynne, C. D. L., Delius, J. D., and Staddon, J. E. R. (1991). Transitive inference formation in pigeons. *Journal of Experimental Psychology: Animal Behavior Processes, 17,* 334-341.

Gillan, D. J. (1981). Reasoning in the chimpanzee: II. Transitive Inference. *Journal of Experimental Psychology: Animal Behavior Processes, 7,* 150-164.

Gorn, G. J. (1982). The effects of music in advertising on choice behavior: A classical conditioning approach. *Journal of Marketing, 46*, 92-101.

Lonon, A. M., & Zentall, T. R. (1996). *The "magical" transfer of value from S+ to S- in simultaneous discriminations.* Manuscript submitted for publication.

McGonigle, B. O., & Chalmers, M. (1977). Are monkeys logical? *Nature, 267,* 694-696.

Piaget, J. (1928). *Judgement and reasoning in the child.* London: Kegan, Paul, Trench, & Trubner.

Reynolds, G. S. (1961). Behavioral Contrast. *Journal of the Experimental Analysis of Behavior, 4,* 57-71.

Smith, G. H., & Engel, R. (1968). Influence of a female model on perceived characteristics of an automobile. *Proceedings of the American Psychological Association, 3*, 681-682.

Spence, K. W. (1937). The differential response in animals to stimuli varying within a single dimension. *Psychological Review, 44*, 430-444.

Steirn, J. N., Weaver, J. E., and Zentall, T. R. (1994). Transitive inference in pigeons: Simplified procedures and a test of value transfer theory. *Animal Learning and Behavior, 23,* 76-82.

Tesser, A. (1988). Toward a self-evaluation maintenance model of social behavior. In L. Berkowitz (Ed.), *Advances in experimental social psychology, (Vol. 21,* pp. 181-227). New York: Academic Press.

Weaver, J. E., Steirn, J. N., & Zentall, T. R. (in press). Transitive Inference in Pigeons: Control for value transfer. *Psychonomic Bulletin and Review.*

Woodruff, G., Premack, D., & Kennel, K. (1978). Conservation of liquid and solid quantity by the chimpanzee. *Science, 202,* 991-994.

Wynne, C. D. L., Fersen, L. von, & Staddon, J. E. R. (1992). Pigeons' Inferences are Transitive and the Outcome of Elementary Conditioning Principles: A Response. *Journal of Experimental Psychology: Animal Behavior Processes, 18,* 313-315.

Zentall, T. R. & Sherburne, L. M. (1994). Transfer of value from S+ to S- in a simultaneous discrimination. *Journal of Experimental Psychology: Animal Behavior Processes, 20,* 176-183.

Zentall, T. R., Sherburne, L. M., Roper, K. L., Kraemer, P. J. (1996). Value transfer in a simultaneous discrimination appears to result from within-event Pavlovian conditioning. *Journal of Experimental Psychology: Animal Behavior Processes, 22,* 68-75.

13

The Experience-dependent Maturation of an Evaluative System in the Cortex

Allan N. Schore
Department of Psychiatry and Biobehavioral Sciences
UCLA School of Medicine
9817 Sylvia Avenue
Northridge, California 91324
Telephone: 818-886-4368 / Fax: 818-349-4404

Abstract

The ability to be continuously informed about external changes in the environment as well as about the current status of internal bodily states is a fundamental function of the brain. In early development this adaptive capacity is essential to the survival of the infant, and its emergence is a product of the incipient interactions between a developing brain and its social environment. The maturation of an evaluative system in the frontolimbic cortex is experience-dependent, and is specifically shaped by face-to-face synchronized interactions in which the caregiver psychobiologically attunes to the child's internal state in order to form a system of mutually regulated arousal. Once this attachment bond is formed, the mother's facially expressed emotional communications can be accessed in order to provide the infant with salient maternal appraisals of interactions and events that regulate his affect, cognition, and behavior. These "social referencing" attachment experiences are imprinted into and directly influence the organization of the orbital cortex, a prefrontal area that undergoes a critical period of growth from the last quarter of the first until the last quarter of the second year. This region hierarchically dominates the limbic system and directly connects into the autonomic nervous system, and it is expanded in the early developing right hemisphere that is dominant for the processing of socioemotional information and for the regulation of body state. Due to the organization of its dense connections with sites in both the cortex and subcortex, this system plays an essential psychobiological role - at the orbitofrontal level cortically processed exteroceptive information concerning the external environment (such as visual and prosodic information emanating from an emotional face) is integrated with subcortically processed interoceptive information regarding the internal visceral environment (such as concurrent changes in bodily states). This cortex, acting in an "evaluative" capacity, thus functions to refine emotions in keeping with current sensory input, and allows for the adaptive switching of internal bodily states in response to changes in the external environment that are appraised to be personally meaningful.

The appraisal of changes in the environment within which an organism exists is a basic adaptive function of the nervous system. For humans, as for all other species, the most salient aspects of the environment are located in not so much the physical as in the social context, the realm of interactions between one individual and another. The transactions within human relationships are both verbal and nonverbal, and thereby contain both cognitive and emotional elements. These communications are not solely "psychological" but more correctly "psychobiological", and as mind-body events they involve activities in both the central and autonomic nervous systems. This means that although evaluative operations have usually referred to purely cognitive assessments of the external environment, they also involve concurrent appraisals of changes in the internal environment that are registered as emotional states. In addition to determining the significance of an environmental stimulus, the brain must also monitor feedback about the internal state in order to make assessments of coping resources, and it must update appropriate response outputs in order to make adaptive adjustments to particular environmental perturbations. Self-organizing appraisals thus involve feedback between social cognition and emotion, and they are a central component of a sequence of psychobiological processes that involve mind and body, cortex and subcortex.

There is now a consensus that appraisal systems assign value to current stimuli based on past experience, according to the accrued developmental history of an individual's preferences and aversions. But there has been some uncertainty about what particular types of early experiences are involved, and a vagueness about the mechanism by which these experiences influence the initial organization of appraisal systems. In light of the developmental principle articulated by Freud, Piaget and others that the beginnings of living systems set the stage for every aspect of an organism's internal and external function for the rest of the lifespan, I suggest that a deeper understanding of the ontogeny of appraisal systems is an essential area of inquiry. This effort involves an integration of current information from a spectrum of disciplines - developmental psychology, developmental psychobiology, and developmental neurobiology.

In this paper I will give a brief overview of my current work in this area (Schore, 1994; 1996), and suggest that certain specific early experiences, in conjunction with genetic factors (Schore, in press), occuring in particular critical periods of the individual's developmental history, are required for the organization of brain systems that come to mediate the appraisal of socioemotional information. These organismic-environmental experiences are embedded in the primordial interactions between a developing human and the first environment, the relationship between the infant and mother. Indeed, it is now clear that development can only be understood in terms of a continuing dialectic between an active and changing organism and an active and changing environment, and that the most important part of that environment are the interactions and relationships the child has with others (Hinde, 1990). The adaptive capacity to automatically and efficiently evaluate changes in the environment and in the self is an important ontogenetic goal, and it's evolution is not only indelibly influenced by the first relationship with another human being, its emergence literally depends upon whether of not the infant has sufficient opportunities to be a co-participant of an open dynamic system of reciprocal interchanges with an emotionally responsive caregiver.

In other words, the establishment of an attachment bond of emotional communication with the mother, the most important environmental object in early infancy, enables the child to receive the mother's affective appraisals of objects in the nonmaternal environment in late infancy. These interactively transmitted, affectively charged external appraisals provide the developing individual with the requisite experiences that ultimately allow for the organization, in the second year, of brain networks that can generate internal evaluations of the personal significance of what is happening in an encounter with the environment (Lazarus, 1991a) and can elicit emotions to actual or expected changes in events that are important to the individual (Frijda, 1988). A growing literature now indicates that the attachment object acts as an external psychobiological regulator of the "experience-dependent" growth of the infant's nervous system, and that these attachment experiences are imprinted into the neurobiological structures that are maturing during the brain growth spurt of the first two years of life, and therefore have far-reaching and long enduring effects.

This model fits well with Eisenberg's (1995) recent writings in the psychiatric literature on "the *social* construction of the human brain", and Tucker's (1992) assertion in the developmental psychological literature that the most important information for the successful development of the human brain is conveyed by the social rather than the physical environment. It has recently been suggested that "the best description of development may come from a careful appreciation of the brain's own self-organizing operations" (Cicchetti & Tucker, 1994, p. 544). There is now widespread agreement that the brain is a self-organizing system, but there is perhaps less of an appreciation of the fact that the self-organization of the developing brain occurs in the context of a relationship with another self, another brain (Schore, 1996).

In the following, I will present material on the psychobiology and neurobiology of attachment experiences, and then I describe how social referencing experiences, a highly visual form of attachment behavior, serve as an interpersonal matrix for the development of the child's evaluative capacities. Next I will offer ideas about how these developmental interactive events influence the experience-dependent maturation of an evaluative system in the infant brain, especially in the right hemisphere which is undergoing a growth spurt in the first year-and-a-half of life. This right brain system is characterized by particular cognitive, attentional, and arousal capacities that make it uniquely suited to respond to environmental events in an adaptive fashion (Heller, 1993). In light of the fact that the organization of this right hemispheric system operates in conjunction with a system localized to the frontal lobes that is involved in modulating the emotional valence of experience, I will emphasize the importance of the development of the paralimbic cortices, especially the orbitofrontal areas, which act, according to Pribram (1987), in an "evaluative" capacity. The orbitofrontal system is intimately invovled in "cognitive-emotional interactions" (Barbas, 1995), a functional role that is relevant to the current emphasis on "cognition-emotion feedback" and self-organizing appraisals (Lewis, 1996).

Mutual Gaze as a Dynamic Mechanism for the Coordination of Facial Appraisals and Changes in Psychobiological State

Although much has been written about cognitive development in infancy, until recently relatively few studies have specifically focused on the ontogeny of social cognition and emotional states. This development is closely tied into the maturation of sensory systems, especially visual systems. In fact, over the first year of life visual experiences play a paramount role in social and emotional development (Hobson, 1993; Preisler, 1995; Wright, 1991). The mother's emotionally expressive face is, by far, the most potent visual stimulus in the infant's environment, and the child's intense interest in her face, especially in her eyes, leads him to track it in space, and to engage in periods of intense mutual gaze. The infant's gaze, in turn, reliably evokes the mother's gaze, thereby acting as a potent interpersonal channel for the transmission of reciprocal mutual influences. These sustained face-to-face transactions are quite common and can be of very long duration, and they mediate the dialogue between mother and child. Indeed, a body of research demonstrates that gaze represents the most salient channel of nonverbal communication and indeed the most intense form of interpersonal communication.

With the onset of increasing myelination of the visual areas of the infant's occipital cortex, mutual gaze interactions increase significantly over the second and third quarter of the first year.

Since these interactions occur within the split second world of the mother and infant, they are therefore not easily visible. This dialogue is best studied by a frame-by-frame analysis of film, and in such work Beebe and Lachmann (1988) observe a "mirroring sequence," synchronous rapid movements and fast changes in affective expressions within the dyad (Fig. 1). In these mother-infant communications, each partner responds to the other extremely rapidly, in latencies ranging from simultaneous to one-half second. This affective mirroring is accomplished by a moment-by-moment matching of affective direction in which both partners increase together their degree of engagement and facially expressed positive affect. The fact that the coordination of responses is so rapid suggests the existence of a bond of unconscious communication. Notice that as the child breaks out in an intensely emotional "full gape smile" its attention is focused on one particular aspect of the environment - the visual image of the mother's face. I suggest that in light of the fact that the human face is a unique stimulus whose features display biologically significant information, these early mutual gaze episodes highlight the central role of emotionally charged facial expressions as a focus of appraisal processes.

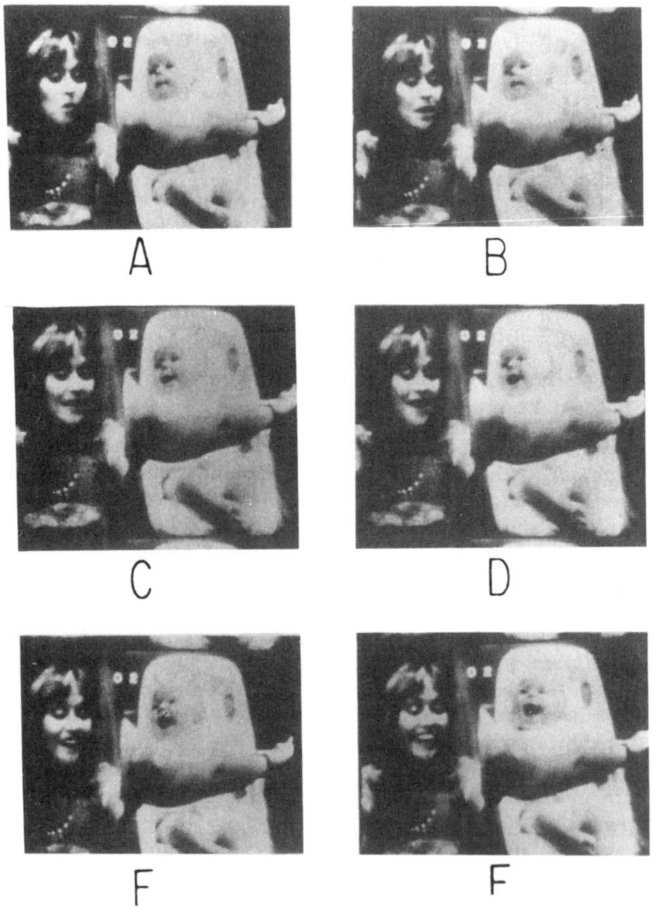

Figure 1. Photographic illustrations of a "mirroring" sequence. Mother and infant are seated face to face, and are looking at each other. At point A, mother shows a "kiss-face," and infant's lips are partially drawn in, resulting in a tight, sober-faced expression. At point B, .54 seconds later, mother's mouth has widened into a slightly positive expression, and infant's face has relaxed with a hint of widening in the mouth, also a slighly positive expression. At point C, .79 seconds later, both mother and infant show a slight smile. At point D, .46 seconds later, both

mother and infant further widen and open their smiles. Again at points E, .46 seconds later, and F, .58 seconds later, both mother and infant further increase their smile display. Points E and F illustrate the infant" "gape smile." At point F the infant has shifted the orientation of his head further to his left, and upward, which heightens the evocativeness of the gape-smile. (From Beebe & Lachman, 1988).

This microregulation continues, as soon after the "heightened affective moment" of an intensely joyful full gape smile the baby will gaze avert in order to regulate the potentially disorganizing effect of this intensifying emotion (Fig. 2). In order to maintain the positive emotion the psychobiologically attuned mother takes her cue and backs off to reduce her stimulation. She then waits for the baby's signals for reengagement. In this process of "contingent responsivity" the more the mother tunes her activity level to the infant during periods of social engagement, the more she allows him to recover quietly in periods of disengagement, and the more she attends to his reinitiating cues for reengagement, the more synchronized their interaction. The caregiver thus facilitates the infant's information processing by adjusting the mode, amount, variability, and timing of stimulation to the infant's actual integrative capacities. Facial mirroring thus illustrates interactions organized by ongoing regulations, and the development of mutually attuned synchronized interactions is fundamental to the ongoing affective development of the infant.

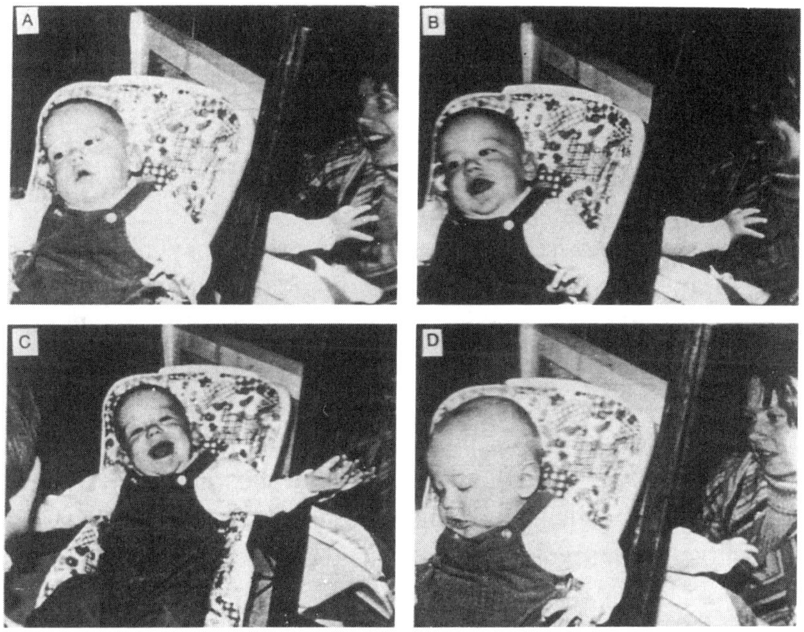

Figure 2. A typical sequence observed during "attuned" interactions of normal infants and their mothers: (a) the infant looks at the mother and the mother shows an exaggerated facial expression (mock surprise); (b) the infant and mother smile; (c) the infant laughs, the mother "relaxes" her smile; and (d) the infant looks away, the mother ceases smiling and watches her infant. (From Field & Fogel, 1982).

These facial mirroring exchanges generate much more than overt facial changes in the dyad; they represent a transformation of inner events. In these episodes of reciprocal facial signalling, as the mother and infant synchronize with each other's temporal and affective patterns, each recreates an inner psychophysiological state similar to the partner's. In this mutually attuned dynamic system the crescendos and decrescendos of the infant's psychobiological state are in resonance with similar states of crescendos and decrescendos, cross-modally, of the mother. Consequently, both experience a state transition as they move together from a state of neutral affect and arousal to one of heightened positive emotion and high arousal. In physics, a property of resonance is sympathetic vibration, the tendency of one resonance system to enlarge and augment through matching the resonance frequency pattern of another resonance system. Infant researchers refer to a particular maternal social behavior which can "blast the infant into the next orbit of positive excitation," and generate "vitality affects" (Stern, 1985). The mother's face has been called the child's "biological" or "emotional" mirror," but I suggest that it is not simply a reflective mirror, but rather an amplifying mirror. A major task of the first year is the evolution of affective tolerance for increasingly higher levels of arousal, and this is facilitated by the mother's modulation of the infant's highly stimulated states (Fogel, 1982).

Interactive transactions are also generating important events in the infant's bodily state. Hofer's (1990) developmental psychobiological research is revealing that in the "symbiotic" state the adult's and infant's individual homeostatic systems are linked together in a superordinate organization which allows for mutual regulation of vital endocrine, autonomic, and central nervous systems of both mother and infant by elements of their interaction with each other. Hofer emphasizes the importance of "hidden" psychobiological regulatory processes by which the caregiver's more mature and differentiated nervous system regulates the infant's "open," immature, internal homeostatic systems. His work reveals that the mother influences the neural substrates for infant emotion by directly regulating the neurochemistry of the infant's maturing brain, including arousal-regulating dopamine and noradrenaline levels, in order to generate high levels of positive affect (Hofer, 1984). The attachment relationship is thus essentially a regulator of arousal (van der Kolk & Fisler, 1994).

Indeed, regulatory processes are the precursors of psychological attachment and its associated emotions (Hofer, 1994), and psychobiological attunement is now thought to be the mechanism that mediates attachment bond formation (Field, 1985). These dialogues act as a crucible for the forging of an attachment bond that allows for the dyadic regulation of emotion (Sroufe, 1996). Infant research now suggests that the baby becomes attached to the modulating caregiver who expands opportunities for positive affect and minimizes negative affect (Demos & Kaplan, 1986). In other words, the affective state underlies and motivates attachment, and the central adaptive function of dyadic attachment dynamics is to interactively generate and maintain optimal levels of positively valenced states.

The Neurobiology of Dyadically Regulated Psychobiological States

According to Bowlby (1969) vision is central to the establishment of a primary attachment to the mother, and imprinting is the learning mechanism that underlies attachment bond formation. Furthermore, attachment is more than overt behavior, it is internal, "being built into the nervous system, in the course and as a result of the infant's experience of his transactions with the mother" (Ainsworth, 1967, p. 429). Imprinting involves a state of mutually entrained central nervous system propensities and a synchrony between sequential infant-maternal stimuli and behavior (Petrovich & Gewirtz, 1985). This points to another level of analysis - the neurobiological level. How are developing systems of the organizing brain influenced by these interactions with the social environment?

The work of Trevarthen (1993) on maternal-infant protoconversations bears directly on this problem (Fig. 3). Coordinated with eye-to-eye messages are auditory vocalizations (tone of voice, Motherese) and tactile and body gestures as a channel of communication. A traffic of visual and prosodic auditory signals induce instant emotional effects, namely excitement and pleasure builds within the dyad. The resonance of the dyad ultimately permits the intercoordination of positive affective brain states. His studies underscore the fundamental principle that the baby's brain is not only affected by these transactions, it's growth literally requires brain-brain interaction and occurs in the context of a positive affective relationship between mother and infant. Trevarthen concludes that "the affective regulations of brain growth" are embedded in the context of an intimate relationship, and that they promote the development of cerebral circuits. This interactive mechanism requires older brains to engage with mental states of awareness, emotion, and interest in younger brains, and involves a coordination between the motivations of the infant and the feelings of adults.

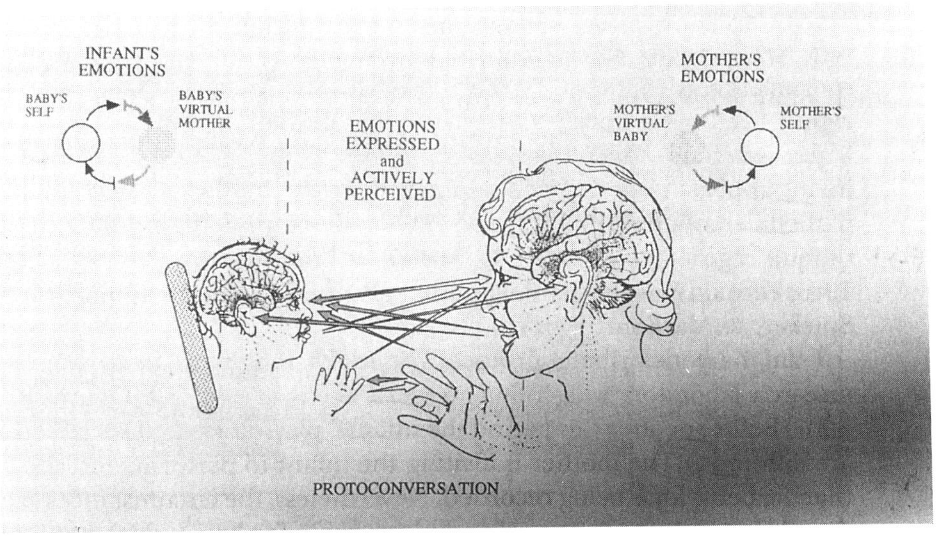

Figure 3. Channels of face-to-face communication in protoconversation. Protoconversation is mediated by eye-to-eye orientations, vocalizations, hand gestures, and movements of the arms and head, all acting in coordination to express interpersonal awareness and emotions. From Trevarthen (1993).

In fact, there is now consensus that interactions with the environment during sensitive periods are necessary for the brain as a whole to mature (Greenough, Black, & Wallace, 1987; Cicchetti & Tucker, 1994). But we know that different regions of the brain mature at different rates. Can we tell what specific parts of the growing brain are affected by these emotion transacting events? In describing the dialogue Trevarthen (1996) notes that the prosody of the voice of the mother is responded to by the infant's right hemisphere, which is "more receptive and self-regulatory". He also points out that the left side of the mother's face has a stronger expression of an attractive smile. It is now thought that "The emotional experience of the infant develops through the sounds, images, and pictures that constitute much of an infant's early learning experience, and are disproportionately stored or processed in the right hemisphere during the formative stages of brain ontogeny" (Semrud-Clikeman & Hynd, 1990, p. 198). The right cortex, which matures before the left (Crowell, Jones, Kapunai, & Nakagawa, 1973; Chi, Dooling, & Gilles, 1977; Geschwind & Galaburda, 1987), is known to be specifically impacted by early social experiences (Denenberg et al., 1978).

I suggest that in these psychobiologically attuned face-to-face transactions the infant's right hemisphere, which is dominant for the infant's recognition of the maternal face (de Schonen, Gil de Diaz, & Mathivet, 1986), and for the perception of arousal-inducing maternal facial affective expressions (Nelson, 1987), visual emotional information (Saxby & Bryden, 1985), and the prosody of the mother's voice (Fernald, 1989), is appraising the output of the mother's right hemisphere, which is dominant for nonverbal communication (Benowitz et al., 1983) and the processing and expression of emotional information (Bradshaw & Nettleton, 1983; Natale, Gur, & Gur, 1983; Hellige, 1990). The right hemisphere is known to be activated in intense states of elation (Tucker, 1981), to show an advantage in the recognition of positive facial expressions (Hugdahl, Iversen, & Ness, 1989), and to contribute to the development of reciprocal interactions within the mother-infant regulatory system (Taylor, 1987). The child is using the output of the mother's right cortex as a template for the imprinting, the hard wiring of circuits in his own right cortex that will come to mediate his expanding cognitive-affective capacities to appraise variations in both external and internal information. It has been said that in early infancy the mother is the child's "auxiliary cortex" (Diamond, Balvin, & Diamond, 1963). There is now solid evidence that the parenting environment influences the developing patterns of neural connections that underlie behavior (Dawson, 1994), including attachment behavior.

Indeed, the right hemisphere is centrally involved in attachment experiences (Henry, 1993; Schore, 1994; 1996). Developmental researchers have held for some time that as a result of attachment experiences the infant develops a schema, a mental image of the mother, especially her face (Mussen, Conger, & Kagan, 1969). Leventhal (1984) notes that the infant's automatic appraisals of the affective aspects of parental interactions, such as smiles, frowns, and vocalizations, are stored in memory along with the memory of emotional experience accompanying these events. These episodes produce prototypes which can then generate automatic and nonconscious affective attributions about environmental interactions. Along this same line of thinking, the internal working model of the attachment relationship is now understood to be a mental representation that guides appraisals of experience (Main, Kaplan, & Cassidy, 1985).

Social Referencing Attachment Experiences and the Origin of the Appraisal of Emotionally Significant Environmental Events

In the last quarter of the first year a critical milestone in human development is reached - with the attainment of upright posture and independent locomotion the infant becomes a toddler. This capacity allows the child to now separate himself from the mother in order to begin to explore the non-maternal physical environment. When the infant locomotes more widely in its environment, vision continues to be the primary mode of connection with the mother. The socioemotional function of gaze emerges at this very time, as the appearance of the infant's new cognitive capacity to "read" mother's face at reunions after separations coincides in time with the increase in motility.

At 10 to 12 months, social referencing experiences, a special form of attachment behavior (Bretherton, 1985), first appear (Walden & Ogan, 1988), in which the child searches the mother's face for emotional information about the physical environment and then follows her gaze (Scaife & Bruner, 1975). In these visuoaffective transactions the mother's facially expressed emotional communications provide the infant with salient maternal appraisals of interactions and events (Hornik, Risenhoover, & Gunnar, 1987), especially of novel persons or objects. Also at this time "joint visual attention" becomes intensified and the child first exhibits communicative pointing (Butterworth, 1991), a social gesture (Leung & Rheingold, 1981) that occurs in the context of a shared activity (Murphy, 1978). In this sequence, the child points to an object in the environment but also looks at the other to check the person's gaze as well (Masur, 1983). This visually-driven dyadic mechanism, so synchronized that it has been described as a "shared visual realiity" (Scaife & Bruner, 1975), allows the infant to see from the adult's point of view, and therefore acts as an interactive matrix for cultural learning (Tomasello, 1993).

The child's first sorties into the world occur under "the watchful eye of the caregiver," in which he uses the mother as a "beacon of orientation". The infant, now at greater distances from the mother yet increasingly sensitive to her gaze, keeps an eye on the feelings expressed on the mother's face (Oatley & Jenkins, 1992), and uses this signal for evaluations of safety and danger in the nonmaternal environment. But even more than this, at the point when he returns to the secure base the mother's attention to the child's emotionally expressive face intensifies. With these facial cues the psychobiologically attuned mother is now able to appraise the child's internal state, and on the basis of this she reengages in synchronized patterns of visuoaffective communication. The toddler, in turn, attentively responds to the visual stimulation emanating from the mother's emotionally expressive face. These visuoaffective transactions that serve as episodes of "microregulation" are thus critical moments of reciprocal signaling that mediate emotional reconnection after separations. The dyad thus evolves an operative practical system for processing high intensity affective transmissions that rapidly maintains psychobiological attunement and sustains the attachment bond without the need for frequent and prolonged physical contact, and these refueling transactions that co-generate high levels of positive affect allow for exploration of the child's expanding world.

It is important to note that these developmental advances depend upon the earlier co-construction of an attachment bond of emotional regulation within the dyad. In social referencing, an affectively charged dialogic process of the communication of "emotional vision" (Bauer, 1982), the mother induces a mood modification in the infant (Feinman, 1982), and is directly influencing the infant's learning of "how to feel," "how much to feel," and "whether to feel" about particular objects in the larger environment. With regard to the development of interest and curiosity, social referencing maternal attention-focusing strategies may also be essential to the caregiver's enduring effect on the infant's learning of "what to feel" about objects in the social environment, and "what to be interested in" amongst the objects in the physical environment. Developmental studies of 12-month-olds support the notion that social referencing accounts for the maternal emotional biasing of infant reactions to novel inanimate objects (Hornik et al., 1987). In classical work, Vygotsky (1978) emphasized that the child's discovery of the physical environment is socially mediated. This seminal idea has been incorporated into and expanded upon in the latest models of emotion. Lazarus (1991b) now contends that it is not the physical properties of the environment but the subjective meanings that count in the emotion process.

The visual-emotional communication of social referencing attachment transactions provides access to the mother's appraisal of objects in the animate and inanimate world, and this influences the development of an internalized system in the infant that can evaluate the personal emotional meaning of any particular environmental event. These critical period events may induce familiarization (Pribram. 1991) or "topographic familiarity" and begin to generate "personally relevant" aspects of the individual's world (Van Lancker, 1991). Emotions are currently understood to "arise in response to the·meaning structures of given situations, to events that are important to the individual, and which importance he or she appraises in some way" (Frijda, 1988, p. 349), and they involve reactions to fundamental relational meanings that have adaptive significance (Lazarus, 1991a). In the current developmental literature, internal working models enable the individual to evaluate information relevant to attachment, and are unconsciously accessed to interpret and act on new experiences (Crowell & Feldman, 1991).

These internal working models are first measured at the end of the first year, and reflect the fact that psychobiological attachment experiences are imprinted into the early developing brain. Indeed, stable attachment bonds are vitally important for the infant's continuing neurobiological development (Trad, 1986). Mary Main (1995, p. 214), perhaps the most influential current attachment researcher, now concludes that "The formation of an attachment to a specified individual signals a quantitative change in infant behavioral (*and no doubt also brain*) organization" [italics added]. Do we now know what parts of the brain begin a critical period of structural growth at 10 to 12 months and are involved in attachment, evaluative functions, and the regulation of emotion?

The Experience-Dependent Maturation of an Evaluative System in the Frontolimbic Cortex Occurs in a Critical Period of Infancy

In my book, *Affect Regulation and the Origin of the Self: The Neurobiology of Emotional Development* (Schore, 1994) I offer evidence to show that high intensity visual and auditory positive affective stimulation provided in face-to-face transactions represent a growth-promoting environment for the prefrontal cortex, an area that is known to undergo a major maturational change at 10 to 12 months (Diamond & Doar, 1989; Huttenlocher, 1979). It is now established that the orbital (as opposed to the later maturing nonlimbic dorsolateral) prefrontal cortex (Fig.4) is critically and directly involved in attachment functions (Steklis & Kling, 1985), appraisal proceses (Pribram, 1987), and directed attention (King, Corwin, & Reep, 1989). This region (along with the anterior temporal area) is a central component of the paralimbic cortex whose structure surrounds the more medial and basal areas of the limbic forebrain and whose function involves the processing of social signals necessary for the initiation of social interactions (Raleigh & Steklis, 1981). Furthermore, this frontolimbic structure which contains neurons that specifically respond to the emotional expressions of faces (Thorpe, Rolls, & Maddison, 1983) is part of a system that has evolved for the rapid and reliable identification of individuals from their faces, because of the importance of this in social behavior" (Rolls, 1986).

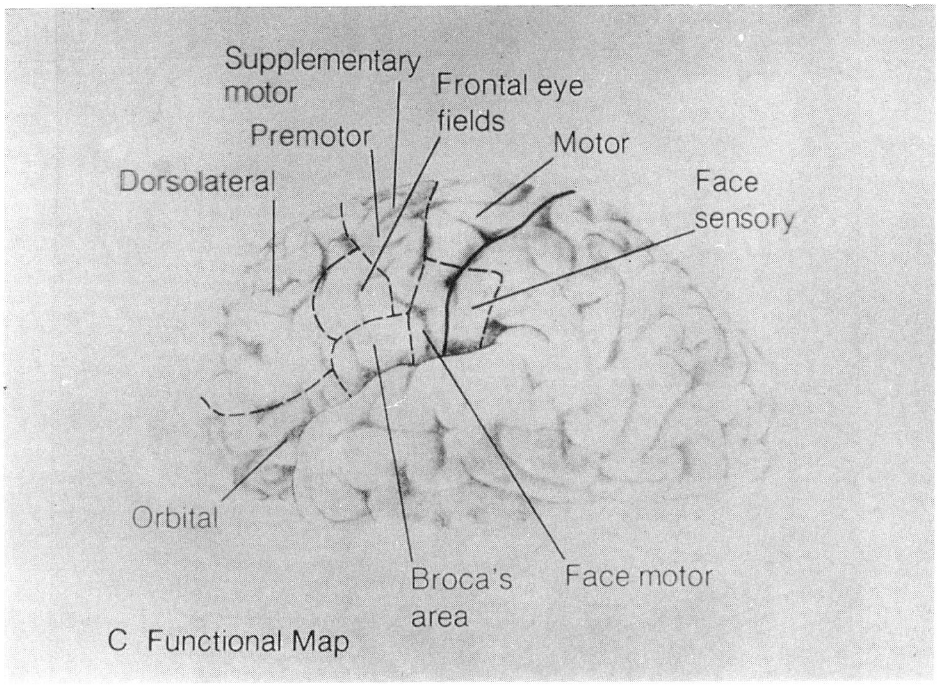

Figure 4. Approximate boundaries of functional zones of the human cerebral cortex, showing the dorsolateral and orbital prefrontal areas. (From Kolb & Whishaw, 1990).

The orbitoinsular frontal cortex is "hidden" in the ventral and medial surfaces of the prefrontal lobe. Due to its location at the interface of cortex and subcortex (Figs. 5-7), it acts as a "convergence zone," and is one of the few brain regions that is "privy to signals about virtually any activity taking place in our beings' mind or body at any given time" (Damasio, 1994, p. 181). In addition to receiving input from all sensory association areas of the posterior cortex (including projections from the face and head region of the somatosensory cortex and from the inferior temporal regions related to central vision), as well as outputs to motor areas in the anterior cortex and ventral striatum, it uniquely projects extensive pathways to limbic areas in the temporal pole and the central nucleus of the amygdala, to glutamate receptors of mesocorticolimbic dopamine neurons in ventral tegmental areas of the anterior reticular formation, and to subcortical drive centers in the paraventricular hypothalamus that are associated with the sympathetic branch of the autonomic nervous system. This excitatory limbic circuit, what Nauta and Domesick (1982) term the ventral tegmental limbic forebrain-midbrain circuit, is involved with the generation of positively valenced states associated with motivational reward and approach behavior. Orbitofrontal regions also send axons onto subcortical targets in parasympathetic autonomic areas of the hypothalamus, and to noradrenergic neurons in the medulla and the vagal complex in the brain stem caudal reticular formation, thereby completing the organization of another limbic circuit, the lateral tegmental limbic forebrain-midbrain circuit that activates the onset of an inhibitory, negatively valenced state associated with avoidance.

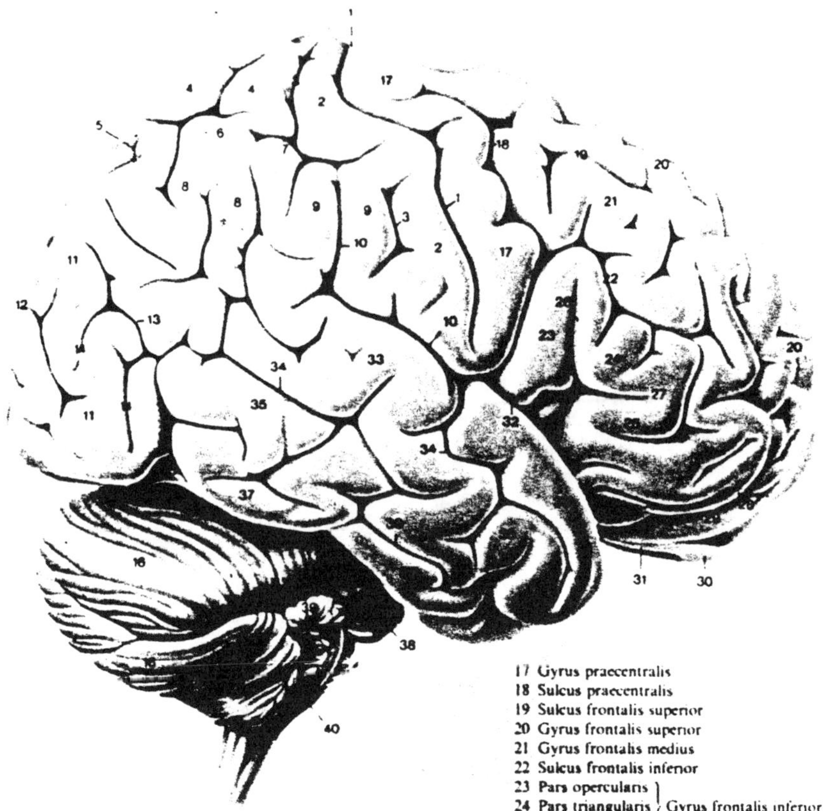

Figure 5. Lateral view of the human right hemisphere. Note the position of the orbital sulci (28) and gyri (29) in the frontal undersurface. (From Nieuwenhuys, Voogd, & van Huijzen, 1981).

Figure 6. Photograph of the base of the human brain showing orbital gyri and sulci at sites lablelled 29, 30, and 31 . (From Watson, 1977).

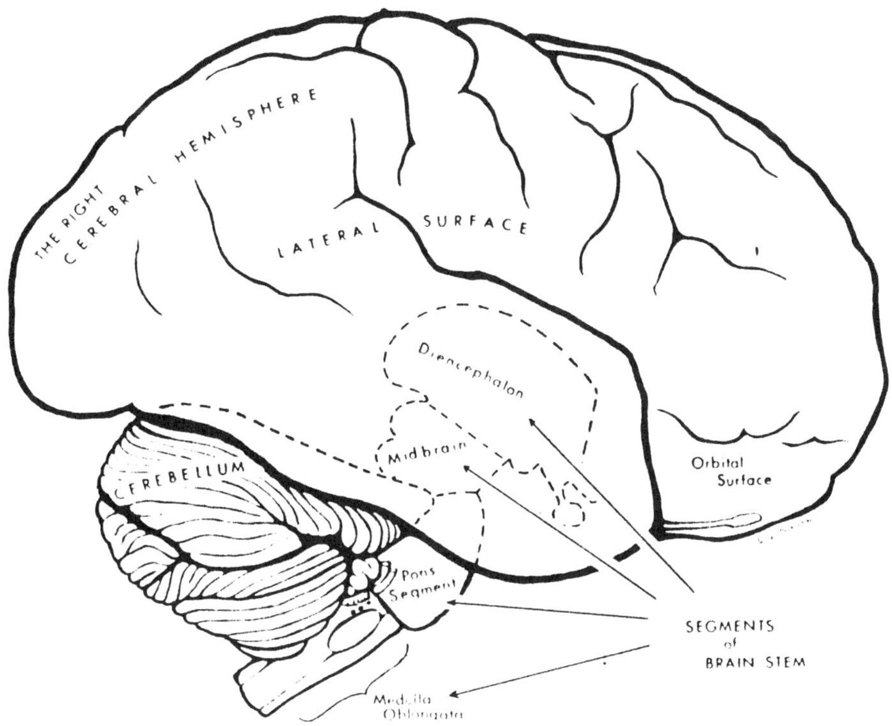

Figure 7. Relationships of brain stem structures to the orbital surface of the right hemisphere. (From Smith, 1981).

This frontolimbic system is so intimately interconnected into limbic areas that it has been conceived of as an "association cortex" (Martin, 1989; Pribram, 1981) for the limbic forebrain (see Fig. 8). Indeed it sits at the apex of the limbic system, the brain system responsible for the rewarding-excitatory and aversive-inhibitory aspects of emotion. This "frontolimbic cortex" acts as a major center of CNS hierarchical control over the energy-expending sympathetic and energy-conserving parasympathetic branches of the ANS, and due to its autonomic connections, it plays an important cortical role in the feedback from bodily systems and in the reception of what Damasio (1994) calls "somatic markers. The orbitofrontal system is involved, according to Luria, in "the regulation of the body state and reflect changes taking place in that state" (1980, p. 262). At the orbitofrontal level cortically processed information concerning the external environment (such as visual and auditory stimuli emanating from the emotional face of the object) is integrated with subcortically processed information regarding the internal visceral environment (such as concurrent changes in the emotional or bodily self state), thereby enabling incoming information to be associated with motivational and emotional states (Pandya & Yeterian, 1985). This far frontolimbic attentional system determines the "regulatory significance" of stimuli that reach the organism (Luria, 1980), and in such manner, "forebrain circuits concerned with the recognition and interpretation of life experiences are capable of influencing virtually all, if not all, regulatory mechanisms in the body" (Wolf, 1995, p. 90).

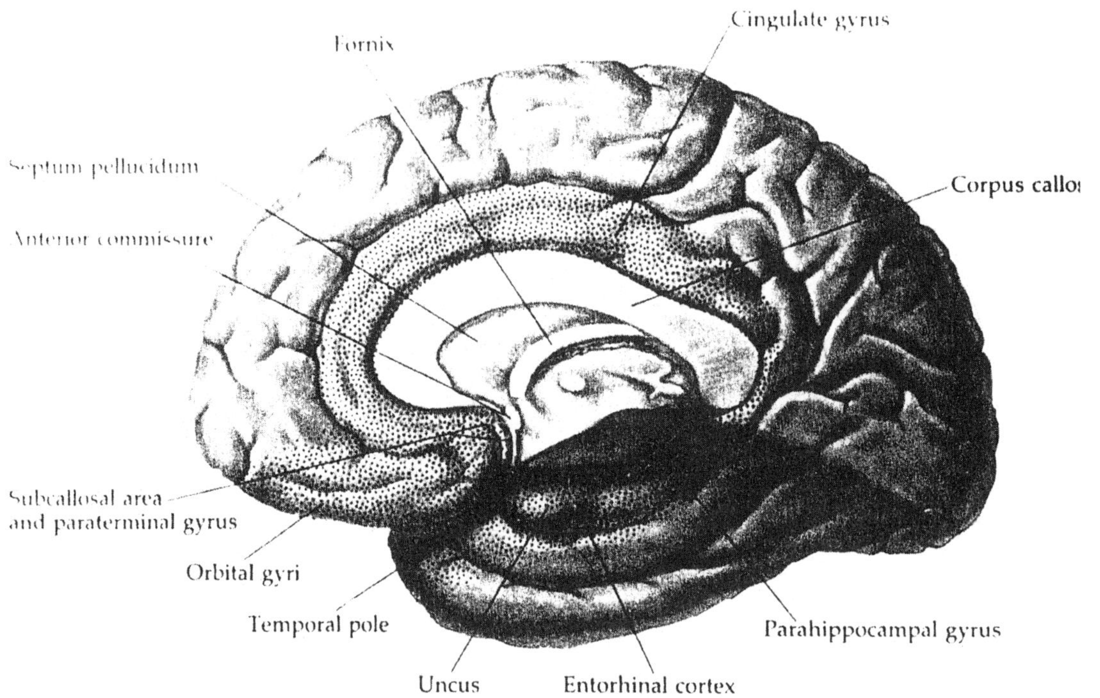

Figure 8. Midsagittal view of the right cerebral hemisphere, with brain stem removed. The limbic association cortex is indicated by the dotted region. Note the orbital gyri. (From Martin, 1989).

The orbital prefrontal region is especially expanded in the right cortex (Falk et al. 1990; Tucker, 1992), and it comes to act in the capacity of an "executive control system" (Pribram, 1991) for the entire right cortex, the "visuospatial" hemisphere that is centrally involved in selective attention to facial expressions (Etcoff, 1984), appraisal (Davidson et al., 1990), and unconscious processes (Galin, 1974; Watt, 1990). The right cortex is dominant for evaluating emotional expressions (Ahern et al., 1991), perceiving briefly presented emotionally expressive faces (Suberi & McKeever, 1977), and eliciting autonomic responses after subliminal presentations of facial expressions (Johnson & Hugdahl, 1991). These operations define the implicit perception of affective information (Niedenthal, 1990), and they reflect the specialization of this hemisphere not only for detecting stimuli faster than the left (Anzola, Bertlolini, Buchtel, & Rizzolatti, 1977), globally directed attention (Fink et al., 1996), processing novel information (Bever, 1983), and holistic analysis (Van Kleeck, 1989), but also for processing the autonomic correlates of emotional arousal (Wittling & Roschmann, 1993). According to Lazarus, rapid unconscious appraisals on the basis of the optic stimulus array are a necessary and sufficient condition of emotion (Lazarus & Smith, 1988). Bear (1983) argues that extensive direct corticolimbic connections of the right hemispheric "ventral pathway," which includes orbitofrontal and anterior temporal components, are responsible for its function in enabling a heightened autonomic response and an immediate, powerful affective reaction to particular stimuli.

Automatic emotional processes are a central element of nonverbal communicaton between humans, and their appraisal is essential to to interpersonal functioning. In light of the fact that emotions can be facially expressed in as little as one-half second (Izard, 1991), an accurate perception of a facial stimulus must be rapidly evaluated in terms of its significance to the particular individual. Ohman (1986) describes a very quick, involuntary, unconscious and holistic "preattentive stimulus analysis" of an emotionally expressive face, Broadbent (1977) refers to a "hidden preattentive process" of analysis of incoming emotional stimuli, and Zajonc (1984) describes an early, fast evaluation which computes the affective significance of an external stimulus. It is now established that evaluative learning occurs without awareness (Baeyens, Eelen, & Van den Bergh, 1990), and that emotional stimuli (Wexler, Warrenburg, Schwartz, & Janer, 1992) and the recognition of familiar faces (De Haan, Young, & Newcombe, 1987) are processed unconsciously.

Socioaffective stimuli, especially patterns of change in visual and auditory stimuli emanating from an emotionally expressive face, are processed in the right posterior occipitoparietal (Bradshaw & Nettleton, 1983) and right parietotemporal (Ross, 1983) association cortices, respectively. The final steps in the discrimination of an incoming stimulus pattern takes place in multimodal or paralimbic association areas in the temporal and frontal cortices which register the motivational significance of the stimulus. In this manner the paralimbic system, via its role in visceroautonomic processing and valuation (Pribram, 1991), performs a "valence tagging" function, in which perceptions receive an affective charge of a positive (pleasurably toned, idealized, good) or negative (unpleasurable, dysphoric, bad) hedonic quality. Ledoux (1989) refers to the "core of the emotional system" that computes the affective significance of environmental stimuli, and Tucker (1992) speaks of frontal networks of the "paralimbic core" that function in the evaluation of information in terms of its motivational significance.

The developmental literature now indicates that the infant's “biologically organized affective core” becomes biased with tendencies toward certain emotional responses, depending on early experiences in the caregiving relationship" (Emde, 1983). Emotion biases in personality, which first appear in the second year of life, may be due to the fact that certain neurophysiological and neuroanatomical emotive circuits become more readily activated than others (Malatesta, Culver, Tesman, & Shepard, 1989). Ultimately, the affective core "biases the infant's evaluation of a new situation and his interactive patterns even before the information arising from the situation has been processed" (Tronick, Cohn, & Shea, 1986, p. 23). These experiences may permanently influence the types of information channels and the specific patterns of input used by a particular personality organization to receive and recognize idiosyncratically meaningful socioaffective signals that trigger particular motivational systems. The early interactional events which provide for the experience-dependent maturation of the orbital cortex thus permanently influence the unique manner by which this system augments and enhances sensitivities as to what to do in a particular environmental context (Deecke, Kornhuber, Long, & Schreiber, 1985; Pribram, 1991). It has been noted that, “With progressively higher organizational levels in evaluative mechanisms, there appears to be a general expansion in the range and relational complexity of contextual controls and in the breadth and flexibility of adaptive response” (Cacioppo & Berntson, 1992, p. 1026).

Limbic areas of the human cerebral cortex show anatomical maturation at 15 months, suggesting that corticolimbic functions expressed in "emotional activities and mechanisms of memory" are operating at this specific time (Rabinowicz, 1979). The early affective experiences that occur during the critical period of growth of this frontoliombic system are imprinted and stored in implicit-procedural memory systems in the orbital prefrontal system and its cortical and subcortical connections as interactive representations. Damasio describes dispositonal representations, stored in the ventromedial cortex, that “embody knowledge pertaining to how certain types of situations have usually been paired with emotional responses in [one's] individual experience" (1995, p. 22). These representations are repositories of stimulus appraisals and memories of somatic value acquired in the individual's experience, and when they are activated they create a bodily state which regulates approach or avoidance towards an object. Such representations also encode an appraisal of the organism's bodily state to generate energy in response to the challenge of a particular external context, that is an estimate of the individual's capacity to cope with a particular environmental situation (Schore, 1994).

The orbitofrontal system plays a major role in the internal state of the organism (Mega & Cummings, 1994), the temporal organization of behavior (Fuster, 1985) and in the adjustment or correction of emotional responses (Rolls, 1986), that is, affect regulation. Pribram (1987) proposes that the ventrolateral cortex, acting in an "evaluative" or "appraisal" capacity, functions to "refine" emotions in keeping with current sensory input and with the consequences of actions. I suggest that the essential activity of this psychic system is thus the adaptive switching of internal bodily states in response to changes in the external environment that are appraised to be personally meaningful. This emergent function, in turn, enables the individual to recover from disruptions of state and to integrate a sense of self across transitions of state, thereby allowing for a continuity of experience in various environmental contexts.

The orbital cortex matures in the middle of the second year, a time when the right hemisphere ends its growth phase and the left hemisphere begins one (Thatcher, 1994). Most intriguingly, infant observers report that in the course of the second year the infant acquires the capacity to generate a "theory of mind," in which an individual imputes mental states to self and to others and predicts behavior on the basis of such states (Bretherton, McNew, & Beeghly, 1981; Meltzoff, 1995). Indeed, there is now evidence that the orbital frontolimbic system functionally mediates the the capacity of empathizing (Mega & Cummings, 1994) and inferring the states of others (Baron-Cohen, 1995), and of reflecting on one's own internal emotional states, as well as others (Povinelli & Preuss, 1995). The activity of the "non-dominant" right hemisphere, and not the later maturing "dominant" verbal-linguistic left, is instrumental to the capacity of empathic cognition and the perception of the emotional states of mind of other human beings (Voeller, 1986). The human appraisal system, from its very beginnings throughout the rest of the lifespan, is not just directed towards evaluating the overt behavior of others, but in attempting to understand the mind of other humans.

Reference

Ahern, G.L., Schomer, D.L., Kleefield, J., Blume, H., Rees Cosgrove, G., Weintraub, S., & Mesulam, M-M. (1991). Right hemisphere advantage for evaluating emotional facial expression. *Cortex*, *27*, 193-202.

Ainsworth, M.D.S. (1967). Infancy in Uganda: Infant care and the growth of love. Baltimore: Johns Hopkins University Press.

Anzola, G.P., Bertolini, G., Buchtel, H.A., & Rizzolatti, G. (1977). Spatial compatibility and anatomical factors in simple and choice reaction times. *Neuropsychologia*, *15*, 292-302.

Baeyens, F., Eelen, P., & Van den Bergh, O. (1990). Contingency awareness in evaluative conditioning: a case for unaware affective-evaluative learning. *Cognition and Emotion*, *4*, 3-18.

Barbas, H. (1995). Anatomic basis of cognitive-emotional interactions in the primate prefrontal cortex. *Neuroscience and Biobehavioral Reviews*, *19*, 499-510.

Baron-Cohen, S., (1995). *Mindblindness: An Essay on Autism and Theory of Mind.* Cambridge, MA: MIT Press.

Bauer, R.M. (1982). Visual hypoemotionality as a symptom of visual-limbic disconnection in man. Archives of Neurology, 39, 702-708.

Bear, D.M. (1983). Hemispheric specialization and the neurology of emotion. *Archives of Neurology*, *40*, 195-202.

Beebe, B. & Lachman, F.M. (1988). Mother-infant mutual influence and precursors of psychic structure. In A. Goldberg (Ed.), *Progress in self psychology, Vol. 3* (pp.3-25). Hillsdale, NJ: The Analytic Press

Benowitz, L.I., Bear, D.M., Rosenthal, R., Mesulam, M-M., Zaidel, E., & Sperry, R.W. (1983). Hemispheric specialization in nonverbal communication. *Cortex*, *19*, 5-11.

Bever, T.G. (1983). Cerebral lateralization, cognitive asymmetry, and human consciousness. In E. Perecman (Ed.), *Cognitive processing in the right hemisphere* (pp. 19-39). New York: Academic Press.

Bowlby, J. (1969). *Attachment and loss Vol 1: Attachment.* New York: Basic Books.

Bradshaw, J. L., & Nettleton, N.C. (1983). *Human cerebral asymmetry.* Englewood Cliffs, NJ: Prentice Hall.

Bretherton, I. (1985). Attachment theory: Retrospect and prospect. *Monographs of the Society for Research in Child Development*, *50*, 3-35.

Bretherton, I., McNew, S., & Beeghly, M. (1981). Early person knowledge in gestural and verbal communication: When do infants acquire a "theory of mind"? In M.Lamb & L. Sherrod (Eds.), *Infant social cognition* (pp. 335-373). Hillsdale, NJ: Lawrence Erlbaum Associates.

Broadbent, D.E. (1977). The hidden preattentive process. *American Psychologist*, *32*, 109-118.

Buck, R. (1993). Spontaneous communication and the foundation of the interpersonal self. In U. Neisser (Ed.), *The Perceived Self: Ecological and Interpersonal Sources of Self-knowledge,* (pp. 216-236). New York: Cambridge University Press.

Butterworth, G.E. (1991). The ontogeny and phylogeny of joint visual attention. In A. Whiten (Ed.), *Natural theories of mind* (pp. 223-232). Oxford: Basil Blackwell.

Cacioppo, J. T., & Berntson, G.G. (1992). Social psychological contributions to the decade of the brain: Doctrine of multilevel analysis. *American Psychologist*, *47*, 1019-1028.

Chi, J. G., Dooling, E.C., & Gilles, (1977). F.H. Gyral development of the human brain. *Annals of Neurology*, *1*, 86-93.

Cicchetti, D. & Tucker, D. (1994). Development and self-regulatory structures of the mind. *Development and Psychopathology*, *6*, 533-549.

Crowell, D.H., Jones, R.H., Kapuniai, L.E., & Nakagawa, J. K. (1973). Unilateral cortical activity in newborn humans: An early index of cerebral dominance? *Science*, *180*, 205-208.

Crowell, J. A., & Feldman, S.S. (1991). Mothers' working models of attachment relationships and mother and child behavior during separation and reunion. *Developmental Psychology*, *27*, 597-605.
Damasio, A.R. (1994). *Descartes' error.* New York: Grosset/Putnam.
Damasio, A.R. (1995). Toward a neurobiology of emotion and feeling: operational concepts and hypotheses. *The Neuroscientist*, *1*, 19-25.
Davidson, R.J., Ekman, P., Friesen, W.V., Saron, C.D., & Senulis, J.A. (1990). Approach-withdrawal and cerebral asymmetry: Emotional expression and brain physiology I. *Journal of Personality and Social Psychology*, *58*, 330-341.
Dawson, G. (1994). Development of emotional expression and emotion regulation in infancy. Contributions of the frontal lobe. In G. Dawson & K.W. Fischer (Eds.), *Human behavior and the developing brain* (pp. 346-379). New York: The Guilford Press.
Deecke, L., Kornhuber, M., Long, M., & Schreiber, H. (1985). Timing function of the frontal cortex in sequential motor and learning tasks. *Human Neurobiology*, *4*, 143-154.
De Haan, E.H.F., Young, A.W., & Newcombe, F. (1987). Face recognition without awareness. *Cognitive Neuropsychology*, *4*, 385-415.
Demos, V. & Kaplan, S. (1986). Motivation and affect reconsidered: Affect biographies of two infants. *Psychoanalysis and Contemporary Thought*, *9*, 147-221.
Denenberg, V.H., Garbanti, J., Sherman, G., Yutzey, D.A., & Kaplan, R. (1978). Infantile stimulation induces brain lateralization in rats. *Science*, *201*, 1150-1152.
de Schonen, S., Gil de Diaz, M., & Mathivet, E. (1986). Hemispheric asymmetry in face processing in infancy. In H.D. Ellis, M.A. Jeeves, F. Newcombe, & A.W. Young (Eds.), *Aspects of face processing* (pp. 199-209). Dordrecht: Nijhoff.
Diamond, A. & Doar, B. (1989). The performance of human infants on a measure of frontal cortex function, the delayed response task. *Developmental Psychobiology*, *22*, 271-294.
Diamond, M.C., Krech, D., & Rosenzweig, M.R. (1963). The effects of an enriched environment on the histology of the rat cerebral cortex. *Journal of Comparative Neurology* 123: 111-120.
Eisenberg, L. (1995). The social construction of the human brain. *American Journal of Psychiatry*, *152*, 1563-1575.
Emde, R.N. (1983). The pre-representational self and its affective core. *Psychoanalytic Study of the Child*, *38*, 165-192.
Etcoff, N.L. (1984). Selective attention to facial identity and facial emotion. *Neuropsychologia*, *22*, 281-295.
Falk, D., Hildebolt, C., Cheverud, J., Vannier, M., Helmkamp, R.C., & Konigsberg, L. (1990). Cortical asymmetries in frontal lobes of Rhesus monkeys (*Macaca mulatta*). *Brain Research*, *512*, 40-45.
Feinman, S. (1982). Social referencing in infancy. *Merrill-Palmer Quarterly*, *28*, 445-470.
Fernald, A. (1989). Intonation and communicative interest in mother's speech to infants: is the melody the message? *Child Development*, *60*, 1497-1510.
Field, T. (1985). Attachment as psychobiological attunement: Being on the same wavelength. In M. Reite & T. Field (Eds.), *The psychobiology of attachment and separation* (pp. 415-454). Orlando, FL: Academic Press.
Field, T., & Fogel, A. (1982). *Emotion and early interaction.* Hillsdale, NJ: Lawrence Erlbaum Associates.
Fink, G.R., Halligan, P.W., Marshall, J.C., Frith, C.D., Frackowiak, R.S.J., & Dolan, R.J. (1996). Where in the brain does visual attention select the forest from the trees? *Nature*, *382*, 626-628.
Fogel, A. (1982). Affect dynamics in early infancy: affective tolerance. In T. Field & A. Fogel (Eds.), *Emotion and early interaction.* Hillsdale, NJ: Erlbaum.
Frijda, N.H. (1988). The laws of emotion. *American Psychologist*, *43*, 349-358.
Fuster, J.M. (1985). The prefrontal cortex and temporal integration. In A. Peters & E.G. Jones (Eds.), *Cerebral cortex. Vol. 4. Association and auditory cortices*, (pp. 151-171). New York: Plenum Press.

Geschwind, N., & Galaburda, A. M. (1987). *Cerebral lateralization: Biological mechanisms, associations, and pathology*. Boston: The MIT Press.
Greenough, W.T., Black, J., & Wallace, C. (1987). Experience and brain development. *Child Development*, *58*, 539-559.
Heller, W. (1993). Neuropsychological mechanisms of indivual differences in emotion, personality, and arousal. *Neuropsychology*, *7*, 476-489.
Hellige, J.B. (1990). Hemispheric asymmetry. *Annual Review of Psychology*, *41*, 55-80.
Henry, J.P. (1993). Psychological and physiological responses to stress: The right hemisphere and the hypothalamo-pituitary-adrenal axis, an inquiry into problems of human bonding. *Integrative Physiological and Behavioral Science*, *28*, 369-387.
Hinde, R. (1990). Causes of social development from the perspective of an integrated developmental science. In G. Butterworth & P. Bryant (Eds.), *Causes of development* (pp. 161-185). Hillsdale, NJ: Erlbaum.
Hobson, R.P. (1993). Through feeling and site through self and symbol. In U. Neisser (Ed.), *The perceived self: Ecological and interpersonal sources of self-knowledge* (pp. 254-279). New York: Cambridge University Press.
Hofer, M. A. (1984). Relationships as regulators: A psychobiologic perspective on bereavement. *Psychosomatic Medicine*, *46*, 183-197.
Hofer, M.A (1990). Early symbiotic processes: Hard evidence from a soft place. In R.A. Glick and S. Bone (Eds.), *Pleasure Beyond the Pleasure Principle*, (pp. 55-78). New Haven: Yale University Press.
Hofer, M.A. (1994). Hidden regulators in attachment, separation, and loss. *Monographs of the Society for Research in Child Development*, *59*, 192-207.
Hornik, R., Risenhoover, N., & Gunnar, M. (1987). The effects of maternal positive, neutral, and negative affective communications on infant responses to new toys. *Child Development*, *58*, 937-944.
Hugdahl, K., Iversen, P.M., & Ness, H-M. (1989). Hemispheric differences in recognition of facial expressions: A VHF-study of negative, positive, and neutral emotions. *Journal of Neuroscience*, *45*, 205-213.
Huttenlocher, P.R. (1979). Synaptic density in human frontal cortex - developmental changes and effects of aging. *Brain Research*, *163*, 195- 205.
Izard, C.E. (1991). *The psychology of emotions*. New York: Plenum Press.
Johnsen, B.H., & Hugdahl, K. (1991). Hemispheric asymmetry in conditioning to facial emotional expressions. *Psychophysiology*, *28*, 154-162.
King, V., Corwin, J. V., & Reep, R.L. (1989). Production and characterization of neglect in rats with unilateral lesions of ventrolateral orbital cortex. *Experimental Neurology*, *105*, 287-299.
Kolb, B., & Whishaw, I.Q. (1990). *Fundamental of human neuropsychology (3rd ed)*. New York: W.H. Freeman.
Lazarus, R.S. (1991a). Progress on a cognitive-motivational-relational theory of emotion. *American Psychologist*, *46*, 819-834.
Lazarus, R.S. (1991b). Cognition and motivation in emotion. *American Psychologist*, *46*, 352-367.
Lazarus, R.S., & Smith, C.A. (1988). Knowledge and appraisal in the cognition-emotion relationship. *Cognition and Emotion*, *2*, 281-300.
LeDoux, J. E. (1989). Cognitive-emotional interactions in the brain. *Cognition and Emotion*, *3*, 267-289.
Leung, E.H.L., & Rheingold, H.L. (1981). Development of pointing as a social gesture. *Developmental Psychology*, *17*, 215-220.
Leventhal, H. (1984). A perceptual-motor theory of emotion. *Advances in Experimental Social Psychology*, *17*, 117-182.
Lewis, M.D. (1996). Self-organizing appraisals. *Cognition and Emotion*, *10*, 1-25.
Luria, A. R. (1980). *Higher cortical functions in man (2nd ed.)*. New York: Basic Books.

Main, M. (1995). Discourse, prediction, and recent studies in attachment: Implications for psychoanalysis. In T. Shapiro & R.N. Emde (Eds.), *Research in psychoanalysis: Process, development, outcome*. Madison, CT: International Universities Press.
Main, M., Kaplan, N., & Cassidy, J. (1985). Security in infancy, childhood and adulthood: A move to the level of representation. *Monographs of the Society for Research in Child Development, 50*, 66-104.
Malatesta, C.Z., Culver, C., Tesman, J.R., & Shepard, B. (1989). The development of emotion expression during the first two years of life. *Monographs of the Society for Research in Child Development, 54*, 1-103.
Martin, J.H. (1989). *Neuroanatomy: Text and atlas*. New York: Elsevier.
Masur, E.F. (1983). Gestural development, dual-directional signaling, and the transition to words. *Journal of Psycholinguistic Research, 12*, 93-109.
Mega, M.S., & Cummings, J.L. (1994). Frontal-subcortical circuits and neuropsychiatric disorders. *Journal of Neuropsychiatric and Clinical Neuroscience , 6*, 358-370.
Meltzoff, A.N. (1995). Understanding the intentions of others: Re-enactment of intended acts by 18-month-old chidren. *Developmental Psychology, 31*, 838-850.
Murphy, C.M. (1978). Pointing in the context of a shared activity. *Child Development, 49*, 371-380.
Mussen, P. H., Conger, J. J., & Kagan, J. (1969). *Child development and personality*. New York: Harper and Row.
Natale, M., Gur, R.E., & Gur, R.C. (1983). Hemispheric asymmetries in processing emotional expressions. *Neuropsychologia, 21*, 555-565.
Nauta, W.J.H., & Domesick, V.B. (1982). Neural associations of the limbic system. In A.L. Beckman (Ed.), *The neural basis of behavior* (pp. 175-206). New York: SP Medical and Scientific Books.
Nelson, C.A. (1987). The recognition of facial expressions in the first two years of life: Mechanisms of development. *Child Development, 58*, 889-909.
Niedenthal, P.M. (1990). Implicit perception of affective information. *Journal of Experimental Social Psychology, 26*, 505-527.
Nieuwenhuys, R., Voogd, J., & van Huijzen, Chr. (1981). *The human central nervous system: A synopsis and atlas, 2nd ed. rev.* New York: Springer-Verlag.
Oatley, K., & Jenkins, J.M. (1992). Human emotions: Function and dysfunction. *Annual Review of Psychology, 43*, 55-85.
Ohman, A. (1986). Face the beast and fear the face: Animal and social fears as prototypes for evolutionary analyses of emotion. *Psychophysiology, 23*, 123-145.
Pandya, D.N., & Yeterian, E.H. (1985). Architecture and connections of cortical association areas. In A. Peters & E.G. Jones (Eds.), *Cerebral cortex. Vol.4. Association and auditory Cortices* (pp. 3-61). New York: Plenum Press.
Petrovich, S. B. & Gewirtz, J.L. (1985). The attachment learning process and its relation to cultural and biological evolution: Proximate and ultimate considerations. In M. Reite & T. Field (Eds.), *The psychobiology of attachment and separation* (pp. 259-291). Orlando: Academic Press.
Povinelli, D., & Preuss, T.M. (1995). Theory of mind: evolutionary history of a cognitive specialization. *Trends in Neuroscience* 18: 418-424.
Preisler, G.M. (1995). The development of communication in blind and in deaf infants - similariites and differences. *Child: Care, Health and Development* 21: 79-110.
Pribram, K.H. (1981). Emotions. In S.B. Filskov & T.J. Boll (Eds.), *Handbook of clinical neuropsychology* (pp. 102-134). New York: Wiley.
Pribram, K.H. (1987). The subdivisions of the frontal cortex revisited. In E. Perecman (Ed.), The frontal lobes revisited (pp. 11-39). Hillsdale, NJ: Lawrence Erlbaum Associates.
Pribram, K.H. (1991). *Brain and perception: Holonomy and structure in figural processing*. Hillsdale, NJ: Lawrence Erlbaum.

Rabinowicz, T. (1979). The differentiate maturation of the human cerebral cortex. In F. Falkner & J.M. Tanner (Eds.), *Human growth, Vol. 3, Neurobiology and nutrition* (pp. 97-123). New York: Plenum Press.

Raleigh, M.J., & Steklis, H.D. (1981). Effects of orbitofrontal and temporal neocortical lesions on the affiliative behavior of vervet monkeys (*Cercopithecus aethiops sabaeus*). *Experimental Neurology*, *73*, 378-389.

Rolls, E.T. (1986). Neural systems involved in emotion in primates. In R. Plutchik & H. Kellerman (Eds.), *Emotion: Theory, research, and practice Vol. 3* (pp. 125-143). Orlando, FL: Academic Press.

Ross, E.D. (1983). Right-hemisphere lesions in disorders of affective language. In A. Kertesz (Ed.), *Localization in neuropsychology* (pp. 493-508). New York: Academic Press.

Saxby, G.P., & Bryden, M.P. (1985). Left visual-field advantage in children for processing visual emotional stimuli. *Developmental Psychology*, *21*, 253-261.

Scaife, M., & Bruner, J.S. (1975). The capacity for joint visual attention in the infant. *Nature* , *253*, 265-266.

Schore, A.N. (1994). *Affect regulation and the origin of the self: The neurobiology of emotional development.* Hillsdale, NJ: Erlbaum.

Schore, A.N. (1996). The experience-dependent maturation of a regulatory system in the orbital prefrontal cortex and the origin of developmental psychopathology. *Development and Psychopathology* 8: 59-87.

Schore, A.N. (in press). The early organization of the nonlinear right brain and the development of a predisposition to psychiatric disorders. *Development and Psychopathology.*

Semrud-Clikeman, M., & Hynd, G.W. (1990). Right hemisphere dysfunction in nonverbal learning disabilities: Social, academic, and adaptive functioning in adults and children. *Psychological Bulletin*, *107*, 196-209.

Smith, C.G. (1981). *Serial dissection of the human brain.* Baltimore-Munich: Urban & Scwarzenberg.

Steklis, H. D., & Kling, A. (1985). Neurobiology of affiliative behavior in nonhuman primates. In M. Reite & T. Field.(Ed.), *The psychobiology of attachment and separation* (pp. 93-134). Orlando: Academic Press.

Stern, D. N. (1985). *The interpersonal world of the infant.* New York: Basic Books.

Suberi, M., & McKeever, W.F. (1977). Differential right hemispheric memory storage of emotional and non-emotional faces. *Neuropsychologia, 15,* 757-768.

Taylor, G. (1987). *Psychosomatic Medicine and Contemporary Psychoanalysis.* Madison, CT: International Universities Press.

Thatcher, R.W. (1994). Cyclical cortical reorganization: Origins of human cognitive development. In G. Dawson & K.W. Fischer (eds.), *Human behavior and the developing brain* (pp. 232-266). New York: Guilford Press.

Thorpe, S.J., Rolls, E.T., & Maddison, S. (1983). The orbitofrontal cortex: Neuronal activity in the behaving monkey. *Experimental Brain Research*, *49*, 93-115.

Tomasello, M. (1993). On the interpersonal origins of self-concept. In U. Neisser (Ed.), *The perceived self: Ecological and interpersonal sources of self-knowledge* (pp. 174-184). New York: Cambridge University Press.

Trad, P.V. (1986). *Infant depression.* New York: Springer-Verlag.

Trevarthen, C. (1993). The self born in intersubjectivity: The psychology of an infant communicating. In U. Neisser (Ed.), *The perceived self: ecological and interpersonal sources of self-knowledge* (pp. 121-173). New York: Cambridge University Press.

Trevarthen, C. (1996). Lateral asymmetries in infancy: Implications for the development of the hemispheres. *Neuroscience and Biobehavioral Reviews*, *20.*

Tronick, E.Z., Cohn, J., & Shea, E. (1986). The transfer of affect between mothers and infants. In T.B. Brazelton & M.W. Yogman (Eds.), *Affective development in infancy* (pp. 11-25). Norwood, NJ: Ablex.

Tucker, D.M. (1981). Lateral brain function, emotion, and conceptualization. *Psychological Bulletin*, *89*, 19-46.

Tucker, D.M. (1992). Developing emotions and cortical networks. In M. R. Gunnar & C.A. Nelson (Eds.), *Minnesota symposium on child psychology. Vol. 24, Developmental behavioral neuroscience* (pp. 75-128). Hillsdale, NJ: Lawrence Erlbaum Associates.

van der Kolk, B.A. & Fisler, R.E. (1994). Childhood abuse and neglect and loss of self-regulation. *Bulletin of the Menninger Clinic*, *58*, 145-168.

Van Kleek, M.H. (1989). Hemispheric differences in global versus local processing of hierarchical visual stimuli by normal subjects: New data and a meta-analysis of previous studies. *Neuropsychologia*, *27*, 1165-1178.

Van Lancker, D. (1991). Personal relevance and the human right hemisphere. *Brain and Cognition*, *17*, 64-92.

Voeller, K.K.S. (1986). Right-hemisphere deficit syndrome in children. *American Journal of Psychiatry*, *143*, 1004-1009.

Vygotsky, L.S. (1978). *Mind in society*. Cambridge, MA: Harvard University Press.

Walden, T.A., & Ogan, T.A. (1988). The development of social referencing. *Child Development*, *59*, 1230-1240.

Watson, C. (1977). *Basic human neuroanatomy, An introductory atlas , 2nd ed.* Boston: Little, Brown.

Watt, D.F. (1990). Higher cortical functions and the ego: Explorations of the boundary between behavioral neurology, neuropsychology, and psychoanalysis. *Psychoanalytic Psychology*, *7*, 487-527.

Wexler, B.E., Warrenburg, S., Schwartz, G.E. & Janer, L.D. (1992). EEG and EMG responses to emotion-evoking stimuli processed without conscious awareness. *Neuropsychologia*, *30*, 1065-1079.

Wittling, W., & Roschmann, R. (1993). Emotion-related hemisphere asymmetry: Subjective emotional responses to laterally presented films. *Cortex*, *29*, 431-448.

Wolf, S. (1995). Psychosocial forces and neural mechanisms in disease: Defining the question and collecting the evidence. *Integrative Physiological and Behavioral Science*, *30*, 85-94.

Wright, K. (1991). *Vision and Separation: Between Mother and Baby.* Northvale NJ: Jason Aronson.

Zajonc, R.B. (1984). On primacy of affect. In K.R. Scherer & P. Ekman (Eds.) *Approaches to emotion* (pp. 259-270). Hillsdale, NJ: Lawrence Erlbaum Associates.

14

The Electricity of Touch: Detection and measurement of cardiac energy exchange between people

Rollin McCraty, MA[1], Mike Atkinson[1] Dana Tomasino[1] and William A. Tiller, PhD[2]
(1) HeartMath Research Center, Institute of HeartMath, 14700 West Park Avenue, Boulder Creek, CA 95006
(2) Department of Materials Science and Engineering, Stanford University, Stanford CA 94305-2205

Correspondence and reprint requests to:
Rollin McCraty, HeartMath Research Center, Institute of HeartMath, 14700 West Park Ave., Boulder Creek, CA 95006.
Phone: (408) 338-8727
FAX: (408) 338-1182
Electronic mail: ihmlab@znet.com

All work performed at HeartMath Research Center, Institute of HeartMath, Boulder Creek, CA.
Key Words: Touch, energy, healing, ECG, EEG, coherence, emotion, stochastic resonance, signal averaging

Abstract

The idea that an energy exchange of some type occurs between individuals is a central theme in many healing techniques. This concept has often been disputed by Western science due to the lack of a plausible mechanism to explain the nature of this energy or how it could affect or facilitate the healing process. The fact that the heart generates the strongest electromagnetic field produced by the body, coupled with the recent discovery that this field becomes more coherent as the individual shifts to a sincerely loving or caring state prompted us to investigate the possibility that the field generated by the heart may significantly contribute to this energy exchange.

We present a sampling of results which provide intriguing evidence that an exchange of electromagnetic energy produced by the heart occurs when people touch or are in proximity. Signal averaging techniques are used to show that one's electrocardiogram (ECG) signal is registered in another person's electroencephalogram (EEG) and elsewhere on the other person's body. While this signal is strongest when people are in contact, it is still detectable when subjects are in proximity without contact.

This study represents one of the first successful attempts to directly measure an energy exchange between people, and provides a solid, testable theory to explain the observed effects of many healing modalities that are based upon the assumption that an energy exchange takes place. Nonlinear stochastic resonance is discussed as a mechanism by which weak, coherent electromagnetic fields, such as those generated by the heart of an individual in a caring state, may be detected and amplified by biological tissue, and potentially produce measurable effects in living systems. One implication is that the effects of therapeutic techniques involving contact or proximity between practitioner and patient could be amplified by practitioners consciously adopting a sincere caring attitude, and thus introducing increased coherence into their cardiac field.

Introduction

The concept of an energy exchange between individuals is culturally a universal belief and is a central theme in many of the healing arts of both Eastern and Western medicine, now often referred to as Energy Medicine. One of the main blocks to the acceptance of these so-called alternative therapies by western science has been the lack of a plausible mechanism that could explain the nature of this energy or how it is exchanged. Nevertheless, numerous of studies of Therapeutic Touch practitioners, healers and other individuals have demonstrated a wide variety of effects on healing rates of wounds,[1, 2] pain,[3, 4] hemoglobin levels,[5, 6] conformational changes of DNA and water structure[7] as well as psychological improvements.[8, 9] If we define energy as the capacity to produce an effect, these experiments suggest that an exchange of energy has occurred. It has also been demonstrated that many of these therapeutic effects occur without physical touch, indicating that energy of some kind is radiated or broadcast between practitioner and patient.[8]

References to the concept of an energy exchange between people can also be found in the psycho-therapeutic field as a sense of energetic interaction between the practitioner and patient. This concept dates back at least to Freud, who proposed in *The Anxiety Neuroses* that an energy exchange between practitioner and patient operated at an unconscious level to bring about changes in the patient's mental, emotional and physical well-being.[9]

Many of the healing professions emphasize the importance of the attitude or intention of the practitioner in order for the greatest facilitation of the healing process to occur.[8, 10, 11] The importance of intention has been demonstrated in several studies,[12] including one at our laboratories.[7] In addition, we have previously shown that a person's inner emotional state directly affects the coherence in the electromagnetic field generated by the heart,[12, 13] and that sincere feelings of appreciation, love or care produce increased coherence in the cardiac field. This is especially significant, as the heart generates the strongest electromagnetic field produced by the body, measurable a number of feet away from the body with SQUID-based magnetometers[13] and sensitive electrostatic detectors.[14]

It has been argued that even if there were an energy exchange between people, the energy contained in the signal would be too weak to produce significant effects in a biological system. However, recent research has established that the noise in biological systems can play a constructive role in the detection of weak periodic signals via a mechanism known as stochastic resonance.[15] In essence, stochastic resonance is a nonlinear cooperative effect in which a weak, normally sub-threshold periodic (coherent) stimulus entrains ambient noise, resulting in the periodic signal becoming greatly enhanced and able to produce large scale effects. The signature of stochastic resonance is that the signal-to-noise ratio in the system rises to a maximum at some optimal noise intensity, corresponding to the maximum cooperation between the signal and the noise. Essentially, the noise acts to boost the sub-threshold signal to a level above the threshold value, enabling it to generate measurable effects. Stochastic resonance is now known to occur in a wide range of systems, including sensory transduction, neural signal processing and oscillating chemical reactions,[15, 16] and is firmly established as a valid and far more general phenomenon than previously thought.

There has been much debate over the capacity of extremely low frequency electromagnetic fields to affect living tissue. Theoretical estimates predict the interaction energies of these fields after penetrating the tissue to be up to three orders of magnitude smaller than the average energy of thermal fluctuations.[16] However, the effect of stochastic resonance operating in the noisy environment of a biological tissue would be to greatly amplify the external field's energy, possibly to the point of enabling it to have significant repercussions in the system. The electromagnetic energy patterns produced by the human heart when an individual is in the internal coherence mode, a state reached when feeling sincere love (Figure 1), is a clear example of a coherent, extremely low frequency electromagnetic field.[17, 18] Recent advances in our understanding of the interaction between coherent signals and noise in nonlinear systems has led to the hypothesis that under certain circumstances these nonthermal, coherent electromagnetic fields are detectable by biological systems at the cellular and sub-cellular level.[15, 19, 20] For example, it has recently been demonstrated that nonthermal, extremely low frequency electromagnetic signals can affect intracellular calcium signaling.[21] In addition, coherent electromagnetic fields have been shown to produce substantially greater effects than incoherent signals on enzymatic pathways, such as the ornithine decarboxylase pathway.[22] This suggests that increased cardiac coherence, and thus one's emotional state, may affect cellular function.

The fact that the heart's electromagnetic field (ECG) can be measured anywhere on the surface of the body and also several feet away from the body, coupled with the recent discovery that this field can become more coherent as the individual shifts to a sincerely loving or caring state, prompted us to investigate the possibility that the field generated by the human heart may be the source of the energy exchanged between practitioner and patient in many healing practices. We therefore set out to develop a method of measuring an electrical exchange between people when they touch or are in proximity. This paper presents a few examples from a number of experiments demonstrating that when individuals touch or are in proximity, one person's electrocardiogram (ECG) signal is registered in the other person's electroencephalogram (EEG) and elsewhere on the other person's body. Simultaneously and independently, Russek and Schwartz conducted similar experiments in which they also showed the registration of one individual's cardiac signal in another's EEG recording when two people sat quietly opposite one another.[23] In a recent publication entitled "Energy Cardiology,"[24] Russek and Schwartz discuss the implications of this finding in the context of what they call a "dynamical energy systems approach" to describing the heart as a prime generator, organizer and integrator of energy in the human body.

The research described here was not designed as a comprehensive, rigorous study to yield results to be subject to statistical analysis, and is not intended to be presented or evaluated as such a study. Rather, we present here a small sampling of results gathered over several years of experimentation that provide intriguing evidence of the exchange of electromagnetic energy produced by the human heart that occurs when two people touch or are in proximity, as well as an experimental protocol that allows such effects to be measured. The results described in this paper are representative examples of the types of data that have been collected from numerous experiments conducted with many different subjects over several years' time. We recognize that these results raise more questions that they answer; and it is our intention that this initial compilation of data might stimulate other interested researchers to pursue the challenge of designing and conducting experiments that further address some of these questions.

Signal Averaging

The measurements presented in this paper were achieved using signal averaging techniques. Signal averaging is a digital procedure for separating a repetitive signal from noise without introducing signal distortion (Figure 2). The superimposition of any number of equal-length epochs, each containing a repeating periodic signal, emphasizes the periodic signal at the expense of irregular variations constituting the noise. The technique was first used in detecting radar signals and was later applied in human physiology to detect and record cerebral cortical responses to sensory stimulation, now known as the cortical evoked potential or event-related potential.[25] The procedure is also used in cardiology to analyze the ECG and is known in this field as micro-potential analysis. In this study, the signal averaging technique was applied to detect signals that were synchronous with the peak of the R-wave of one subject's ECG in recordings of another subject's EEG or body surface.

Method

Subjects were either seated in comfortable, high-back chairs to minimize postural changes or were lying down on a massage table. Prior to each session, subjects were informed of the tasks they were to perform and asked to refrain from talking, falling asleep or engaging in exaggerated body movements. The subjects were carefully monitored to ensure that there were no exaggerated respiratory or postural changes during the session.

Disposable silver/silver chloride electrodes were used for all bipolar ECG measurements. The positive electrode was located on the left side at the sixth rib and the reference was placed in the right supraclavicular fossa. Grass model 7P4 amplifiers were used for ECG amplification and Grass model P5 amplifiers were used for EEG and body surface measurements. The low frequency filters were set at the 1 Hertz setting and the high frequency filters at 35 Hertz. EEG electrodes were attached according to the International 10-20 system; the various recording sites and referencing are specified in each experimental session. Electrode resistance was measured with a UFI model 1089 electrode tester. Electrode to electrode resistance was typically in the range of 2 to 5 K . All data were digitized by a Bio Pac 16 bit digitizer and software system. The sample rate was 256 Hz. All post analysis was done with DADiSP/32 digital signal processing software.

All of the experiments monitored various recording sites on 2 subjects simultaneously. In all experiments, both subjects were wired with ECG electrodes as described above. To clarify the direction in which the signals were analyzed, the subject whose ECG R-wave peak was used as the signal time reference for the signal averaging is referred to as the "signal source," or simply "source." It should be emphasized that the subject designated as the source did not consciously intend to send or transmit a signal. The subject whose EEG or body surface recordings were analyzed for the registration of the source's ECG signal is referred to as the "signal receiver," or simply "receiver." Signal averaging techniques were used to detect the appearance of the source's ECG signal on the surface of the receiver's body at various electrode locations. The resulting waveform appearing on the receiver is referred to as the signal-averaged waveform (SAW). The signal-averaged waveforms were triggered by the peak of the source's ECG R-wave. The number of averages used in the majority of the experiments was 250 ECG cycles or roughly 4 minutes of data.

It is well known that the electrical potential generated by one's heartbeat can be recorded from any site on the body, including the sites recorded by the EEG.[23, 26, 27] (Figure 3). Therefore, in each of our experiments, the possibility had to be considered that the signal appearing in the receiving subject's recordings was the receiver's own ECG rather than that of the other subject designated as the source. Given the signal averaging procedure employed, this would only be possible if the ECG of the source was continually and precisely synchronized with the receiver's ECG. To definitively rule out this unlikely possibility, in all experiments both the source and the receiver's ECG were recorded.

Experimental Examples

Example 1: Holding hands

The purpose of these experiments was to test the hypothesis that when 2 people touch, an exchange of electrical energy produced by their hearts occurs. In the experiments reported on here, 6 subjects were paired in groups of 2. Each pair was monitored on a separate day. The experiment was designed to test for the appearance of the source's ECG signal in the receiver's EEG recording when the subjects were sitting several feet apart and when they held hands but made no other contact. Data were also analyzed to check for the transfer of energy in the reverse direction.

The subjects were seated and fitted with ECG and EEG electrodes. The EEG electrodes were attached to the CZ, C3 and C4 locations on both subjects. The reference for both the C3 and C4 electrode was at the CZ location. The 2 subjects were simultaneously monitored using a 10-minute baseline period during which they were separated by 4 feet, followed by a 5-minute hand holding period. In this experiment subjects were instructed to hold hands and were not instructed to have any specific intention or feeling state.

Signal averaging was used to detect the appearance of the source's ECG signal in the receiver's signal-averaged waveform (SAW) at the various electrode locations. The SAW in the receiver was triggered by the R-wave (250 ECG cycles) of the source's ECG.

Results:

When the subjects were seated 4 feet apart, there was no indication of a transfer of energy between them from the 250 averages used in these experiments. However, when they held hands, the source's ECG could be clearly detected in the receiver's SAW at both the C4 and C3 locations. Figure 4 shows the data from one set of subjects. In this particular subject pair, we were not able to detect an energy transfer in the reverse direction (*i.e.* the receiver's ECG did not appear in the source's SAW).

In all 3 sets of subjects, the ECG of one of the subjects was easily detected in the other's SAW. However, in only one set of these 3 experiments were we able simultaneously to see the effect in both directions.

The data were also analyzed to see if the ECG of the source was synchronized with the receiver's ECG. It was determined that there was no synchronization between the two ECG's, thus confirming that the ECG signal appearing in the receiver's recordings was indeed transmitted from the source's heart rather than the receiver's own. This was true in all the experimental examples which follow.

Example 2: Hand holding orientation

This experiment was designed to determine whether the transfer of cardiac energy, as observed in Example 1, would be affected by changes in the orientation of the subjects' hand holding (*i.e.* source's left hand holding receiver's right hand *vs* source's right hand holding receiver's left hand, etc.). Subjects were seated and fitted with ECG and EEG electrodes. Electrode placement was the same as in Example 1, with the exception that the EEG electrodes were referenced to linked ears. The subjects held hands for 5 minutes in each of the four possible orientations (source's left hand holding receiver's left hand; source's right hand holding receiver's left hand; source's left hand holding receiver's right hand; source's right hand holding receiver's right hand). The recordings were analyzed as in Example 1.

Results:

In the four different hand holding orientations tested, measurable differences were observed in the transfer of cardiac energy between subjects, as measured by the amplitude of the source's ECG signal appearing in the receiver's EEG recording. As seen in Figure 5, the source's ECG appeared with the largest amplitude in the receiver's SAW at the CZ location when the receiver's right hand was held by either the source's left or right hand (top right and bottom right panels). When the receiver's left hand was held by the source's right hand, the source's ECG was could still be detected in the receiver's SAW, but at a somewhat lower amplitude. Finally, when the receiver's left hand was held by the source's left hand, the source's ECG signal was not detected in the receiver's SAW.

Example 3: Wearing glove

This experiment was designed to see if the source's ECG could be picked up on the receiver's arms and to determine whether the signal was being transferred by means of electrical conduction or by radiation. Electrodes were placed 8 inches apart on the reciever's right upper arm and in the standard locations for ECG measurement on both subjects. Once a 5-minute baseline period was recorded, the subjects joined hands and recording continued for the next 5 minutes. The experiment was then repeated with the source wearing a form-fitting full-length latex lab glove.

Results:

The lefthand panel in Figure 6 shows that the source's ECG could be clearly detected on the receiver's right arm when neither subject was wearing a glove. The righthand panel depicts the results when the source was wearing the latex glove. In this case, the source's ECG signal was still present in the receiver's SAW; however it was approximately tenfold lower in signal strength.

Example 4: Light touch

This purpose of this experiment was to determine whether the signal could be transferred by the source lightly touching the receiver's body at different locations. In this experiment, the receiver was lying supine on a padded massage table while the source stood next to the table. Three separate trials were performed: In the first, the source lightly placed his right hand on the receiver's forehead; in the second, he placed his right hand lightly on the receiver's stomach; in the third trial, the source placed one hand on the receiver's forehead and the other on his stomach. Electrodes were placed 4 inches apart on the receiver's left and right lower arms and on the standard locations for ECG measurements on both subjects.

Results:

In all three trials, the source's ECG signal was clearly detectable in the receiver's SAW on both arms; however, the signal measured across the receiver's right arm was consistently 5 times greater in amplitude than the signal picked up on the left arm (Figure 7; note scales).

Example 5: Wired together

This experiment was designed to determine whether the cardiac energy transfer could be increased through forming a hard wire connection between the subjects. The subjects were seated side by side with 18 inches between them. After baseline recordings were established, 5 minutes of data were collected with the subjects wired together. A hard wire connection between subjects was created by placing ECG electrodes on the right side of each subject's rib cage and connecting the electrodes with a 36-inch ECG lead wire. Electrodes were placed 4 inches apart on the receiver's left and right lower arms, 2 inches apart on the receiver's forehead and at the standard locations for ECG measurements on both subjects.

Results:

The source's ECG signal was detected in the receiver's SAW on both arms and on his forehead; however, the amplitude of the transferred signal was not increased with respect to the hand holding or light touch experiments. (Figure 8). As was also observed in the light touch experiments (Example 4), the signal measured across the receiver's right forearm was approximately 5 times greater than that picked up on the left forearm.

Example 6: Proximity without contact

As the cardiac signal is known to be radiated outside the body, in this experiment we sought to determine whether the signal would be detected by the receiver when subjects were not touching. The subjects were seated side by side with 18 inches between them at the closest point. Electrodes were placed 4 inches apart on the receiver's left and right lower arms and on the standard locations for ECG measurements on both subjects. Two thousand averages were used for this experiment (approximately 30 minutes recording time).

Results:

Figure 9 is an overlay plot showing the readings from the electrodes on the receiver's arms and the source's ECG. We were able to detect a signal on the receiver's arms; however, there was a phase shift of 10 ms between the source's ECG and the appearance of the signal across the electrodes on the receiver's arms.

Discussion

The data presented here clearly show that when people touch or are in proximity, a transference of the electromagnetic energy produced by the heart occurs. This energy exchange was evidenced by the registration of one individual's electrocardiogram R-wave peak at different sites on another person's body surface. The transference of the signal appears to depend on the distance between individuals, as would be expected if the signal transferred is electromagnetic in nature. The effect was evident when people were touching or positioned 18 inches apart, but it was not detectable when subjects were separated by a distance of 4 feet and 250 averages were used in the signal averaging process. However, it is quite possible that by measuring longer time periods and using more averages, signal transfer could be detected at greater distances. Russek and Schwartz's measurement of an exchange of cardiac energy between subjects separated by 3 feet certainly supports this possibility.[23] The observation that the signal was still transferred when subjects were not in contact demonstrates that the transference occurs at least to some degree through radiation. However, the tenfold reduction in the amplitude of the transferred signal observed in both the non-contact experiment and in the hand holding trial in which one subject wore an insulated glove suggests that skin-to-skin contact plays an important role in facilitating the signal transfer. Interestingly, forming a hard wire connection between subjects did not increase the amplitude of the transferred signal with respect to the experiments in which subjects simply held hands or touched lightly. The signal amplitude was also unaffected in other experiments (data not shown) in which electrode gel was used to decrease skin-to-skin contact resistance.

There were a number of interesting observations made for which we feel there is not yet sufficient data to attempt to offer an explanation at this point. These include: (1) While in all cases a signal transfer between two subjects was measurable at least in one direction, a transfer was sometimes, but not always, detectable in both directions (*i.e.* In some cases the designated "receiver's" ECG was not observed in the "source's" recordings). From other experiments we have done, this does not appear to be related to the gender of the subjects. (2) Significant differences were observed in the amplitude of the transferred signal depending on the hand holding orientation adopted. The amplitude was highest when the receiver's right hand was held by the source's left hand, and the transfer was not detected at all when subjects held hands left hand to left hand. (3) In the light touch and wired together trials reported on here (Examples 4 and 5), the signal picked up on the receiver's right forearm was consistently 5 times greater in amplitude than the signal registered on the left forearm. This difference was observed in some, but not all, similar experiments performed. (4) In the non-contact experiment (Example 6), but in none of the other trials, a phase shift of 10 ms between the sender's ECG and the appearance of the signal across the receiver's arms was observed. All of these observations pose intriguing research questions and invite additional experimentation to determine whether they do, in fact, represent significant trends to consider in further characterizing this energy exchange.

It should be noted that the appearance of the source's ECG signal in the receiver's EEG does not necessarily indicate that the signal has produced an alteration in the receiver's brainwaves. These data simply indicate that the source's ECG signal can be measured on the receiver's scalp as well as at other sites on the receiver's body surface, such as the forearms and legs. The fact that the signal is indeed registered, however, together with the recent demonstration of nonlinear stochastic resonance effects in several biological systems, certainly raises the possibility that it may exert some effect on the receiving subject's brain and/or other components of the receiver's physiology.

This possibility is in fact supported by experiments conducted by Schandry and co-workers which demonstrated that cortically generated potentials are affected by one's own ECG. These experiments have shown that the registration of one's own ECG R-wave in the EEG is modulated by psychological factors such as attention and motivation, in a fashion analogous to the cortical processing of external stimuli.[26, 28-30] This is also supported by work in our laboratory which has shown that when individuals focus their attention in the area of the heart and consciously generate a positive emotion, the heart rate variability patterns become more orderly and coherent.[17] When a person is in this more coherent state, the portion of the heartbeat evoked potential which reflects cortical processes[28] is dramatically changed.[27] The idea that the registration of another person's ECG across the scalp could also give rise to characteristic cortical potentials is certainly a possibility that deserves further investigation.

A biological response to an externally applied field implies that the field has caused changes in the system greater than those due to random fluctuating events, or "noise." Traditional linear theory predicted that weak, extremely low frequency electromagnetic fields, such as that radiated from the human heart, could not generate enough energy to overcome the thermal noise limit and thus to affect biological tissue. However, a number of experiments have revealed cellular responses to electric field magnitudes far smaller than the theoretical estimates for the minimum field strength required to overcome the thermal noise limit in these systems.[31-33] (cited in [34]). It has been proposed that this discrepancy can in part be accounted for by biological cells' capacity to rectify and essentially signal average weak oscillating electric fields through field-induced variation in the catalytic activity of membrane-associated enzymes or in the conformation of membrane channel proteins.[20, 34] Signal rectification and averaging provide a mechanism by which a signal from an external periodic electric field could be accumulated over time by a cell, and would significantly lower theoretical estimates of the system's threshold of response to external fields, though still not enough to fully explain all the experimental data.

Theoretical estimates of the limitations on the detection of very small signals by sensory systems imposed by the presence of thermal noise (thermal noise limit) were traditionally made using linear approximation under the assumption that the system is in a state of equilibrium.[35] More recently, it has been recognized that a linear and equilibrium approach is not appropriate for biological systems, which are intrinsically nonlinear, nonequilibrium and noisy. The recent advent of the nonlinear stochastic resonance concept[15] has caused further revisions of the theoretical estimates for the minimum field strengths required to affect biological systems. The concept of stochastic resonance was first used in a theoretical study of the ion binding model for the explanation of weak EMF effects on biological systems.[19] The effect of very weak, coherent electromagnetic signals as small as one hundred to one thousand times smaller than the amplitude of the surrounding random noise was studied using numerical simulation. It was shown that coherent signals having an amplitude substantially below that of the background thermal noise could change the mean time it takes for a biological ion to escape from the binding site of a regulatory protein, and thus influence cellular response.[19] Remarkably, in subsequent experimental studies[36-38] the effect of subthermal, coherent signals was observed in different biological systems for signal amplitudes as small as one-tenth or even one-hundredth the amplitude of the random noise component. Whereas initial studies of stochastic resonance in biological systems dealt exclusively with single-frequency signals embedded in a broadband noise background, recent experimental work has shown that stochastic resonance can also be observed with broadband stimuli,[37] thus further generalizing this phenomenon. In addition, a voltage-dependent ion channel system has recently been shown to exhibit stochastic resonance with no detectable response threshold.[38] These data confirm that biological systems under certain circumstances are able to detect arbitrarily small coherent signals. Theory, simulation and experimental data all suggest that nonlinear stochastic resonance may play an important role in the dynamics of sensory neurons,[15, 37, 39] and the demonstration of over a thousand-fold increase in signal transduction across voltage-dependent ion channels induced by the addition of external noise provides evidence that stochastic resonance may also be operative at a sub-cellular level.[36, 38]

Many healing modalities involving contact or proximity between practitioner and patient, including Therapeutic Touch, holoenergetic healing, healing touch, Chi Gong, Reiki, Shiatsu, the Trager technique and polarity therapy, are based upon the assumption that an exchange of energy occurs to facilitate healing. While there exists scientific evidence to substantiate the physiological and psychological effects of many of these treatments, science has as yet not been able to describe a mechanism by which this putative energy exchange between individuals takes place. This study, together with the work of Russek and Schwartz, represents one of the first successful attempts to directly measure an exchange of energy between people. As such, it provides a foundation for a solid, testable theory to explain the observed effects of these healing modalities. We propose that through cellular signal averaging and nonlinear stochastic resonance, a therapist's cardiac field, registered by the patient, may be amplified so as to produce significant effects. As a weak field signal becomes more coherent, the greater its capacity becomes to entrain ambient noise and thus to produce effects in biological tissue. Recent research has shown that the heart's electromagnetic field decreases in electrical coherence as an individual becomes angry or frustrated and increases in coherence as a person shifts to such positive emotional states as sincere love, care or appreciation.[17] Preliminary results indicate, further, that individuals who intentionally increase their cardiac coherence by maintaining a focused state of sincere love or appreciation can induce changes in the structure of water[7] and in the conformational state of DNA.[40] An obvious implication, if the stochastic resonance model is valid, is that the effects of therapeutic techniques involving contact or proximity between practitioner and patient could be amplified by practitioners adopting a sincere caring attitude, and thus introducing increased coherence into their cardiac field.

This may explain why many healing practices have as a core tenet that the therapeutic effects of the treatment are dependent upon the intention of the practitioner to help or heal the patient. The Therapeutic Touch literature describes the role of the practitioner of this technique as attempting "to focus completely on the well-being of the recipient in an act of unconditional love and compassion.[41] It has been demonstrated that hospitalized cardiovascular patients treated with Non-Contact Therapeutic Touch experienced a significantly greater decrease in post-treatment state anxiety than did patients who were administered a control intervention in which nurses mimicked the movements of the Therapeutic Touch technique but did not focus their intention on helping the patients.[8] Of particular relevance to the work described in the present study is Russek and Schwartz's finding that people more accustomed to receiving love and care appear to be better receivers of others' cardiac signals.[23] In a group of subjects in late adulthood, those who in college had rated themselves as having been raised by loving parents exhibited significantly greater registration of an experimenter's cardiac signal in their EEG in a non-contact experiment than those who had rated their parents low in loving. This implies that the exchange of cardiac energy described here may be influenced not only by the degree of coherence of the transmitted signal (which, in turn, can depend on the source's emotional state and intention), but also by the degree of the receiver's receptivity to the signal. Individuals raised in an environment which they perceive to be loving are not only more accustomed to receiving others' love, but also often tend to be more loving themselves. Thus, it is possible that signal registration may be enhanced by increased coherence in the receiver's system. It is not surprising that many of the healing modalities mentioned above emphasize not only that the practitioner have the intention to heal but also that there be a mutually caring relationship between practitioner and patient.

It should also be mentioned that there is an extensive literature concerning nonlocal effects, prayer and distance healing. Larry Dossey has pointed out that the term "energy" as it is used in this paper may not be the appropriate term to describe nonlocal effects, which cannot be explained by conventional electromagnetic theory.[42] We use the term "energy" here, as we believe that the results described in this paper can be explained by conventional electromagnetic theory. This paper does not attempt to explain nonlocal effects; however, it would be interesting to determine whether the effectiveness of nonlocal forms of healing is related to the degree of coherence in the practitioner's cardiac field. Gough and Shacklett[43] as well as Tiller[44] have proposed models which expand and connect conventional electromagnetic theory with an inherently nonlocal and multidimensional realm. Paddison has also written at length concerning the coupling between the electricity generated by the heart and more subtle levels of reality.[45] According to these models, increased coherence in conventional electromagnetic fields would serve to enhance nonlocal effects.

If the electromagnetic field generated by our heart indeed has the capacity to significantly affect those around us, the implications of this would of course extend far beyond healer-patient interactions. It has long been observed that our emotions have the capacity to affect those in our proximity. Evidence that the cardiac field changes with different emotions experienced, combined with the finding that this field is registered physiologically by those around us provides the foundation of one possible mechanism to describe the impact of our emotions on others at a basic physiological level. In addition, if touch, as we have shown, serves to facilitate this exchange of cardiac energy between individuals, this would give new and more precise meaning to the concept of touch as the first and most fundamental means of communication[46] and facilitator of human interactions. Future study of the effects of the electrical exchange that occurs when individuals are in contact or proximity may eventually foster increased awareness of our inner feeling states both in therapeutic interventions and in the broader context of our daily interactions with those in our immediate environment.

Future Directions

These experiments represent an initial attempt to identify and objectively measure an exchange of energy between individuals. The phenomenon highlighted by the results presented here is an intriguing one that has many potential implications and certainly invites further characterization. It is our hope that these data will serve to stimulate critical discussion and encourage interested researchers to pursue further the investigation of the many unanswered questions that have been raised by this work. The repetition of the experiments discussed in this paper with expanded sample sizes will help to distinguish anecdotal observations from real trends and also begin to paint a picture of the variability that exists among individuals with regard to this phenomenon. To continue to characterize this energy exchange, it will be important to refine our understanding of how it varies with distance. More precisely mapping out how transmission of the signal decays with distance will allow us to determine whether there exists an effective "cut-off" point and whether this varies among individuals.

We feel that individual variability in both the transmission and reception of cardiac energy is an important area of investigation that raises a number of questions: Future research might seek to increase our understanding of how one's emotional state affects both energy transmission and reception as well as investigate the role that intention may play in facilitating the energy exchange. In particular, does consciously shifting to a state such as sincere love or appreciation, in which the heart's energy field becomes measurably more coherent, affect signal transference? Also along these lines, does the exchange vary according to the type of relationship people share? Would the signal transference be measurably different in subjects who did not know each other as compared to people who shared a close personal relationship? Finally, studies analyzing the exchange of cardiac energy between individuals in conjunction with the practice of various therapeutic techniques may serve to elucidate any relationships that may exist between this type of energy exchange and the physiological effects of these treatments.

Reference

1. Wirth, D. P. The effect of non-contact therapeutic touch on the healing rate of full thickness dermal wounds. *Subtle Energies* 1990; **1**(1):1-20.
2. Grad, B. Some biological effects of the laying on of hands: review of experiments with animals and plants. *J. Am. Soc. Psychical Res.* 1965; **59**:95-171.
3. Keller, E. Effects of therapeutic touch on tension headache pain. *Nur. Res.* 1986; **35**(2):101-105.
4. Redner, R., Briner, B. and Snellman, L. Effects of a bioenergy healing technique on chronic pain. *Subtle Energies* 1991; **2**(3):43-68.
5. Krieger, D. The response of in vivo human hemoglobin to an active healing therapy by direct laying on of hands. *Human Dimensions* 1972; **1**:12-15.
6. Krieger, D. Healing by the laying on of hands as a facilitator of bio-energetic change: the response of in vivo human hemoglobin. *Psychoener. Sys.* 1974; **1**:121-129.
7. Rein, G. and McCraty, R. Structural changes in water and DNA associated with new physiologically measurable states. *J. Sci. Explor.* 1994; **8**(3):438-439.
8. Quinn, J. Therapeutic touch as an energy exchange: testing the theory. *Adv. Nursing Sci.* 1984; (January):42-49.
9. Freud, S. *The Standard Edition of the Psychological Works of Sigmund Freud.* London: Hogarth Press; 1962: 107-111
10. Krieger, D. *The relationship of touch, with the intent to help or heal, to subjects' in-vivo hemoglobin values: A study in personalized interaction.* In:*Proceedings of the Ninth American Nurses' Association Research Conference.* 1973. New York: American Nurses' Association.
11. Laskow, L. *Healing with love.* New York: HarperCollins; 1992.
12. Tiller, W. A. A gas discharge device for investigating focussed human attention. *J. Sci. Explor.* 1990; **4**(2):255-271.
13. Stroink, G. Principles of cardiomagnetism. In: Williamson S. J., Hoke M., Stroink G., and Kotani M., *Advances in Biomagnetism.* New York: Plenum Press; 1989: 47-57.
14. Green, E., Parks, P., Guyer, P. M., Fahrion, S. and Coyne, L. Anomalous electrostatic phenomena in exceptional subjects. *Subtle Energies* 1991; **2**(3):69-97.

15. Wiesenfeld, K. and Moss, F. Stochastic resonance and the benefits of noise: from ice ages to crayfish and SQUIDs. *Nature* 1995; **373**:33-36.
16. Bulsara, A. R. and Gammaitoni, L. Tuning into noise. *Physics Today* 1996; **March**:39-45.
17. Tiller, W., McCraty, R. and Atkinson, M. Cardiac coherence; A new non-invasive measure of autonomic system order. *Alternative Therapies* 1996; **2**(1):52-65.
18. McCraty, R., Atkinson, M. and Tiller, W. A. New electrophysiological correlates associated with intentional heart focus. *Subtle Energies* 1995; **4**(3):251-268.
19. Poponin, V. Nonlinear stochastic resonance in weak EMF interactions with diamagnetic ions bound within proteins. In: Allen M. J., Cleary S. F., and Sower A. E., *Charge and Field Effects in Biosystems*. New Jersey: World Scientific; 1994: 306-319.
20. Astumian, R. D., Weaver, J. C. and Adair, R. K. Rectification and signal averaging of weak electric fields by biological cells. *Proc. Natl. Acad. Sci. USA* 1995; **92**(3):740-743.
21. Walleczek, J. Field effects on cells of the immune system: the role of calcium signaling. *Fed. Amer. Soc. Exp. Biology* 1992; **6**:3177-3185.
22. Liovitz, T. A., Krause, D. and Mullins, J. M. Effect of coherence time of the applied magnetic field on ornithine decarboxylase activity. *Biochem. Biophys. Res. Com.* 1991; **178**:262-265.
23. Russek, L. and Schwartz, G. Interpersonal Heart-Brain Registration and the Perception of Parental Love: A 42 Year Follow-Up of the Harvard Mastery of Stress Study. *Subtle Energies* 1994; **5**(3):195-208.
24. Russek, L. and Schwartz, G. Energy Cardiology: A Dynamical Energy Systems Approach for Integrating Conventional and Alternative Medicine. *Advances* 1996; **12**(4):4-24.
25. Coles, M., Gratton, G. and Fabini, M. Event-Related Brain Potentials. In: Cacioppo J. and Tassinary L., *Principles of Psychophysiology: Physical, Social and Inferential Elements*. New York: Cambridge University Press; 1990.
26. Schandry, R., Sparrer, B. and Weitkunat, R. From the heart to the brain: a study of heartbeat contingent scalp potentials. *Intern. J. Neuroscience* 1986; **30**:261-275.
27. McCraty, R., Tiller, W. A. and Atkinson, M. *Head-Heart Entrainment: A Preliminary survey.* In:*Integrating the Science and Art of Energy Medicine*. 1995. Boulder, Colorado: ISSSEEM.
28. Schandry, R. and Montoya, P. Event-related brain potentials and the processing of cardiac activity. *Bio. Psychol.* 1996; **42**:75-85.
29. Weitkanut, R. and Schandry, R. Motivation and heartbeat evoked potentials. *J. Psychophysiol.* 1990; **4**:33-40.
30. Montoya, P., Schandry, R. and Muller, A. Heartbeat evoked potentials (HEP): topography and influence of cardiac awareness and focus of attention. *Electroenceph. Clin. Neurophy.* 1993; **88**:163-172.
31. Kalmijn, A. J. Electric and magnetic field detection in elasmobranch fishes. *Science* 1982; **218**(4575):916-918.
32. McLeod, K. J., Lee, R. C. and Ehrlich, H. P. Frequency dependence of electric field modulation of fibroblast protein synthesis. *Science* 1987; **236**(4807):1465-1469.
33. Cleary, S. F., Liu, L. M., Graham, R. and Diegelmann, R. F. Modulation of tendon fibroplasia by exogenous electric currents. *Bioelectromagnetics* 1988; **9**(2):183-194.
34. Weaver, J. C. and Astumian, R. D. The response of living cells to very weak electric fields: the thermal noise limit. *Science* 1990; **247**:459-462.
35. Bialek, W. *Annu. Rev. Biophys. Biophys. Chem.* 1987; **16**:455.
36. Bezrukov, S. M. and Vodyanoy, I. Noise-induced enhancement of signal transduction across voltage-dependent ion channels. *Nature* 1995; **378**:362-364.
37. Levin, J. E. and Miller, J. P. Broadband neural encoding in the cricket cercal sensory system enhanced by stochastic resonance. *Nature* 1996; **380**:165-168.
38. Bezrukov, S. and Vodyanoy, I. Stochastic resonance in non-dynamical systems without repsonse thresholds. *Nature* 1996; **December,** In press.
39. Douglass, J. K., Wilkens, L., Pantazelou, E. and Moss, F. Noise enhancement of information transfer in crayfish mechanoreceptors by stochastic resonance. *Nature* 1993; **365**(6444):337-340.

40. Rein, G. and McCraty, R. *Modulation of DNA by coherent heart frequencies*. In: *Proceedings of the Third Annual Conference of the International Society for the Study of Subtle Energy and Energy Medicine*. 1993. Monterey, CA.
41. Quinn, J. F. and Strelkauskas, A. J. Psychoimmunologic effects of Therapeutic Touch on practitioners and recently bereaved recipients: A pilot study. *Adv. Nursing Sci.* 1993; **15**(4):13-26.
42. Dossey, L. But is it energy? Reflections on consciousness, healing and the new paradigm. *Subtle Energies* 1992; **3**(3):69-82.
43. Gough, W. C. and Shacklett, R. L. The science of connectiveness, Part III: The human experience. *Subtle Energies* 1993; **4**(3):187-214.
44. Tiller, W. A. What are subtle energies? *J. Sci. Explor.* 1993; **7**(3):293-304.
45. Paddison, S. *The Hidden Power of the Heart*. Boulder Creek, CA: Planetary Publications; 1992.
46. Barnett, K. A theoretical construct of the concepts of touch as they relate to nursing. *Nurs Res* 1972; **21**(2):102-110.

Figure 1. Coherent and incoherent ECG spectra. Both of the above graphs are amplitude spectra of 10-second epochs of ECG data. The lefthand graph is an example of the internal coherence mode of heart function. This coherence is associated with sustained, sincere feelings of love and other positive emotions. The graph on the righthand side depicts an incoherent spectrum and is typical of feelings of anger or frustration.

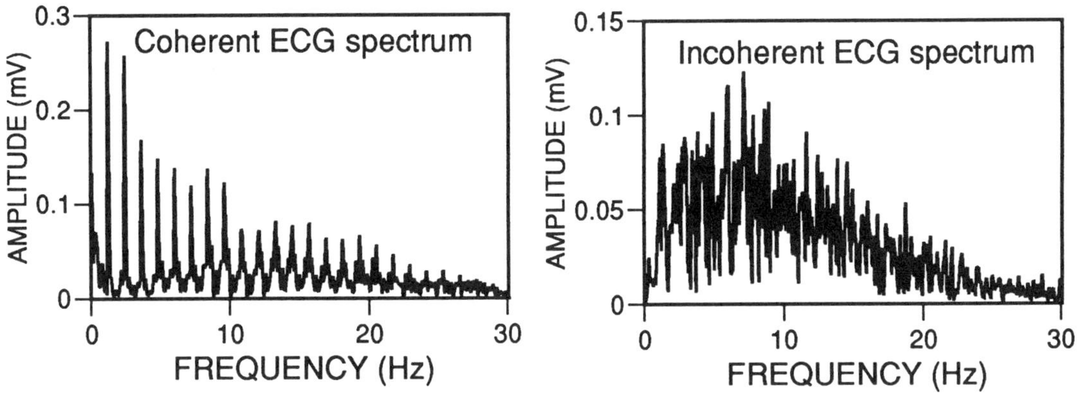

Figure 2. The signal averaging technique. The sequence of the signal averaging procedure is shown above. First, the signals recorded from two subjects are digitized and recorded in a computer. The R-wave (peak) of the ECG recorded from the individual designated as the "signal source" is used as the time reference for cutting the two signals into individual segments. The individual segments are then averaged together to produce the resultant waveforms. Only signals that are repeatedly synchronous with the signal source's ECG are present in the resulting waveform. The signals which are not related to the signal source are eliminated by the process.

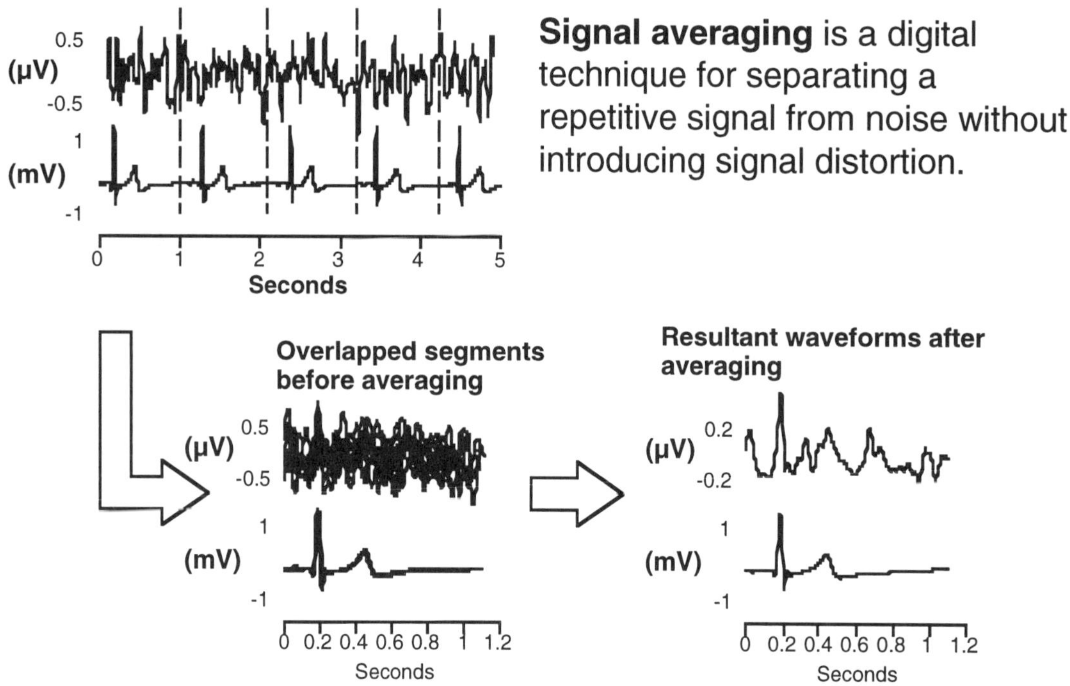

Figure 3. Heartbeat evoked potential. Illustrates an example of the heartbeat evoked potential when a subject's own ECG is used as the signal source. The top trace is the EEG recorded at the CZ location and the middle trace is the blood pressure wave, which was recorded at the earlobe. The signal from the heart arrives at the CZ location around 10 milliseconds after the ECG R-wave and the blood pressure wave arrives around 240 milliseconds latter.

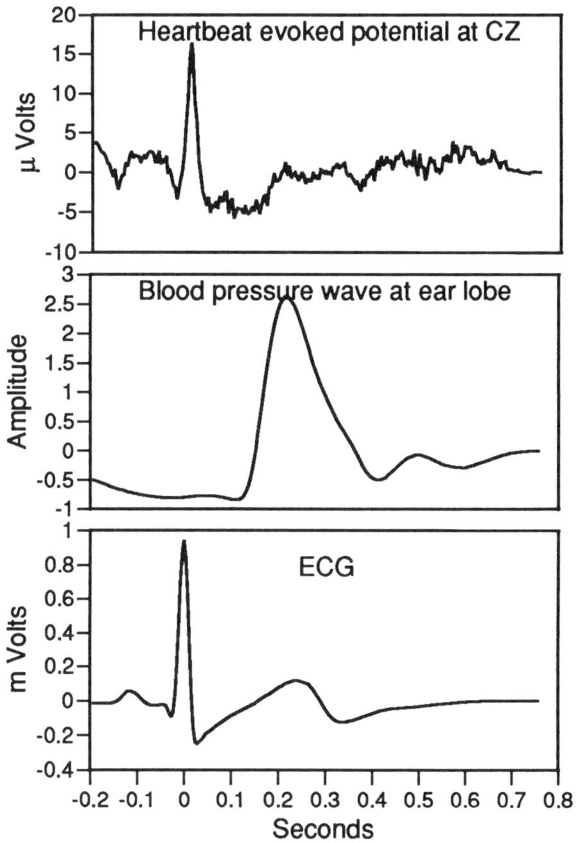

Figure 4. Cardiac signal averaged waveforms before and while holding hands. Signal averaged waveforms showing a transference of the electrical energy generated by the source's heart to the receiving subject's head. The baseline recording (lefthand column) was from a 10-minute period during which the subjects were seated 4 feet apart. The righthand column of panels shows the recording from a 5-minute period during which the subjects held hands. The EEG electrodes on the receiver were placed at the C3 and C4 locations.

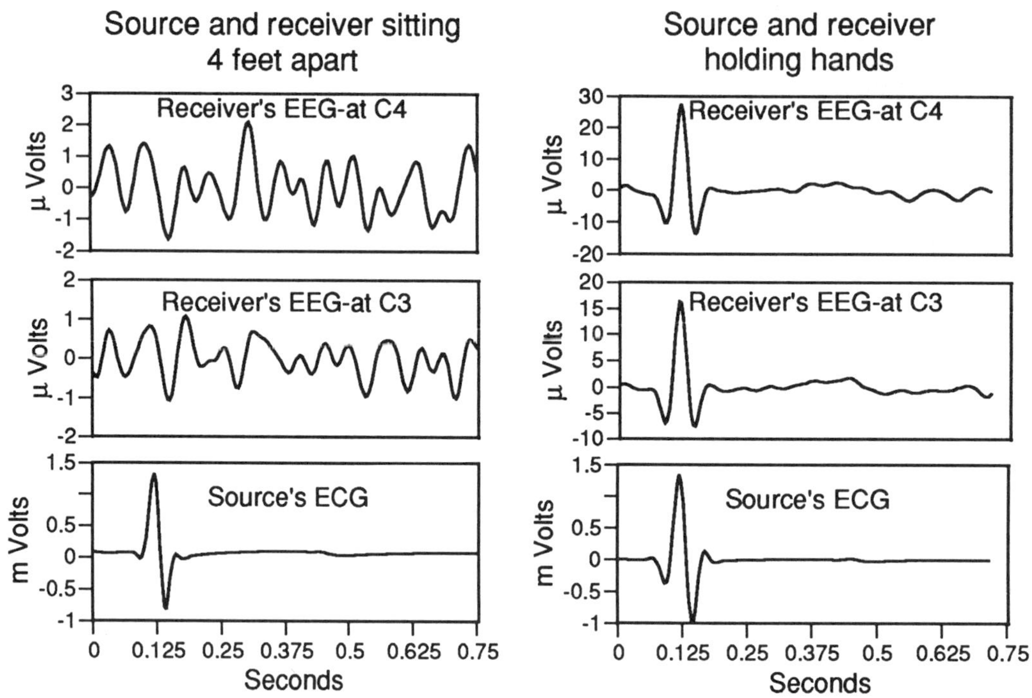

Figure 5. Cardiac signal averaged waveforms with different hand holding orientations. Signal averaged waveforms showing differences in the transference of the electrical energy generated by the source's heart to the receiving subject's head depending on which hand holding orientation was adopted. Subjects held hands for 5 minutes in each of the four orientations shown. Data shown are from the CZ location on the receiver.

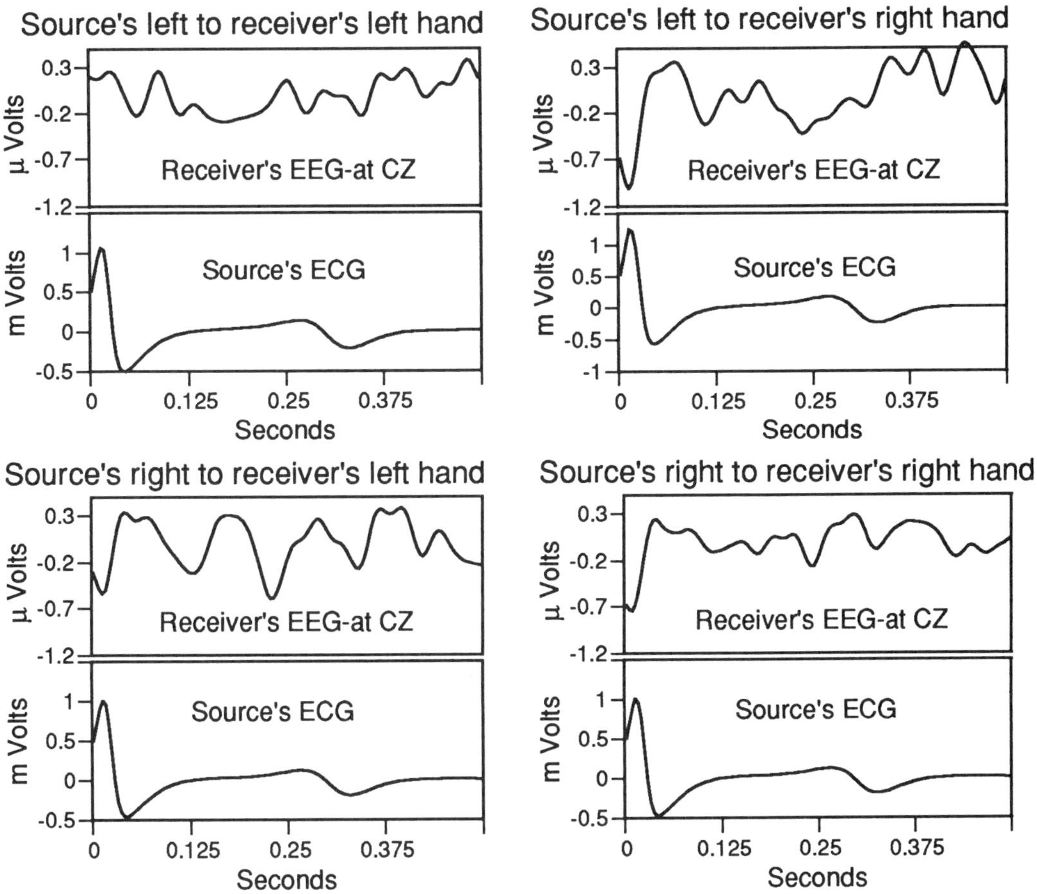

Figure 6. Cardiac signal averaged waveforms: holding hands with and without glove. Illustrates the difference in signal strength measured on the receiver's arm with (righthand panels) and without (lefthand panels) the source wearing a latex glove when they held hands. Note that the source's ECG signal is still present in the receiver's SAW when the glove is worn; however, its amplitude is reduced by a factor of 10 (note scales).

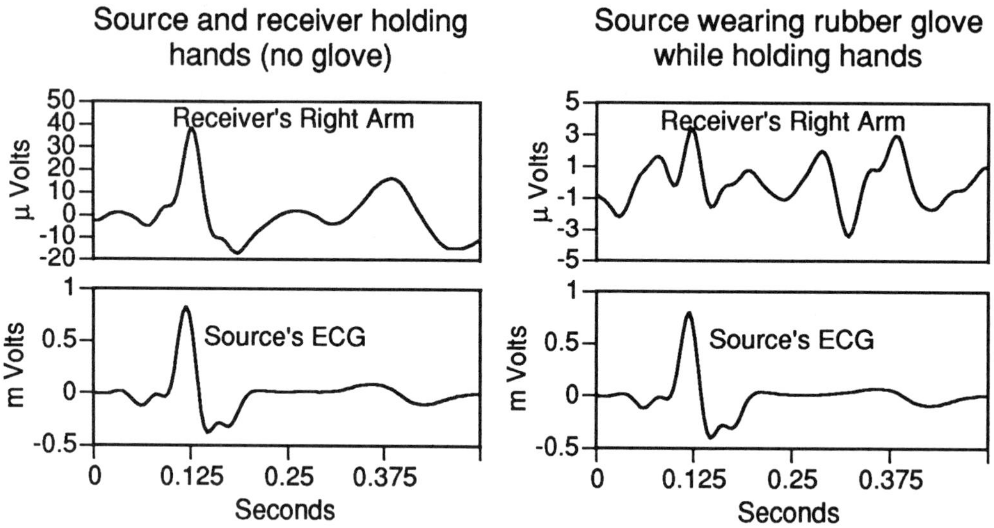

Figure 7. Cardiac signal averaged waveforms: light touch. Signal averaged waveforms showing the transference of the electrical energy generated by the source's heart to the receiving subject's forearms when the source lightly touched the receiver's forehead (lefthand panels), the receiver's stomach (middle panels), or both stomach and forehead (righthand panels).

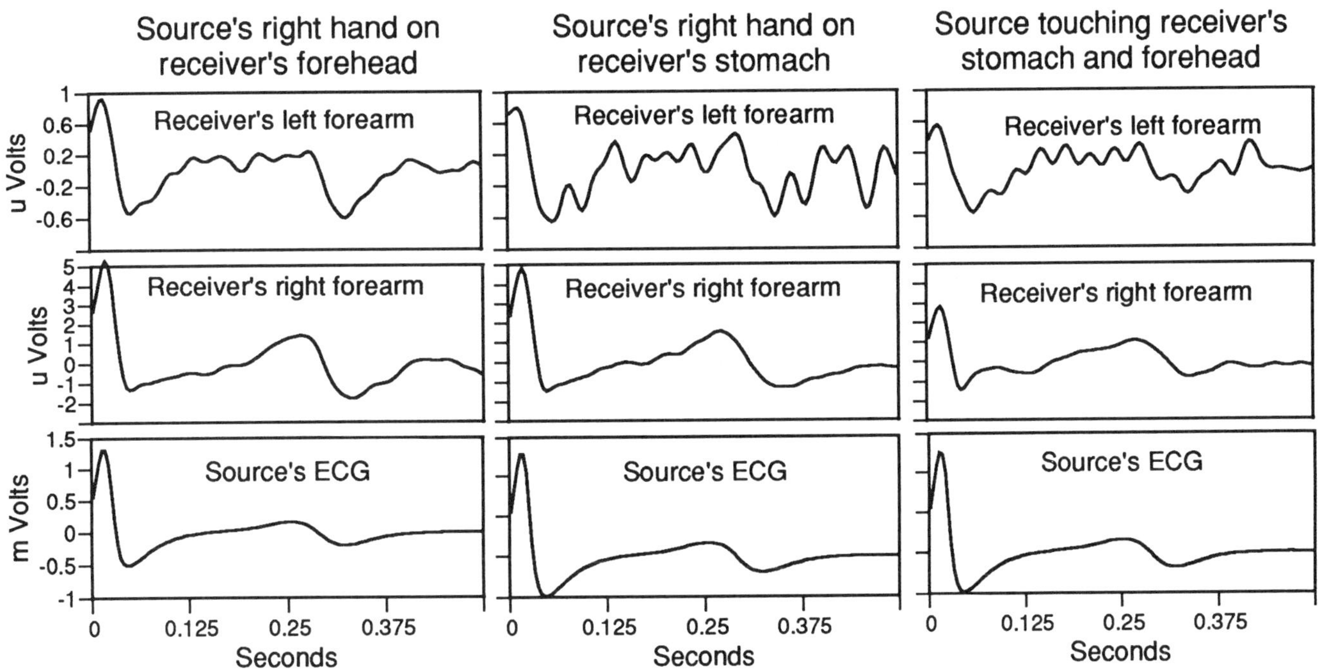

Figure 8. Cardiac signal averaged waveforms: subjects wired together. Signal averaged waveforms showing the transference of energy generated by the source's heart to the receiving subject's forearms and forehead when subjects were wired together right rib to right rib. No increase in the amplitude of the transferred signal was observed with respect to the experiments in which subjects held hands or touched lightly.

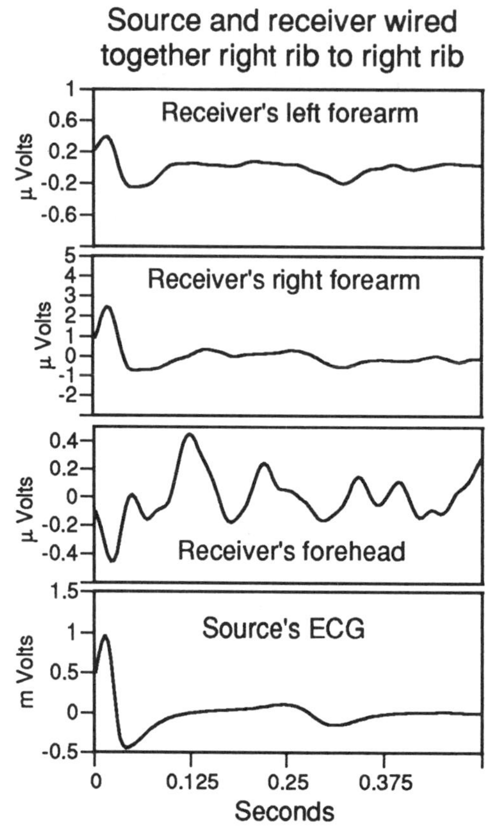

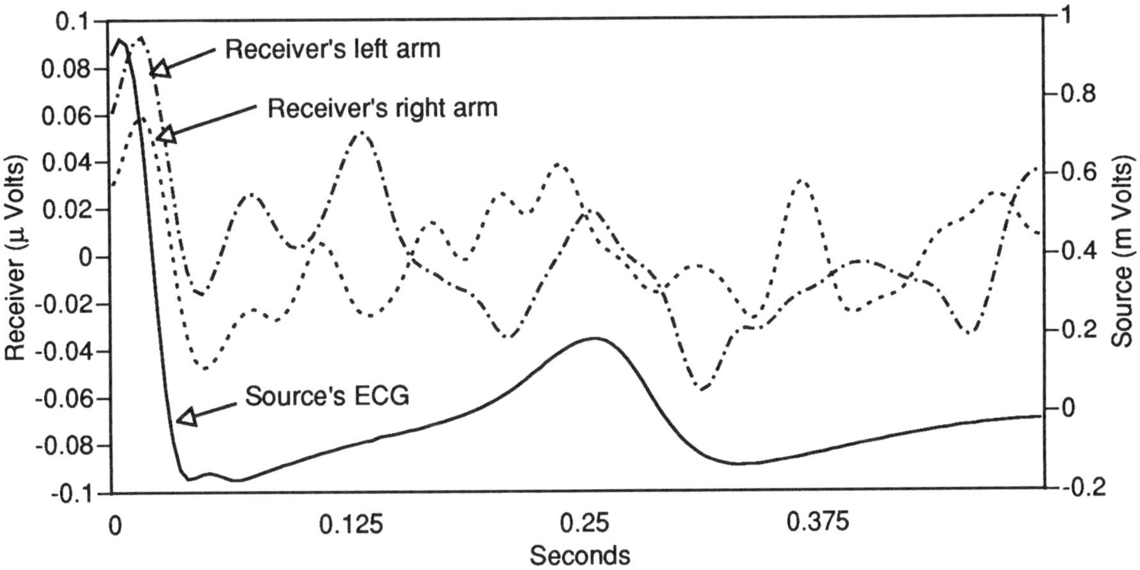

Receiver's left arm
Receiver's right arm
Source's ECG
Receiver (µ Volts)
Source (m Volts)
Seconds
0.1
0.08
0.06
0.04
0.02
0
-0.02
-0.04
-0.06
-0.08
-0.1
1
0.8
0.6
0.4
0.2
0
-0.2
0
0.125
0.25
0.375

15

Readiness For Action

by Lüder Deecke and Wilfried Lang
University Clinic of Neurology,
Währinger Gürtel 18-20, A-1090 Vienna, Austria

Abstract

Readiness for action is expressed in the brain by means of a slow negative DC-potential which is a preparatory facilitation process (synaptic drive) over certain cortical areas that will be involved in the intended volitional act. This 'volition-related potential' has been termed *Bereitschaftspotential* (BP) or readiness potential (Kornhuber & Deecke 1964, 1965). The importance of the availability of the BP is the possibility of making philosophical values such as volition or will accessible to neurophysiological experimentation. As an important result of such research, we postulated from our studies that the supplementary motor area (SMA) participates in the preparation of willful, self-initiated action (Deecke & Kornhuber, 1978). By now this postulate that a certain brain potential is generated in a certain brain area has long been confirmed. The BP or more precisely its early component BP1 was found to be generated by the SMA. Our notion of two components of the BP, the bilaterally-symmetrical BP1 (1.5 to 0.5 sec prior to the onset of voluntary action) and the asymmetrical, i.e. contralaterally dominant BP2 (0,5 to 0 sec prior to movement onset) was as well confirmed by other groups. Our 1978 publication appeared in the same year as Lassen, Ingvar & Skinhøj (1978), who found that not only the primary motor area (MI) but also SMA showed increased rCBF (regional cerebral blood flow) in conjunction with hand movements. With the rCBF method - the subjects are continuously making movements (similar as with the emission CTs or FMRI) - one cannot distinguish whether the SMA is activated before, during or after the movement. The BP, being an entirely pre-movement phenomenon proofs that the SMA activation starts already prior to the onset of voluntary action. The BP amplitude is a function of psychological values such as level of motivation, strength if imagination, intentional engagement, directed attention, inertial as well as emotional load, proactive interference, etc. Using magnetoencephalography (MEG), 3 principal generators of the Bereitschaftsfeld (BF) or readiness field (RF) preceding unilateral movement can be distinguished: (1) the supplementary motor area (SMA) for BF1, (2) the contralateral motor cortex (MI) for BF2 and (3) the ipsilateral MI (Lang, Cheyne, Kristeva, Beisteiner, Lindinger & Deecke 1991). According to recent evidence the cingulate motor area (CMA) may also be involved in BP1 generation. Philosophy of volition had some tradition in the past, at present will is not particularly 'in'. It is hoped that experimental access by natural sciences tools may change the present neglect. Present day philosophy mostly deals with 'retrorolandic matters' (perception, cognition, etc.) but it is important to focus more interest on 'prerolandic' matters, such as motor, motivation, volition, will, planning, providence, precaution, emotionally-engaged anticipation of action, cortically monitored vs. automatic (subconscious) actions, etc.

Introduction

Movement is the essence of life. Without movement there is no survival, without movement literally nothing can be achieved. Quiet thinking of a genius is of limited use if it cannot be communicated to others. Consequently, movement has been optimized to highest levels of precision during phylogenesis. This has been achieved by assigning more and more brain for motor. A carnivore´s motor cortex is just a 'dimple' in its brain supported by merely a minimal frontal cortex. In the primate, the dimple evolved into a motor strip and moved from an anterior to a central position, in order to accommodate more and more frontal brain. Neandertal man with his sloping forehead still had less frontal cortex than Cro Magnon man. In the latter including ourselves, more than half the cortex is pre-rolandic, i.e. assigned to motor in the broader sence, i.e. including areas that supply to motor such as premotor, prefrontal, etc. areas. These *frontal supplier areas to motor* are involved in functions such as motivation, preparation, volition, will, intention, planning, providence, active anticipation, precaution, etc. i.e. all these functions are frontal in location, a special role may be assigned to the frontomesial cortex including the SMA. In short, the brain apparatus supporting motor is among the most highly organized cerebral structures, and the *corticalisation of movement* developed in a saltatory way in primate phylogenesis. There are many movements man can perform, but basically there are only *two* primordial categories of movements: Self-initiated voluntary movements = *actions* and responsive (externally triggered) movements = *re-actions*.

Re-Actions and Actions

Re-Actions

The experimental paradigm for typical re-actions is the CNV paradigm. Walter, Cooper, Aldridge, McCallum & Winter (1964) introduced this classic *reaction time experiment* into slow cerebral potential research, which features a warning stimulus S1 and an imperative stimulus S2 after a second or so which the subject has to *re-act* to. In this condition another slow DC-potential shift occurs which in laboratory terminology was called the 'expectancy wave'. In the publication the less implicating term 'contingent negative variation' (CNV) was preferred, surely inviting associations with the classic pavlovian heritage of the contingent reflex in animal conditioning. Thus, the CNV paradigm is the prototype for movements that are not willful in the strict sense, since they do not require will for initiating them: The starting command comes from the outer world. Certainly, the CNV setting may also require will. For example in a choice reaction time experiment, the decision making process undoubtedly needs will - however, will in decision, not in initiation. Will in initiation, so to say 'will in the time domain', is required in the BP paradigm only. In the CNV paradigm, the individual does not need to care about 'when to do' (i.e. about the right moment to start the movement), the stimulus from the outer world takes care of that. The 'when to do' is the last of three questions related to frontal motor and premotor function, and in the following we will understand, why our group attributes this question to the SMA (Deecke and Kornhuber 1978):

Actions

Our volitional self-initiated acts are preceded by the Bereitschaftspotential (BP) or readiness potential (Kornhuber and Deecke, 1964, 1965). The BP paradigm in contrast to the above CNV paradigm is the voluntary movement paradigm. The subject performs movements or actions in a self-initiated way, i.e. out of his or her own volition. The BP has two components, an early component (BP1) and a late component (BP2, see Fig. 1). The early component BP1 lasts from the very beginning of the BP (1-2 sec or more prior to movement onset depending on the complexity of the movement) to app. ½ sec prior to movement onset and the late component BP2 lasts from ½ sec before to the onset of movement itself (sec 0 in Fig. 1). BP1 is symmetrical even for unilateral movement, while BP2 is larger over the contralateral hemisphere.

Many experiments have been carried out designed at investigating human voluntary, self-initiated movements. Although the time period prior to movement onset (Bereitschaftspotential, BP) can only be investigated using (DC-) EEG and MEG, Emission Tomography (SPECT and PET) and recently FMRI can add topographical information. It must, however, be born in mind that the 'neural signal' recordable by DC-EEG and MEG at an instantaneous time resolution, must not necessarily coincide with the 'cerebrovascular signal' recorded by Emission Tomography and FMRI, which is an activity-dependent vasomotor reflex and accordingly has too poor time resolution than necessary for real *event-related* recording conditions. It is, therefore, of advantage to use a multimethodological approach. Typical examples for experiments in movement using such combination of tools are incorporated in this paper.

DC-EEG

Electroencephalography originally was the only method fast enough to record movement-related cortical events. Attempts to record movement-related potentials date back as early as 1951, when Bates probed into the pre-movement period using a cathode ray oscilloscope on a photographic plate superimposing several movement-related epochs. He did not find any pre-movement potential, for the BP escaped him due to the short time constant (t) necessary for the superposition method. Kornhuber and Deecke (1964, 1965) were lucky enough to select a t of 1.2 sec in order to enable discovery of the BP. Certainly for modern standards this still is too short - attenuating BP amplitude by more than 1/3 - and subsequently we used longer and longer time constants (t = 3, 5, 10, 20, etc.) until since 1988 we are employing only recordings with t = ∞, i.e. true DC-EEG (64-channel computer-assisted DC-EEG amplifier; Lindinger, Svasek, Lang & Deecke 1991).

Fig. 1 gives an example of the BP preceding voluntary right index finger flexions. We see the slow negativity of the BP starting already 1½ sec prior to the onset of movement (movement onset is at sec 0, vertical line). We also see the two components, the BP has. As indicated at the top of Fig. 1, in precentral leads and the vertex we see an early, more shallow negative slope, BP1, and a late, steeper negative slope, BP2. Barrett, Shibasaki & Neshige (1986) reported on three components of the BP, introducing an 'intermediate component' between BP1 and BP2. However, we never found an 'intermediate conponent' and believe that the pre-voluntary movement period is best described postulating simply two principal BP components corresponding to two principal generators. It is our hypothesis supported by many pieces of experimental evidence that the BP1 principal generator is the mesial prefrontal cortex including the supplementary motor area (SMA) and probably also the cingulate motor area (CMA), while the BP2 principal generator is the primary motor cortex (MI, area 4, precentral gyrus, 'motor strip'). The BP as such has an early start of 1 to 2 sec prior to movement onset, for complex movements even up to 3 sec. For simple movements as in Fig. 1, the BP is positive in frontal leads, negative in central and parietal leads as well as at the vertex. According to theory of EEG, negativity can be related to activity of the cortical areas under study, while positivity is related to inactivity (Creutzfeldt 1983).

What is the BP? Certainly a slow negative cortical potential, a negative-going slope of the cortical steady potential, it is a cortical facilitation process, i.e. synaptic drive engaging those cortical areas that are or will be involved in the intended volitional act. Perhaps it could also be described as a functional dendritic field (Pribram 1971, 1977, 1982), albeit a 'macroprocess' rather than the dendritic microprocess described by Pribram (1997). This means the BP is activity from the vast dendritic connection web (Pribram) that imbibes the entire cortex (summation of EPSPs). However, the BP is not the non-specific arousal process known from sleep/wakefulness research to be of reticular origin; it is rather specific and, what is more, *selective* that is restricted to the areas which will be needed for the action under study. Maybe the BP also represents a temporal 'hold', for which Pribram (1997) picks the expectancy wave (CNV) as the example. The BP has some elements in common with the CNV (the polarity and the slowness, i.e. time characteristics). However, while the expectancy wave is *evoked* by stimuli from the outer world, the BP paradigm (voluntary movement paradigm) investigates internally *produced* movement, i.e. performed out of our free will (Deecke, Scheid & Kornhuber 1969, Deecke 1974, Deecke, Grözinger &Kornhuber 1976, Deecke 1987, 1990). For these volitional movements we postulate an 'act of will' necessary to self-initiate the movement (cf. Kornhuber, Deecke, Lang, Lang & Kornhuber 1989). This certainly is a 'frontal act', and the decision of 'when to move' is thought to be elaborated by the mesial fronto-central cortex, containing the SMA. Our hypothesis that SMA has something to do with the start of actions rather than re-actions has been confirmed by PET experiments: Deiber, Passingham, Colebatch, Friston Nixon & Frackowiak (1991) found that SMA activity is greater during the performance of internally-generated rather than externally-cued movements. We now think that not only SMA but also adjacent parts of the cingulate gyrus are activated prior to the onset of voluntary movements and actions ('cingulate motor area', CMA; also cf. Shima, Aya, Mushiake, Inase, Aizawa & Tanji 1991).

This is the situation with unilateral voluntary movements. How is it with bilateral movements? Experiments investigating unilateral versus bilateral movements can tell us something about the temporal order of the moments *when* activation starts in the two principal generators that produce the BP, SMA as the principal generator for BP1 and MI as the principal generator for BP2. When does activation start? There are three possibilities: Do both generators start activation at the same time, as the Cleveland group believes (Ikeda, Lüders, Burgess & Shibasaki 1993), or is MI earlier than SMA, which nobody believes or is SMA activation earlier than MI activation, which we believe (Deecke 1987, Deecke & Lang 1996). From animal experiments we know that in SMA - and markedly less in MI - there exist neurons finring with long latency (app. 1.5 sec) prior to the onset of a self-initiated movement of the monkey (Okano & Tanji 1987). Our hypothesis that SMA is upstream in the final motor cascade when it comes to channeling motivation, intention or the act of will into motor execution is demonstrated in Fig. 2.

In A, the BP preceding simple monophasic bilateral index finger flexions is shown. At the fronto-central midline (vertex, Cz), negativity starts already at -1.1 sec, while at the primary motor cortex (MI, C'3, C'4), the BP starts significantly later, at -0.7 sec (movement onset at 0 sec, vertical line). In B a nice confirmation of such thinking can be seen: Experiments in movement involving *bilateral* simultaneous voluntary self-initiated index finger extensions with different loads were carried out by Cui, Huter, Lindinger, Lang and Deecke (1996). We see an early BP onset over the supplementary motor area (SMA, fronto-central midline, FCz and Cz) at -1.8 sec, and a significantly later BP onset over primary motor cortex (MI, C3 and C4) at -0.8 sec. The generally earlier BP onset times are due to the fact that the loaded movements of Fig. 2B are complex movements as opposed to the simple movements of Fig. 2A. We have shown that BP onset times ('latencies') increase with the level of complexity of the movement, the longest onset times having been observed with speaking (Deecke, Engel, Lang & Kornhuber 1986) or writing and drawing (Schreiber, Lang, Lang, Kornhuber, Heise Keidel, Deecke & Kornhuber 1983).

MEG

Experiments in movement using the magnetoencephalogram (MEG) started in 1982. Generally, all graphoelements of the EEG can also be found in the MEG. Thus, the equivalent of the Bereitschaftspotential BP has been recorded in the MEG and has been termed Bereitschaftsfeld (BF) or readiness field (RF, Deecke, Weinberg & Brickett 1982). The advantage of MEG over EEG is better localization. The BF late component (BF2) has been localized in MI (area 4, Deecke, Boschert, Weinberg & Brickett 1983). It reveals the typical somatotopic organization known from Penfield & Rasmussen (1950), i.e.for movements of different parts of the body, the equivalent current dipoles corresponding to BF2 were distributed in a homuncular pattern (Cheyne, Kristeva & Deecke 1991, from where Fig. 3 is taken).

Fig. 3A shows three-dimensional dipole source locations for 7 different movement conditions foot, wrist, forefinger, thumb, little finger, face and tongue. Sources of neural activity - identified using the non-invasive measurements of cerebral magnetic fields (MEG) - were found to confirm the somatotopic organization of primary motor cortex for movements of different parts of the body in normal human subjects. The somatotopic map produced with this technique revealed slight differences to the classic homunculus obtained from studies using invasive cortical stimulation in epileptic patients (Penfield and Rasmussen, 1950). In Fig. 3B these source locations were projected onto a coronal magnetic resonance image (MRI) taken from the same subject. The typical homuncular distribution of foot-, wrist-, digit-, face- and tongue movements indicated that the generators of BF2/RF2 are indeed localized in area 4. An exception is the dipole for voluntary tongue movements. It showed more variability due to mechanical triggering and other reasons.

So far so good for the MI generator (BF2). MEG localization of the BF1 generator was less readily achieved. In the healthy subject, both SMA generators are active also in case of unilateral movement. Due to the anatomical localization of the SMA on the mesial surface of the hemisphere, the two SMAs face eachother and partially cancel one another. Therefore, we made experiments in movement in a patient with a right SMA lesion (Lang, Cheyne, Kristeva, Beisteiner, Lindinger & Deecke 1991). Unilateral SMA lesion was selected in order to overcome the cancellation problem in MEG of the two opposing SMA dipoles in the normal subject. The patient having only one remaining SMA (the left), performed voluntary flexions with his right index finger. The results were quite convincing: In the early phase of the readiness magnetic field (1.2 to 0.8 sec prior to the onset of movement, corresponding to BP1) field lines were going out of the head at the vertex (Cz) and were going into the head at a frontal position between F3 and Fz (Fig 4, far left). They thus envelopped an electric dipole on the mesial surface of the left hemisphere in the left (intact) SMA. A similar dipole was found during the period 0.8 to 0.6 sec prior to the onset of movement, still corresponding to the BP1-SMA system. However, in the late phase of the readiness field (during BP2), in this case measured between 0.2 and 0 sec prior to movement onset, activity had shifted: Magnetic field lines now came out of the head at FC3 and went into the head at FCz. This indicated an electrical dipole in the left area 4 hand representation (MI). This again supports our hypothesis that the SMA leads the MI motor cortex activation prior to human voluntary movement. We believe that any self-initiated movement is preceded by SMA activity (cf. Deecke, Grözinger & Kornhuber. 1976 and - largely founding on our BP results - Eccles 1982).

Thus, we can state: Activation of SMA/CMA leads activation of MI in time. For this and other reasons we attribute the *starting function* of voluntary action to the SMA. Patients with chronic unilateral lesions of one SMA were examined using a voluntary movement paradigm involving different kinds of bimanual motor sequences (Asenbaum, Oldenkott, Lang, Lindinger & Deecke. 1990).

Clinically, these patients presented with bradykinesia contralateral to the lesion, deceleration of initial movement, switching from sequential performance into simultaneity and frequent failure to initiate or inhibit a movement on one side or the other. The clinical syndrome of SMA lesion is termed *motor dysrhythmia* or *dyschronokinesia.* BP (prior to movement) and negativity of performance (N-P, post movement onset) were larger over the lesioned hemisphere. This seemingly paradoxical result is due to the orientation of the dipole of the remaining SMA on the mesial surface of the hemisphere.

The bradykinesia of these frontal patients resembled the one seen in basal ganglia disorder (Parkinson's Disease, PD). In the latter, the BPs are normal over the SMA but reduced and delayed over MI. On closer look, in PD BP1 is reduced, while a normal BP2 helps the patient to pulling up and at movement onset reaching same amplitudes as the healthy controls - even overshooting after movement onset. The BP changes in PD as compared to healthy controls reveal that postsynaptic changes occur as well as presynaptic changes (dopamine depletion) in PD: Dysfunction of *cortical* structures as a 'hodological' consequence of the dopaminergic deficiency. That is why PD patients have trouble in self-initiation (action) while still being able to respond to external stimuli (re-action; Deecke, Englitz, Kornhuber & Schmitt 1977, Deecke & Kornhuber, 1978, Dick, Rothwell, Day, Cantello, Buruma, Gioux, Benecke, Berardelli, Thompson & Marsden 1989; Harasko-van der Meer, Gerschlager, Lalouschek, Lindinger, Deecke & Lang 1996, Simpson & Khuraibet, 1986, 87). These postsynaptic dysfunctions seem to be partially reversible with L-Dopa substitution (Dick, Rothwell, Day, Cantello, Buruma, Gioux, Benecke, Berardelli, Thompson & Marsden 1989; Feve, Bathien, Rondot 1992). Analyses of BP topography revealed that the contribution of the SMA to the BP was reduced in PD (Dick, Rothwell, Day, Cantello, Buruma, Gioux, Benecke, Berardelli, Thompson & Marsden 1989; Harasko-van der Meer, Gerschlager, Lalouschek, Lindinger, Deecke & Lang 1996; Cunnington, Iansek, Bradshaw & Phillips 1995). PET studies confirmed a diminution of movement-related SMA activation (Playford, Jenkins. Passingham, Nutt, Frackowiak & Brooks 1992; Jenkins, Fernandez, Playford, Lees, Frackowiak, Passingham & Brooks 1992; Rascol, Sabatini, Chollet, Celis, Montastruc, Marc-Vergnes, Rascol 1992). In primate models of PD, it became now clear that the feedback circuits between cortex, basal ganglia, and thalamus are disturbed in this disease (Alexander 1994, Alexander DeLong & Strick, 1986, DeLong 1990). In detail it is the lack of inhibition of the inhibitory action of the Globus pallidum internum (Gpi) onto the excitatory (glutamatergic) thalamo-cortical pathway. This dopaminergic lack of inhibition is the intrinsic pathophysiology of the disease: Parkinson patients are akinetic because they are caught in the state of this enormous Gpi-'hyperinhibition', which in the untreated severe cases downregulates their motor capacity to a complete '*rien ne va plus.*' Internal pallidotomy can markedly improve akinesia (DeLong 1990, Laitinen, Bergenheim & Hariz 1992). Why is it effective? Now we can understand, why: It removes this 'all paralyzing' hyperinhibition from the thalamocortical output circuits. The SMA is one of the major projection areas of these thalamo-cortical circuits receiving even stronger input than MI. It has been shown that post pallidotomy, movement-related SMA activity as assessed by PET is restored in PD patients (Ceballos-Baumann, Obeso, Vitek, DeLong, Bakay, Linazasoro & Brooks 1994; Grafton, Waters, Sutton, Lew & Coudwell 1995). It is likely that the Parkinson-typical changes of the Bereitschaftspotential will also normalize post pallidotomy, which has to be shown yet. However, it is clear that Parkinson akinesia is the result of dysfunction of a complicated feedback circuit system, in which not only the basal ganglia but also the SMA plays an important role.

Emission CT (SPECT)

Experiments in movement have also been carried out using emission CT. Out of the two methods of this family (PET and SPECT), SPECT has been employed. Using the compound HMPAO that shows a trapping effect in brain tissue, rCBF as the one physiological parameter and movement-related DC-potentials as the other were measured in the same 17 normal Ss. A visuomotor tracking (T) paradigm was employed (Lang, Lang, Podreka, Steiner, Uhl, Suess, Müller & Deecke 1988). Trajectorial learning was required in a conflicting situation: a visual target moved on a screen and had to be tracked by moving a light stylus in the right hand on a photo detector plate in an inverted fashion (IT), e.g. movements of the target to the right side required hand movement to the left and vice versa. Compared to a normal non-inverted control task (T), IT required the development of a novel motor program and the prevention of returning to routine direct persuit. These additional demands in IT caused a relative hyperperfusion in regions including the middle frontal gyri, frontomedial cortex (containing the SMA), right basal ganglia, and left cerebellum. Correlations of rCBF values between the middle frontal gyrus and basal ganglia indicated a functional relation between these two brain structures. Trajectorial performance was accompanied by slow negative DC-potential shifts *before* tracking (BP) and *during* tracking (negativity of performance, N-P). In frontolateral and frontomedial (containing the SMA) recordings, amplitudes of DC-negativity were significantly higher in IT than they were in T. What was most striking in the experiment of Lang, Lang, Podreka, Steiner, Uhl, Suess, Müller & Deecke 1988, however, was that this additional frontal negativity covaried with the success in trajectorial learning. These results substantiated, using a dual approach, our previous suggestion that the frontal lobe plays an important role in trajectorial learning in providing a learning memory for novel motor skills. Cognitive skills as relevant for performing music were investigated as well, also emphasizing a role of the SMA in timing.

Other studies dealt with the investigation of motor imagery, that is we are able to see something 'in our mind's eye.' For example athletes make use of motor imagery ('mental rehearsal') before the trial in order to improve performance. Their 'mental representation of motor acts' is so good that they know aforehand whether a stroke for instance in golf will be a hit or a miss. We conducted a topographical study on the mental imagery of motor and other tasks (Goldenberg, Podreka, Uhl, Steiner, Willmes & Deecke 1989; Uhl, Goldenberg, Lang, Lindinger, Steiner & Deecke 1990).

The Imagery Paradigm

An experiment was carried out in which human mental activity was recorded and mapped through scalp electrodes using 12-channel DC-amplifiers (Fig. 4; Uhl, Goldenberg, Lang, Lindinger, Steiner & Deecke 1990).

The Problem

Mental activity usually occurs spontaneously and unpredictably. The recording of cortical DC-potentials requires averaging. How can the averaging process be triggered by an unpredictable event?

The Solution

The S starts the trial by a self-initiated voluntary movement (preceded by a BP), i.e. the S presses buttons of a slide projector presenting himself with a slide. This initiating movement provides the trigger for time-locked potential averaging (at sec 0 in Fig. 2). Instantaneously, a slide appears on the screen in front of the S for 200 msec (ending at sec 0.2). By the contents of the slide, the S was instructed which of the 3 categories - colours, faces or a spatial map - would follow. After 3 sec, a recorded voice spoke out the item that was to be imagined by mental imagery for a duration of 0.2 sec (from sec 3.2 to 4.4 in Fig. 2). Ss were instructed to create a visual image of the item, i.e. to see it in their mind's eye. They were asked to generate this image as soon as possible after they had heard the word and to hold it until the experimenter announced the end of the trial. This was done at irregular intervals in order to prevent the occurence of an expectancy wave (CNV; Walter, Cooper, Aldridge, McCallum & Winter 1964). When taking the measurement of the cortical DC-level between sec 7 and 9, one has a high probability that the cooperative S actually performs the mental task of imagery, the cortical DC-negativity of which can then be recorded and mapped over the scalp. The results showed that the bilateral movement of pressing buttons, occuring at sec 0, was preceded by a BP. This showed the usual central maximum and was bilaterally symmetrical. It was frontally absent, since no learning was involved in this paradigm. However, it was also present in occipital leads, where the BP is usually absent preceding simple finger movement tasks. Thus, already in the distribution of the BP preceding it, the present visual task displayed itself in a modality-specific manner. After the onset of movement, a visual evoked potential in response to the self-presentation of the slide occured. Thereafter, we see a typical expectancy wave (CNV) in anticipation of the loudspeaker stimulus announcing the item to be imagined, followed by the auditory (verbal) evoked potential in response to the item presentation. While the early potential complex seemed to be rather normal, the late components appeared to be quite remarkable: The peak that appeared at sec 4.4 (1.2 sec after the onset of the loudspeaker signal) showed reversal; it was negative over frontal areas (F3 to C4) and positive over posterior areas (P3) at sec 2. Thus, it pointed towards a generator process in deeper structures that is perhaps related to word comprehension and memory storage.

DC-Shifts During Imagery

Since Ss were instructed to immediately start imagining the item announced and maintain the image, the subsequent DC-potential shifts are undoubtedly related to creating and maintaining mental imagery. While over frontal and central areas, negative shifts declined, there was sustained DC-negatvity over the retrorolandic brain, that is only at areas showing sustained negativity we assumed imagery to actually occur. Since frontal and central areas showed large initial DC-negativity it seems reasonable to assume that these areas, in particular frontal ones, participate in the generation of imagery. The maintainance of the imagery finally seems to occur in retrorolandic brain regions only, measurements being taken between sec 7 and 9 (hatched area in Fig. 2). And here we found interesting topographical differences between the 3 categories of imagined items. The level of negativity was almost lowest for faces, medium for colours and highest for the map. It is surprising that imagination of images that complex as faces generated less negativity than did imagining of colours (plain unicoloured sheets). The explanation might be that faces are more imagined in infero-temporobasal areas than are colours and - in particular - the map. Negative DC-shifts in basal cortical areas were not seen by our electrodes arranged over the convexity of the skull, and, what is more, these areas can even inject a positive bias on the DC-shift recorded via the linked ears reference electrodes. The fact that the map caused maximum overall DC-negativity also fits this assumption because now imagery particularly involved areas of the convexity, namely the parietal lobes. The cortical DC-potential (prevailing between sec 7 and sec 9 of the epoch - imagery period) was mapped using current source density mapping algorithms (Laplacian transformation and spline interpolation; Lindinger, Baumgartner, Burgess, Lüders & Deecke 1994). With this display, the 2 experimental conditions involving *visual imagery* (faces and colours) differed from the condition *spatial imagery* (map). The faces′ and colours′ regional DC-negativities were distributed over the occipital cortex and over temporal areas, showing a preponderance over the left hemisphere. For the spatial map the topography was different, regional DC-negativity being distributed occipitally (but somewhat more anterior) and parietally (not temporally). A slight left preponderance was observed also here. We conclude from these observations that visual imagery and spatial imagery differ in their regional DC-negativity pattern, the first having an occipital and temporal distribution the latter having an occipital and parietal topography. Since the same difference was found for visual and spatial perception, we considered our data as subserving further evidence for the postulate that imagery takes place at the same areas that elaborate the *percepts*.

HMPAO-SPECT and Mental Imagery

30 Ss -28 of whom were identical with those of the DC-potential study- participated in the experiments. The distribution of rCBF was assessed during a resting state and during imagining either colours or faces -visual imagery- or a route on a map -spatial imagery- similar as in the DC-potential experiment. Ss were in a supine position, and were blindfolded. An intravenous line was laid to a cubital vein before the experiment. Ss wore earphones connected to a tape recorder, which announced the items to be imagined at a rate of 15 sec. Imagery experiments were compared to resting state sessions during which the tape was silent and Ss were advised to lie relaxed and 'to think of nothing'.

The results showed that - as compared to the resting state - the imagination of faces caused distinctly more regional activation than with any of the other 2 imagery conditions. Statistically, imagining faces was also the only condition that led to an increase of activity in - particularly the left - inferior occipital region which has been suggested by previous studies as being a crucial area for visual imagery (Basso, Bisiach & Luzatti 1980, Goldenberg, Podreka, Uhl, Steiner, Willmes & Deecke 1989; Farah 1984; Goldenberg 1987; Uhl, Goldenberg, Lang, Lindinger, Steiner & Deecke 1990). Also for colours, basal temporal and occipital regions are activated but less so than for faces.

The imagery of routes on a map, on the contrary, showed a tendency towards more activation in visual association areas on the convexity of the brain, particularly in parietal areas. This is another piece of evidence for a 'visual/spatial dichotomy' as suggested by Mishkin, Ungerleider & Macko 1983. We now know that the visual modality has many submodalities such as form, disparity, colour, motion perception and others that are envisaged of being represented in different visual subsystems, some of which occupying distinct areas of the visual association cortex V1 to V7. Mishkin's distinction between *object vision* and *spatial vision* being represented in even 2 separate visual pathways is a particularly intriguing concept which is corroborated partially by the 2 imagery studies reported here. Furthermore, the finding in our SPECT study that imagining faces involves basal cortical areas to a greater extent than does colour and map imagery - the latter rather activating the parietal convexity - can explain the above paradoxical result of our DC-potential study that the imagination of faces yielded less overall negativity than did colours. Since the information content of faces is by orders of magnitude higher than the one of plain unicoloured sheets, this paradox can be explained by the greater degree of 'basality' (inferotemporal, inferooccipital) of faces as compared to colours. It is important to consider that there are large cortical regions which the electroencephalographer does not see. This again calls for the conduction of joint studies. From the SPECT study on imagery we can conclude as well that *imagery* occurs in those brain areas where *perception* takes place.

Experiments in Movement Using FMRI and MEG

Since a few years, magnetic resonance imaging (MRI) goes functional (FMRI, Belliveau, Kennedy, McKinstry, Buchbinder, Weisskoff, Cohen, Vevea, Brady & Rosen 1991). The FMRI technique is another way - in addition to the emission tomography methods - that allows the measurement of activation-related cerebral blood flow changes occuring with specific tasks. However, the spatial relationship between neuronal activity and functional cerebral blood flow changes is not known yet. Therefore, again the combination of tools is required: A study was carried out in order to compare the center of neuronal activation (measured by MEG) with that of the blood flow responses (measured by FMRI). 8 Ss participated in a typical BP paradigm (i.e. voluntary movement paradigm) of tapping with their right index finger (Beisteiner, Gomiscek, Erdler, Teichtmeister, Moser & Deecke 1995a,b). The results show that MEG helps to improve FMRI (Beisteiner, Erdler, Teichtmeister, Diemling, Moser, Edward & Deecke 1997) As the common center of neuronal activation, the mean of the motor and sensory cortex MEG centers was taken. The center of the blood flow response has been incorporated in the images as well. It occured that the 'neuronal signal' (MEG) and the 'cerebrovascular signal' (FMRI) were not at the same location. Rather a difference in topography between the 2 signals was observed. On the average, the two signals were about 1½ cm separate from eachother, which demands for methodological improvements. That means both signals lay in the left MI area but 1½ cm apart from eachother. One source of localization error may be pixels with large signal amplitudes, since these pixels may be expected to stem from larger vessels that may be even remote from the center of neuronal activation (large vessel effect). Using MEG - thought to be very close to the 'neuronal signal' - in addition to the 'cerebrovascular signal' (FMRI) in a simple finger tapping task, such multimethodological approach helped to improve FMRI brain mapping results. It was in favour of this hypothesis that results showed, indeed, a deterioration of FMRI localization quality with increasing signal amplitudes indicating increased contribution of the large vessel effect. Thus, for FMRI evaluation it is recommended to exclude pixels with large signal amplitudes in order to reduce the large vessel effect and to get closer to the neuronal signal which is the important parameter in functional brain topography.

SMA: Supervision of Acquisition of Novel Motor Skills

In order to investigate the cerebral potentials accompanying the performance of voluntary movement sequences over a longer period, Lang , Beisteiner, Lindinger & Deecke 1992 investigated changes of cortical activity when executing learned motor sequences over an epoch of 20 sec. Ss had trained the sequences before. The quality of performance remained constant over the epoch and the duration of the entire experiment. However, the topographies of the DC potentials changed considerably: In parietal recordings the potential remained stable over the 20 sec epoch, over the contralateral MI it declined only slighly. However, over the frontocentral midline (SMA) it declined significanty, reaching baseline at the end of the epoch. This may have to do with a transition from consciously-controlled to more automatic movement execution and it appears that the SMA activity is not immediately associated with processes such as the programming and execution of movement but rather with a placed-over or *supervising control* of these processes. Thus, SMA activity is not only necessary for the intention and initiation of the voluntary act but also for its *supervising control* during execution - if necessary, i.e. mainly during the acquisition of novel tasks. Once the task, albeit temporally complicated, has been learned to the extent that performance is more 'automatic' rather than 'conscious', SMA participation seems to be less and less required. This again shows that SMA is not simply involved in performance control but rather has a role in *supervising control.* Regarding a hierarchy in control for movement, SMA seems to range over MI also in this respect. However, some authors think that the proof of this hierarchical concept is still lacking (cf. Wiesendanger & Wise 1992), although also these authors state that 'in the region of the medial-frontal cortex there is a gradual change from caudal to rostral with the posterior portion being more 'motor', the anterior more 'complex'. We think this is in good agreement with our thinking and would mean that our interpretation of SMA function would mainly belong to the anterior SMA (or pre-SMA of Luppino, Matelli, Camarda & Rizzolatti, 1993). In Wiesendanger's intra cortical motor stimulation experiments, the anterior SMA was 'clearly less excitable, although more sluggish responses may be obtained, dependent on whether the animal is about to move.' Also the cortico-cortical connections were more widespread in the anterior SMA (Wiesendanger, Rouiller, Kazennikov & Perrig 1996, Hummelsheim, Bianchetti, Wiesendanger & Wiesendanger 1988). Luppino, Matelli, Camarda & Rizzolatti (1993) in a detailed hodological analysis also described fundamental differences in connectivity between the two SMA portions, the pre-SMA receiving rich input from areas 46, F5, *anterior* cingulate, and only modest input from PG, PFG and disgranular insula. However, the SMA proper is richly linked with the MI, *posterior* cingulate and receives modest inputs from the PM, and postrolandic areas PC and PE and the granular insula. Again it would be rather the pre-SMA that has the properties which we attribute to the SMA on the basis of our experiments. However, it appears that the cingulate motor area (CMA) also is activated prior to voluntary movement.

SMA's role in bimanual and other interlimb coordination is well documented by our experiments, and there are little objections against the view that the SMA is important for interlimb coordination. Participation of the SMA in motor learning is also well documented. The SMA is not the only area engaged in motor learning, the lateral premotor cortex is as well (electrodes F3 and F4 in EEG, MFG [middle frontal gyrus] of both sides in SPECT) as are basal ganglia and cerebellum.

Conclusion

Readiness for action expresses itself in the brain by the Bereitschaftspotential, whose early component BP1 appears to be largely generated by the SMA. It still remains difficult to define *the* function of the SMA. Neither lesion studies in man and monkey, nor DC-potentials and fields in EEG and MEG in man, nor Emission-CT studies in man nor single unit recordings in monkeys nor direct cortical recordings in epileptic patients have yielded really unequivocal results. The reason is that the function of the SMA is so extremely task-dependent. Already Foerster (1936) has reported on the enormous complexity of this area. On the other hand, our hypothesis that SMA - as a prime function in voluntary self-initiated action - has to do with intention (for review see Kornhuber, Deecke, Lang, Lang & Kornhuber (1989) has never been falsified. On the contrary, piece-by-piece, direct cortical recordings by several groups (Neshige, Lüders & Shibasaki 1988; Ikeda, Lüders, Burgess & Shibasaki 1993, and Rektor, Feve, Buser, Bathien & Lamarche 1994) have surely confirmed our notion. This is now established. While in the beginning, pre-movement activity in the SMA had been denied on the basis of intracranial recordings, more recent studies have confirmed our hypothesis and have ascertained that a BP can, in fact, be recorded in the SMA (Ikeda, Lüders, Burgess & Shibasaki 1993; Rektor, Feve, Buser, Bathien & Lamarche 1994).

Does SMA activity lead in time over MI activity? On the basis of our data this is clearly so: In all our BP studies since the beginning (Kornhuber & Deecke 1964, 1965), BP latency measurements clearly have shown that it is always at the midline (vertex Cz or FCz, at times Pz) where the BP has its earliest onset. As an example for bilateral movement Fig 2B is given here. In agreement with Barrett, Shibasaki & Neshige (1986) , the scalp recorded BP has two principal components, we call them BP1 and BP2 (see above). Barrett, Shibasaki & Neshige (1986) label BP1 as BP and BP2 as NS'. We have reported from the outset that BP1 is bilateral, and what is more: almost bilaterally symmetrical, reflecting the activity of *both* SMAs even prior to unilateral finger or hand movement. The later component, BP2, is strongly lateralized with a *contralateral* preponderance for movements of the upper extremity and an ipsilateral preponderance for movements of the lower extremity, reflecting mainly activity of the MI areas.

The cancellation problem: In view of the fact (now proven by intracranial recordings, Rektor, Feve, Buser, Bathien & Lamarche 1994) that both SMAs are active even preceding unilateral movement and given the stereogeometrical arrangement of the dipoles on the mesial cortical surface between the hemispheres, there should be a large amount of mutual cancellation of dipole strength of the two SMA activations. The cancellation problem may not be a problem in the EEG but it is a problem in the MEG. There are 2 solutions to the cancellation problem. Solution 1: Selecting a patient with only one SMA left, as Lang, Cheyne, Kristeva, Beisteiner, Lindinger & Deecke 1991 did (see above, Fig. 4), which unequivocally showed that SMA mainly contributes to BP1 (or BF1, respectively), while BP2 (BF2) or NS' is mainly generated by the (contralateral) MI. Solution 2: In praxis, cancellation should not be 100% in the healthy subject (Rektor, Feve, Buser, Bathien & Lamarche 1994 in intracerebral recordings report: 'Regarding SMA recording, the amplitudes of RPs to both ipsilateral and contralateral movements were in general similar although the RP tended to be smaller on the side of the performed movement'). Also Emission CT investigations show a little more activation of the *contralateral* SMA over the ipsilateral one. This resultant contralateral dipole moment should be detectable in MEG. Recently, we were able to show this, using the 65 channel whole head MEG system at Vancouver (Deecke, Cheyne, Lang and Weinberg, in preparation). Prior to a self-initiated finger tapping task, BF1 (corresponding to BP1) was best fitted assuming a dipole in the (contralateral) SMA, the *resultant dipole* so to say, while BF2 was best fitted by assuming two dipoles, a strong contralateral and a weak ipsilateral MI dipole. *During* the tapping, two dipoles were prevailing: SMA *and* contralateral MI. These important tapping movements are presently investigated using our new 143-channel MEG system in Vienna.

In the two important invasive investigations (Ikeda, Lüders, Burgess & Shibasaki 1993 using subdural electrodes in 3 epileptic patients and Rektor, Feve, Buser, Bathien & Lamarche 1994 using intracerebral recordings in 13 epileptic patients), both studies attempted to test our hypothesis whether the BP in SMA leads in time over the BP in MI: In Ikeda & coworkers' (1993) paper, this question can only be answered when recordings of both, SMA *and* MI, responsible for the movement under study were carried out at the same time in the same patient. This was only the case in patient 1 of Ikeda & coworkers, in whom foot movements were investigated. Of particular importance are the recordings over the anterior SMA (pre-SMA, according to Luppino, Matelli, Camarda & Rizzolatti 1993). In electrodes A5, B4 and B5, a very early (about 1.8 sec prior to movement onset) BP was recorded for the repetitive task and the BP over MI started much later (around 1 or 0.9 sec). The BP in anterior SMA reached (as the authors measured it, 300 msec prior to movement onset) -18.5 μV! The authors deliberately write that 'subsequently no NS' and MP developed and the negative potentials were sustained after the EMG onset', which can nicely be seen in their Fig. 2. This notion exactly describes what we found in Cz of our DC scalp EEG with repetitive movements, as explained in the present paper, and now confirmed using the current source density technique (Cui, Huter, Lang, Lindinger, Beisteiner & Deecke 1996): 'Earlier onset than MI, no change in slope (i.e.

no NS' or MP), and sustained during the performance of the task ' (because, as we speculate, the timing control properties of the [anterior] SMA are further needed, i.e. as well *during* the repetitive movements for certain tasks that are not 'automatic'.

Of particular interest is the publication by Rektor, Feve, Buser, Bathien & Lamarche 1994. These authors found considerably higher BP amplitudes than Ikeda & coworkers, obviously because they used intracerebral recordings. Rektor & coworkers showed the scalp recordings at Cz, C3 and C4 together with their intracerebral recordings. They also found electrically positive intracranial BPs. Since they recorded with intracerebral electrodes, their positive BPs have to be explained differently from that of Ikeda & coworkers. We know from Sasaki, Gemba & Hashimoto (1981) with their 'transcortical' recordings of the BP in monkeys that the negative potential is restricted to the 1.5 mm gray matter of the cortex itself. In the white matter immediately below it, the potentials are positive. This means for Rektor & coworkers's positive BPs that the recording lead of the electrode was just outside the cortex, i.e. in the white matter. The essentials of Rektor & coworkers' paper are the large BPs in both SMAs for unilateral movement with a trend towards a slightly higher amplitude contralaterally, the large BPs in the contralateral MI, and none or very small ones in the ipsilateral MI (a definite ipsilateral MI-BP can be seen in their Fig. 4 III electrode P'3-2.) We have always stressed from our MEG recordings that the ipsilateral MI dipole is small but it is definitely there (Deecke, Weinberg & Brickett 1982; Kristeva, Cheyne & Deecke 1991), and Ikeda & coworkers 1993 also found small ipsilateral MI sources. Rektor & coworkers (1994) make no distinction between the posterior and the anterior SMA, probably they recorded rather in the posterior SMA. In most recordings our impression is that the BP starts earlier in SMA than in MI. The authors themselves described in their patient 12 that 'the SMA (electrode F) preceded the motor cortex (electrode P) in the case of foot movements by about 600 msec'.

Maybe the activity of the *posterior* SMA is more in parallel with that of MI (more 'motor' according to Wiesendanger & Wise 1992, while anterior SMA is more 'complex') The activity of the *anterior* SMA is as we postulate *upstream* in the motivational cascade with respect to MI. This does not mean a 'supramotor' but simply a premotor function of the anterior or pre-SMA. In PET investigations, the activation in the midline with voluntary movement often appears in a rather deep location. Often it is not clear whether it is above or rather below the cingulate sulcus. If, as Tanji (1994) describes in his review, the dorsal bank of the cingulate sulcus is already the cingulate motor area (CMA), it is very likely that this area is active as well in voluntary movement (cf. also Lüders, personal communication). This would create an upward radial dipole, which cannot be seen by the MEG but could well contribute to the large vertex negative BP in EEG and current source density maps (cf. Cui, Huter, Lang, Lindinger, Beisteiner & Deecke 1996). Thus, it might be that it is the anterior SMA *and* the CMA that are the sources for our DC-EEG findings. Both areas are limbic or iuxtalimbic and therefore may well include functions such as motivation, intention or preparation, i.e. the SMA has - among others - the function of channelling motivation into execution of action.

References

Alexander GE, DeLong MR, Strick PL: Parallel organization of functionally aggregated circuits linking basal ganglia and cortex. Ann Rev Neurosci 9: 357-381 (1986)

Alexander G: Basal ganglia - thalamocortical cortical circuits: Their role in control of movements. J Clin Neurophysiol 11: 420-431 (1994)

Asenbaum S, Oldenkott B, Lang W, Lindinger G, Deecke L: Motor sequences in patients with unilateral SMA lesions: Movement-related potentials and motor disorders. In: Brunia CHM, Gaillard AWK, Kok A (Eds.): Psychophysiological Brain Research. Vol. I, Tilburg University Press, pp. 119-123 (1990)

Barrett G, Shibasaki H,. Neshige R: Cortical potentials preceding voluntary movement: evidence for three periods of preparation in man. Electroenceph Clin Neurophysiol 63(4): 327-339 (1986)

Basso A, Bisiach E, Luzatti C: Loss of mental imagery: a case study. Neuropsychologia 18: 435-442 (1980)

Bates JAV: Electrical activity of the cortex accompanying movements. J Physiol (Lond) 113: 240 (1951)

Beisteiner R, Gomiscek G, Erdler M, Teichtmeister C, Moser E, Deecke L: Funktionelles Magnet Resonanz Imaging (FMRI) - Ergebnisse und Kombinationsmöglichkeiten mit der Magneto- und Elektroenzaphalographie. In: W Lang, L Deecke, HC Hopf (Eds): Topographische Diagnostik des Gehirns. Springer Wien Vh Dt Ges Neurol 9: 189-192 (1995a)

Beisteiner R, Gomiscek G, Erdler M, Teichtmeister C, Moser E, Deecke L: Comparing localization of conventional functional magnetic resonance imaging and Magnetoencephalography. Eur J Neurosci 7:1221-1224 (1995b)

Beisteiner R, Erdler M, Teichtmeister C, Diemling M, Moser E, Edward V, Deecke L: Magnetoencephalography may help to improve functional MR brain mapping. Eur J Neurosci 9: 1072-1077 (1997)

Belliveau JW, Kennedy DN, McKinstry RC, Buchbinder BR, Weisskoff RM, Cohen MS, Vevea JM, Brady TJ, Rosen BR: Functional mapping of the human visual cortex by magnetic resonance imaging. Science 254: 716-719 (1991)

Ceballos-Baumann AO, Obeso JA, Vitek JL, DeLong MR, Bakay R, Linazasoro G, Brooks DJ: Restoration of thalamocortical activity after postero-ventral pallidotomy in Parkinson´s disease. Lancet 344: 814 (1994)

Cheyne D, Kristeva R, Deecke L: Homuncular organization of human motor cortex as indicated by neuromagnetic recordings. Neurosci Lett 122: 17-20 (1991)

Creutzfeldt OD: Cortex cerebri. Springer Berlin, Heidelberg New York, Tokyo pp 145-149 (1983)

Cunnington R, Iansek R, Bradshaw JL, Phillips JG: Movement-related potentials in Parkinson´s disease: Presence and predictability of temporal and spatial cues. Brain 118: 935-950 (1995)

Cui RQ, Huter D, Lang W; Lindinger G, Beisteiner R, Deecke L: Multichannel DC current source density mapping of the Bereitschaftspotential in the supplementary and primary motor area preceding differently loaded movements. Brain Topography 9: 83-94 (1996)

Deecke L, Scheid P, Kornhuber HH: Distribution of readiness potential, pre-motion positivity, and motor potential of the human cerebral cortex preceding voluntary finger movements. Exp Brain Res 7: 158-168 (1969)

Deecke L: Die corticalen Potentiale des Menschen vor raschen willkürlichen Fingerbewegungen. Habilitationsschrift. University of Ulm Library (1974)

Deecke L, Grözinger B, Kornhuber HH: Voluntary finger movement in man: Cerebral potentials and theory. Biol Cybern 23: 99-119 (1976)

Deecke L, Englitz H-G, Kornhuber HH, Schmitt G: Cerebral potentials preceding voluntary movement in patients with bilateral or unilateral Parkinson akinesia. In: Desmedt JE. (Ed): Attention, voluntary contraction, and event-related cerebral potentials. Prog Clin Neurophysiol Vol 1, pp 151-163 (1977)

Deecke L, Kornhuber HH: An electrical sign of participation of the mesial „supplementary" motor cortex in human voluntary finger movement. Brain Res 159: 473-476 (1978)

Deecke L, Weinberg H, Brickett P: Magnetic fields of the human brain accompanying voluntary movement. Bereitschaftsmagnetfeld. Exp Brain Res 48: 144-148 (1982)

Deecke L, Boschert J, Weinberg H, Brickett P: Magnetic fields of the human brain (Bereitschaftsfeld) preceding voluntary foot and toe movements. Exp Brain Res 52: 81-86 (1983)

Deecke L, Engel M, Lang W, Kornhuber HH: Bereitschaftspotential preceding speech after holding breath. Exp Brain Res 65: 219-223 (1986)

Deecke L, Bereitschaftspotential as an indicator of movement preparation in supplementary motor are and motor cortex. In: R. Porter (Chair), Motor Areas of the Cerebral Cortex. Wiley. Chichester, Ciba Found Symp 132: pp 231-250 (1987)

Deecke L: Electrophysiological correlates of movement initiation. Rev Neurol 146/10: 612-619 (1990)

Deecke L, Lang W: Generation of movement-related potentials and fields in the supplementary sensorimotor area and the primary motor area. In: Lüders HO (Ed.) Supplementary Sensorimotor Area - Advances in Neurology. Vol 70 Lippincott/Raven Publishers pp 127-146 (1996)

Deecke L, Cheyne D, Lang W, Weinberg H: Cerebral magnetic fields accompanying voluntary finger tapping. In preparation

Deiber M-P, Passingham RE, Colebatch JG, Friston KJ, Nixon PD, Frackowiak RSJ: Cortical areas and the selection of movement: a study with positron emission tomography. Exp Brain Res 84: 393-402 (1991)

DeLong MR: Primate models of movement disorders of basal ganglia origin. Trends Neurosci 13: 281-285 (1990)

Dick JPR, Rothwell JC, Day BL, Cantello R, Buruma O, Gioux M, Benecke R, Berardelli A, Thompson PD, Marsden CD: The Bereitschaftspotential is abnormal in Parkinson´s disease. Brain 112: 233-244 (1989)

Eccles JC: The initiation of voluntary movements by the supplementary motor area. Arch Psychiat Nervenkr 231: 423-441 (1982)

Farah MJ: The neurological basis of mental imagery: a componential analysis. Cognition 18: 245-272 (1984)

Feve AP, Bathien N, Rondot P: Chronic administration of L-Dopa affects the movement-related cortical potentials of patients with Parkinson´s disease. Clin Neuropharmacol 15: 100-108 (1992)

Foerster O: Motorische Felder und Bahnen. In: O. Bumke, O. Foerster (Hrsg.) Handbuch der Neurologie IV, Berlin Springer (1936)

Grafton ST, Waters C, Sutton J, Lew MF, Coudwell W: Pallidotomy increases activity of motor association cortex in Parkinson´s disease: A positron emission tomographic study. Ann Neurol 37: 776-783 (1995)

Goldenberg G: Neurologische Grundlagen bildlicher Vorstellungen, Springer Wien, New York (1987)

Goldenberg G, Podreka I, Uhl F, Steiner M, Willmes K, Deecke L: Cerebral correlates of imagining colours, faces and a map - I. SPECT of regional cerebral blood flow. Neuropsychologia 27: 1315-1328 (1989)

Harasko-van der Meer C, Gerschlager W, Lalouschek W, Lindinger G, Deecke L, Lang W: Breitschaftspotential preceding onset and termination of a movement is abnormal in Parkinson´s disease. Movement Disorders 11 Suppl 1: 84 (1996)

Hummelsheim H, Bianchetti M, Wiesendanger M, Wiesendanger R: Sensory inputs to the agranular motor fields: a comparison between precentral , supplementary motor and premotor areas in the monkey. Exp Brain Res 69: 289-298 (1988)

Ikeda A, Lüders HO, Burgess RC, Shibasaki H: Movement-related potentials associated with single and repetitive movements recorded from human supplementary motor area. Electroenceph Clin Neurophysiol 89: 269-277 (1993)

Jenkins IH, Fernandez W, Playford ED, Lees AJ, Frackowiak REJ, Passingham RE, Brooks DJ: Impaired activation of the supplementary motor area in Parkinson´s disease is reversed when akinesia is treated with apomorphine. Ann Neurol 32: 749-767 (1992)

Kornhuber HH, Deecke L: Hirnpotentialänderungen beim Menschen vor und nach Willkürbewegungen, dargestellt mit Magnetbandspeicherung und Rückwärtsanalyse. Pflügers Arch 281: 52 (1964)

Kornhuber HH, Deecke L: Hirnpotentialänderungen bei Willkürbewegungen und passiven Bewegungen des Menschen: Bereitschaftspotential und reafferente Potentiale. Pflügers Arch 284: 1-17 (1965) "Citation Classic"

Kornhuber HH, Deecke L, Lang W, Lang M, Kornhuber A: Will, volitional action, attention and cerebral potentials in man: Bereitschaftspotential, performance-related potentials, directed attention potential, EEG spectrum changes. Chapter 6 in: WA Hershberger (Ed) Volitional action, Elsevier (North Holland), Amsterdam, pp 107-168 (1989)

Kristeva R, Deecke L: Cerebral potentials preceding unilateral and simultaneous bilateral finger movements. In: HH Kornhuber, L Deecke (Eds): Motivation, motor and sensory processes of the brain: Electrical potentials, behaviour and clinical use. Progr Brain Res Vol 54 Elsevier Amsterdam pp 748-754 (1980)

Kristeva R, Cheyne D, Deecke L: Neuromagnetic fields accompanying unilateral and bilateral voluntary movements: topography and analysis of cortical sources. Electroenceph Clin Neurophysiol 81: 284-298 (1991)

Laitinen LV, Bergenheim AT, Hariz MI: Leksell´s posteroventral pallidotomy in the treatment of Parkinson´s disease. J Neurosurg 76: 53-61 (1992)

Lang W, Lang M, Podreka I, Steiner M, Uhl F, Suess E, Müller C, Deecke L: DC-potential shifts and regional cerebral blood flow reveal frontal cortex involvement in human visuomotor learning. Exp Brain Res 71: 353-364 (1988)

Lang W, Cheyne D, Kristeva R, Beisteiner R, Lindinger G, Deecke L: Three-dimensional localization of SMA activity preceding voluntary movement. A study of electric and magnetic fields in a patient with infarction of the right supplementary motor area. Exp Brain Res 87: 688-695 (1991)

Lang W, Beisteiner R, Lindinger G, Deecke L: Changes of cortical activity when executing learned motor sequences. Exp Brain Res 89: 435-440 (1992)

Lassen NA, Ingvar DH, Skinhøj, E: Brain function and blood flow. Sci Amer 239: 50-59 (1978)

Lindinger G, Svasek P, Lang W, Deecke L: PC-supported 64-channel DC-EEG amplifier. In: Adlassnig KP, Grabner G, Bengtsson S, Hansen R (Eds) Lecture notes in medical informatics 45. Springer Berlin pp 1005-1009 (1991)

Lindinger G, Baumgartner C, Burgess R, Lüders H, Deecke L: Topographic analysis of epileptic spikes using spherical splines and spline laplacian (CSD). J Neurol 241: 276-277 (1994)

Luppino G, Matelli M, Camarda R, Rizzolatti G: Corticocortical connections of area F3 (SMA proper) and area F6 (preSMA) in the macaque monkey. J Comp Neurol 338: 114-140 (1993)

Mishkin M, Ungerleider LG, Macko KA: Object vision and spatial vision: the two cortical pathways. Trends Neurosci 6: 414-417 (1983)

Neshige R, Lüders H, Shibasaki H: Recording of movement-related potentials from scalp and cortex in man. Brain 111: 719-736 (1988)

Pribram KH: Languages of the brain: Experimental paradoxes and principles in neuropsychology. Englewood Cliffs NJ: Prentice -Hall, 1971; Monterey CA: Brooks/Cole, 1977; New York: Brandon House, 1982

Pribram KH: Thoughts on the meaning of brain electrical activity. Int J Psychol in press 1997 (Proceed Symp 'Neuropsychology of consciousness' Int Congr Psychol Montreal 1996)

Okano K, Tanji J: Neuronal activities in the primate motor fields of the agranular frontal cortex preceding visually triggered and self-paced movement Exp Brain Res 66: 155-166 (1987)

Penfield W, Rasmussen AI: Cerebral cortex of man. A clinical study of localization of function. Mac Millan, New York (1950)

Playford ED, Jenkins IH, Passingham RE, Nutt J, Frackowiak RSJ, Brooks DJ: Impaired mesial frontal and putamen activation in Parkinson´s disease: a positron emission tomography study. Ann Neurol 32: 151-161 (1992)

Rascol O, Sabatini U, Chollet F, Celis P, Montastruc JC, Marc-Vergnes JP, Rascol A: Supplementary and primary sensory motor area activity in Parkinson´s disease. Regional cerebral blood flow changes during finger movements and effects of apomorphine. Arch Neurol 49: 144-148 (1992)

Rektor I, Feve A, Buser P, Bathien N, Lamarche M: Intracerebral recordings of movement-related readiness potentials: an exploration in epileptic patients. Eletroenceph Clin Neurophysiol 90: 273-283 (1994)

Sasaki K, Gemba H, Hashimoto S: Premovement recordings of slow potentials on self-paced hand movements and thalamocortical and corticocortical responses in the monkey. Exp Neurol 72: 41-50 (1981)

Schreiber H, Lang M, Lang W, Kornhuber A, Heise B, Keidel M, Deecke L, Kornhuber HH: Frontal hemispheric differences of the Bereitschaftspotential associated with writing and drawing. Hum Neurobiol 2: 197-202 (1983)

Shima K, Aya K, Mushiake H, Inase M, Aizawa H, Tanji J: Two movement-related foci in the primate cingulate cortex observed in signal-triggered and self-paced forelimb movements. J Neurophysiol 65: 188-202 (1991)

Simpson JA, Khuraibet AJ: Readiness potential of cortical area 6 in Parkinson´s disease. Evidence for a dopaminergic striatal control of postural set involving supplementary motor area. J Neurol Neurosurg Psychiat 49: 475 (1986)

Simpson JA, Khuraibet AJ: Readiness potential of cortical area 6 preceding self paced movement in Parkinson´s disease. J Neurol Neurosurg Psychiat 50: 1184-1191 (1987)

Tanji J: The supplementary motor area in the cerebral cortex. Neurosci Res 19: 251-268 (1994)

Uhl F, Goldenberg G, Lang W, Lindinger G, Steiner M, Deecke L: Cerebral correlates of imanining colours, faces and a map - II. Negative cortical DC-potentials. Neuropsychologia 28: 81-93 (1990)

Walter WG, Cooper R, Aldridge VJ, McCallum WC, Winter AI (1964) Contingent negative variation: An electric sign of sensori-motor association and expectancy in the human brain. Nature 203: 380

Wiesendanger M, Wise SP: Current issues concerning the functional organization of motor cortical areas in nonhuman primates. Adv Neurol 57: 117-134 (1992)

Wiesendanger M, Rouiller EM, Kazennikov O, Perrig S: Is the supplementary motor area a bilaterally organized system? In: Lüders HO (Ed.) Supplementary Sensorimotor Area - Advances in Neurology. Vol 70 Lippincott/Raven Publishers pp 8593 (1996)

Fig. 1: Movement-related potentials in different recording locations of a subject performing flexions of the right index finger (N=1100 self-initiated movements). Bereitschaftspotential (BP) precentro-parietally negative; frontally positive. BP1 (early component of the BP) app. 1½ to ½ sec prior to movement onset. BP2 (late component of the BP) ½ sec before until movement onset (sec 0, vertical line). Pre-motion positivity (PMP, from culmination point 90 to 80 msec prior to movement onset in all centro-parietal leads, type C-subject). Motor potential (MP) in bipolar recording L/R precentral 60 to 50 msec prior to movement onset. Reafferent potentials (RAP) after movement onset. From Deecke 1974

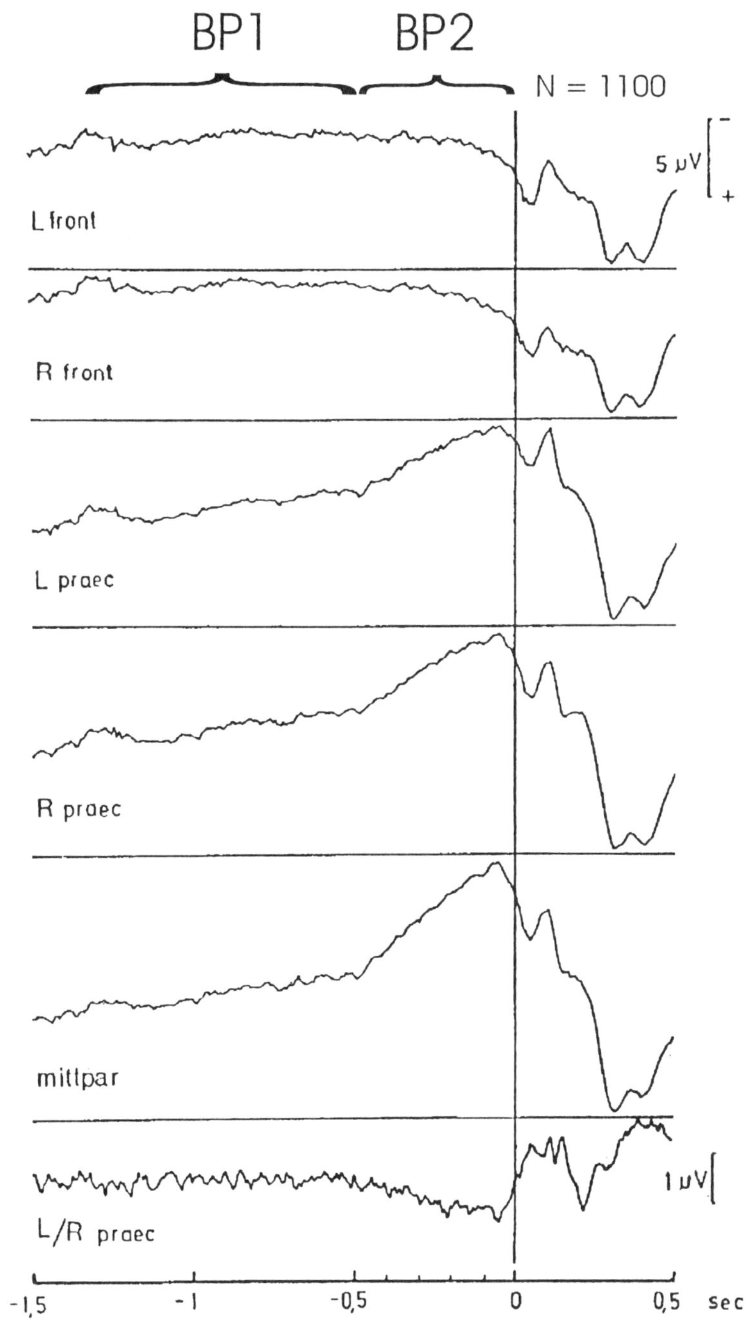

Fig. 2: Movement-related potentials preceding *bilateral* voluntary movements.

A: Cerebral potentials preceding *bilateral* simultaneous simple monophasic index finger flexions. Monophasic means as in all our experiments in movement that the subject performed self-initiated brisk voluntary flexions of the two index fingers that remained in flexed position returning to indifferent position only during ITI (inter trial interval). Bereitschaftspotential (BP) earliest and largest at vertex. Note that at the fronto-central midline (vertex, Cz), negativity starts already at -1.1 sec, while at the primary motor cortex (MI, C'3, C'4), the BP starts significantly later, at -0.7 sec (movement onset as usual at 0 sec, vertical line).128 artifact-freee trials. Monopolar recordings against linked mastoids. C'3 or C'4 1cm anterior of C3 or C4. Detail of Fig. 1 from Kristeva and Deecke, 1980

B: Cerebral potentials preceding *bilateral* simultaneous voluntary self-initiated index finger extensions with different loads. Note similar finding as in A: Early BP onset over the supplementary motor area (SMA, fronto-central midline, FCz and Cz) at -1.8 sec; significantly later BP onset over primary motor cortex (MI, C3 and C4) at -0.8 sec. The generally earlier BP onset times are due to the fact that the loaded movements are complex movements as opposed to the simple movements of A. From experiments of Cui, Huter, Lindinger, Lang and Deecke (1996)

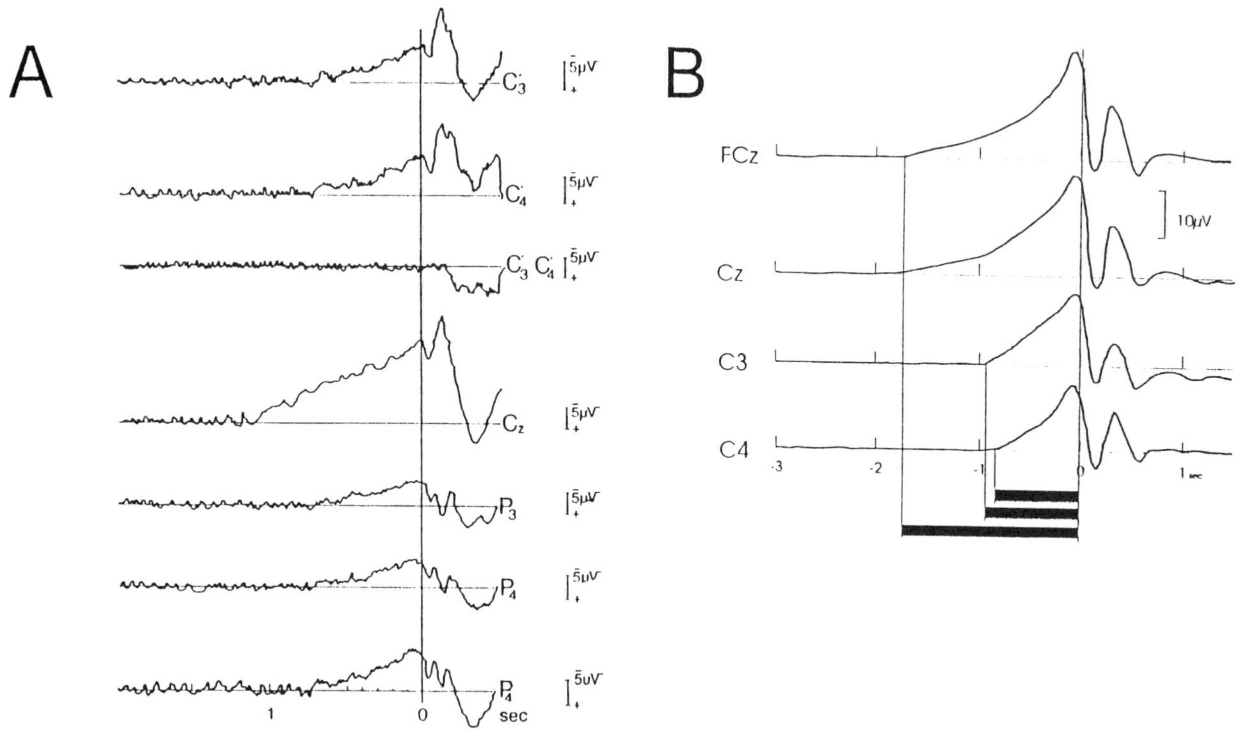

Fig. 3: Homuncular organization of the Bereitschaftsfeld's (BF) or readiness magnetic field's (RF) late component (BF2/RF2).
A: Three-dimensional dipole source locations for 7 different voluntary movement conditions projected onto lateral and posterior views of the subject's head. Head outlines are drawn from digitized points taken over the sagittal and coronal midlines using the 3-D digitizing system also used to produce the the head coordinate system. Dipole locations are indicated by filled circles and are numbered with respect to the voluntary movement performed, and the muscle from which the EMG trigger was recorded as indicated in the legend below (tibialis anterior, etc.).
B: Source locations projected onto a coronal magnetic resonance image (MRI) taken from the same subject. A thin tube filled with contrast agent (gadolinium) was placed over the anatomical landmarks used to establish the head coordinate system (pre-auricular points and nasion) in order to align the dipole coordinate system to that of the MR images. The MR plane shown here lies parallel to the vertical-lateral plane of the head coordinate system at the level indicated by the thin vertical line labelled 'MRI' anterior of the vertical line of the head coordinate system. Note homuncular distribution of foot-, wrist-, digit-, face- and tongue movements indicating that the generators of BF2/RF2 lie in primary motor cortex (area 4). From Cheyne, Kristeva & Deecke, 1991

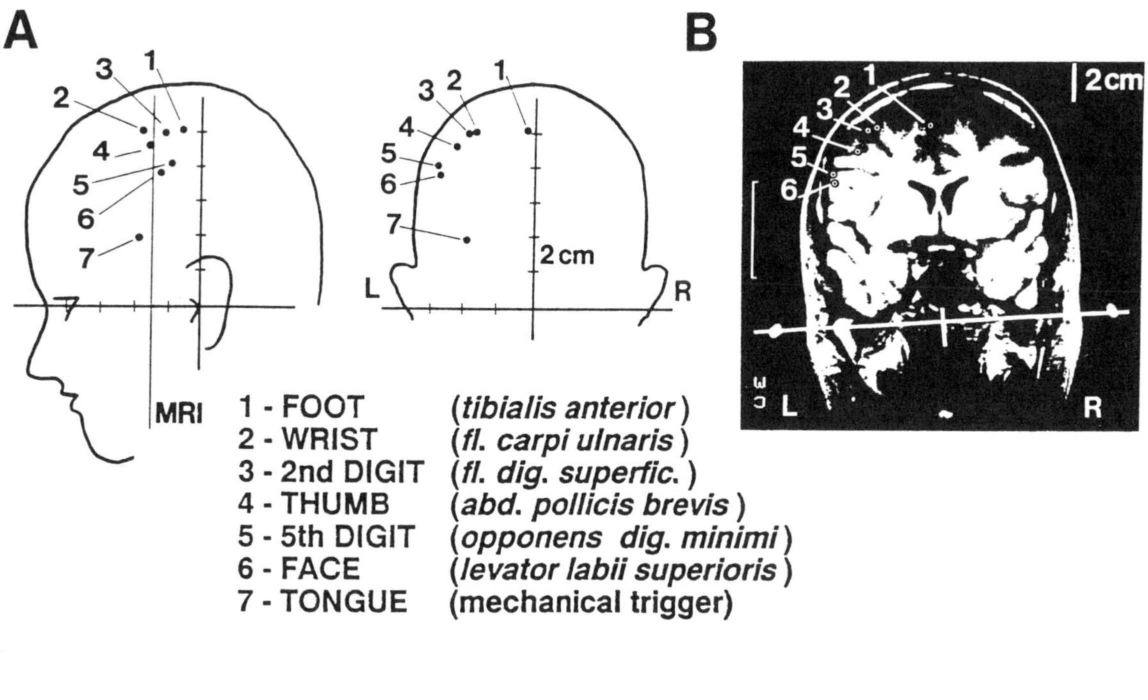

Fig. 4: MEG recordings in three different 200 msec time windows of the foreperiod in a patient with a right hemisphere SMA lesion (cerebral infarction in the territory of the right anterior cerebral artery); he performed brisk flexions of his right index finger in a volitional self-initiated manner.

A: Early phase of the readiness field (BP1 or rather BF1) 1200 to 1000 msec prior to the first EMG activity related to the voluntary contraction. Field lines come out of the head (solid isofield lines) at the vertex Cz, they go into the head (stippled) at the frontal region corresponding to a position F1 (between F3 and Fz). Magnetic field lines envelop an electric dipole situated on the mesial surface of the left hemisphere (see projections of the dipole into the head coordinate system at the bottom). In the center (coronal) head sketch the dipole points to the left, i.e. negative *intra*cellular currents point to the left, which means that negative *extra*cellular currents point to the right. This is consistent with negative activity in superficial cortical layers of the left SMA. Note dipole in head coordinate system at the bottom.

B: Same situation a little later in the foreperiod -800 to -600 msec prior to the onset of movement, still BP1/BF1 period. SMA dipole still active and well-pronounced (very little residual variance). Note dipole remaining in same place as in A in head coordinate system at the bottom.

C: In the late phase of the readiness field BP2/BF2 between -200 and 0 msec prior to the onset of movement, field lines come now out of the head at FC3 (solid isofield lines) and go into the head at FC3 (stippled), corresponding to an electric dipole in the left precentral area (MI hand field). Note completely different dipole location as in A and B in head coordinate system at the bottom. This late dipole is less unequivocal, which probably is due to the fact that the early SMA dipole is not completely silent but still shows some activity. Modified from Lang, Cheyne, Kristeva, Beisteiner, Lindinger & Deecke 1991

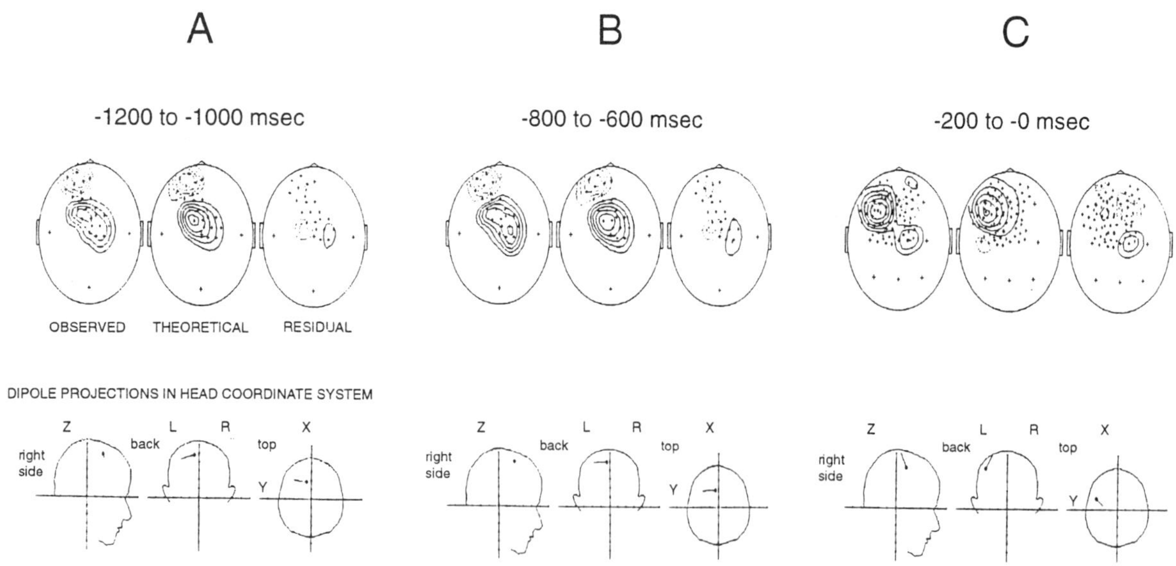

Fig. 5: Experiments in movement imagery.

A: *Imagery paradigm.* Negative DC-shifts (upper trace) were recorded from 4 sec before to 9 sec after the volitional initiation of the trial (subjects presented themselves with a slide by self-initiated button presses with both index fingers simultanously). Center trace, time marker indicating presentation of slide revealing the condition (COLOUR, FACE or MAP). The lower trace depicts the moment of the auditory item presentation (particular politician, or particular colour or route from 1 to 2 on the spatial map). The hatched area marks the time range from sec 7 to 9 of epoch for calculating the mean amplitude of DC-negativity during mental imagery.

B: DC potential waveforms averaged across all 28 subjects for imagining COLOURS (left), FACES (right) and the spatial MAP (bottom). Negativity up. The vertical line represents the moment when subjects initiated the stimulus programme by pressing buttons with both index fingers simultaneously. Note that COLOURS created higher amplitudes (generators on the convexity) than did FACES (generators at the brain base, infratemporal and infraoccipital locations). Highest amplitudes with MAP due to occipito-parietal generators (separate visual pathways for object vision and spatial vision according to Mishkin. Modified from Uhl, Goldenberg, Lang, Lindinger, Steiner & Deecke 1990.

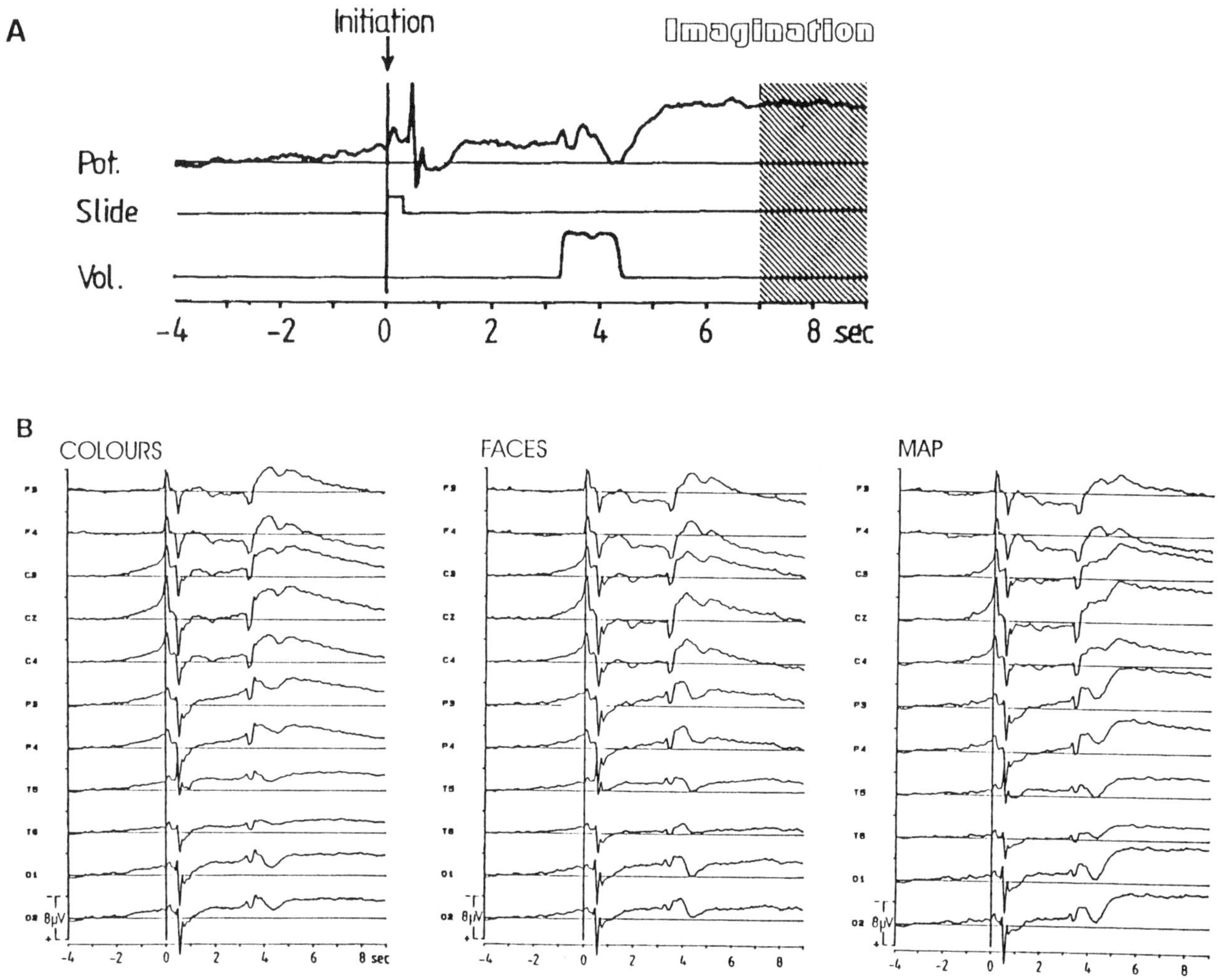

CREATIVITY

16

A Larmarckian Model of Creativity

Robert J. Sternberg
Department of Psychology
Yale University
Box 208205
New Haven, CT 06520

Lamarckian Model of Creativity

People talk about creativity; they care about creativity. But what is creativity? Creativity is the ability to produce work that is both novel (i.e., original, unexpected) and appropriate (i.e., useful, meets task constraints) (Lubart, 1994; Ochse, 1990; Sternberg, 1988a; Sternberg & Lubart 1991, 1995). Creativity is a topic of wide scope that is important at both the individual and societal levels for a wide range of task domains. At an individual level, creativity is relevant, for example, when solving problems on the job and in daily life. At a societal level, creativity can lead to new scientific findings, new movements in art, new inventions, and new social programs. The economic importance of creativity is clear because new products or services create jobs. Furthermore, individuals, organizations, and societies must adapt existing resources to changing task demands to remain competitive.

Creativity Comes into its Own

As the first half of the twentieth century gave way to the second half, Guilford (1950), in his APA Presidential Address, challenged psychologists to pay attention to what he found to be a neglected but extremely important attribute, namely, creativity. Guilford reported that less than two-tenths of one percent of the entries in Psychological Abstracts up to 1950 focused on creativity.

Interest in creativity research began to grow somewhat in the 1950s and a few research institutes concerned with creativity were founded. Today, there is widespread and dynamic interest in creativity, and it is this current interest that motivates this essay.

In this essay, I will briefly review seven approaches to understanding creativity--mystical, pragmatic, psychoanalytic, psychometric, cognitive, social-personality, and confluence. I will then review in more detail an eighth, biological-evolutionary approach. This approach can actually be viewed as consisting of two subapproaches--Darwinian and Lamarckian. Of course, these approaches do not exhaustively cover all possible approaches to creativity, nor is it possible in a chapter even to do full justice to the eight approaches considered. But I believe that the survey of the eight approaches covers some of the major highlights. I will argue that our own approach, the investment theory of creativity--a confluence model--is Lamarckian. Creativity can be Darwinian as well--but typically only when applied to natural evolution.

Mystical Approaches

The study of creativity has always been tinged--some might say tainted--with associations to mystical beliefs. Perhaps the earliest accounts of creativity were based on divine intervention. The creative person was seen as an empty vessel that a divine being would fill with inspiration. The individual would then pour out the inspired ideas, forming an other-worldly product.

In this vein, Plato argued that a poet is able to create only that which the Muse dictates, and even today, people sometimes refer to their own Muse as a source of inspiration. In Plato's view, one person might be inspired to create choral songs, another, epic poems (Rothenberg & Hausman, 1976). Often, mystical sources have been suggested in creators' introspective reports (Ghiselin, 1985).

The mystical approaches to the study of creativity have probably made it harder for scientists to be heard. Many people seem to believe, as they believe for love (see Sternberg, 1988a, 1988b), that creativity is something that just doesn't lend itself to scientific study, because it is a more spiritual process. Pragmatic approaches, like mystical approaches, steer away from science. Instead, they try to provide paths people might take toward enhancing their own creativity.

Pragmatic Approaches

Equally damaging for the scientific study of creativity, in my view, has been the takeover of the field, in the popular mind, by those who follow what might be referred to as a pragmatic approach. Those taking this approach have been concerned primarily with developing creativity, secondarily with understanding it, but almost not at all with testing the validity of their ideas about it.

Perhaps the foremost proponent of this approach is Edward De Bono, whose work on lateral thinking and other aspects of creativity has had what appears to be considerable commercial success (e.g., De Bono, 1971, 1985, 1992). De Bono's concern is not with theory, but with practice. Thus, for example, he suggests using a tool such as PMI to focus on the aspects of an idea that are pluses, minuses, and interesting. Or he suggests using the word po, derived from hypothesis, suppose, possible, and poetry, to provoke rather than judge ideas. Another tool, that of "thinking hats," has individuals metaphorically wear different hats, such as a white hat for data-based thinking, a red hat for intuitive thinking, a black hat for critical thinking, and a green hat for generative thinking, in order to stimulate seeing things from different points of view.

DeBono is not alone in this enterprise. Osborn (1953), based on his experiences in advertising agencies, developed the technique of brainstorming to encourage people to solve problems creatively by seeking many possible solutions in an atmosphere that is constructive rather than critical and inhibitory. Gordon (1961) also attempted to stimulate creative thinking by a method called synectics, which involves primarily analogies.

More recently, authors such as Adams (1974/1986) and von Oech (1983) have suggested that people often construct a series of false beliefs that interfere with creative functioning. For example, some people believe that there is only one «right» answer and that ambiguity must be avoided whenever possible. People can become creative by identifying and removing these mental blocks. Also, von Oech (1986) has suggested that we need to adopt the roles of explorer, artist, judge, and warrior in order to foster our creative productivity.

These approaches have had considerable public visibility, in much the way that Leo Buscaglia has given visibility to the study of love. And they may well be useful. From the present point of view as psychologists, however, the approaches lack any basis in serious psychological theory as well as serious empirical attempts to validate them. Of course, techniques can work in the absence of psychological theory or validation. But the effect of such approaches is often to leave people associating a phenomenon with commercialization, and to see it as less than a serious endeavor for psychological study.

Pragmatic approaches may well have been useful to some, but they have been divorced from psychological theory. In contrast, psychoanalytic approaches are deeply grounded in a general theory of the mind, and especially, the unconscious mind.

Psychodynamic Approaches

The psychodynamic approach can be considered the first of the major twentieth-century theoretical approaches to the study of creativity Based on the idea that creativity arises from the tension between conscious reality and unconscious drives, Freud (1908/1959) proposed that writers and artists produce creative work as a way to express their unconscious wishes in a publicly acceptable fashion. These unconscious wishes may concern power, riches, fame, honor, or love (Vernon, 1970). Case studies of eminent creators, such as Leonardo da Vinci (Freud, 1910/1964), were used to support these ideas.

Later, the psychoanalytic approach introduced the concepts of adaptive regression and elaboration for creativity (Kris, 1952). Adaptive regression, the primary process, refers to the intrusion of unmodulated thoughts in consciousness. Unmodulated thoughts can occur during active problem solving, but often occur during sleep, intoxication from drugs, fantasies or daydreams, or psychoses. Elaboration, the secondary process, refers to the reworking and transformation of primary-process material through reality-oriented, ego-controlled thinking. Other theorists (e.g., Kubie, 1958) have emphasized that the preconscious, which falls between conscious reality and the encrypted unconscious, is the true source of creativity because thoughts are loose and vague but interpretable. In contrast to Freud, Kubie claimed that unconscious conflicts actually have a negative effect on creativity because they lead to fixated, repetitive thoughts. Recent work has recognized the importance of both primary and secondary process (Noy, 1969; Rothenberg, 1979; Suler, 1980; Werner & Kaplan, 1963).

Psychoanalytic approaches sought to highlight the deep-seated unconscious factors that may have given rise to creativity. But they never provided any satisfactory way of measuring the creative phenomena or people they were studying. Psychometric approaches, in contrast, did provide such measurement.

Psychometric Approaches

When we think of creativity, eminent artists or scientists such as Michelangelo or Einstein immediately come to mind. However, these highly creative people are rare and difficult to study in the psychological laboratory. In his APA address, Guilford (1950) noted that the rarity of these individuals had limited research on creativity. He proposed that creativity could be studied in everyday subjects with a psychometric approach, using paper-and-pencil tasks. One of these was the Unusual Uses Test, in which an examinee thinks of as many uses for a common object (such as a brick) as possible. Many researchers adopted Guilford's suggestion and "divergent-thinking" tasks quickly became the main instruments for measuring creative thinking. The tests were a convenient way of comparing people on a standard "creativity" scale.

Building on Guilford's work, Torrance (1974) developed the Torrance Tests of Creative Thinking. These tests consist of several relatively simple verbal and figural tasks that involve divergent thinking plus other problem-solving skills. The tests can be scored for fluency (total number of relevant responses), flexibility (number of different categories of relevant responses), originality (the statistical rarity of the responses), and elaboration (amount of detail in the responses).

An example of a subtest from the Torrance battery is Asking Questions, in which the examinee writes out all the questions he or she can think of, based on a drawing of a scene.

This psychometric revolution for measuring creativity had both positive and negative effects on the field. On the positive side, the tests facilitated research by providing a brief, easy-to-administer, objectively scorable assessment device. Furthermore, research was now possible with "everyday" people (i.e., non-eminent samples). However, there were some negative effects as well. First, some researchers criticized brief paper-and-pencil tests as trivial, inadequate measures of creativity (see essays in Sternberg, 1986); more significant productions, such as actual drawings or writing samples, should be used in addition, or better, instead. Second, other critics suggested that fluency, flexibility, originality, and elaboration scores failed to capture the concept of creativity (see Amabile, 1983). In fact, the definition and criteria for creativity are a matter of ongoing debate, and relying on the objectively-defined statistical rarity of a response with regard to all the responses of a subject population is only one of many options. Other possibilities include using a consensus of judges regarding a product's creativity. Third, some researchers rejected the assumption that non-eminent samples could shed light on eminent levels of creativity, which was the ultimate goal of many studies of creativity. Thus a certain malaise developed and continues to accompany the paper-and-pencil assessment of creativity. Some psychologists, at least, avoided this measurement quagmire in favor of less problematic research topics.

Psychometric approaches provide measurement of creativity, but little understanding of the cognitive processes that give rise to measured individual differences in creativity. Cognitive approaches seek to illuminate such processes.

Cognitive Approaches

Cognitive approaches look at the mental representations and processes underlying creativity. One such approach to creativity has been to view creativity as <u>merely</u> a special case of a more general cognitive phenomenon (e.g., Langley et al., 1987; Weisberg, 1986). By studying, say, perception or memory, one would already be studying the bases of creativity; thus, the study of creativity would merely represent an extension, and perhaps not a very large one, of work that is already being done under another guise. For example, in the cognitive area, creativity was often subsumed under the study of intelligence. I do not argue with the idea that creativity and intelligence are related to each other. However, the subsumption has often been so powerful that researchers such as Wallach and Kogan (1965), among others, have had to write at length on why creativity and intelligence should be viewed as distinct entities.

In more recent cognitive work, Weisberg (1993) proposes that creativity involves essentially ordinary cognitive processes yielding extraordinary products. Using case studies of eminent creators and laboratory research, such as studies with Duncker's (1945) candle problem, which requires subjects to attach a candle to a wall using only objects available in a picture (candle, box of tacks, and book of matches), Weisberg attempts to show that the insights depend on subjects using conventional cognitive processes (such as analogical transfer) applied to knowledge already stored in memory.

Not all cognitive ventures view creativity as ordinary. Approaches that steer away from this view and that are based on the study of human subjects are perhaps prototypically exemplified by the work of Finke, Ward, and Smith (1992) (see also contributions to Smith, Ward, & Finke, 1995; Sternberg & Davidson, 1995). Finke and his colleagues have proposed what they call the Geneplore model, according to which there are two main processing phases in creative thought: a generative phase and an exploratory phase. In the generative phase, an individual constructs mental representations referred to as preinventive structures, which have properties promoting creative discoveries. In the exploratory phase, these properties are used to come up with creative ideas. A number of mental processes may enter into these phases of creative invention, such as retrieval, association, synthesis, transformation, analogical transfer, and categorical reduction (i.e., mentally reducing objects or elements to more primitive categorical descriptions). In a typical experimental test based on the model (see, e.g., Finke, 1990), subjects will be shown parts of objects, such as a circle, a cube, a parallelogram, and a cylinder. On a given trial, three parts will be named, and subjects will be asked to imagine combining the parts to produce a practical object or device. For example, subjects might imagine a tool, a weapon, or a piece of furniture. The objects thus produced are then rated by judges for their practicality and originality.

Computer-simulation approaches, reviewed by Boden (1992), have as their goal the production of creative thought by a computer in a manner that simulates what people do. Langley, Simon, Bradshaw, and Zytgow (1987), for example, developed a set of programs that rediscover basic scientific laws. These computational models rely on heuristics--problem-solving guidelines--for searching a data set or conceptual space and finding hidden relationships between input variables. The initial program, called BACON, uses heuristics such as "if the value of two numerical terms increase together, consider their ratio" to search data for patterns. One of BACON's accomplishments has been to examine observational data available to Kepler on the orbits of planets and to rediscover Kepler's third law of planetary motion. Further programs have extended the search heuristics, the ability to transform data sets, and the ability to reason with qualitative data and scientific concepts. There are also models concerning an artistic domain. For example, Johnson-Laird (1988) developed a jazz improvisation program in which novel deviations from the basic jazz chord sequences are guided by harmonic constraints (or tacit principles of jazz) and random choice when several allowable directions for the improvisation exist.

The cognitive approaches look primarily inside the individual to understand creativity. Social-personality approaches, in contrast, look outside.

Social-Personality Approaches

Developing in parallel with the cognitive approach, work in the social-personality approach has focused on personality variables, motivational variables, and the sociocultural environment as sources of creativity. Researchers such as Amabile (1983), Barron (1968, 1969), Eysenck (1993), Gough (1979), MacKinnon (1965) and others have noted that certain personality traits often characterize creative people. Through correlational studies and research contrasting high- and low-creative samples (at both eminent and everyday levels), a large set of potentially relevant traits has been identified (Barron & Harrington, 1981). These traits include independence of judgment, self-confidence, attraction to complexity, aesthetic orientation, and risk taking.

Proposals regarding self-actualization and creativity can also be considered within the personality tradition. According to Maslow (1968), boldness, courage, freedom, spontaneity, self-acceptance, and other traits lead a person to realize his or her full potential. Rogers (1954) described the tendency toward self-actualization as having motivational force and being promoted by a supportive, evaluation-free environment.

Focusing on motivation for creativity, a number of theorists have hypothesized the relevance of intrinsic motivation (Amabile, 1983; Crutchfield, 1962; Golann, 1962), need for order (Barron, 1963), need for achievement (McClelland, Atkinson, Clark, & Lowell, 1953), and other motives. Amabile (1983; Hennessey & Amabile, 1988) and her colleagues have conducted seminal research on the importance of intrinsic motivation. Studies using motivational training and other techniques have manipulated motivation and observed effects on creative-performance tasks, such as writing poems and making collages.

Finally, the relevance of the social environment to creativity has also been an active area of research. At the societal level, Simonton (1984, 1988, 1994a, 1994b) has conducted numerous studies in which eminent levels of creativity over large spans of time in diverse cultures have been statistically linked to environmental variables such as cultural diversity, war, availability of role models, availability of resources (such as financial support), and number of competitors in a domain. Cross-cultural comparisons (e.g., Lubart, 1990) and anthropological case studies (e.g., Maduro, 1976; Silver, 1981) have demonstrated cultural variability in the expression of creativity. Moreover, they have shown that cultures differ simply in the amount that they value the creative enterprise.

The cognitive and social-personality approaches have each provided valuable insights into creativity. However, if you look for research that investigates both cognitive and social-personality variables at the same time, you will find only a handful of studies. The cognitive work on creativity has tended to ignore or downplay the personality and social system, and the social-personality approaches have tended to have little or nothing to say about the mental representations and processes underlying creativity. Confluence approaches, in contrast, look at cognition in interaction with the sociocultural system.

Confluence Approaches

Many recent works on creativity hypothesize that multiple components must converge for creativity to occur (Amabile, 1983; Csikszentmihalyi, 1988; Gardner, 1993; Gruber, 1989; Lubart, 1994; Mumford & Gustafson, 1988; Perkins, 1981; Simonton, 1988; Sternberg, 1985; Sternberg & Lubart, 1991, 1995; Weisberg, 1993; Woodman & Schoenfeldt, 1989). Sternberg (1985), for example, examined laypersons' and experts' conceptions of the creative person. People's implicit theories contain a combination of cognitive and personality elements, such as "connects ideas," "sees similarities and differences," "has flexibility," "has aesthetic taste," "is unorthodox," "is motivated," "is inquisitive," and "questions societal norms."

At the level of explicit theories, Amabile (1983) describes creativity as the confluence of intrinsic motivation, domain-relevant knowledge and abilities, and creativity-relevant skills. The creativity-relevant skills include (a) a cognitive style that involves coping with complexities and breaking one's mental set during problem solving, (b) knowledge of heuristics for generating novel ideas, such as trying a counterintuitive approach, and (c) a work style characterized by concentrated effort, an ability to set aside problems, and high energy.

Gruber and his colleagues (1981, 1988; Gruber & Davis, 1988) have proposed a developmental evolving-systems model for understanding creativity. A person's purpose, knowledge, and affect grow over time, amplify deviations that an individual encounters, and lead to creative products. Developmental changes in the knowledge system have been documented in cases such as Charles Darwin's thoughts on evolution. Purpose refers to a set of interrelated goals, which also develop and guide an individual's behavior. Finally, the affect or mood system encompasses the influence of joy or frustration on the projects undertaken.

Csikszentmihalyi (1988) takes a different "systems" approach and highlights the interaction of the individual, domain, and field. An individual draws upon information in a domain and transforms or extends it via cognitive processes, personality traits, and motivation. The field, consisting of people who control or influence a domain (e.g., art critics and gallery owners), evaluates and selects new ideas. The domain, a culturally defined symbol system, preserves and transmits creative products to other individuals and future generations. Gardner (1993) has conducted case studies that suggest that the development of creative projects may stem from an anomaly within a system (e.g., tension between competing critics in a field) or moderate asynchronies between the individual, domain, and field (e.g., unusual individual talent for a domain).

All of these approaches have looked at "the system" as social-cultural. But humans are also embedded in a complex biological system, which itself can be used as a basis for understanding creativity.

Biological Approaches

No doubt about it: When it comes to the evolution of organisms, Darwin had it right and Lamarck had it wrong: The evolution of organisms is largely a passive rather than an active process. Darwin believed that acquired traits could not be passed on genetically; Lamarck believed they could be. The evidence is clear: Generations of organisms have to wait around passively until a random mutation just happens to come along that renders one or more of these organisms a better fit to the environment; this mutation can then be passed on to subsequent generations.

Even if one or more organisms manage through active interaction with the environment to find a way to become more adaptive, the change will not be passed on through that organism's genetic material to the organism's offspring. Only genetic mutations can be passed on. Unfortunately, most random mutations decrease rather than increase adaptiveness.

On the one hand, we might bemoan the Darwinian nature of genetic evolution. If acquired traits could be passed on genetically, just think of all the benefits we might be able to behold--people saving on college costs because they are born knowing what their parents learned in college; people being born with cosmetically attractive noses, courtesy of a parent's nose job; some people realizing once and for all that unprotected sex led, well, to their own birth, and could lead to the birth of others like them.

On the other hand, the Darwinian rather than Lamarckian nature of evolution saves us from passing on to our ancestors the genital mutilation of women that is still common in many parts of the world, or the physical deformities that accidents, torture, and careless treatment of the fetus in utero can produce. When all is said and done, we are probably fortunate that genetic evolution is Darwinian. We do not get the benefits, but we are spared the pain, of changes that result from the environment being passed on to subsequent generations.

The ideas of Charles Darwin are so powerful that they have spread into many and diverse fields, including psychology. A whole branch of psychology has, shall we say, "evolved." This field is called evolutionary psychology (see, e.g., Barkow, Cosmides, & Tooby, 1992; Buss, 1995a; Wright, 1994). Much of the initial impetus for this development resulted from the work of Dawkins (1976), who argued that evolutionary theory could be used to explain psychological adaptations. Dawkins was not the first to make this argument (see, e.g., Trivers, 1971), but was perhaps the first to capture the public imagination with his arguments.

The goal of evolutionary psychology is to apply ideas regarding natural evolution--most of them tracing back to Charles Darwin (1859, 1871)--to understanding why humans feel, think, and behave the way they do. Many psychologists--myself included--believe that these ideas have been successful in helping us understand certain aspects of human behavior, such as aspects of physical attraction and mating behavior (Buss, 1994).

Although evolutionary models are difficult to test, they can be tested at least to some extent (e.g.,

Buss, 1995b; Cosmides, 1989), albeit, not always as conclusively as we would like.

When a set of ideas proves useful in understanding or even generating various kinds of phenomena, it is tempting to apply the set of ideas as broadly as possible. Eventually, a field, or at least a major portion of it, may be captured by a new paradigm that looks at old phenomena in a new way (Kuhn, 1962). This evolution of paradigms may itself be seen as Darwinian in a certain sense, in that only the "fittest" of the new ideas survive. Of course, such as statement comes dangerously close to tautology, in which the fitness of ideas may be used, more or less, to explain why the ideas are fit.

The danger of any paradigm is that it will be applied beyond the domain in which it best applies. Many psychologists, including myself, believe that this is what happened with concepts of instrumental conditioning, such as those expressed in the ideas of Skinner (1974). Ideas that well explained simple learning phenomena were less than adequate when applied to, say, complex language learning (Chomsky, 1959). Darwinian ideas, like any other ideas, may be overextended, leading to poorer rather than improved understanding of certain phenomena. One such phenomenon may be human creativity. In particular, the random trial-and-error type of model that does indeed work well for certain aspects of the simple learning that takes place in instrumental conditioning does not work well for human creativity.

A Darwinian Approach to Human Creativity

Several investigators have suggested evolutionary interpretations of human creativity, whereby Darwinian ideas about natural selection are applied to the creative process. The pioneer in the application of evolutionary ideas to human creativity was Donald Campbell (1960), who in a classic article in the Psychological Review, suggested that creative thought could be understood in large part as the result of mechanisms of blind variation and selective retention.

Other investigators, such as Findlay and Lumsden (1988), Perkins (1995), and Simonton (1995), have greatly expanded upon Campbell's original work, but have heeded Campbell's call to apply Darwinian thinking to the understanding of human creativity. Of course, all of these recent investigators have had much more to say about human creativity than is expressed in their neo-Darwinian work. However, the focus of this chapter will be on that particular aspect of theirs and others' work.

The basic idea underlying all of these theories was expressed by Campbell (1960) in three postulates (p. 380), which are paraphrased here. First, a blind-variation-and-selective-retention process is fundamental to all inductive achievements (including creative ones). Second, even additional processes that shortcut a full blind-variation-and-selective-retention process are in themselves inductive achievements. Third, these additional shortcut processes therefore contain in their own operation a blind-variation-and-selective-retention process at some level.

The basic idea in such a model is that creative ideas come about through a process of selective retention of ideas originated through random generation. Such a process seems, on its face, to be implausible, and hence much of Campbell's essay (as well as the essays of some others) is devoted to explanations of how such a model could generate the phenomena we observe in the world that pertain to human creativity. It is not possible nor is it appropriate to reiterate here all of the objections and replies. But consider just one, that regarding individual differences.

How could a blind-variation-and-selective-retention model account for the large, some might even say, stunning individual differences we observe across individuals in their levels of expressed creativity? If we compare highly creative thinkers such as Einstein, Picasso, and Darwin himself to ordinary thinkers, how could the observed differences between the highly creative and the apparently ordinary thinkers be explained?

For one thing, we may have, as humans, "a tendency to see marvelous achievements rooted in equally marvelous antecedents" (Campbell, 1960, p. 390). In other words, we may be guilty of inappropriately utilizing the representativeness heuristic (Kahneman & Tversky, 1971), whereby we expect causes to resemble their effects. Whether or not we are guilty of such potentially maladaptive heuristics, however, does not directly bear on whether human creativity can be explained, in part, in terms of blind variation and selective retention. Campbell suggests several mechanisms by which individual differences might arise.

First, thinkers may differ in the accuracy and detail of their mental representations of the external world, or of possible locomotions through these mental representations. Second, thinkers may differ in the number and range of variations in thoughts that they produce. More numerous and varied thoughts would be associated with greater creative success. Third, thinkers may differ in the selective criteria they apply in weeding out ideas that, for one reason or another, are inadequate.

Neither Campbell nor anyone else brings any data directly to bear upon the question of whether the Darwinian models they propose actually fit whatever data may be available about the nature of human creativity. Most probably, they do not discuss specific data because there are none that they see as directly testing the model. Rather, their arguments are more of the kind found in somewhat speculative thought pieces, where a point of view is proposed, and existing data are shown to be consistent with, although not necessarily directly supportive of, the proposed point of view.

It may be, however, that there are data that could be brought to bear on these models. I believe there are, and that they are overwhelmingly at odds with the proposed Darwinian models. These data derive from the study of expert-novice differences in performance in many domains of inquiry (as summarized, for example, by Chi, Glaser, & Farr, 1988; and by Ericsson & Smith, 1991).

There are many differences between experts and novices that have emerged from multiple studies of expertise (Sternberg, 1996a), among which is the strong trend among experts to try no more paths to solution of a problem than do novices, but to try better paths (Holyoak, 1990). This difference is particularly obvious in the realm of chess, where the number of alternative paths is unforgivingly large, and where human working memory will allow only the exploration of a few of these paths in the actual playing of a game (Charness, 1991). In few situations is there a more clear need to balance knowledge and search. Moreover, in chess, as in other competitive games, winning means coming up with a solution that is sufficiently novel and high in quality (which is to say, creative--Sternberg & Lubart, 1995, 1996) that the opponent is at a loss as to how appropriately to respond to the challenge of the move that has been made.

How can we tell whether expert chess players, or scientists or writers, for that matter, really generate possible courses of action at random, or whether instead they use a guided search that is anything but random? The problem is that, when one writes in a speculative mode, almost any behavior could be "explained" post hoc as the result of a very extensive random search. In the case of chess, however, we may quite by accident have a test of whether the blind-variation model works.

Most strong AI programs that play chess use a method called "brute force," which means that they generate moves at random from all of the potential moves that are possible, and then are highly selective about which of the generated moves they actually make. In other words, they operate in a "blind-variation-and-selective-retention" mode. The strongest programs are strong largely through their having available extremely powerful algorithms that enable them to test many millions of possible moves in the time it would take a human to test one or two. Brute-force program have now reached the point where, through sheer speed of working through an extremely large solution space (Newell & Simon, 1972), they can beat all but the very best human players.

A case in point is the recent matchup between "Deep Thought," an AI program constructed by computer scientists at IBM and generally considered to be the most powerful chess-playing program available, and Gary Kasparov, generally considered to be the leading human chess player in the world today. Deep Thought actually won the first game, and drew several other games. It eventually lost the match, however. Kasparov explained after the tournament that he used the first game to figure out Deep Thought's strengths and weaknesses, and then devised strategies that would be novel to Deep Thought and thwart, to the extent possible, its programming. Thus, no standard path of play was likely to be able to beat Deep Thought, which could take into account through its programming these familiar paths from past games. Kasparov had to play in a highly creative way to win.

The point is not whether Kasparov won, but how his play differed from that of Deep Thought. Kasparov and other creative experts generate no more ideas than do less creative individuals. The ideas they generate, however, are better than are the ideas of the less creative individuals. They do not generate lots of moves at random and then pick better moves; they just generate better moves from the start.

In models of human creativity, as in models of so many other things, psychologists (as well as other scientists) may tend to resort to "random-process" arguments when they do not well understand a phenomenon. The heuristic seems to be that if the scientist does not understand it, the lack of understanding is because the thing is not understandable--it is random. This heuristic is understandable, but nonoptimal. It does not really explain the phenomenon in question, but rather, explains it away.

There are occasional cases in which the random-generation theory may be correct. Those who are creative once or twice in their life, such as Engelbert Humperdinck (who wrote one successful opera, Hansel and Gretel, and nothing else of any significance)--may very well have been successful through random generation that just happened to pass the tests of the environment for adaptive fit. The random-generation theories lack plausibility, however, for individuals who are repeatedly creative. Such theories seem patently inadequate, if not ridiculous, in explaining the creativity of a Mozart, a Picasso, or a Shakespeare, all of whom were more creative many, many times than most people ever are even once in their lives. Monkeys typing random strings of letters may produce a play comparable to one of Shakespeare's once in several billion years, but they are unlikely to repeat the performance!

Of course, "parameters" of the kind referred to by Campbell can be added to a Darwinian type of model in order better to account for the stunning success of highly creative individuals. The problem is that it is never made explicit just how these parameters enter into a process model of human creativity. Indeed, when computer programs have been written that actually simulate creative information processing (see, e.g., Johnson-Laird, 1988; Langley, Simon, Bradshaw, & Zytkow, 1987; or the review of such programs in Boden, 1990), rather than trying to achieve it by brute force as do most of the successful chess programs, their performance is most decidedly nonrandom.

The task facing these AI programs is often quite a bit easier than that facing the creative individual, because in many cases, the programs rediscover something rather than discovering something originally, raising the question as to just how creative they are. These models use nonrandom processing. I would argue that there is not even one case in which it has been shown convincingly that a random-processing model has given rise even to modestly creative work.

Why might a Darwinian process work for natural evolution when, if the present argument is correct, it does not work at all for the evolution of ideas? The difference is in the time scale. Evolution works on a time scale of eons. A millennium is a drop in the bucket of evolutionary time: It makes no discernible difference. Humans, for example, have not changed genetically, so far as we know, in a millennium, although they have changed phenotypically through the effects of improved environments (at least, for some people) (Gould, 1996). But as a result of creative endeavors, society has changed radically from the last millennium--the year 1097, or thereabouts. The difference is that human creativity is an active process that imposes itself systematically upon the environment, rather than a passive one that waits for adaptive mutations to come along at random.

A Lamarckian Approach to Human Creativity

If Darwin doesn't describe how human creativity typically functions, perhaps Lamarck will. Gould (1996) has argued that cultural evolution, unlike the evolution of organisms, is a Lamarckian process, and the argument here is that the same can be said for human creativity, in general. Mozart, Picasso, Freud, and Einstein did not "luck out" again and again, nor has anyone shown how a random-generation model could replicate their creative achievements. Rather, they used their extraordinary knowledge of their respective domains, combined with generous measures of creative-thinking skills, to produce their superlative achievements.

We who are scholars do the same as Einstein and other geniuses, although on a much more modest scale. For example, when we write an article, we generate ideas that we believe have a chance of having a creative impact. We use what knowledge we have to predict what such ideas might be. We then receive feedback from referees, and systematically use this feedback to improve our ideas and the communication of them. We neither generate creative ideas, nor process feedback about these ideas, at random. But just what do we do? An extension of the triarchic theory of human intelligence (Sternberg, 1985, 1996) may contain a possible answer.

According to the triarchic theory of human intelligence (Sternberg, 1985a), we can understand human intelligence in terms of three main aspects: creative, analytical, and practical. These three aspects all play a role in creativity, broadly defined (Sternberg & Lubart, 1995, 1996b). The creative aspect of intelligence is used to generate ideas in an active, systematic fashion; the analytical aspect to determine which ideas are worth pursuing; and the practical aspect to decide how people are likely to react to the ideas, how therefore to present the ideas effectively, and how to persuade people of their validity.

Far from generating ideas at random, many creative people use both formal and informal knowledge bases in order to generate ideas that they believe or at least hope will be well received in the marketplace for these ideas, although often only after substantial effort on the part of the originator of the ideas to persuade people of their value. Formal knowledge corresponds to what Csikszentmihalyi (1988) refers to as knowledge of a domain (or discipline), and informal knowledge corresponds to what he refers to as knowledge of a field (how the social system of knowledge works in the discipline). As both Gardner (1993) and Gruber (1981; Gruber & Davis, 1988) have shown in case studies, creative individuals have a large store of both kinds of knowledge.

Creative people may also discard ideas that seem to have little or no value in this marketplace. For example, an artist might recognize that no matter how unusual her Impressionist painting, the time for Impressionist paintings to be most valued as creative is not the present one.

Dawkins (1976) and, more recently, Lynch (1996), have suggested that just as genes may pass from one generation to another, so may memes, which are what Lynch (1996) refers to as "actively contagious ideas" (p. 2). From a triarchic point of view, creative intelligence is involved in generating new ideas. Analytic intelligence is involved in judging which of these are high-quality ideas. But practical intelligence is involved in having a basis for making this judgment on the basis of informal knowledge about how ideas spread--the so-called science of "memetics."

What are the characteristics of ideas that tend to be "actively contagious"? Probably they are ideas that make connections with what people already believe, but that, in some sense, defy these beliefs. It is probably for this reason that ideas that, on the surface, seem like oxymorons--such as "multiple intelligences" (Gardner, 1983) or "emotional intelligence" (Goleman, 1995)--have particular potential: They are both familiar and unfamiliar, in a way that extends what we know, but that does not totally leave it behind. In terms of our investment theory of human creativity, they are ideas that "defy the crowd" (Sternberg & Lubart, 1995), yet can be presented in a way that will make sense to the crowd, and perhaps encourage them to change the direction of their thinking.

People who are creative actively defy existing beliefs, while at the same time using them as a springboard for their own beliefs. Less creative people will accept, largely in a passive fashion, the ideas that are extant at a given time. The creative person, far from waiting passively for good ideas to occur to him or her, actively seeks out ideas that he or she will potentially be able to "sell" to the largely passive crowd.

What kinds of informal (or tacit) knowledge might an individual have that he or she could bring to bear in deciding on whether an idea is likely to become actively contagious? Having such knowledge goes well beyond practical intelligence. Practical intelligence is involved in knowing what ideas are likely to be well-received in the present (Sternberg, Wagner, Williams, & Horvath, 1995), whereas human creativity also involves knowing what ideas are likely to be well-received in the future--in other words, knowing not just where things are, but where they are going.

What are the kinds of informal procedural knowledge would creative individuals need in order to assess for themselves whether their ideas are likely to have a fresh impact? There are a variety of kinds of knowledge. For example, Lynch (1996) suggests that ideas have a better chance of being widely propagated if, among other things, they are subject to transmission from parents to children, if they are conveyed by active proselytizing, and if they effectively sabotage their competition. Sternberg, Wagner, Williams, and Horvath (1995) have provided a production-system formalism for informal knowledge, and suggested that it spreads despite the resistance of organizations to its diffusing through these organizations.

Although various kinds of knowledge can help in generating creative ideas, there is no magic formula. If there was, the formula would cease to work as soon as multiple people started using it, because what once was creative no longer would be. Thus, creative people can make guesses as to which of their ideas will become actively contagious, but as Simonton (1994b) has shown, not only do creative people not know for sure which ideas will succeed; they do not even get better at predicting the success of their ideas with age. Nevertheless, creative people continue over the lifespan to attempt to generate creative ideas, and to pass them on to their intellectual (as well as physical!) descendants.

Once creative ideas are formed, how do they pass on from one generation of thinkers to the next so as to ensure their propagation? As Kuhn (1962) pointed out, we probably need not worry. Humans have a tendency to follow existing paradigms rather than to create new ones.

In terms of our own investment theory of creativity (Sternberg & Lubart, 1991, 1992, 1995), once a creative idea has been proposed and accepted, there are many forces that support the following of that creative idea rather than the generation of new creative ideas: People and the educational systems they develop often tend to accept existing definitions of problems rather than to seek out new ones; to emphasize recall rather than expansion of knowledge; to foster an "executive" style of thinking whereby people like to follow existing ways of doing things rather than formulating their own (Sternberg, 1988c, in press); to be risk-adverse and to succumb in the face of obstacles to new ideas; and to provide environments that discourage innovative thinking (Amabile, 1983; Hennessey & Amabile, 1988). At the same time, creative innovation never seems to stop entirely, precisely because there have always been people who are willing and able to "defy the crowd."

One of the main mechanisms whereby ideas continue to evolve in creative ways is through the so-called Hegelian dialectic, whereby a new thesis is proposed, followed later by an antithesis that opposes the original thesis, and followed still later by a synthesis that incorporates the best elements of both the thesis and the antithesis. Eventually, the synthesis becomes the new thesis to be opposed (Kalmar & Sternberg, 1988). This model works quite well in describing the evolution of ideas in many fields, including psychology (Sternberg, 1995). For example, the Gestalt movement was in many respects an antithesis to the behaviorist movement, and the cognitive movement is in many respects a synthesis of both.

Thus, once humans form creative ideas, they pass them down to the next intellectual generation. In a sense, the success of a creative idea is probably best measured not by its longevity in its original form, but in its contribution to an evolutionary process. Probably the mark of a truly successful idea is not that it stays around in its original form forever, but rather that it gives birth, in dialectical fashion, to its antithesis and eventually to a synthesis. That may well be the greatest value of the original article by Campbell (1960) on the evolutionary approach to creativity.

Darwinian Creativity Strikes Again

I have argued that human creativity is an active, nonrandom process whereby "acquired mutations" in ideas can be passed down through generations by dialectical processes--in other words, that human creativity is Lamarckian in nature. But I wish to point out that Darwinian systems can give rise to creative products, even though their blind generation and selective retention seems itself to be anything but creative. The model I believe to be appropriate, however, is not humans' creative generation of ideas, but rather, evolution itself. Evolution is a process that, through blind generation and selective retention, has produced highly adaptive organisms that are quite different from the organisms that lived earlier in the evolutionary process.

We humans can best see how evolution, as a system, produces creative products, by looking at an evolutionary process that is substantially accelerated relative to our own, namely, that of microbes. Microbes go through thousands and thousands of generations in what for us is just a single life time. It is therefore possible to see evolutionary processes at work in microbes in a way that is not possible to see in humans.

We can see the creativity of the evolutionary system by observing that, in the creative contest between humans and microbes, the results have been, for us, at best a draw. The days when doctors are talking about the end of the age of infectious microbes are themselves at an end; no matter what clever medications we invent, microbes have shown an ability to outwit us, or at least to bring us to a draw (Garrett, 1994; Karlen, 1995; Nesse & Williams, 1994). Successive generations of microbes evolve that produce mutations that are immune to antibiotics or other agents that killed earlier generations of microbes. Often through our own human stupidity--such as when a prescribed course of antibiotics is not completed--we actually foster the propagation of these mutations. Eventually, many of us (unlike Gary Kasparov!) lose the contest, and expire in the face of microbes with which even modern medicine is unable successfully to compete.

Certain microbes even have other, Lamarckian techniques for fighting the agents we use to attempt to overcome them. For example, bacteria can literally transmit genes to each other by exchanging chromosomal DNA or small bundles of genetic material referred to as plasmids (1995). In this way, they can confer immunity upon each other, and even (in Lamarckian fashion) pass it along to future generations. This process, again, however, is systemic to bacteria, rather than individually determined.

When we speak of the "creativity" of microbes, we need to be careful to make clear that we are referring to the creativity of the evolutionary system in which they are embedded, not to the creativity of the individual microbes, which do not even have brains, much less "creative thought processes." The individual microbes have no minds at all; it is the evolutionary system, rather than they themselves, that is demonstrating creativity. Individual organisms begin to show creativity in the nonblind, actively originated, Lamarckian sense only with the evolutionary emergence of brains, and creativity of the kind we value in humans further requires sufficient amounts of cerebral cortex to generate it.

The Darwinian creative process is an inadequate model of human creativity because the research on expertise cited earlier shows that humans do not generate hundreds or even thousands of candidates for propagation--or mutations--for every one that succeeds. We just do not have the time for what, for us, would be such an inefficient system. With their tens of thousands of generations taking place in short time, infectious microbes can afford to be profligate. In their case, millions can die off within the body while the system waits for the mutation that will succeed in outwitting the human or other body's natural and artificial defenses.

Over the course of eons, evolution in general can afford the same kind of profligacy. As humans, we cannot. We need a rapid, nonrandom form of evolution that does not require so much time and so much loss. We have found it in a Lamarckian form of creativity. It is how cultural, as opposed to biological evolution, takes place (Gould, 1996).

Conclusion

The view of human creativity favored here is a Lamarckian, not a Darwinian one. Human creativity is a highly active process rather than a passive one. It is systematic rather than random. New ideas are contagious--they are memetic--and can be carried over from one generation to another, forming the bases of the ideas of the next generation. According to the present view, the tendency to view human creativity as Darwinian results, in part, from a desire to integrate the study of human creativity with Darwinian theory, in general, and with evolutionary psychology, in particular. But it results in an overextension of the ideas of Darwin to a domain in which they do not well apply. Underlying this overextension may be an assumption that what we do not understand is, at some level, not understandable, and hence random. The system that works well to explain the creativity of evolution as a system does not work well to explain the creative of individual humans.

Repeated creativity in people is anything but random. At the same time, occasional creativity, as shown in the work of authors, composers, scientists, or others, who "hit it big" once and then disappear, may well result from the good luck that a normal distribution of random processes will generate. Of course, even highly creative people will, at times, produce work that, despite their best efforts, fails to have much of an impact upon the field in which they work.

To the extent we wish to develop creativity in our students and in our children, a random model of idea generation does also does not seem to serve either us or them well from a pragmatic point of view. Rather, we may be better off taking existing theories, such as our own investment theory (Sternberg & Lubart, 1995), and showing how these theories can be applied to developing creativity in the students we teach, as well as in ourselves (Sternberg & Williams, 1996). We can also teach subject matter to students in ways that encourage and value their creativity, and that can benefit creative as well as other kinds of learners (Sternberg, Ferrari, Clinkenbeard, & Grigorenko, 1996).

Darwinian models have been very helpful in understanding many aspects of human functioning, as noted above. But these models seem to have been most helpful when they have been used in a way that explains present behavior in terms of what kinds of behavior would have been adaptive for our ancestors in the distant past, and may even be adaptive in the present. For example, Buss (1994) as well as others have pointed out how men and women's ideal mating strategies might have evolved in different ways to reflect how men versus women would maximize the replication of their genes. An analogous model of human creativity might point out--probably correctly--that human creativity would be selected for because of the advantage at which it would place humans in an environment that changes at a very rapid rate.

I have not proposed a precise mechanism by which the generation of creative thought in humans occurs. Nor am I convinced that anyone else has yet suggested a compelling set of mechanisms. Computer simulations have provided a valuable step in this direction, although I suspect that too much is given to the computer, relative to what is available to the creative human. Finke, Ward, and Smith (1992) have also made valuable contributions in this direction in their theory-based experiments on creative cognition. In any case, the field will be better served by trying to discover what these processes are than by assuming they are merely random permutations in a Darwinian landscape.

Author Note

Todd I. Lubart contributed to the review of literature of all except the biological approaches, and has been a wonderful collaborator in much of the research on creativity in which I have been involved. Preparation of this chapter was supported under the Javits Act program (Grant R206R5001) as administered by the Office of Educational Research and Improvement, U.S. Department of Education. The findings and opinions expressed in this chapter do not necessarily reflect the positions or policies of the Office of Educational Research and Improvement or the U.S. Department of Education. Requests for reprints should be sent to Robert J. Sternberg, Department of Psychology, Yale University, Box 208205, New Haven, CT 06520-8205.

References

Adams, J. L. (1986). Conceptual blockbusting (3rd Ed., original work published 1974). New York: Addison-Wesley.

Amabile, T. M. (1983). The social psychology of creativity. New York: Springer-Verlag.

Barkow, J. H., Cosmides, L., & Tooby, J. (Eds.) (1992). The adapted mind: Evolutionary psychology and the generation of culture. New York: Oxford University Press.

Barron, F. (1963). Creativity and psychological health. New York: Van Nostrand.

Barron, F. (1968). Creativity and personal freedom. New York: Van Nostrand.

Barron, F. (1969). Creative person and creative process. New York: Holt, Rinehart, and Winston.

Barron, F., & Harrington, D. M. (1981). Creativity, intelligence, and personality. Annual Review of Psychology, 32, 439-476.

Boden, M. A. (1990). The creative mind: Myths and mechanisms. New York: Basic Books.

Boden, M. (1992). The creative mind: Myths and mechanisms. New York: Basic Books.

Buss, D. M. (1994). The evolution of desire: Strategies of human mating. New York: Basic Books.

Buss, D. M. (1995a). Evolutionary psychology: A new paradigm for psychological science. Psychological Inquiry, 6, 1-30.

Buss, D. M. (1995b). The future of evolutionary psychology. Psychological Inquiry, 6, 81-87.

Campbell, D. T. (1960). Blind variation and selective retention in creative thought as in other knowledge processes. Psychological Review, 67, 380-400.

Charness, N. (1991). Expertise in chess: the balance between knowledge and search. In K. A. Ericsson & J. Smith (Eds.), Toward a general theory of expertise: Prospects and limits (pp. 39-63). New York: Cambridge University Press.

Chi, M., Glaser, R., & Farr, M. (Eds.) (1988). The nature of expertise. Hillsdale, NJ: Lawrence Erlbaum Associates.

Chomsky, N. (1959). A review of Skinner's Verbal Behavior. Language, 35, 26-58.

Cosmides, L. (1989). The logic of social exchange: Has natural selection shaped how humans reason? Cognition, 31, 187-276.

Crutchfield, R. (1962). Conformity and creative thinking. In H. Gruber, G. Terrell, & M. Wertheimer (Eds.), Contemporary approaches to creative thinking (pp. 120-140). New York: Atherton.

Csikszentmihalyi, M. (1988). Society, culture, and person: A systems view of creativity. In R. J. Sternberg (Ed.), The nature of creativity (pp. 325-339). New York: Cambridge University Press.

Darwin, C. (1859). On the origin of the species by means of natural selection, or, Preservation of favoured races in the struggle for life. London: Murray.

Darwin, C. (1871). The descent of man and selection in relation to sex. London: Murray.

Dawkins, R. (1976). The selfish gene. New York: Oxford University Press.

De Bono, E. (1971). Lateral thinking for management. New York: McGraw-Hill.

De Bono, E. (1985). Six thinking hats. Boston: Little, Brown and Company.

De Bono, E. (1992). Serious creativity: Using the power of lateral thinking to create new ideas. New York: HarperCollins.

Duncker, K. (1945). On problem solving. Psychological Monographs, 68(5), whole no. 270.

Ericsson, K. A., & Smith, J. (Eds.) (1991). Toward a general theory of expertise: Prospects and limits. New York: Cambridge University Press.

Eysenck, H. J. (1993). Creativity and personality: A theoretical perspective. Psychological Inquiry, 4, 147-178.

Findlay, C. S., & Lumsden, C. J. (1988). The creative mind: toward an evolutionary theory of discovery and innovation. San Diego: Academic Press.

Finke, R. A,, Ward, T. B., & Smith, T. M. (1992). Creative cognition: Theory, research, and applications. Cambridge, MA: MIT Press.

Finke, R. (1990). Creative imagery: Discoveries and inventions in visualization. Hillsdale, NJ: Erlbaum.

Finke, R. A., Ward, T. B., Smith, S. M. (1992). Creative cognition: Theory, research, and applications. Cambridge, MA: MIT Press.

Frensch, P. A. & Sternberg, R. J. (1989). Expertise and intelligent thinking: When is it worse to know better? In R.J. Sternberg (Ed.) Advances in the psychology of human intelligence (Vol. 5, pp. 157-158). Hillsdale, NJ: Erlbaum.

Freud, S. (1908/1964). Leonardo da Vinci and a memory of his childhood. New York: Norton. (Original work published in 1910).

Gardner, H. (1983). Frames of mind: The theory of multiple intelligences. New York: Basic Books.

Gardner, H. (1993). Creating minds. New York: Basic Books.

Garrett, L. (1994). The coming plague: Newly emerging diseases in a world out of balance. New York: Farrar Straus Giroux.

Ghiselin, B. (Ed.). (1985). The creative process: A symposium. Berkeley, CA: University of California Press.

Golann, S. E. (1962). The creativity motive. Journal of Personality, 30, 588-600.

Goleman, D. (1995). Emotional intelligence. New York: Bantam.

Gordon, W. J. J. (1961). Synectics: The development of creative capacity. New York: Harper & Row.

Gough, H. G. (1979). A creativity scale for the Adjective Check List. Journal of Personality and Social Psychology, 37, Gould, S. J. (1996). Full house: The spread of excellence from Plato to Darwin. New York: Harmony Books.

Gruber, H. E. (1981). Darwin on man: A psychological study of scientific creativity (2nd ed.). Chicago: University of Chicago Press.

Gruber, H. E., & Davis, S. N. (1988). Inching our way up Mount Olympus: the evolving-systems approach to creative thinking. In R. J. Sternberg (Ed.), The nature of creativity: Contemporary psychological perspectives (pp. 243-270). New York: Cambridge University Press. 1398-1405.

Gruber, H. (1981). Darwin on man: A psychological study of scientific creativity (2nd ed.). Chicago: University of Chicago Press. (Original work published 1974)

Gruber, H. E. (1988). The evolving systems approach to creative work. Creativity Research Journal, 1, 27-51.

Gruber, H. E. (1989). The evolving systems approach to creative work. In D. B. Wallace & H. E. Gruber (Eds.), Creative people at work: Twelve cognitive case studies (pp. 3-24). New York: Oxford university Press.

Gruber, H. E. & Davis, S. N. (1988). Inching our way up Mount Olympus: the evolving-systems approach to creative thinking. In R. J. Sternberg (Ed.), The nature of creativity (pp. 243-270). New York: Cambridge University Press.

Guilford, J. P. (1950). Creativity. American Psychologist, 5, 444-454.

Hennessey, B. A. & Amabile, T. M. (1988). The conditions of creativity. In R. J. Sternberg (Ed.), The nature of creativity (pp. 11-38). New York: Cambridge University Press.

Holyoak, K.J. (1990). Problem solving. In D. N. Osherson & E.E. Smith (Eds.), An invitation to cognitive science: Vol. 3. Thinking (pp. 116-146). Cambridge, MA: MIT Press.

Johnson-Laird, P. N. (1988). Freedom and constrain in creativity. In R. J. Sternberg (Ed.), The nature of creativity: Contemporary psychological perspectives (pp. 202-219). New York: Cambridge University Press.

Kahneman, D., & Tversky, A. (1971). Subjective probability: A judgment of representativeness. Cognitive Psychology, 3, 430-454.

Kalmar, D. A., & Sternberg, R. J. (1988). Theory knitting: An integrative approach to theory development. Philosophical Psychology, 1, 153-170.

Karlen, A. (1995). Man and microbes: Disease and plagues in history and modern times. New York: Tarcher Putnam.

Kris, E. (1952). Psychoanalytic exploration in art. New York: International Universities Press.

Kubie, L. S. (1958). The neurotic distortion of the creative process. Lawrence: University of Kansas Press.

Kuhn, T. (1962). The structure of scientific revolutions. Chicago: University of Chicago Press.

Langley P., Simon, H. A., Bradshaw, G. L., & Zytkow, J. M. (1987). Scientific discovery: Computational explorations of the creative process. Cambridge, MA: MIT Press.

Lubart, T. I. (1990). Creativity and cross-cultural variation. International Journal of Psychology, 25, 39-59.

Lubart, T. I. (1994). Product-centered self-evaluation and the creative process. Ph.D. Dissertation. New Haven, CT: Yale University.

Lubart, T. I. & Sternberg, R. J. (1995). An investment approach to creativity: Theory and data. In S. M. Smith, T. B. Ward, & R. A. Finke (Eds.), The creative cognition approach. Cambridge, MA: MIT Press.

Lynch, A. (1996). Thought contagion: How belief spreads through society: The new science of memes. New York: Basic Books.

MacKinnon, D. W. (1965). Personality and the realization of creative potential. American Psychologist, 20, 273-281.

Maduro, R. (1976). Artistic creativity in a Brahmin painter community. Research monograph 14, Berkeley, CA: Center for South and Southeast Asia Studies, University of California.

Maslow, A. (1968). Toward a psychology of being. New York: Van Nostrand.

McClelland, D. C., Atkinson, J. W., Clark, R. A., & Lowell, E. L. (1953). The achievement motive. New York: Appleton-Century-Crofts, Inc.

Mumford, M. D., & Gustafson, S. B. (1988). Creativity syndrome: Integration, application, and innovation. Psychological Bulletin, 103, 27-43.

Nesse, R. M., & Williams, G. C. (1994). Why we get sick: The new science of Darwinian medicine. New York: Times Books.

Newell, A., & Simon, H. A. (1972). Human problem solving. Englewood Cliffs, NJ: Prentice-Hall.

Noy, P. (1969). A revision of the psychoanalytic theory of the primary process. International Journal of Psychoanalysis, 50, 155-178.

Ochse, R. (1990) Before the gates of excellence: The determinants of creative genius. New York: Cambridge University Press.

Osborn, A. F. (1953). Applied imagination (rev. ed.). New York: Charles Scribner's Sons.

Perkins, D. N. (1981). The mind's best work. New York: Harvard University Press.

Perkins, D. N. (1995). Insight in minds and genes. In R. J. Sternberg & J. E. Davidson (Eds.), The nature of insight (pp. 495-533). Cambridge, MA: MIT Press.

Rogers, C. R. (1954). Toward a theory of creativity. ETC: A Review of General Semantics, 11, 249-260.

Rothenberg, A. (1979). The emerging goddess. Chicago: University of Chicago Press.

Rothenberg, A. & Hausman, C. R. (Eds.). (1976) The creativity question. Durham, NC: Duke University Press.

Rubenson, D. L. & Runco, M. A. (1992). The psychoeconomic approach to creativity. New Ideas in Psychology, 10(2), 131-147.

Silver, H. R. (1981). Calculating risks: The socioeconomic foundations of aesthetic innovation in an Ashanti carving community. Ethnology, 20(2), 101-114.

Simonton, D. K. (1984). Genius, creativity, and leadership. Cambridge, MA: Harvard University Press.

Simonton, D. K. (1988). Age and outstanding achievement: What do we know after a century of research? Psychological Bulletin, 104, 251-267.

Simonton, D. K. (1994a). Greatness: Who makes history and why. New York: Guilford Press.

Simonton, D. K. (1994b). Individual differences, developmental changes, and social context. Behavioral and Brain Sciences, 17, 552-553.

Simonton, D. K. (1995). Foresight in insight? A Darwinian answer. In R. J. Sternberg (Ed.), The nature of insight (pp. 466-494). Cambridge, MA: MIT Press.

Skinner, B. F. (1974). About behaviorism. New York: Knopf.

Smith, S. M., Ward, T. B., & Finke, R. A. (Eds.) (1995). The creative cognition approach. Cambridge, MA: MIT Press.

Sternberg, R. J. (1985a). Beyond IQ: A triarchic theory of human intelligence. New York: Cambridge University Press.

Sternberg, R. J. (1985b). Implicit theories of intelligence, creativity, and wisdom. Journal of Personality and Social Psychology, 49, 607-627.

Sternberg, R. J. (1986). Intelligence applied: Understanding and increasing your intellectual skills. San Diego: Harcourt Brace Jovanovich.

Sternberg, R. J. (Ed.) (1988a). The nature of creativity: Contemporary psychological perspectives. New York: Cambridge University Press.

Sternberg, R. J. (1988b). The triangle of love. New York: Basic Books.

Sternberg, R. J. (1988c). The triarchic mind: A new theory of human intelligence. New York: Viking.

Sternberg, R. J. (1995). In search of the human mind. Orlando: Harcourt Brace College Publishers.

Sternberg, R. J. (1996a). Cognitive psychology. Orlando: Harcourt Brace College Publishers.

Sternberg, R. J. (1996b). Successful intelligence: How practical and creative intelligence determine success in life. New York: Simon & Schuster.

Sternberg, R. J. (in press). Thinking styles. New York: Cambridge University Press.

Sternberg, R. J., & Davidson, J. E. (Eds.) (1994). The nature of insight. Cambridge, MA: MIT Press.

Sternberg, R. J., Ferrari, M., Clinkenbeard, P. R., & Grigorenko, E. L. (1996). Identification, instruction, and assessment of gifted children. Gifted Child Quarterly, 40, 129-137.

Sternberg, R. J., Gordeeva, T. (1996). The anatomy of impact: What makes an article influential? Psychological Science, 7, 69-75.

Sternberg, R. J., & Lubart, T. I. (1991). An investment theory of creativity and its development. Human Development, 34, 1-31.

Sternberg, R. J., & Lubart, T. I. (1992). Buy low and sell high: An investment approach to creativity. Current Directions in Psychological Science, 1, 1-5.

Sternberg, R. J., & Lubart, T. I. (1995). Defying the crowd: Cultivating creativity in a culture of conformity. New York: Free Press.

Sternberg, R. J., & Lubart, T. I. (1996). Investing in creativity. American Psychologist, 51, 677-688.

Sternberg, R. J., Wagner, R. K., Williams, W. M., & Horvath, J. A. (1995). Testing common sense. American Psychologist, 50, 912-927.

Sternberg, R. J., & Williams, W. M. (1996). How to develop student creativity. Alexandria, VA: Association for Supervision and Curriculum Development.

Suler, J. R. (1980). Primary process thinking and creativity. Psychological Bulletin, 88, 144-165.

Torrance, E. P. (1974). Torrance tests of creative thinking. Lexington, MA: Personnel Press.

Trivers, R. (1971). The evolution of reciprocal altruism. Quarterly Review of Biology, 46, 35-56.

Vernon, P. E. (Ed.) (1970). Creativity: Selected readings. Baltimore, MD: Penguin Books.

von Oech, R. (1983). A whack on the side of the head. New York: Warner.

von Oech, R. (1986). A kick in the seat of the pants. New York: Harper & Row.

Wallach, M. A., & Kogan, N. (1965). Modes of thinking in young children: A study of the creativity-intelligence distinction. New York: Holt, Rinehart & Winston.

Weisberg, R. W. (1986). Creativity, genius, and other myths. New York: W.H. Freeman.

Weisberg, R. W. (1993). Creativity: Beyond the myth of genius. New York: Freeman.

Werner, H., & Kaplan, B. (1963). Symbol formation. Hillsdale, NJ: Erlbaum.

Woodman, R. W., & Schoenfeldt, L. F. (1989). Individual differences in creativity: An interactionist perspective. In J. A. Glover, R. R. Ronning, & C. R. Reynolds (Eds.), Handbook of creativity. New York: Plenum.

Wright, R. (1994). The moral animal: Why we are the way we are: The new science of evolutionary psychology. New York: Vintage.

17

Creativity: Reframed As A Biological Process

Sam Leven
Scientific Cybernetics, Inc.
For a New Social Science
4701 Falstone
Chevy Chase, Maryland 20815

Thanks are due to: *FOR A NEW SOCIAL SCIENCE* ® for a decade of support; the B.R.A.I.N.S. Center, created by Karl Pribram and Al Harris, for the hard work of tolerance and wisdom they have displayed, particularly to me; Joe King, Bill Hudspeth, and Peter Kugler, senior members of the Center, for many hours of discussion and education; Greg Marsella, who loaned me a clinical and academic perspective to the new psychopharmacology; Dan Levine, whose intellect and good spirits have always left me wiser and more comforted; Paul Werbos, outside of Karl the most dangerous man I know, for penetrating queries I struggled to answer; and Rod Smith, who never let me forget there's a real world out there, filled with difficult problems (and people). Gratitude would not suffice; only excellence could settle these debts (and all of them know how hard to come by excellence is).

My family deserves special recognition. My wife, Nina, has tolerated a decade of constant befuddlement and frustration, inattention and despair. Perhaps this, and other recent work, is a talisman of my gratitude and affection. My daughter, Rachel, is proof that genetics skip male generations; some day, she will be as grateful as I am already that her sturdiness and generosity of spirit are curried from her mother and from my antecedents she remains a reminder that grace, intellect, and beauty sometimes emerge and remain, ineradicable. Last, for my son Joshua, whose heart is one with mine (and so, keeps my aging vessel vibrant if tremulous) and whose spirit is luminous in the ruins, this is the first step in keeping my promise to you; be patient, not only with the Lord, but with me, too. I have promised (foolishly, as fathers do) not to fail. I pledge myself, step by step, to reach out, with you, to that "daring joy, but safe -/ Are they not *all* the seas of God?"

Abstract

We find in the creative act the highest expression of human mental achievement. Models of creativity have, however, not specified the underlying brain-behavior processes that give rise to it.

Human creativity is characterized as a psychobiological phenomenon. Its common function in everyday and sublime action is described in an integrated model. This view draws from literature on autism (Baron-Cohen, 1994), creative language use (Turner, 1996; Turner & Fauconnier, 1996), dysfunctions of memory and planning (Pribram, 1996), affect's role in cognition (Redfield, 1993) and creativity (Rothenburg, 1990).

Broad models of brain function are employed to explain the integrated nature of creativity (Pribram, 1991; MacLean, 1990). Convergent zones of brain function (Damasio, 1996) are tied to mental space models of metaphor comprehension (Fauconnier, 1994); these are tied to Pribram's (1996) model of frontal function and Sternberg's (1994, 1995) view of complex problem-solving and creativity. Cloninger's (1986, *et. seq.*) view of temperament serving as substrate – tendencies – and character as determinant of outcome is related to Sternberg's (1996) view of "successful intelligence". Pribram's "surface structure" of memory is demonstrated to comprise the higher-order substrate for the characterological dimensions of processing; this surface structure constitutes the developed store and constructed values of a lifetime – it is "consequences of behavior that stand in relation to each other " (Pribram, 1996*b*). The tendencies to actively *experience*, which underlie behavior lay the groundwork for mental instantiations of sensorimotor and internal state data to object form and immediate (unintentional response) represent Pribram's deep surface – and underpin Cloninger's temperament and the early processes of Baron-Cohen's (1995) and Turner's (1996) "being-in-the world".

Shore's (1995) work emphasizes – and delineates the processes of – intersubjectivity which allow us to construct intention, develop attention, and suspend disbelief (in a complex world) to generate effort. The bonds between self and other that represent the grounding of experience, likewise represent the basis for self- and other-regulation – which produce the creation of "mental spaces" (Fauconnier, 1994) and the grounding of episodes that lead from primitive, syntactic-level experience to episode-building into parable (Pribram, 1991 and 1996*a* and *b*; Turner, 1996).

It is argued that models of brain dysfunction (Redfield, 1993; Leven, 1996) can be used to comprehend the links between affect, context, and creativity. Yet, most "creative" acts (Csikszimenthalyi, 1996) are not dysfunctional

This apparent conflict between creativity as a "special, private" act and as a "common, public" experience is resolved in presentation of a physiological and computational model which allows us to test its implications – and offers a means to learn from its failings.

Metaphor, Healing, and Creativity:. Re-Framing as a Biological Phenomenon[2]

So much time to make up, everywhere you turn,
Time we have wasted on the way.
So much more than you need, underneath the bridge;
Let the water come and carry us away. – Graham Nash, *Wasted on the Way*

The drive towards the formation of metaphors is the fundamental human drive, which one cannot for a single instance dispense with in thought, for one would thereby dispense with man himself. – F. Nietzsche, quoted in Fiumara (1995)

Introduction

For Brahms, . . . Mozart was a fellow modernist [who] had stretched the power of music . . ., making it capable of wrestling with metaphysical questions: desire and morality, life and death. . . .
Sigmund Freud confronted similar issues . . . . Maturation, he suggested, involves the internalization of the disciplining presence of the father, but he could never work out a model for internalization of the mother. Modern cultural roles would not allow such a thing. (Steinberg, 1996, pp. 31-2).

The sense Brahms encountered in his fellow composer -- the individual's reckoning with mortality, sensibility, and virtue – was, indeed the modern experience of *difference* (Derrida, 1981), the lack of inevitable ends promised by enlightenment values. What is remarkable about Mozart's premature discovery of Camus' (1952) dilemma of desiring determinacy in an open world (Eco, 1989) is that it emerged from his personal experience, rather than some determined philosophical search (Einstein, 1945). Mozart's mother died; his sense of stability and assured portion of joy were desiccated; he found himself unable to express his despair in words; he reached to the metaphoric experience he trusted most, piano music, and stepped out of the "Classical" schema and created Romantic Music and the beginnings of Modernity (Davies, 1987).

This experience of expressing the apparently ineffable is at the heart of what we consider creativity (Boden, 1992), what we value most. It is at the heart of the *multi-directed partiality* (Goldenthal, 1996) that the psychological novels of Dostoevsky embody (Corson, 1996) – and has led to the development of narrative therapy (Freedman & Combs, 1996). It is also the basis of the recognition that shared creative commitment underlies business innovation (Leven & Smith, 1995; Kao, 1996).

---

[2] For Karl Pribram: role model for three generations of scholars and humanists -- who taught me to honor my persistence and other endowments and demanded that higher values guide my efforts, regardless of consequence. With gratitude and affection. And for my son, Josh: inheritor of an unsought position in an "invisible college". May he persist.

Perhaps most difficult for current commentators to grasp is that Freud's vision has laid the groundwork for modern developments, from neuropsychiatry (Schore), to psychological models (Sternberg, 1988, *et. al.*) to linguistic theory (Turner, 1996; Fauconnier, 1994) and philosophical models of creativity (Holyoak & Thagard, 1995). As Karl Pribram (Pribram & Gill, 1976, 168-9) grasped, Freud's work laid the groundwork "for investigators and practitioners working at the interface between disciplines" and "welcomes psychoanalysis back into the natural sciences."

We shall find that it is only in grasping the foundations of human experience Freud recognized that we can explain the most common and the most sublime of our expressions. As Pribram predicted, we require tools of linguistic theory, biopsychology, organization theory, and philosophy to encounter human creativity. And we shall find that, as Freud and Pribram have predicted, the ability to reach out for the basis of relativity (Holton, 1996) is no more complex than the ability to reach out for one's mother (Schore, 1995) and seek to communicate (Baron-Cohen, 1995).

It is the "figural processing" Pribram (1991) has defined as our fundamental mode of communication – "dynamical cognition", as Horgan and Tienson (1996) term it – that underlies all perception, action, and anticipation. And, as all of these functions operate within characteristic "executive" processes (Sternberg & Lubart, 1995), we experience the same sense of "sense-making" we do crossing an unfamiliar street, cooperatively arriving at budget estimates or hearing a a rarely-performed Haydn string quartet.

Creativity as Phenomenon and Artifact

What is common between Mozart's creation of the Romantic musical framework, the therapeutic process, and modern business product development? Each is, in its own way, an act of great creativity. Mozart's work has received attention from the psychological community (Wallace & Gruber, 1985; Czikismanthalyi, 1996; *et. al.*); the efforts of ordinary people achieving extraordinary ends has been less-celebrated (but see Argyris & Schon, 1996; and, generally, Bateson, 1972). Each violates norms and belief-systems, each changes some piece of the human condition forever. Each, then, is a creative act. Each is, as Pribram (1991) terms it, an act of "figural processing."

Students of creativity have employed laboratory studies (Sternberg, 1988a, *et. al.*) and historical analysis (*e.g.*, Simonton, 1984, *et. seq.*). Pribram (1991, *et. al.*), Bateson (1972), and others have argued that the only appropriate cauldron into which a theory of complex mental function can be poured is that of a biological model. I have suggested that preliminary models of behavior are not only plausible, but necessary (Leven, 1988 and 1992): only when we see the implications of our models can we grasp their inherent strengths and failings.

The creative act stands at the apogee of human experience. It is here we must test the fitness of our theories.

Human creativity has been the subject of two raging debates. First, Is ordinary people's analogical and metaphoric use related to the creative acts of "great men"? Second, Is disturbed affect useful in creative acts and "conceptual revolutions"?

Simonton (1988, *et al.*) and others have suggested that special capabilities are required for high creativity – which is entirely separate from normal linguistic invention and problem-solving. Writers such as Zuckerman (1994) have maintained that qualities like heightened scanning and higher-speed processing combine to create a unique class of "special minds". But Turner (1996) has emphasized the universal nature of creative acts, from visualization to story-telling. Pribram (1991) tells us that "figural processing" is inherent in acts from perception to creativity; the brain as an ensemble creates and recreates perception and meaning from moment to moment – even in the most modest mind.

Recent writers have divided on the second question: Redfield (1994, *et. refs.*) has supported the proposition that mental disorders can be critical to creative processes, while Rothenburg (1990) and Csikszimenthalyi (1996) have dissented, but noted that special qualities distinguish creative people.

Others have "democratized" the debate, arguing that all people engage in private acts that are as revelatory to themselves as any scientific or literary discovery are to scholarly communities (Harre and Stearns, 1995). The notion that each person "composes a life" (Eco, 1989) out of fragmentary objective events derives from the Gestalt tradition (Lewin, 1940). And the notion that creativity is inherent in normal human function is one of the Gestalt movement's testaments (Wertheimer, 1945/1990).

I shall argue that these "opposite" accounts are commensurable, given new models of mind (Pribram, 1991; Benson, 1995; Cloninger,1986 *et seq.*) and the wealth of new research (*e.g.*, Damasio, 1996). The argument does not derive from "teachable" creativity models (*i.e.*, Perkins, 1994). Rather, I follow Sternberg's (1994, 1995a) argument that two pairs of three sets of "component" psychological systems allow construction of complex mental processes – analogous to the biological ones I have identified in discussing depressive mentation (Leven, 1992, and *in press*), creativity (*e.g.*, Leven, 1995), and interpersonal/bargaining behaviors (Elsberry & Leven, 1989; Smith & Leven, 1996).

Viewing the brain as a dynamical system allows us to resolve these views into a single approach that allows psychobiological insights (Pribram, 1990; Schore, 1995) to be leavened with advances in developmental theory (Baron-Cohen, 1995), linguistics (Turner, 1996), and literary criticism (Morson, 1996). Healing a psychic wound is, arguably, as creative an act as composing a symphony (Fiumara, 1995); creating a new business product may well be as generative as writing a poem (Kao, 1996). Cloninger (esp. 1993) has provided the skeleton key; it is all a continuum.

First, we shall examine the diffuse literature on creativity. We shall find that many views are "assumed" by other participants; there is less conflict than varying foci of attention. Then, we shall present a psychobiological account that comprehends the main lines of thought.

Finally, we shall examine three cases of creativity. First, we shall examine one of the hallmarks of Western music and its genesis in disturbed affect. Second, we shall study the resolution of a reported case of psychological trauma. Third, we shall examine the process of introducing a radically new business technique to a conservative firm.

We shall find that each of these fits a single model of behavior. This model, which has been approximated in various literatures, has commonalities noted by Bateson (1972) and Pribram (1991), among others. We shall trace the biological substrate which most plausibly replicates the process we review and present a putative neural network model which allows us to comprehend human creativity from its simplest (grasping communicative intent), through its standard implementation (grasping and creating an ordinary metaphor), through its most sublime.

From Communicative Competence to Invention

Turner (1996) and his partner Fauconnier (1994) argue that creativity with language is inherent in the process of making sense of a complex environment. We must *tell ourselves a story* to provide continuity from multifarious moment to moment – in the face of changes in our internal physiological and affective states and in the environment we encounter. How we tell these stories ("parables"), he argues, reveals the operation of mind.

First, we shall follow his argument in order to pose questions for non-linguistic approaches, focusing on the work of Sternberg and Lubart. We shall, then, seek biological bases for the operations which Turner and Sternberg specify, from Pribram, Schore, and Cloninger – and propose a neural network model for the creative process.

The traditional account of language (Chomsky, 1957) suggests that language is built hierarchically, of given basic components. From syntactic phrase structures emerge languages; out of language grows narrative, then literary narrative. Finally, at the apex of his structure, is the capacity to generate complex metaphor and parables.

Turner (1996, p. 168) argues for modification, in light of modern findings:

> With story, projection, and their powerful combination in parable, we have a cognitive basis from which language can originate. The dynamic processes of parable are basic to the construction of meaning and . . . language. Story precedes grammar. Projection precedes grammar. Parable precedes grammar. Language follows from these mental capacities as a consequence.

How does parable emerge, whole, from the formative human mind? Turner specifies the processes. We begin, he argues, by "projecting", predicting from our current situation important events that will follow. We evaluate whether this future is pleasing by "narrative imagining". We construct an imaginative narrative path of action to achieve the goal. We "explain" this plan by reference to our reconstruction of the past to match these values and chosen actions. We manipulate "objects", "events", and "actors" (categories which emerge from our frame of reference) to construct our account, using them as "emblems" from related memorable episodes and match the course of action to an "image schema" while describes the choice of frame and action. We perform this magic by "conceptual blending".

Conceptual blending is the process of mapping representations of motor/readiness, sensory/affective, and semantic data into "mental spaces" (Fauconnier, 1994). Fauconnier argues that language emerges from the dynamic process of representing the integration of these states ("making sense" of them) by mapping them into a virtual space or image. These mental spaces take input data, transform them (based on prior state and expectation) onto a "generic space", which is a flexible frame suited to current experience. These are then manipulated into "middle spaces", in which the images are transformed into coherent, aesthetically pleasing integrated representations.

While the entire process takes place instantaneously (and continuously), the transformation of past and complex present into a coherent image allows us to "parse" experience into sensible categories and make effective predictions. The mapping involved is an effort to achieve sensibility.

Invoking Damasio's (1996) "convergence zone hypothesis", Turner argues that cross-modal integration is effectively identical to merging relevant facets of one image of experience with another. "Input spaces" represent the initial merging of motor/readiness, sensory/affective, and semantic data into many simultaneous integrated mental states. "Generic spaces" are comparable to Baars' (1988) "global workspace" model, in which current states evoke recalled states and are eligible to be mapped ("projected") into a contextualized image schema (Figure 1, from Fauconnier & Turner, 1995, pp. 5*ff*). They derive from underlying metaphors which are based on our physical and social experience, such as "grasping" a problem represents a mapping from the physical experience of the child in having to work at enveloping and having the capacity to manipulate an object (*cf.* Johnson, 1987).

Recent results by Vandenberghe, *et. al.* (1996) shows that the Lakoff-Johnson-Fauconnier-Turner model is supported by studies of neural activity involved in processing the meaning of words. Using positron-emission tomography, the London scholars discovered that subjects involved in iconic and semantic processes (visual and verbal) employed heavily-overlapping regions, including portions of left frontal, temporal, parietal, and occipital areas. In fact, only where activation was in response to pictures or to words, the operations were not modality-specific. If there were some core syntactic/semantic region (such as posited by Chomsky, 1957), then the broad interregional co-operation would be unnecessary: the "noun" area would call a simple store of names. Instead, as Vandenberghe and associates (1996, p. 255) say there is "a distributed semantic system shared" by vision and language. As Pribram (1996) has suggested, this "deep memory" store probably includes all sensorimotor modalities, activated by cues from different modes but representing multi-dimensional, multi-sensory maps.

The collective store we build from sensorimotor, affective, and semantic orderings allows us to select distributed representations from recollections and current perceptions, feelings, and thoughts. As Turner (1996) has maintained, the simplest components of expression emerge from experience – and are "complexified" by the blending process.

Blends constitute the elaboration of the mapping process through three devices:

- selective projection applies to inputs that share generic structure and can be partially mapped homomorphically onto each other;
- emergent structure is generated in three ways: composition of projections from the inputs, completion based on independently recruited frames and scenarios, and elaboration ("running the Blend");
- revision of the inputs can apply by reverse mapping from the Blend.

These processes are qualitatively similar to the mental functions defined by Sternberg (1988, 1995). Sternberg has argued that "cognitive components" underlie creative and effective intelligent processes: selective encoding, selective comparison, and selective combination. He has contended, further, that "metacomponents" are the analytics used to plan, monitor, and evaluate progress (the "white collar" tasks); "performance components" are inductive tools employed to encode a map between perception and known information structures, infer a directional relationship, and apply the "vector" to the data, then apply it to make a match; "knowledge-acquisition components" entail the use of insight, "the ability to detect and use clues embedded in the context" (Sternberg, 1989 , p. 176).

Sternberg notes other critical elements in insight processes. Knowledge of the domain is essential when a problem is embedded in a body of learning. He notes (Sternberg, 1988, p. 181) that specific cognitive styles are most effective: "The most successful . . . are those who manage to combine well both impulsive and reflective styles. At certain points . . ., they act on impulse; at other times, the act only after great reflection." And motivation is required in challenging tasks: "we found that the better problem solvers were more persevering than the poorer ones."

Can Turner's model serve as a template for the processes Sternberg describes? We turn to the model introduced, studied, and verified by Cloninger (1986, *et. seq.*). As Pribram (Pribram, 1971 and 1991) has maintained, behavior is based in a biological substrate of processes. We shall find Cloninger presents opportunities to explicate the processes Turner and Sternberg describe – and lays the groundwork for a broader understanding of human action as figural and creative.

First, we shall briefly review other elements of creative processes, as presented by a variety of researchers. Starting from Lubart (1995), we shall essay work by a variety of psychological researchers. These shall be followed by models of creative process by Thagard and Holyoak (1995) and Finke, *et. al.* (1992, 1995). We shall then integrate these approaches in a single model, based on Cloninger and Pribram.

Psychological and Computational Models of Creative Process

Lubart (1994) provides, perhaps, the most direct summary of approaches to creative process and functioning. Creativity is novel and appropriate work, unexpected in content and directed at some unfulfilled need. To be fully appreciated it must be executed with a high level of skill and be directed at a pressing need; further, it must be the product of unique combinations of knowledge and persistent efforts. It should pull together disparate ideas and evoke surprise and admiration in its audience. Thus, its social setting (Amabile, 1983) and its audience's preferences (Csikszentmihalyi, 1988) are central to its success. Finally, creativity is a human attribute which may be characterized by a unique cognitive style or character – which tend (Eysenck, 1993) to be the product of a combination of normal characteristics, felicitously combined.

Lubart defines four approaches to the sources of creativity. "Mystical" sources may underlie a creative process (Ghiselin, 1985). Kris' (1952) combination of unmodulated and intrusive thoughts (adaptive regression) with rational extension of the products (elaboration) is comparable to Kubie's (1958) "Psychodynamic" approach in which the preconscious makes available vague but accessible thoughts which struggle against the unconscious conflicts that Freud (*e.g.*, 1964) emphasized – they tend, he said, to lead to repetitive, almost compulsive ideation. Rothenberg (1990) stresses the centrality of both primary and secondary processes. The "Cognitive" approach derives from Guilford's (1977) characterization of divergent thinking, exploiting a set of abilities which included figural examination of multiple levels of problems. The recognition of analogies developed either through insightful (Finke, *et. al.*, 1992) or semirandom efforts (Simonton, 1988) led to a new set of approaches. Finally, the Simon group (Langley, *et. al.*, 1987) produced a heuristic-exploiting program to combine existing patterns of explanation with new data. The boundaries of normality are tested by creatives, claim Prentky (1989), who stresses lability of sense-data input among high creatives – along with the mentally-ill. Eysenck's (1993) work suggests a biological model which, again, suggests that creatives are on the edges of pathology. Finally, the "Social-Psychological" group emphasizes personality traits – independence of judgment, self-confidence, attraction to complexity, aesthetic sensitivity, and risk-preference – distinguish creative people (Barron & Harrington, 1981). Environmental variables and developmental influences are emphasized (Amabile (1983).

Finally, a "Systems" or "Confluence" view has emerged. Csikszentmihalyi (1996) emphasizes cognitive processes interacting with personality traits and motivation in the context of a socially-governed field (Kuhn, 1962), a culturally-defined symbol system whose "holes" create opportunities for creative developments. Amabile (1983) stresses flexible and complexity-tolerant cognitive style, use of creative heuristics, and the ability to employ high energy in concentrated effort. Finally, Sternberg and Lubart's (1995) own "investment approach" emphasizes Sternberg's typology, motivation, and sensitivity to the environment – leading to a "buy low-sell high" approach to recognizing problems strategically and devoting appropriate effort levels, based on potential productivity of outcome.

This strategy dimension is emphasized by Sternberg and Davidson (1995). They propose that insight represents a "cognitive transition". The problem is restructured; elements are selectively combined or compared recursively, until Metcalfe's (1986) "warm feeling" is evoked.

Metaphoric thinking has dominated a large group of writers, including modelers we shall consider below. Barron (1969) and Gruber (1985) emphasize scientific breakthroughs which were based on direct analogies and visualization. Boden (1992) suggests that computational models, along the lines of connectionist models, might define conceptual spaces which would allow recognition of similarities among disparate entities (*cf.*, Rumelhart, McClelland, & the PDP Research Group, 1986, esp. Ch. 14). Related to these approaches is Rothenberg's (1990) combination of "homospatial" and "janusian" thinking. Homospatial approaches superimpose one problem on another that is well-understood; janusian techniques emphasize recognition of antipodal qualities as relevant.

Finally, as Sternberg (1996) has emphasized, there are motivational and personological issues. Sternberg and Lubart (1991) emphasize five personality traits: tolerance of ambiguity, perseverance, openness to new experiences and new roles, willingness to take risks, and, finally, high belief in one's own capacity (and locus of control). These qualities must co-occur with substantial intelligence and motivation. Sternberg and Lubart (1995) suggest that there is a dynamical nature to the interaction of these qualities.

`Lubart (1994, p. 314) distinguishes among four approaches to "the creative process": "the sequence of thoughts and actions that lead to a creative product [as opposed to] a routine problem-solving process." Weisberg (1993) and his group see little difference between creative and routine problem-solving; more effective use of information, more practice at problem-solving, higher standards for a satisfactory solution. Persistence and increased investment at each step are emphasized by Getzels and Csikszentmihalyi (1976). Following particular routines (*e.g.*, allowing unconstrained consideration of the problem, with general review and recursion, only at the beginning of problem-solving) is emphasized by Lubart. Finally, the notion that *special* processing is involved is exemplified by Rothenberg's (1990) janusian/homospatial approach.

We shall emphasize, below, the great commonality in these views (Figure 5). But the framework on which virtually all writers seem to rest has been discussed frequently (Hadamard, 1945, *et. al.*). Four stages appear to take place in all views. *Preparation* involves systematic scene-setting for solving the problem – gathering materials, consulting informed opinion, even finding an hospitable environment. *Incubation* involves gradual development of the problem definition and evocation of related memories and impressions. *Illumination* is the discovery of a provisionally promising solution. It may well be Sternberg and Davidson's (1995) warm moment of insight. *Verification* is the evaluative step, in which the solution is applied to sample problems and checked for logical validity. If the proposed solution fails at Verification, the process begins anew.

So we have a shared vision of the step-wise process, by most authors. Most agree that there is a confluence of store of information, environment, personality, motivation, and intelligence. We present the similarities and differences in Figure 4.

We turn, now to two leading models of the creative process, those of Holyoak and Thagard (1995) and Finke, *et. al.* (1992, 1995). We find, in each, sophisticated efforts to embody complex processes in a Dynamical Cognition (Horgan & Tienson, 1996) framework. Each model has components which correspond to experimentally-observed steps in the creative process. Each demonstrates a sophisticated approach to the task Pribram laid down in (Pribram & Gill, 1976) and work towards the progress psychologists have made (Pribram, 1994, p. 98):

> (1) . . . a group of investigators can begin with a general metaphor – a broad and somewhat undefined sense of the similarities [in their results], (2) how they can "trim" this metaphor into more . . . precise shape [by] reasoning by analogy . . ., and (3) how, once they have gone far enough, the original metaphor is transformed into a precise scientific model . . . that can be shared with and tested by the larger scientific community.

The Holyoak and Thagard (1995, p. 242) effort to build an "analogical computer" began with the recognition of an *analogy*: Holyoak read the account of Rumelhart, *et. al.* (1986, esp. Chs. 2 and 3) on the back-propagation neural network representing the Necker Cube illusion (Figure 6), in which the network learned to "see" the reversing faces perceived by humans:

> we had vague ideas that analogy depended on structure, similarity, and purpose, ... Holyoak realized that the two alternative interpretations of the Necker-cube drawing can be seen as two alternative mappings from the two-dimensional

> picture to a [cube] schema . . . . we worked together to figure out how a network could be used to perform analogical mapping by simultaneously satisfying constraints of similarity, structure, and purpose. We quickly succeeded . . . .

The two combined in designing three networks. *ACME* (Analogical Constraint Mapping Engine) was designed to model inferences according to the constraints the two had emphasized. *ECHO* (Explanatory Coherence by Harmony Optimization) seeks to model evaluation of hypotheses on the basis of their explanatory coherence; this was succeeded by a decision-analysis analog call *DECO*[3]. Finally, they designed a new model, exploiting the availability of several different sources, *ARCS* (Analog Retrieval by Constrain Satisfaction).

The account of parallel constraint satisfaction Holyoak and Thagard (1995, pp. 244-5) is clear. They define a symmetrically-connected neural network, trained by traditional Back-Propagation methods (Werbos, 1973/1995). Elements in retrieval or decision making or recall are represented by excitatory connections (if positive from *A* to *B*, then equally negative from *B* to *A*). An external constraint (Master or teacher) provides a special unit which continuously affects satisfactory units, until some units are activated and others deactivated (have no effect on the network). They analogize (p. 245):

> The elements of a decision are various actions and goals. The positive internal constraints come from facilitation relations: if an action facilitates a goal, then the action and goal tend to go together. The negative internal constraints come from incompatibility relations, when two actions or goals cannot be performed or satisfied together. . . . Once the elements and constraints have been specified for a particular decision problem, a constraint network can be formed . . . .

The problem with modeling creative thought in this manner may be evident: many different elements enter into the creative process, by common consent. Environment, motivation, sensitivity to ambiguity, and others all must be represented by the parallel constraint network. Is each of these sufficiently similar to be represented in a single network? I have argued (Leven, 1988, *et. seq.*) that theories ought to match data as closely as possible.

In a simple decision exercise (modeling why a consumer would prefer New Coke in a test environment, then reject it at the supermarket), Dan Levine and I designed a biologically-inspired network that captured the Nonisomorphism Problem (Holbrook, *et. al.*, 1985). Preferences among artificial experimental objects are not the same as choices in actual settings – because affect, heuristics, environment, and individual differences play critical roles in decision behavior. As we wrote (Leven & Levine, 1996, p. 289), "some of the most creative accomplishments . . . have been made by individuals who do not 'play the percentages,' . . . but rely on creative intuitions that violate easily measurable criteria for success." The network we implemented allowed affective attachment to Old Coke ("the *Real* thing") and high arousal levels for novelty in testing to be represented in a seven neural network model that utilized different structures to embody qualitatively different phenomena.

---

[3] It appears that DECO represents Decision Explanation by Coherent Optimization, although Holyoak and Thagard (pp. 245-7, the only reference to the network) are silent.

While Holyoak and Thagard admirably demonstrated that parallel constraint systems have impressive computational power, they failed to emulate the nonisomorphic elements in the decision environment. Like the mechanic trying to fix an automobile during lunch, using a can-opener for all functions is as vain an effort as representing all elements of a decision as if they were similar.

A superior effort is presented by the ongoing research program of Finke, *et. al.* (1992, 1995). Taking a rigorous approach to the psychological literature, they produce two networks in the Geneplore Cycle Model (Figure 6, Finke, *et. al.*, 1992, p. 18). The initial phase is Generation of Preinventive Structures. This is the combination (often randomly-produced) of retrieved forms which satisfy pre-existing Product Constraints (rules about purpose, function, or costliness) or freely generated from past experience. The second structure is Preinventive Exploration and Interpretation. The emergent (novel) properties generated by the combinations are elaborated and tested for aesthetic fitness. Should the properties be too "narrow", the next Generative stage seeks more extreme results; if they are too "broad", the next stage imposes deeper constraints.

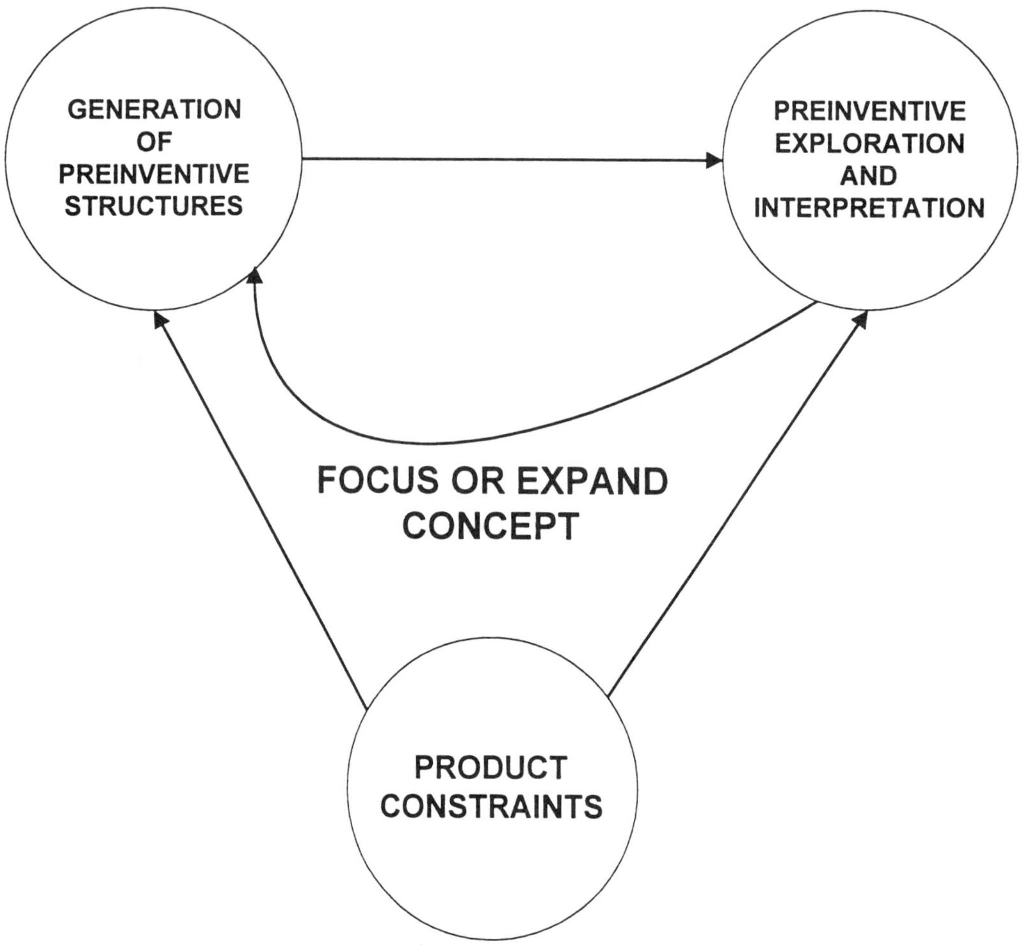

Figure 5. The Geneplore Model (adapted from Finke, *et. al.*, 1992, p. 18).

Rather than suggesting that Geneplore is biologically plausible, Finke, *et. al.* (Table 1, from 1992, p. 20, 2.1) provide a number of functions that the general technique could perform. If one assumes that cognition is dynamic and recursive (as in Horgan & Tienson, 1996), then such functions could take place "anywhere" in the system. But the Creative Cognitionists are not so naïve. The series of functions they suggest all require rearranging and reassembling – cognitive remapping which, as we shall see, is not so implausible.

Generative Processes	Preinventive Structures	Preinventive Properties	Exploratory Processes	Product Constraints
Retrieval	Visual patterns	Novelty	Attribute finding	Product type
Association	Object forms	Ambiguity	Conceptual interpretation	Category
Synthesis	Mental blends	Meaning	Functional inference	Features
Transformation	Category exemplars	Emergence	Contextual shifting	Functions
Analogical transfer	Mental models	Incongruity	Hypothesis testing	Components
Categorical reduction	Verbal combinations	Divergence	Searching for limitations	Resources

Table 1. Adapted from Finke, *et. al.* (1992, p. 20).

The main distinction between the Creative Cognition effort and *ECHO* is that Geneplore is an *exploratory* effort. No definitive architecture has been proposed – (although several have been used "experimentally").[4] It stands as a suggestion for both basic (Ward, *et. al.*, 1995, Ch. 3) and higher-order (Ward, *et. al.*, 1995, Ch. 8) models.

And the notion of recursive invention, recombination, and tuning is attractive. It is, plainly, in the tradition of Sternberg (1988, *et. seq.*) and his variations of selective encoding, selective comparison, and selective combination. It also recalls Ebert's (1994) "Cognitive Spiral" model, involving the recursive cycle Perceptual Thought ⇒ Creative Thought ⇒ Inventive Thought ⇒ Metacognitive Thought ⇒ Performance Thought. Based on Piagetan theory, Ebert (1994, p. 288) recalls Creative Cognition when he writes of "a perspective which accommodates the uniqueness of stimuli and the dynamic nature of the individual's knowledge base, while acknowledging the specialized contribution of creative thinking."

Geneplore represents a superior architectural proposal to the one offered by Holyoak and Thagard (1995)[5], although not as fully realized. It remains as a basis for advancing discussion – as we shall employ it, below.

The remaining outstanding differences, as we detailed at the start, the "great man" argument and the "disturbed affect" debate, are both based on the degree of individual difference between average problem-solvers and those able to resolve very difficult issues

[4] D. S. Levine, personal communication.

[5] For mathematical plausibility of the approach, see Wolpert & Macready (1996).

(*e.g.*, Einstein, Beethoven). There is agreement that culture, setting, knowledge-store, motivation, and tolerance for ambiguity are factors (in differing degrees).

Our argument renders the main differences moot. We shall discover that people who can achieve *special states* are capable of creative solutions. We shall find, further, that the nature of the representation process is central to these special states – which are experienced by people with varying genetic tendencies, early experiences, intellectual preparation, and current social environments. Sternberg's (1996) notion of Successful Intelligence will make much greater sense – once we have presented a biological model which supports his observations.

A Biology of Successful Intelligence?

We begin with a puzzle, presented to us by Csikszentmihihalyi (1996, pp. 55-7), who demands, "when are we going to deal with the *real* characteristics of creative persons? . . . [You] cannot assume the mantle of creativity just by assuming a certain personality style. . . .If I had to express in one word [their unique quality], it would be *complexity*. . . . They contain contradictory extremes – . . . each of them is a 'multitude.'" [*sic*]

The author, first reviewing the literature on personalities of the creative and then studying individuals himself, arrives at a set of "contradictions". Creative personalities express "dialectics" of traits. They contain greater *range* across a number of dimensions. Csikszentmihalyi (pp. 58-76) presents ten polar tendencies, each attained by the creative persons he has assayed (Figure 6).

ENERGY<-------------------->REST

SMART<-------------------->NAIVE

PLAYFUL<-------------------->DISCIPLINED

FANTASTIC<-------------------->CONCRETE

EXTROVERT<-------------------->INTROVERT

ARROGANT<-------------------->SHY

ANDROGYNOUS<-------------------->SEXUAL

REBELLIOUS<-------------------->CONFORMIST

PASSIONATE<-------------------->OBJECTIVE

SUFFERER<-------------------->JOYFUL

Figure 6. Csikszentmihalyi's typology of creative people's qualities.

Ostensibly, this set of astonishing contrasts suggests underlying psychopathology. Redfield (1993, p. 6) points out that most people with manic-depression are never diagnosed, let alone treated. Still, she points to the compelling qualities bipolar illness can confer[6]:

> It is the interaction, tension, and transition between changing mood states, as well as the sustenance and discipline drawn from periods of health, that is critically important; and it is these same tensions and transitions that ultimately give such power to [creative products].

Like the Geneplore model, Redfield (1993, p. 104) points out that creativity requires "a dipping into pre-rational or irrational sources [and] integration of these deeper . . . sources with more logical processes . . . ." Certainly, she suggests, in all creative process the fluency, rapidity, and flexibility of thought required for insight (Davidson, 1995) must be coupled with combinations of ideas or categories of thought. These are, as her research has attested, the hallmarks of manic-depression illness (Redfield, 1989).

Can such antipodal states be native to "normal" people? We may resort to analysis of language creativity for support. As Mark Turner has told us, everyday speech involves creative production of "parables", making sense of events we only partly understand or imagine. As we "fill in" visual scenes we only sample (Pribram, 1991, Ch. 3*ff.*), we employ "figural processing" commonly.

Turner (1996, p. 83) suggests that blending in parable (story-telling) follows patterns:

- The blend exploits and develops counterpart connections between input spaces.
- Counterparts may or may not both be brought [or fused] into the blend.
- The projection from the input spaces is selective.
- Blends recruit a great range of conceptual structure and knowledge . . . .
- [Blends] can be inputs to other blends.
- Blends develop structure not provided by the inputs. . . .
- The recruitment of a conventional metaphor to the blend is . . . partial, selective, and transforming.

Parable, then, "involves dynamic construction of input spaces, generic spaces, and blended spaces, multiply linked, . . . Meaning is a complex operation of projecting, blending, and integrating over multiple spaces. Meaning never settles down . . . ." (Turner, 1996, p. 106), just as Geneplore suggests. We *make sense* by combining appropriate components of our experience that seem related (proximally or distally) to recent events and important historical events. We *draw conclusions*, writing in space a story over time that explains the complexity we encounter (Figure 2, adapted from Fauconnier & Turner, 1996).

Turner (1996, p. 142*ff.*) suggests that grammar itself emerges from parable and intersubjectivity. As Lakoff and Johnson (1984) have observed, central experiences structure our notions of action, time, and others' feelings and beliefs. We learn from our own experience developing physical and mental mastery and from others who cue us and practice us in norms and common knowledge. Syntax and grammar *emerge*, then, from the complex experience of action over time and the process of blending experiences into parables.

---

[6] See her Chapters 3 and 4 (and references, *infra*) for review of supporting literature.

But is such a model biologically plausible? Work with autistic children suggests it is. Baron-Cohen (1995) and others have demonstrated that specific deficits in the ability to recognize others' support or suggestions – or to identify with their affects – produce profound failures to construct either experience or language. Failures to develop certain faculties prevent development of components of other-orientation which leave children autistic (Baron-Cohen, 1995, Ch. 5; see Figure 7).

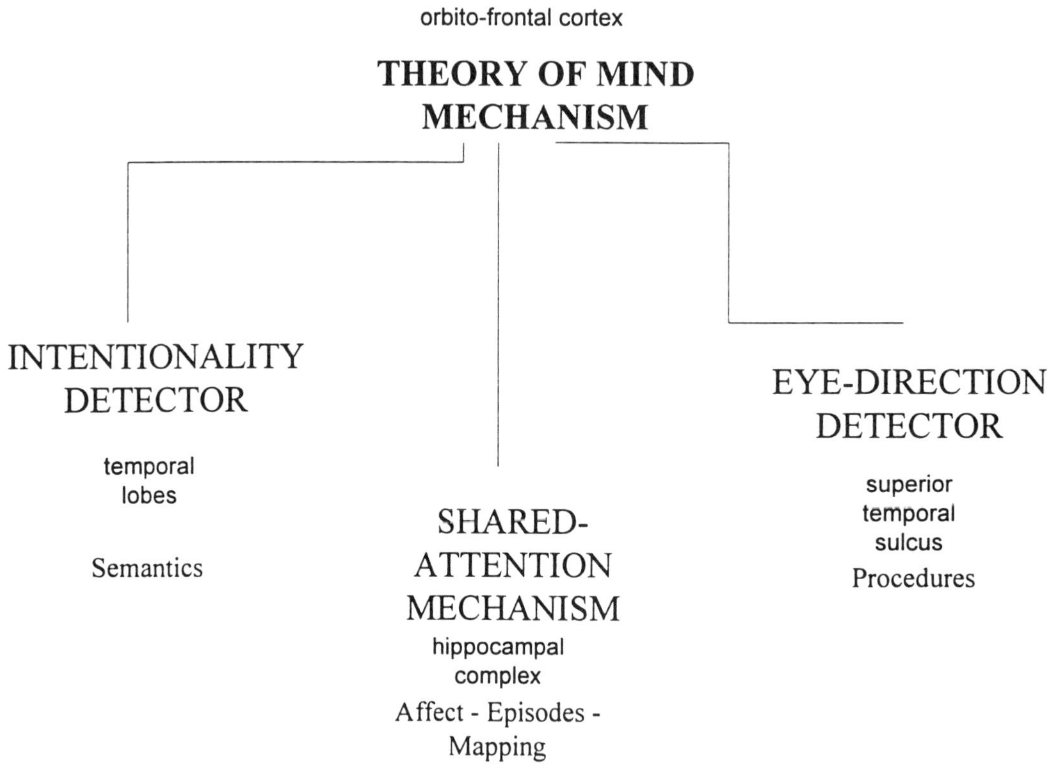

Figure 7. Baron-Cohen's (1995) theory: development of communicative competence.

Baron-Cohen (1995, p. 31) argues that disturbances with four fundamental processes in the infant prevents specific capabilities which allow most of us to recognize our own – and, specifically, others' – mental states. The mechanisms "reflect four properties of the world: volition, perception, shared attention, and epistemic states." While these mechanisms do not emerge *ab initio*, they are fundamental components of the first two critical periods of brain/behavior development and integration (the first, taking place within the first twenty-six weeks following birth, the next during the fourteenth to eighteenth months; Schore, 1995, Ch. 3).

The first, *Intentionality Detection*, allows us to interpret motion stimuli in terms of primitive volitional states, approach and avoidance (or in Baron-Cohen's, p. 32, terms "goal and desire"). Infants attribute goal-direction to movement and force (sometimes misinterpreting mechanistic actions as being goal-oriented). This finding is supportive of the "object-relations" vision of development; the child learns to recognize an external source of support and supply of needs.

The second mechanism Baron-Cohen identifies is the *Eye-Direction Detector*, a visually-based mechanism which allows locating others' eyes, finding their focus (and produces a pleasurable experience when mutual contact arises), and attributes mind-like capabilities to the looking other. These are dyadic mechanisms, allowing only the representation of object and goal. Without the sense that another and the person are perceiving and having intentions toward some object or goal, "your universe would be . . . an 'autistic' one. You would have sensations [and] images of people doing [and] wanting and seeing things, but you would have no way of knowing that [your goal] was the very same thing." (Baron-Cohen, p. 44).

The ability to produce "triadic representations" is produced by the *Shared Attention Mechanism*, which receives information about another's perceptual state, then computes shared attention by comparing one's own state. This process, he suggests, is a "comparator" which has the capacity to share information among *ID* and *EDD*.

Finally, Baron-Cohen posits a mechanism that allows a means of representing the set of epistemic mental states and tying the volitional, perceptual, and epistemic data into a unified framework, his *Theory of Mind Mechanism*. This permits use of Leslie's (1994) *M-Representations* which allow for locality of reference (*e.g.*, "We met a guy named Bill Clinton on our tour of the White House" is *not* necessarily equivalent to "We met the President on our tour of the White House"). This function, grasped by normal 4-5-year-olds, makes possible the deceptions children enjoy (the Prince fits the glass slipper on Cinderella's foot – and only then recognizes his true love).

Baron-Cohen is able to tie *ID* to Superior Temporal Sulcus, *EDD* to Amygdala, and *ToMM* to Orbital Prefrontal Cortex. He confesses puzzlement about the location of *SAM*. We shall demonstrate, below, the plausibility of the Hippocampal-Frontal region for this function.

Let us review what Baron-Cohen's work teaches. A child who cannot attribute goals, intentions, or shared interaction is closed to the world about him or her. It is only in the intersubjective environment children experience even before birth that we can orient ourselves to a world filled with others, make sense of their behavior, and express ourselves effectively. This world, co-constructed with others, is the only one in which creative action is possible – or, even, sensible.

Schore (1995) ties this cooperative construction to the object relations approach in psychotherapy. The self, formed through development of biological regulators, produces control of physiological, representational, and affective responses based on internalization of the interaction with the caregiver's psychobiological regulation of internal arousal. Energy dissipation is facilitated by learned responses, as Freud proposed in his Project (Pribram & Gill, 1975). Imagery is mobilized as internal working models to produce affect regulation and generate coping strategies. Schore (1995, p. 314) integrates the arousal-control, novelty-detection, object-recognition, and self-theory-building as, "mental representations that contain affective as well as cognitive components." He emphasizes that this Theory of Mind Model underlies both the recognition of others and *the internalization of an other*, quoting Greenberg and Mitchell (1983, p. 10), "a psychic representation of a person which in itself has the power to influence both the individual's affective state and his overt behavioral reactions."

Other functions than generalized theories of self and others are established. Appropriate attachments allow expectancies to develop that disruptions of arousal and dyadic coordination are subject to interactive repair and self-remediation. Negative affect can be transformed into positive experience; these expectancies, developing from the middle of the second year, allow the availability of complex symbolic working models for regulating and recovering from stressful states. These models are also guides for further interactions; they are the basis for working memory (Pribram, in press).

At the core of this process, Schore reports, are the development of orbitofrontal regulation of emotion-generating structures to hypothalamus and lower structures; sympathetic processing begins to dominate as the person's characteristic style of empotional expression emerges. The person develops the capacity to utilize reciprocal autonomic modes in response to challenge. It is also, at this point that heightened unconscious affect and dysregulation lay the groundwork for susceptibility to psychological disorder.

So the process Baron-Cohen reports has other implications. The brain develops pathways and capacities (physical sensitivities, degrees of connectivity, and capacities for representation) which allow the development of the successor to Theory of Mind – personality itself. Schore emphasizes the dual noradrenergic and dopaminergic limbic pathways as bases for affect and autonomic regulation, based orbitofrontally. He argues, "The confluence of their dissociable outputs is phenomenologically experienced as discrete emotions and enduring affect states. . . that are externally expressed as discrete behavioral states" (Schore, 1995, p. 366). He couples exploratory play and shame as bases for developing states, based on responses to hypostimulating (deprivation) and hyperstimulating (social) stresses which must be internalized as controllable behaviors. He emphasizes the centrality of the ability to maintain stable "tone" in the developing infant – up to age three, he reports serotonergic activity dominates activity in orbitofrontal cortex and, so, in the development of interaction with the world (Schore, pp. 525-7).

This other, associated form of early learning provides "prototypic presymbolic representations", stored as emotion-evoking non-verbal stimuli in the "early maturing visuospatial right cerebral cortex. . . which stores and processes self-and-object images" (Schore, 1995, p. 447). As Freud suggested, this nonverbal component of object relations is older and allows orbitofrontal regions to selectively process experience-appropriate affective information.

The notion that some forms of relations between self and self and self and other are preverbal and frame expectancies is the basis for the empathic work of therapy, Schore suggests. The ability of the therapist to recognize and resonate responses not available to a client's verbal processes creates an *intersubjective field*, "a system of differently organized, interacting subjective worlds" in which the therapist establishes positive affects within the implicit Theory of Mind of the patient, allowing for "dyadic merger" (Schore, p. 449). The therapist is able to "excite an emotional resonance within enduring internal object images (excitatory reverberations within the ventral and lateral tegmental forebrain-midbrain limbic circuits)" (Schore, p. 450).

Ultimately, the psychotherapeutic bond engages physiological stress inducing and stress regulating activities that are echoic of the infant experiences. Schore (1995, p. 471) argues that "these episodes of nonlinear psychodynamics are triggered by transferential communications . . . . affective interactions [which] represent critical nonverbal phenomena. . . ." Phase transitions in mood states are induced by engaging reward, inhibition, and (presumably) novel facilities that permit transitions into states which allow joint experience and exploration of negative experiences. Verbal self-regulatory require the replaying of the process of inhibition that allows development of language during the second year – and the establishment of object constancy (see discussion of Pribram and objects, below). It is only then that a "dialogical self" may be engaged – and the full Theory of Mind mechanism is engaged, as empathic function occurs in the elaboration of orbitofrontal cortex.

Ironically, as Turner would emphasize, our access to these experiences requires the use of related metaphors (parables). As Fiumara (1995, p. 62) writes, "The force of certain metaphoric presentations [is] the correlate of our active search for symbolic equivalents of inchoate transformational agents [just as] in our early experiences parents *can* transform our inner world through their behaviour [sic]." Mental objects exist in multidimensional *mental spaces* which need not take on verbal forms, once evoked. As Arbib and Hesse (1986, p. 38) emphasize, evoking these generic spaces is a product of

> the notion of an *essentailly embodied subject* [which] aspires to break the daulism of mind/body, . . ., self/other, and fact/value; it holds these dualisms to be untenable.

Fiumara (1995, pp. 78-9) suggests that effective therapy is, inherently, creative. Its course takes on a "plot", a space of discovery, and operates by the "myth of transformation". She writes that invoking metaphor in therapy makes it "natural to direct attention to constructs that involve isomorphic analogies moving across different domains. . . . A 'symbolic' game is proposed, leaving in the background the multiple reasons for playing it."

So Turner and Fauconnier's generic spaces are the places where empathic (non- or pre-verbal) therapeutic interaction takes place. Schore provides a biological substrate for this evocation.

Simply because such a model is biologically plausible and supported by behavioral research does not mean we have fulfilled Pribram's (1994) command that theories be demonstrable and testable as models. Further, we must find a means of connecting Geneplore with the six-fold Sternberg model. But work by Cloninger (1986 *et. seq.*) provides us a model that is based on biopsychology, has produced a testing modality which seems to confirm it, and represents a basis for producing a calculable neural network model that is as demonstrable as Holyoak and Thagard's. Its strengths (and inevitable weaknesses) shall serve as the foundation for our inquiry into normal and pathological creative process.

The project Cloninger initially set for himself, representing a theory of personality which could explain and classify anxiety-based disorders, has grown into a deeper creative project. It became, next, an effort to develop a theory and test to classify the most profound characterological disorders and comprise a theory of the genesis of healthy personality. Finally, as Geneplore and Sternberg's aggressive Successful Intelligence would predict, it has become an inquiry into the genesis and growth of human temperament and personality – a subtle effort to understand, classify, and treat inherited weakness and to support inherited strengths through the development of adult character.

Cloninger (1986, p. 170) presented a "unified biosocial theory of personality and its role in the development of anxiety states." He had distinguished two chronic anxiety syndromes, which he labeled Cognitive and Somatic Anxiety. These had different patterns of information processing, learning, and drug responses:

> Cognitive anxiety is prominent in patients . . . who have a high threshold for sedation and a low tolerance for stimulants or physiological stress. In contrast, patients with high somatic anxiety tend to have low pain and sensation thresholds, low sedation thresholds and improved performance on some perceptual tasks in response to stimulants or novelty. Individuals with high cognitive anxiety learn to avoid aversive stimuli passively more readily than others, but do not learn to maintain rewarded behavior more easily.

Cloninger noted a host of global differences in among these patients (Table 2, from C., 170). From this early observation, he was led to a review of the animal literature on anxiety and learned helplessness (reviewed in Leven, 1992 and 1997) and to studies by Eysenck (*e.g.*, 1993) and others on arousability and character. These induced an analogy to the typology devised by Eysenck; he adapted the Eysenck scale (neuroticism, introversion, psychoticism). Re-analyzing the factor-analytic data, he produced a three-dimensional typology involving reward dependence, harm avoidance, and novelty seeking (Figure 7).

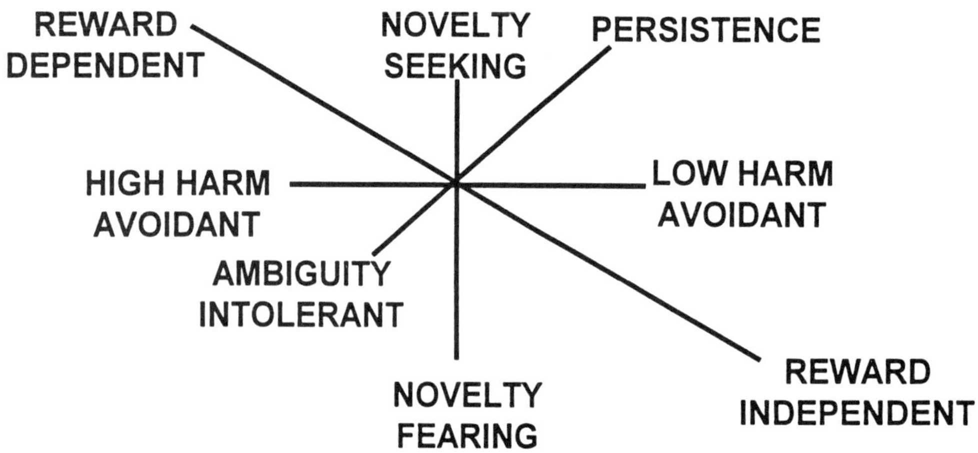

Figure 8. Adapted from (Cloninger, 1986, pp. 177-9, Figs. 2-4).

Drawing heavily on Gray's (1982) theory of anxiety, he posited comparability between his Harm Avoidance system and Gray's Behavioral Inhibition system. He emphasized the roles of serotonergic responses to punishment in the septo-hippocampal-frontal circuit and the finding that frustration produced a reduction in GABA-ergic processes in frontal and hippocampal regions. Serotonergic inhibitions from dorsal raphe to substantia nigra inhibit reward-signaling dopaminergic neurons based in striatum.

Cloninger tied his Novelty-Seeking (Incentive) system closely to Gray's Behavioral Activation system, ascending dopaminergic projections from ventral tegmentum to nucleus accumbens, and fronto-limbic regions. The dopaminergic pathways offered an "incentive function" , rewarding exploration.

Lastly, he tied Reward Dependence to Gray's Behavioral Maintenance system. The process is driven by supplies of noradrenaline, rising from locus coeruleus and projecting to amygdala, septum, hippocampus, and frontal regions. NE antagonists have produced anxious withdrawal (depression); reversal with agonists provides reduction in acetylcholine precursors and calms and rewards.

Cloninger produced a typology of personality type and susceptibility to anxiety and depression, based on biological and psychological measures of patient populations (Table 3), including longitudinal twin studies, questionnaire-behavioral studies, and pre-typed-drug response studies. He saw these as based on heritable tendencies, affected by sensitization in real-life experience.

Personality Type	Somatic Anxiety	Cognitive Anxiety	Reactive Dysphoria
Histrionic	++++	-	+++
Antisocial	+++	-	-
Passive-aggressive	++	+	++
Explosive-schizoid	+	++	-
Passive-dependent	-	+++	-
Obsessional	-	++++	-
Cyclothymic	-	-	++++
Imperturbable schiz.	-	-	-

Table 3. Cloninger's (1986, p. 193) typology of personality/susceptibility, adapted.

Cloninger (1986, pp. 217-8) emphasized that genetic tendencies were not solely determinative, only tendencies to respond to life events:

> stimulus and response become inextricably confounded when individuals vary in their response predisposition and their expectancies of the salience and significance of the same events. . . .
>
> Likewise genes do not determine the phenotype directly. . . . Relatively specific genes may be involved in susceptibility to specific neuropsychiatric complications . . . . However, even for such specific phenomena it is valuable to take into account the multifactorial systems that influences development . . . .

Cloninger (1987, p. 575) clarified his approach in a specifically general theory of personality, which followed his initial effort (Table 4). These, he suggested, were subject to construction of a systematic scale for hypothesis-testing.

Brain System Personality Style	Principal Neuromodulator	Relevant Stimuli	Behavioral Response
Behavioral activation novelty-seeking	Dopamine	Novelty Potential reward	Exploratory pursuit Appetitive approach
Behavioral inhibition harm avoidance	Serotonin	Conditioned signals for punishment, novelty, or frustrative reward	Active avoidance Escape
Behavioral maintenance reward dependence	Noradrenaline	Conditioned signals for reward or relief of punishment	Resistance to extinction

Table 3. Three Major Brain-Behavior Systems (adapted from Cloninger, 1987).

Cloninger followed this theoretical statement with development of the Tridimensional Personality Questionnaire, whose effective predictions were supported in normative (Cloninger, *et. al.*, 1991 and affective disorder populations (Svrakic, *et. al.*, 1992). The tests confirmed elicited self-report data from the questionnaire with behavioral and observational reports.

Support for Cloninger's genetic typing has come from studies of two types of alcoholics (Tihonen, 1996). The dopaminergic pathways from ventral tegmentum to accumbens and frontal cortex which are central to reward processes showed somewhat higher density (and receptor populations for D_2) in violent alcoholics – and even greater differentiation from normals in basal ganglia. Meanwhile, non-violent alcoholics showed extremely low densities. These results, Cloninger (1996, p. 623*ff.*) asserts are in keeping with his characterization of the two very different temperament and character types he has predicted. The impulsive (high novelty-seeking), risk-taking (low harm avoidant) drinker takes on the *hostile* profile; he is adventurous and socially-hostile, due to serotonergic insensitivity that tends to produce intentional isolation and high dopamine-induced aggressive exploration. This group, in experimental groups, has tended to commit violent crimes and be diagnosed with Antisocial Personality Disorder. The *passive* type begins drinking later; he is anxiety-prone (low dopaminergic response) and cautious (high serotonergic-deficits). This is the more heritable type, showing low novelty seeking (and the low dopaminergic intake predicted) and rapid losses of self-regulation.

Further confirmation of Cloninger's dopamine/reward/genetic predicted relation has flooded in. Manki, *et. al.* (1996, 8) note that "the dopamine D_2 receptor is the major site of action of neuroleptics. . . . Due to similarities in structure and function, two other D_2-like receptors, D_3 and D_4, [show evidence of involvement] in the pathophysiology of mood disorders." In fact, two variants of the gene for receptor four seem to provide a protective factor for mood disorders (four-repeat on DRD4) and a risk factor (five repeat on DRD4).

In a pointed editorial in *Molecular Psychiatry*, Licino (1996) emphasized the plausibility of Cloninger's dopamine prediction. Without citing two leading studies that emphasized the role of novelty seeking played by the specific complex (Ebstein, *et. al.*, 1996; Benjamin, *et. al.*, 1996), he wrote,

> data suggest that polymorphic variation in the gene encoding the D_4 dopamine receptor may contribute to the expression of . . . ADHD. The data are consistent with two recent independent studies . . . showing that . . . 7-fold repeat form of the DRD4 gene is elevated in individuals who rate high on a 'novelty seeking scale' [which includes] impulsiveness, exploratory behavior, and hyperactivity. The conclusions fit nicely with the work of Cloninger . . . .

As Cloninger predicted, Accili and Fuchs (1996, p. 93) have observed

> A physiologic response to dopamine requires a balance between activation of D_1-like. . . and activation of D_2-like receptors. . . . Mice harboring the mutant D_3 receptor gene [which cannot mediate dopamine action] are considerably more active than their litter mates with normal. . . genes. Our study. . . rekindles the notion of a specific role of D_3 receptors to regulate behavior [and that dysfunctions] may be responsible for behavioral disorders associated with hyperactivity.

This approach derived support not only from its own research program, but from experimental observations ordered by MacLean (1990), Benson (1995), and Leven (1992, *et. al.*). MacLean's evolutionary approach had motivated research by many writers (including Gray and Leven). Benson noted the connections between the three systems MacLean had projected and the frontal region's three areas (Pribram, 1987). Leven observed the similarity between the learned helplessness paradigm and its three classes of induction and elimination, mediated by similar neurotransmitters to those specified by Cloninger.

But further research efforts have shown that the three dimensions failed to differentiate individuals with personality disorders or poor social adjustment from well-adapted people with extreme personality profiles (*e. g.*, Cloninger, *et. al.*, 1988). Other personality disorders, neglected in the focus on anxiety disorders, were unmeasured by the TPQ. Further, Cloninger sought to embrace broader issues, such as autonomy, maturity, and self-actualization (Cloninger, , *et. al.*, 1993). He sought to extend his reach to prediction of diagnoses of all standard (DSM) disorders[7], designing the Temperament and Character Inventory (Svrakic, *et. al.*, 1993).

Cloninger's group believed "it would be useful to know whether the characteristics of personality disorders can be decomposed to distinguish between [their] core features [and] permit discrimination . . . ." (Svrakic, *et. al.*, 1993, p. 992). Further, it could be essential to distinguish biological from social causes of behavioral development, making it possible to distinguish adult growth which is obscured by current instruments.

Cloninger, *et. al.* (1993, p. 976) learned, "the presence of an interview diagnosis of personality disorder by *DSM-III-R* criteria is consistently predicted by low scores on two personality dimensions: self-directedness and cooperativeness" – two dimensions absent from the earlier typology. The five other dimensions they isolated permitted isolated sub-typing of all DSM disorders.

Memory research had demonstrated to Cloninger that a central determinant of recall and response was the bifurcation between working memory and long-term, temperament-specific memory. As Pribram (1996) has noted, the frontal sections of the brain encode working memory processes and do not fully develop until adulthood.

Urgency was provided by another researcher's developing perspective. Akiskal (1989) had expressed frustration with diagnostic dichotomies based on a confusion of biological and psychological causation. The classifications failed to capture behavioral complexes. He proposed, in a series of papers focused on affective disorder (reviewed in Akiskal, 1995), a new classification based on trait-state spectrum models. Genetics and experience led to three fundamental *temperament* types.

Akiskal (1995, pp. 100-2) argued that *temperament* "refers to the individual's activity level, basal affective tone, . . . intensity, reactivity, and variability [based on] underlying biologic dysregulation." *Personality*, Akiskal suggested, had to be distinguished as "more distal phenotypes, and refer primarily to interpersonal operations. . . ." Only by distinguishing genetic predisposition (temperament), developmental, and gender-related factors can use be made of those who "carried the same gene(s) as the . . . more severe phenotypes, but had somehow escaped the fate of being clinically ill." Akiskal emphasized the continuity between inherited tendencies and behavioral ranges; developmental and environmental factors either precipitated illness or enabled the same traits to become strengths in stable personalities – as Goodwin and Jamison (1990) have argued.

Perhaps the greatest incentive to reconceptualize, however, has been the emergence of a *fourth* factor in analyses of the Cloninger instruments (Stallings, *et. al.*, 1996[8]). *Persistence* was originally considered a component of reward dependence. When defined as "perseverance despite frustration and fatigue", it had higher heritability than all but one other dimension of personality (Cloninger, *et. al.*, 1993, p. 978). It showed no correlation with the other aspects originally conceptualized as reward dependence (sentimentality, social attachment, and dependence on approval). Cloninger makes no biological argument concerning this dimension (or genetic or chemical); while we anticipate, we note

---

[7] No self-report or dimensional method had been validated for DSM diagnostics.

[8] But see Heath, *et. al.*, 1994, for a more skeptical evaluation.

that he had earlier noted a distinctive combination of processes inducing Harm Avoidance – Serotonin only reduced impact of pain, while frustration seemed susceptible to the "downstream" impact of GABA at frontal and hippocampal regions (Cloninger, 1986, p. 181).

This finding led Cloninger to reconsider the social-developmental dimensions he had observed as exogenous to his earlier "personality" correlates. These biologically-based tendencies, he termed "Temperament" dimensions. He sought to extend his theory of personality to allow character development – "insight learning or reorganization of self-concepts." Cloninger was reaching out to Geneplore and Turner – people begin with a heritage (temperament), then construct a world from their experiences, efforts, and satisfactions.

People, he discerned, develop throughout life along three dimensions. These Cloninger termed "Character" dimensions. *Self-directedness* is the ability "to control, regulate, and adapt behavior to fit the situation in accord with individually chosen goals . . . ." This dimension is related to such concepts as self-efficacy and locus of control. *Cooperativeness* is "identification with and acceptance of other people." The scale was especially sensitive to social acceptance *vs.* Intolerance, empathy *vs.* social disinterest, helpfulness *vs.* unhelpfulness, compassion *vs.* vengefulness, other-considering *vs.* self-advantage. Finally, *Self-Transcendence* emerged as self-forgetful *vs.* self-conscious experience, identification with environment *vs.* self-differentiation, and spiritual acceptance *vs.* rational materialism (Cloninger, *et. al.*, pp. 979-82).

When the seven variables (Table 4) were factor-analyzed, the Persistence factor proved central to the three Character dimensions. Whether adults developed autonomy, selflessness, and concern for the whole was more dependent on their native tendency to persist than their attachment to novelty, their concern for punishment, or their search for rewards. The more people were *determined*, the more they matured (as Kohlberg, 1980, would suggest).

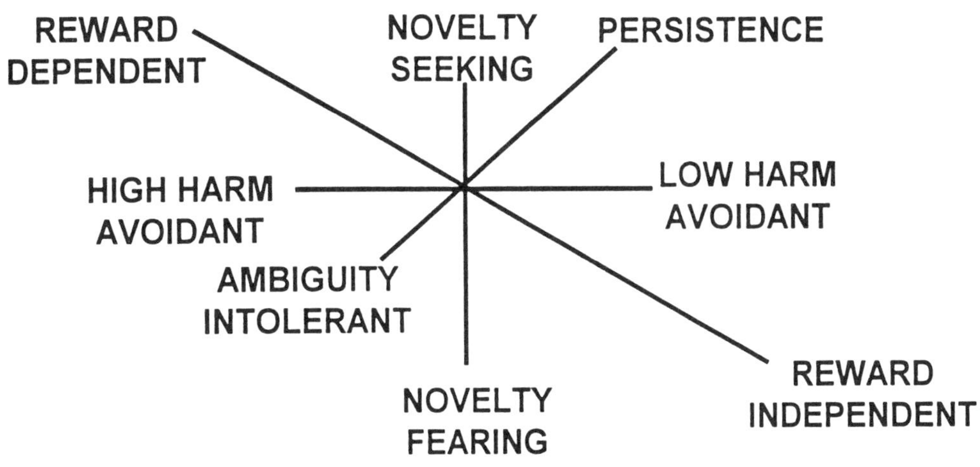

Figure 9. Cloninger's Temperament Dimensions: heritable characteristics.

Meanwhile, what is inherited is malleable, as he had emphasized in his early work. Sensitization and habituation were the original devices he suggested for modifying inherited tendencies. Now, the developing person can take control, establish goals, consider the needs of others, envision the long-term impact of his work – he can engage in Bateson's (1972, 1979) "deutero-learning", discovering through recursive examination the import of his experience.

All of which leads me to suggest the centrality of GABAergic descending systems as the basis for Cloninger's Persistence dimension. He has suggested that it is necessary to suspend the functioning of standard incentive systems (punishment and reward) in order to achieve long-term goals. The result of this suspension is ambiguity, whose tolerance his own data suggest underlies Persistence. One of the central functions of the GABA-benzodiazapine complex is to mediate frustration in the absence of reward (Ballenger, 1995; Petty, *et. al.*, 1993). It reduces anxiety. Those with high-availability and -sensitivity to GABA should be able to persevere, focusing on longer-term issues.

So, Cloninger proposes, the seven-dimensional framework is supported by experiment and commonsense. Whether it is biologically-sound remains for additional analysis – for the contributions of Pribram and Schore. Whether it is supportable as a theory of intelligence and tractable as a model for creative process requires analysis in light of Sternberg.

The solution proposed presents a puzzle (Table 4). Which is where creativity begins.

DIMENSION (Temp./Char.)	BRAIN SYSTEM	RELEVANT STIMULI	BEHAVIORAL RESPONSE
Novelty-Seeking	**Dopaminergic**	**Novelty**	**Pursuit**
Reward-Dependent	**Noradrenergic**	**Reward**	**Work**
Harm-Avoidant	**Serotonergic**	**Punishment**	**Avoidance**
Persistent	?	?	**Frustration resistant**
Self-Directed	?	?	**Internal locus**
Cooperative	?	?	**Empathy**
Self-Transcendent	?	?	**Selflessness**

Table 4. Cloninger's Seven-Sided Puzzle.

Toward a Biopsychological Model of Creativity

We begin where Sternberg (1996) advises: we employ the best mentors we can find, internalize their values, extend their work to a new problem. We start with one of Sternberg's mentors (Karl Pribram), integrate the work of the incisive Allen Schore, apply the results to Sternberg's own work – utilizing the insights of Pribram's inheritors, Stephen Grossberg and Paul Werbos.[9]

Pribram's work has presented another puzzle: he has insisted on the centrality of *four* primary brain-behavior functions (Pribram, 1991) and the equal importance of fron-

[9] Let me, gratefully, acknowledge another mentor and friend, Dan Levine, whose publications are annotated – but whose support and insights cannot be catalogued.

tal functions which, like Caesar's Gaul, are divided in three functional areas (Pribram, 1987). Reconciling the processes involved will present us with seven systems to underlie Cloninger's seven dimensions. We shall, then, require Schore to explicate the nature-nurture chemical-behavioral development processes – and Sternberg's old (1988) and new (1996) work to fill a framework for creative life. Finally, drawing on past modeling efforts (Leven, 1992, *et. seq.*, Levine, *et. al.*, 1992), we shall provide Pribram's (1994) demanded *objective-correlative* – for scientists to challenge, disprove, and remake in the image of superior understanding.

Although Pribram had spawned the widely-adopted "triune brain" hypothesis which served as the basis for two generations' research,[10] he reached out for the results he had encouraged – and specified four fonts of behavioral tendencies and an image-based substrate[11]. First, Pribram (1991, p. 134) cites the somato-motor system of pre- and post-Rolandic cortex as the source of *Images of Achievement*. He writes, "somatic motor processing is . . . the mechanism by which the organism manipulates and acts on the environment. Thus an independent set of action systems has developed." We image action as the integration in motor cortex of thalamically-mediated sensory processes which enable us to "figure out" or imagine and execute actions. This structure mediates skills (writing a bicycle, performing a memorized piano piece). It also allows us to organize our world into invariant forms in an object space (enabling us to move around a room without having to reorient to new *relative* furniture locations). We define ourselves, similarly, in the context of these object spaces: "The corporeal self thus becomes defined as that which remains invariant across all targets, that is, all achievements" (Pribram, 1991, p. 155).

This extraction of constancy from the world lies at the basis of Baron-Cohen's Eye-Detection process. It is the basis of routine action on the world (grasping, releasing) that is the basis for Turner's hypothesis that syntax derives from constancies we experience – and is the basis for the metaphoric foundation of language Lakoff and Johnson stress. Hitting a tennis ball and imagining ourselves hitting it are equivalent in the somato-motor *Images* world of skill and routine. "It is within this action space," Pribram (1991, p. 161) writes, "that the self achieves."

Pribram (1991, p. 197) labels the functions "Proprieties", "Practicalities", "Priorities", and "Envisioning", emphasizing that the brain functions across four competencies in two ways, common to all four (Figure 10):

> (a) a heterarchical parallel-*distributed* process that occurs at each level of a neural system, and (b) a *localized* hierarchy of relationships between levels of the system. . . . By this means, as well as by movement, the "dismembered," distributed processing "core" . . . becomes "re-membered".

[Insert Fig. 10, from Pribram, 1991, p. 3)

The basis for proper thinking is placed, in Pribram's holonomic framework, in the system comprised of striate, inferotemporal, and precentral systems, regions suffused

---

10 MacLean (1990, p. 421) implicitly acknowledges this, *e.g.*., at Refs. 63.

11 My construction of this material differs from the presentation in Pribram (1991); it is in keeping, however, with lectures by Pribram, Radford University, 1990-2, and with extensive discussions at The B.R.A.I.N.S. Center, Radford, 1990-3. Many thanks for the help.

with noradrenaline. The somato-motor systems respond to reward signals (from amygdala), learning to produce vectors of output representing correct (rewarded) motion for the situation. Inferotemporal cortex takes on a locally-executive role. Pribram (1991, pp. 180-1) explains the IT's role by telling of experiments that showed IT signaling the non-reinforced action (while the other systems responded appropriately). His team was puzzled until a member went to England and got stuck while crossing streets:

> He exclaimed, I now know what was going on . . . . It's just as when I cross a street here. I have learnt to think: don't look left, look right." . . . The correct response was made as much on the basis of comprehending "not red" as on the basis of [the correct criterion].

Priorities (including Pribram's now-famous "four F's) are ignored when the almond-shaped amygdala is removed. Damage to the amygdala, a system inundated with serotonergic input, eliminates recognition of social rank, physical danger (such as swallowing a match), fullness of eating and sexual desires. Punished behaviors are not recognized (and, so, avoided). The valuation process, the perception of comfort are disrupted. Lacking the sense of reward or punishment, we fail to become familiar with events – every event remains novel. The ability to recognize thresholds for pleasure and satiety also provide the ability to detect episodes, Pribram (1991, p. 217) asserts:

> An episode is thus marked by the achievement of a temporary stability [which] is characterized by . . . the structure of redundancy. The [representation of stimuli sets] connecting equi-valent polarizations in the network . . . is unique to the episode. . . .

Valuation, facilitated by the amygdalar complex, provides frames to experience. It allows us to attach worth, based on genetic tendency and experience. It connects events to values – allowing us to establish priorities. Content becomes significant context.

The hippocampus and limbic forebrain, driven by ascending dopaminergic systems (Kalivas, *et. al.*, 1993), allow us to distinguish innovation and practicality. We can distinguish the reduction of uncertainty from the new: novelty does not change the amount of uncertainty, but rather rearranges what is familiar. Three functions are combined in the complex. First is a buffer which holds stimuli for a period, then "flushes them" – unless there is a significant related change. Such a change then produces fluctuating activity in granular and CA1 regions, especially high during REM sleep and exploration; this may constitute pattern-fixing behavior. The second is irregularly distributed through the system, especially active when presented with a novel object; this system responds to approach-appetitive-consummatory cues *even after amygdalar removal.* This system produces cognitive effort in the absence of external reward; here, it seems, novelty is its own reward.

The hippocampus and limbic cortex, it seems, produce a *reference context*. Pribram (1991, p. 237) hypothesizes, "this context is formed by the . . . interference [computed] on the basis of mismatch signals. These result when the output layer generates something other than what the input layer is programmed to accept." Efficiency of effort seems to define the topography of a cognitive map. Like ancient mariners, whose renderings of experience was vague and frequently inaccurate, this practical system helps us navigate the new and the surreal.

Pribram (1996, p. 3) suggests that frontal functions allow us to know *what* to do, *when* to do, and *how* to do: "attentional and intentional processes necessary to the processing of 'one time learning' tasks . . . ." The three sections find, in the world as we experience it immediately, the proper, valuable, practical events to attend to and tasks to perform.

The orbitofrontal sector is connected primarily with amygdalar and temporal areas. It facilitates processing novel events, finding appropriate contexts to integrate them with, and is sensitive to autonomic responses to experience. It guides behavior to appropriate ends. Pribram has analogized this process to a database in the form of a "look-up table"; a particular cue requires a specific known response which is automatically located.

The dorsal segment utilizes cingulate cortex to gain access to the hippocampal complex. It monitors spatial and temporal processing that produce context. It establishes valuation (priorities) for actions. Pribram (see, also, Watanabe, 1996, p. 629) has analogized this process to a declarative (rule-based) search program, with capabilities for restructuring the data by semantic logic (classical, modal, non-monotonic).

The middle of the far frontal cortex serves as the integrator. With extensive access to sensory processing and input from the other two frontal regions, it assures the practicality of action. Pribram compares this process to a context-sensitive parser that retains partial matches and multiple clusters; it must be flexible enough to instantiate the learned structures immediately. Dias, *et. al* (1996, p. 69., suggest that this center may facilitate "response inhibition," shifts in motivation and attention that promote "cognitive flexibility."

Pribram suggests that the output of all of these "programs" resides in working memory, where it can be sampled and structured. The process of re-ordering this mass seems to be the function of each of the frontal areas as they develop "memory of the future" (Ingvar, 1985).

The most striking recent finding of Pribram, though, is that frontal areas continue development deep into maturation. Rapid frontal development has been documented by Pribram (1996) through the age of 21. This is supportive of the role of frontal cortices as skilled organizers of experience.

The six-fold model of Pribram is strikingly similar to the proposal of Cloninger – and, in fact, may be commensurable. While the map between the brain systems they describe is not homomorphic, the similarities between the behavioral descriptions they present is striking. The limbic novelty systems, the amygdalar reward systems, and the hippocampus-based inhibition systems take on striking similarity. The higher order systems (Pribram's Proprieties, Priorities, and Practicalities) *may* map well (to Cloninger's Self-Directed, Cooperative, and Self-Transcendent). Ironically, the Persistence dimension may be easiest to match, considering the Pribram-McGuinness (1975) work on effort (see Pribram, p. 237*ff.*) In fact, one could consider the extensive test support of Cloninger's to be supportive generally of Pribram's, as well.

The portrait Pribram paints is strikingly similar in function to the one described by Sternberg. Similar processes are postulated by both (see Figure 11). The recursive nature of search, evidenced in Geneplore and in Pribram's account of working memory, is evident. The Investment Perspective seems to take into account the Persistence dimension identified by Cloninger.[12]

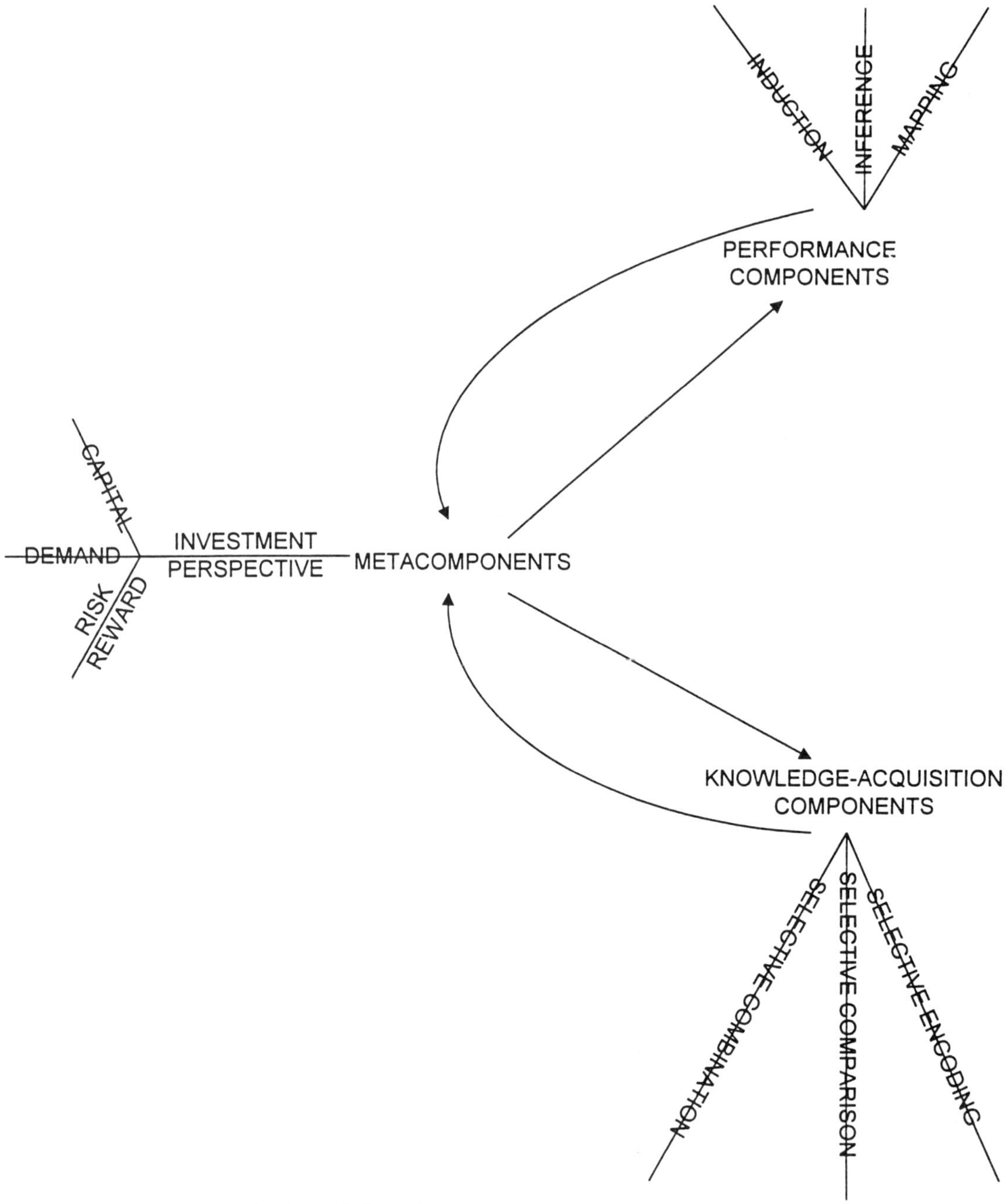

Figure 11. Sternberg's Creativity (from Sternberg, 1988; Sternberg & Lubart, 1995).

Whether the Sternberg model meets the Cloninger criteria is another matter. Recall the role of adult development. Clearly, the Investment Perspective meets the standard of self-directed behavior. In fact, Sternberg and Lubart (1995, p. 556) note that co-

---

[12] Pribram & Bradley (1996, p. 19*ff.*) note Sternberg's (1986) representation of the *decision/commitment* dimension in love relations. The long-term commitment may be comparable to Persistence, as well; factor analytic research may be indicated.

operation from associates can only take place "after the insightful thinker has laid the . . . groundwork." But, as Csikszentmihalyi and Sawyer (1995, p. 331) have observed, "The periods of hard work [surrounding] a creative insight are fundamentally social, deeply rooted in interaction with colleagues . . . ." There is a rhythm of intense interaction and withdrawal required – and a sense of commitment to the group and the domain.

What may be true is that the Sternberg group and the Csikszentmihalyi group have found what they were trained to find. Sternberg, as a personality psychologist, sought internal motivation. Csikszentmihalyi, as a social psychologist, sought environmental influences.

If both are correct, then the Cloninger approach is essential to building an effective understanding. Our model provides, as Pribram (1994) demanded, a first test.

A Provisional Model of Creative Process

Building a meso-level model of this body of work presents a substantial problem. We shall revise earlier work (Leven, 1995) and suggest a continuing research program. Plainly, however, the presentation must be impressionistic at this point. We begin with a case study.

In 1778, Mozart was earning a modest living as Europe's first composer-entrepreneur. His dazzling improvisations and light-hearted inventions provided him with a unique independence among the leading musicians of his time. While his music had grown in complexity and power, he remained true to the public's demand for cleverness and intricacy. Then his mother died.

She had been his main source of comfort and support (while his father provided the somewhat distasteful critical drive). He experienced her as unalloyedly happy and uncritically supportive. He experienced his first major loss, expressing despair and hopelessness in his correspondence.

It was then that Mozart resorted, for the first time, to the minor keys for a piano sonata. In K. 310, the A-Minor Piano Sonata, he begins with uncharacteristic sharp sadness. New shifts in the sad theme occur, as the angry voice in the treble register is uncharacteristically answered in the bass with a profoundly sad firmness. Voices lead his alternately plaintive and bitter argument from register to register, always advancing the argument from sadness to despair. Finally, after echoing her happy voice in an otherwise unremarkable second movement, Mozart arrives at confusion, reorganization, and

While there is nothing new in "voice leading" (a theme being repeated and transformed in a new register or instrument, as Bach had perfected), the emergence of unique characters in the two registers is astonishing. Mozart's music has become *dialogic*, the interplay of separately motivated voices expressing themselves differently with the same themes.

Dissonance, a tool in the composer's repertoire for adding interest in the return to tonality ("equilibrium"), becomes a staple of the angry voices of Mozart. Long sections of the attack in the last movement revel in irresolution ("disequilibrium") until, at the very last instant Mozart provides an angry, minor-driven resolution that is so sudden it shocks the listener.

It is not surprising that the A-Minor Sonata was not among Mozart's most popular works during his lifetime. It was disturbing, passionate, and personal in a new way that he would revisit years later. His D-Minor Piano Concerto (K. 466) represents his exposition of this new form, of voices not merely echoing each other but rather taking independent and assertive roles (with "minds of their own"). The ironic conclusion of K. 466, the piano quietly dissonant while flutes "laugh" at the conventional required resolution, derives from the twisted expression of despair at the close of K. 310.

We should not be surprised that the D-Minor, composed a dozen years later, was not popular with audiences or critics either. It did, however, engage the attention of a young composer who determined to study with Mozart and re-invent music in this new spirit. Although Beethoven never achieved his goal of learning from the now-unpopular composer, he composed a tribute to the new beginning his idol had made: he composed two *cadenzas* for K-466, which Beethoven played frequently as he created Romantic Music.

There are few revolutionary elements in K. 310, aside from the relentless sadness of its opening and the advance of development by various voices. Mozart combined many existing techniques, leavened by his desperate sense of loss, to give birth to Romantic music. His easy access to the many devices developed by the Bachs, the formal innovations of Haydn, and his own radically changed affect made possible one of the most profound creative acts of Western art.[13]

Yet the resolution of Mozart's crisis – and those of other creative people (Wallace & Gruber, 1989, *et. al.*) – is not dissimilar to the resolution of individual crises of "normal" people facing depression, panic or despair. As Emmy Gut (1989, p. 36-7) has written that the emergence of change

> is connected with significant thoughts and memories or new information now having become linked in a new way. Some of this linkage has also reached awareness. . . . It may take time and intense effort to make them available for linking with other data. . . . Actually, the linkage that resolves the depression does not even need to become conscious and elicit emotion. At times a depression subsides upon a burst of meaningful action without conscious recognition of why we became depressed or why we now are happily active again.

We shall argue that the process of "re-framing" which is embodied in modern psychotherapies (Harre, 1995; Gergen, 1982) revolves around the manipulation of linkages, as Gut has remarked. We shall demonstrate that the analysis of metaphor by Turner (1996) and Fauconnier (1994, *et. al.*) can explain how the evolution of internal conflict can allow us to understand how we overcome a conflict whose "intrinsic nature . . . stands in the way of its resolution" (Shapiro, 1989, p. 7).

Just as is language, our lives are "open works" (Eco, 1989), subject to reinterpretation as new insights connect mental spaces (Fauconnier, 1994) and acceptance of the

---

[13] Weisberg (1993, 255), discussing Mozart, maintains that "These processes – continuity based on near analogies, and discontinuities based on reasoning and sensitivity to external events – are components of ordinary thinking." Approximately, our argument.

many implicit possible ways of interpreting the past (Morson, 1996). The healing process and the creative process share the capacity for – and need for – revision and reconciliation.

But it is the process of language that opens our door. We shall begin by describing the manner in which readers grasp metaphors by breaking down assumptions and fixed "walls of belief". As Fiumara (1995) has suggested, the sharing of "the metaphoric process" is fundamental to the sharing of pre- and non-verbal expressive internal states. We shall proceed to find the connection between this process and the creation of ordinary metaphors (Lakoff & Johnson, 1987). Finally, we shall find that these ordinary mental processes are the genesis of the creative process of "geniuses" like Einstein and Mozart – and the source of healing for those suffering grief.

As Baron-Cohen and Schore have suggested, the evolving bi-personal field cannot be conceptualized in semantic terms. Fiumara (1995, p. 106) suggests such a "developing interactive field cannot be entirely representable even though a variety of fantasies may 'represent' our inner experience. . . ." Ironically, therapy is dependent on language, description, and a level of cognitive mastery. But, once empathic relations are established, by creating "consensual metaphors. . ., then one. . . may choose to respond *within* the same metaphoric agreement, . . . or else within an *intermediate* area extending between the metaphor which creates intimacy. . . , and the surrounding epistemology."

The creation of empathic mental spaces can be a *cooperative* venture, in which modes of experience are introduced into a **common mental space** (Figure 12). Fiumara (1995, p. 107) suggests, "The germinative area for metaphorical processes may then extend between human instincts. . . and a propositionally structured space. . . ." In this affective shared space, non-verbal experience may take expression by gesture and tone or by verbal allusion. The vulnerability (experience of pain, frustration, or confusion) can be reframed by the insight, cooperation, and persistence of the other – as if the temperamental features and deep memories of the afflicted could be processed, parsed, and represented by the higher-order functions of the other. Fiumara suggests that trauma may have to be segmented and re-viewed (in order to avoid falling into false, "local minima"), then reconstructed. She (Fiumara, 1995, p. 110) suggests,

> A dynamic process is constituted by its own past and by the way it generates linguistic attractors which direct its own future behavior; . . . When two different approaches are metaphorically connected [to] support one another, the link between them is strengthened. [The] subsequent relational activity will be likely to utilize. . . that same connection; it may thus even expand its field of attraction to interact with yet more thoughts that are simultaneously active.

The therapist shares, then, not only the supportive affect which enables the envisioning or re-experiencing of the trauma, but also the cooperative and self-transcendent role of resolving painful parables into useful, constructive ones. Langs' (1975) bipolar field is not simply a plain on which images are replayed and viewed sympathetically by another. Rather, they are cooperatively revisioned, by a therapist with "a constructive ear" who finds commonality with the "stories of others" and externalizes his/her intentions (Parry & Doan, Ch. 3).

The notions of "constructive ears" and germinative common mental spaces find resonance in Kao's (1996, p. 81-2) suggestion that business creativity and jazz music performance are similar. Personal stories (parables) must continually by revised:

> A creative [business] culture values give-and-take, friction and argument. A jam session is a polemical interaction of instruments, a conversation. That's the kind of tension [a] corporate culture [should] encourage. . . . How else are we to test the resilience of the creative proposal.

Sounding like a multi-person Geneplore, Kao (1996, p. 7) says creativity emerges from a resonant process – "Contradiction and divergence create puzzlement, tension, and stress. And the almost instinctive desire to overcome or encompass [it requires] creativity [and is] likely to increase on networks of information exchange." It is the empathic ability to merge a group "working memory" with the "deep memory" representations of each member, modulated by the individuals' characteristic responses. The resonance has become multidimensional, as have harm avoidance, reward dependence – and even the group's self-reinforcing persistence.

Further, each member of a business team has a different vision of its goals, a different set of values for each part of its duties, and a different orientation toward the tools available for achieving them (Leven & Smith, 1995; Smith & Leven, 1996). These views are not myopic – there is no unitary "economic reality" but only a dynamic interactive process by which organizations arrive at choices which are more or less efficient (Leven & Smith, 1996). The ability of a team to capture many views of reality and the many options they imply, by shared parables in the common mental space, empowers effective decisions.

The creation of a quasi-secret team to create the IBM PC, allowing the group to violate company policy of employing internal resources and standards, permitted engineers to envision an assembly-line metaphor constructed of licensed and purchased common parts; it allowed creative responses by many employees with differing expertise and acquaintance with technologies. The group created a re-visioned computer (Parry & Doan, 1994), made of replaceable parts built to simple specifications.

The regnant metaphor of a computer as a system made of complex systems was replaced with the parable of a system composed of simple systems which could be upgraded without changing the broad definition of the "computer". This exchange of an image for a story allowed the ease of transition from 8088-based machines capable of word-processing to "80686" multi-processor machines, capable of performing trillions of computations a second (launching space vehicles or responding to thousands of complex requests simultaneously; see Leven, 1996).

The decision to free the engineers to "jam", to contribute to the construction of a new story that

While not all of us can resolve deep sadness by revolutionizing a field of art or science, as have Mozart and many others, we can create productive metaphors for our lives where repression or ignorance had lain before. And these acts of personal re-creation may be just as important, to each of us, as the inception of Romantic Music or the development of the Theory of Relativity.

Our model takes Mozart's creative process as he experienced it: a crushing loss, inaccessible to easy expression or purging. Figure 12 demonstrates the faculties brought to bear, structurally and chemically. We can assume that the keen sense of loss produced a high state of arousal. His attentional resources must, initially, have been reduced; we know from the result he achieved, that he quickly recovered focus and the capacity for cognitive effort. Still, the pain he suffered must have resulted from a shortage of serotonergic resources.

Instead of engaging in Harm Avoidance, however, Mozart accepted his need to express this most personal loss. As he persisted, we must believe that his system compensated by supplying massive GABAergic resources which sensitized serotonin and noradrenergic receptors (Petty, *et. al.*, 1993), provided immediate tolerance for his disorientation by binding hippocampally and distally, and promoted release of endogenous opiates (Ballenger, 1995).

Mozart must have engaged resources he had developed over many years of practice and study. Using a Geneplore-like technique, he utilized his Knowledge-Acquisition Components. His choice of key, the unlikely A-Minor (to which he would return rarely), was surely initiated by selectively encoding his pain and mapping the frustration he felt into the repeated staccato beginning: Alfred Einstein (1962) tells us A-Minor is, forever more, his "key of despair." After re-stating the theme, he must have engaged in selective comparison with the many pieces he knew; only the Bach fugues he had practiced could capture the sense of urgency he sought. He would not introduce a second theme, but only provide a figure he could beckon without developing. Restlessly, as Geneplore would suggest, Mozart returns to the single theme again and again.

In the slow movement, a listless first theme is followed by a brief reprieve (as if a single recalling of his mother's laughter); by now, the goal focusing would not allow him to withdraw from his statement. He would return to the minimal second movement theme.

Finally, as the endorphins must have suffused his system, he could see the global aspect of his loss: his cry for his lost consolation could give voice for thousands of others (as we know now, of course, millions. Selflessly, he threw away the spectacle Schuster's fireworks offered and turned the final movement into a transformational moment. Einstein (1962, p. 244) says, "one may look in vain in all the works . . . of the period for anything similar." The driving irony of the conclusion, surely the expression of supreme cognitive effort (a mapping across culture and time), would return as he "looked back in anger" in the magisterial D-Minor Piano Concerto. Beethoven would listen and learn.

As Emmy Gut (1989) has emphasized, personal grief can be transformed into purposeful self-fulfillment. The expression of primary process, through a grieving mother's creation of Mothers Against Drunk Driving or a devastated adult testifying against the sexual predator who violated him years before, is the escape through cognitive process Freud promised in his Project (Pribram & Gill, p. 127).

Mozart had disciplined himself into transcendence; his music would never be the same. He would four years later reach further into the future, finding a C Minor voice (in K. 457 and 475) that Einstein (1962, p. 247) would assert, "contributed a great deal towards making 'Beethovenisme' possible."

This music did not enjoy the popularity of his lighter pieces (the A-Minor was ignored when it was published in Paris) and rarely performed publicly during his lifetime. But, even as Mozart turned to variations on a children's tune and a "lovable" sonata (K. 330), echoes of his themes recur in odd corners – as if the genius, himself, were amazed at his own achievement.

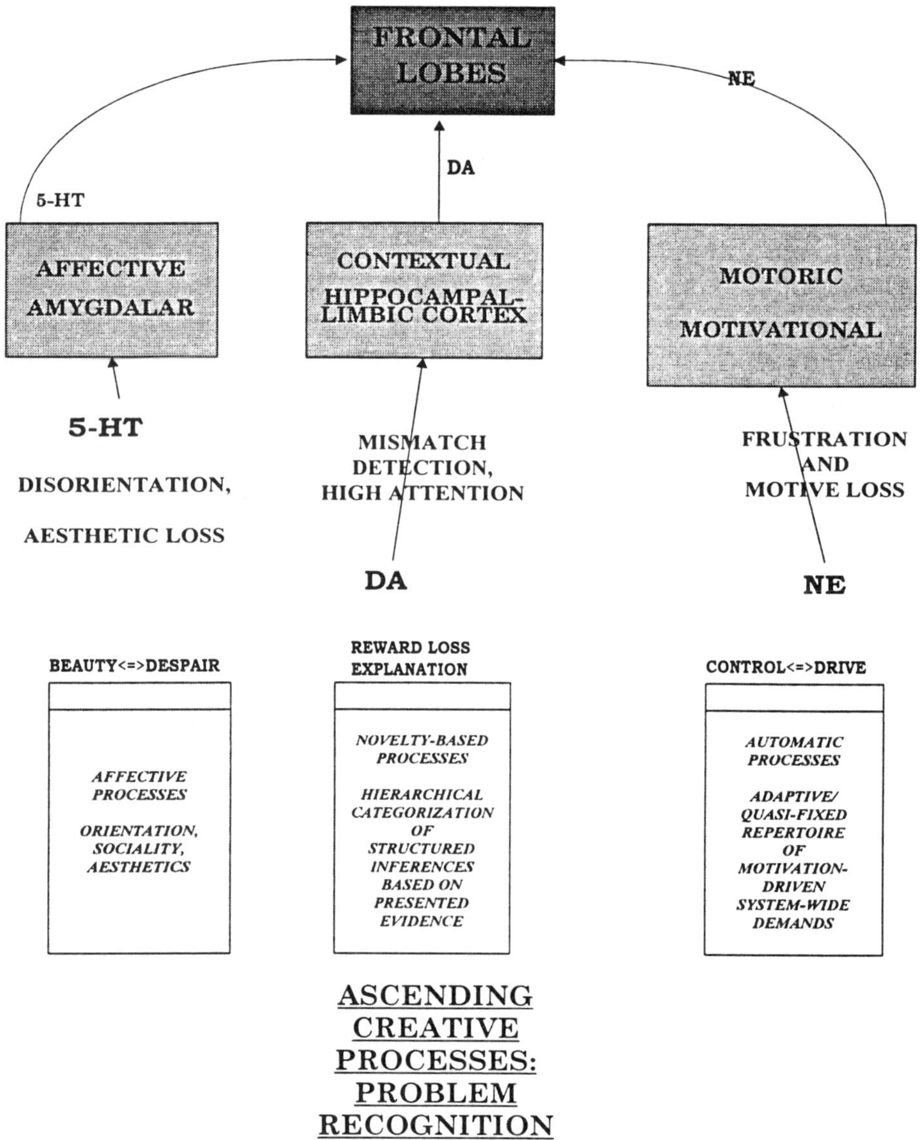

Figure 11. Beginning of the creative response to crisis.

Suffering from bipolar illness (Davies, 1987, Ch. 16), he reached out to family in the midst of composition. His brother in law wrote (in Davies, p. 150):

> when he was busied with an important work[,] he not only spoke confusedly and disconnectedly, but occasionally made jests of a nature which one did not expect of him . . . .

He gambled irresponsibly, drank to physically-dangerous excess, and offered extravagant gifts to strangers. He would jump on the piano as students played well, sing along with the music, and then "meow".[14]

As Wallace and Gruber (1989) and Csikszentmihalyi (1996) emphasize, however, most creative people are not disabled by their gift. Which leads us to specify a model of creativity, following the observational leads of Sternberg and Davidson, *et. al.*

Plainly, creative people have extraordinary lability. Both D2 and D1 families of receptor must be susceptible to rapid sensitization, along with a_2 receptor up-regulability. They must have ready access to GABA and endogenous opiate supplies; if they are high on Cloninger's Cooperative dimension, they must have high oxytocin sensitivity.

DIMENSION (Temp./Char.)	BRAIN SYSTEM	RELEVANT STIMULI	BEHAVIORAL RESPONSE
Novelty-Seeking	**Dopaminergic**	**Novelty**	**Pursuit**
Reward-Dependent	**Noradrenergic**	**Reward**	**Work**
Harm-Avoidant	**Serotonergic**	**Punishment**	**Avoidance**
Persistent	**GABA/FC-Hippo**	**Ambiguity tolerance**	**Frustration-resistant**
Self-Directed	**OrbitoFrontal/D1**	**Goal focus**	**Internal Locus**
Cooperative	**Dorsal-F./oxytocin**	**Other's affect**	**Empathy**
Self-Transcendent	**Mid-F./endorphin**	**Global good**	**Selflessness**

Table 4. Cloninger's Seven-Sided Puzzle Provisionally Solved.

Generative Processes	Preinventive Structures	Preinventive Properties	Exploratory Processes	Product Constraints
Retrieval	**Visual patterns**	**Novelty**	**Attribute finding**	**Product type**
Association	**Object forms**	**Ambiguity**	**Conceptual interpretation**	**Category**
Synthesis	**Mental blends**	**Meaning**	**Functional inference**	**Features**
Transformation	**Category exemplars**	**Emergence**	**Contextual shifting**	**Functions**
Analogical transfer	**Mental models**	**Incongruity**	**Hypothesis testing**	**Components**
Categorical reduction	**Verbal combinations**	**Divergence**	**Searching for limitations**	**Resources**

Table 5. Geneplore re-presented, for comparison's sake.

We compare Geneplore to the provisional solution we have offered for a purpose: clearly, Geneplore captures critical creative qualities and represents a validation check for any model purporting to be a vehicle for evaluation. We see that novelty in Geneplore is tied to visual patterns, as Schore would suggest fits with exploratory pursuits into pre- and non-verbal object domains. Clearly, attribute finding is a pursuit-oriented process.

---

[14] Jamison (1993, p. 298, n. 63) is agnostic.

We may match Association with the Persistent style; here ambiguity tolerance and ambiguity meet. Conceptual integration is, plainly, a frontohippocampal "mapping" function, in Pribram's terms. Synthesis seems tied to Cooperative empathic behavior, reaching across experiential domains (as in metaphor and in therapeutic alliance); affective qualities seem essential to any such process.

Transformation involves emergence from one conceptual domain to another, across generic spaces; it seems similar to Reward Dependent behavior of repetitive effort and attention focusing. Analogical transfer seems analogous to Self Transcendence, the reaching across incongruous spaces, patiently seeking a global solution. And Categorical reduction seems to match Harm Avoidance: each is based on awareness of limitations and careful mindfulness of resources.

If this reading is accurate, Geneplore seems to encompass most of the Cloninger dimensions. The remaining one, Self Directedness, appears to match the Sternberg "executive" model – producing a practical calculus of problem selection and time management.

We shall argue that the ingredients for a successful Pribram metaphor are present: data has produced models that have been shown to be commensurate. We now seek demonstrability.

Finally, we shall outline a neural network approach to creativity in seven dimensions (Table 4). First, we define neural network designs which capture the flavor of processing of each of the seven dimensions. Then, we shall propose limited linking structures among them. This effort, different from any we have made before (*e.g.*, Leven & Levine, 1996*a* and *b*, Smith & Leven, 1996, Leven, 1997), will certainly have unique qualities. We shall continue to seek to base them on the biology we understand.

The pre-training of these networks (prior to "real-life experience" data) is intended to confer on them tendencies that are in keeping with the biological and early-life model of Cloninger and Schore. Different pre-training, of course, is intended to represent the relative sensitivity and reactivity to characteristic personal exposures; the responses elicited should be in keeping with these tendencies.

The Novelty-Seeking network (Figure 13) is constituted of a set of Adaptive Resonance networks (representing the CA1-CA3 interaction), filtered by a READ field reading input somatosensory data and by a dipole field (representing Dentate Gyrus and Cingulate Cortex). The READ field is motivated (modulated) by output from Amygdala. The output of the dipole field is read directly by a multi-dimensional ART layer (representing middle Frontal Cortex): this layer also accepts input from the other two frontal areas.

The Harm-Avoidant network (Figure 14) is a partially self-regulating hysteretic tensor network (Mayergoyz, 1989, *et. al.*). This network meets the several requirements laid out in Pribram (1991, Ch. 7): begin with clear tendencies (genetic endowment), have the capacity for many simultaneous minima states of many dimensions, have substantial learning capacity and significant stability of memory states affected by outside criteria (sensory-motor control response), and have the capacity – in the extreme – to be "wiped out" and self-train with the reference signals from amygdala for reward/punishment and episode beginning/end. The layers of the network "predict" future states of higher layers, allowing for preparatory set and elaborate procedural responses.

The Reward-Dependent (Figure 15) network is an embedded hierarchy of dynamic back-propagation networks, as discussed in Leven (1995, *et. al.*). The hierarchy is capable of developing new vertical representations, based on error feedback from amygdala. Three hierarchies co-exist at each level: a classical inference network, a modal logic network, and a non-monotonic logic network. The fit of incoming data determines which network receives data (through mutually inhibitory connections based on goodness of fit). The Amygdalar portion is a vector hysteretic network, capable of relative stability and incentive and punishment learning; it provides the hierarchy with cues, based on a one-to-one-level mapping.

The Persistent (Figure 16) network is a descending Adaptive Resonance network, taking its activation level from output of the activation level of the central novelty-seeking ART. Its chemical extensions are generated by a spreading activation (Anderson, 1973) network, which spreads ambiguity tolerance in response to frontal states.

The Self-Directed (Figure 17) network is a set of three layers of hysteretic networks, connected to the central mid-Frontal network. The flexibility of hysteresis (learning "flux") is based on read-out of the current ascending NE-representing nets.

The Cooperative network (Figure 18) is a set of self-training back-propagation layers (taking activation levels from the Reward-Dependent output and mid-Frontal output). This network outputs to a set of outstars, representing basally-released oxytocin. This output, in turn, affects sensory-motor inputs to the above-noted networks; in other words, "warmth" expressed by this network can filter patterns earlier inducing hostility or fear.

The Self-Transcendent network (Figure 19) is an inherently recursive five-dimensional Adaptive Resonance network with high gain and low activation levels, ensuring rapid convergence. The network receives direct input from Novelty-Seeking, Persistent, Self-Directed, and Cooperative networks and from the output to the sensory-motor inputs to Reward-Dependent and mid-Frontal networks. Thus, the network is "grounded in reality" and is capable of driving the Geneplore-like process which must be the font of change and sense-making in the environment. This is the "response inhibition" and change in affective association in real-time identified by Dias, *et al.* (1996).

The Quasi-Biological Birth of Creativity as Parable

We are left, then, with the puzzles first presented: is Turner's account of language and thought plausible? are Turner and Pribram and Cloninger, *et. al.*, commensurable? do the described networks "fill in" Cloninger's puzzle satisfactorily? can we implement the simple blending technology (p. 7) and the deeper problem of parable (p. 15) it promises? does any of this tell us anything about creativity? can we provide a meso-level model sufficiently robust that it is both descriptive and defeasibly? and, even if it is, does it advance the field, offering a useful objective-correlative, as Pribram (1994) demanded?

Even these questions are inadequate – in the same way Pribram acknowledges his own deficiencies (Pribram, 1991, Epilogue). Still, is there any alternative to response?

It appears that there are strong mappings between Pribram's account of fronto-limbic interaction, the Geneplore model, and Turner's broad notion of blends as recursive, partial, selective, and transforming. The more basic notion finds resonance with both Geneplore and Sternberg: is selective projection not similar to selective combination and selective encoding? Emergent structure, almost a paraphrase of the Geneplore principle, is at the heart of the recursive nature of Sternberg's Metacomponents. And revision by reverse mapping simply represents re-entrant connectivity, built into the conceptual and computational models.

The proposed networks should not fit Cloninger precisely. Instead, they represent the first gun-shot in a long battle to reach common ground.

What have we learned about creativity? Especially, what have we learned about the intersubjective element Csikszentmihalyi emphasizes? As our networks have been designed (see Smith & Leven, 1996, *et. al.*), they seek to embrace symbolic, gestural, and affective communication. Certainly, the input data of the real world would overwhelm such a model quickly; neural network modelers like to snicker that "that's just a *scale* problem." It is fair to recognize that many models of the Space Shuttle worked flawlessly in the days before *Challenger* destroyed seven lives. Still, that is the strength of models: we can test our ideas and *nobody gets killed*!

Finally, does our model meet muster? We wait for the inevitable response.

Lastly, it is important to represent what was *not* said (though not for lack of attention). Little was said about the roles of metaphor and parable in therapeutic relationships (Langs' "bipolar fields"). Little was established about the conceptual revolution applying an other-regarding model, such as Cloninger's, to policy analysis, as Donald Schon would like.

Finally, there remains a Buberian silence. The assumed over persists. Who is he or she: internalized, intimate, or obscure? And, does it matter? As Camus (1955) might have insisted, Sisyphus would have thought so.

References

Accili, D., and S. Fuchs. A new look at dopamine D3 receptors. Molecular Psychiatry 1. 1996. Pp. 93-4.

Akiskal, H. Toward a Temperament-Based Approach to Depression: Implications for Neurobiologic Research. In Gessa, B., W. Fratta, L. Pani, and G. Serra, eds.. Depression and Mania. NYC: Raven, 1995. Pp. 99-111.

Akiskal, H. Validating affective personality types. In Robins, L., and J. Barrett. The validity of psychiatric diagnosis. NYC: Raven. 1989. Pp. 215-235.

Amabile, T. The Social Psychology of Creativity. Berlin: Springer-Verlag. 1983.

Baron-Cohen, S. Mindblindness. Cambridge: MIT. 1995.

Ballenger, J. Benzodiazepines. In Schatzberg, A., and C. Nemeroff. The American Psychiatric Association Textbook of Pharmacology. Washington: American Psychiatric. 1995. Pp. 215-30.

Bateson, G. Toward an Ecology of Mind. NYC: Ballentine.

Benjamin, J., L. Li, C. Patterson, B. Greenberg, D. Murphy, and D. Hamer. Population and familial association between the D4 receptor gene and measure of Novelty Seeking. Nature Genetics 12. 1996. Pp. 81-4.

Boden, M. The Creative Mind. NYC: Basic. 1992.

Camus, A. The Myth of Sisyphus and Other Essays. NYC: Anchor. 1952.

Cloninger, C. A Systematic Method for Clinical Description and Classification of Personality Variants. Archives of General Psychiatry 44, pp. 573-88. 1987.

C. Cloninger. The psychobiological regulation of social cooperation. Nature Medicine 1, pp. 623-5. 1996.

Cloninger, C. A Unified Biosocial Theory of Personality and its Role in the Development of Anxiety States. Psychiatric Developments 3, pp. 167-226. 1986.

Cloninger, C., T. Przybeck, and D. Svrakic. The Tridimensional Personality Questionnaire: US normative data. Psychological Reports 69, 1047-57. 1991.

Cloninger, C., S. Sigvardsson, and M. Bohman. Childhood personality predicts alcohol abuse in young adults. Alcoholism 12, pp. 494-505. 1988.

Cloninger, C., D. Svrakic, and T. Przybeck. A Psychobiological Model of Temperament and Character. Archives of General Psychiatry 50, pp. 975-89. 1993.

Csikszentmihalyi, M. Creativity. NYC: Harper-Collins, 1996.

Damasio, A. Descartes' Error. NYC: Grosset/Putnam. 1996.

Davidson, J. The Suddenness of Insight. In Sternberg, R., and J. Davidson. The Nature of Insight. Cambridge: MIT. 1995. Pp. 125-56.

Davies, P. Mozart in Person. NYC: Greenwood. 1987.

Dias, R. T. Robbins, and A. Roberts. Dissociation in prefrontal cortex of affective and attentional shifts. Nature 380, pp. 69-72. 1996.

Ebert, E. The Cognitive Spiral. Journal of Creative Behavior 28, pp. 275-91. 1994.

Ebstein, R., O. Novick, R. Umnasky, B. Priel, Y. Osher, D. Blaine, E. Bennett, L. Nemanov, M. Katz, and R. Belmaker. Population and familial association between the D4 receptor (D4DR) exon III polymorphism associated with the human personality trait of Novelty Seeking. Nature Genetics 12. 1996. Pp. 78-88.

Eco, U. The Open Work. Cambridge: Harvard. 1989.

Einstein, A. Mozart. NYC: Norton. 1945.

Fauconnier, G. Mental Spaces. NYC: Cambridge. 1994.

Finke, R. Creative Insight and Preinventive Forms. In Sternberg, R., and J. Davidson. The Nature of Insight. Cambridge: MIT. 1995. Pp. 255-80.

Finke, R., T. Ward, and S. Smith. Creative Cognition. Cambridge: MIT, 1992.

Goldenthal, P. Doing Contextual Therapy. NYC: Norton, 1995.

Goldman, D. Dopamine transporter, alcoholism and other diseases. Nature Medicine 1, pp. 624-5. 1196.

Goodwin, F., and K. Jamison. Manic-Depressive Illness. NYC: Oxford. 1990.

Guilford, J. Way beyond the IQ. Buffalo: Creative Education Foundation, 1977.

Greenberg, J., and S. Mitchell. Object relations in psychoanalytic theory. Cambridge: Harvard. 1983.

Gruber, H. The evolving systems approach to creative work. In Wallace , D., and H. Gruber, eds. Creative people at work. NYC: Oxford, 1989.

Hadamard, J. The psychology of invention in the mathematical field. Princeton, N.J.: Princeton, 1945.

Heath, A., C. Cloninger, and N. Martin. Testing a Model for the Genetic Structure of Personality. Journal of Personality and Social Psychology, 66, 762-75. 1996.

Holyoak, K., and P. Thagard. Mental Leaps. Cambridge: MIT. 1995.

Horgan, T., and J. Tienson. Connectionism and the Philosophy of Psychology. Cambridge: MIT. 1996.

Hudspeth, W., and K. Pribram. Stages of brain and cognitive maturation. Journal of Educational Psychology 82, pp. 881-4. 1990.

Ingvar, D. "Memory of the future." Human Neurobiology 4, 127-36. 1985.

Kao, J. Jamming. NYC: Harper-Collins. 1996.

Kalivas, P., L. Churchill, and M. Klitenick. The Circuitry Mediating the Translation of Motivational Stimuli into Adaptive Motor Responses. In Kalivas, P., and C. Barnes, eds. Limbic motor circuits and neuropsychiatry. Boca Raton: CRC. 1993. Pp. 237-88.

Knight, R. Contribution of human hippocampal region to novelty detection. Nature 383, pp. 256-9.

Langs, R. The Bipersonal Field. NYC: Jason Aronson. 1975.

Leven, S. Choice and Neural Process. Unpublished dissertation. UT-Arlington. 1988.

Leven, S. Modeling Creative Process. Presented at M.I.N.D. Conference on Models of Higher Mental Function. College Station, TX. 1995.

Leven, S. Learned Helplessness, Memory, and the Dynamics of Hope. In Leven, S., and D. Levine, eds. Motivation, Emotion, and Goal Direction in Neural Networks. Hillsdale, N.J.: Erlbaum. 1992. Pp. 259-99.

Levine, D., S. Leven, and P. Prueitt. Integration, Disintegration, and the Frontal Lobes. In Leven, S., and D. Levine, eds. Motivation, Emotion, and Goal Direction in Neural Networks. Hillsdale, N.J.: Erlbaum. 1992. Pp. 301-35.

Licino, J. Editorial. Molecular Psychiatry 1. 1996. Pp. 83-4.

Lubart, T. Creativity. In Sternberg, R., ed. Thinking and Problem-Solving. NYC: Academic. 1994. Pp. 289-332.

Manki, H., S. Kanba, T. Muramatsu, S. Higuchi, E. Suzuki, S. Matsushita, Y. Ono, H. Chiba, F. Shintani, M. Nakamura, G. Yagi, and M. Asai. Dopamine D_2, D_3 and D_4 receptor and transporter gene polymorphisms and mood disorders. Journal of Affective Disorders 40. 1996. Pp. 7-13.

Mayergoyz, I. Mathematical Models of Hysteresis. Berlin: Springer-Verlag. 1989.

Morson, G. Narrative and Freedom. New Haven: Yale. 1996.

Parry, A., and R. Doan. Story Re-visions. NYC: Guilford. 1994.

Petty, F., G. Kramer, and W. Hendrickse. GABA and Depression. In Mann, J., and D. Kupfer. Biology of Depressive Disorders: A. NYC: Plenum. 1993. Pp. 79-108.

Pribram, K. Brain and Perception. Hillsdale, N.J.: Erlbaum. 1991.

Pribram, K. From metaphors to models. In Leary, D., ed. Metaphors in the History of Psychology. NYC: Cambridge. 1994. Pp. 79-103.

Pribram, K. Subdivisions of the frontal cortex revisited. In Brown, E., and E. Perecman. The frontal lobes revisited. NYC: IRBN. 1987. Pp. 11-39.

Pribram, K. The hippocampal system and recombinant processing. In Isaacson, R., and K. Pribram, eds. The hippocampus, Vol. 4. NYC: Plenum. Pp. 329-370.

Pribram, K. The Deep and Surface structure of Memory and Conscious Learning: Toward a 21st Century Model. In Solso, R. The Science of the Mind: the 21st Century. Cambridge: MIT. To appear.

Pribram, K. The Work in Working Memory. Presented at N.I.H.C.H.H.D. Conference, Prefrontal Cortex. Bethesda, Md. 1996.

Pribram, K., and D. McGuinness. Arousal, activation and effort in the control of attention. Psychological 82, pp. 116-49. 1975.

Pribram, K., and D. McGuinness. Attention and para-attentional processing. In Friedman, D., and G. Bruder, eds. Annals of the New York Academy of Sciences, 658. NYC: NYAS. 1992. Pp. 65-92.

Pribram, K., and R. Bradley. The Brain, the Me and the I. Manuscript. 1996.

Pribram, K., and M. Gill. Freud's Project Revisited. NYC: Basic. 1976.

Redfield, K. Mood disorders and patterns of creativity. Psychiatry, 52, 125-34, 1989.

Redfield, K. Touched with Fire. NYC: Free Press, 1993.

Schore, A. Affect Regulation and the Origin of the Self. Hillsdale, N.J.: Erlbaum. 1995.

Simonton, D. Foresight in Insight? A Darwinian Answer. In Sternberg, R., and J. Davidson. The Nature of Insight, 465-94.

Simonton, D. Genius, creativity, and leadership. Cambridge: Harvard. 1984.

Smith, S., T. Ward, and R. Finke, eds. The Creative Cognition Approach. Cambridge: MIT. 1995.

Stallings, M., J. Hewitt, C. Cloninger, A. Heath, and L. Eaves. Genetic and Environmental Structure of the Tridimensional Personality Questionnaire. Journal of Personality and Social Psychology, 70, 127-40. 1996.

Steinberg, M. Old Brahms a Modernist? New York Times, September 24, 1996. Pp. H-31-2.

Sternberg, R. Successful Intelligence. NYC: Simon & Schuster. 1996.

Sternberg, R. A triangular theory of love. Psychological Review 93. 1986. Pp. 119-35.

Sternberg, R. The Triarchic Mind. NYC: Penguin. 1988.

Sternberg, R., and J. Davidson. Preface. In Sternberg, R., and J. Davidson, The Nature of Insight. Cambridge: MIT. 1995. PP. xv-xviii.

Sternberg, R., and T. Lubart. An Investment theory of creativity and its development. Human Development, 34, 1-31. 1991.

Sternberg, R., and T. Lubart. An Investment Perspective on Creative Insight. In Sternberg, R., and J. Davidson. The Nature of Insight. Cambridge: MIT. 1995. Pp 535-58.

Sternberg, R., and T. Lubart, Creative giftedness: A multivariate investment approach. Gifted Child Quarterly, 37(1), 7-15. 1993.

Svrakic, D., T. Przybeck, and C. Cloninger. Mood states and personality traits. Journal of Affective Disorders 24, 217-26. 1992.

Svrakic, D., C. Whitehead, T. Przybeck, and C. Cloninger. Differential Diagnosis of Personality Disorders by the Seven-Factor Model of Temperament and Character. Archives of General Psychiatry 50, pp. 991-9. 1993.

Tihonen, J., J. Kuikka, K. Bergstrom, P. Hakola, J. Karhu, O-P. Ryynanen, and J. Fohr. Altered striatal dopamine re-uptake site densities in habitually violent and non-violent alcoholics. Nature Medicine 1, pp. 654-7. 1996.

Turner, M. The Literary Mind. New York: Oxford. 1986.

Vandenberghe, R., C. Price, R. Wise, O. Josephs, and R. Fackowiak. Functional anatomy of a common semantic system for words and pictures. Nature 383, pp. 254-6. 1996.

Ward, T., R. Finke, and S. Smith. Creativity and the Mind. NYC: Plenum. 1995.

Watanabe, M. Reward expectancy in primate prefrontal neurons. Nature 382, pp. 629-32. 1996.

Werbos, P. The Roots of Backpropagation. NYC: Academic. 1994.

Werbos, P. Presentation at Fifth Appalachian Conference. 1995.

Wolpert, D., and W. Macready. Combining Stacking with Bagging to Improve a Learning Algorithm. SFI-TR-96-03-123. Santa Fe Institute. 1996.

18

Values, Agency, and the Theory of Quantum Vacuum Interaction

Raymond Trevor Bradley
Institute for Whole Social Science, Carmel, CA
and
Center for Brain Research and Informational Sciences
Radford University
September, 1997

Send All Correspondence Regarding This Manuscript To:
Dr. Raymond Trevor Bradley
Institute for Whole Social Science
25400 Telarana Way
Carmel, CA 93923
Tel: (408) 626-8057; FAX: (408) 622-9423

Abstract

Laszlo's theory of quantum vacuum interaction describes the evolution of all order in the universe as a result of the interaction between the holofield, a subquantum field of potential energy, and the multi-level hierarchy of matter and living systems. A review of the scheme reveals two problems which question its application to psycho-social interaction. One is the use of classical holography (*image* processing), which is inherently deterministic. The second is Laszlo's view of the brain as a passive processor of information. Both preclude human agency in the generation of psycho-social organization. An alternative approach is developed here that solves both of these problems. First, I draw on Piaget's work to show how thought and reason, and thus agency, is produced by the logic of cooperative interaction. And second, I use both classical holography and quantum holography (*information* processing) to develop a nondeter-minist account of communication in social collectives. The principles of classical holography are used to describe how purposeful action is imaged and processed; and the principles of quantum holography are used to describe how a stable order of endogenous organization is generated by the processing of information about the interactions among members. By describing how the collective's normative order (the system of shared values, beliefs, and norms) operates to effect the construction and regulation of the collective's communicative structure, I show how indeterminacy is introduced and, hence, the potential for an active human agent in social life.

Introduction

In his recent book, *The Interconnected Universe*, Ervin Laszlo (1995) outlines a general theory of nature, the "theory of quantum vacuum interaction" (QVI). The theory aims to describe the basic processes "whereby the universe evolves in an unbroken (though not necessarily linear) sweep from basic physical entities to the complex open systems that are physicochemical, biophysical, biochemical, biological, ecological, and even psychological in nature." It therefore "encompasses physics and biology, and ultimately the human and the social sciences, as special aspects within a general scheme" (Laszlo, 1995, page xiv).[1]

In brief, the theory views the evolution of *all* order in the universe as the result of the interaction between a subquantum field of potential energy (the "holofield") and the multi-level hierarchy of matter and living systems. The holofield is a holographic order in which information from all physical, biological, and psycho-social interaction is enfolded into the energy wave-forms of the "empty" space of the quantum vacuum. The quantum vacuum, therefore, is the medium for constant universal interconnection--a field of potential energy that interrelates all orders and scales of organization in the universe. Action in the material world is guided by a holographically generated "prompt." Because the holographic prompt contains information about a system's past states as well as information about its position within a hierarchy of systems, the prompt is the means by which novelty and change are generated so that self-organizing evolution results.

Laszlo demonstrates the theory's utility by showing how it provides a unified description of order at the physical and the biological levels. As a social scientist, though, my main interest is in how QVI accounts for order at the psycho-social level. Although Laszlo shows how he believes QVI explains such extraordinary psychic and socio-cultural interaction at-a-distance phenomena as extra-sensory perception and collective mind, he leaves unanswered the critical question: the role of an *active (willful) human agent* in the generation of psycho-social organization.

Agency requires an active brain: a self-conscious brain that can assess the significance of and assign meaning to sensory input; a brain that can evaluate and deliberately reorganize the contents of memory; a brain that can establish priorities and plans; and a brain that can implement, and coordinate with others, purposeful programs of social action (Miller, Galanter, and Pribram, 1960). In short, agency requires an intentional brain with the capacity to make decisions and take action about *what is valued* by the individual or the collective.

In relation to this question, though, there are some fundamental scientific difficulties with Laszlo's scheme. In addition to the nonmaterialist basis of QVI and the determinism resulting from the use of classical (or *image* processing) holography, there is also the view of the brain as simply a passive processor of information. Thus in this scheme, the values that in-form behavior are predetermined and, like all information, are merely transduced by a passive brain. Given these elements, how can interaction between the quantum vacuum's holofield and the psycho-social order involve a willful agent? And how is purposeful collective organization is generated? In short, as the scheme is presently conceived it appears to preclude human agency.

Following a review of the theory of QVI in Part I, my goal in the remainder of this essay is to show how these difficulties can be resolved, while remaining within the context of a holographic approach. In Part II, I draw on Piaget's recently translated analysis of the genesis of thought and reason in cooperative systems (Piaget, 1965/1995a & b) to show how the logic of cooperative interaction produces a rational actor in social life, and how this logic, therefore, is the basis of purposeful forms of collective organization. Then, in Part III, I distinguish between classical and quantum holography (information processing), and present evidence of holographic-like social organization.

In Part IV, I develop a nondeterminist holographic account of communication in social

---

[1]Unless otherwise indicated, all page references are to Laszlo, 1995.

collectives that incorporates human agency. I use the principles of classical holography to describe how the collective constructs and communicates an image of the object of its purposeful activity, and the principles of quantum-holography to describe how the collective processes and communicates information from the interactions among members to generate stable endogenous organization. To resolve the problem of holographic determinism, I show how the construction of the communicative structure in social collectives opens the system to new input. By describing how the normative order (the collective's system of shared values, beliefs, and norms) sets and regulates the parameters that govern the construction and operation of the communicative structure, I show how indeterminacy is introduced and, hence, the potential for an active agent in social life.

PART I. The Theory of Quantum Vacuum Interaction

Theoretical Framework

The basic axiom upon which the theory of QVI rests is the "reasonable assumption" that "*preferential interconnections* must exist in nature."

> When a set of parts or elements are constantly and systematically interconnected, the set generates order spontaneously. ... Connections must be selective, though not exclusive of other connections; they must be of the kind where each event is preferentially cor-related with a specific set of other events. [If] The system's preferential correlations ... are specific yet sufficiently broad-ranging, the order that evolves does not merely repeat the order that has already emerged, but creates possibilities for attaining genuine novelty within the range of possibility.
> (Pages 4 and 5; addition in parenthesis mine.)

To elaborate this precept into the "simplest possible scheme of thought capable of uniting the *prima facie* disparate domains of physical and of living nature" (page 25), Laszlo develops three concepts: constant universal interconnection among all elements, information storage and retrieval according to holographic principles, and action directed by the probabilistic dynamics of chaotic-state large-scale Poincare systems.

QVI's fundamental concept is the postulate of *constant universal interconnection*: that everything in the universe--every entity and every event--is fundamentally and inseparably interconnected in a constant state of interaction. Like Bohm's concept of the implicate order (Bohm, 1980), the physical basis for universal interconnection resides in the quantum vacuum, a nonmaterial subquantum field filled with zero-point energies that define the ground state of the universe. A basic property of the quantum vacuum is its enormous energy density, which, based on Wheeler's calculations, is equivalent to 10^{94} g/cm^3, and, according to Bohm, exceeds the energy in all known matter by a factor of 10^{40}. Another property is the complex nature of quantum interactions, which Laszlo attributes to the electromagnetic field in which quanta are embedded.[2] The interaction between electrically charged quanta and the zero-point electromagnetic field (ZPF) produces secondary electromagnetic fields within the vacuum which "must be universally extended" (page 30). The vacuum's enormous energy density and the production of universal secondary electromagnetic

[2]This is a significant point of difference with Bohm who, in his theory of the quantum potential *Q*, views quanta as highly complex systems capable of responding to a pilot wave of "active information" (see Bohm and Hiley, 1993).

fields enable it to be conceptualized as a cosmically extended field, a medium with virtually unlimited information-storage capacity. This field is the "memory" into which and from which information about *all* interactions in the universe is stored and retrieved.

To describe the physical processes by which the vacuum processes information, Laszlo conceptualizes it as a "universal field with holographic properties" (page 13). Holographic organization is an order in which information about the features of an object is encoded in the interference patterns created by intersecting wave-forms of energy bouncing off the object. Since information about the organization of the whole object is distributed and stored (enfolded) throughout the holographic field, an image of the entire object can be reconstructed from any location. Thus as a field "endowed with memory," the "hologram provides the kind of memory that can connect phenomena at a universal scale" (page 12). Envisaging the structure of the ZPF to be mediated by secondary scalar fields created by the motion of quanta, Laszlo postulates that the scalars approximate Schrodinger waves. Produced throughout an entire medium, Schrodinger-wave holograms store vast amounts of information. They also allow for storage and retrieval of time-varying information.

QVI and the Organization of Matter

To produce an effect on "observable phenomena" of the material world, the scalar-mediated ZPF of the vacuum must interact with quanta. This interaction between scalars and quanta approximates a two-way Fourier transformation in that "...the vacuum encodes the coefficients of the interfering scalarwave fronts produced by the motion of quanta. In doing so the vacuum carries out the equivalent of the forward Fourier transform: it translates a pattern from the spatiotemporal domain to the spectral domain. In the inverse transform--from the spectral to the spatiotemporal domain--the interference patterns encoded in the vacuum 'in-form' the motion of quanta in space and in time" (page 33).

For the vacuum to affect the action path-ways of matter beyond the quantum level, Laszlo employs the macro-statistical dynamics of large-scale Poincare systems.[3] Because large-scale Poincare systems are in constant interaction with their environments, the selection of the action path is influenced by a "subtle but statistically significant [contextual] constraint" (page 9; my addition). At the level of microscale systems--of ensembles of quanta--the constraints on action paths appear in the form of probabilistic irreversabilities, as minute but significant biases in direction generated, thus, by the statistical dynamics of populations of systems. For macroscale systems the interaction-effects from fluctuations in the vacuum, when amplified by chaos-dynamics, appear as Poincare resonances in ensembles of quanta. Because "these interactions involve ... the $3n$-dimensional configuration-space of the whole multiquantal system ... in dynamically indeterminate ('chaotic') states, macroscale systems can be 'in-formed' with the ZPF-conveyed Fourier-transform of their $3n$-dimensional configuration-space" (page 36).

It is through the wave-generating properties of large-scale Poincare systems (viz, continuous spectra and resonance) that the evolutionary direction of action in the material world is selected. Because holographic information can be stored as superposed wave-patterns in multiple dimensions, a holographically generated "prompt" can therefore connect each system to its own evolutionary

[3]He argues against Prigogine's dissipative system dynamics (Nicolis and Prigogine, 1977; Prigogine and Stengers, 1984) on the grounds that the individual system is treated as if it is separated from its environment during the bifurcation process, so that, in Prigogine's scheme, chance determines the system's evolutionary trajectory.

history and also connect entire hierarchies of multi-level systems to one another. This constant holographic exchange of information between parts and levels within a given system, and between individual systems and the complex multi-level structure in which they are embedded, results in self-generated evolution of increasingly more complex yet coherent forms of order (pages 13-15).

QVI and Biological and Psycho-Social Organization

In applying the scheme to living systems, Laszlo draws on his concept of the quantum vacuum as an information-rich holofield and on his concept of chaotic-state large-scale Poincare systems. Living organisms are viewed as far-from-thermodynamic equilibrium systems whose form is holographically encoded into the ZPF of the subquantum domain so that the vacuum's holofield operates as a multidimensional morphogenetic field. The morphogenetic field not only connects the individual organism (via a species-specific frequency resonance) to the evolutionary morphology of its own species, but connects it, as well, to the space-time structure and dynamics of its environment. As a complex multi-level ensemble of populations of macroscale systems, the organism is inherently unstable and ultrasensitive; it is affected by "unmeasurably fine fluctuations" in the form of minute internal and external variations which, when amplified by chaos dynamics, produce "measurable--in fact, decisive--effects" on the organism's evolution (pages 81-82):

> The scalar-mediated vacuum holofield encodes the species-specific imprints of all organisms, and those that 'match' these imprints are constantly affected by them. ... Whenever the ultrasensitive dynamics of chaos come into play, the interaction-effect guides not only processes of [organic] generation, but also processes of regeneration. ... the orienting biofield [is] the scalar-mediated ZPF. The Fourier transforms of the state-space of particles and the configuration-space of living systems are encoded and conveyed to systems in isomorphic spaces, in-forming them both with their own prior states, and with that of the larger systems in which they are integrated.
> (Pages 83 and 84, respectively; my addition.)

At the psychological and sociological levels, QVI plays a "functionally analogous" role. First, the quantum vacuum is the medium of transmission for a wide variety of "space-leaping" spontaneous communications among human "brain/minds:" extrasensory perception, anomalous recall of near-death and past-life experiences, altered states of consciousness (Chapter Five), creativity, the sharing of sociocultural 'archetypes,' and the 'collective unconscious' (pp. 129-136). Second, the vacuum's holofield functions as an extrasomata memory for storing patterns of psychic and also of socio-cultural "transpersonal experience."

Contrary to present understanding of the brain,[4] these patterns and the other data that make up long-term memory are *not*, according to Laszlo, physically located and stored in the neural structures of the brain. Instead, QVI suggests that the contents of long-term memory are stored *outside* the brain; they are encoded as interference patterns *in* the vacuum's holofield. Thus the brain acts like a radio or a TV receiver in that it operates "***only** as the transducer of signals* received from the vacuum holofield" (page 100; my italics and emphasis). The pattern encoded in the signals is multi-dimensional and includes information on the individual's position in the multi-level ensemble of systems in which s/he is embedded: "in one direction the pattern would include the Fourier

[4] See for example, Freeman, 1995, and Pribram, 1991, and the articles in Gazzaniga, 1995.

transform of the configuration of cells and molecules that constitute the human body; while in the other it would include the dynamic $3n$-dimensional configuration-space of the social and ecological milieu of the individual" (page 134).

Retrieval from the vacuum's holofield follows holographic principles: the brain accesses the holofield, and by performing a reverse Fourier transformation it converts the interference pattern into an image of the original object or event that recorded the pattern. Thus recall of the contents of psychic experience, or of the interactional schemata that in-form socio-cultural life, is a "holographic process occurring in an interaction between the brain of the subject and an ambient holographic field" (page101).

The Difficulty

Laszlo's insistence that the brain is merely a *passive* processor of information seems opposed to any concept of a willful human agent and creates a basic difficulty for two aspects of his account of psycho-social organization. It creates a problem for his hypothesis of "interactive creativity" (pages 129-132). This he defines as "the *elaboration* of an idea or a pattern in two or more minds in interaction" by way of interconnection through the vacuum's holofield (page 130; my emphasis). By its very definition, 'elaboration' connotes an active, purposeful brain--a brain able 'to produce by labor' or 'work out in detail' (*Conc. Ox. Dict.*, 1964: 391). Such a brain is at odds with Laszlo's passive brain, a non-active processor of through-put.

The same difficulty arises in Laszlo's account of the collective level. Focusing on the sharing of collective experiences (e.g., "feelings of oneness," "the collective unconscious," or *communion*), Laszlo views QVI-brain/mind interaction as the means by which individuals are integrated into the sociocultural order. But this focus on communion takes what the collective values as given and leaves aside the most important aspect of human social experience, namely, the phenomenon of purposeful organization. Whether in the form of a family, a corporation, a market economy, or a nation state, purposeful organization requires an order of social *communication* that is constructed by an *active* brain--a thinking, reasoning brain that evaluates information, establishes priorities, makes decisions and plans, and implements complex programs of social action to accomplish a multiplicity of objectives in the material world.

As noted at the outset, these difficulties have their roots in the inherent determinism of classical holography. And it is clear from his reference to Gabor's paper in *Nature*, "A New Microscopic Principle" (Gabor, 1948), that this is the kind of holographic organization that Laszlo has in mind in the theory of QVI.

But before we take up this problem, I want to resolve the question of an active brain by drawing on Piaget's analysis of the genesis of rationality in cooperative social systems (Piaget, 1965/1995a & b). This will inform the understanding of purposeful interaction in collaborative social collectives that is developed in Part IV.

PART II. The Basis of Agency

In natural systems, optimal action--that is, action that is both internally efficient and externally effective--results when the principles of least action and self-organization combine to hold the system in an adaptive relationship of co-evolution with its environment (Jantsch, 1980). In human collectives, however, to the degree that the brain is an *active* processor of information, a new alternative is presented: namely, rational choice of action in relation to a purpose or plan (Miller,

Galanter, and Pribram, 1960). Piaget shows that the conditions that generate thought and reason--an active brain--occur *only* in systems of cooperative interaction.

According to Freud (1895/1966), Piaget (1965/1995a), and Pribram (1991), among others, *thoughts are implicit acts*: they are images of planned or possible sequences of actions--*virtual actions*--that an individual visualizes mentally as objects. As such, the generation of thought is a prerequisite to rational or purposeful action.

The Genesis of Thought and Reason

In his little-known essay, "Logical Operations and Social Life," Piaget (1965/1995a) uses group theory from mathematics to show that the requirements for the development of logic in individual actions are essentially the same as the conditions that characterize cooperation as a system of social exchanges. He defines cooperation as a "system of operations carried out in common" involving reciprocal interactions within the context of a common system of language, values, and social norms (page 153).[5] Piaget demonstrates that stability in both systems (that is, *psycho*-logical action and *socio*-logical [inter-]action) presupposes certain basic conditions which hold for formal systems composed of logical or qualitative operations.

Table:
Showing Requirements for a Group, Mathematical Functions, and Piaget's "Mental Function" Equivalents

Requirement for a Group	**Mathematical Function**	**Mental Function Equivalent**
Composability	A new operation is composable from the application of two operations of the set. A + A^ = B	Coordination of two actions by union or serial order etc.
Invertibility	Each operation has a corresponding inverse operation. +A, -A	Psychological reversibility, which allows the reconstitution of a sequence of mental actions.
Substitutional Equivalence	Operations that produce equivalent compositions by substitutions of different elements in the set. (+A +(A^ + B^)) = ((A + A^) + B^)	Ability to use different paths to achieve the same result.
General Identity	Operation resulting from a direct operation and its inverse. +A -A =0	Assures the identity of the objects of thought.

[5]Unless otherwise indicated, the page references in this section are to Piaget (1965/1995).

The mathematical basis for a precise conception of the "group," was developed in the nineteenth century by the work of Helmholtz, Lie, and Poincare. As reviewed by Pribram (1991, Lecture 5), the relevance of group theory to psychology was only rediscovered in the mid-twentieth century with a body of important work on perceptual constancy that began with Cassirer.

Mathematically, a stable system of logical operations is a *group* comprised of elements (or classes of elements), operations acting on the elements, and outputs resulting from application of the operations. Four requirements must be met for a system of logical operations to be considered a "group:" the system of operations must be composable, invertible, possess substitutional equivalence, and preserve identity. The table below defines these requirements and the mathematical logic that meets each requirement, and presents the corresponding "mental function" Piaget equates to that mathematical operation.

Unlike mathematical groups, stable systems of qualitative or simply logical operations are tautological in that an operation composed of elements of itself remains identical (e.g., A + A = A). Therefore, for systems composed of *psycho*-logical operations, Piaget adds a fifth requirement: "***groupement.***" This he defines as the *final form of logic that generates a stable system of human actions* (pages 141-142). To reason, thus, is "to *effect* on objects, materially or mentally, the most general actions possible while 'grouping' those actions according to a principal of reversible composition" (page 139; my emphasis).

Piaget begins his analysis by treating the individual as a closed system involving exchanges only with the physical environment:

> ... it is necessary to conceive the development of logic as a progressive transition from irreversible effective actions [in the child] to reversible virtual actions that are [adult] operations. One can, therefore, interpret logic as the final equilibrated form of actions toward which all sensory-motor and mental evolution tends.
> ...[Thus] considered from the point of their psychological development, logical operations constitute the final equilibrial form of actions when they are grouped into mobile systems that are both indefinitely composable and rigorously reversible.
> (Pages 142 and 145; additions in parentheses, mine).

Piaget suggests that logic, at this level, is simply the "final equilibrial [stable] form"[6] of a system of actions, of movements or sensory-motor processes that have become completely coordinated with one another and thus constitute a system of reversible compositions. Reversibility enables conservation of the whole system of logic--its identity and structural integrity--for it provides the means for returning to the initial condition by way of an inverse function. Because an operation is never conducted in an isolated state, the evolution of operations involves reworking operations as new situations are confronted, which progressively reconstitutes the whole system of operations. By successive iterations of this restructuring process, a final point is reached where action is stable and the configuration (the *groupement*) of the system of logical operations associated with this condition is the final form of logic that will generate this stability. In essence, the importance of Piaget's

---

[6]It should be noted that in a personal communication to Pribram (see Pribram and Bradley, 1996) from Inhelder, she stated that Piaget, after becoming acquainted with the ideas proposed by Prigogine (Prigogine and Stengers, 1984), agreed completely with a change in terminology for his (Piaget's) concepts from "equilibrium" to stability far-from-thermodynamic-equilibrium.

concept of *groupement* is that it *provides the means to get from thought--mental images of virtual actions--to effective, purposeful action in the material world.*

Piaget goes on to examine the conditions required for the development of a stable system of intellectual exchange. He identifies three requirements: a common system of signs and definitions (language); conservation of valid propositions and obligations; and reciprocity of thought among the individuals involved.

Finally, he analyzes three different social contexts within which interpersonal interactions occur--egocentrism, where the individuals involved do not succeed in coordinating their points of view; constraint, where unified collective thought is externally imposed by authority or tradition; and cooper-ation, where the exchange of thought is organized as a system of reciprocal operations. He shows that it is *only* in cooperation that the *socio*-logical organization of thought corresponds to a *groupement*.

> ... cooperation is opposed to both autism and constraint. It progressively eliminates the processes of autistic or egocentric thought ... . Discussion produces internal reflection; mutual verification produces the need for proof and objectivity. The exchange of thought presupposes the principles of contradiction and identity conceived as regulative of discourse, etc. As for constraint, cooperation destroys it to the extent that there is a differentiation of individuals and free discussion. Hence cooperation provides for the mind the psychological conditions necessary for the attainment of truth.
>
> Only cooperation assures mental equilibrium, which allows one to distinguish the *de facto* state of psychological operations from the *de jure* state of the rational ideal. (Piaget, 1965/1995b, pages 208 and 210.)

Piaget's analysis is consistent with the cooperative logic that Searle (1995) identifies in his account of the "construction of social reality." Searle (1995) argues that "genuine cooperative behavior" is the basis for a nonreductionist order of social life that he calls "collective intentionality": "The crucial element in collective intentionality is a sense of doing (wanting, believing, etc.) something together, and the individual intentionality that each person has is derived *from* the collective intentionality that they share" (Searle, 1995: 24-25; emphasis in original). Moreover, Piaget's conclusion of the critical role that cooperation plays in the psycho-social development of rationality is supported by Schore's (1994) extensive review of the multi-disciplinary research on infant development. Schore shows that the requisite neurological organization for the development of a psychologically stable and effective social self--in our terms, an infant with an active brain--is prompted by cooperative interaction (touching, holding, feeding and especially mutual eye gazing) along two dimensions in the mother-infant dyad: one involving the stimulation of positive affect and a second entailing modulation and regulation of the infant's behavior.[7]

In short, not only is cooperation generative of thought and reason, but because intentionality presupposes logical thought--that is, in the form of plans, means-ends schemata etc., (Freeman, 1995; Miller, Galanter, and Pribram, 1960; Searle, 1983)--cooperation also is the basis for purposeful social

---

[7]Schore shows how a breakdown in this cooperative mother-infant relationship, resulting in a prolonged exposure to heightened negative affect during a critical period (approximately the first year of life), can affect the growth and organization of the infant's developing frontal cortex with enduring pathological consequences for subsequent social behavior (see Schore, 1994: 159-167).

action and hence the source of the active agent in psycho-social life.

PART III. Holographic Organization: Concept and Evidence

With an active brain now back in the picture, we move on to the second difficulty in Laszlo's account of psycho-social organization, namely, the inherent determinism of classical holography. In laying the foundation to resolve this problem in the alternative approach developed in Part IV, we begin by distinguishing between classical and quantum holography, and go on to consider evidence of holographic organization in social collectives.

Classical Holography: Image Processing

Classical holography was developed to understand the physics of *image processing* (Gabor, 1948). It is based on a field concept of order in which information about the organization of an object as a whole is encoded as an interference pattern in energy waveforms distributed throughout the field. This makes it possible to retrieve an image of the whole object--a *holograph*--from any location within the field.

To construct a hologram, two or more Fourier transformed records must be linearly superposed. Mathematically, encoding the features of an object onto a holographic record and retrieving an image is achieved by an invertible linear process involving a Fourier transformation (FT). A forward FT enfolds an image of the object from the spatio-temporal domain into the spectral domain. It does this by decomposing the image into a set of waveform components, the elements of which are completely regular sine and cosine waves. The waveform components are thus the interference pattern of these waveforms, and it is these that are stored in the medium of the holographic field. By representing the entire interference pattern, each location in the field contains information about the whole image.

An inverse FT unfolds the encoded image from the spectral domain back into its form as a three-dimensional image of the object in the spatio-temporal domain. It is important to note that there are *no* degrees of freedom in the transformation relations here; chance and probability are not involved. Both in terms of object identity and structural integrity, the image that is retrieved from the spectral domain is the same as the object's form in space-time.

Thus classical holography is a principle of system organization in which constituent parts are enfolded into, and whose behavior is in-formed and, therefore, *determined* by the global organization of the system as a whole. According to this principle, then, the generation of all order would be the product of the predetermined harmony of the universe. Spontaneous or purposeful behavior (novelty or intentionality) *within* the system would *not* be possible because everything is enfolded in a predetermined order. Therefore, without an exogenous means for breaking the endogenous determinism of this form of holography, *human agency*--be it planned or spontaneous--***within** a holographic [imaging] system* simply ***cannot** occur*. This is true whether we are talking about a hologram or a metaphorical social group.[8]

We now introduce the concept of quantum holography, the second kind of holographic organization.

---

[8]This problem of holographic determinism also confounds the use of holographic analogies in social science. For examples, see El Sawy, 1985; Hutchins, 1991; MacKenzie, 1991; Morgan and Ramierez, 1984; Ravn, 1985.

Quantum Holography: Information Processing

Known as *holonomy* or a *quantum holographic* process, quantum holography is used in the physics of signal (information) processing (see Pribram, 1991, Chap. 2). Virtually unknown in the social and psychological sciences, quantum holography is based on Gabor's (1946) energy-based concept of *information*, the ***logon***, which provides the foundation for a non-determinist kind of holographic organization.

In his classic article, "Theory of Communication," Gabor defines a unit of information as *the minimum uncertainty with which a signal can be encoded as a pattern of energy oscillations across a waveband of frequencies,* as in the encoding and transmission of vocal utterances for telephonic communication. Gabor's concept is radically different from, though related to, the more commonly used measure of information developed by Claude Shannon (1949).[9] While Shannon dealt with a *reduction in uncertainty*, Gabor designates the *minimum uncertainty* beyond which a message cannot be compressed.[10]

Gabor determined that there exists a restriction to the efficient processing and communication of a set of telephone signals. The restriction is due to the limit of precision that can be achieved in concurrent measurements of the signal's spectral components (frequency, amplitude, and phase) and its (space)time epoch. This restriction is illustrated in Figure 1a in which time and frequency are treated as orthogonal coordinates. Although the energy frequency of a signal, represented by a dashed vertical line, is exactly defined, its duration in time is totally undefined. Conversely, a sudden surge or change in the signal--a unit impulse function--the solid horizontal line, is sharply defined in time, but its energy is distributed evenly throughout the whole frequency spectrum. Since, at the limit, accurate measurement of the signal can be made only in time **or** in frequency, *it **cannot** be simultaneously made in **both** beyond this point* (Gabor, 1946: 431-432).

Gabor was able to show, mathematically, that this limit could be given formal expression by Heisenberg's uncertainty principle.[11] In its rigorous form the uncertainty relation is given as D*t* D*f* $\geq$ ½ (where D = *delta*), which states that time (*t*) and frequency (*f*) cannot be simultaneously defined in exact terms, but only with a latitude of greater than or equal to one-half in the product of the uncertainties. Since certainty can be obtained only by minimizing uncertainty on both ordinates, the *minimum measurement* of the signal in time **and** frequency is D*t* D*f* = ½, which defines an ***elementary unit of information*** (Gabor, 1946: 431-437). This unit of information both minimizes uncertainty and provides the maximally efficient compression of communication, that is, it provides the minimum space or time of transmission for a signal to maintain fidelity in telecommunication.

Gabor called his unit of information a ***logon***, or a *quantum of information*, and showed that the signal that occupies this minimum area "is the modulation product of a harmonic oscillation [of energy] of any frequency with a pulse in the form of a probability function" (Gabor, 1946: 435; my addition). Mathematically, the logon is a sinusoid variably constrained by space-time coordinates--

---

[9]Shannon's measure is that of information as a reduction of uncertainty through choice among alternatives. His smallest unit of information is the **BIT**, the **BI**nary digi**T**--nowadays corresponding to the smallest standard unit of information in computational information systems. In such systems each unit of information in a sequence contributes to resolution of the signal's message by reducing the probability of alternative meanings.

[10]See Cherry (1966) for an excellent review of these ideas, and Kaiser (1994) for a readable introduction to the physics of signal processing.

[11]Heisenberg had developed his mathematical formulation of uncertainty to define the discrete units of energy, *quanta*, emitted by subatomic radiation.

essentially a ***space-time constrained hologram.***[12]

An important point, which bears directly on the question of determinism, is that logons are *not* discrete units but occur as space-time constrained sinusoids each wrapped in a Gaussian probability envelope, a series of overlapping units in which the "heads" and "tails" of adjacent units interpenetrate one another (this is illustrated in Figure 1b). This 'overlap' among logons introduces indeterminacy in that the data in each unit are spectrally enfolded, to some degree, into the data of adjoining logons. This means, as Gabor puts it, that each logon contains an "*overlap (with) the future*" (Gabor, 1946: 437;

Figure 1. Representation of Limits of Measurement, and the Overlap of Logons (Adapted from Gabor, 1946, Figures 1.3 and 1.7).

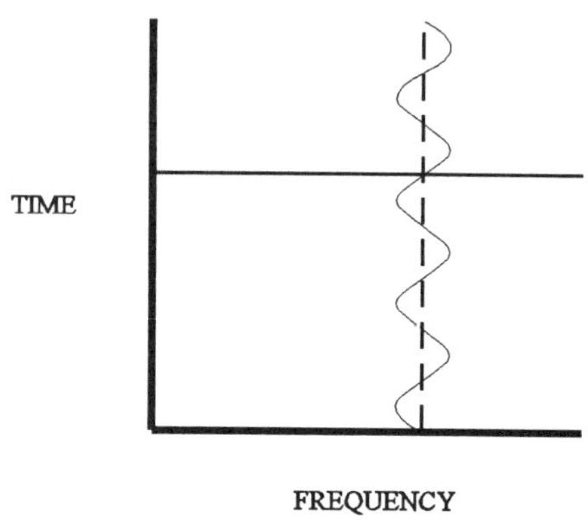

Fig. 1a. Limits of Concurrent Measurement of Time and Frequency of a Signal.

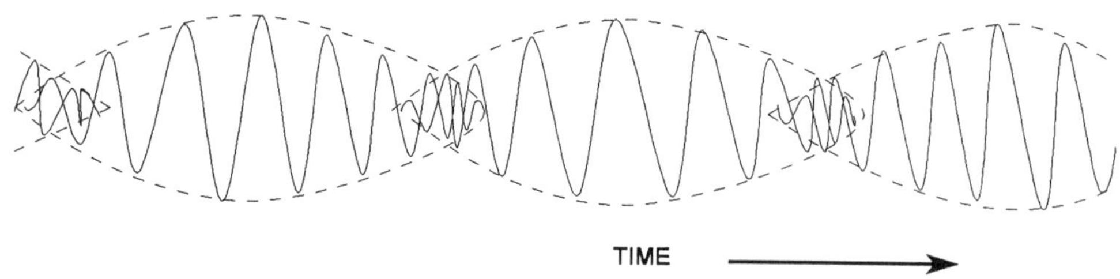

Fig. 1b. Representation of the Overlap of Logons.

[12]This unit differs from Shannon's unit of information, the binary digit, which is the Boolean choice between alternatives.

emphasis added).[13] In other words, that each unit of information, by virtue of its spectral nature, "anticipates" the information of the unit that succeeds it.

However, the overlap among logons is not the only feature of this kind of holography that enhances indeterminacy. The degree of overlap itself has been shown to be a function of the contextual properties of the signal generating apparatus which affect the energy frequency and the time(space) components of the signal (Xie, 1995). The maintenance of an optimal overlap among logons for efficient information processing requires adjusting the communicative apparatus to set and maintain an optimal relationship between the signal's two components.

In sum, quantum holography is the second kind of holography, and it provides a non-determinist energy-based model of information processing. Quantum holography thus defines two distinct dimensions: an energy dimension (measured in terms of frequency times Planck's constant), and a space-time constraint (either Gaussian or rectangular).

Evidence of Holographic Social Organization

Pribram has pioneered the now extensive research on how holographic-like processing operates in the brain with respect to sensory perception and memory (Pribram, 1971 & 1991). In his holonomic brain theory, Pribram shows how Gabor's concept of information can be used to characterize perceptual processing in the neural microstructure (that is, the interaction between horizontal dendritic networks and vertical axonal transmission pathways) for several sensory systems in the cerebral cortex (see Pribram, 1991, Lectures 1-5, for a review of the evidence).[14] Here, in this essay, I add to the small beginning I have made in my own research (see Bradley, 1987, Chaps. 8 & 9; Bradley, 1997; Bradley and Roberts, 1989b), and to the joint work that Pribram and I have undertaken (Bradley, 1996; Bradley and Pribram, 1995, 1996, 1997a, 1997b; Pribram and Bradley, 1997), to describe how holographic-like processing is accomplished in social collectives.

For the reader who is unacquainted with this work, however, this begs the question of the utility of a holographic approach to social organization. As just described, the basic principle of holographic organization is that of a field concept of order in which information about the organization of a system as a whole is enfolded and distributed to all parts so that a description of the system's global organization can be obtained from any part or location within the field. Is there any compelling empirical evidence that this kind of system organization exists in social collectives?

Although there has been little interest in holography among social scientists, except as a descriptive metaphor (e.g., El Sawy, 1985; Hutchins, 1991; MacKenzie, 1991; Morgan, 1986; Morgan and Ramierez, 1984; Ravn, 1985), there is strong evidence from my own study of 57 urban communes suggesting that the processing of communication in social collectives may be structured according to holographic principles (Bradley, 1987; see Chaps. 8 & 9 for the details).

One of the study's instruments was a sociometric questionnaire which was used to measure a commune's social structure by mapping all possible dyadic relations (the relationship between each

---

[13]This overlap is a result of using a measure of time as one of the (measurement) dimensions. "The principle of causality requires that any quantity at an epoch *t* can depend *only* on data belonging to epochs earlier than *t*. .... In fact, *strict causality exists only in the 'time language'*" (Gabor, 1946: 437). Thus, in the non-constrained spectral domain, since time is totally enfolded, causality (in the sense of Aristotle's efficient causation) becomes meaningless.

[14]For example, in a series of recent studies on the barrel cortex of the rat involving temporal and spatial stimulation of the rat's whiskers, Pribram and his collaborators have shown that the response activity of receptive fields could be described in terms of spectral manifolds, and that each of these manifolds could be derived from Gabor-like functions (King et al., 1994; Santa Maria et al., 1995).

pair of members) in the group.[15] Operationally, this entailed asking each member to complete a set of standardized questions about his/her perception of the dyadic relationship between themselves and each of the other members, considered one at a time.[16] Among the questions asked, was one concerning the flow of "power" in the dyadic *relationship between each pair of individuals*.[17] When the data were aggregated for all dyads in the group and converted into the circles and arrows of a sociogram, what emerged for most communes, as shown in Figure 2, was a view of the *global organization* of power in the group--a dense network of transitive relations interconnecting virtually all members in a single multi-level hierarchy. The presence of such coherent global organization was found to be invariant across communes of different size, culture, degree of formal organization, length of existence, time of founding, and member characteristics (see Bradley, 1987, Chap. 8).

Another example from the study is presented in Figure 3, which shows the sociometric structure of "power" (Figure 3a) and "sexual" relations (Figure 3b) for a commune of ten members. Again, a single multi-level hierarchy of "power" relations connecting all members is evident. Although, for "sexual" relations, there was no overt social norm prescribing or favoring a nonmonogamous order,[18] it is clearly apparent that not only are most individuals involved with multiple sexual partners, but that there is a global logic to the arrangement of sexual relationships in this commune.[19] Moreover, when the sociometric structure of "power" AND "sexual" relations is examined (Figure 3c), it is evident that the two kinds of relations are *not* discrete systems but appear to be intertwined at the global level.

However, given the measurement procedures employed, these results are a paradox. The data from which these images of global organization were constructed were *not* measured at the collective level; they were generated by measuring the *discrete* relationship between each pair of members. The commune members were asked ***only*** about dyadic relations--pairwise bonds: *they were **not** asked at all about the global organization of relations for the commune as a whole*. How is it, then, from measurements of the properties of parts (perceptions of discrete dyads) that we can obtain an image of the organization of the whole (the social structure of the commune)?

It is important to point out that these patterns are unlikely to be an artifact either of the data gathering procedures or of aggregating the dyads together for a commune. In regard to the former, each individual's responses to the sociometric instrument were gathered in a way to ensure that their answers were independent; strict supervision of the data gathering by field workers made collusion between respondents impossible. And for the latter to be plausible, that each dyad reflects a discrete

---

[15]In a group of size N, there are N(N-1) dyadic relations. Thus, for a group of 10 members there are 90 different dyadic relations.

[16]A copy of the sociometric instrument is provided in Bradley, 1987, "Appendix B."

[17]The "power" question read: "Even the most equal of relationships sometimes has a power element involved. However insignificant it may be in your relationship with this person, which of you do you think holds the greater amount of power in your relationship?"

[18]Unless there is a group norm sanctioning or requiring promiscuity, normally sexual relations are monogamously structured in communes (Zablocki, 1980).

[19]Four months of field observation in this commune found no evidence of a social norm favoring multiple sexual relations nor overt recognition by members that such relations existed in the group. However, snippets of information revealed during interviews with individual members suggested that beneath the veneer of an informal "public" monogamy lay a complex web of covert, multi-party sexual relationships.

reality, one would expect much less global coherence in order; more cliques and disjointed subsets, more intransitivity--patterns that reflect idiosyncratic individual perceptions or specific interpersonal realities, rather than the forms we see which are more consistent with an emergent order of collective organization. If anything, the data gathering procedures--mapping the properties of dyads--have actually worked *against* the discovery of such coherent images of global organization.

Therefore, to the degree that such measurement artifacts are ruled out, the answer is given in the paradox itself: *it is possible to construct an image of global organization from measurements of discrete dyads because* ***information*** *about the organization of the whole (the commune) is distributed throughout the group and* ***enfolded into all of the parts*** *(the dyads).* Such correspondence *in information* between parts and the whole is consistent with holographic organization. In short, there is compelling evidence that the organization of information in social collectives appears to be structured as a holographic order.

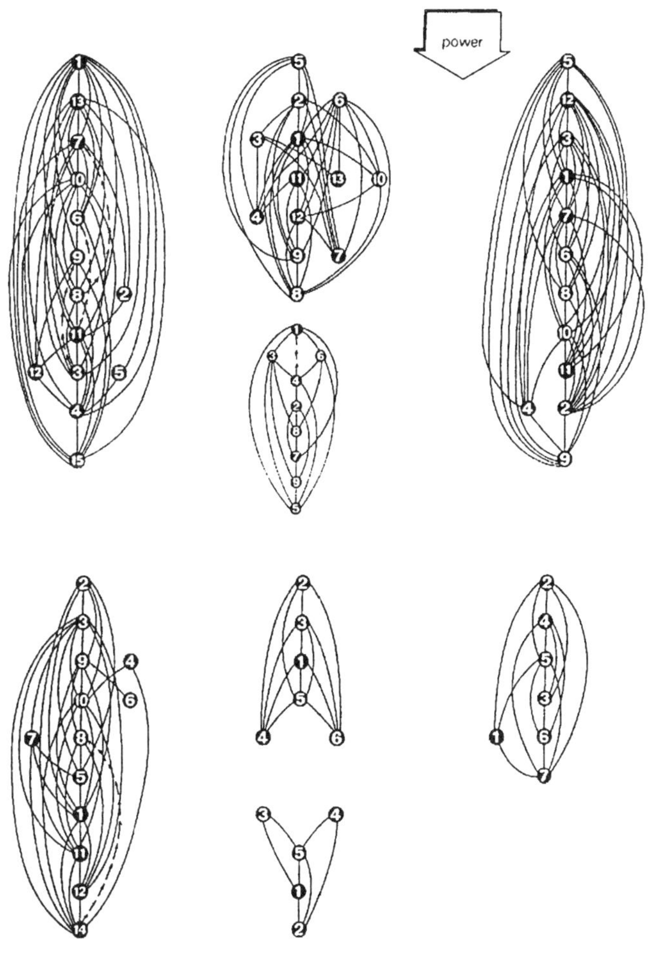

Figure 2. Examples of the Global Organization of "Power" from Mappings of Dyadic Relations in Selected Communes.

Fig. 3a. " Power" Relations

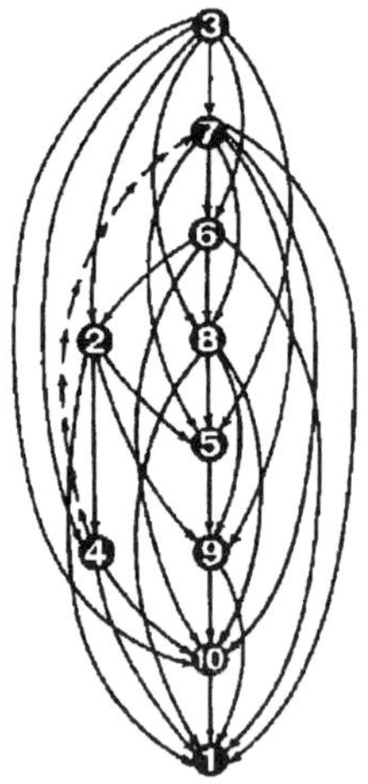

Fig. 3b. " Sexual" Relations

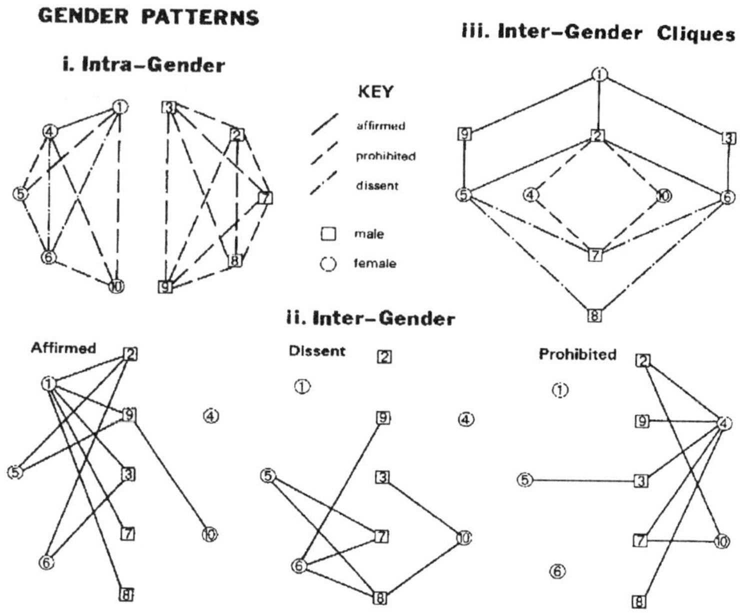

Fig. 3c. "Power AND Sexual" Relations

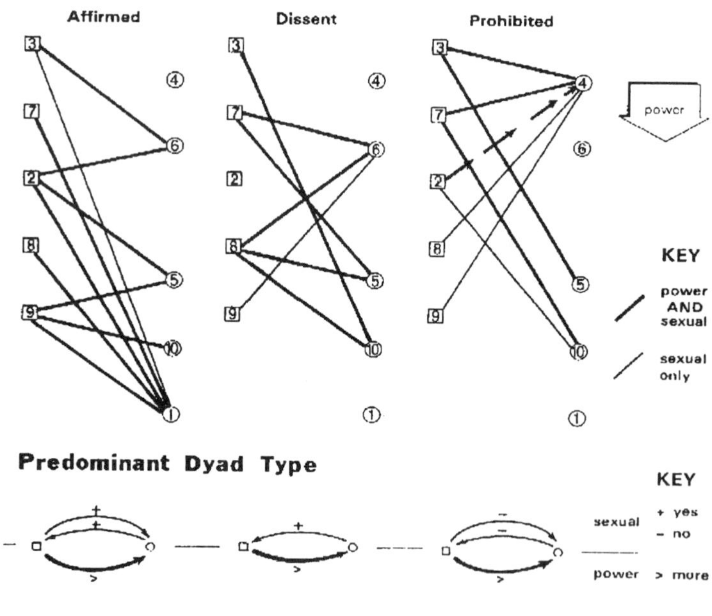

Figure 3. Global Organization of "Power," "Sexual," and "Power AND Sexual" Relations from Mappings of Dyadic Relations in a Commune.

PART IV. Communication in Social Collectives: A Non-determinist Holographic Approach

Indeterminacy and Biosocial Order

At the physical level, the quantum domain is naturally organized as a nonlocal order with the properties of a holographic field. However, to create a hologram in a laboratory or to send a signal encoded as a pattern of energy oscillations through a telephone wire, requires the construction and operation of an "artificial" signal processing device. Because the generation and maintenance of order in living systems requires active constructive endogenous processes (Kauffman, 1993; Prigogine, 1997), the structures that produce and process communication in biological and social collectives are *not* given. Thus, to the degree that holographic-like processes operate in biological and sociological systems, the communicative structures that generate such processes must be constructed and maintained by the biology of the organism or by the sociology of the social collective. It is because such communicative structures *must* be constructed and then actively regulated for effective operation in an organism or a social collective that the holographic determinism of the quantum physical world is broken in living systems.

A *social collective* is defined as a durable arrangement of individuals that is distinguished by shared membership and collaborative interaction relative to a common purpose or goal. In social collectives, both the construction and the regulation of the communicative structure is governed by the normative order: the collective's system of common language, values, beliefs, and social norms. By motivating individuals to join and participate, the normative order of the collective establishes the boundary of the communicative structure--members only--and the elements from which the communicative structure is constructed--a particular set of individuals (the members) and the network of social relations through which their participation in the collective is channeled. The normative order defines the nature and scope of the relations among members, thereby establishing the channels for communication (including the dimensions of reciprocity and asymmetry) and, in broad terms, the extent and direction of information flow. Thus, the basis for the members' sharing of experience and meanings is established through the collective's system of shared symbols, language and beliefs.

Additionally, the normative system regulates the collective's communicative structure by calibrating the relations among the collective's members. This it achieves through the framework of common values and beliefs. Acting as a tuning device, this framework establishes a distinctive psycho-social bandwidth for social communication. The bandwidth is defined by the emotional resonance of the social relations within the collective which, as described below, has its basis in the affectively-charged attachments among members. In short, the operation of normative order opens the communicative structure to the indeterminacy of new input.

In what follows, I draw on both classical and quantum holography to construct a non-determinist holographic model of social organization that incorporates active brains and describes communication processing in purposeful social collectives. The principles of classical holography are used to describe how the collective images the objects towards which its purposeful activity is directed. The principles of quantum holography are used to show how the collective processes information from the on-going interactions among members to generate a stable order of endogenous organization. Because a stable platform of organization is a prerequisite for any kind of effective collective action, I begin with the latter.

Communication and Endogenous Organization[20]

To appropriately apply the principles of quantum holography to the operation of the communicative structure in social collectives, we must first establish that social interaction has a dimensionality that is equivalent to Gabor's energy and space-time ordinates.

Dimensionality of Social Organization

While referred to by different terms, social science has long recognized the importance of two basic patterns of social organization: *gemeinschaft* and *gesellschaft* organization (Toennies, 1957), organic and mechanical solidarity (Durkheim, 1949), informal and formal organization (Roethlisberger and Dickson, 1939), markets and hierarchies (Williamson, 1975), flat networks and hierarchical networks (Burt, 1992) among others. Underlying these conceptualizations is a distinction between ***field-like*** and ***hierarchical*** forms of organization.[21] In field-like organization the pattern of social relations is fluid and transitory, and is based on an equivalence among individuals. It is an order of social connection that, like the structure of a market economy, is distributed over the whole region of a social space. Such an order is essentially holographic-like in organization in that each interaction enfolds the operations of the collective as a whole in much the same way that each transaction in a market economy enfolds the activity of the entire market place (Pribram, 1982). In hierarchical organization, by contrast, the pattern is ordered and stratified by relations of authority and social control. As a system of social constraints, the hierarchy locates each individual in a position that is spatially and temporally localized.

My own previous studies of a variety of types of urban communes, community organizations, and small groups have shown that these two patterns of interconnection form the *communicative structure* in stable social collectives.[22] As revealed for stable communes (communes surviving at least 24 months beyond measurement of their social structure), one pattern is a dense web of relations of positive affect (mutual "loving" ties) interconnecting virtually all members (see Figure 4). This measurement is viewed as an index of the members' potential energy in that it reflects the emotion excited by affectively-charged attachments. The web of these mutual attachments can be interpreted as a social field--i.e., as a distributed, massively parallel network of reciprocated relations through which there is an unrestricted flow of information about virtually all interactions throughout the web. The second pattern is a densely interlocking order of "power" relations which also extends to interconnect virtually all individuals. This is a hierarchical order, a highly stratified structure of asymmetrical relations which operates as a system of spatial and temporal constraints on the behavior of each individual.

In following up on these earlier findings that describe *what* the communicative structure was composed of, Pribram and I sought to understand *how* the interaction between field and hierarchy operates as an information processing system to in-form organization in social collectives.

---

[20]This section draws from my collaborative work with Karl Pribram (Bradley and Pribram, 1995, 1996, 1997a, 1997b; Bradley, 1996).

[21]There is much empirical evidence that these two dimensions of order are the basis for stable organization at the neurobiological, neuropsychological, psychological, and sociological levels of behavior (see Pribram and Bradley, 1997).

[22]See Bradley (1987), Bradley and Roberts (1989a, 1989b), Roberts and Bradley (1988, 1991); also see the studies by Carlton-Ford (1993), Roberts and King (1996), and Zablocki (1971, 1980).

A Quantum-Holographic Model

Ontologically, a rigorous concept of energy, or its equivalent (e.g., Rosenstein, 1996), is fundamental to an understanding of collective organization. Energy is a measure of the means--the fuel--for maintaining order in the face of challenge (novelty) or changing an order in the face of inertia. As individual biological organisms, a collective's members possess the potential for work, measured as energy. To exist as an entity, a social collective *must* mobilize and appropriate the members' potential for work, their biological capacity for physical behavior and activity, and direct it toward collective ends. "Energy," so defined, also is the medium for information processing, the medium for encoding and relaying communications as signals back and forth among the elements of a system. In what follows we focus primarily on the role of energy in social communication. Elsewhere, we have begun work on the dynamics of energy expenditure in the collective's action (see Bradley and Pribram, 1997b; Bradley, 1997).

Elements of the Communicative Structure.

To develop our understanding, we begin with the premise that collective organization is, first and foremost, a relationship of ***collaboration***--of individuals *working together* to achieve a shared end.[23] This concept is equivalent to Piaget's concept of cooperation, a "system of operations carried out in common" (Piaget, 1965/1995a: 153), as described in Part II above. To collaborate entails work, and work requires a supply of energy. As just noted, we assume that, as biological organisms, the members of a social collective are the source of this energy, and that they expend this energy as they work and interact in relation to realizing a shared goal. In addition to the availability of a pool of potential energy, collaboration also requires that each individual's expenditure of this energy be coordinated and directed toward the collective's objective.

Collaboration, therefore, involves two complementary processes. The first is ***flux***,[24] the constant movement of energy throughout the collective as it is activated and expended by the members in physical behavior and social interaction. The second is ***control***, the construction of a system of social constraints which directs the expenditure of energy (at specific locations in space and particular moments in time) into collective action. As described below, the interpenetration between flux and control operates as a communication processing network that *in*-forms the transformation of potential energy into collective work.

Flux, the movement of energy, occurs as a distributed, massively parallel process of *equivalent* (of equal value) relations interconnecting all members. This field is established by membership in the collective which creates a sociocultural boundary separating members from nonmembers. Membership thus effects a nominal bond of connection by which members are socially attuned to one another. As an undifferentiated web of psycho-social connection extending throughout the collective, this field is the intermediary for *all* interactions among individuals; this includes verbal communication and also nonverbal communication as signaled by bodily position and movement and by the symbolic use of artifacts and objects. Thus the field is the order through which all movement of the collective's energy is processed.

---

[23] As discussed elsewhere (Roberts and Bradley, 1991), the word "collaboration" is derived from the French verb *collaborer* and means "working" (*laborer*) "together" (*col*) to achieve a common objective or outcome.

[24] Interestingly enough, flux is defined in *The Concise Oxford Dictionary* as the "flood of talk" and denotes a "continuous succession of changes" (Fowler and Fowler, 1964: 469).

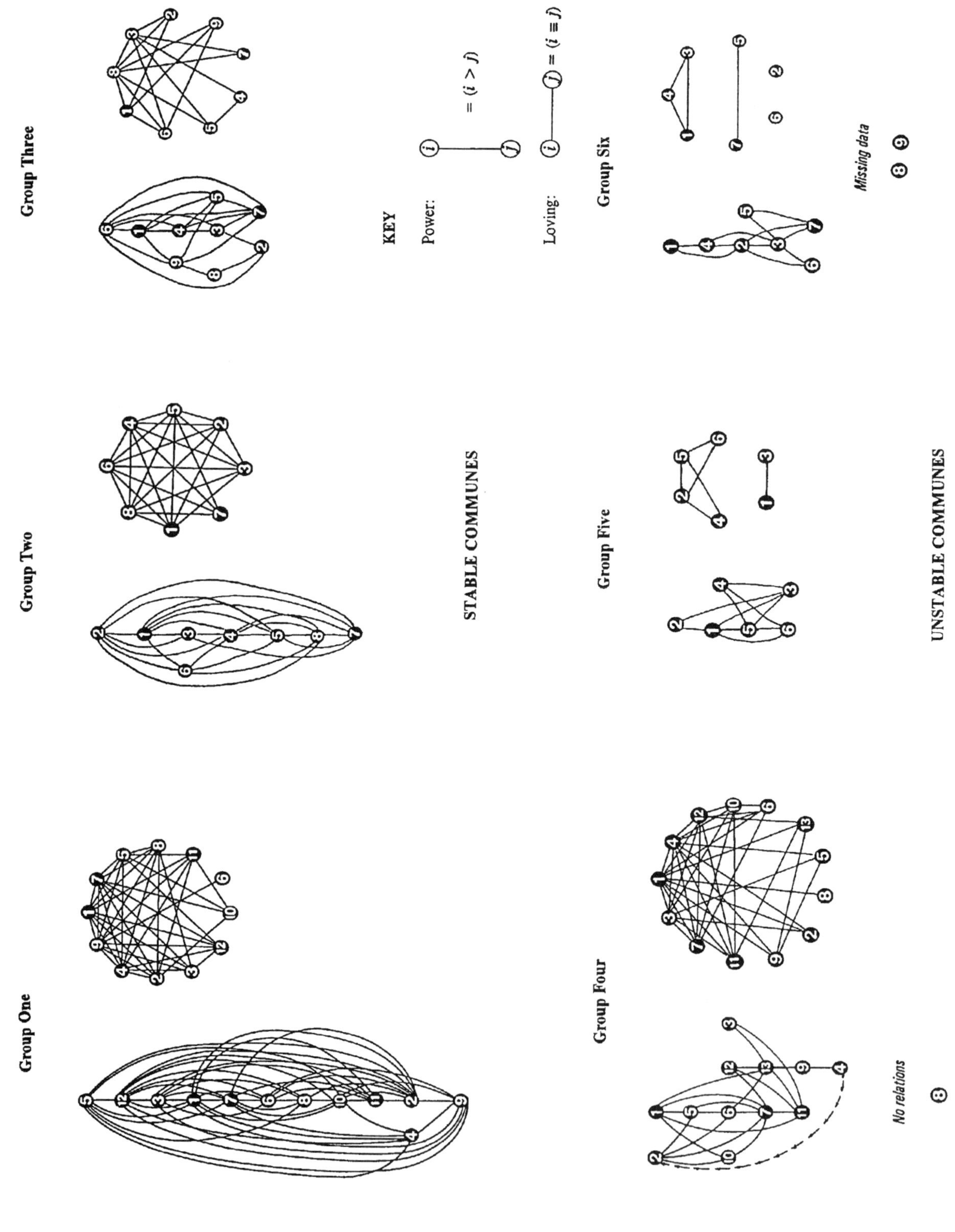

Figure 4. Sociometric Structure of "Loving" (Field-Like) and "Power" (Hierachy) Relations in Selected Stable and Unstable Communes.

The collective operates on this field of relations to *activate* individuals to action by *arousing* affective attachments among members. Arousal of affective bonds *excites* emotions, which in turn mobilizes the individual's propensity for action (Pribram and McGuinnes, 1975; 1992; Schore, 1994), and thus the potential for expending energy (Pribram and Bradley, 1997). Thus, the *level of aroused affect* reflects the degree to which members are activated to action. It is a *measure of the amount of potential biological energy that has been mobilized and is available for collective or individual use*.

Also at work here, as just noted, is control--the system of social constraints that operates to channel the members' energy toward collective ends thereby preventing the energy's dissipation on other social activity and physical behavior. The controls influence the movement of energy by constraining the spatial and temporal direction of the paths of flux, thereby in-forming the collaborative interactions among individuals. By differentially constraining the paths by which individuals expend their energy, both with respect to specific locations in space and with respect to particular moments in time, the controls *in-form* (give shape to)[25] the patterning of collective organization.

Operation of the Communicative Structure.

We should now be in position to apply Gabor's concept of information and show how the interaction between flux and control operates as a communicative structure to in-form the collective's expenditure of energy. The symmetric bonds of the distribution of flux indicate that members are essentially interchangeable and that, among members, the patterning of flux throughout this field is more or less equivalent. As this field is an energy field, it lies within the spectral domain and is related to space and time by way of a transformation (the Fourier transform function). By contrast, because individuals are asymmetrically connected in the hierarchical order, the system of controls operates differentially on the collective's members, both with respect to their energy expenditure in a particular location in space as well as with respect to their energy expenditure at particular moments in time.

Drawing on Gabor, it is expected, therefore, that the operation of hierarchical controls on the distribution of flux generates information as a moment-by-moment--*quantized*--description of the collective. Encoded in terms of both structure (spatial-temporal position) and flux (distribution of energy), each unit of information is essentially a space-time constrained hologram that contains a description of the collective's internal organization. By processing and distributing, via a holographic-like process, a series of successive descriptions within space-time and spectral coordinates, the interaction between flux and control operates as a communication system. It *in-forms* the continuous stream of *inter*-actions among members to produce stable organization.

However, whenever there is an imbalance between the amount of distribution of flux and the amount of control, quantization breaks down resulting in a lowering of information transmission. The reduction in information transmission impairs the efficient operation of the collective which, in turn, increases the likelihood of instability. This impairment is due to what Ashby (1956) has characterized as the need for "requisite variety" in cybernetic (information and control) systems.

A point of particular relevance is an important implication of Gabor's concept of information for the role of causality in communication. As described above, logons are not discrete units but, as a result of their spectral nature, occur as overlapping units in which the data in each logon are

---

[25]This conception is similar to Bohm and Hileys' notion of "active information" (see Bohm and Hiley, 1993: 35-42, 59-71).

enfolded into the data of adjoining logons. In addition to the indeterminant order that results, what is of special interest here is that this spectral enfoldment among logons yields an information processing system in which data about the "future" (potential) order are enfolded into the units that are processed in the "present." In this way, the communicative structure *anticipates*, to some extent, the next moment of collective organization (Bradley, 1996).

The generation of such anticipated order by the communicative structure requires optimizing the degree of overlap among logons. This entails adjusting the contextual properties of the communicative apparatus that regulate the frequency and time(space) components of the signal. In complex systems, like biological and social collectives, the maintenance of an optimal overlap among logons for efficient information processing requires constant exogenous adjustments to the communicative structure to regulate communication. This is a primary function of the normative order in human social collectives. In addition to defining social objectives and proscribing the means for achieving them, the normative order sets and regulates the social parameters that affect the energetic and space-time aspects of communication.[26]

Communication and Purposeful Action

Having described the processes of communication by which a collective generates a stable order of endogenous organization, we turn our attention to the matter of how the collective visualizes and communicates plans for purposeful social action. We begin by conceptualizing a plan, in Pribram's (1991: 148) terms, as an "image of achievement"-- in this case, as the target for a program of collective action.

Plans as Images of Achievement

While action is the environmental consequence of an individual's movement or behavior, plans are mental constructions or *thoughts*. As already noted, thoughts are implicit acts in that they are *images* of planned or possible sequences of actions that the individual visualizes mentally as objects. Thus thoughts precede purposeful acts.

In *Brain and Perception*, Pribram (1991, Lecture 6) describes how the execution of an individual's action in the material world is controlled by an internal representation of a "target." The target is a neurological image of achievement, the anticipated or imagined pattern of environmental consequences that result from a sequence of bodily movements. According to the holonomic brain theory, targets are "objects" processed in the frequency domain (Pribram, 1991: 134).

Although targets can be unconscious or conscious neurological constructs, for our purposes here I am limiting them to conscious images. Accordingly, a *plan* is defined as a *way* to achieve. It is a mentally constructed target *object* to which the consequences of a program of action are directed and against which the results of actions in the program are compared. In short, the plan is a control system.

In translating this concept to the sociological level, we draw on our earlier discussion of Piaget's concept of "*groupement*."

[26]See Bradley and Pribram (1997) for an example showing the limits of such normative regulation of an Gabor-like energy-based communication system in musical social collectives.

***Groupement*: the Object of Collective Achievement**

As discussed above in Part II, Piaget (1965/1995a) uses the mathematical theory of groups to show that the conditions for the development of a stable system of logic in individual actions are essentially the same as the conditions that hold for cooperative social actions. He shows that stability in both kinds of systems depends on five conditions: the four basic requirements for stable systems of purely logical operations (formal groups), composability, invertibility, associativity, and identity; and his fifth requirement for stable systems of *psycho*-logical or *socio*-logical actions, "*groupement*." A *groupement* is the final form of logic in a system of operations that generates a stable order of human actions. As emphasized above, the importance of this concept is that it provides the means to get from mental images of virtual actions to effective intentional action in the material world. At the sociological level, such images are embodied and shared among the collective's members by its system of collective expressions and representations, its shared language and beliefs.

In applying Piaget's concept of *groupement* to social collectives, we use the term to mean an order of endogenous organization that generates a stable system of effective, purposeful collective action. In essence, *groupements* are *images of objects of collective achievement*. Examples are the musical score for a tune, a "play" for a football team, or the script for a theatrical play. The deliberate recording and preservation of such information enables the collective to recall and recreate or modify orders of organization required for its purposeful activity. The recording and retrieval of *groupements* is thus vital to the functional viability of a purposeful collective.

According to Piaget (1965/1995a: 144-145), *groupements* are "conserved" by the operation of the normative order (what he calls the "common system of signs and references"), the collective's system of common language, values, and social norms. Such a "collective form of thought" functions to "conserve" the operatory logic and outputs--the "propositions"--of prior interactions, thereby ensuring logical consistency in the development and evolution of subsequent *co*-operations.

> An operatory *groupement* is a system of operations with compositions exempt from contradiction, reversible, and leading to the conservation of the totalities envisioned. Now it is clear that thinking jointly with others facilitates noncontradiction. It is much easier to contradict oneself when one thinks only for himself (egocentrism) than when partners are present to recall what one has said previously and what one has agreed upon. Reversibility and conservation are, on the other hand, contrary to how things appear and only become rigorous on the condition of replacing objects by signs, which is to say by a system of collective expressions. In a still more general way, a *groupement* is a system of concepts (classes or relations) implying a coordination of points of view and a pooling of thought. This is even clearer on the formal plane ... since hypothetico-deductive thought is, more than anything else, thought based on language (common or mathematical) and, therefore, on a collective form of thought (Piaget, 1965/1995a: 144-145).

In short, the normative system acts to select certain orders of endogenous organization and key them as socially significant *groupements* for conscious incorporation into the collective's "memory." In doing so the normative system functions to preserve plans or strategies for action that are *valued* by the collective.

While Piaget does not provide a description of the process, an understanding of how *groupements* are processed for storage and retrieval in social collectives can be developed by complementing holographic image processing with quantum-holographic information processing.

Outline of a Holographic Model

Viewing the recording and retrieval of *groupements*--or, for that matter, any configural order of information processed by the collective--in the terms of the *image processing* principles of classical holography, has certain processing advantages (speed and efficiency, robustness of the image due to information redundancy, preservation of object integrity and identity, etc.) not offered by traditional sociological approaches to the question of how social order is preserved.[27] These features are described in summary form in the table below in terms of the six properties of a hologram.

At the sociological level, the plan is a shared image of the pattern of environmental consequences resulting from a program of intentional collective action. To the degree that the processing is similar across the neuropsychological and sociological levels, it is the *outcome*--the pattern of environmental consequences--that is imaged by the collective as the target object. Using our conception, thus, of a plan as a shared image of achievement, we can translate this framework to the sociological level and outline a holographic model of the communicative processes involved in purposeful collective organization. As shown schematically in the flow chart in Figure 5, there are four aspects to this process.

Holographic Image Processing.

The first involves imaging the plan as an object for collective achievement and communicating its features to all members. As the target for the collective's action in the material world, the plan must be rendered as an *ontological* description of the intended outcome if it is to effectively control the collective's achievement process. This means construction of a shared description of the *groupement* defined by the plan in terms of its spatial and temporal features as a set of action consequences.

The construction of the shared description is mediated in social interaction by the collective's system of collective expressions and representations--its language and culture. Through a process of symbol-based interaction (i.e., symbolic interaction), which often includes rehearsing the plan's implementation, a shared conscious image of the *groupement's* object-form is constructed. This is done by encoding its spatial and temporal features, and associated pattern of energy expenditure, to particular words, signs, or gestures. This creates an enfolded order in which its set of features as an object-form are encoded and represented in social interaction by the symbols. The interaction thus enfolds the set of features, and the set's encoding to particular symbols, into the relations of flux and distributes them throughout the collective to all members.

At the cognitive level, the individual member remembers, recalls, and images the plan's *groupement* as it is encoded in the symbols embodying its shared collective representation. At the neurological level, it is stored in the brain as a set of Fourier descriptors. The Fourier descriptors encode its object-form features and can be regarded, therefore, as a hologram.

Retrieval of the plan by the collective from the brains of the individual members requires activating the codes or symbols associated with the plan. This recalls the set of Fourier descriptors that encode the *groupement's* object-form. Spectrally enfolded as a sequence of harmonic frequencies, the Fourier descriptors also encode its ordering as a movement. Recall therefore enables

---

[27]See a rview of the standard sociological approaches to this question in Turner (1986).

the sequences of action outcomes the *groupement* embodies to be *re*-presented as the image of a dynamic process in the brains of the members.

Table Showing the Properties of a Hologram and their Sociological Equivalents*

Property	**Description**	**Sociological Equivalent**
Independence from Physical Waves	No necessity for actual physical waves; only that waveform description fits data.	Wave-like, periodic behavior of social phenomena such as business cycles in stock market, political swings in voter behavior.
Distributed Information Store	Information re order of whole encoded, via a FT, into all parts/locations; creates redundancy of information and ability to back-transform, via a FT, to retrieve image of whole from any part or location.	Evidence that information re global organization of social structure is enfolded into sociometric data collected on discrete dyads.
Gigantic Storage Capacity	Encoding and concentration of huge quantities of information into small holographic record store, while still preserving object integrity and identity.	Creation of social situations requires storage and processing of enormous quantities of information for on-going immediate recall by social actors to re-construct coherent, familiar patterns of social organization.
Image Constancy & Highly Textured 3D, Parallax Reconstruction	Image has constancies in object-form invariant with angle of view. Hidden features of object revealed with change angle of view.	Social reality imaged and experienced by social actors as emergent object with an invariant dimensionality; new features revealed from different social positions and roles.
Image Projection Away from Holographic Record's Surface	Object's retrieved image not coextensive with holographic record but projected away from the surface.	Construction of social organization as a shared reality, projected outside the brains of individuals and off the surface of social relationships as an emergent form.
Associative Image Retrieval	Enfolded order of information enables associative retrieval of image of a missing object from information on another object.	Projection of an alternative social situation while ontologically present in the shared construction and experience of a different situation; e.g., projection of a familial order of relationships into a business or professional context.

---

*Adapted from Bradley (1987: 241-246).

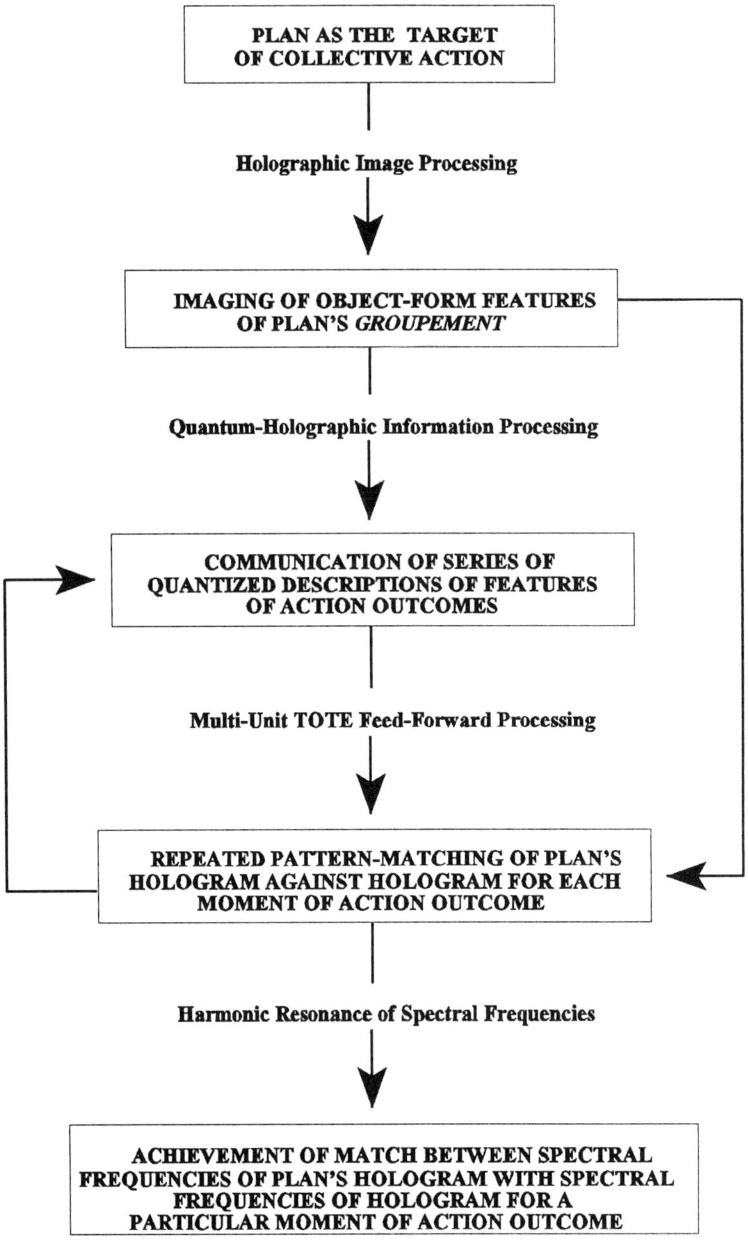

Figure 5. Flow Chart Showing Communication Processing of Purposeful Collective Action.

The image of the plan's object-form is projected away from the member's brains and onto the flux field which, to the degree the relations are in phase so that it is organized as a harmonious order, provides a uniform surface from which a highly detailed, coherent collective image is *re*-constructed. Thus the action of the normative order in regulating and synchronizing the relations of flux, through its system of shared beliefs and norms, affects the degree to which a coherent collective image of the plan is reconstructed. Because of the enormous redundancy in information in this holographic order, an image of the plan's global features can still be obtained, with some loss in resolution, even if memory of its *groupement* was not retained by all members or was destroyed by member turnover or loss.

Quantum-Holographic Information Processing. The second aspect of the process involves the production of a continuous succession of descriptions of the action outcomes, and the communication of the features of these descriptions to all members on a moment-by-moment basis. This can be understood as a quantum-holographic process in the terms described above.

Each outcome of the collective's action, as a pattern of spatial consequences at a given moment in time, is reflected in the collective's pattern of energy expenditure at that moment. As previously described, the relations of flux, as a field of energy, are the medium through which the processing of all interactions--both verbal and nonverbal communication--within the collective occurs. This is where I agree with Laszlo's idea that the distribution of communication in psycho-social interaction is mediated by an energy field. However, rather than the nonmaterial ether-like medium Laslzo has in mind, for me the energy field has a tangible basis and properties that are amenable, in some degree, to scientific measurement. So that in addition to encoding and transmitting verbal signals as acoustical waves of energy, the flux field also encodes and conveys nonverbal signals such as eye-contact, body posture, bodily movements including gestures and physical contact, the use of symbolic objects and artifacts, etc., in the energy waves of radiant light. In short, the energy in the flux field provides the ontological means by which members "read" each other's actions.[28]

Thus *all* movement or action in or by the collective interacts with the energy wave-forms of the flux field and encodes the features of the collective's evolving endogenous organization in the field. This creates, at any given moment, an interference pattern which records in the field the set of energy and spatial frequencies that characterize the collective's pattern of distribution of energy expenditure. In effect, a space-time constrained hologram is created in the flux field that contains an a spectral description of the collective's organization along energy frequency and spatial coordinates at that moment. This spectral description is enfolded by the on-going movement of the energy waveforms and distributed, via a holographic-like process, throughout the relations of flux to all members. The spectral description is *re*-presented in the individual's brain as a set of Fourier descriptors that encode the energy and spatial frequencies of the object-form of the outcome of collective action at a particular moment in time.

---

[28]Based on his extensive review of the evidence, Schore's (1994) finding that a relation of positive affect is the channel through which communication is processed in mother-infant dyad, and his description of the signaling media involved (mutual eye-gazing, touching and holding, cooing, etc.), are consistent with this idea. There is also sociological evidence, from the work on speech convergence and accommodation, of an energic system of nonverbal information transmission operative in human social interaction (see the review by Giles and Coupland, 1991). For instance, in a recent study of 25 dyadic interviews between a talk show host and his guests, Gregory and Webster (1996) conducted a spectral (fast Fourier transform) analysis of vocal spectra of speech samples from the interviews, and found a nonverbal signal in the low [energy] frequency band (beneath 0.5 kHz) that appears to carry encoded information about the relative social status of the individuals involved.

Pattern-Matching TOTE Control. The third aspect of the process is essentially a pattern-matching operation in which successive quantum-holographic descriptions, each representing a different moment of the energy and spatial spectra for the outcome of the collective's action, are compared against the hologram for the plan's *groupement* until a match is achieved. The logic of this type of control has been described by the Test-Operate-Test-Exit (TOTE) model (Miller, Galanter, and Pribram, 1960) in which a test matches an input (the action consequence) against a setpoint (the target) in the same way that control operates in a thermostat. Viewed this way, planned or purposeful action involves a feed-forward process in which the brain compares the mentally-projected image of the achievement target (the plan as an imaged outcome of action, *imagined* as a discernable *pattern* of anticipated environmental consequences) against successive perceptions of the consequences of the individual's behavior (the sensory-perceptual *re*-presentation of the *actual pattern* of behavioral consequences) until a "match" is achieved. This 'pattern-matching' process is conducted in the frequency domain and involves the brain comparing the Fourier descriptors of the object-form of the achievement target with the Fourier descriptors of the object-form of the perception of the consequences of action (Pribram, 1991, Lecture 5).[29]

Applying this logic to the communication involved in purposeful collective action, the process can be modeled as a multi-unit TOTE scheme with individual TOTE units linked hierarchically to create a *feed-forward* control system.[30] In the terms of this logic, the input is the quantum-holographic description of the collective's outcome at a given moment in time, which is compared against the setpoint of the hologram for the collective's plan. The operation of the control system calls for further rounds of collective action until a match between the two holograms is produced.

Match of Spectral Frequencies. Achievement of a match between the two holograms is the fourth aspect of the processing. This occurs by harmonic resonance and entails matching the spectral frequencies of the object-form of the plan encoded in the member's brain with the spectral frequencies of the object-form of successive action outcomes encoded in the flux field. Because matching takes place in the frequency domain, it involves the equivalent of the simple linear mathematical operation of convolution or superposition. This provides for fast efficient image processing of the object-form of the plan in relation to the object-form of each moment of action outcomes.

Empirical Support

There is space here for only the briefest review of empirical evidence that supports the holographic model I have outlined above. We will start with a summary of the evidence for the quantum-holographic model of communication.

The basic proposition of the quantum-holographic model is that the interaction between flux

---

[29]In an ingenious application of this kind of approach, Glazer (1987) uses the principles of optical (classical) holography to develop a holographic model of "individual decision-making," as an alternative to traditional 'rational-choice' models. In Glazer's model, individual choice is viewed as a process of "recognition" whereby the features of a set of external objects (the choice set) are matched against the features of a preferred object (the target) that the individual has "in mind." Viewed this way, "recognition" is a pattern-matching process which involves the brain matching the Fourier transformed spatial features of the external objects to the Fourier encoded spatial features of the preferred object.

[30]The logic of a single TOTE unit is a feed-*back* control system (see Miller, Galanter, and Pribram, 1960).

and control forms a communicative structure which, by in-forming the collective's expenditure of energy, generates stable organization. In a test of the theory using sociometric measures of flux (networks of positive affect) and control (networks of power relations) from data collected in a longitudinal study of 60 urban communes (Bradley, 1987; Zablocki, 1980), Pribram and I found a strong pattern of results consistent with the theory.[31] Of particular interest here is our finding, based on multivariate discriminant function analysis, that flux and control are predictive of stability: that the relationship between flux and control at a given point in time was found to predict a commune's survival status twenty-four months in the future (Bradley and Pribram, 1996 and 1997a).

Turning from communication to purposeful organization, the basic proposition of this holographic model is that the collective forms an "image" of the object of its purposeful activity which is communicated and stored, via a holographic-like process, throughout the system. While I know of no direct evidence of holographic processing and storage of the target object of purposeful activity in social collectives,[32] there is evidence of holographic-like processing of collectively-held "images" of global organization in the communes data (Bradley, 1987). I tested the degree to which the communes were able to project and maintain a collective "memory" of their global organization over time, and in particular, whether the persistence of such "memory" was affected by membership turnover or loss. I used graph-theoretic measures of network organization (Holland and Leinhardt, 1976) to provide systematic structural indices of coherent global organization in the sociogrametric image of "power" for each commune, as constructed from the dyadic measurements described in Part III, above. I found that the persistence (over a 12 month measurement period) of a coherent image of global organization was independent of the degree of change in a group's membership (Bradley, 1987, Chap. 9). This suggests that not only does the collective encode and distribute an "image" of its global organization by a holographic-like process, as the evidence presented in Part III shows, but that it also records the "image" in a holographic-like manner that is robust and resilient in the face of disturbance or destruction to the memory store. Thus, there is good reason to suggest that the collective uses comparable communicative processes to image and record the objects of its purposeful activity.

Summary

Laszlo's theory of quantum vacuum interaction is organized around two sets of theoretical principles. The first set is derived from his postulate of universal interconnection: that everything in the universe is fundamentally related by a process of constant universal interaction. This process is mediated by the virtually unlimited potential energy of the quantum vacuum. The wave-forms of energy in this field encode information on all interactions, interactions both across and between the physical, biological, and psycho-social levels of organization. The information describing these interactions is stored and retrieved according to the principles of classical holography.

---

[31] Details of the research procedures and results are available elsewhere (see Bradley, 1987, Chaps. 8 & 9; Bradley and Pribram, 1996 and 1997a).

[32] Evidence of an underlying common cognitive reality, from a "cognitive map" technique developed by Bougon (1983) to measure "implicit knowledge" held by individuals who share certain social positions or social categories (e.g., "expert" teachers versus "novice" teachers, Bougon and Baird, 1987), displays many of the characteristics of holographic organization--an encoded enfolded order, distributed information store, part-whole correspondence in information, associative recall, etc.

The second set of postulates rests on the axiom that all interactions are guided by preferential correlations. This introduces an order-generating bias in the interactions between ensembles of systems. These interactions occur by way of a holographic prompt which in-forms the pathways of action. Because the holographic prompt encodes information about the specific system and the multi-level hierarchy to which it belongs, the order-generating process results in self-organized evolution of increasingly complex forms of organization.

However, in considering this framework as an account for interaction at the psycho-social level, we found two basic difficulties, each of which appears to rule out any possibility of an active human agent in social life. One is Laszlo's use of classical holography, the holography of image processing. Based on the linear mathematics of a two-way Fourier transformation, this type of holography renders a predetermined order which by definition precludes chance and choice, and, therefore, agency. The second difficulty is Laszlo's view of the human brain as merely a passive receiver of information. This view makes it virtually impossible to account for purposeful forms of social organization within his quantum vacuum interaction framework. Thus, as the scheme is presently conceived, it is insufficient: it leaves the key question of an active human agent unaddressed.

This is the basic difference with the alternative understanding of psycho-social organization that I have outlined here. Human agency is provided for in two respects. The first stems from my assumption that the human brain *is* an active processor of information; this seems the most reasonable interpretation of the evidence (Freeman, 1995; Gazzaniga, 1995; Pribram, 1991; Schore, 1994; Searle, 1983). Our description of social collaboration, following Piaget's analysis, as generative of a purposeful human agent, rests upon this neuropsychological assumption. The second way that agency enters is through the operation of the collective's normative order. In my account, the construction and regulation of the communicative structure by the normative order introduces indeterminacy by opening the system to new input. The normative order not only defines the nature

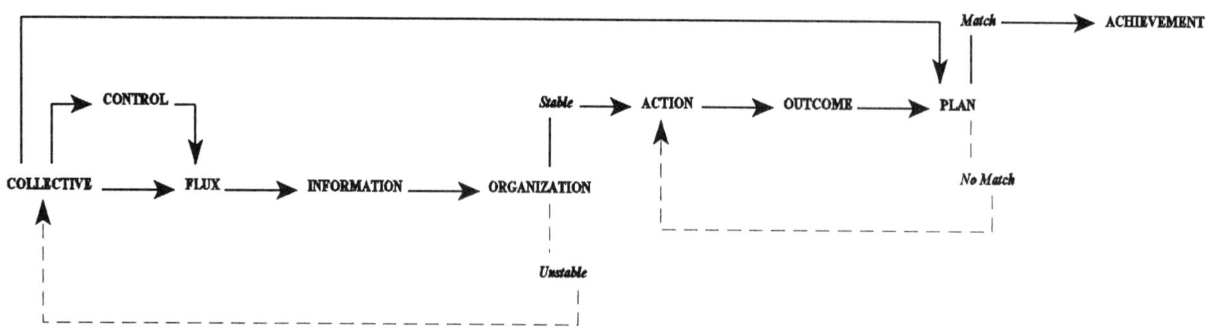

Figure 6. Logic of Communication for Endogenous Organization and Purposeful Action.

and scope of relations among members, and hence the potentials for information flow, but it also establishes a distinctive psycho-social bandwidth for communication by attuning members to one another through its framework of shared values and beliefs. However, I am *not* substituting cultural determinism for holographic determinism here. In my account, the normative order is not a causal agent but plays a much more subtle role. It sets only the broad parameters within which the actual form and direction of collective organization evolves as a dynamic, self-organizing, multi-level process.

It is from this foundation that I develop an alternative non-determinist holographic account which uses the principles of both classical and quantum holography. The principles of quantum holography are used to describe how the interaction between flux and control creates a communicative structure that processes information on the interactions among members to in-form the collective's endogenous organization. And the principles of classical holography are used to describe the image processing required for the collective's purposeful action. Achievement of the collective's plan is modeled as a pattern-matching process in which the hologram for the plan is compared against successive quantum-holographic descriptions of the consequences of collective's action until a match is achieved. Figure 6 summarizes the combined logic of these processes of communication in the terms of a single substantive model.

A final point concerns a tantalizing question of connection between Laszlo's scheme and ours: namely, the relation between his concept of quantum vacuum interaction and our concept of flux. Could it be that flux, the distribution of energy in psycho-social fields, is actually mediated by the quantum vacuum? In discussing this question for future research with Karl Pribram, we believe that the answer may not lie in the already extensively studied properties of electromagnetic fields but must await improvements in the measurement of magnetic phenomena and gravity.

References

Ashby, W. R. (1956). *An Introduction to Cybernetics.* London: Chapman & Hall.

Bohm, D. (1980). *Wholeness and the Implicate Order.* London: Routledge.

Bohm, D., & B. J. Hiley (1993). *The Undivided Universe*. London: Routledge.

Bougon, M. G. (1983). Uncovering cognitive maps: the self-Q technique. In G. Morgan (ed.), *Beyond Method: Strategies for Social Research*, pp. 173-188. New York: Sage Publications.

Bougon, M. G., & N. Baird (1987). Unfolding the wholistic social knowledge of novices and experts. Unpublished paper, presented to Annual Meeting of the Academy of Management, New Orleans.

Bradley, R. T. (1987). *Charisma and Social Structure: A Study of Love and Power, Wholeness and Transformation*. New York: Paragon House.

Bradley, R. T. (1996). The anticipation of order in biosocial collectives. *World Futures: The Journal of General Evolution*, Vol. 49: 93-116.

Bradley, R.T (1997; in press). Quantum vacuum interaction and psycho-social organization. In D. Loye (ed.), *The Evolutionary Outrider: The Impact of the Human Agent on Evolution*, Chap. 10. New York: Gordon and Breach.

Bradley, R. T., & K. H. Pribram (1995). Communication and stability in social collectives. Unpublished manuscript. Carmel, CA: Institute for Whole Social Science.

Bradley, R. T., & K. H. Pribram (1996). Self-Organization and the social collective. In K. H. Pribram & J. King (eds.), *Learning as Self-Organization*, pp. 478-506. Mahwah, NJ: Lawrence Erlbaum Associates.

Bradley, R. T., & K. H. Pribram (1997a). Optimality in biosocial collectives. In D.S. Levine & W. R. Elsberry (eds.), *Optimality in Biological and Artificial Networks*, pp. 449-488. Mahwah, NJ: Lawrence Erlbaum Associates.

Bradley, R.T., & K.H. Pribram (1997b; in press). Generation of collective order. In A.A. Montuori & R.E. Purser (eds.), *The Social Dimensions of Creativity*, Volume 3. New York: Hampton Press.

Bradley, Raymond T., & N. C. Roberts (1989a). Relational dynamics of charismatic organization: The complementarity of love and power. *World Futures: The Journal of General Evolution*, Vol. 27: 87-123.

Bradley, R. T., & N. C. Roberts (1989b). Network structure from relational data: Measurement and inference in four operational models. *Social Networks,* 11, 89-134.

Burt, R. S. (1992). *Structural Holes: The Social Structure of Competition*. Cambridge, MA, & London: Harvard University Press.

Carlton-Ford, S. (1993). *The Effects of Ritual and Charisma.* New York: Garland Publishing.

Cherry, C. (1966). *On Human Communication: A Review, a Survey, and a Criticism*. Cambridge, MA: The MIT Press.

Durkheim, E. (1949). *The Division of Labor in Society* (trans. George Simpson). Glencoe, Ill.: The Free Press.

El Sawy, O. A. (1985). From separatism to holographic unfolding: the evolution of the techno-structure of organizations. Unpublished paper, ORSA/TIM Conference, Boston.

Fowler, H., & F.G. Fowler (eds.) (1964). *The Concise Oxford Dictionary* (5th ed.). Oxford, England:Oxford University Press.

Freeman, W. J. (1995). *Societies of Brains: A Study in the Neuroscience of Love and Hate*. Hillsdale, NJ: Lawrence Erlbaum Associates.

Freud, S. (1895/1966). *Project for a Scientific Psychology*. (Standard Ed., Vol. 1). London: Hogarth.

Gabor, D. (1946). Theory of communication. *Journal of the Institute of Electrical Engineers*, 93, 429-457.

Gabor, D. (1948). A new microscopic principle. *Nature*, 161, 777-778.

Gazzaniga, M. S. (ed.) (1995). *The Cognitive Neurosciences*. Cambridge, Mass. & London: The MIT Press.

Giles, H., & N. Coupland (1991). *Language: Contexts and Consequences.* Pacific Grove, CA: Brooks/Cole.

Glazer, R. (1987). Holographic decision-making: An example of the analysis of interdependence in individual choice. Unpublished paper, presented to Annual Meeting of the Academy of Management, New Orleans.

Gregory, S. W., & S. Webster (1996). A Nonverbal signal in voices of interview partners effectively predicts communication accommodation and social status perceptions. *Journal of Personality and Social Psychology,* 70, 6: 1231-1240.

Holland, P. W., & S. Leinhardt (1976). Local structure in social networks. In D. R. Heise (ed.) *Sociological Methodology 1976*, pp. 1-45. San Francisco: Jossey-Bass.

Hutchins, E. (1991). The social organization of distributed cognition. In Lauren B. Resnick, John M. Levine, and Stephanie D. Teasley (eds.), *Perspectives on Socially Shared Cognition*. Washing-ton, DC: American Psychological Association.

Jantsch, E. (1980). *The Self-organizing Universe: Scientific and Human Implications of the Emerging Paradigm of Evolution*. Oxford: Pergamon Press.

Kaiser, Gerald (1994). *A Friendly Guide to Wavelets*. Boston: Birkhauser.

Kauffman, S. A. (1993). *The Origins of Order: Self-Organization and Selection in Evolution*. New York: Oxford University Press.

King, J. S., M. Xie, B. Zheng, & K. H. Pribram (1994). Spectral density maps of receptive fields in the rat's somatosensory cortex. In *Proceedings of the Second Appalachian Conference on Behavioral Neurodynamics*, pp. 557-571. Hillsdale, NJ: Lawrence Erlbaum Associates.

Laszlo, E. (1995). *The Interconnected Universe: Conceptual Foundations of Transdisciplinary Unified Theory*. Singapore: World Scientific.

MacKenzie, Kenneth D. (1991). *The Organizational Hologram: The Effective Management of Organizational Change*. Norwell, MA: Kluwer Academic Publishers.

Miller, G. A., E. H. Galanter, & K. H. Pribram (1960). *Plans and the Structure of Behavior*. New York: Holt, Rinehart & Winston.

Morgan, G. (1986). *Images of Organization*. Beverly Hills, CA: Sage Publications, Inc.

Morgan, G., & R. Ramirez (1984). Action learning: a holographic metaphor for guiding social change. *Human Relations*, 37: 101-106.

Nicolis, G., & Prigogine, I. (1977). *Self-Organization in Nonequilibrium Systems: From DissipativeStructures to Order Through Fluctuation*. New York: Wiley-Interscience.

Piaget, J. (1965/1995a). Logical operations and social life. In Leslie Smith (ed.), *Sociological Studies*, pp. 134-157. London and New York: Routledge.

Piaget, J. (1965/1995b). Genetic logic and sociology. In Leslie Smith (Ed.), *Sociological Studies*, pp. 184-214. London and New York: Routledge.

Pribram, K. H. (1971). *Languages of the Brain: Experimental Paradoxes and Principles in Neuropsychology*. New York: Brandon House.

Pribram, K. H. (1982). The place of brain in the ecology of mind. In W. Weimer & D. Palermo (eds.), *Cognition and the Symbolic Process, Vol. 2*. Hillsdale, NJ: Lawrence Earlbaum Associates.

Pribram, K. H. (1991). *Brain and Perception: Holonomy and Structure in Figural Processing*. Hillsdale, NJ: Lawrence Erlbaum Associates.

Pribram, K. H., & R. T. Bradley (1997; in press). The brain, the Me and the I. In R. Sternberg & M.Ferrari (Eds.), *The Development of the Self*. New Haven, CT: Yale University Press.

Pribram, K. H., & D. McGuinness (1975). Arousal, activation and effort in the control of attention. *Psychological Review*, 82: 116-149.

Pribram, K. H., & D. McGuinness (1992). Attention and para-attentional processing: Event related brain potentials as tests of a model. In D. Friedman & G, Bruder (eds.), *Annals of the New York Academy of Sciences*, 658: pp. 65-92. New York: New York Academy of Sciences.

Prigogine, I., & I. Stengers (1984). *Order out of Chaos: Man's New Dialogue with Nature*. New York:Bantam Books.

Prigogine, I. (1997). *The End of Certainty: Time, Chaos, and the New Laws of Nature*. New York: The Free Press.

Ravn, I. (1985). Creating futures, constructing realities. *General Systems Yearbook.*

Roberts, N. C., & R. T. Bradley (1988). Limits to charisma. In J. Conger & R. Kanungo (Eds.), *Charismatic Leadership: The Elusive Factor in Organizational Effectiveness* (pp. 253-275). San Francisco: Jossey-Bass.

Roberts, N. C., & R. T. Bradley (1991). Stakeholder collaboration and innovation: A study of public policy initiation at the state level. *Journal of Applied Behavioral Science*, 27, 2: 209-227.

Roberts, N. C., & P. J. King (1996). *Transforming Public Policy: Dynamics of Policy Entrepeneurship and Innovation.* San Francisco: Jossey-Bass.

Rosenstein, G. (1996). The income-choice approach and some unsolved problems of psychology--a "bridge over time." In D. S. Levine & W. R. Elsberry (eds.), *Optimality in Biological and Artificial Networks*, pp. 396-423. Mahwah, NJ: Lawrence Erlbaum Associates.

Roethlisberger, F. J., & W. J. Dickson (1939). *Management and the Worker*. Cambridge, MA: Harvard University Press.

Santa Maria, M., J. King, M. Xie, B. Zheng, K. Pribram, & D. Doherty (1995). Responses of somato-sensory cortical neurons to spatial frequency and orientation: A progress report. In J. King & K. H. Pribram (eds.), *Scale in Conscious Experience: Is the Brain too Important to Left to Specialists to Study?* Pp. 157-168. Mahwah, NJ: Lawrence Erlbaum Associates.

Schore, A. N. (1994). *Affect Regulation and the Origin of the Self: The Neurobiology of Emotional Development.* Hillsdale, NJ: Lawrence Erlbaum Associates.

Searle, J. R. (1983). *Intentionality: An Essay on the Philosophy of Mind.* Cambridge, Eng.: Cambridge University Press.

Searle, J. R. (1995). *The Construction of Social Reality.* New York: The Free Press.

Shannon, C. E. (1949). The mathematical theory of communication. In C. E. Shannon & W. Weaver, *The Mathematical Theory of Communication* (pp. 3-91). Urbana, IL: The University of Illinois Press.

Toennies, F. (1957). *Community and Society* (trans. Charles P. Loomis), Book I. East Lansing, MI: Michigan State University Press.

Turner, J. H. (1986). *The Structure of Sociological Theory.* Chicago: The Dorsey Press.

Williamson, Oliver E. (1975). *Markets and Hierarchies.* New York: Free Press.

Xie, M. (1995). Signal decomposition for non-stationary processes. Unpublished doctoral dissertation, Virginia Polytechnic and State University.

Zablocki, B. D. (1971). *The Joyful Community.* New York: Penguin Books.

Zablocki, B. D. (1980). *Alienation and Charisma: A Study of Contemporary Communes.* New York: The Free Press.

19
On Cognitive Maps, Vicarious Trial-and-Error, and Impulsivity

Abram Amsel
Department of Psychology and Institute for Neuroscience
University of Texas at Austin
Austin, TX 78712
Telephone: 512-471-3827/Fax: 512-471-6175
E-mail: amsel@psy.utexas.edu

Abstract

The most prominent hypothesis of hippocampal function likens the hippocampus to a cognitive map, a term used by a famous learning theorist, E. C. Tolman, to explain maze learning. The usual current application of this concept of cognitive map, as it applies to the hippocampus, is to what is called spatial learning, mainly in the radial-arm maze of Olton and the Morris water maze. In a recent Hippocampus Forum in the journal Hippocampus, evidence for the cognitive map hypothesis was reviewed in a lead article by L. Nadel, followed by a series of commentaries by leading investigators of hippocampal function. In a later issue of Hippocampus, I offered an alternative not represented in the forum--that the function of the hippocampus in spatial learning is not as a cognitive map, but that it subserves another function also proposed by Tolman in his work on simple discrimination learning, vicarious trial-and-error (VTE), based on incipient, conflicting dispositions to approach and avoid.

Experimental data are presented to show that reduced hippocampal function affects learning--single alternation and the partial reinforcement extinction effect in a straight runway--that is in no way spatial. They support the hypothesis that intact VTE may itself be a function of the hippocampus, and that there may be connections among deficits in VTE, impulsivity, and attentional disorders. Some recent experiments on this latter conjecture are also presented. These experiments, also in infant and preweanling rats, involve reducing hippocampal function pharmacologically, following pre-and postnatal exposure to alcohol, and via x-irradiation, and restoring some of this loss of function with d-amphetamine.

The Hullians of my generation, and, particularly, my mentor, Kenneth W. Spence, would probably not be amused by my emphasis, in this presentation, on Edward Tolman and some of his followers, particularly in the opening part of my remarks; but, except for the S's and R's and the little r_g's in a later allusion to my own theorizing, the work of Tolman, a famous learning theorist of the 1930's and 1940's, will be my main point of departure.

Introduction

The Hippocampus as a Cognitive Map

A major focus of this presentation is on the hippocampus, and the most prominent hypothesis of hippocampal function, originated in a book by O'Keefe and Nadel (1978), likens the hippocampus to a "cognitive map," a term coined by E. C. Tolman (1948) to explain, among other things, complex maze learning. The more recent application of this concept of cognitive map, particularly as it applies to functions of the hippocampus, is to what is called "spatial learning," mainly in the radial-arm maze of Olton and the Morris water maze.

In a Hippocampus Forum "Hippocampus and Space Revisited," in the journal Hippocampus, in 1991, evidence for the cognitive map hypothesis was reviewed in a lead article by L. Nadel, followed by a series of commentaries by leading investigators of hippocampal function. In one of the next issues of Hippocampus, I offered, as an epilogue to this love feast-like commentary, an hypothesis that had not been represented in the Forum (Amsel, 1993).

My hypothesis was that the function of the hippocampus in spatial learning is not as a cognitive map, but that it subserves another function proposed many years ago by Tolman: A decade before his original allusion to the cognitive map (Tolman, 1948), in an article titled "Cognitive Map in Rats and Men" Tolman (1938), along with his colleague, Muenzinger (1938), had proposed a mechanism they called vicarious trial-and-error (VTE), which referred to a rat's head movements from one discriminandum to the other on a Lashley Jumping Stand. These head movements were taken to represent incipient "evaluations" of the stimuli, resulting in conflicting pre-discrimination dispositions to approach and avoid both discriminanda. (More on head-movements later.) Figure 1 is a drawing of a Lashley jumping stand, named for Karl Lashley, who invented it, and which was the classical apparatus for demonstrating VTE.

In the old jumping-stand experiments, the rat jumps from the stand over a gap of a few inches toward one of two stimulus doors (the visual discriminanda), behind one of which there is food. The left-right positions of the positive and negative stimuli are randomized from trial to trial, so that the rat cannot discriminate purely on the basis of left-right position. This results in a lot of VTE'ing.

In 1979, in a multiple review in Behavioral and Brain Sciences of the then-recent book by O'Keefe and Nadel (1978), The Hippocampus as a Cognitive Map, I was one of a small number of dissenters--perhaps the strongest--from the implication of the title--that the rat hippocampus functioned (exclusively?) in a cognitive, mainly "spatial" mode. One of the things that troubled me was that "cognitive" and "spatial" had become functionally synonymous. In my brief 1979 review of O'Keefe & Nadel, I cited all the evidence I could get into a small space suggesting that a variety of noncognitive (nonspatial, in their sense) kinds of learning were clearly affected by damage to the hippocampus. Among the decidedly nonspatial learning (and memory) I cited was a kind that Thomas (1984) and I (Amsel, 1986) had called "dispositional" (as opposed to "declarative"). Included in my definition were patterned single alternation (PSA: go/no-go memory-based learning); the partial-reinforcement extinction effect (PREE: learned persistence); successive negative contrast (SNC: learned suppression); and discontinuously negatively correlated reinforcement (DNC: learned, idiosyncratic, time-taking), all studied in a straight-alley runway. In fact, at about the time of appearance of the O'Keefe and Nadel book, we were in the process of working out, at Texas, in a program of developmental research in the rat, the order of first appearance of a family of these reward-schedule effects. Table 1 lists nine of these in order of their first appearance in the ontogeny of the rat (Amsel 1986, 1992).

Of course, none of these learned dispositions--such as running or not running in runway acquisition and extinction; running fast, then slow (SNC); running persistently or not persistently (the PREE); running idiosyncratically, or, in human terms, superstitiously (DNC)--none of these effects, based on the specific intermixture of rewards and nonrewards, requires for its demonstration a radial-arm maze or a Morris water maze, or has an obvious spatial or cognitive-map component. And, starting in the mid-'seventies, Jeffrey Gray's laboratory at Oxford and my own, among others, had begun to show that dispositions, as nonspatial as the PREE and SNC, were clearly affected by hippocampal interventions. We were influenced in this work by existing nonspatial conceptualizations of hippocampal function in terms such as novelty, discrepancy, "hesitation and doubt" (e.g., Simonov, 1974; Vinogradova, 1975); suppression and inhibition (e.g., Altman et al., 1973; Douglas, 1967; Kimble, 1968); neuromodulatory effects (Isaacson, 1980); and, in the spirit of Vinogradova and Simonov, frustration, suppression, and persistence (e.g., Amsel, 1992; Amsel et al., 1973; Glazer, 1974a, b). Later, Gray (1982) referred to hippocampal function, broadly and inclusively, as the "behavioral inhibition system."

Angevine and Cotman (1981), in what could stand as a summary of a kind of pre-O'Keefe & Nadel characterization of hippocampal function, wrote the following: "[T]he hippocampus holds the secret to limbic system function, or at least a large part of it....Lesions of the hippocampus in experimental animals make it more difficult for these animals to change an ingrained response to a new one" (p. 261, italics added).

The term "cognitive map," in O'Keefe & Nadel's title, was, of course, borrowed from a paper in the Psychological Review by Edward Tolman (1948). In this paper, titled "Cognitive Maps in Rats and Men," Tolman had, obviously, taken a new course, and had set aside the power of his decade-earlier VTE concept. Interestingly, exactly bounding this 1948 date, from 1947 to 1949, Hugh Blodgett, an early student of Tolman, then at the University of Texas at Austin, was the principal author of the three most analytical papers ever written, in my opinion, on movement of rats in space. I do not say "spatial" here, but rather "movement of rats in space," because these experiments examine the rat's movements, in space, in terms of the relative strengths of three possible components: response, direction and place--as it turned out, in that order of importance. In the last of these papers (Blodgett, McCutchan, & Mathews, 1949), they used an arrangement of elevated T-mazes reproduced as Figure 2.

I won't go into detail here on the intricacies of all of these experiments, but with the help of Figure 2 we can follow the basis of Blodgett et al.'s results in one of the last in this series of experiments, whose title was "Spatial learning in the T-maze: The influence of direction, turn, and food location." The following is a summary of their definitions of responding on the basis of place (location), direction, and response (turn), respectively. Their general procedure was to train animals on (say) S_1-F_2 then test on S_3-F_2 as an indication of learning on the basis of "place." And so on with direction and response, as follows:

Gp I: Place	Train on S_1-F_2 test on S_3-F_2
Gp II: Direction	Train on S_1-F_1 test on S_4-F_2
	or
	Train on S_1-F_2 test on S_4-F_3
Gp III: Response	Train on S_1-F_1 test on S_2-F_2
	or
	Train on S_1-F_2 test on S_2-F_1

From this and other experiments published from 1947 to 1949, Blodgett et al. concluded "that in learning the elevated T-maze, in which, of course, distal cues were available, the rat acquires a direction disposition and, of lesser strength, a response disposition ... [but there is] no evidence that the rat acquires a place disposition" (Blodgett, McCutchan, & Mathews, 1949, p. 809).

One of the very interesting, and, I think, instructive things about this last paper by Blodgett et al. is that the word "spatial," which actually appears in the title of all three papers, and is so frequently used (in their sense, misused) today, has no specific, functional or theoretical meaning: it is simply a general term covering everything the animal does in space. There is no reference in these experiments to "spatial learning," or to "cognitive maps." To repeat, "spatial," for Blodgett et al., is an overall term that includes "direction" or general orientation, "turn or response" (left or right), and "place" (food location)--important in that order: In fact, they found in the rat no evidence for a place or location disposition in what they, in overall terms, called spatial learning. The whole recent (since O'Keefe & Nadel, 1978) goings-on about "spatial learning," mainly in relation to "The Hippocampus as a Cognitive Map," is covered in the work of Blodgett et al. by the terms direction, response, and place--again, with no real evidence for a "place disposition."

As I pointed out earlier, the term VTE was coined some 60 years ago by Tolman (1938, 1939) and Muenzinger (1938); and if one thinks of Tolman as the father of the cognitive-map concept, as he surely was some 50 years ago, in his paper "Cognitive maps in rats and men" (Tolman, 1948), his concept of VTE, a decade earlier, would seem very "unTolman-like." There is nothing particularly cognitive about VTE. In fact, in its use, even then, it was much more like opposing (or conflicting) stimulus-response direction dispositions, represented, specifically and in fact, by head-movements, testing the "reward value" of the two stimuli.

A Digression

Well before my 1993 paper on VTE in the journal Hippocampus, there was, I recently discovered, a literature on head movements (but not on VTE) in the rat, mainly related to recording from subiculum (pre and post) and from "other hippocampal areas," though some of these studies alluded also to thalamus and thalamocortical circuits. A paper by Blair & Sharp (1996) cites a

1984 Society for Neuroscience Abstract, by J.B. Ranck, Jr., on "head-direction cells" as perhaps the first of these. I was followed by two papers by Taube, Muller, & Ranck (1990a, b), and then several others in 1994 and 1995 (cited in the Blair & Sharp, 1996, paper). One very recent paper by Taube and Burton (J. Neuroscience, 1997, in press), seems particularly relevant for the concept of VTE: its title actually includes the phrase "head-direction activity ... during a cue conflict situation," almost a paraphrase of Tolman's VTE.

VTE in its Various Manifestations

It is clear, then, that VTE has been regarded as the external manifestation of conflicting tendencies to approach and avoid at a choice-point. It can also be regarded as a behavioral actualization of Hull's (1931) anticipatory goal reaction with its feedback stimulation (r_G-s_G) or, more to the exact point, to the external manifestation of conflict in the third stage of my four-stage theories of the partial reinforcement extinction effect and, particularly, in simultaneous discrimination learning (Amsel 1958, Figure 3); The third stage in each of these two cases involves behavioral variability stemming from conflict between approach and avoidance. In the case of runway behavior, on which most of this theorizing was based, this involves going forward to the goal, then retracing toward the startbox--conflict between anticipatory reward and anticipatory frustration is a runway version of VTE.

Support for my preference for VTE over cognitive map as a manifestation of hippocampal function comes from several sources. One is an observation by M. F. Brown (1992) that, in the familiar radial-arm maze, "door investigations" and "microchoices" represent go-no-go decisions, and that such decisions do not require "the assumption that choice behavior is guided to baited arms by a cognitive map" (p. 57).

In another very important case, Whishaw (1991) has pointed out that, even in the Morris water maze, an apparatus in which the absolute importance of externally placed cues may not seem so obvious as a basis for VTE behavior as in a radial-arm maze, "rats use associative learning processes rather than cognitive mapping to solve place problems" (p. 83). Whishaw and Tomie (1987) had shown earlier that central cholinergic blockade in rats treated with atropine sulfate (but not atropine methylnitrate) reduced the number of "searches" in learning in the Morris water maze, searches defined as pauses during swimming, raising the head from the water, and head scans made at the wall of the pool.

The evidence that animals acquire distal cue information en route to the platform has also seemed quite clear to Devan, Blank, & Petri (1992), who alluded to Morris's own early recognition that, whereas in his apparatus no local (intramaze) cues were available to the rat, the animals were "able to generate accurate trajectories to the submerged platform on the basis of its fixed spatial relation with distal (extramaze) room cues" (p. 120).

Finally, in the words of Leaton and Borszcz (1990, p. 81), "one could develop a nonspatial analog of [radial-arm] maze learning as a sequence of conditioned stimuli (CSs) culminating in a positive or negative reinforcement." In short, the argument is that cognitive mapping, the mechanism ascribed to the rat in the major current theory of hippocampal function, can be derived from the kind of incipient sampling of cues and the obvious momentary hesitation and conflict that defines VTE, and therefore that VTE, itself, can be derived from what can be regarded as behavioral first principles: conflicting tendencies to approach and to avoid, in the terms of the present symposium, a most fundamental representation of values, both positive and negative.

Tolman, Blodgett, and VTE in "Spatial" Learning

In Tolman's experiment on VTE, there were three groups of rats, representing three levels of discrimination difficulty on the Lashley jumping stand. For all groups the rewarded door was white; however, the nonrewarded door was black for one group, medium gray for another, and light gray for the third. (Recall that the discriminanda were moved quasi-randomly from side to side.) Tolman's data show a very clear positive relationship between number of VTEs per day and the level of discrimination performance: The group for which black was negative was at 100 percent correct choice after 20 days, medium gray at about 75 percent, and light gray still at chance performance. Average number of VTEs per day increased in the two groups that learned, reaching about 18 in the black-negative group and about 10 in the medium gray-negative group; the light-gray-negative VTEs per day stayed at about four. Clearly at odds with his own concept of cognitive map, which, as we have seen, came about a decade later, Tolman writes about VTE as follows:

The theory ... consists in the assumption that all higher organisms are in a certain underlying sense tropistic. This, not literally but figuratively speaking, tropistic or taxic character of higher organisms provides, I believe, the necessary model for stating how in this experiment with rats different strengths of hunger, different goodnesses of perceptual differentiations, and different degrees of learning, combine to produce the actual amounts of correct choosing and of VTE-ing that they do. (Tolman, 1939, p. 323)

Tolman went on in this paper to reemphasize the taxic nature of this kind of learning with a model he called the "schematic sowbug," a rather complicated analysis of the perceptual-orienting factors comprising VTE. Put simply, the schematic sowbug is actually averaging and resolving, from moment to moment, the strengths of its tendencies to approach and to avoid. As an old Hullian, I am more than comfortable with this Tolman of 1939 (as opposed to his later "cognitive-map" phase), because the VTE analysis can readily be expressed in conflicting or opposing habit strengths or, even better, r_g's and s_g's. Nevertheless, in the case at hand, as I pointed out earlier, Tolman's, and not Hull's, analysis is central to my argument.

VTE or Cognitive Maps in Spatial Learning?

What I have proposed (Amsel, 1993) is that what is now called cognitive mapping or spatial (locale) learning can be reduced to the operation of Muenzinger's and Tolman's VTE, which has been detected not only in the rat's head-movements on a Lashley jumping stand, but also at choice-points on an elevated maze, such as Olton's radial-arm maze, or in the water maze--in fact, in any situation in which the rat responds on the basis of <u>surrounding</u> external cues. As I pointed out in the 1993 <u>Hippocampus</u> commentary, O'Keefe and Nadel (p. 73) cite Blodgett and McCutchan (1947) to the effect that in a constant heterogeneous environment place learning is faster than response learning, but that this superiority disappears in a homogeneous environment. However, they do not cite an article a year later, subtitled, "Anticipation as a failure of place discrimination," in which Blodgett and McCutchan (1948) provided data to support the position that "a response disposition is...stronger than a place disposition" (p. 23), or an article two years later, to which I have referred in some detail, titled "Spatial learning in the T-maze: The influence of direction, turn, and food location." Here they report, as we have seen, that "the rat acquires a direction disposition and, of lesser strength, a response disposition [but there is] no evidence [for] a place disposition" (Blodgett, McCutchan, & Mathews, 1949, p. 809).

In fact, this order of strength: first, direction; second, turn or response; and third, place--<u>with no real evidence for place learning</u>, is very much in line "with the hypothesis that VTE and the underpinnings on which it is based, approach and avoidance, are the first principles from which the general construct of spatial learning can be derived" (Amsel, 1993, p. 253).

The Concept of VTE in Our Own Recent Work

In some of our recent work, which is based on a developmental paradigm--and, therefore, involves hippocampal immaturity as well as hippocampal insult--the emphasis has been mainly on the two earliest-appearing reward-schedule effects: The first, PSA, a form of memory-based discrimination learning (e.g. Diaz-Granados et al., 1992, 1994; Greene et al., 1992; Lobaugh et al., 1989, 1995), involves inhibition or suppression on trials following reward (the "braking" system); the second, the PREE, involves partial--compared with continuous--reinforcement and resistance to extinction (relative learned persistence). The other effects (Table 1) also--and in other ways--involve reaction to discrepancy, novelty, and frustration (see Amsel, 1986, 1992, for reviews). Figure 4 is a schematic of our typical runway apparatus for experiments on PSA and PREE (and other reward-schedule effects). Figure 5 presents PSA results in four treatment conditions: hippocampal lesions, (Lobaugh et al., 1989), high-peak blood ethanol (Greene et al.,

1994), infantile x-irradiation (Diaz-Granados et al., 1994), and an untreated control. In all cases the experiments were conducted in preweanling rats--17 to 18 days of age at 60s ITI--and in all cases the effects were the same: the treatments eliminated PSA over a 2-day period involving 240 trials. Figure 6 presents results showing the negative effect on PSA of atropine, a hippocampal muscarinic receptor blocker, in four conditions of administration (Lilliquist et al., 1993).

A study, recently published, adds to our work on PSA the effect of the NMDA antagonist, MK-801, again in preweanling rats, after administration of 0.05 mg/kg 30 minutes before each of 3 daily sessions on successive days (Figure 7, Highfield, Nixon, & Amsel, 1996). As you can see, the effect increases with ITI, from 8 to 30 to 60 seconds.

None of these effects in memory-based (PSA) learning, observed in a simple, homogeneous, straight runway (or, as we shall see, in a simple simultaneous visual discrimination) has an obvious spatial component. In all of them we see "VTE-like" behavior. For example, in the third stage of the PSA discrimination (or in the PREE experiment) VTE-like behavior takes the form of conflict between approaching and avoiding the goalbox. The reward-schedule effects, outlined earlier, emerge in the rat as the hippocampus matures in the first four postnatal weeks. The results just reviewed show that the earliest-appearing of these effects, PSA--a form of go/no-go discrimination based on memory--is attenuated, or even eliminated, by a variety of hippocampal interventions in the preweanling rat. These deficits fit Simonov's (1969) characterization of the hippocampus as "an organ of hesitation and doubt," or, in Angevine & Cotman's phrase, that hippocampal damage (or, we would add, hippocampal immaturity) "makes it more difficult for [the animal] to change an ingrained response to a new one."

The VTE mechanism, as originally characterized, however, was designed to cover cases of conflict related to approach and avoidance of differential, external, spatially arranged cues, e.g., the Lashley Jumping Stand. If VTE is absent, an infant rat, or a rat with extensive damage to the hippocampus, or a preweanling rat with damage from x-irradiation, is more likely (a) to go into the arm of a radial maze in which it has already found food, or (b) to swim in a tub of water toward external landmarks (usually including the experimenter) that do not lead to the submerged platform. There is, in such an infant or damaged, preweanling experimental animal, less "hesitation," less "doubt," less conflict between approach and avoidance, and less VTE in the service of learning and memory, than in an animal whose hippocampus, broadly taken to include input and output structures, is mature and intact.

In some recent work (Hu & Amsel, 1995) that involved a simpler to-choice arrangement of external cues, we showed that bilateral dorsal hippocampal lesions reduced the appearance of VTE and retarded black-white discrimination learning in what has been referred to as a Grice Box, a startbox and alley leading to a choice area in which the two cues can be quasi-randomly arranged from side-to-side. Figure 8 shows the extent and location of the lesions; Figure 9A mean rewards per day, and 9B mean VTEs per day in sham lesions and hippocampal lesions. Figure 10 is a comparison of VTE performance in sham and lesion conditions in relation to rewards. It is clear from these data that VTEs are strongly related to rewards: that hippocampal lesions retard VTE onset and onset of discrimination learning.

Additional experiments (Griesbach, Hu, & Amsel, unpublished) indicates that pre-administration of MK-801 retards the acquisition of an odor discrimination and reduces VTE in a Y-maze, with no effect on the reversal of that discrimination. However, in a second experiment, if administration of MK-801 does not precede acquisition, but is delayed until the reversal phase, both rate of reversal and VTE are affected.

Approach and Withdrawal, Seeking and Avoidance: The Role of Expectancy and VTE in Mature Hippocampal Function

Some years ago, Schneirla (1959) proposed a very important distinction which has not had the broad attention it has deserved, particularly in those interested in developmental processes. Schneirla's is actually a distinction between distinctions. He used the terms approach and avoidance somewhat differently, and in a sense more definitively, than, say, N. E. Miller did. Referring, particularly, to their application to ontogeny and phylogeny, he suggested two separate distinctions: The first was between "approach and withdrawal," this more appropriate for less developed or "advanced" animals, ontogenetically and phylogenetically. The second distinction was between "seeking and avoidance," this more appropriate for more mature or higher-level animals. He expanded on the latter distinction as follows: "Psychologists who emphasize disproportions, reversals, and exceptions between stimulus magnitude and response properties are therefore talking of adult stages at higher psychological levels" (p. 22).

In this terminology Schneirla was acknowledging the greater role of expectancy in higher-level than in lower-level organisms. Speaking ontogenetically, in the case of the rat, the distinction would describe the appearance of seeking and avoidance (rather than simple approach and withdrawal), and therefore of the appearance of VTE in the 17-18-day-old, preweanling rat, but not in the infant rat, the rat being a species much less precocial at birth than the human infant (Figure 11). (I must remind you, however, that, generally, in the literature of learning and memory, and for example, in my own theorizing, including in the present paper, the tendency of virtually every investigator of conflict in animal learning has been to follow N. E. Miller's style [e.g., Miller, 1959], using the terms "approach and avoidance," as the descriptors of conflict, and, as in Tolman's case and the present one, its visible accompaniment: VTE.) Adopting Schneirla's terminology for the moment, however, it seems reasonable to suppose that, not only very young (infant) mammals and lower-level organisms, but also older ones with hippocampal damage, are operating at the lower levels of "approach and withdrawal" rather than at the higher ones of "seeking and avoidance," since seeking and avoidance connote positive and negative expectancy, whereas approach and withdrawal do not.

It follows also that, in the ontogeny of the intact rat, VTE behavior is present more and more as the hippocampus and, particularly, the dentate gyrus (Figure 12), which arises postnatally, continues to mature from birth to as late as 25-30 days of age (Altman & Das, 1965). It would then also be the case that VTE behavior would be present less and less as the nature and extent of hippocampal damage increases in severity, for example, as a consequence of extended focal x-irradiation of the dentate gyrus, which, applied very early in its development, from postnatal days 2 through 15, reduces hippocampal cell number by about 80 percent (Figure 13, Diaz-Granados et al., 1993).

Seeking and avoidance (rather than approach and withdrawal) and mature hippocampal function exist in what Schneirla referred to as "more advanced organisms" that act "purposively" and at "adult stages at higher psychological levels." Seeking and avoidance also govern reactions to "disproportions, reversals, and exceptions... " (p. 22). That is to say, seeking and avoidance, in contrast to approach and withdrawal, connote expectancy, and the competing expectancies of seeking and avoidance (approach and avoidance, in the terms we have come to employ) result in conflict.

According to earlier theorists (Douglas, 1967; Kimble, 1968), and later ones (e.g., Gray, 1982), the hippocampus is part of a "behavioral inhibition system" that governs, particularly, the negative expectancy. According to frustration theory (most recently, Amsel, 1992), the inhibitory effect of anticipatory frustration, the negative reward expectancy, is overridden in the prediscrimination phase of discrimination learning, because response to either stimulus is partially reinforced, which amounts to training to persist. Following extensive training, this persistence breaks down. This breakdown is marked by an <u>increase</u> in VTE, as avoidance is reinstated to the negative stimulus. All of this involves relatively mature hippocampal function, which, in the rat, involves maturation of the dentate gyrus, substantial at about preweanling age.

As I showed earlier, neither in learned persistence (the PREE) nor in simple memory-based discrimination learning (PSA) are differential external cues involved; yet there are, in both cases, clear instances of a kind of VTE which takes the form of approach-avoidance conflict. In runway work we call the VTE-like behavior retraces: the preweanling rat--in our case--turns in the alley and runs back and forth from goalbox to startbox. In these cases, experiments described earlier show that hippocampal damage from electrolytic and ibotenic acid lesions (Lobaugh et al., 1989, 1995); from fetal or early postnatal exposure to ethanol (e.g., Greene et al., 1992; Lobaugh, et al., 1991); from early postnatal exposure to X-irradiation (Diaz-Granados et al., 1992, 1994); from intrahippocampal injections of atropine (Lilliquist et al., 1993); or from treatment with MK-801 (Highfield et al., 1996)--all of these treatments produce deficits in these cue-bereft situations in preweanling rats, which survive into adulthood.

On the other hand, the radial-arm or water maze would not be apparatuses of choice in which to demonstrate deficits in extinction and discrimination reversal; or in response suppression; or on DRL or DNC schedules that reward time-taking or slow responding. In both the radial-arm and water-maze apparatuses, some spatial arrangement of objects or cues external to the apparatus is necessary for learning: The animal must form a "cognitive map." To be fair, then, it is no doubt the case that in alleys and boxes, elevated mazes and swimming pools--in this kind of apparatus--we have examples of how the conceptualization of a brain function suggests the behavioral test situation which best "illuminates" that conceptualization, to the likely neglect of others.

Nevertheless, there is in the end a certain lack of symmetry to be dealt with: I have argued here, as have others, that, in the words of Whishaw (1991, p. 83), "rats use associative learning processes rather than cognitive mapping to solve place problems in a swimming pool." It is more difficult to argue that rats require cognitive maps to learn to run in simple runways or to press levers in Skinner boxes. However, both these latter kinds of apparatus, over many years, have been in the service of assessing hippocampal integrity. Is it simply the case that the work in simple runways and boxes--some earlier and some later than 1978--has been neglected, because there is not in these simpler cases a close fit to the "cognitive map and spatial learning" explanations? Or is it, even more simply, that--unknowingly, at first--we returned to the concept of "spatial learning," which Blodgett et al. investigated some 60 years ago, in which they included direction, response, and place, and concluded that there was no evidence for place learning in the rat?

Cognitive Maps Versus VTE

It is perfectly reasonable for a theory to include both taxon and locale systems. The question is only whether both are required to account for what has been called "spatial" or "place" learning in the rat: We have seen that there are very different understandings of the meaning of spatial, Blodgett's all-inclusive one and the current one derived from O'Keefe and Nadel. If Tolman's cognitive map is replaced by his VTE, a construct with observable referents, theorizing about hippocampal function can become more analytical: In the words of Leaton and Borszcz, "more progress will be made in unraveling the mysteries of hippocampal function if the effects are analyzed in terms of the well-established and reliably measured phenomena of conditioning" (1990, pp. 87-88).

It may be the case that the mazes of Olton and Morris, which are in such wide use in university laboratories and in drug companies around the world, are indeed diagnostic for general hippocampal impairment, but are relatively non-analytical instruments for comprehending the subtleties of hippocampal function. Neglected though it has been for many years, the VTE concept, based as it is on conflicting tendencies to approach and avoid, appears the more analytical.

To complete this part of my over-long presentation, the point is that VTE--and head-movements research, generally--represent an approach that is very close to the learned behavior itself, in rats, overt, and in humans, usually--but not always--implicit. On the other hand, cognitive maps are "representations" that don't really represent anything. As Walter Freeman and Christine Skarda (1990) asked--and answered in the negative--"Representation: Who needs them?"

Dispositions, VTE, and Impulsivity

Finally, there is a possible tie-in (at least, as I see it) among the concept of "disposition," the conflict-related concept of VTE--and its opposite, impulsivity--and the symptoms and the most common, current treatment of attention deficit-hyperactivity disorder (ADHD), which, in the theme of the present symposium, are all value-laden concepts.

In an earlier paper (Amsel, 1990), I suggested that the sequence of first appearance of reward-schedule effects I had proposed in 1986 (reproduced here as Table 1) represents stages of development, and that, along with the action of pharmacological agents and other treatments (e.g., fetal and postnatal alcohol, focal x-irradiation, atropine, MK-801), they began to suggest an animal model for some of the symptoms normally associated with the diagnosis of ADHD in children. This leap to the concept of ADHD from simple animal experiments was emboldened by the earlier work of Virginia Douglas and her students (e.g., Douglas, 1980; Douglas & Parry, 1983) who have pointed to the importance "of consistent contingent reinforcers" (Douglas & Parry, 1983, p. 324) in guiding and controlling the behavior of hyperactive children. The essence of the 1990 proposal was that, whereas "normal" children (and experimental animals) learn to be persistent (the PREE) on the basis of the inconsistent presentations of rewards (Rs) and frustrative nonrewards (Ns), and learn on the basis of short-term memory to alternate their behavior to single alternations of Rs and Ns (PSA), animals with compromised hippocampal function do not learn as well, or at all. And it has been the position of Douglas and her students (and my own more second-hand view) that children with ADHD, because of their own strong responses to rewards and anticipated frustrations and, consequently, to "the activating component of r_F [anticipatory frustration] by raising ... [the child's] level of arousal beyond an optimum limit" (Freibergs & Douglas, 1969, p. 394)--that such children would be impaired in learning normal persistence. It is as though in moving through the normal four stages required for the formation of the PREE, as postulated by frustration theory (Amsel, 1958), ADHD children, in generally identifiable situations, are prone to be locked into the third stage, which leaves them in conflict.

Further, as the right-hand side of Figure 14, in what I have called A More General Theory of Persistence (Amsel, 1972), suggests, these ADHD children, in the context of "ongoing" stimulation and responding (S_O-R_O) are prone to be disrupted by intrusive external stimulation (S_X), and to be left in generalized conflict and agitation. On the other hand, "normal" children countercondition R_O's to S_X's: that is to say, they habituate more readily to such novel or unexpected external (or internal) stimulation. The point is, that whether from frustration and conflict, on the left-hand side of Figure 14, or from lack of generalized persistence--the failure to countercondition R_O to S_X--on the right-hand side, these children are less likely, in given situations, to learn persistence, and more likely to be in a VTE-like, conflict state than "normal" children. Note--again--that in the case of Figure 14, my definitions of persistence and habituation, and of conflict and generalized disruption of behavior by S_X's are, in each case, the same.

A program of research, some time ago, by Robbins and others (e.g., Robbins et al., 1983), at Cambridge University, suggested that one very important differential effect of d-amphetamine was to increase the strength of positive (approach) tendencies, but not of negative (avoidance) tendencies. As I applied this to ADHD children (Amsel, 1990), it was clear, in this context, that this differential action of amphetamine--to raise the approach gradient relative to the avoidance gradient (top left panel of Figure 15)--would, in the terms of Miller's (1944) analysis of conflict, prevent the interaction of--the crossing of--these gradients, thereby preventing or relieving conflict.

In non-medical terms, ADHD is characterized by irregularities in behavior such as fidgeting and squirming, and by learning-related attentional deficits such as forgetfulness, accompanied in some cases by hyperactivity and/or impulsivity. ADHD is widely diagnosed, particularly in boys in their childhood and early adolescence. The emphasis has been on treatment with a family of drugs, namely, d-amphetamine, methylphenidate (Ritalin) and clonidine, which are sometimes effective in treating some of the symptoms of ADHD. The areas of the brain affected by these drugs include the dopaminergic and noradrenergic systems, which are known to be related to attention and reward.

Speculation on the causes of ADHD have ranged from hypoxia during childbirth (when oxygen is unavailable for a short period of time to the newborn brain); to excessive maternal intake of alcohol, which may also cause hypoxia and brain damage in the fetus, resulting in learning deficits short of the more recognizable fetal alcohol syndrome; to other possible causes such as high levels of lead or other toxic substances that also affect brain development in the growing fetus. There is, apparently, no easy-to-identify or easy-to-diagnose brain damage in these children. Altman (1986)--in an out of the way chapter, but one that in my opinion is "must reading" for those involved in ADHD--suggested the concept, "microneuronal hypoplasia," as representing a condition in which the number or the smallest neurons (the microneurons) in the brain, found mainly in the cerebellum, the olfactory bulbs, and the dentate gyrus of the hippocampus, are, for one or more of the above-mentioned reasons, or for other reasons, severely reduced in development. The main factors in this granule-cell hypoplasia (reduced mitosis) in humans would have their effect (possibly late) in the third trimester of pregnancy. Referring to our earlier discussion on relative human and rat development (Figure 13), it is clear that in the rat the equivalent factors producing microneuronal hypoplasia would operate in early postnatal development, mainly in the first 10 to 15 days, when 85-90% of granule-cell mitosis occurs.

To repeat, Altman suggested that these deficits in mitosis, in the human fetus, as in the rat infant, might result from teratogens such as lead and alcohol, or from drugs, or other factors and, certainly, from x-irradiation. As described earlier, we have, like Altman, produced 80-85% reductions, specifically in hippocampal granule cells, by preventing any possible effects on granule-cell development in cerebellum or olfactory bulbs with appropriate lead shielding. And in every case, the effects of our x-ray treatments take the form of learning deficits, specifically, in short-term memory (PSA, Diaz-Granados et al., 1992) and in persistence (the PREE, e.g., Diaz-Granadoset al., 1994). In these experiments the subjects are preweanling rats (17-18-day-olds) in which the radiation protocol had been executed in infancy.

In addition to the clear deficits in PSA, particularly at 60-s ITI, and in the PREE, these preweanling rat pups were more hyperactive than their controls. And it is interesting to note that they appeared to show this "hyperactivity" or "impulsivity" (as we might call it in the VTE experiments) only in situations in which rewards and frustrative nonrewards were involved (Diaz-Granados et al., 1994). The strongest evidence here is that, after the PREE experiment, when they were tested in automated running wheels (in which rewards and nonrewards are not involved), x-ray-treated pups and controls did not differ in running speed or pattern. Douglas and her associates (e.g., Douglas & Parry, 1983; Parry & Douglas, 1983) hold that children diagnosed with ADHD can also be said to show unusually strong emotional reactions and sensitivity in response to conflict and frustration. In these reports from McGill University, the hyperactivity and impulsivity of these children is said to result from their unusual sensitivity to rewards and nonrewards and to the "activating component of r_F [anticipated frustration] ..." (Freibergs & Douglas, 1969, p. 394). We are, therefore, cautious in speculating, following Altman (1986), that, in the case of our preweanling rats, hippocampal microneuronal hypoplasia may be a factor in deficits in learning and in hyperactivity (or impulsivity) related to frustration and conflict.

If indeed microneuronal hypoplasia is a possible--but very difficult to diagnose--basis for ADHD, it would follow that treatments that improve learning in children with ADHD should also be beneficial to rats subjected to hippocampal x-ray treatment in infancy. To test this hypothesis, Highfield, Hu, & Amsel (in preparation), following Altman's protocol, x-irradiated infant rats on postnatal days 2, 3, 5, 7, 9, 11, 13, and 15, the period of greatest granule-cell proliferation, along with the appropriate untreated controls. The pups, at preweanling age (postnatal day 17-18) received intraperitoneal injections (0.3 mg/kg) of d-amphetamine (the controls, the saline vehicle) and were tested on PSA, the short term memory-based learning procedure. (As we have seen, normal preweanling pups learn this discrimination by 240 trials over 2 days, but x-irradiated rat pups do not.) The result of the Highfield et al. experiment, was that the x-irradiated rats that received d-amphetamine injections before training learned the PSA discrimination as well as the controls that had not received the treatment.

As an additional test of this model, we are putting alcohol in the artificial diet of rat pups from postnatal days 2 to 11, and testing them as preweanlings with and without d-amphetamine. The results in the animals we have tested are that amphetamine restores persistence in the PRF relative to the CRF condition in the alcohol-treated weanling rats. The test of alcohol's effect on memory-based (PSA) learning after amphetamine treatment is underway. [I have not covered our work on pre- and postnatal alcohol-treatment work in detail; however, the basic findings of these earlier experiments are that pups undergoing this form of early postnatal, intragastric ("pup-in-cup") alcohol treatment are deficient in persistence (the PREE), and in short-term memory (PSA) (e.g., Diaz-Granados, Greene, & Amsel, 1993; Greene, Diaz-Granados, & Amsel, 1992).]

This paper, in many ways, has been a recasting of a combination of much-earlier and relatively recent experimentally-based concepts into a more modern synthesis. The concepts are Muenzinger's (1938) and Tolman's (1938) VTE and their modern counterparts; the insights of Blodgett and others (1947, 1948, 1949) into the identification of the roles of response, direction, and place in spatial learning; Altman's (1986) brilliant discussions of minimal brain dysfunction, summarized in his concept of microneuronal hypoplasia; and Amsel's version of frustration theory, both specific (1958) and general (1972), and of discrimination learning (1962), summarized and redirected to suggest an animal model for ADHD (Amsel, 1990, 1992). Of course, the notions about ADHD are conceptual and merely suggestive, and should be so considered.

References

Altman, J. (1986). An animal model of minimal brain dysfunction. In M. Lewis (Ed.), Learning disabilities and prenatal risk. Urbana, Illinois: University of Illinois Press.

Altman, J., R. L. Brunner, and S. A. Bayer (1973) The hippocampus and behavioral maturation. Behav. Biol. 8:557-596.

Altman, J., and G. D. Das (1965) Autoradiographic and histological evidence for postnatal hippocampal neurogenesis in rats. J. Comp. Neurol. 124:319-336.

Amsel A (1958) The role of frustrative nonreward in noncontinuous reward situations. Psychol Bul 55:102-119.

Amsel, A. (1972). Behavioral habituation, counterconditioning, and a general theory of persistence. In A. H. Black & W. F. Prokasy (Eds.), Classical conditioning II: Current research and theory (409-426). New York: Appleton-Century-Crofts.

Amsel, A. (1979) Hippocampus, memory and movement. Behav. and Brain Sci. 2:494-495.

Amsel A (1986) Daniel Berlyne Memorial Lecture. Developmental psychobiology and behaviour theory: Reciprocating influences. Canadian J Psychol 40:311-342.

Amsel, A. (1990). Arousal, suppression, and persistence: Frustration theory, attention, and its disorders. Cognition and Emotion, 4:239-268.

Amsel, A. (1992) Frustration theory: An analysis of dispositional learning and memory. Cambridge University Press, New York.

Amsel, A. (1993). Hippocampal function in the rat: Cognitive mapping or vicarious trial-and-error? Hippocampus 3:251-256.

Amsel, A., H. Glazer, J. R. Lakey, T. McCuller, and P. T. P. Wong (1973) Introduction of acoustic stimulation during acquisition and resistance to extinction in the normal and hippocampally-damaged rat. J. Comp. Physiol. Psychol. 84:176-186.

Angevine, J. B., Jr., and C. W. Cotman (1981) Principles of neuroanatomy. Oxford University Press, Oxford.

Blair, H. T. & Sharp, P. E. (1996). Visual and Vestibular Influences on Head-Direction Cells in the Anterior Thalamus of the Rat. Behav. Neurosci. 110:643-660.

Blodgett, H. C., and K. McCutchan (1947) Place versus response learning in the simple T maze. J. Exp. Psychol. 37:412-422.

Blodgett, H. C., and K. McCutchan (1948) Relative strength of place and response learning in the T maze. J. Comp. Physiol. Psychol. 41:17-24.

Blodgett, H. C., K. McCutchan, and R. Mathews (1949) Spatial learning in the T maze: The influence of direction, turn, and food location. J. Exp. Psychol. 39:800-809.

Brown, M. F. (1992) Does a cognitive map guide choices in the radial-arm maze? J. Exp. Psychol. [Anim. Behav.] 18:56-66.

Devan, B. D., G. S. Blank, and H. L. Petri (1992) Place navigation in the Morris water task: Effects of reduced platform interval lighting and pseudorandom platform positioning. Psychobiology 20:120-126.

Diaz-Granados, J. L., Greene, P. L., & Amsel, A. (1992). Memory-based learning in preweanling and adult rats after infantile X-irradiation-induced hippocampal granule-cell hypoplasia. Behav. Neurosci. 106:940-946.

Diaz-Granados, J. L., Greene, P. L., & Amsel, A. (1993) Mitigating effects of combined prenatal and postnatal exposure to ethanol on learned persistence in the weanling rat: A replication under high-peak conditions. Behav. Neurosci. 107:1059-1066.

Diaz-Granados, J. L., Greene, P. L., & Amsel, A. (1994). Selective activity enhancement and persistence in weanling rats after hippocampal x-irradiation in infancy: Possible relevance for ADHD, Behavioral and Neural Biology, 61:251-259.

Douglas, R. J. (1967) The hippocampus and behavior. Psychol. Bul. 67:416-442.

Douglas, V. I., & Parry, P. A. (1983). Effects of reward on delayed reaction time task performance of hyperactive children. J. Abnorm. Child Psychol. 11:313-326.

Freibergs, V., & Douglas, V. I. (1969). Concept learning in hyperactive and normal children. J. Abnorm. Psychol., 74:388-395.

Glazer, H. I. (1974a) Instrumental conditioning of hippocampal theta and subsequent response persistence. J. Comp. Physiol. Psychol. 86:267-273.

Glazer, H. I. (1974b) Instrumental response persistence following induction of hippocampal theta frequency during fixed-ratio responding in rats. J. Comp. Physiol. Psychol. 86:1156-1162.

Gray, J. A. (1982) The neuropsychology of anxiety: An enquiry into the functions of the septo-hippocampal system. Oxford University Press, New York.

Greene, P. L., J. L. Diaz-Granados, and A. Amsel (1992) Blood ethanol concentration from early postnatal exposure: effects on memory-based learning and hippocampal neuroanatomy in infant and adult rats. Behav. Neurosci. 106:51-61.

Greene, P. L., Diaz-Granados, J. L., & Amsel, A. (1994). Postnatal high-peak blood ethanol concentration and external cue-based discrimination learning and reversal in the preweanling rat: Comparison with memory-based discrimination learning. Behav. Neurosci. 108:333-339.

Griesbach, G. S., Hu, D., & Amsel, A. (in preparation). Effects of the NMDA receptor antagonist MK-801 on vicarious trial-and-error and on olfactory discrimination learning and its reversal.

Highfield, D. A., Nixon, K., & Amsel, A. (1996). The NMDA antagonist MK-801 affects non-spatial learning in preweanling rats. Behav. Neurosci. 110:300-304.

Hu, D. & Amsel, A. (1995). A simple test of the vicarious trial-and-error (VTE) hypothesis of hippocampal function. Proc. Nat. Acad. Sci. 92:5506-5509.

Hull C. L. (1931) Goal attraction and directing ideas conceived as habit phenomena. Psychol. Rev. 38:487-506.

Isaacson, R. L. (1980) A perspective for the interpretation of limbic system function. Physiol. Psychol. 8:183-188.

Kimble, D. P. (1968) Hippocampus and internal inhibition. Psychol. Bul. 70:285-295.

Leaton, R. N., and G. S. Borszcz (1990) Hippocampal lesions and temporally chained conditioned stimuli in a conditioned suppression paradigm. Psychobiology 18:81-88.

Lilliquist, M. W., Burkhalter, E. C., Lobaugh, N. J., & Amsel, A. (1993). Age-dependent effects of hippocampal muscarinic receptor blockade on memory-based learning in the developing rat. Behavioral Brain Research, 53:119-125.

Lobaugh, N. J., Greene, P. L., Grant, M., Nick, T., & Amsel, A. (1989). Patterned (single) alternation in infant rats following combined or separate lesions of hippocampus and amygdala. Behav. Neurosci. 103:1159-1167.

Lobaugh, N.J., Norton, D., Novak, K. & Amsel, A. (1995). Age and dose-related effects of hippocampal ibotenic acid on lesion size, mortality, and non-spatial working memory in infant rats. Neurobiology of Learning and Memory, 63:94-106.

Lobaugh, N. J., Wigal, T., Greene, P. L., Diaz-Granados, J. L., & Amsel, A. (1991). Effects of prenatal ethanol exposure on learned persistence and hippocampal neuroanatomy in infant, weanling and adult rats. Behavioural Brain Research, 44:81-86.

Miller, N. E. (1944). Experimental studies of conflict. In J. McV. Hunt (Ed.), Personality and the behavior disorders: Vol. 1. New York: Ronald.

Miller, N. E. (1959) Liberalization of basic S-R concepts: Extensions to conflict behavior, motivation, and social learning. In Psychology: A study of a science (Vol. 2), S. Koch, ed., McGraw-Hill, New York.

Muenzinger, K. F. (1938) Vicarious trial and error at a point of choice: I. A general survey of its relation to learning efficiency. J. Genet. Psychol. 53:75-86.

Nadel, L. (1991) The hippocampus and space revisited. Hippocampus 1:221-229.

O'Keefe, J., and L. Nadel (1978) The hippocampus as cognitive map. Oxford University Press.

Parry, P. A., & Douglas, V. I. (1983). Effects of reinforcement on concept identification in hyperactive children. J. Abnorm. Child Psychol. 11:327-340.

Robbins, T. W., Watson, B. A., Gaskin, M., & Ennis, C. (1983). Contrasting interactions of pipradrol, d-amphetamine, cocaine, cocaine analogues, apomorphine and other drugs with conditioned reinforcement. Psychopharmacology (Berlin), 80:113-119.

Schneirla, T. C. (1959) An evolutionary and developmental theory of biphasic processes underlying approach and withdrawal. In Nebraska Symposium on Motivation, M. R. Jones, ed., University of Nebraska Press, Lincoln, Nebraska.

Simonov, P. V. (1969) Studies of emotional behavior of humans and animals by Soviet physiologists. Annals of the New York Academy of Science 159:1112-1121.

Simonov, P. V. (1974) On the role of the hippocampus in the integrative activity of the brain. Acta Neurobiologicae Experimentalis 34:33-41.

Taube, J. S. & Burton, H. L. (in press). Head-direction cell activity monitored in a novel environment and during a cue conflict situation. J. Neurosci.

Taube, J. S., Muller, , R. U., & Ranck, J. B.,Jr. (1990a). Head-direction cells recorded from the postsubiculum in freely moving rats: I. Description and quantitative analysis. J. Neurosci. 10:420-435.

Taube, J. S., Muller, , R. U., & Ranck, J. B.,Jr. (1990b). Head-direction cells recorded from the postsubiculum in freely moving rats: II. Effects of environmental manipulations. J. Neurosci. 10:436-447.

Thomas, G. J. (1984) Memory: Time finding inn organisms. In Neuropsychology of memory, L. R. Squire and N. Butlers, eds., Guilford: New York.

Tolman, E. C. (1938). The determiners of behavior at a choice point. Psychol. Rev. 45:1-41.

Tolman, E. C. (1939) Prediction of vicarious trial and error by means of the schematic sowbug. Psychol. Rev. 46:318-336.

Tolman, E. C. (1948) Cognitive maps in rats and men. Psychol. Rev. 55:1-4.

Vinogradova, O. S. (1975) Functional organization of the limbic system in the process of registration of information: Facts and hypotheses. In The hippocampus: Volume 2. Neurophysiology and behavior, R. L. Isaacson and K. H. Pribram, eds., Plenum: New York.

Whishaw, I. Q. (1991) Latent learning in a swimming pool place task by rats: Evidence for the use of associative and not cognitive mapping processes. Quarterly J. Exp. Psychol. 43b: 83-103.

Whishaw, I. Q., & Tomie, J.-A. (1987). Cholinergic receptor blockade produces impairments in a sensorimotor subsystem for place navigation in the rat: Evidence from sensory, motor, and acquisition tests in a swimming pool. Behav. Neurosci. 101: 603-616.

Table 1: Ontogeny of reward schedule effects. First appearance in the ontogeny of the rat of a number of effects based on a variety of arrangements of reward and nonreward. (See text and Amsel, 1986, 1992.)

Effect	Age(days)
Successive acquisition and extinction (Amsel, 1986)	0-10
Single patterned alternation (PSA) (Stanton, Dailey, & Amsel, 1980)	10-11
Partial reinforcement extinction effect (PREE) (Letz, Burdette, Gregg, Kittrell, & Amsel, 1978; Chen & Amsel, 1980a, b)	12-14
Variable magnitude of reinforcement (VMREE) (Chen, Gross, & Amsel, 1981)	16-18
Partial delay of reinforcement (PDREE) (Chen, Gross, & Amsel, 1981)	16-18
Partial reinforcement acquisition (PRAE) (Chen, Gross, Stanton, & Amsel, 1980)	18-20
Magnitude reinforcement extinction (MREE) (Burdette, Brake, Chen, & Amsel, 1976; Chen, Gross, & Amsel, 1981)	20-21
Successive negative contrast (SNC) (Chen, Gross, & Amsel, 1981; Stanton & Amsel, 1980)	25-26
Slow Responding (DNC) (Chen, Gross, Stanton, & Amsel, 1981)	21-63

Figure 1: Schematic of a Lashley Jumping Stand used in simultaneous discrimination experiments.

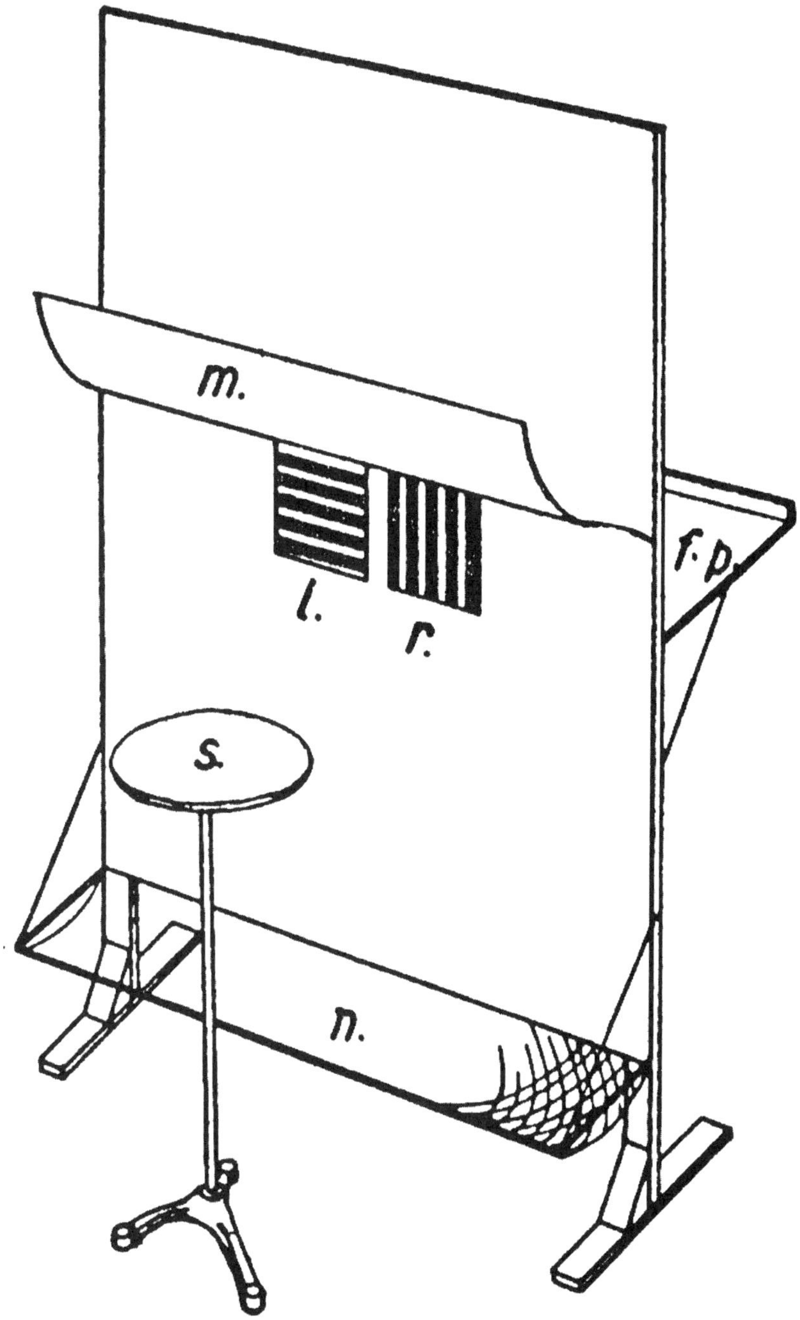

Figure 2: Elevated mazes in experiments by Blodgett, McCutchan, and Mathews (1949). See text for description of how experiments determined the influence of response, direction and place in spatial learning.

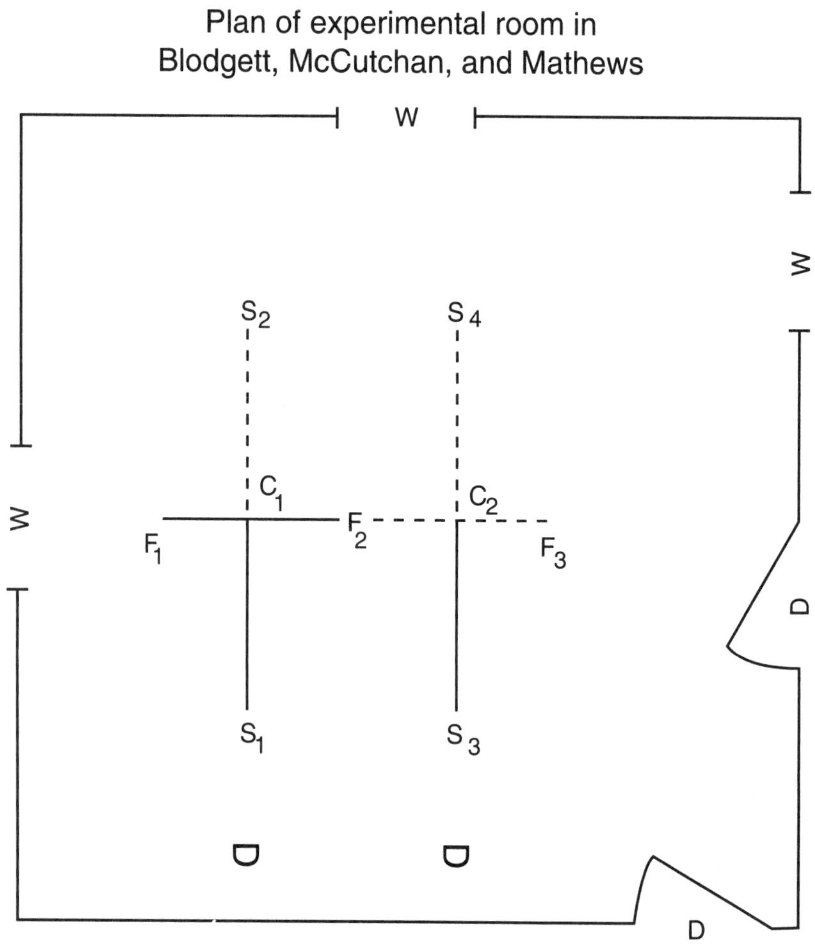

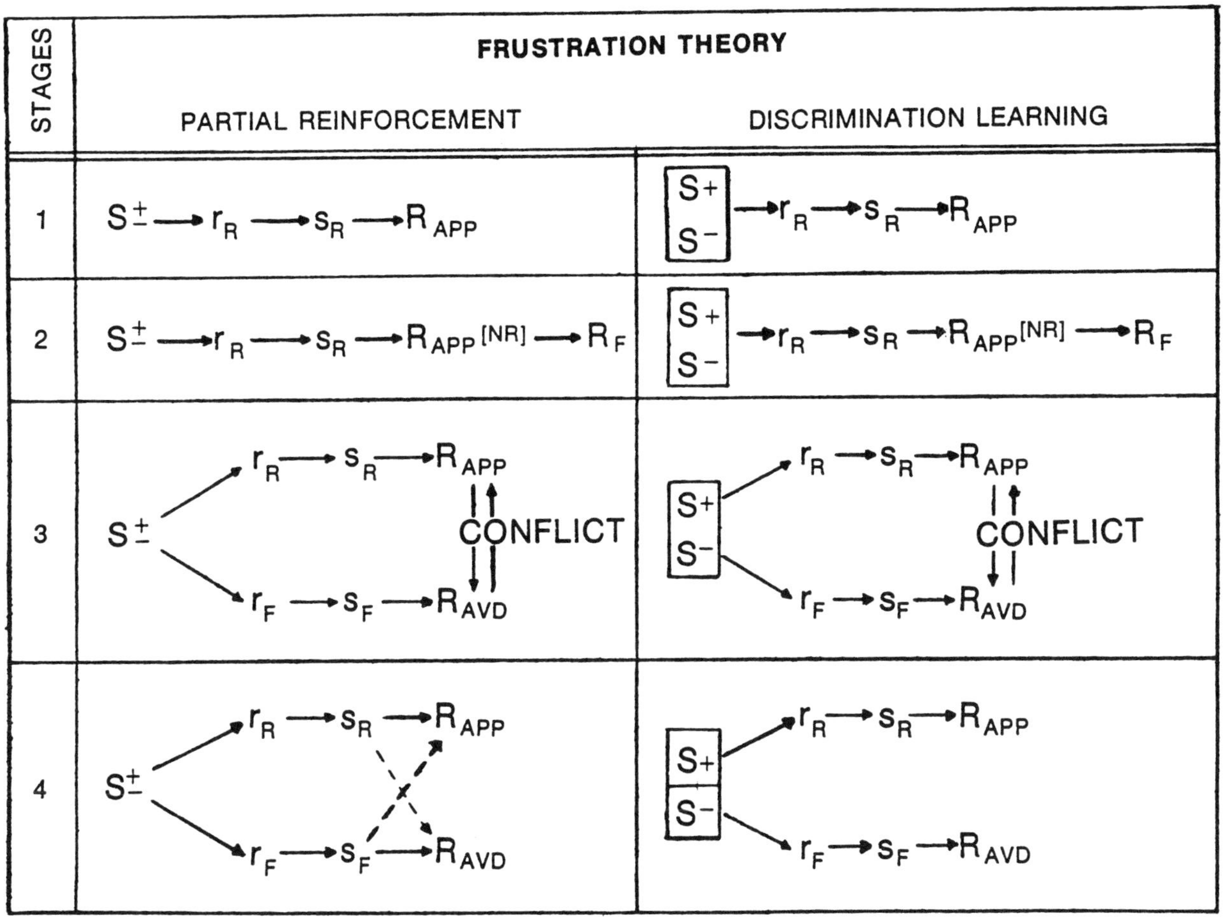

Figure 3: Four-stage theories of the PREE and discrimination learning (Amsel, 1958). See text for elaboration.

Figure 4: Schematic of runway apparatus for pups from about 10-11 days of age to preweanlings, 17-18 days of age. The pup is trained to approach a two-sections goal box. On rewarded (R) trials the door between these sections opens automatically and the pup enters, attaches to the nipple of an anesthetized dam, and is rewarded with milk injected onto its tongue via implanted cannula. (See any of referenced articles.)

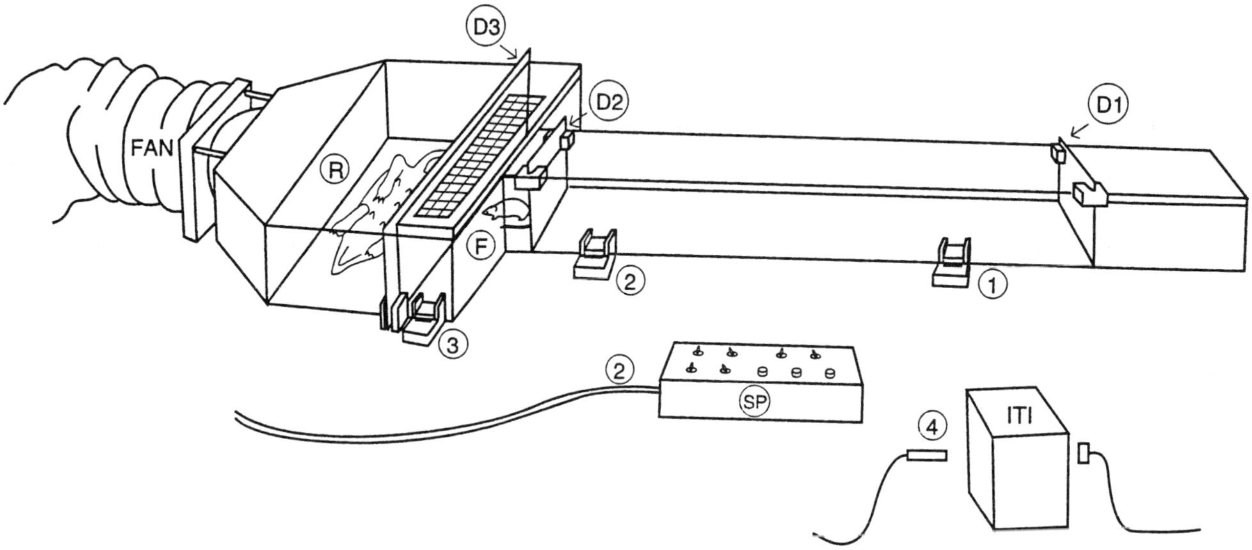

PUP RUNWAY

Figure 5: Results from experiments in runway apparatus described in Figure 4 caption under the four experimental conditions described in the text. PSA training at 60s ITI after lesions of hippocampus; a high-peak blood-ethanol condition; infantile x-irradiation to selectively affect mitosis of granule cells; and a control condition.

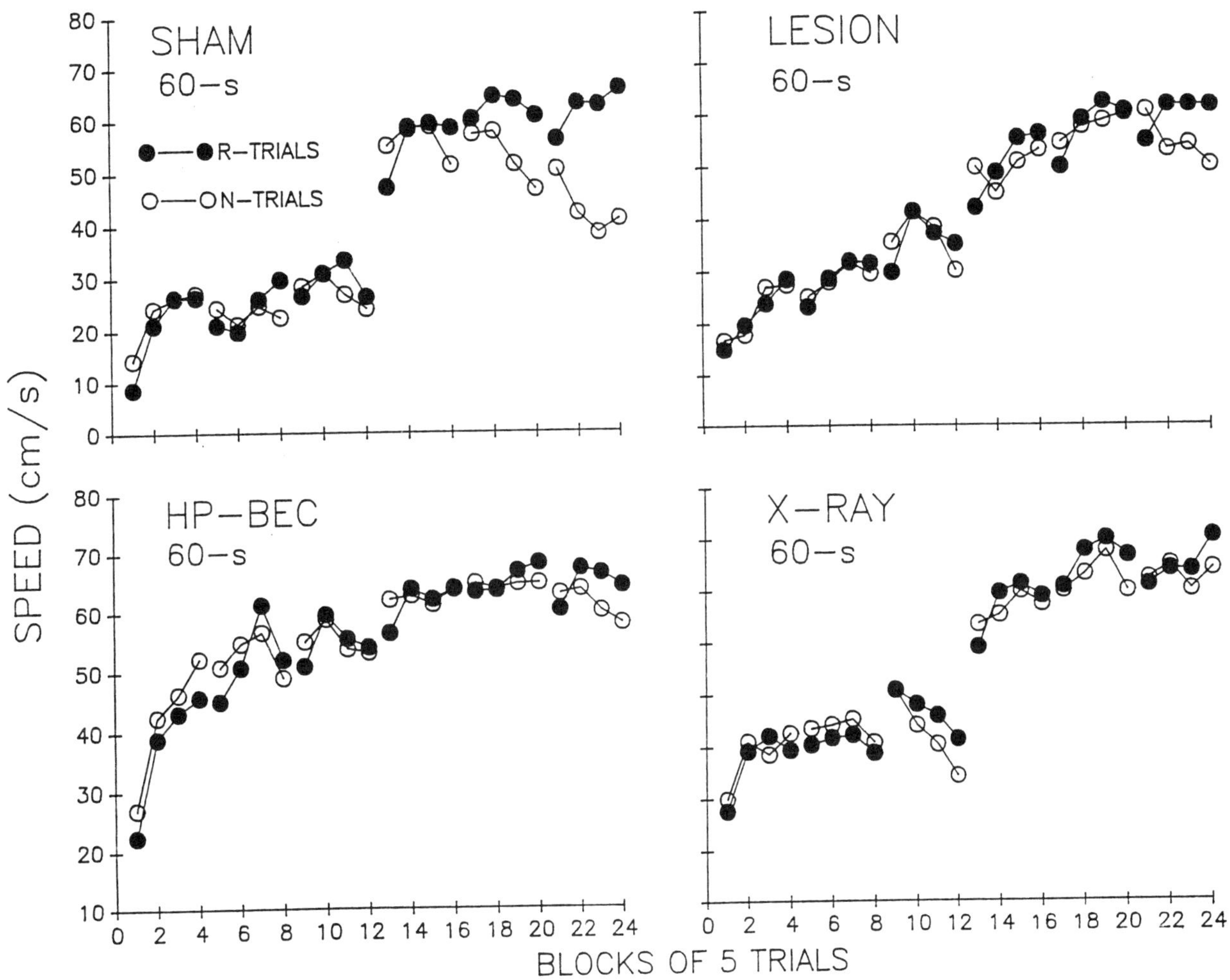

Figure 6: Effect on PSA of atropine.

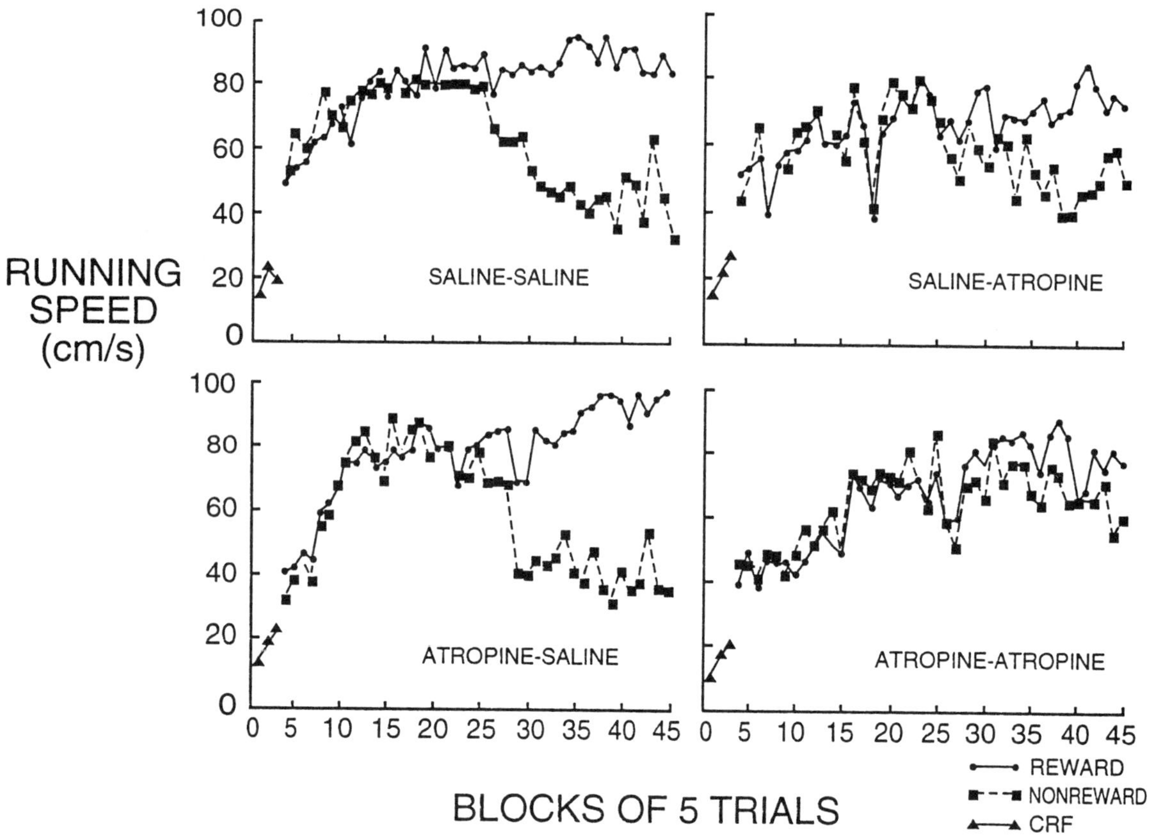

Figure 7: Effects on PSA of NMDA antagonist, MK-801, as a function of ITI--8, 30, or 60 seconds.

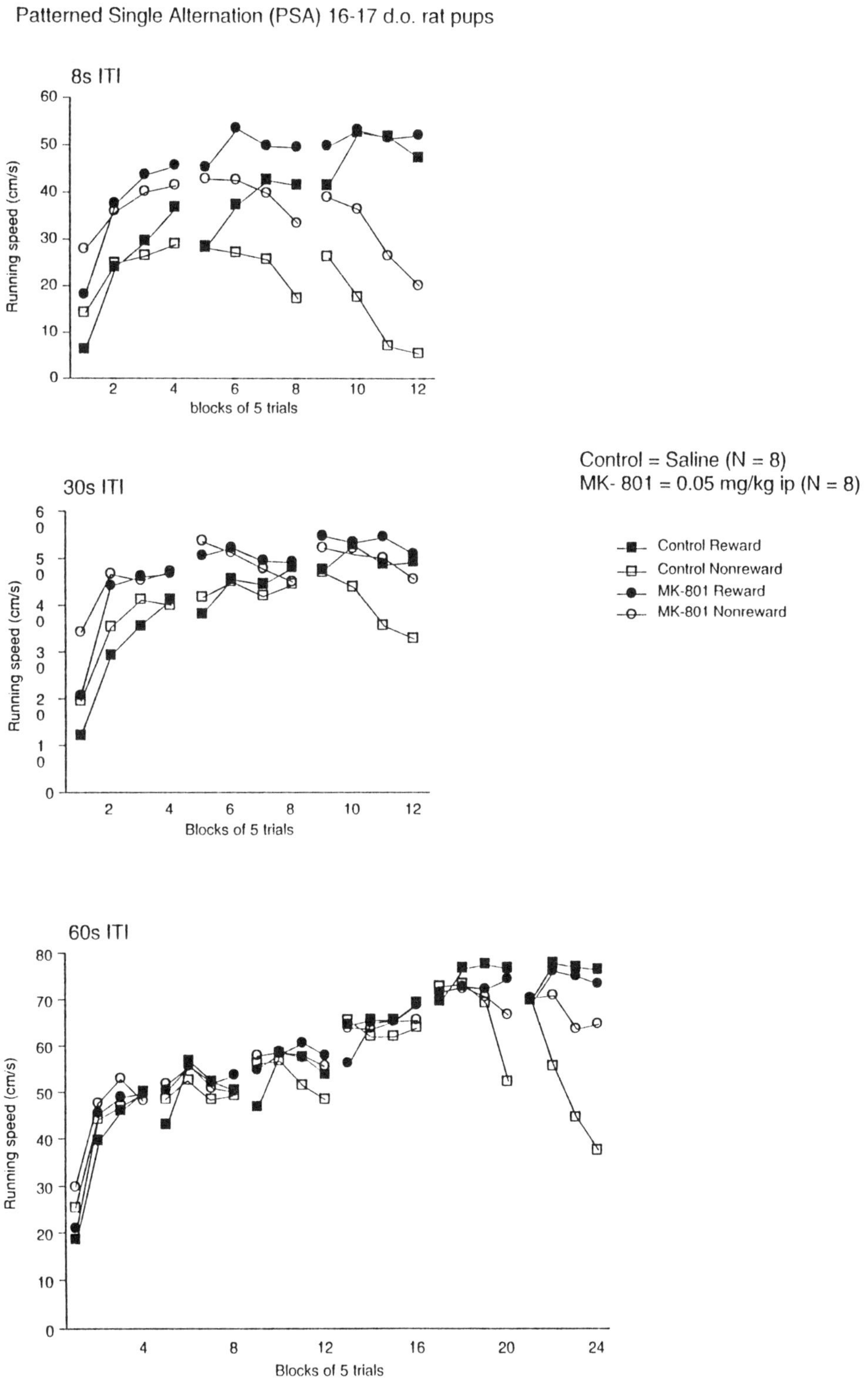

Figure 8: Examples of hippocampal damage and extent of cortical damage in sham and lesion conditions.

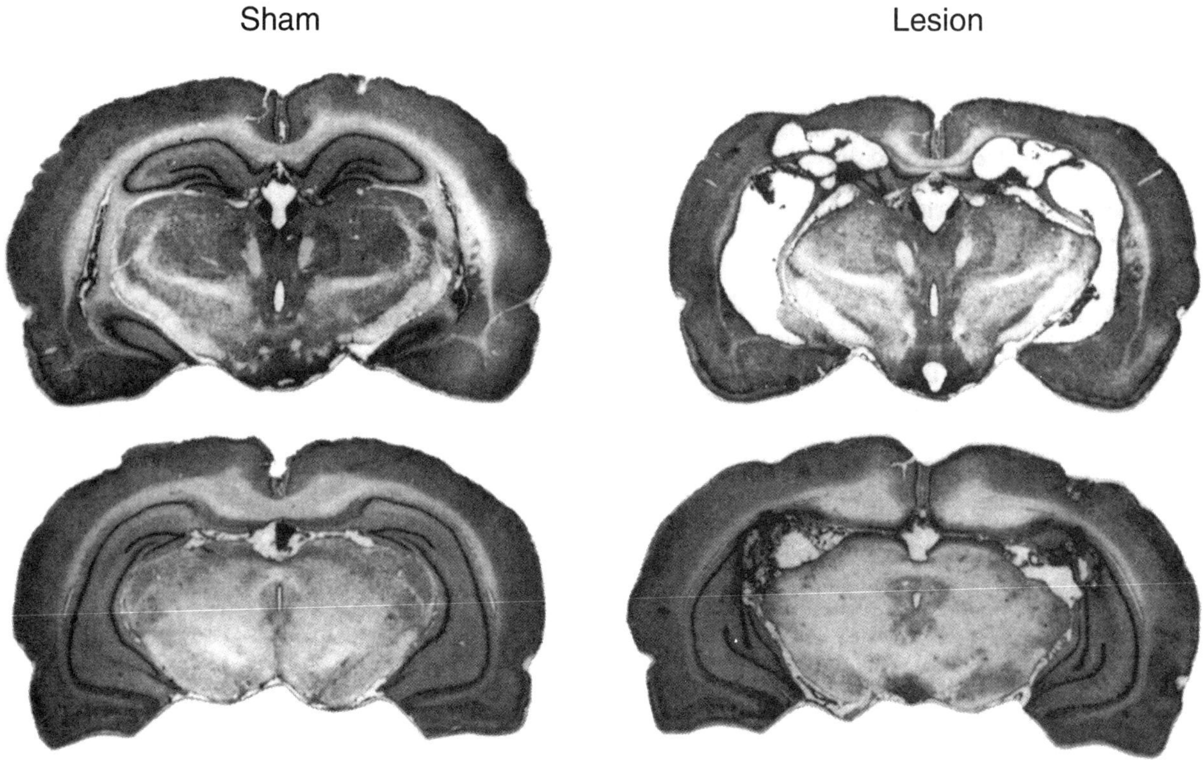

Figure 9: (A) Mean number of rewards each training day. (B) Number of VTEs each training day in sham and lestion conditions.

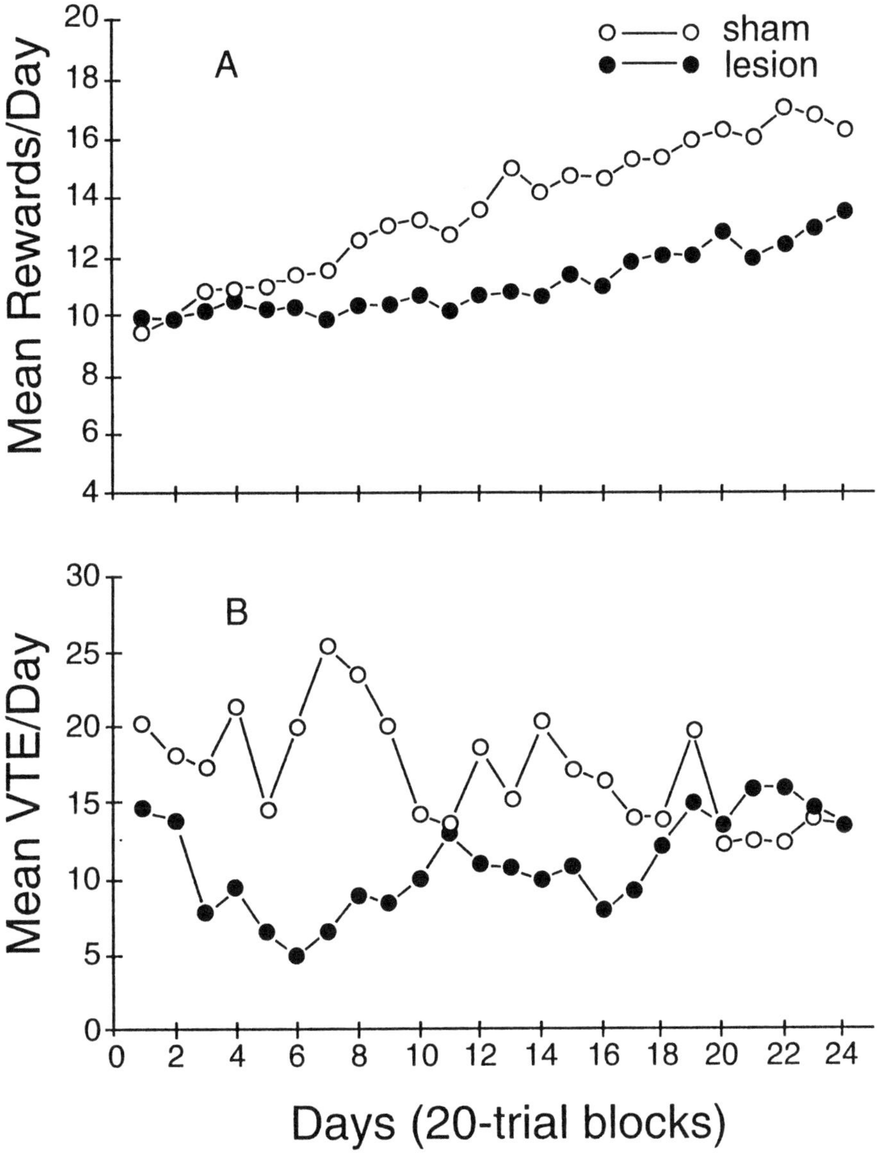

Figure 10: Comparison of VTE performance patterns in relation to rewards in sham and lesion conditions.

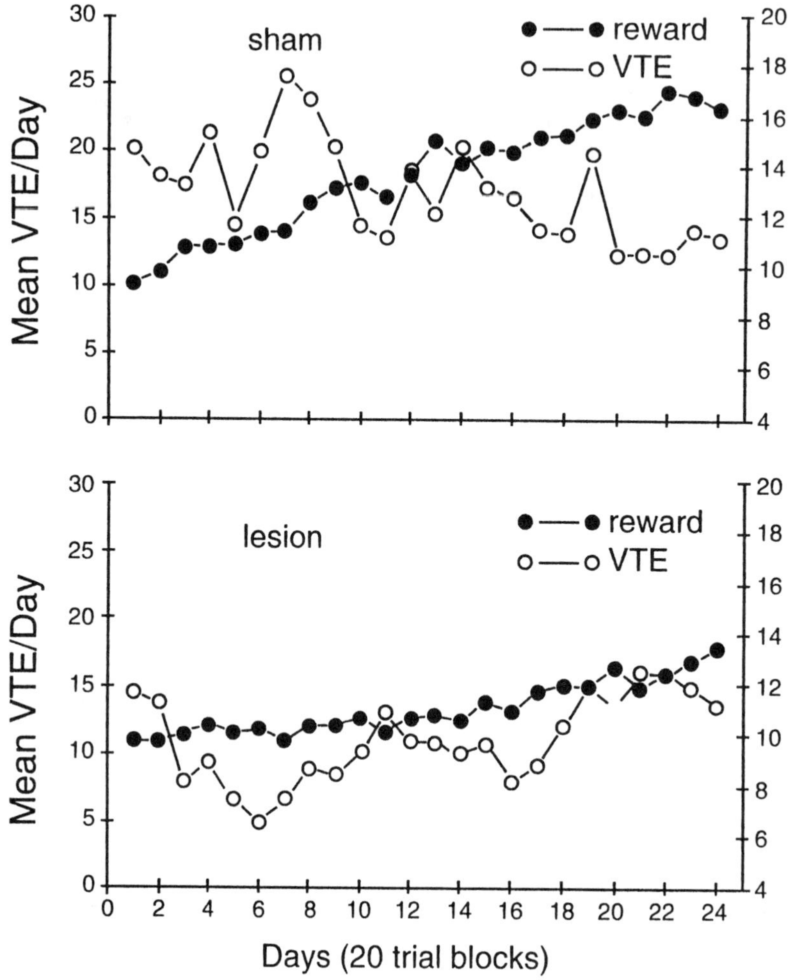

Figure 11: Comparison of developmental stages in rats and humans, showing rats relatively altricial at birth and in early development.

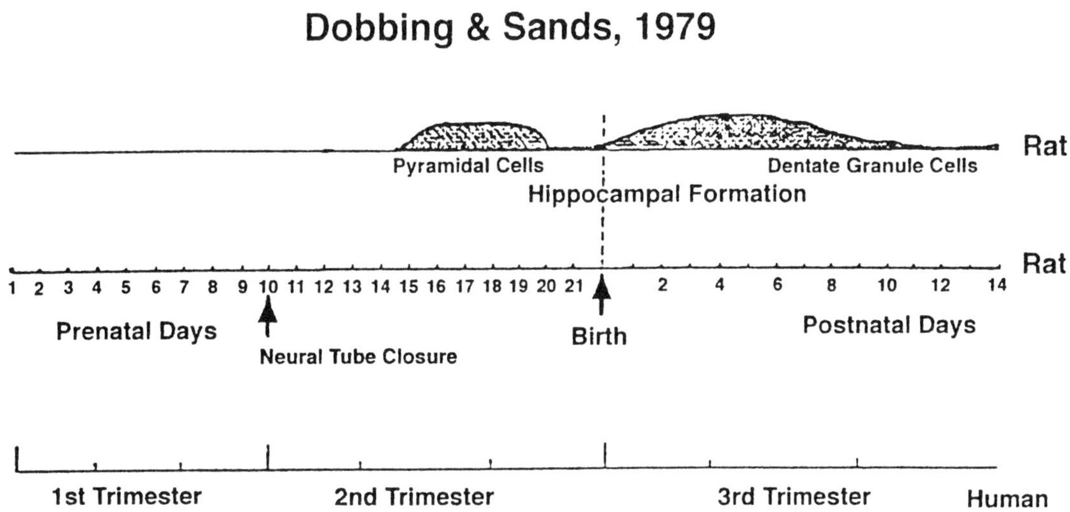

Figure 12: Maturation (mitosis) of hippocampus granule cells from birth to postnatal days 25-30 in the rat (Altman & Das, 1965).

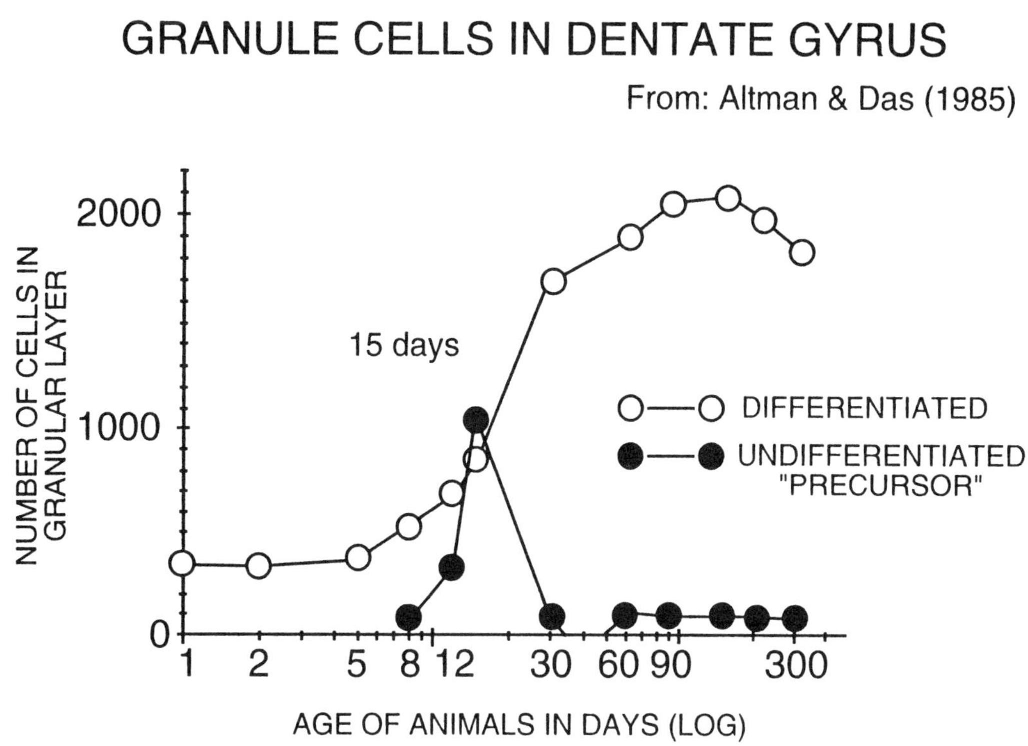

Figure 13: Granule cell density in control and x-irradiated dentate gyrus in the rat (Diaz-Granados et al., 1992).

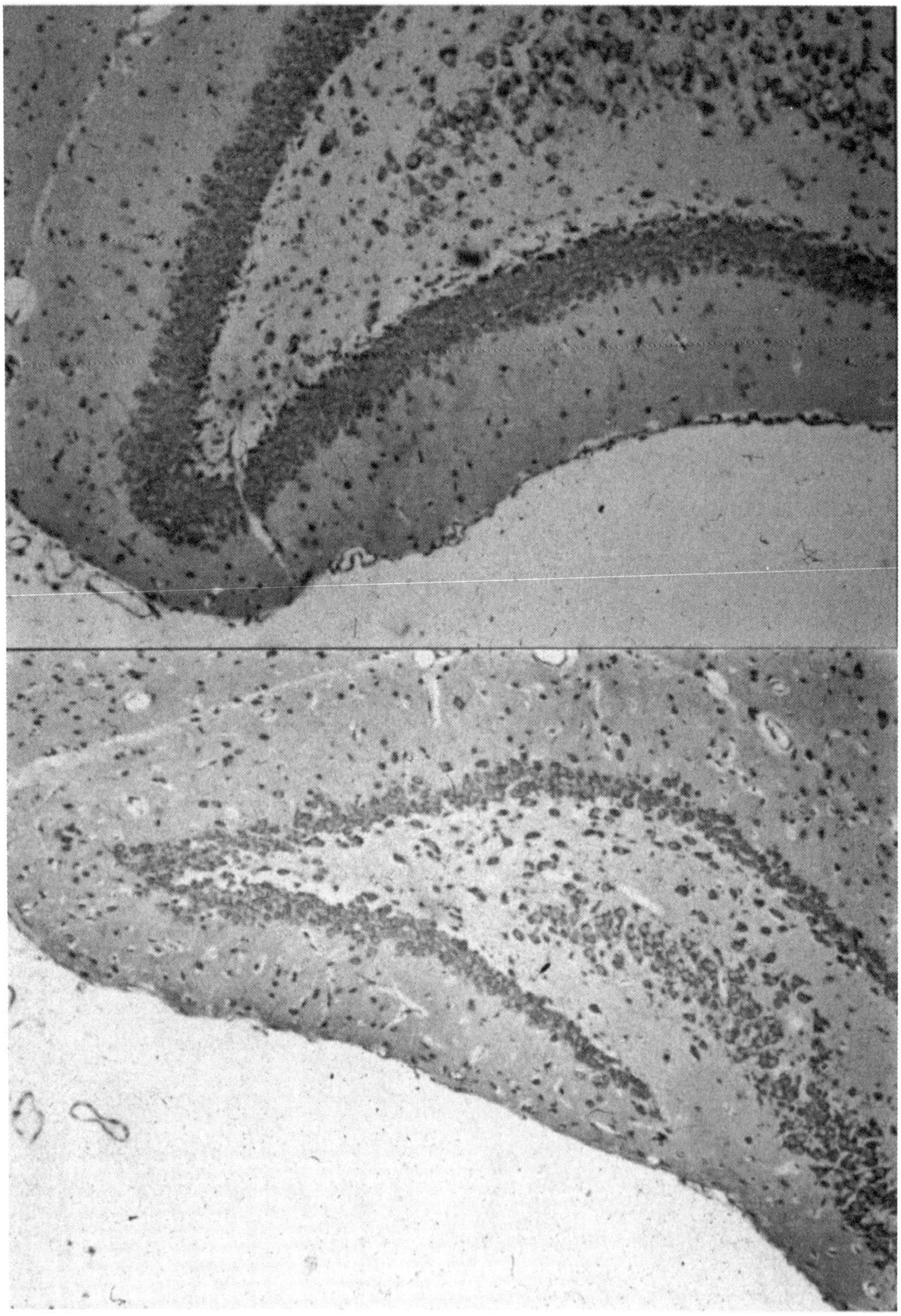

Figure 14: Specific and general theories of persistence (Amsel, 1992).

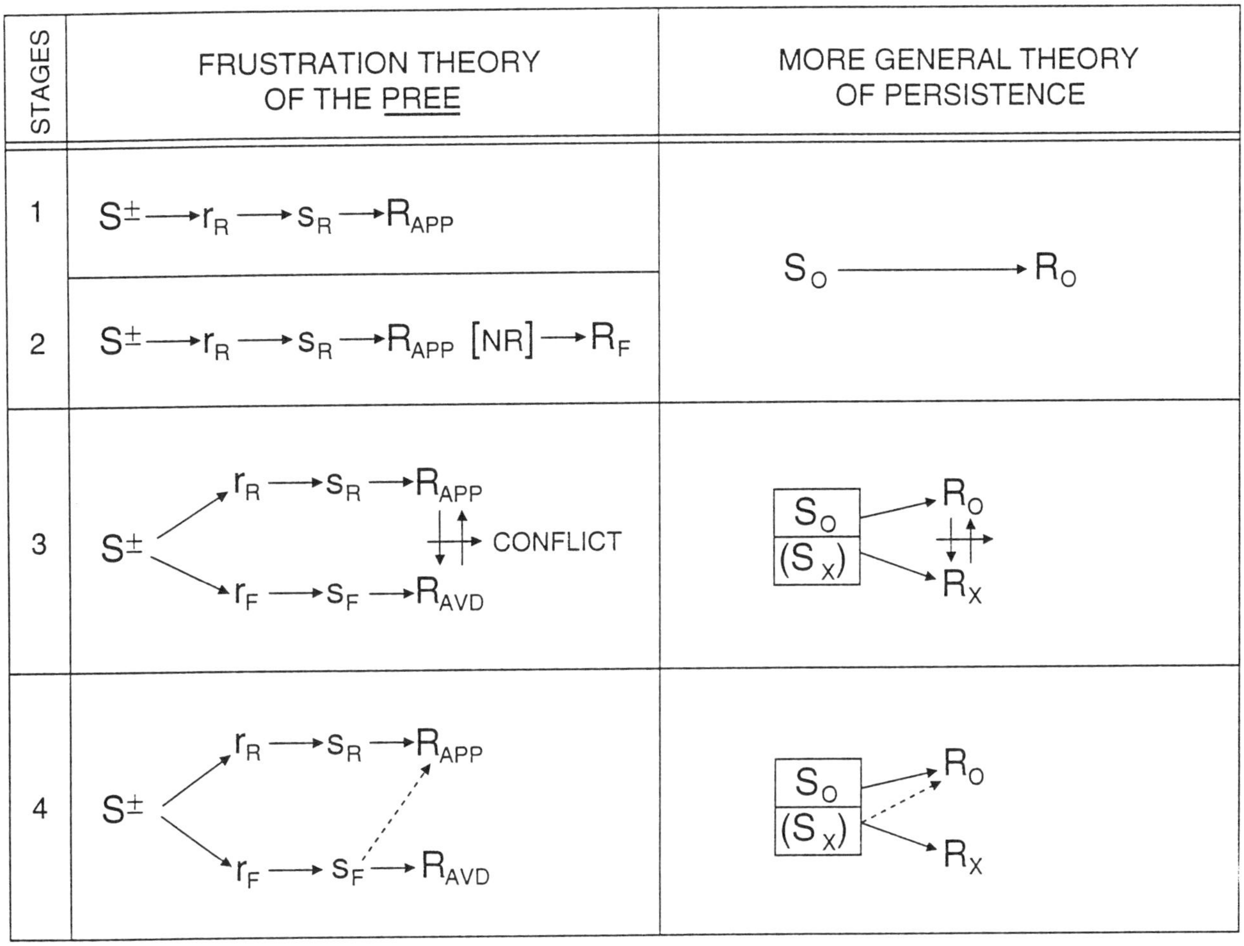

Figure 15: Hypothetical approach and avoidance gradients with and without administration of d-amphetamine and resulting effects on conflict (Amsel, 1990).

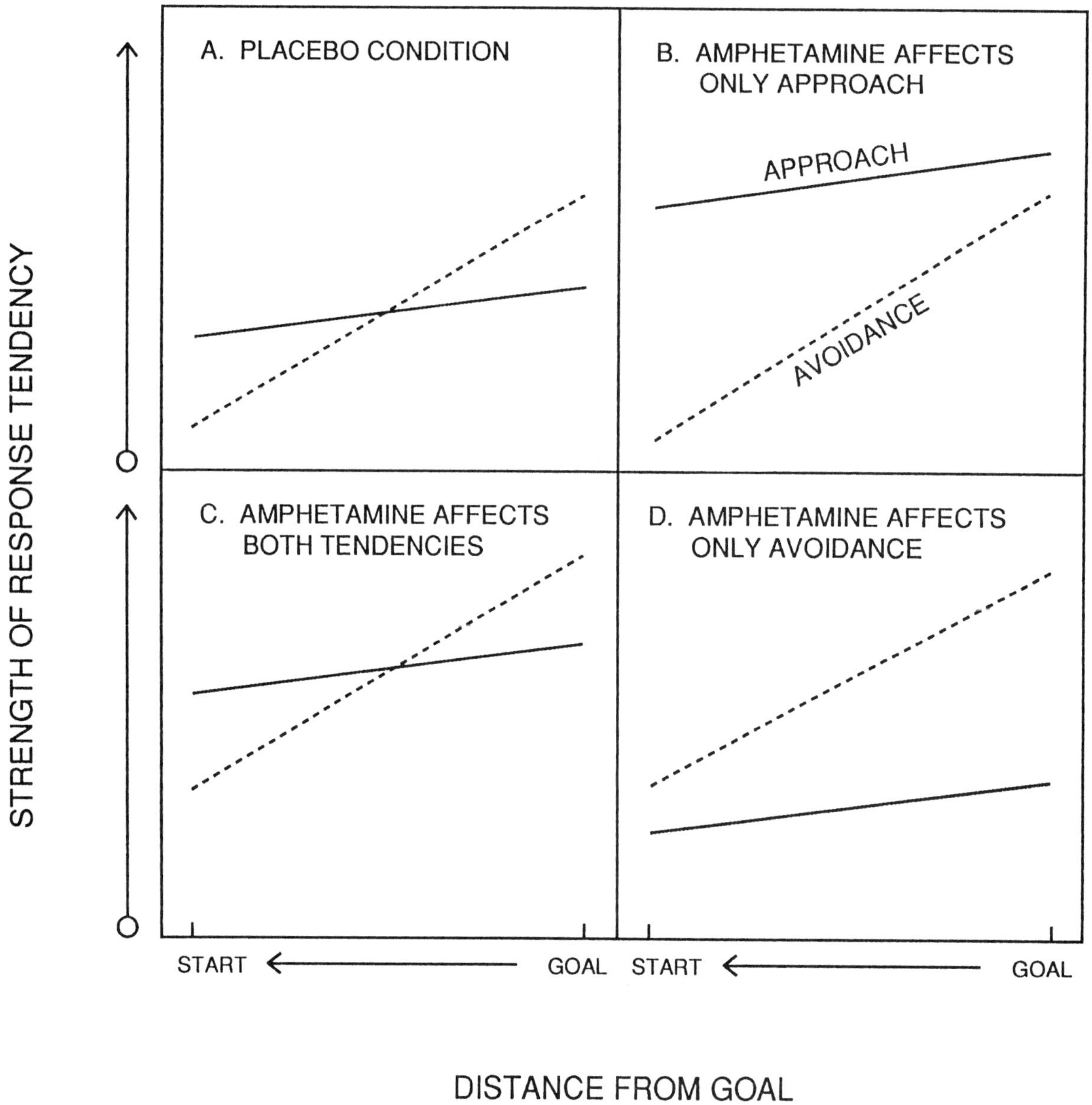

Figure 16: Rat pups treated with x-irradiation do not learn the PSA discrimination, but a separate group, so treated, learns if d-amphetamine is injected before training (Highfield et al., in preparation).

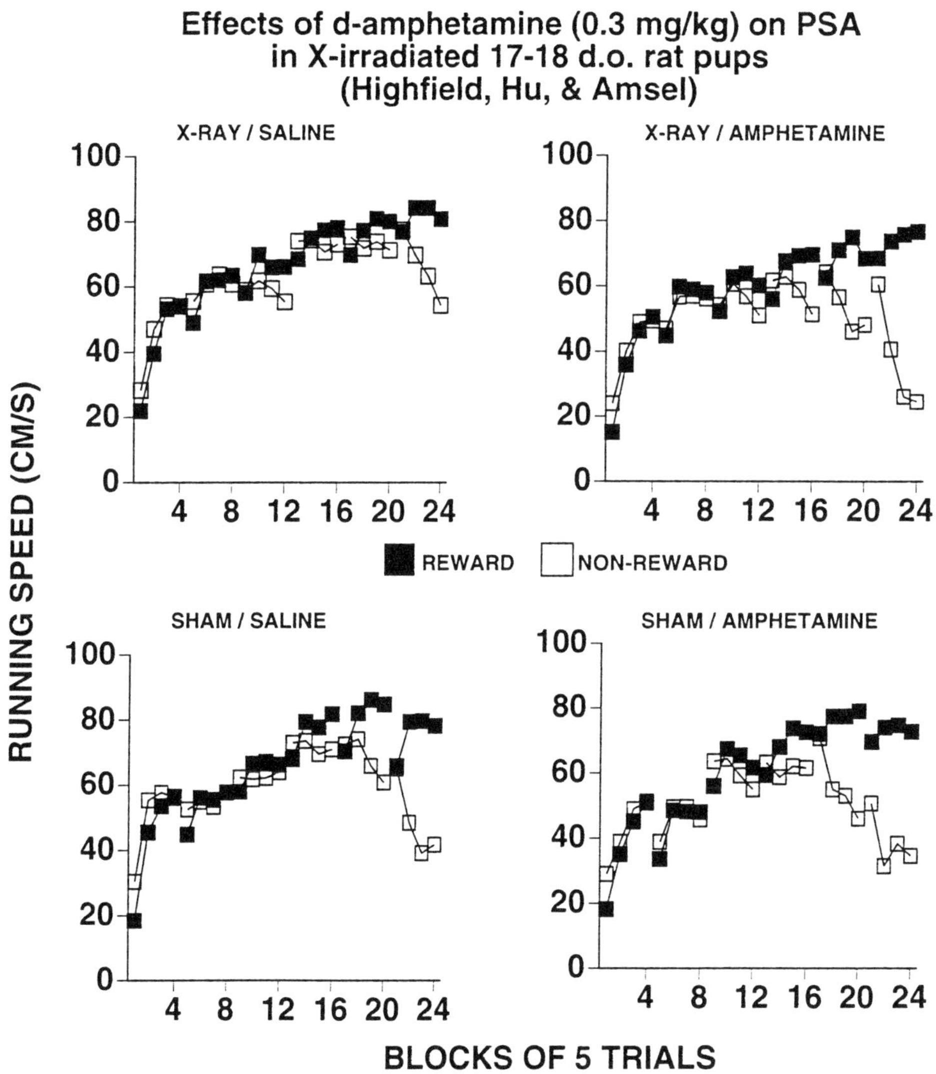

20

Role of the Hippocampus in Learning and Memory: A Computational Analysis

J. L. McClelland
Center for the Neural Basis of Cognition and Carnegie Mellon University
Department of Psychology
115 Mellon Institute
4400 Fifth Avenue
Pittsburgh, PA 15213

Human amnesia

A striking finding in the literature on human memory is the profound but selective deficit seen after damage to the hippocampus and related structures in the temporal lobes of the human brain. This pattern of deficits was first seen in patient H.M. (Scoville and Milner, 1957), and has now been replicated in the data of a large number of other patients with similar lesions. These patients show a dramatic deficit in the ability to acquire new explicit memories for the contents of specific episodes and events. For example, the ability to learn arbitrary word-pairs such as "LOCOMOTIVEDISHTOWEL" is drastically impaired. Normal subjects will learn such lists after a few presentations, so that whenever the first word of a pair is presented they will he able to recall the second word of each pair. But, H.M. and other similar amnesics may tail to learn even one pair after repeated presentations. There is also profound impairment in everyday memory abilities, such as the ability to learn names of new people, or the ability to remember important personal events and experiences.

At the same time these amnesics are completely normal in the use of their existing semantic and procedural knowledge, and in fact their acquisition of new skills appears to be completely intact. They also show after-effects from processing particular items that are identical to the after-effects seen in normal people, in a wide range of tasks. *Ex*, if amnesic subjects read a word such as "WINDOW" on a list, they show a normal amount of perceptual facilitation in the identification of visually presented words due to a prior presentation of the word. Amnesics also show a severe temporally graded retrograde amnesia for episodic information. Figure 1. shows data from four

1996.In T.Ono, B.L. McNaughton, S. Molotchnikoff, E. T. Rolls, and H. Nichijo (Eds.), Perception, Memory and Emotion: Frontier in Neuroscience. Oxford: Elsevier Science, Ltd. 601-613.

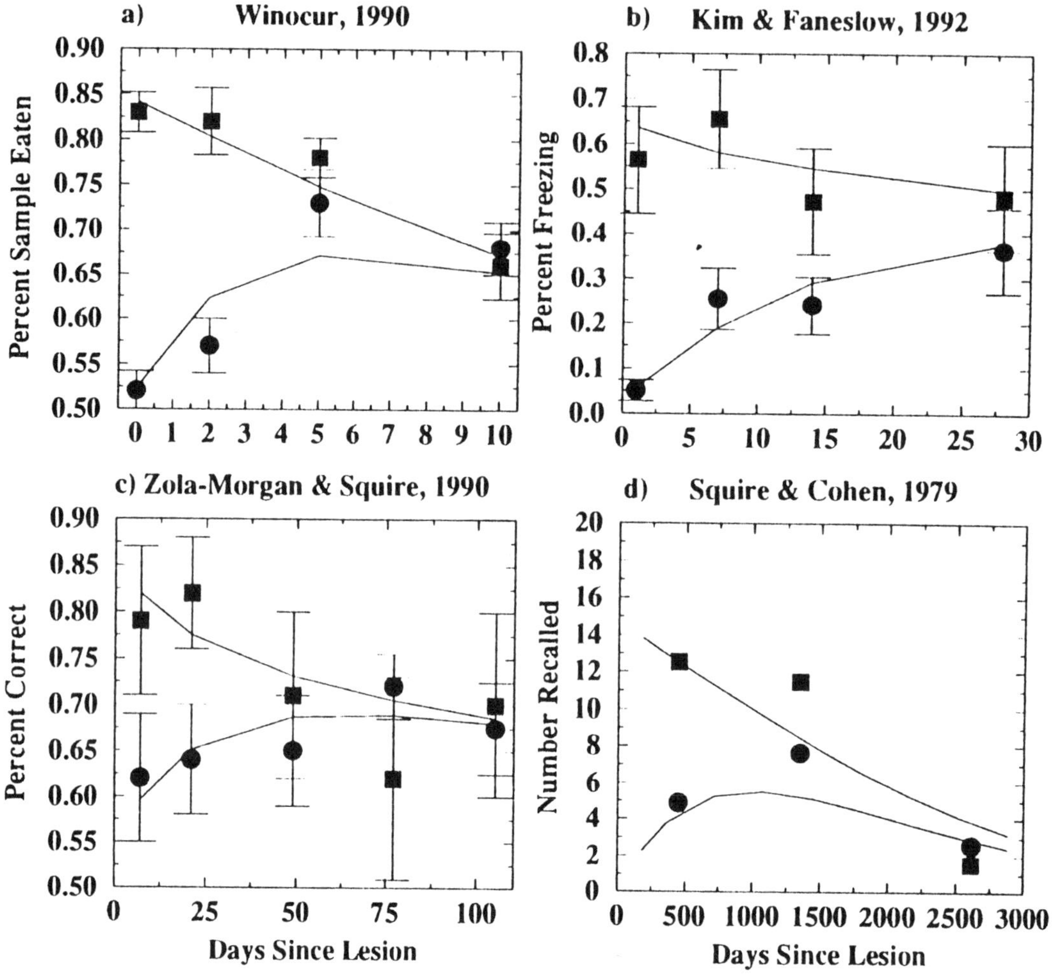

FIG. 1. Panels (a)–(c) show behavioral responses of animals lesioned different numbers of days after exposure to relevant experiences, while panel (d) shows a comparable pattern from an experiment with human subjects. (a) Food preferences exhibited by rats exposed to a conspecific (Winocur, 1990). (b) Fear behavior shown by rats exposed to a contingency between tone and foot shock in a novel environment (Kim and Fanselow, 1992). (c) Choices of reinforced objects by monkeys exposed to 14 training trials with each of 20 object pairs (Zola-Morgan and Squire, 1990). Panel (d) shows recall by depressed human subjects of details of television shows aired different numbers of years prior to the time of test, either before of after ECT (Squire and Cohen, 1979). Note that we have translated years into days to allow comparison with the results from the animal studies. Reprinted from McClelland *et al.* (1995).

studies of retrograde amnesia in rats, primates, and humans. In all cases we see that there is a selective loss of memory for material occurring shortly before the lesion, relative to more remote time periods. In humans, the period of selective loss can extend over several years, though it is much shorter in rats and monkeys.

Organization of the human memory system

One view of the organization of mammalian memory that is consistent with these facts is illustrated in Fig. 2. According to this view (McClelland *et al.*, 1995), all kinds of knowledge can ultimately be stored in the connections among neurons in the cortex and other nonhippocampal structures such as the basal ganglia (for brevity I will refer

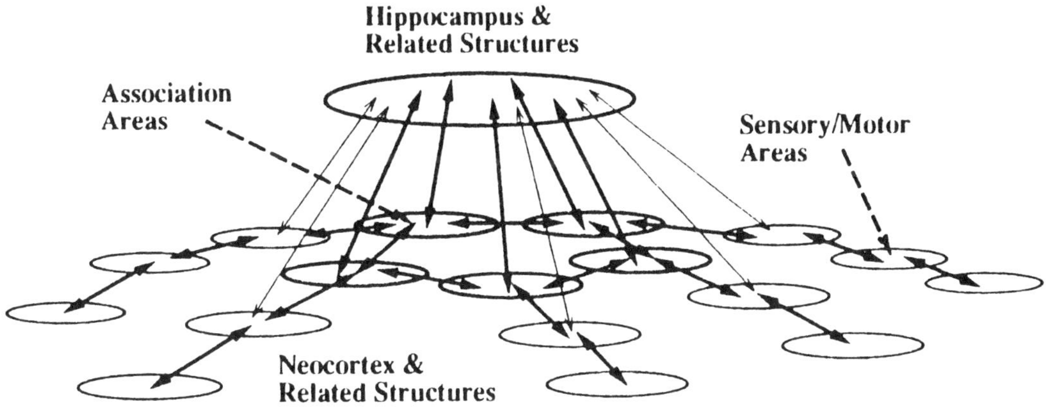

FIG. 2. Conceptual sketch of the neocortical system and its connections with the hippocampal system.

to this as the cortical system). We imagine that this cortical system learns gradually, with each experience producing very small changes to connection weights. These small changes, we assume, are the basis of repetition priming effects, and they gradually give rise to cognitive skills. However, in addition to this cortical system, we assume that there is also rapid storage of traces of specific episodes within the hippocampus. On this view, when an event or experience produces a pattern of activation in the neocortex, it produces at the same time a pattern of activation at the input to the hippocampus. So for example, if I am introduced to someone, I form a pattern of activation including the appearance of the person, the name, and other aspects of the situation. Synaptic modifications within the hippocampus itself then autoassociate the parts of the pattern. Later, when a retrieval cue is presented (let us say the person reappears and I wish to recall his name), this then produces a partial reinstatement on the hippocampal input pattern. This is then completed with the aid of the modified synapses in the hippocampus, and then reinstated in the neocortex via return projections. The assumption is that these changes are large enough so that with one or a few trials they may sustain accurate recall, while the changes that take place in the cortex are initially too small for that. Gradually, though, through repeated reinstatement of the same trace, the cortex may receive enough trials with the same association to learn it in the neocortical connections. This reinstatement can occur, we suggest, either through repetition of the association over many different events; or through repeated reinstatement from the hippocampus. Thus on this view the hippocampus can serve both as the initial cite of storage and also as teacher to the neocortex.

Why is the system organized this way?

The above description provides an account of the data, and although not everyone accepts it, it is not really totally new. Many investigators have suggested some or most of these same points. What I do not know of, however, is any clear explanation of why the system might be organized in this way. We can phrase the questions as follows: First, why is it that we need a special system for rapid storage of memories, if knowledge of all types is eventually stored in connectionist within the neocortex? Second, why is consolidation of information into the neocortex so slow? Why does it require many repeated reinstatements apparently spanning years in some cases?

The answers to these questions require us to consider the acquisition of both procedural and semantic representations in connectionist networks. In such networks, the effective discovery of the shared structure that underlies an entire domain—either a procedural domain such as learning a skill or a semantic domain—requires gradual learning, in which the learning of any one association is interleaved with learning about other associations.

First we consider procedural learning. The network—a fairly standard feedforward network trained with backpropagation—was developed by Plaut and McClelland (1993) to learn to translate alphabetic patterns representing the spellings of words into phonological patterns representing their sounds. The training corpus consisted of a set of 3000 monosyllabic words. Training proceeded very gradually, capturing the gradual nature of the acquisition of reading skill; in fact the network was exposed to the entire corpus of training examples several hundred times, with presentations of more-frequent words given greater weight than presentations of less-frequent words. After training, the network could read 99.7% of the words in the corpus correctly, missing only ten words that were low in frequency and very exceptional either in their spelling or in their pronunciation given the spelling—two examples are the French words "sioux" and "bas", which are very rare and completely inconsistent with English pronunciation. For the vast majority of words, including exceptions such as "HAVE", "GIVE" and "PINT", the network was correct. The network was also tested with many pronounceable nonwords, and scored as well as normals in providing plausible pronunciations—as high as 98% correct on some sets of items. Based on comparisons of the networks performance with typical adult English speakers, we can say that the network has gradually learned both the regular mapping between spelling and sound, and the commonly occurring exceptions to it in a way that corresponds to native English speakers' capabilities in this regard. It has mastered, in short, the procedural skill of translating from spelling to sound—through gradual, incremental learning.

Let us now consider a model that learns semantic information. This model was constructed by Rumelhart (1990) to demonstrate how structured semantic representations can arise in PDP systems from experience with propositions about objects and their properties. The network was trained to capture the information standardly stored in old-fashioned semantic networks of the kind illustrated in the Fig. 3. The goal is to allow the network to store information about concepts in such a way as to permit it to generalize from what it has learned about some concepts to other related concepts. In the old approach, this was done by storing information true of a class or subclass of concepts at the highest possible level of the tree. In the new approach, this is done by gradually learning to assign each concept a distributed representation, capturing its similarity relations to other concepts.

Rumelhart's network is shown in Fig. 4. It consists of a set of input units, one for each concept and one for each of several relations: "ISA", "HAS", "IS", and "CAN". At the output it consists of a set of units for various completions of simple propositions, such as "ROBIN ISA BIRD", "ROBIN CAN FLY", etc. In between there are two layers of hidden units, one to represent the semantic relations of the concepts and another to combine them with the relations to activate the correct outputs.

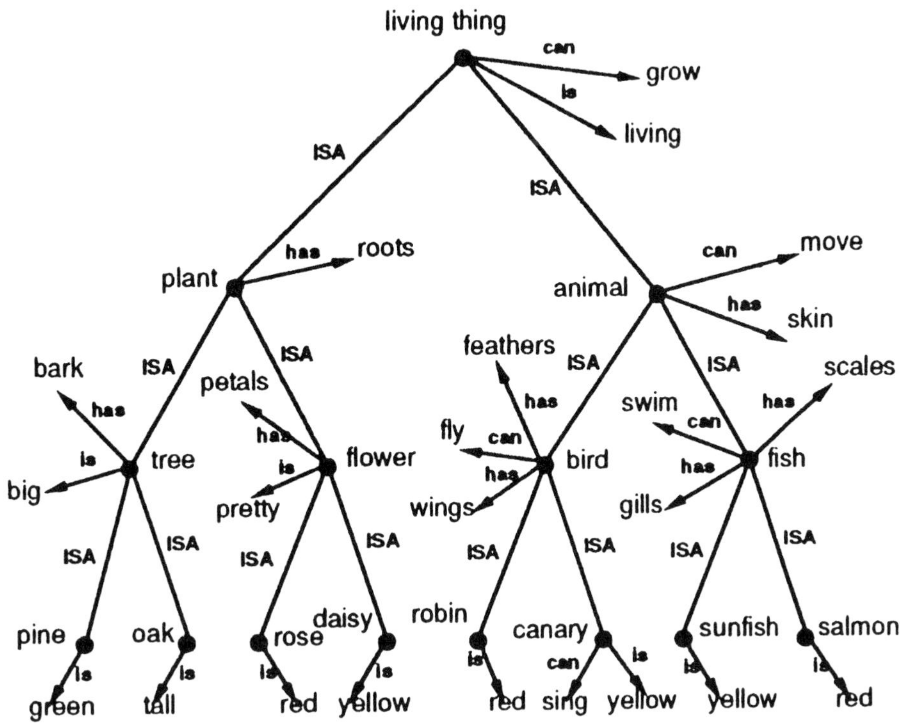

FIG. 3. A semantic network of the type formerly used in models of the organization of knowledge in memory. After Rumelhart and Todd (1993). Reprinted from McClelland *et al.* (1995).

Rumelhart trained this network with a set of propositions involving the concepts shown in the previous figure. There was one input pattern for each concept-relation pair, and the task was to turn on all the output units representing correct completions of the proposition. For example, when the input is "ROBIN ISA" the correct output is "BIRD", "ANIMAL", "LIVING THING". The entire set of patterns was presented to the network many times, making small adjustments to the strengths of the connections after the presentation of each pattern.

After the network has mastered the training set, it is possible to examine the representations that the network has learned to assign to the words in the concepts. We can look either at the patterns themselves, as well as their similarity relations. McClelland *et al.* (1995) repeated this simulation, and the results are shown in Fig. 5. The figure demonstrates that the network has assigned similar patterns to similar concepts—so for example, the pattern for oak is similar to the pattern for pine, the pattern for daisy is similar to the pattern for rose, etc.

Now we may consider how we may achieve generalization in this network. We may do so if we can assign to a new pattern a representation that is similar to the representation of other similar concepts. To illustrate this, Rumelhart trained the network to produce the correct class inclusion output for the concept sparrow. He simply presented "Sparrow isa" as input, trained the network to produce the correct response = "BIRD, ANIMAL, LIVING THING". This caused the network to assign a pattern to sparrow very close to the patterns for robin and canary. As a result the network was able to answer reasonably when probed with other propositions involving sparrows. It said the sparrow can grow and can fly, it has wings, features and skin, and it is small. It was unsure about whether the sparrow could sing and about its color.

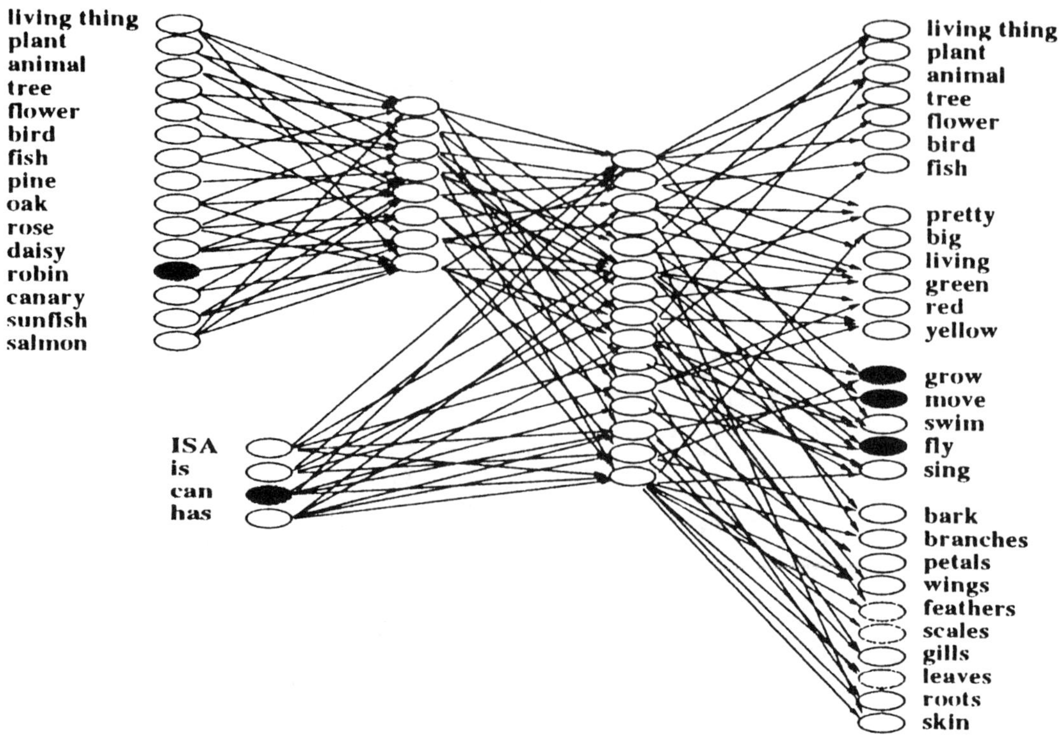

FIG. 4. The connectionist network used by Rumelhart to learn propositions about the concepts shown in the Fig. 3. The entire set of units used in the actual network is shown. Inputs are presented on the left, and activation propagates from left to right. Where connections are indicated, every unit in the pool on the left (sending) side projects to every unit in the right (receiving) side. An input consists of a concept-relation pair; the input *robin can* is illustrated here by darkening the active input units. The network is trained to turn on all those output units that represent correct completions of the input pattern. In this case, the correct units to activate are *grow*, *move* and *fly*; the units for these outputs are darkened as well. Subsequent analysis focuses on the concept representation units, the group of eight units to the right of the concept input units. Reprinted from McClelland *et al.* (1995).

Thus the network learned to inherit knowledge from what it had learned about other birds without direct instruction.

The point here is that PDP models, trained slowly via interleaved presentation on a representative sample of an entire domain of knowledge, can gradually acquire knowledge of structured procedures such as spelling–sound correspondence and structured semantic domains such as the domain of living things.

To make the point even more clear, let us consider what happens in Rumelhart's semantic network if we try to teach it some new information in either of two different ways. The first way, which we call focused learning, involves teaching the network the new information all at once, without interleaving it with ongoing exposure to the structure of the entire domain. The second way, which we call interleaved learning, involves simply introducing the new information into the mix of experiences that characterize the entire domain. As our example, we will consider the case of the penguin. Now as we all know, the penguin is a bird, but it can swim, and it cannot fly. We therefore took Rumelhart's semantic network and taught it these two things in the two different ways just described. The focused case involved simply presenting the two items repeatedly and watching the network learn. The interleaved case involved adding the two new items to the same training set used previously and continuing training as

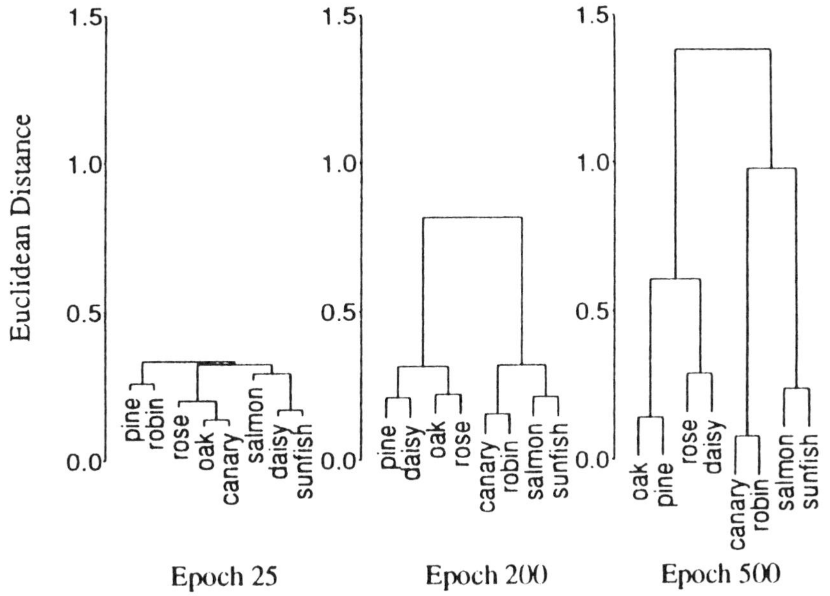

FIG. 5. Similarity structure discovered in our replication of Rumelhart's learning experiment. Initially, the patterns are all quite similar, and the weak similarity structure that exists is random. The concepts become progressively differentiated as learning progresses. Reprinted from McClelland *et al.* (1995).

before. The results of these experiments, for the case of "penguin can grow–swim–fly", are shown in Fig. 6. Here if we look at the number of presentations required for acquisition, in the left panel of the figure, we see that focused learning is better, since acquisition is considerably faster with focused learning than it is with interleaved learning. But it turns out that this slight advantage has been purchased at a considerable cost. For it turns out that with focused learning, the training on the case of the penguin has strongly corrupted the models understanding of other concepts. Indeed if we consider the networks knowledge of the concept robin, in the right panel, we see that focused learning has produced a dramatic interference. In fact, the model now thinks that all animals can swim, and even some plants. But in the case of interleaved learning, we see very little interference with what is already known. To be sure there is a slight effect, but as interleaved learning proceeds this is even reduced gradually over trials. Thus with focused learning new knowledge corrupts the structured database in the network; with interleaved learning new knowledge is gradually added with no disruption.

This demonstration recapitulates a phenomenon that has previously been reported by researchers who have tried to use networks of the type used in Rumelhart's semantic network to model episodic memory (McCloskey and Cohen, 1989). They termed the phenomenon catastrophic interference, and rejected networks of the type that gradually discover structure through incremental learning as models of episodic memory, since storage in episodic memory clearly involves what we have called focused learning. Backpropagation networks are poor models for this kind of learning. However, the kind of learning that they are good at is just the kind of learning we think the cortex is specialized for—gradual discovery of skills from the overall structure of experience.

For the discovery of overall structure, gradual, interleaved learning is necessary, so that the connection weights in the network come to reflect influences of all of the

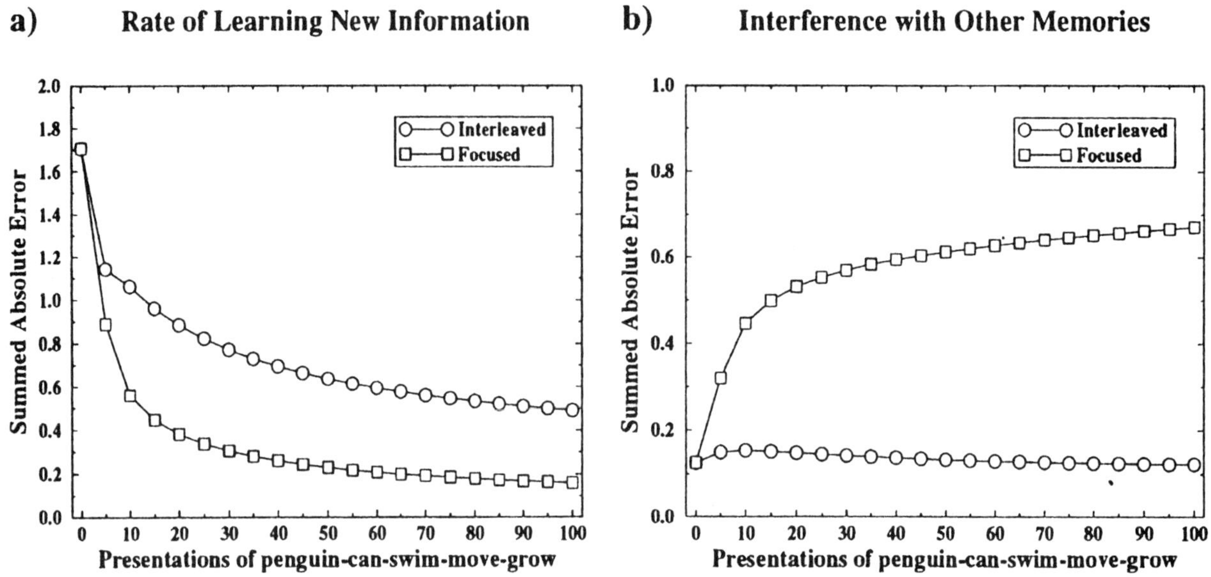

FIG. 6. Effects of focused and interleaved learning on the acquisition of new knowledge and on interference with existing knowledge. Simulations were carried out using Rumelhart's network, using the connection weights resulting from the initial 500 epochs of training with the base corpus. Reprinted from McClelland *et al.* (1995).

examples in the domain. In fact, it can be shown mathematically that in order to converge on a set of connection weights that best captures the structure of a particular domain, it is necessary in the limit to reduce the learning rate asymptotically to 0; the closer it gets to zero, the better we will approximate correct connection weights. This point leads to the observation that if the brain is to be able to extract the structure, it is going to have to learn slowly.

With these ideas in mind, we can now return to the two questions asked above.

First, why is it that we need a special system for rapid learning of the contents of specific episodes and events, if knowledge of all types is eventually stored in connectionist within the neocortex?

The answer begins with the idea that the cortical system is specialized for the extraction of shared structure of events and experiences. In order to do this, and to avoid catastrophic interference with that structure, it is necessary for the cortical system to use a very small learning rate. In this context, the role of the hippocampus is to provide a system in which the contents of specific episodes and events can be stored rapidly, without at the same time interfering with the structure that has been extracted by the cortical system.

Second, why is consolidation of information into the neocortex so slow? Why does it require many repeated reinstatements apparently spanning years in some cases?

Our answer is that consolidation is slow precisely because it would be disruptive if information stored in the hippocampus were incorporated in the neocortex all at once. It is only when new information is gradually interleaved in this way that we are able to incorporate new knowledge into the neocortical system without interfering with what we already know.

Simulation of retrograde amnesia

This observation has now lead to simulations of retrograde amnesia. Due to space limitations I will describe only the simulation of the experiment of Zola-Morgan and Squire (1990). They taught monkeys a series of conditional discriminations. Each conditional discrimination involved two junk objects (for example, a blue plastic cup and a yellow toy dump truck). Under one object the animal could find a reward. The monkeys received 14 trials with each pair of junk objects. Training was arranged so that each monkey learned 20 different discriminations at 16 weeks before surgery, another 20 at 12 weeks before surgery, and 20 each at 8, 4, and 2 weeks before surgery. Some animals received lesions removing the hippocampus and related structures, and others were controls with sham surgery. After time for recovery from surgery the animals were tested and the hippocampal group showed a selective deficit for items learned 2–4 weeks before surgery compared to items learned 12–16 weeks earlier. To simulate this, we imagined that the memory system consists of two parts, a slow-learning cortical network—simulated using a standard backpropagation network—and a more rapidly learning hippocampal network. We do not actually implement a hippocampal network, but instead we treat is at a black box with certain storage, decay, and retrieval properties, and we imagine that it can serve as a source of training patterns for the neocortex. The basic idea (illustrated in Fig. 7) is that when an experience occurs it leaves a trace both in the hippocampus and in the neocortex, but initially the trace in the hippocampus is much stronger. The trace decays relatively rapidly from

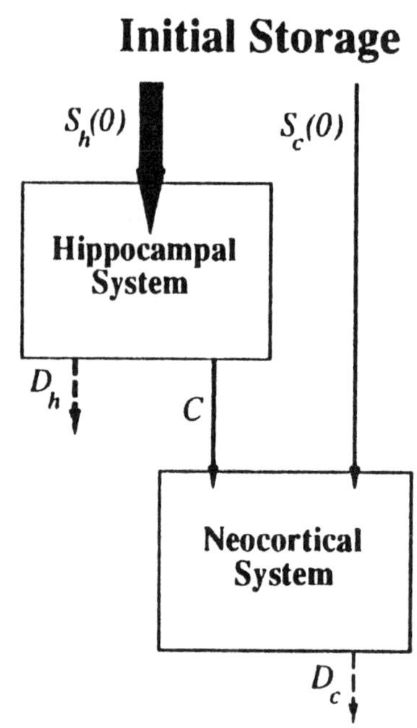

FIG. 7. A simple two-compartment model that characterizes memory storage and decay in the hippocampal system, together with consolidation and subsequent decay in the neocortical system. Arrows are labeled with the parameters of the simple model: $S_h(0)$ and $S_c(0)$ refer to the strength of the hippocampal and neocortical traces due to the initial exposure to the event, D_h and D_c refer to the rate of decay from the hippocampal system and the neocortical system respectively, and C refers to the rate of consolidation. Reprinted from McClelland *et al.* (1995).

TABLE 1. *Parameter values used in fitting the simplified two-memory model to data from four consolidation experiments*

	Parameter				
Experiment	D_h	C	D_c	$S_h(0)$	$S_c(0)$
Winocur (1990)	0.250	0.400	0.075	0.900	0.100
Kim and Fanselow (1992)	0.050	0.040	0.011	0.800	0.030
Zola-Morgan and Squire (1990)	0.035	0.020	0.003	1.000	0.100
Squire and Cohen (1979)	0.001	0.001	0.001	0.500	0.000

D_h represents the rate of hippocampal decay; C represents rate of consolidation in offline contexts; D_c represents rate of decay from neocortex; $S_h(0)$ represents initial strength of the hippocampal trace; $S_c(0)$ represents initial strength of the neocortical trace.

the hippocampal system, but while it remains it has the chance to be reinstated in the neocortex. Meanwhile in the neocortex other events and experiences are coming along and changing the connection weights as well so the learning of the new experience is interleaved with ongoing exposure to other events and experiences. The simulation results are shown along with the data from Zola-Morgan and Squire (1990) in Fig. 8.

This simulation has led us to consider the parameters of the consolidation process. As noted previously there are vast differences between the rate of consolidation in different studies. Indeed, we were able to fit parameters to the simple consolidation model shown in Fig. 7—the results are shown in Table 1. The fits indicate that the rate

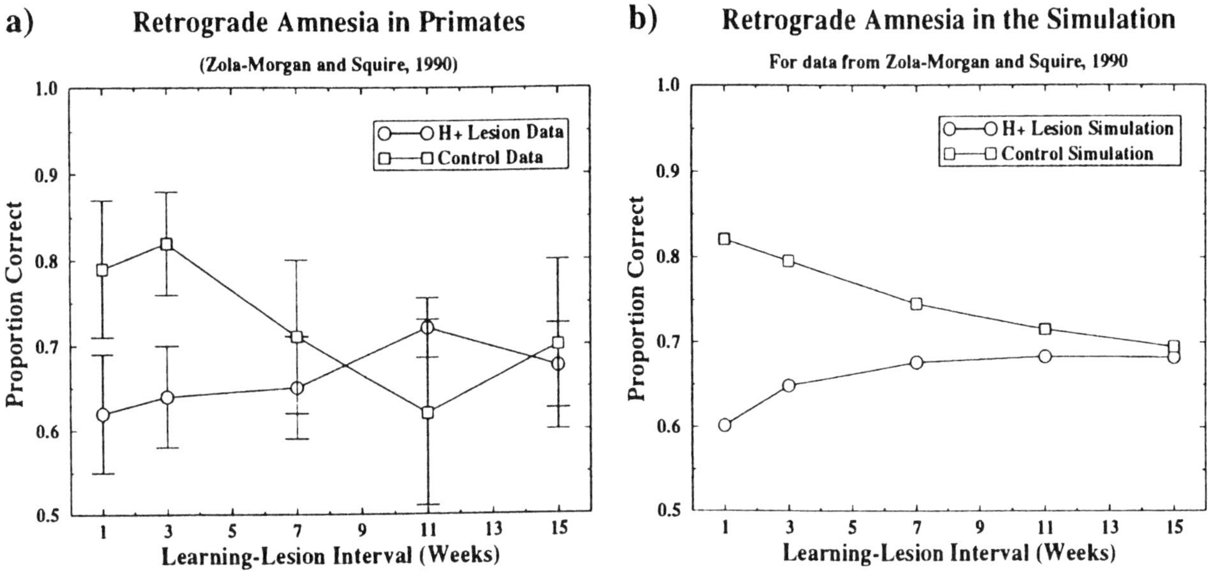

FIG. 8. Experimental data (a) and simulation (b) of the experiment of Zola-Morgan and Squire (1990). Reprinted from McClelland *et al.* (1995).

of decay from the hippocampal system, the rate of consolidation into the neocortical system, and the rate of decay from the neocortical system tend to covary across studies. We consider possible reasons for the differences between species below. Here we consider the difference between the studies of Winocur (1990) and Kim and Fanselow (1992). In the former, the rate of hippocampal decay is apparently far greater than in the latter. The presentation of the shock used by Kim and Fanselow (1992) apparently led to effects that persist for far longer than the experience used by Winocur (1990), who simply exposed rats to a conspecific who had eaten food flavored with cinnamon or chocolate. In the latter case the animals acquired a preference for the food eaten by the conspecific, but even in normals this preference wanes over several days. One possible explanation for this may be that highly emotionally charged experiences such as those involving the kind of intense shock used by Kim and Fanselow (1992) produce much longer lasting hippocampal traces than less emotionally charged experiences like exposure to a conspecific who has eaten a particular food. Interestingly there is evidence (Barnes, 1979; Abraham and Otani, 1991) that strong inducing stimulation produces a longer lasting form of long-term potentiation in hippocampal slices than is produced by weaker stimulation. The decay constants of the weak and strong forms of potentiation in the rat hippocampus are remarkably similar to the decay constants found in our fits to the data from Winocur (1990) and Kim and Fanselow (1992), respectively.

Implications

The ideas presented here have many important implications. I can only mention two implications briefly here. First, they suggest possible explanations for the large species differences between rats, primates and humans. It may be that the human neocortex must learn much more slowly than in less advanced animals, so that humans can take advantage of the much more complex, culturally transmitted, structure present in their environment. Also, it may be noted that the optimal procedure for learning structured environments may be to start with relatively large weight adjustments and then gradually reduce the size; in this way it is possible to guarantee convergence to the best possible set of weights (White, 1989; Darken and Moody, 1991). In this case, we would expect more rapid cortical learning early in life compared to later time periods. Second, these ideas provide one possible account of the phenomenon of infantile amnesia—the fact that we forget our early childhood experiences (Howe and Courage, 1993). According to the present theory, infantile amnesia could be due to the fact that the neocortical representations are changing relatively rapidly early in life, so that even if they are consolidated initially into the neocortical system they would tend to be overwritten by subsequent changes in the representations and connections. Later in life, as the neocortical representations became relatively stabilized, consolidated knowledge would tend to become less susceptible to interference.

Acknowledgments

The work reported here was supported by Grants MH00385 and MH47566, by a collaborative grant from the Human Frontiers Program, and by a center grant from the McDonnell–Pew Program in Cognitive Neuroscience to the University of Arizona. This article is adapted from McClelland (1994).

References

Abraham, W. C. and Otani, S. (1991). Macromolecules and the maintenance of long-term potentiation. In: *Kindling and Synaptic Plasticity*, pp. 92–109, Morrell, F. (ed.). Birkhauser: Boston, MA.

Barnes, C. A. (1979). Memory deficits associated with senescence: A neurophysiological and behavioral study in the rat. *J. Comp. Physiol. Psychol.* **93,** 74–104.

Darken, C. and Moody, J. (1991). Note on learning rate schedules for stochastic optimization. In: *Advances in Neural Information Processing Systems 3*, pp. 832–838, Lippman, R. P., Moody, J. E. and Touretzky, D. S. (eds). Morgan Kaufmann: Palo Alto, CA.

Howe, M. L. and Courage, M. L. (1993). On resolving the enigma of infantile amnesia. *Psychol. Bull.* **113,** 305–326.

Kim, J. J. and Fanselow, M. S. (1992). Modality-specific retrograde amnesia of fear. *Science* **256,** 675–677.

McClelland, J. L. (1994). Computational insights into the organization of human memory. In *Synapse-94*.

McClelland, J. L., McNaughton, B. L. and O'Reilly, R. C. (1995). Why there are complementary learning systems in the hippocampus and neocortex: insights from the successes and failures of connectionist models of learning and memory. *Psychol. Rev.* **102,** 419–457.

McCloskey, M. and Cohen, N. J. (1989). Catastrophic interference in connectionist networks: the sequential learning problem. In: *The Psychology of Learning and Motivation*, Vol. 24, pp. 109–165, Bower, G. H. (ed.) Academic Press: New York.

Plaut, D. C. and McClelland, J. L. (1993). Generalization with componential attractors: Word and nonword reading in an attractor network. *Proceedings of the 15th Annual Conference of the Cognitive Science Society*, pp. 824–829. Lawrence Erlbaum: Hillsdale, NJ.

Rumelhart, D. E. (1990). Brain style computation: learning and generalization. In: *An Introduction to Neural and Electronic networks*, pp. 405–420, Zornetzer, S. F., Davis, J. L. and Lau, C. (eds). Academic Press: San Diego, CA.

Rumelhart, D. E. and Todd, P. M. (1993). Learning and connectionist representations. In: *Attention and Performance XIV: Synergies in Experimental Psychology, Artificial Intelligence, and Cognitive Neuroscience*, pp. 3–30, Meyer, D. E. and Kornblum, S. (eds). MIT Press: Cambridge, MA.

Scoville, W. B. and Milner, B. (1957). Loss of recent memory after bilateral hippocampal lesions. *J. Neurol., Neurosurg. Psychiat.* **20,** 11–21.

Squire, L. R. and Cohen, N. (1979). Memory and amnesia: resistance to disruption develops for years after learning. *Behav. Neural Biol.* **25,** 115–125.

White, H. (1989). Learning in artificial neural networks: a statistical perspective. *Neural Comput.* **1,** 425–464.

Winocur, G. (1990). Anterograde and retrograde amnesia in rats with dorsal hippocampal or dorsomedial thalamic lesions. *Behav. Brain Res.* **38,** 145–154.

Zola-Morgan, S. and Squire, L. R. (1990). The primate hippocampal formation: evidence for a time-limited role in memory storage. *Science* **250,** 288–290.

AFTERWORD

Afterword

Karl H. Pribram
Center for Brain Research and Informational Sciences
Radford University
Peters Hall Laboratory/Adams Street
Radford, VA 24142
Telephone: 540-831-6108 / Fax: 540-831-6236
E-mail: kpribram@runet.edu

As a result of the conference, are we any more enlightened as to how brain function influences human values than we were before? In a strange sense, I believe we are. The strangeness arises from the fact that many, if not most, of the speakers did not directly address the relationship between brain and values. At the same time, they brought intuitions to bear that can be tested by making them explicit -- and this has great value in science because it makes possible the asking of precise questions. As I indicated in my introductory statements, there seems to be some implicit understanding that memory and value are closely related; and if so, learning must also be integral to the process of valuation both in setting up the process and running it off. At another pole, values appear to be set by biological factors: drive stimuli and resulting emotional and motivational states as predispositions and dispositions to value one or another perception or action.

The papers by McClelland and by Ahern and Wood bring new dimensions to these issues, dimensions worth pursuing a bit further.

Commentary on Role of Hippocampus in Learning and Memory

From my standpoint, this paper is a seminal contribution to our understanding of the functions of the hippocampal system. In the final chapter of Robert Isaacson's and my four volume compendium on the hippocampal system, I summarized evidence that indicates this system makes possible "recombinant processing". In Isaacson's book on the limbic systems (1982) and in Lecture 9 of mine on perception (Pribram, 1991) we noted that hippocampal function makes innovation possible. What was lacking in these summaries and proposals was a precise model of just how hippocampal processes would accomplish recombination and innovation.

McClelland, in this contribution to Appalachian V, presents such a precise model. Furthermore, his model indicates how hippocampal intervention – which is present whenever a stimulus is novel to the organism (Thatcher & John, 1977) – can lead to dysfunction (catastrophic interference) as well as to creative innovation. McClelland shows how hippocampal input to the cortex can, at other times, lead to interleaved learning. According to McClelland, non-limbic learning is slow and is produced "via interleaved presentation on a representative sample of an entire domain of knowledge." (See Page 535. Also, see Amsel's chapter, Page 505.)

However, Douglas and I (1966) showed that such interleaved learning can also occur in the absence of the hippocampal formation (and thus independently of episodic processing). Could this be due to a difference in brain organization between rodent and primate such as the massive increase in the area of the frontal cortex? In an earlier report (Pribram, 1958), I had reviewed evidence that animals with more complex nervous systems actually learned more slowly than animals with simpler nervous systems but that the range of what can be learned increases with an increase in brain complexity. McClelland's model shows precisely how an increase in the complexity of the brain can accomplish this enhanced range.

As an addition to the overall model presented by McClelland, there are elaborations of its neurological underpinnings that can fill out the particulars of the "how."

McClelland's current formulation was developed in collaboration with Bruce McNaughton and takes into account the latter's finding of representation in hipppocampal neurons of a path in space. Thus, their model directly matches hippocampal activity with the activity of the cortical convexity (as would be expected of a comparator). On the input side such a model is plausible. However, their model also demands such a comparator process on the output side. This is implausible in view of results obtained by Paul MacLean and myself (Pribram & MacLean, 1953) when mapping cortical connectivity by strychnine neuronography While we were able to readily show multiple inputs to the hippocampal formation, we were totally unable to activate <u>any</u> isocortical region by stimulating the hippocampal cortex. The finding was so striking that MacLean (1990) developed the theme of a schizophysiology of cortical function.

On the other hand outputs are plentiful to the amygdala, to the n. accumbens septi and other subcortical structures via the fornix. Confirmation of the difference between input (encoding) and output (decoding) operations involving the hippocampal formation has recently come from studies in humans using *f*MRI (Gabrieli, Brewer, Desmond & Glover 1997). Encoding into memory was found to activate the para-hippocampal cortex (including the entorhinal cortex which receives input from the remainder of the isocortex), whereas decoding (retrieval) was found to activate the subiculum which "provides the major <u>subcortical</u> output of the hippocampal region via the fornix" (p. 265).

The subcortical nuclei do not have the laminar structure of cortex and so are poor candidates for the point to point match a computer would ordinarily be conceived to implement. On the other hand, a match could readily be achieved if the comparison would involve a stage during which processing entailed a distributed stage, much as when a holographic memory is used to store and retrieve information (for instance with holofishe). It is the evidence that a distributed store is, in fact, built up in the hippocampal formation during learning that makes this sort of model plausible.

Landfield (1976) and O'Keefe (1986) have developed this sort of model. O'Keefe reviews the evidence and describes the model as follows:

> Attempts to gain an idea of the way in which an environment is represented in the hippocampus strongly suggest the absence of any topographic isomorphism between the map and the environment. Furthermore, it appears that a small cluster of neighboring pyramidal cells would map, albeit crudely, the entire environment. This observation, taken together with the ease that many experimenters have had in finding place cells with arbitrarily located electrodes in the hippocampus, suggests that each environment is represented many times over in the hippocampus, in a manner similar to a holographic plate. In both representation systems the effect of increasing the area of the storage which is activated is to increase the definition of the representation.

A second major similarity between the way in which information can be stored on a holographic plate and the way environments can be represented in the hippocampus is that the same hippocampal cells can participate in the representation of several environments (O'Keefe & Conway, 1978; Kubie & Ranck, 1983). In the Kubie and Ranck study the same place cell was recorded from the hippocampus of female rats in three different environments: All of the 28 non-theta cells had a place field in at least one of the environments, and 12 had a field in all three environments. There was no systematic relationship amongst the fields of the same neurone in the different environments. One can conclude that each hippocampal place cell can enter into the representation of a large number of environments, and conversely, that the representation of any given environment is dependent on the activity of a reasonably large group of place neurones.

The third major similarity between the holographic recording technique and the construction of environmental maps in the hippocampus is the use of interference patterns between sinusoidal waves to determine the pattern of activity in the recording substrate (see Landfield, 1976). In optical holography this is done by splitting a beam of monochromatic light into two, reflecting one beam off the sconce to be encoded and then interacting the two beams at the plane of the substrate. In the hippocampus something similar might be happening. . . . The beams are formed by the activity in the fibers projecting to the hippocampus from the medial septal nucleus (MS) and the nucleus of the diagonal band of Broca (DBB).

Pioneering work by Petsche, Stumpf and their colleagues (Stumpf, 1965) showed that the function of the MS and DBB nuclei was to translate the amount of activity ascending from various brainstem nuclei into a frequency moduled code. Neurons in the MS/DBB complex fire in bursts, with a burst frequency which varies from 4-12 Hz. Increases in the strength of brainstem stimulation produce increases in the frequency of the bursts but not necessarily in the number of spikes within each burst (Petsche, Gogolak and van Zweiten, 1965). It is now widely accepted that this bursting activity in the MS/DBB is responsible for the synchronization of the hippocampal theta rhythm (O'Keefe, 1986, pp. 82-84).

The November 1995 issue of *Scientific American* shows how such a holographic matching process could work. Of course, in this quotation, the matching process works by way of illuminating crystals, and the neural substitutes for this would be surface distributions of dendritic polarizations in somatosensory cortex (King et. al, 1994; Pribram, 1995):

Given a hologram, either one of the two beams that interfered to create it can be used to reconstruct the other. What this means, in a holographic memory, is that it is possible not only to orient a reference beam into the crystal at a certain angle to select an individual holographic page but also to accomplish the reverse, illuminating a crystal with one of the stored images gives rise to an approximation of the associated reference beam, reproduced as a plane wave emanating from the crystal at the appropriate angle.

A lens can focus this wave to a small spot whose lateral position is determined by the angle and therefore reveals the identity of the input image. If the crystal is illuminated with a hologram that is not among the stored patterns, multiple reference beams -- and therefore multiple focused spots, are the result. The brightness of each spot is proportional to the degree of similarity between the input image and each of the stored patterns. In other words, the array of spots [weights in a layer of a PDP network] is an encoding of the input image, in terms of its similarity with the stored database of images. (Psaltis & Mok, 1995, p. 76.)

Putting this together with the McNaughton's data showing that the hippocampus represents the path taken by an animal down an alley maze, encoding in the hippocampus may be both holographic-like and patterned in space and time. The hypothesis is that as multiple paths become represented in the hippocampal formation, a transformation into holographic-like surface distribution in the spectral domain is effected. When subsequently a particular path is sought, the buffer operates much as does the holographic memory described above. Initially, actual paths construct the holographic memory and during retrieval, scan paths activate the comparator to access a particularly appropriate actual path. Essentially, the process implements a shift in coordinates from space-time (configurational) to spectral and back to configurational.

The shift of coordinates is suggested to take place by way of scanning, that is, constructing a particular scan path. Computational models such as those proposed by Harth, Unnikrishnan and Pandya (1987) and by Yasue, Jibu and Pribram (Pribram, 1991) have been developed for vision to account for the selection of coordinates as a result of internal scanning. The shift from spectral to the configuration coordinates has been demonstrated in the visual system both at the thalamic and cortical level. Electrical stimulation of temporal or frontal lobe cortex enhances or diminishes the extent of the inhibitory surrounds and flanks of dendritic receptive fields in thalamus and cortex so that the sensory channels can either become multiplexed or fused. As the dendritic fields can be described in terms of a space-time constraint on a sinusoid -- such as the Gabor elementary function, the constraint is embodied in the inhibitory surround of the field. Enhancing the surround enhances processing in configuration coordinates; diminishing the surround enhances the sinusoidal (spectral domain) aspects of processing. Thus, the development of scanpaths operates on the inhibitory process that characterize the fluctuations of the polarizations of the dendritic connection web.

References

Douglas, R. J. & Pribram, K. H. (1966) Learning and limbic lesions. *Neuropsychologia, 4,* pp. 197-220.

Gabrieli, J.D.E., Brewer, J.B., Desmond, J.E. & Glover, G.H. (1997) Separate neural bases of two fundamental memory processes in the human medial temporal lobe. *Science, Vol. 276,* pp. 264-266.

Harth, E., Unnikrishnan, P., & Pandya, A.S. (1987) The inversion of sensory processing by feedback pathways: Model of visual cognitive functions. *Science, 237,* pp. 184-187.

Issacson, R.L. (1982) *The Limbic System.* New York: Plenum Press, 2nd Ed.

King, J.K., Xie, M., Zheng, B., Pribram, K. (1994) Spectral density maps of receptive fields in the rat's somatosensory cortex. In K.H. Pribram (Ed.) *Origins: Brain and Self Organization.* Mahwah, NJ: Lawrence Erlbaum Associates, pp. 556-571.

Kubie, J.L. & Ranck, J.B. (1983) Sensory-behavioral correlates in individual hippocampus neurones in three situations: space and context. In W. Seifert (Ed.), *Neurobiology of the Hippocampus,* (pp. 433-447). London: Academic Press.

Landfield, P.W. (1976) Synchronous EEG rhythms: Their nature and their possible functions in memory, information transmission and behavior. In E.H. Gispen (Ed.) *Molecular and Functional Neurobiology*. Amsterdam: Elsevier.

MacLean, P. (1990) *The Triune Brain in Evolution: Role in Paleocerebral Functions*. New York: Plenum Press.

O'Keefe, J. (1986) Is consciousness the gateway to the hippocampal cognitive map? A speculative essay on the neural basis of mind. *Brain and Mind*k, vol. 15, pp. 59-98.

O'Keefe, J. & Conway, D.H. (1978) Hippocampal place units in the freely moving rat: why they fire where they fire. *Experimental Brain Research 31*, 573-590.

Petsche, H., Gogolak, G. & van Zweiten, PA (1965) Rhythmicity of septal cell discharges at various levels of reticular excitation. Electroencephalography and Clinical Neurophysiology, 19, pp. 25-33.

Pribram, K.H. (1995) The Enigma of Reinforcement. *Neurobehavioral Plasticity: Learning, Development and Response to Brain Insults*. Proceedings of the Bob Isaacson Symposium in Clearwater, FL, Lawrence Erlbaum Associates, NJ, pp. 381-403.

Pribram, K.H. (1991) *Brain and Perception: Holonomy and Structure in Figural Processing*. Mahwah, NJ: Lawrence Erlbaum Associates, Inc.

Pribram, K. H. (1958) Comparative neurology and the evolution of behavior. In A. Roe & G. G. Simpson (Eds.), *Behavior and evolution.* New Haven: Yale University Press, pp. 140-164.

Pribram, K.H. & MacLean, P. (1953) Neuronographic analysis of medial and basal cerebral cortex. II. *Journal of Neurophysiology, 16*, pp. 324-340.

Psaltis, D. & Mok, F. (1995) Holographic memories. In *Scientific American*, Vol 273, 5, pp. 70-76.

Stumpf, C. (1965) Drug action on the electrical activity of the hippocampus. *International Review of Neurobiology, 8*, pp. 77-138.

Thatcher, R.W. & John, E.R. (1977) *Functional Neuroscience, Vol. 1.* Hillsdale, NJ: Lawrence Erlbaum Associates.

McClelland Reactions to Pribram's Commentary

J.L. McClelland
Center for the Neural Basis of Cognition
Carnegie Mellon University
Pittsburgh, PA 15213
August 1997

I thank Pribram for his kind words about our model of the role of the hippocampus in learning and memory, and agree with several aspects of his commentary. I will comment on two points: (1) the necessity of hippocampal playback for normal cortical learning, and (2) the relationship between hippocampal memory storage and holographic forms of memory. Is hippocampal playback necessary for normal cortical learning?

Pribram mentions data suggesting that in the monkey, the hippocampus is not necessary for normal acquisition of memories in the neocortex, and suggests that this may be a difference between monkey and man. In fact, recent data suggests the same in humans. Vargha-Khadem, (1997), and Mishkin have recently reported three cases of young adults who sustained apparently selective damage to the hippocampus early in life. At least one of these cases shows an above-average IQ, with good performance on various tests that assess adequacy of 'semantic' memory, including vocabulary and encyclopedia knowledge. For example, the patient, who received his education in England, gave a very cogent and accurate answer to the question, 'Who was Martin Luther King?' Yet the patient's episodic memory is grossly deficient, as one would expect from the extensive apparent damage to the hippocampus. Since the hippocampal damage is most likely to have occurred during anoxic periods around the time of his birth, it would seem clear that acquisition of normal semantic memory does not depend on an intact hippocampal system.

This finding is actually completely consistent with my views of the role of the hippocampus, which I fear my writings on the subject have not made totally clear. There is evidence of replay of memories stored in the hippocampus, and the finding of temporally graded retrograde amnesia after a single punctate exposure seems best interpreted as reflecting the neocortical learning that results from the use of this replay mechanism. However, I did not mean to imply that replay of memories stored in the hippocampus is necessary for cortical learning. Day to day exposure to structured information embodied in the events and experiences we have in our interactions with the world are in my view the main source of the experience that drives neocortical learning.

Similarities and Differences Between Hippocampal Memory Storage and Holographic Forms of Memory

Pribram is quite right to note that there are several striking similarities between hippocampal memory and holographic memory. The spatial structure of the world is completely dispersed across the hippocampus, just as it is in a hologram, and small patches of hippocampus do indeed provide degraded representations of entire spatial environments. However, I do not believe the mechanism that gives rise to this is holographic in the sense of relying on interference patterns and linear summation. Instead, as articulated in some detail in a recent paper by McClelland and Goddard (1996), I believe that this is accomplished by a mechanism we call 'sparse, random, conjunctive coding'. Essentially, the idea is this. First, neurons within hippocampal field CA3 recieve projections from a small fraction (perhaps 1-5%) of neurons in entorhinal cortex (EC). While there may be some topographic organization within EC, the fibers leading from EC to field CA3 are quite dispersed, so that a given CA3 neuron can receive direct input from up to 1/2 of the EC. Collareral interactions among CA3 neurons likewise spread very widely, leading to the idealization that each CA3 neuron receives inputs selected completely at random from the full population of EC. Second, neurons in CA3 project to and recieve projections back from a pool of inhibitory interneurons, whose function is to strictly regulate the number of CA3 neurons active at a particular time. The hippocampal theta rhythm modulates the activity of these interneurons, allowing only a very small fraction to fire in the 'trough' of the theta cycle, and a larger fraction to fire at the peak. At the trough, then, the hippocampal representation is extremely sparse (few neurons are active) and the ones that are active are the ones with the strongest total synaptic input from entorhinal cortex. They thus represent a sparse, distributed code, e.g., of the animal's location in space. Because they are distributed approximately uniformly over the hippocampus, the link between external position and position in the hippocampus is broken. Even though the code is sparse, it is still redundant, in the following sense. There are 300,000 CA3 neurons in the rat, and even if there are 2% active at any given point in the environment, this still comes to 6000 neurons. Each neuron encodes a random subset of the elements that define the particular location in the environment, but each element is represented several times, so that any given element is likely still to be represented if only a fraction of the CA3 is preserved. Tsodyks, Skaggs, McNaughton, and Sejnowski have recently suggested what may happen during the peak of the theta cycle: here, as more and more hippocampal neurons are allowed to become active, the influence of collateral connections among CA3 neurons, rather than bottom-up connections from EC, may become more and more important in determining which CA3 neurons fire. Thus the neurons activated in the later stages of the theta cycle may reflect an associative pattern completion process. During spatial navigation, this process is thought to be responsible for that fact that neurons active in the peak of one theta cycle represent locations in space toward which this animal is moving.

As a more general comment, I have always found Pribram's perspective, emphasizing the distributed nature of memory storage, to be an extremely valuable and indeed largely valid perspective. It seems to me that the brain has found ways of capturing the positive features of this idea. Whether it does so by making explicit use of interference patterns remains a controvertial point, but the more fundamental insight is that representation and processing are indeed highly distributed, and this is especially true within the hippocampus.

References

McClelland, J.L., & Goddard, N.H. (1997). Considerations arising from a complementary learning systems perspective on hippocampus and neocortex. *Hippocampus,* 6, 654-665.

Vaegha-Khadem, F., Gadian, D.G., Watkins, K.E., Connelly, A., Van Paesschen, W., & Mishkin, M. (1997). Different effects of early hippocampal pathology on episodic and semantic memory. *Science,* 277, 376-379.

Commentary on the Role of the Frontolimbic System on Episodic Processing

In this scheme of things, valuation is also influenced by more momentary utilitarian considerations provided by desires and the probability that these desires will be fulfilled.

Critical to this aspect of valuation is the case of T.J. presented by Charles Ahern and Frank Wood. Ahern and Wood framed their analysis in the terms used in memory research: the distinction between episodic, skill (procedural), and semantic (reference) types of processing. For reasons that, I hope, are apparent, I equate episodic processing with the current, momentary utilities (desirabilities and probabilities) that influence valuation; and I equate semantic processing with discriminations and preferences.

Given these equations, T.J. is important in that his semantic learning proceeded within normal limits and that his semantic memory is unimpaired, as is his skill in using it. As Zajonc (1966) has claimed on the basis of his studies, episodic processing appears to utilize a brain system independent of procedural and semantic processing.

Furthermore, episodic processing is not a necessary antecedent to skill or semantic (reference) learning or remembering. In these respects T.J. confirms the findings obtained for decades, with non-human primates. (For review see Pribram 1991, Lectures 7, 8, 9, and especially10). Also, T.J.'s lesions involve the mediobasal structures of the limbic system and the related anterior frontal cortex. However, in monkeys the resections that produced the defects were symmetrical; in T.J. the lesion was primarily, though not exclusively, in the right hemisphere. What is striking about T.J. is how well he can traverse the social scene despite his catastrophically severe deficit in episodic processing. What remains to be investigated is just how his judgements, his values are affected. So far he manages on what he has learned; in the jargon adopted in my papers in this volume he has obtained his preferences from his social environment.

But there is also considerable intactness in T.J.'s emotional processing. Are drive stimuli and their influence on emotions and values influenced by a system separate from that which influences episodic processing? If so, is the drive-emotional system centered on the amygdala and the orbitofrontal cortex (as reviewed by Schore in this volume, while episodic processing involves the hippocampal-dorsal-frontal-cingulate cortex? And if this is so, how do we reconcile the data and models of hippocampal function presented in this conference (and in my discussion thereof) with the fact that semantic learning can proceed without hippocampal intervention? An answer to this last question may be the one given by Vindogradova (1970) (and by Douglas and myself, 1966; 1969) that in learning, an ounce of "emotion" is the equivalent of a ton of repetition. Perhaps the hippocampal models presented here apply to those circumstances when something novel catches our interest -- and not to the rote learning of the alphabet, and other prerequisites to semantic processing.

I have pursued only two of the possible trains of thought brought up during the proceeding of this conference. Of equal interest is the model of visual processing developed by Vogl, Blackwell and Alkon. And what about Sternberg's demonstration of the importance of values in shaping our testing procedures and therefore the formation of their values during the educational process in youths whose far frontal cortex is still developing (Hudspeth and Pribram 1990).

Each of these trains of thought deserves a whole conference of its own. For now, I do feel this conference has added a great deal to our understanding of the role of the brain in shaping our values. But one does have to dig a bit in order to extract the gold.

References

Douglas, R. J. & Pribram, K. H. (1969) Distraction and habituation in monkeys with limbic lesions. *J. Comp. Physiol. Psychol., 69,* pp. 473-480.

Douglas, R. J. & Pribram, K. H. (1966) Learning and limbic lesions. *Neuropsychologia, 4,* pp. 197-220.

Hudspeth, W. J. & Pribram, K. H. (1990) Stages of brain and cognitive maturation. *Journal of Educational Psychology*, 82(4), pp. 881-884.

Pribram, K.H. (1991) *Brain and Perception: Holonomy and Structure in Figural Processing.* New Jersey: Lawrence Erlbaum Associates, Inc.

Vindogradova, O. (1970) *Short-term Changes in Nueral Activity & Behavior*. Cambridge: Cambridge University Press.

Zajonc, R.B. (1966) *An Experimental Approach.* Basic Concepts in Psychology Series. Belmont, CA: Wadsworth Publishing Company, Inc.

List of Authors Cited

W

X,Y

Z